A Graduate Course in Probability

Other World Scientific Titles by the Author

Lectures on the Geometry of Manifolds
ISBN: 978-981-02-2836-1

Lectures on the Geometry of Manifolds
Second Edition
ISBN: 978-981-270-853-3
ISBN: 978-981-277-862-8 (pbk)

Introduction to Real Analysis
ISBN: 978-981-121-038-9
ISBN: 978-981-121-075-4 (pbk)

Lectures on the Geometry of Manifolds
Third Edition
ISBN: 978-981-121-481-3
ISBN: 978-981-121-595-7 (pbk)

A Graduate Course in Probability
ISBN: 978-981-125-508-3

A Graduate Course in Probability

LIVIU I NICOLAESCU
University of Notre Dame, USA

NEW JERSEY · LONDON · SINGAPORE · BEIJING · SHANGHAI · HONG KONG · TAIPEI · CHENNAI · TOKYO

Published by

World Scientific Publishing Co. Pte. Ltd.
5 Toh Tuck Link, Singapore 596224
USA office: 27 Warren Street, Suite 401-402, Hackensack, NJ 07601
UK office: 57 Shelton Street, Covent Garden, London WC2H 9HE

Library of Congress Control Number: 2022033524

British Library Cataloguing-in-Publication Data
A catalogue record for this book is available from the British Library.

A GRADUATE COURSE IN PROBABILITY

Copyright © 2023 by World Scientific Publishing Co. Pte. Ltd.

All rights reserved. This book, or parts thereof, may not be reproduced in any form or by any means, electronic or mechanical, including photocopying, recording or any information storage and retrieval system now known or to be invented, without written permission from the publisher.

For photocopying of material in this volume, please pay a copying fee through the Copyright Clearance Center, Inc., 222 Rosewood Drive, Danvers, MA 01923, USA. In this case permission to photocopy is not required from the publisher.

ISBN 978-981-125-508-3 (hardcover)
ISBN 978-981-125-509-0 (ebook for institutions)
ISBN 978-981-125-510-6 (ebook for individuals)

For any available supplementary material, please visit
https://www.worldscientific.com/worldscibooks/10.1142/12800#t=suppl

Printed in Singapore

To my mom.

Introduction

> *In no other branch of mathematics is it so easy for experts to blunder as in probability theory.*
> Martin Gardner

I have to confess that my mathematical formation is not that of a probabilist. I am a geometer/analyst by training. About fifteen years ago I stumbled on some probabilistic geometry questions. The ad-hoc methods I used were producing encouraging but unsatisfactory answers. A chance encounter with a trained probabilist led me to a pretty advanced monograph dealing with related problems from a probabilistic view point. I spent a sabbatical year learning probability so I could understand that book.

I eventually did understand that book, I was able to phrase the original questions in a better language and I even offered answers to questions I could not conceive before. A "side effect" of this effort was that I got a taste of probability.

To the geometer in me, the probabilistic thinking looked (and still looks) like mathematics with a bit more, somewhat similar to classical mechanics, that is mathematics with a sprinkle of physical intuition. I find this subject fresh, full of interesting and enticing questions. This is how my probabilistic journey began and I have been enjoying it since. In the meantime I matured a bit more by teaching probability, both at undergraduate and graduate level. This book partially reflects this personal journey.

Probability theory has grown out of many concrete examples and questions and I firmly believe that probabilistic thinking can only be grasped through examples. Compared to other mathematical areas I am familiar with, probability contains an unusually large number of counterintuitive results. To me, these represent one of the attractive features of the subject. So a substantial part of this book is devoted to examples, some truly fundamental and quite a few more esoteric but which are aesthetically very pleasing and pedagogically very revealing. Some of these examples are recurring, appearing in many places in the text and, as we develop more and more sophisticated technology, we dig a deeper and deeper into them.

While teaching probability I discovered that probabilistic simulations enhance the understanding of probabilistic thinking. That is why I have included a brief introduction to R and a few of the simple codes that allows one to do basic Monte-Carlo simulations. I hope I can tempt the reader to try a few of these and be amazed, like myself and my students, of the remarkable agreement between practice and theory.

I have divided the book into five chapters. The first one concentrates on the measure theoretic foundations of probability and its theoretical part it is essentially the content of Kolmogorov's foundational monograph. I assume that the reader is familiar with the measure theory and integration. I survey this subject and I present complete proofs only of results that have important probabilistic applications or significance.

The first genuinely probabilistic concept is that of independence and I prove early on Kolmogorov's zero-one theorem. It is a striking all-or-nothing result and its deeper implications are gradually revealed in the later parts of the book. The ubiquitous concept of random variable and its numerical characteristics are discussed in detail. Along the way I discuss the various modes of convergence of random variables. I made sure the reader has the opportunity to see these ideas at work so I present many classical random variables and some of their probabilistic occurrences. Among the classical problems/themes I discuss I should mention, the inclusion-exclusion principle, sieves and Poissonization, Poisson processes, the coupon collector problem, the longest common subsequence problem.

Section 4, one of the largest of this chapter, is devoted to the concept of conditional expectation, a central probabilistic concept that takes some getting used to. Analytically, the existence of conditional expectation is a simple consequence of the Radon-Nicodym theorem. This however hides its probabilistic significance. I opted for the more involved approach that reveals the meaning of this object as the best predictor given certain information.

To get to the heart of the rather subtle concept of conditional expectation I tried to present many examples, from simple computations to more sophisticated applications to stochastic optimization problems such as the classical secretary problem. I spend considerable time on the concept of kernels a.k.a. random measures, regular conditional distributions and disintegration of measure describing the various connections between them. I opted to only sketch the proof of the existence of regular conditional distributions since I felt that the missing details add little to the understanding of this important concept. Instead, I have included a large and varied number of concrete examples to give the reader a better feel of this concept.

The last section of this chapter is an introduction to stochastic processes. The central result of this section is Kolmogorov's existence/consistency theorem that guarantees that various objects discussed in the previous sections do indeed have a mathematical existence. I decided to present a complete proof of this result so the reader can see the source of this existence, namely Tikhonov's compactness theorem, a result that is deeply rooted in the foundations of mathematics.

Chapter 2 is devoted to a major theme in probability, the law of large numbers and its relatives. The first section is devoted to the Strong Law of Large Numbers. I present Kolmogorov's proof that reduces this result to the convergence of random series with independent summands. I find the Law of Large Numbers philosophically surprising since it extracts order out of chaos. The Monte Carlo method is one convincing manifestation of the order-out-chaos phenomenon. I could not pass the opportunity to introduce the concept of entropy and its application via the law of large numbers to coding/compression of data. The second section is devoted to the central limit theorem.

The third section is devoted to concentration inequalities. We describe the basics of Chernoff's estimates and produce a few fundamental concentration inequalities. As an application we discuss Lindenstrauss-Johnson lemma stating that the geometry of a cloud of points in a high-dimensional vector space is, with high confidence, little disturbed by a orthogonal projection onto a random subspace of much smaller dimension.

Section 4 is devoted to more modern considerations, namely uniform limits of empirical processes. The Glivenko-Cantelli is the pioneering result in this direction. I also discuss more recent results showing how this uniform convergence can be obtained by combining the concentration results in the previous section and the concept of VC-families/dimension. I briefly describe the significance of such results to PAC-learning, a concept central in machine learning.

The last section of this chapter is a brief introduction to the theory of Brownian motion. I used it as an opportunity to discuss more concepts and results involving stochastic processes such as Gaussian processes and Kolmogorov's continuity theorem.

Chapter 3 is devoted to the castle that J. L. Doob built, namely the theory of (sub)martingales, discrete and continuous. I present in detail the theoretical pillars of this edifice: stopping/sampling, asymptotic behavior, maximal inequalities and I discuss a large and diverse collection of examples: occurrence of patterns, Galston-Watson processes, optimal gambling strategies, Azuma and McDiarmid inequalities and their application to combinatorial optimization problems, backwards martingales, exchangeable sequences, de Finetti's theorem, and asymptotics in Polya's urn problem, Brownian motion.

Chapter 4 is an introduction to Markov chains. This beautiful and rich subject is still actual, growing, and has many applications and ramifications. The first three sections are devoted to the "classical" part of this subject and culminates with the law of large numbers for such stochastic processes. Section 4 is devoted to a more recent (1950's) point of view, namely the connection between reversible Markov chains and electrical networks. I adopt a more geometric approach based on the old observations of H. Weyl and R. Bott (see [16]) that Kirckhoff's laws have a Hodge theoretic description. The last section is devoted to finite Markov chains I describe various ways of estimating the rate of convergence of irreducible recurrent Markov chains. The chapter ends with brief discussion of the Markov Chain Monte Carlo methods.

The last chapter of the book is the shortest and is devoted to the classical ergodic theorems. I have included it because I felt I owed it to the reader to highlight a principle that unifies and clarifies the main limit theorems in Chapters 2 and 4.

As the title indicates, this book is meant as an introduction to the modern, i.e., post Kolmogorov's axiomatization, theory of probability. The reader is assumed to have some familiarity with measure theory and integration and be comfortable with the basic objects and concepts of modern analysis: metric/topological spaces, convergence, compactness. In a few places, familiarity with basic concepts of functional analysis is assumed. It could serve as a textbook for a year-long basic graduate course in probability. With this purpose in mind I have a included a relatively large number of exercises, many of them nontrivial and highlighting aspects I did not include in the main body of the text.

The book grew up from notes of a one-semester graduate course in probability that I taught at the University of Notre Dame. That course covered Chapter 1, the classical limit theorems (Secs. 2.1–2.3) and discrete time martingales (Secs. 3.1–3.2). Some of the proofs appear in fine print as a suggestion to the potential student/instructor that they can be skipped at a first encounter with this subject.

Work on this book has been my constant happy companion during these improbable times. I hope I was able to convey my curiosity, fascination and enthusiasm about probability and convince some readers to dig deeper into this intellectually rewarding subject.

I want to thank World Scientific for a most professional, helpful and pleasant collaboration over the years.

Notre Dame, May 2022

Notation and conventions

- We set $\mathbb{N} := \mathbb{Z}_{>0}$, $\mathbb{N}_0 := \mathbb{Z}_{\geq 0}$.
- For $n \in \mathbb{N}$ we set $\mathbb{I}_n := \{1, 2, \ldots, n\}$.
- For $n \in \mathbb{N}$ we denote by $\mathfrak{S}_n$ the group of permutations of $\mathbb{I}_n$.
- We set $\mathbb{R}_+ := [0, \infty)$.
- For $x \in \mathbb{R}$ we set $\lfloor x \rfloor := \max \mathbb{Z} \cap (-\infty, x]$, $\lceil x \rceil := \min \mathbb{Z} \cap [x, \infty)$.
- $x \wedge y := \min(x, y)$, $x \vee y := \max(x, y)$.
- $\boldsymbol{i} := \sqrt{-1}$.
- Given a subset A of a set X we denote by A^c its complement (in X).
- For any set X we denote by 2^X the collection of all the subsets of X.
- For any set X we denote by 2_0^X the collection of all the *finite* subsets of X.
- We will denote by $|S|$ or $\#S$ the cardinality of s set S.
- If T is a topological space, then we denote by $\mathcal{B}_T$ the σ-algebra of Borel subsets of T.
- We denote by $\boldsymbol{\lambda}$ the standard Lebesgue measure on $\mathbb{R}$ and by $\boldsymbol{\lambda}_n$ the standard Lebesgue measure on $\mathbb{R}^n$.
- If $(\Omega, \mathcal{F})$ is a measurable space and $(\mathcal{A}_i)_{i \in I}$ is a collection of subsets of $\mathcal{F}$, then $\sigma(\mathcal{A}_i, i \in I)$ is the smallest sub-σ-algebra of $\mathcal{F}$ containing all the collections $\mathcal{A}_i$.
- For a collection $(X_i)_{i \in I}$ of random variables defined on the same probability space we denote by $\sigma(X_i; i \in I)$ the sub-σ-algebra generated by these variables.
- Given an ambient set Ω and a subset $A \subset \Omega$ we denote by $\boldsymbol{I}_A : \Omega \to \{0, 1\}$ the *indicator function* of A,

$$\boldsymbol{I}_A(\omega) = \begin{cases} 1, & \omega \in A, \\ 0, & \omega \notin A. \end{cases}$$

- We denote by $\boldsymbol{\omega}_n$ the volume of the unit ball in $\mathbb{R}^n$ and by $\boldsymbol{\sigma}_{n-1}$ the "area" of the unit sphere in $\mathbb{R}^n$.

$$\boldsymbol{\omega}_n = \frac{1}{n} \boldsymbol{\sigma}_{n-1}, \quad \boldsymbol{\sigma}_{n-1} = \frac{2\Gamma(1/2)^n}{\Gamma(n/2)}.$$

Contents

Introduction   vii

Notation and conventions   xi

1. Foundations   1
    - 1.1 Measurable spaces   2
        - 1.1.1 Sigma-algebras   2
        - 1.1.2 Measurable maps   6
    - 1.2 Measures and integration   13
        - 1.2.1 Measures   13
        - 1.2.2 Independence and conditional probability   21
        - 1.2.3 Integration of measurable functions   31
        - 1.2.4 L^p spaces   37
        - 1.2.5 Measures on compact metric spaces   39
    - 1.3 Invariants of random variables   41
        - 1.3.1 The distribution and the expectation of a random variable   41
        - 1.3.2 Higher order integral invariants of random variables   48
        - 1.3.3 Classical examples of discrete random variables   53
        - 1.3.4 Classical examples of continuous probability distributions   64
        - 1.3.5 Product probability spaces and independence   69
        - 1.3.6 Convolutions of Borel measures on the real axis   77
        - 1.3.7 Modes of convergence of random variables   83
    - 1.4 Conditional expectation   91
        - 1.4.1 Conditioning on a sigma sub-algebra   92
        - 1.4.2 Some applications of conditioning   102
        - 1.4.3 Conditional independence   110
        - 1.4.4 Kernels and regular conditional distributions   111
        - 1.4.5 Disintegration of measures   120
    - 1.5 What are stochastic processes?   123
        - 1.5.1 Definition and examples   123
        - 1.5.2 Kolmogorov's existence theorem   127

		1.6	Exercises	133

2. Limit theorems 151
   - 2.1 The Law of Large Numbers 152
     - 2.1.1 Random series 152
     - 2.1.2 The Law of Large Numbers 156
     - 2.1.3 Entropy and compression 163
   - 2.2 The Central Limit Theorem 169
     - 2.2.1 Weak and vague convergence 170
     - 2.2.2 The characteristic function 179
     - 2.2.3 The Central Limit Theorem 186
   - 2.3 Concentration inequalities 188
     - 2.3.1 The Chernoff bound 189
     - 2.3.2 Some applications 194
   - 2.4 Uniform laws of large numbers 200
     - 2.4.1 The Glivenko-Cantelli theorem 201
     - 2.4.2 VC-theory 203
     - 2.4.3 PAC learning 212
   - 2.5 The Brownian motion 214
     - 2.5.1 Heuristics 214
     - 2.5.2 Gaussian measures and processes 217
     - 2.5.3 The Brownian motion 225
   - 2.6 Exercises 236

3. Martingales 259
   - 3.1 Basic facts about martingales 260
     - 3.1.1 Definition and examples 260
     - 3.1.2 Discrete stochastic integrals 266
     - 3.1.3 Stopping and sampling: discrete time 269
     - 3.1.4 Applications of the optional sampling theorem . 274
     - 3.1.5 Concentration inequalities: martingale techniques . 280
   - 3.2 Limit theorems: discrete time 286
     - 3.2.1 Almost sure convergence 286
     - 3.2.2 Uniform integrability 293
     - 3.2.3 Uniformly integrable martingales 298
     - 3.2.4 Applications of the optional sampling theorem . 304
     - 3.2.5 Uniformly integrable submartingales 310
     - 3.2.6 Maximal inequalities and L^p-convergence 317
     - 3.2.7 Backwards martingales 322
     - 3.2.8 Exchangeable sequences of random variables 325
   - 3.3 Continuous time martingales 333
     - 3.3.1 Generalities about filtered processes 333

		3.3.2	The Brownian motion as a filtered process	337

         3.3.2   The Brownian motion as a filtered process 337
         3.3.3   Definition and examples of continuous time martingales . . 345
         3.3.4   Limit theorems 347
         3.3.5   Sampling and stopping 349
   3.4   Exercises 354

4. Markov chains   367

   4.1   Markov chains 368
         4.1.1   Definition and basic concepts 368
         4.1.2   Examples 375
   4.2   The dynamics of homogeneous Markov chains 378
         4.2.1   Classification of states 378
         4.2.2   The strong Markov property 384
         4.2.3   Transience and recurrence 386
         4.2.4   Invariant measures 392
   4.3   Asymptotic behavior 400
         4.3.1   The ergodic theorem 401
         4.3.2   Aperiodic chains 403
         4.3.3   Martingale techniques 407
   4.4   Electric networks 413
         4.4.1   Reversible Markov chains as electric networks 413
         4.4.2   Sources, currents and chains 414
         4.4.3   Kirkhoff's laws and Hodge theory 416
         4.4.4   A probabilistic perspective on Kirchoff laws 420
         4.4.5   Degenerations 424
         4.4.6   Applications 432
   4.5   Finite Markov chains 438
         4.5.1   The Perron-Frobenius theory 438
         4.5.2   Variational methods 450
         4.5.3   Markov Chain Monte Carlo 458
   4.6   Exercises 463

5. Elements of Ergodic Theory   475

   5.1   The ergodic theorem 475
         5.1.1   Measure preserving maps and invariant sets 475
         5.1.2   Ergodic theorems 482
   5.2   Applications 492
         5.2.1   Limit theorems 492
         5.2.2   Mixing 495
   5.3   Exercises 501

Appendix A A few useful facts   507
   A.1 The Gamma function 507
   A.2 Basic invariants of frequently used probability distributions 509
   A.3 A glimpse at R 510

Bibliography   525

Index   533

Chapter 1

Foundations

At the beginning of the twentieth century probability was in a fluid state. There was no clear mathematical concept of probability, and ad-hoc methods were used to rigorously formulate classical questions. Probability at that stage was a collection of interesting problems in search of a coherent setup. According to Jean Ville, a PhD student of M. Fréchet, in Paris probability was viewed among mathematicians as "an honorable pastime for those who distinguished themselves in pure mathematics".

The whole enterprise seemed to be concerned with concepts that lie outside mathematics. Henri Poincaré himself wrote that "one can hardly give a satisfactory definition of probability". As Richard von Misses pointed out in 1928, the German word for probability, "wahrscheinlich", translates literally as "truth resembling"; see [155]. Bertrand Russel was quoted as saying in 1929 that "Probability is the most important concept in modern science, especially as nobody has the slightest notion of what it means". The philosophical underpinnings of this concept are discussed even today. For more on this aspect we refer to the recent delightful book [45].

In his influential 1900 International Congress address in Paris D. Hilbert recognized this state of affairs and the importance of the subject. In the sixth problem of his famous list of 23 he asks, among other things, for rigorous foundations of probability. These were laid by A. N. Kolmogorov in his famous 1933 monograph [94]. According to Kolmogorov himself, this was not a research work, but a work of synthesis. A brilliant synthesis I might add. His point of view was universally adopted and modern probability theory was born. The theory of probability can now be informally divided into two eras: before and after Kolmogorov.

The present chapter is devoted to this foundational work of Kolmogorov. The pillars of probability theory are the concept of probability or sample space, random variables, independence, conditional expectations, and consistency, i.e., the existence of random variables or processes with prescribed statistics.

So efficient is his axiomatization that to the untrained eye, probability, as envisaged by Kolmogorov, may seem like a slice of measure theory. In a 1963 interview Kolmogorov complained that his axioms have been so successful on the theoretic side that many mathematicians lost interest in the problems and applications that were and are the main engines of growth of this subject. I understand his criticism since I too was one of those mathematicians that was not interested in these applications. Now I know better.

In this chapter present these pillars of probability theory and prove their main properties. I have included a large number of detailed examples meant to convey the subtleties, depth, power and richness of these concepts. No abstract theorem can capture this richness.

I want to close with a personal anecdote that I find revealing. A few years ago, at a conference, I had a conversation with J. M. Bismut, a known probabilist whose mathematical interests were becoming more and more geometric. He noticed that I was in the middle of a mathematical transition in the opposite direction and asked me what prompted it. I explained my motivation, how I discovered that probability is not just a glorious part of measure theory and how much I struggled to truly understand the concept of conditional expectation, a concept eminently probabilistic. He smiled and said: "Probability theory is measure theory plus conditional expectation". I know it is an oversimplification, but it contains a lot of truth.

1.1 Measurable spaces

1.1.1 *Sigma-algebras*

Fix a nonempty set Ω.

Definition 1.1. (a) A collection $\mathcal{A}$ of subsets of Ω is called an *algebra* of Ω if it satisfies the following conditions

(i) $\emptyset, \Omega \in \mathcal{A}$.
(ii) $\forall A, B \in \mathcal{A},\ A \cup B \in \mathcal{A}$.
(iii) $\forall A \in \mathcal{A},\ A^c \in \mathcal{A}$.

(b) A collection $\mathcal{S}$ of subsets of Ω is called a *σ-algebra* (or *sigma-algebra*) of Ω if it is an algebra of Ω and the union of any countable subfamily of $\mathcal{S}$ is a set in $\mathcal{S}$, i.e.,

$$\forall (A_n)_{n \in \mathbb{N}} \in \mathcal{S}^{\mathbb{N}},\ \bigcup_{n \geq 1} A_n \in \mathcal{S}. \tag{1.1.1}$$

(c) A *measurable space* is a pair $(\Omega, \mathcal{S})$, where $\mathcal{S}$ is a sigma-algebra of subsets of Ω. The subsets $S \in \mathcal{S}$ are called ($\mathcal{S}$-)*measurable*. □

Remark 1.2. To prove that an algebra $\mathcal{S}$ is a σ-algebra is suffices to verify (1.1.1) *only for increasing* sequence of subsets $B_n \in \mathcal{S}$. Indeed, if $(A_n)_{n \in \mathbb{N}}$ is an arbitrary family in $\mathcal{S}$ the new family of sets in $\mathcal{S}$

$$B_n = \bigcup_{k=1}^{n} A_n,\ n \in \mathbb{N},$$

is increasing and its union coincides with the union of the family $(A_n)_{n \in \mathbb{N}}$. □

Example 1.3. (a) The collection 2^Ω of all subsets of Ω is obviously a σ-algebra.
(b) Suppose that $\mathcal{S}$ is a (σ-)algebra of a set Ω and $F : \widehat{\Omega} \to \Omega$ is a map. Then the preimage
$$F^{-1}(\mathcal{S}) = \{ F^{-1}(S);\ S \in \mathcal{S} \}$$
is a (σ-)algebra of subsets of $\widehat{\Omega}$. The σ-algebra $F^{-1}(\mathcal{S})$ is denoted by $\sigma(F)$ and it is called the *σ-algebra generated by F* or the *pullback of $\mathcal{S}$ via F*. We will often use the more suggestive notation
$$\{F \in S\} := F^{-1}(S) = \{\,\hat{\omega} \in \widehat{\Omega};\ F(\hat{\omega}) \in S\,\}.$$
(c) Given $A \in \Omega$ we denote by $\mathcal{S}_A$ the *σ-algebra generated by A*, i.e.,
$$\mathcal{S}_A = \{\,\emptyset, A, A^c, \Omega\,\}.$$
We will refer to it as the *Bernoulli algebra* with success A. Note that $\mathcal{S}_A$ is the pullback of $2^{\{0,1\}}$ via the indicator function $\boldsymbol{I}_A : \Omega \to \{0,1\}$.
(d) If $\mathcal{C} \subset 2^\Omega$ is a family of subsets of Ω, then we denote by $\sigma(\mathcal{C})$ the σ-algebra generated by $\mathcal{C}$, i.e., the intersection of all σ-algebras that contain $\mathcal{C}$. In particular, if $\mathcal{S}_1, \mathcal{S}_2$ are σ-algebras of Ω, then we set
$$\mathcal{S}_1 \vee \mathcal{S}_2 := \sigma(\mathcal{S}_1 \cup \mathcal{S}_2).$$
More generally, for any family $(\mathcal{S}_i)_{i \in I}$ of σ-algebras we set
$$\bigvee_{i \in I} \mathcal{S}_i := \sigma\left(\bigcup_{i \in I} \mathcal{S}_i \right).$$
(e) Suppose that we are given a countable partition $\{A_n\}_{n \in \mathbb{N}}$ of Ω
$$\Omega = \bigsqcup_{n \in \mathbb{N}} A_n.$$
The sets A_n are called the *chambers* of the partition. Then the σ-algebra generated by this partition is the σ-algebra consisting of all the subsets of Ω who are unions of chambers. This σ-algebra can be viewed as the σ-algebra generated by the map
$$X : \Omega \to \mathbb{N},\quad X = \sum_{n \in \mathbb{N}} n \boldsymbol{I}_{A_n},$$
so that $A_n = X^{-1}(\{n\})$.
(f) If $(\mathcal{S}_i)_{i \in I}$ is a family of (σ-)algebras of Ω, then their intersection
$$\bigcap_{i \in I} \mathcal{S}_i \subset 2^\Omega$$
is a (σ-)algebra of Ω.
(g) If $(\Omega_1, \mathcal{S}_1)$ and $(\Omega_2, \mathcal{S}_2)$ are two measurable spaces, then we denote by $\mathcal{S}_1 \otimes \mathcal{S}_2$ the sigma algebra of $\Omega_1 \times \Omega_2$ generated by the collection
$$\{S_1 \times S_2 :\ S_1 \in \mathcal{S}_1,\ S_2 \in \mathcal{S}_2\,\} \subset 2^{\Omega_1 \times \Omega_2}.$$

(h) If X is a topological space and $\mathcal{T} \subset 2^X$ denotes the family of open subset, then the *Borel σ-algebra* of X, denotes by $\mathcal{B}_X$, is the σ-algebra generated by $\mathcal{T}_X$. The sets in $\mathcal{B}_X$ are called the *Borel subsets of X*. Note that since any open set in $\mathbb{R}^n$ is a countable union of open cubes we have

$$\mathcal{B}_{\mathbb{R}^n} = \mathcal{B}_{\mathbb{R}}^{\otimes n}. \tag{1.1.2}$$

Any *finite dimensional* real vector space V can be equipped with a topology by choosing a linear isomorphism $L : V \to \mathbb{R}^{\dim V}$. This topology is independent of the choice of the isomorphism L. It can be alternatively identified as the smallest topology on V such that all the linear maps $V \to \mathbb{R}$ are continuous. We denote by $\mathcal{B}_V$ the sigma-algebra of Borel subsets determined by this topology.

We set $\bar{\mathbb{R}} = [-\infty, \infty]$. As a topological space it is homeomorphic to $[-1, 1]$. For simplicity we will refer to the Borel subsets of $\bar{\mathbb{R}}$ simply as *Borel sets*.

(i) If $(\Omega, \mathcal{S})$ is a measurable space and $X \subset \Omega$, then the collection

$$\mathcal{S}|_X := \{ S \cap X : S \in \mathcal{S} \} \subset 2^X$$

is a σ-algebra of X called *the trace of $\mathcal{S}$ on X*. $\square$

Remark 1.4. In measure theory and analysis, sigma-algebras lie in the background and rarely come to the forefront. In probability they play a more important role having to do with how they are perceived.

One should think of Ω as the collection of all the possible outcomes of a random experiment. A σ-algebra of Ω can be viewed as the totality of information we can collect using certain measurements about the outcomes $\omega \in \Omega$. Let us explain this vague statement on a simple example.

Suppose we are given a function $X : \Omega \to \mathbb{R}$ and the only thing that we can absolutely confirm about the outcome ω of an experiment is whether $X(\omega) \leq x$ for any given and $x \in \mathbb{R}$. In other words, we can detect by measurements the collection of sets

$$\{X \leq x\} := X^{-1}\big((-\infty, x] \big), \quad x \in \mathbb{R}.$$

In particular, we can detect whether $X(\omega) > x$, i.e., we can detect the sets $\{X > x\} = \{X \leq x\}^c$. More generally, we can determine the sets

$$\{a < X \leq b\} = \{X > a\} \cap \{X \leq b\}.$$

Indeed, we can do this using two experiments: on experiment to decide if $X \leq a$ and one to decide if $X \leq b$.

We say that a set S is X-measurable if given $\omega \in \Omega$ we can decide by doing countably many measurements on X whether $\omega \in S$. If $S_1, \ldots, S_n, \ldots \subset \Omega$ are known to be X-measurable, then their union is X-measurable. Indeed,

$$\omega \in \bigcup_{n \in \mathbb{N}} S_n \Longleftrightarrow \exists n \in \mathbb{N} : \omega \in S_n.$$

Let us observe that the set theoretic conditions imposed on a sigma-algebra have logical/linguistic counterparts. Thus, the statement

$$\omega \in \bigcap_{i \in I} S_i$$

translates into the formula $\forall i \in I,\ \omega \in S_i$, while the statement

$$\omega \in \bigcup_{i \in I} S_i$$

translates into the formula $\exists i \in I,\ \omega \in S_i$.

Conversely, statements involving the quantifiers $\exists, \forall$ can be translated into set theoretic statements.

The information we can collect by doing such measurements of the function X is collected into the sigma-algebra $\sigma(X) = X^{-1}(\mathcal{B}_{\mathbb{R}})$ generated by the map $X: \Omega \to (\mathbb{R}, \mathcal{B}_{\mathbb{R}})$. □

Definition 1.5. Let $\mathcal{C}$ be a collection of subsets of a set Ω. We say that $\mathcal{C}$ is a π-*system* if it is closed under finite intersections, i.e.,

$$\forall A, B \in \mathcal{C}:\ A \cap B \in \mathcal{C}.$$

The collection $\mathcal{C}$ is called a λ-*system* if it satisfies the following conditions.

(i) $\emptyset, \Omega \in \mathcal{C}$.
(ii) if $A, B \in \mathcal{C}$ and $A \subset B$, then $B \setminus A \in \mathcal{C}$.
(iii) If $A_1 \subset A_2 \subset \cdots$ belong to $\mathcal{C}$, then so does their union.

□

Note that a collection $\mathcal{C}$ is a σ-algebra if it is simultaneously a π and a λ-system. Since the intersection of any family of λ-systems is a λ-system we deduce that for any collection $\mathcal{C} \subset 2^{\Omega}$ there exists a smallest λ-system containing $\mathcal{C}$. We denote this system by $\Lambda(\mathcal{C})$ and we will refer to it as the λ-system generated by $\mathcal{C}$.

Example 1.6. Suppose that $\mathcal{H}$ is the collection of half-infinite intervals

$$(-\infty, x],\ x \in \mathbb{R}.$$

Then $\mathcal{H}$ is π-system of $\mathbb{R}$. The λ-system generated by $\mathcal{H}$ contains all the open intervals. Since any open subset of $\mathbb{R}$ is a countable union of open intervals we deduce that $\Lambda(\mathcal{P})$ coincides with the Borel σ-algebra $\mathcal{B}_{\mathbb{R}}$. □

Theorem 1.7 (Dynkin's $\pi - \lambda$ theorem). *Suppose that $\mathcal{P}$ is a π-system. Then*

$$\Lambda(\mathcal{P}) = \sigma(\mathcal{P}).$$

In other words, any λ-system that contains $\mathcal{P}$, also contains the σ-algebra generated by $\mathcal{P}$.

Proof. Since any σ-algebra is a λ-system we deduce $\Lambda(\mathcal{P}) \subset \sigma(\mathcal{P})$. Thus it suffices to show that

$$\sigma(\mathcal{P}) \subset \Lambda(\mathcal{P}). \tag{1.1.3}$$

Equivalently, it suffices to show that $\Lambda(\mathcal{P})$ is a σ-algebra. This happens if and only if the λ-system $\Lambda(\mathcal{P})$ is also a π-system. Hence it suffices to show that $\Lambda(\mathcal{P})$ is closed under (finite) intersections.

Fix $A \in \Lambda(\mathcal{P})$ and set

$$\mathcal{L}_A := \{ B \in 2^\Omega : A \cap B \in \Lambda(\mathcal{P}) \}.$$

It suffices to show that

$$\Lambda(\mathcal{P}) \subset \mathcal{L}_A, \ \ \forall A \in \Lambda(\mathcal{P}). \tag{1.1.4}$$

Observe that $\mathcal{L}_A$ is a λ-system. Indeed, $\Omega \in \mathcal{L}_A$ since $A \in \Lambda(\mathcal{P})$. The properties (ii) and (iii) in the definition of a λ-system are clearly satisfied since $\Lambda(\mathcal{P})$ is a λ-system. Thus, to prove (1.1.4), it suffices to show that

$$\mathcal{P} \subset \mathcal{L}_A, \ \ \forall A \in \Lambda(\mathcal{P}). \tag{1.1.5}$$

Note that since $\mathcal{P}$ is a π-system

$$\mathcal{P} \subset \mathcal{L}_A, \ \ \forall A \in \mathcal{P}.$$

In particular

$$\Lambda(\mathcal{P}) \subset \mathcal{L}_A, \ \ \forall A \in \mathcal{P}.$$

Thus, if $A \in \mathcal{P}$ and $B \in \Lambda(\mathcal{P})$, then $B \in \mathcal{L}_A$, i.e., $A \cap B \in \Lambda(\mathcal{P})$. Hence

$$\mathcal{P} \subset \mathcal{L}_B, \ \ \forall B \in \Lambda(\mathcal{P}).$$

This proves (1.1.5) and completes the proof of the $\pi - \lambda$-theorem. $\square$

1.1.2 Measurable maps

Definition 1.8. A map $F : \Omega_1 \to \Omega_2$ called *measurable* with respect to the σ-algebras $\mathcal{S}_i$ on Ω_i, $i = 1, 2$ or $(\mathcal{S}_1, \mathcal{S}_2)$-*measurable* if $F^{-1}(\mathcal{S}_2) \subset \mathcal{S}_1$, i.e.,

$$F^{-1}(S_2) \in \mathcal{S}_1, \ \ \forall S_2 \in \mathcal{S}_2.$$

Two measurable spaces $(\Omega_i, \mathcal{S}_i)$, $i = 1, 2$, are called *isomorphic* if there exists a bijection $F : \Omega_1 \longrightarrow \Omega_2$ such that $F^{-1}(\mathcal{S}_2) = \mathcal{S}_1$ or, equivalently, both F and its inverse F^{-1} are measurable. $\square$

Definition 1.9. Suppose that $(\Omega, \mathcal{S})$ is a measurable space. A function $f : \Omega \to \bar{\mathbb{R}}$ is called $\mathcal{S}$-*measurable* if, for any *Borel* subset $B \subset \bar{\mathbb{R}}$ we have $f^{-1}(B) \in \mathcal{S}$. $\square$

Example 1.10. (a) The composition of two measurable maps is a measurable map.
(b) A subset $S \subset \Omega$ is $\mathcal{S}$-measurable if and only if the indicator function $\boldsymbol{I}_S$ is a measurable function.
(c) If $\mathcal{A}$ is the σ-algebra generated by a finite or countable partition

$$\Omega = \bigsqcup_{i \in I} A_i, \ I \subset \mathbb{N},$$

then a function $f : \Omega \to (\mathbb{R}, \mathcal{B}_{\mathbb{R}})$ is $\mathcal{A}$-measurable if and only if it is constant in the chambers A_i of this partition. $\square$

Proposition 1.11. *Consider a map $F : (\Omega_1, \mathcal{S}_1) \to (\Omega_2, \mathcal{S}_2)$ between two measurable spaces. Suppose that $\mathcal{C}_2$ is a π-system of Ω_2 such that $\sigma(\mathcal{C}_2) = \mathcal{S}_2$. Then the following statements are equivalent.*

(i) The map F is measurable.
(ii) $F^{-1}(C) \in \mathcal{S}_1$, $\forall C \in \mathcal{C}_2$.

Proof. Clearly (i) $\Rightarrow$ (ii). The opposite implication follows from the $\pi - \lambda$ theorem since the set

$$\left\{ C \in \mathcal{S}_2; \ F^{-1}(C) \in \mathcal{S}_1 \right\}$$

is a λ-system containing the π-system $\mathcal{C}_2$ that generates $\mathcal{S}_2$. $\square$

Corollary 1.12. *If $F : X \to Y$ is a continuous map between topological spaces, then it is $(\mathcal{B}_X, \mathcal{B}_Y)$ measurable.*

Proof. Denote by $\mathcal{T}_Y$ the collection of open subsets of Y. Then $\mathcal{T}_Y$ is a π-system and, by definition, it generates $\mathcal{B}_Y$. Since F is continuous, for any $U \in \mathcal{T}_Y$ the set $F^{-1}(U)$ is open in X and thus belongs to $\mathcal{B}_X$. $\square$

Corollary 1.13. *Let $(\Omega, \mathcal{S})$ be a measurable space. A function $X : \Omega \to \mathbb{R}$ is $(\mathcal{S}, \mathcal{B}_{\mathbb{R}})$-measurable if and only if the sets $X^{-1}((-\infty, x])$ are $\mathcal{S}$-measurable for any $x \in \mathbb{R}$.*

Proof. It follows from the previous corollary by observing that the collection

$$\left\{ (-\infty, x]; \ x \in \mathbb{R} \right\} \subset 2^{\mathbb{R}}$$

is a π-system and the σ-algebra it generates is $\mathcal{B}_{\mathbb{R}}$. $\square$

Corollary 1.14. *Consider a pair of maps between measurable spaces*

$$F_i : (\Omega, \mathcal{S}) \to (\Omega_i, \mathcal{S}_i), \ i = 1, 2.$$

Then the following statements are equivalent.

(i) The maps F_i are measurable.

(ii) The map
$$F_1 \times F_2 : \Omega \to \Omega_1 \times \Omega_2, \quad \omega \mapsto \big(F_1(\omega), F_2(\omega)\big)$$
is $(\mathcal{S}, \mathcal{S}_1 \otimes \mathcal{S}_2)$-measurable.

Proof. (i) $\Rightarrow$ (ii) Observe that if the maps F_1, F_2 are measurable then
$$F_1^{-1}(S_1),\ F_2^{-1}(S_2) \in \mathcal{S}, \quad \forall S_1 \in \mathcal{S}_1,\ S_2 \in \mathcal{S}_2$$
$$\Rightarrow (F_1 \times F_2)^{-1}(S_1 \times S_2) = F_1^{-1}(S_1) \cap F_2^{-1}(S_2) \in \mathcal{S}, \quad \forall S_1 \in \mathcal{S}_1,\ S_2 \in \mathcal{S}_2.$$
Since the collection $S_1 \times S_2$, $S_i \in \mathcal{S}_i$, $i = 1, 2$, is a π-system that, by definition, generates $\mathcal{S}_1 \otimes \mathcal{S}_2$ we see that the last statement is equivalent with the measurability of $F_1 \times F_2$.

(ii) $\Rightarrow$ (i) For $i = 1, 2$ we denote by π the natural projection $\Omega_1 \times \Omega_2 \to \Omega_i$, $(\omega_1, \omega_2) \mapsto \omega_i$. The maps π_i are $(\mathcal{S}_1 \otimes \mathcal{S}_2, \mathcal{S}_i)$ measurable and $F_i = \pi_i \circ (F_1 \times F_2)$. □

Definition 1.15. For any measurable space $(\Omega, \mathcal{S})$ we denote by $\mathcal{L}^0(\mathcal{S}) = \mathcal{L}^0(\Omega, \mathcal{S})$ the space of $\mathcal{S}$-measurable random variables, i.e., $(\mathcal{S}, \mathcal{B}_{\bar{\mathbb{R}}})$-measurable functions $\Omega \to \bar{\mathbb{R}}$.

The subset of $\mathcal{L}^0(\Omega, \mathcal{S})$ consisting of nonnegative functions is denoted by $\mathcal{L}^0_+(\Omega, \mathcal{S})$, while the subspace of $\mathcal{L}^0(\Omega, \mathcal{S})$ consisting of bounded measurable functions is denoted $\mathcal{L}^\infty(\Omega, \mathcal{S})$. □

Remark 1.16. The algebraic operations on $\mathbb{R}$ admit (partial) extensions to $\bar{\mathbb{R}}$.
$$c + \pm\infty = \pm\infty, \infty + \infty = \infty, \ c \cdot \infty = \infty, \ \forall c > 0.$$
As we know, there are a few "illegal" operations
$$\infty - \infty, \ 0 \cdot \infty, \ \frac{0}{0} \ \text{etc.}$$
□

Proposition 1.17. *Fix a measurable space $(\Omega, \mathcal{S})$. Then the following hold.*

(i) For any $X, Y \in \mathcal{L}^0(\Omega, \mathcal{S})$ and any $c \in \mathbb{R}$ we have
$$X + Y,\ XY,\ cX \in \mathcal{L}^0(\Omega, \mathcal{S}),$$
whenever these functions are well defined.

(ii) If $(X_n)_{n \in \mathbb{N}}$ is a sequence in $\mathcal{L}^0(\Omega, \mathcal{S})$ such that, for any $\omega \in \Omega$ the limit
$$X_\infty(\omega) = \lim_{n \to \infty} X_n(\omega)$$
exists and is finite. Then $X_\infty : \Omega \to \bar{\mathbb{R}}$ is also $\mathcal{S}$-measurable.

(iii) If $(X_n)_{n \in \mathbb{N}}$ is a sequence in $\mathcal{L}^0(\Omega, \mathcal{S})$ such that, for any $\omega \in \Omega$ the quantities
$$Y_\infty(\omega) = \inf_{n \in \mathbb{N}} X_n(\omega), \quad Z_n(\omega) = \sup_{n \in \mathbb{N}} X_n(\omega)$$
are finite. Then $Y_\infty, Z_\infty \in \mathcal{L}^0(\Omega, \mathcal{S})$.

Proof. (i) Denote by $\mathcal{D}$ the subset of $\bar{\mathbb{R}}^2$ consisting of the pairs (x, y) for which $x + y$ is well defined. Observe that $X + Y$ is the composition of two measurable maps

$$\Omega \to \mathcal{D} \subset \bar{\mathbb{R}}^2, \quad \omega \mapsto \big(X(\omega), Y(\omega)\big), \quad \mathcal{D} \to \bar{\mathbb{R}}, \quad (x, y) \mapsto x + y.$$

Above, the first map is measurable according to Corollary 1.14 and the second map is Borel measurable since it is continuous. The measurability of XY and cX is established in a similar fashion.

(ii) We will show that for any $x \in \mathbb{R}$ the set $\{X_\infty(\omega) > x\}$ is $\mathcal{S}$-measurable. Note that

$$X_\infty(\omega) > x \iff \forall \nu \in \mathbb{N}, \ \exists N = N(\omega) \in \mathbb{N}: \ \forall n \geq N: \ X_n(\omega) > x + 1/\nu.$$

Equivalently

$$\big\{X_\infty(\omega) > x\big\} = \bigcap_{\nu \in \mathbb{N}} \bigcup_{N \in \mathbb{N}} \bigcap_{n \geq N} \big\{X_n > x + 1/\nu\big\} \in \mathcal{S}.$$

(iii) The proof is very similar to the proof of (ii) so we leave the details to the reader. □

Corollary 1.18. *For any function $f \in \mathcal{L}_+^0(\Omega, \mathcal{S})$, its positive and negative parts,*

$$f^+ := \max(f, 0), \quad f^- := \max(-f, 0)$$

belong to $\mathcal{L}_+^0(\Omega, \mathcal{S})$ as well.

Proof. The function f^+ is the composition of the continuous function $x^+ = \max(x, 0)$ with f. □

Definition 1.19. A function $f \in \mathcal{L}^0(\Omega, \mathcal{S})$ is called *elementary* or *step function* if its range is a *finite* subset of $\mathbb{R}$. We denote by $\mathrm{Elem}(\Omega, \mathcal{S})$ the set of elementary functions. □

More concretely, a function $f : \Omega \to \mathbb{R}$ is elementary if there exist *finitely many disjoint measurable sets* $A_1, \ldots, A_N \in \mathcal{S}$, and constants $c_1, \ldots, c_N \in \mathbb{R}$ such that

$$f(\omega) = \sum_{k=1}^N c_k \boldsymbol{I}_{A_k}(\omega), \ \forall \omega \in \Omega. \tag{1.1.6}$$

The decomposition (1.1.6) of an elementary function f is not unique. Among the various decompositions there is a canonical one

$$f = \sum_{r \in \mathbb{R}} r \boldsymbol{I}_{f^{-1}(r)}.$$

The above sum is finite since $f^{-1}(r)$ is empty for all but finitely many r's.

Let us also observe that $\mathrm{Elem}(\Omega, \mathcal{S})$ is a vector space. Indeed if f_0, f_1 are elementary functions with ranges R_0 and respectively R_1, then their sum is measurable and its range is contained in $R_0 + R_1$. This is a finite set since R_0, R_1 are finite. Clearly the multiplication of an elementary function by a scalar also produces an elementary function.

Let us observe that any nonnegative measurable function is the limit of an increasing sequence of elementary functions. For $n \in \mathbb{N}$ we define

$$D_n : [0, \infty) \to [0, \infty), \quad D_n(r) := \sum_{k=1}^{n2^n} \frac{k-1}{2^n} \boldsymbol{I}_{[(k-1)2^{-n}, k2^{-n})}(r).$$

Let us observe that if $r \in [0, n]$, then $D_n(r)$ truncates the binary expansion of r after n digits. E.g., if $r \in [0, 1)$ and

$$r = 0.\epsilon_1 \epsilon_2 \ldots \epsilon_n \ldots := \sum_{k=1}^{\infty} \frac{\epsilon_k}{2^k}, \quad \epsilon_k \in \{0, 1\},$$

then

$$D_n(r) = 0.\epsilon_1 \ldots \epsilon_n.$$

This shows that $(D_n)_{n \in \mathbb{N}}$ is a nondecreasing sequence of functions and

$$\lim_{n \to \infty} D_n(r) = r, \quad \forall r \geq 0.$$

For $f \in \mathcal{L}_+^0(\Omega, \mathcal{S})$ and $n \in \mathbb{N}$ we define $D_n[f] : (\Omega, \mathcal{S}) \to [0, \infty)$

$$D_n[f](\omega) := D_n\big(f(\omega)\big) = \sum_{k=1}^{n2^n} \frac{k-1}{2^n} \boldsymbol{I}_{[(k-1)2^{-n}, k2^{-n})}\big(f(\omega)\big) + n\boldsymbol{I}_{[n,\infty)}\big(f(\omega)\big). \tag{1.1.7}$$

We deduce that the sequence of nonnegative elementary functions $D_n[f]$ converges increasingly to f.

Definition 1.20. Let $(\Omega, \mathcal{S})$ be a measurable. A collection $\mathcal{M}$ of $\mathcal{S}$-measurable random variables is called a *monotone class* of $(\Omega, \mathcal{S})$ if it satisfies the following conditions.

(i) $\boldsymbol{I}_\Omega \in \mathcal{M}$.
(ii) If $f, g \in \mathcal{M}$ are bounded and $a, b \in \mathbb{R}$, then $af + bg \in \mathcal{M}$.
(iii) If (f_n) is an increasing sequence of nonnegative random variables in $\mathcal{M}$ with finite limit f_∞, then $f_\infty \in \mathcal{M}$.

□

Theorem 1.21 (Monotone Class Theorem). *Suppose that $\mathcal{M}$ is a monotone class of the measurable space $(\Omega, \mathcal{S})$ and $\mathcal{C}$ is a π-system that generates $\mathcal{S}$ and such that $\boldsymbol{I}_C \in \mathcal{M}, \forall C \in \mathcal{C}$. Then $\mathcal{M}$ contains $\mathcal{L}^\infty(\Omega, \mathcal{S})$ and all the nonnegative $\mathcal{S}$-measurable functions.*

Proof. Observe that the collection
$$\mathcal{A} := \{A \in \mathcal{S} : \boldsymbol{I}_A \in \mathcal{M}\}$$
is a λ-system so $\mathcal{A} = \mathcal{S}$, by the $\pi - \lambda$ theorem. Thus $\mathcal{M}$ contains all the elementary functions. Since any nonnegative measurable function is an increasing pointwise limit of elementary functions we deduce that $\mathcal{M}$ contains all the nonnegative measurable functions. Finally, if f is a bounded measurable function, then f^+, f^- are nonnegative and bounded measurable functions so $f^+, f^- \in \mathcal{M}$ and thus
$$f = f^+ - f^- \in \mathcal{M}.$$

$\square$

Definition 1.22. The σ-algebra generated by a collection $(X_i)_{i \in I}$ of real-valued functions on a set Ω is
$$\sigma(X_i, i \in I) := \bigvee_{i \in I} X_i^{-1}(\mathcal{B}_{\mathbb{R}}).$$
$\square$

The next result provides an interpretation of the concept of measurability along the lines of Remark 1.4.

Theorem 1.23 (Dynkin). *Suppose that $F : (\Omega, \mathcal{S}) \to (\Omega', \mathcal{S}')$ is a measurable map. Let $X : \Omega \to \mathbb{R}$ be an $\mathcal{S}$-measurable function. Then the following are equivalent.*

(i) The function X is $(\sigma(F), \mathcal{B}_{\mathbb{R}})$-measurable.
(ii) There exists an $(\mathcal{S}', \mathcal{B}_{\mathbb{R}})$-measurable function $X' : \Omega' \to \mathbb{R}$ such that $X = X' \circ F$.

Proof. Clearly, (ii) $\Rightarrow$ (i). To prove that (i) $\Rightarrow$ (ii) consider the family $\mathcal{M}$ of $\sigma(\mathcal{F})$-measurable functions of the form $X' \circ F$, $X' \in \mathcal{L}^0(\Omega', \mathcal{S}')$. We will prove that $\mathcal{M} = \mathcal{L}^0(\Omega, \sigma(\mathcal{F}))$. We will achieve using the monotone class theorem.
Step 1. $I_\Omega \in \mathcal{M}$.
Step 2. $\mathcal{M}$ is a vector space. Indeed if $X, Y \in \mathcal{M}$ and $a, b \in \mathbb{R}$, then there exist $\mathcal{S}'$-measurable functions X', Y' such that
$$X = X' \circ F, \ Y = Y' \circ F, \ aX + bY = (aX' + bY') \circ F.$$
Hence $aX + bY \in \mathcal{M}$.
Step 3. $\boldsymbol{I}_A \in \mathcal{M}$, $\forall A \in \sigma(F)$. Indeed, since $A \in \sigma(F)$ there exists $A' \in \mathcal{S}'$ such that
$$A = F^{-1}(A')$$
so $\boldsymbol{I}_A = \boldsymbol{I}_{A'} \circ F$. Hence $\mathcal{M}$ contains all the $\sigma(F)$-measurable elementary functions.

Step 4. Suppose now that $X \in \mathcal{L}^0\big(\Omega, \sigma(F)\big)$ is nonnegative. Then there exists an increasing sequence $(X_n)_{n\in\mathbb{N}}$ of $\sigma(F)$-measurable nonnegative elementary functions that converges pointwise to X. For every $n \in \mathbb{N}$ there exists an $\mathcal{S}$-measurable elementary function $X'_n : \Omega' \to \mathbb{R}$ such that
$$X_n(\omega) = X'_n\big(F(\omega)\big), \quad \forall \omega \in \Omega.$$
Define
$$\Omega'_0 := \big\{\, \omega' \in \Omega';\ \text{the limit } \lim_{n\to\infty} X'_n(\omega') \text{ exists and it is finite}\,\big\}.$$
Let us observe that Ω'_0 is $\mathcal{S}'$-measurable because
$$\omega' \in \Omega'_0 \Longleftrightarrow \forall \nu \geq 1,\ \exists N \geq 1,\ \forall m, n \geq N:\ |X'_n(\omega') - X'_m(\omega')| < 1/\nu,$$
i.e.,
$$\Omega'_0 = \bigcap_{\nu \in \mathbb{N}} \bigcup_{N \geq 1} \bigcap_{m,n > N} \big\{\,|X'_n(\omega') - X'_m(\omega')| < 1/\nu\,\big\}.$$
Clearly, $F(\Omega) \subset \Omega'_0$. For any $\omega' \in \Omega'$ we set
$$X'_\infty(\omega') := \begin{cases} \lim_{n\to\infty} X'_n(\omega'), & \omega' \in \Omega'_0, \\ 0, & \omega' \in \Omega' \setminus \Omega'_0. \end{cases}$$
Arguing as in the proof of Proposition 1.17(ii) we deduce that X'_∞ is $\mathcal{S}'$-measurable. For any $\omega \in \Omega$ the sequence $X'_n\big(F(\omega)\big) = X_n(\omega)$ is increasing and the limit
$$\lim_{n\to\infty} X'_n\big(F(\omega)\big)$$
exists and it is finite. Hence
$$X'_\infty\big(F(\omega)\big) = X(\omega), \quad \forall \omega \in \Omega.$$
This proves that $\mathcal{M}$ is a monotone class in $\mathcal{L}^0\big(\Omega, \sigma(F)\big)$ that is also a vector space so it coincides with $\mathcal{L}^0\big(\Omega, \sigma(F)\big)$. $\square$

Corollary 1.24. *Suppose that $X_1, \ldots, X_n : (\Omega, \mathcal{S}) \to \mathbb{R}$ are $\mathcal{S}$-measurable random variables. The the function $X : \Omega \to \mathbb{R}$ is $\sigma(X_1, \ldots, X_n)$-measurable if and only if there exists an $\mathcal{B}_{\mathbb{R}^n}$-measurable function $u : \mathbb{R}^n \to \mathbb{R}$ such that*
$$X = u(X_1, \ldots, X_n).$$

Proof. Apply the above theorem with $(\Omega', \mathcal{S}') = (\mathbb{R}^n, \mathcal{B}_{\mathbb{R}^n})$ and
$$F(\omega) = (X_1(\omega), \ldots, X_n(\omega)).$$
$\square$

Remark 1.25. We see that, in its simplest form, Corollary 1.24 describes a measure theoretic form of functional dependence. Thus, if in a given experiment we can measure the quantities $X_1, \ldots, X_n$ and we know that the information $X \leq c$ can be decided only by measuring the quantities $X_1, \ldots, X_n$, then X is in fact a (measurable) function of $X_1, \ldots, X_n$. In plain English this sounds tautological. In particular, this justifies the choice of term "measurable". $\square$

1.2 Measures and integration

1.2.1 *Measures*

Throughout this section $(\Omega, \mathcal{S})$ will denote a measurable space. Given a function $f : X \to \mathbb{R}$ we will use the notation $\{f \leq c\}$ to denote the subset $f^{-1}((-\infty, c])$. The sets $\{a \leq f \leq b\}$ etc. are defined in a similar fashion.

Definition 1.26. A *measure* on $(\Omega, \mathcal{S})$ is a function $\mu : \mathcal{S} \to [0, \infty]$, $S \mapsto \mu[S]$ such that the following hold.

- $\mu[\emptyset] = 0$, and
- it is *σ-additive*, i.e., for any sequence of pairwise disjoint $\mathcal{S}$-measurable sets $(A_n)_{n \in \mathbb{N}}$ we have

$$\mu\left[\bigcup_{n \in \mathbb{N}} A_n\right] = \sum_{n \geq 1} \mu[A_n]. \qquad (1.2.1)$$

The measure is called *σ-finite* if there exists an increasing sequence of $\mathcal{S}$-measurable sets

$$A_1 \subset A_2 \subset \cdots$$

such that

$$\bigcup_{n \in \mathbb{N}} A_n = \Omega \text{ and } \mu[A_n] < \infty, \ \forall n \in \mathbb{N}.$$

The measure is called *finite* if $\mu[\Omega] < \infty$. A *probability measure* is a measure $\mathbb{P}$ such that $\mathbb{P}[\Omega] = 1$. We will denote by $\mathrm{Prob}(\Omega, \mathcal{S})$ the set of probability measures on $(\Omega, \mathcal{S})$. $\square$

Remark 1.27. The σ-additivity condition (1.2.1) is equivalent to a pair of conditions that are more convenient to verify in concrete situations.

(i) μ is *finitely additive*, i.e., for any finite collection of $\mathcal{S}$-measurable sets $A_1, \ldots, A_n$ we have

$$\mu\left[\bigcup_{k=1}^{n} A_k\right] = \sum_{k=1}^{n} \mu[A_k].$$

(ii) μ is *increasingly continuous* i.e., for any increasing sequence of $\mathcal{S}$-measurable sets $A_1 \subset A_2 \subset \cdots$

$$\mu\left[\bigcup_{n \in \mathbb{N}} A_n\right] = \lim_{n \to \infty} \mu[A_n]. \qquad (1.2.2)$$

If $\mu[\Omega] < \infty$ and μ is finitely additive, then the increasing continuity condition (ii) is equivalent with the *decreasing continuity* condition, i.e., for any decreasing sequence of $\mathcal{S}$-measurable sets $B_1 \supset B_2 \supset \cdots$

$$\mu\left[\bigcap_{n \in \mathbf{n}} B_n\right] = \lim_{n \to \infty} \mu[B_n]. \qquad (1.2.3)$$

Indeed, the sequence $B_n^c = \Omega \setminus B_n$ is increasing and $\mu[B_n^c] = \mu[\Omega] - \mu[B_n]$. This last equality could be meaningless if $\mu[\Omega] = \infty$. □

Definition 1.28. (a) A *measured space* is a triplet $(\Omega, \mathcal{S}, \mu)$, where $(\Omega, \mathcal{S})$ is a measurable space and $\mu : \mathcal{S} \to [0, \infty]$ is a measure. □

Our next result shows that a *finite* measure is uniquely determined by its restriction to an algebra generating the sigma-algebra where it is defined.

Proposition 1.29. *Consider a measurable space* $(\Omega, \mathcal{S})$ *and two* finite *measures* $\mu_1, \mu_2 : \mathcal{S} \to [0, \infty]$ *such that* $\mu_1[\Omega] = \mu_2[\Omega] < \infty$, *then the collection*

$$\mathcal{E} := \{ S \in \mathcal{S}; \ \mu_1[S] = \mu_2[S] \}$$

is a λ-system. In particular, if $\mu_1[C] = \mu_2[C]$ for any set C that belongs to a π-system $\mathcal{C}$, then μ_1 and μ_2 coincide on the σ-algebra generated by $\mathcal{C}$.

Proof. Clearly $\emptyset, \Omega \in \mathcal{E}$. If $A, B \in \mathcal{E}$ and $A \subset B$, then

$$\mu_1[A] = \mu_2[A] < \infty, \ \mu_1[B] = \mu_2[B] < \infty$$

so

$$\mu_1[B \setminus A] = \mu_1[B] - \mu_1[A] = \mu_2[B] - \mu_2[A] = \mu_2[B \setminus A],$$

so $B \setminus A \in \mathcal{C}$. The condition (iii) in the Definition 1.5 of a λ-system follows from the σ-additivity of the measures μ_1, μ_2. □

Definition 1.30. A *probability space*, or *sample space*, is a measured space $(\Omega, \mathcal{S}, \mathbb{P})$, where $\mathbb{P}$ is a probability measure. In this case we use the following terminology.

- The subsets $S \in \mathcal{S}$ care called the *events* of the sample space.
- An event $S \in \mathcal{S}$ is called *almost sure* (or a.s.) if $\mathbb{P}[S] = 1$. An event S is called *improbable* if $\mathbb{P}[S] = 0$.
- The measurable functions $X : (\Omega, \mathcal{S}, \mathbb{P}) \to \overline{\mathbb{R}}$ are called *random variables*.
- A random variable $X : (\Omega, \mathcal{S}, \mathbb{P}) \to \overline{\mathbb{R}}$ is called a.s. *finite* if

$$\mathbb{P}[|X| < \infty] = 1.$$

- A random variable on $(\Omega, \mathcal{S}, \mathbb{P})$ is called *deterministic* if there exists $c \in \mathbb{R}$ such that $X = c$ a.s.

□

✎ *Traditionally the random variables have capitalized names X, Y, Z etc. to distinguish them from deterministic quantities that are indicated in small caps. We will try to adhere to this convention throughout this book.*

Example 1.31. (a) If $(\Omega, \mathcal{S})$ is a measurable space, then for any $\omega_0 \in \Omega$, the *Dirac measure* concentrated at ω_0 is the probability measure

$$\delta_{\omega_0} : \mathcal{S} \to [0, \infty), \quad \delta_{\omega_0}[S] = \begin{cases} 1, & \omega_0 \in S, \\ 0, & \omega_0 \notin S. \end{cases}$$

(b) Suppose that S is a finite or countable set. A measure on $(S, 2^S)$ is uniquely determined by the function

$$w : S \to [0, \infty], \quad w(s) = \mu[\{s\}].$$

We say that $\mu[\{s\}]$ is the mass of s with respect to μ. The function w is referred to as the *weight function* of the measure. Often, for simplicity we will write

$$\mu[s] := \mu[\{s\}].$$

The associated measure μ_w is a probability measure if

$$\sum_{s \in S} w(s) = 1.$$

When S is finite and

$$w(s) = \frac{1}{|S|}, \quad \forall s \in S,$$

then the associated probability measure μ_w is called the *uniform probability measure* on the finite set S.

(c) Suppose that $F : (\Omega, \mathcal{S}) \to (\Omega', \mathcal{S}')$ is a measurable map between measurable spaces. Then any measure μ on Ω induces a measure $F_\# \mu$ on Ω' according to the rule

$$F_\# \mu[S'] := \mu[F^{-1}(S')].$$

The measure $F_\# \mu$ is called the *pushforward of μ via F*.

(d) Fix a set T with two elements, $T = \{0, 1\}$. For any $p \in (0, 1)$ the probability measure $\beta_p : 2^T \to [0, \infty)$ defined by

$$\beta_p[1] = p, \quad \beta_p[0] = q := 1 - p$$

is called the *Bernoulli distribution* with success probability p. We abbreviate it by Ber(p).

(e) Given finite or countable sets $\Omega_1, \ldots, \Omega_n$, and probability measures $\mu_i : 2^{\Omega_i} \to [0, 1]$, we obtain a probability measure

$$\mu := \mu_1 \otimes \cdots \otimes \mu_n : 2^{\Omega_1 \times \cdots \times \Omega_n} \to [0, 1]$$

by setting
$$\mu\bigl[(\omega_1,\ldots,\omega_n)\bigr] = \mu_1[\omega_1]\cdots\mu_n[\omega_n], \quad \forall(\omega_1,\ldots,\omega_n)\in\Omega_1\times\cdots\times\Omega_n.$$
In particular, there exists a probability measure $\beta_p^{\otimes n}$ on $\{0,1\}^n$.

Note that we have a random variable
$$N:\{0,1\}^n\to\mathbb{N}_0,\quad N\bigl((\epsilon_1,\ldots,\epsilon_n)\bigr) = \epsilon_1+\cdots+\epsilon_n, \quad \forall \epsilon_1,\ldots,\epsilon_n\in\{0,1\}.$$
The push-forward $\mathbb{P} = \mathbb{P}_{n,p} := N_\#\beta_p^{\otimes n}$ is a probability measure on $\{0,1,\ldots,n\}$ called the *binomial distribution* corresponding to n independent trials with success probability p and failure probability $q = 1-p$. It is abbreviated $\mathrm{Bin}(n,p)$. Note that $\mathrm{Bin}(1,p) = \mathrm{Ber}(p)$. For any $k\in\{0,1,\ldots,n\}$ we have
$$\mathbb{P} = \sum_{k=0}^n \mathbb{P}[k]\delta_k,$$
where
$$\mathbb{P}[k] = \beta_p^{\otimes n}[N=k] = \sum_{\epsilon_1+\cdots+\epsilon_n=k}\beta_p^{\otimes n}\bigl[(\epsilon_1,\ldots,\epsilon_n)\bigr]$$
$$= \sum_{\epsilon_1+\cdots+\epsilon_n=k} p^k q^{n-k} = \binom{n}{k} p^k q^{n-k}.$$

(f) The Lebesgue measure $\boldsymbol{\lambda}$ defines a measure on $\mathcal{B}_\mathbb{R}$. For any compact interval $[a,b]$ the *uniform probability measure* on $[a,b]$ is
$$\frac{1}{b-a}\boldsymbol{I}_{[a,b]}\boldsymbol{\lambda}.\qquad\square$$

Definition 1.32. Let X be a topological space. As usual $\mathcal{B}_X$ denotes the σ-algebra of Borel subsets of X. A measure on X is called *Borel* if it is defined on $\mathcal{B}_X$. $\quad\square$

The Lebesgue measure on $\mathbb{R}$ is a Borel measure.

Definition 1.33. Suppose that $X\in\mathcal{L}^0(\Omega,\mathcal{S},\mathbb{P})$. Its distribution is the Borel probability measure $\mathbb{P}_X$ on $\bar{\mathbb{R}}$ defined by
$$\mathbb{P}_X[B] = \mathbb{P}[X\in B], \quad \forall B\in\mathcal{B}_{\bar{\mathbb{R}}}.$$
In other words, $\mathbb{P}_X$ is the pushforward of $\mathbb{P}$ by X, $\mathbb{P}_X = X_\#\mathbb{P}$. $\quad\square$

Definition 1.34. Suppose that μ is a measure on the measurable space $(\Omega,\mathcal{S})$.

(i) A set $N\subset\Omega$ is called μ-*negligible* if there exists a set $S\in\mathcal{S}$ such that
$$N\subset S \text{ and } \mu[S] = 0.$$
We denote by $\mathcal{N}_\mu$ the collection of μ-negligible sets.

(ii) The σ-algebra $\mathcal{S}$ is said to be *complete* with respect to μ (or μ-*complete*) if it contains all the μ-negligible subsets.

(iii) The μ-*completion* of $\mathcal{S}$ is the σ-algebra $\mathcal{S}^\mu := \sigma(\mathcal{S},\mathcal{N}_\mu)$. $\quad\square$

Remark 1.35. (a) It may be helpful to think of a sample space $(\Omega, \mathcal{S}, \mathbb{P})$ as the collection of all possible outcomes ω of an experiment with unpredictable results. The observer may not be able to distinguish through measurements all the possible outcomes, but she is able to distinguish some features or properties of various outcomes. An event can be understood as the collection of the all outcomes having an observable or measurable property. The probability $\mathbb{P}$ associates a likelihood of a certain property to be observed at the end of such a random experiment.

Take for example the experiment of flipping n times a coin with 0/1 faces. One natural sample space for this experiment is based on the set $\Omega = \{0, 1\}^n$.

If we assume that the coin is fair, then it is natural to conclude that each outcome $\omega \in \Omega$ is equally likely. Suppose that we can distinguish all the outcomes. In this case

$$\mathcal{S} = 2^\Omega.$$

Since there are 2^n outcomes that are equally likely to occur we obtain a probability measure $\mathbb{P}$ given by

$$\mathbb{P}[S] = \frac{|S|}{2^n}, \quad \forall S \in \mathcal{S}.$$

A random variable on a sample space is a numerical attribute X that we can assign to each outcome ω of a random experiment with the following feature: for any $c \in \mathbb{R}$ the property $X(\omega) \leq c$ is observable, i.e., the set $X^{-1}\big((-\infty, c]\big)$ belongs to the collection $\mathcal{S}$ of observable properties. For example, in the situation of n fair coin tosses, the number N of 1's observed at the end of n tosses is a random variable.

(b) Often one speaks of sampling a probability distribution on $\mathbb{R}$. Modern computer systems can sample many distributions. More concretely, we say that a probability measure μ on $(\mathbb{R}, \mathcal{B}_\mathbb{R})$ can be sampled by a computer system if that computer can produce a random[1] experiment whose outcome is a random number X so that, when we run the experiment a large number of times n, it generates numbers $x_1, \ldots, x_n$ and, for any $c \in \mathbb{R}$, the fraction of these numbers that is $\leq c$ is very close to $\mu\big[(-\infty, c]\big]$.

When we speak of sampling a random variable X, we really mean sampling its probability distribution $\mathbb{P}_X$. $\square$

Clearly $\mathcal{S}^\mu$ is the smallest μ-complete σ-algebra containing $\mathcal{S}$. The proof of the following result can be safely left to the reader.

Proposition 1.36. *Suppose that μ is a measure on the σ-algebra $\mathcal{S} \subset 2^\Omega$.*

---

[1] The precise term is pseudo-random since one cannot really simulate randomness.

(i) The completion $\mathcal{S}^\mu$ has the alternate description
$$\mathcal{S}^\mu = \{ S \cup N;\ S \in \mathcal{S},\ N \in \mathcal{N}_\mu \} \subset 2^\Omega.$$

(ii) The measure μ admits a unique extension to a probability measure $\bar{\mu} : \mathcal{S}^\mu \to [0, \infty)$. More precisely
$$\forall S \in \mathcal{S},\ N \in \mathcal{N}_\mu\ \ \bar{\mu}[S \cup N] = \mu[S].$$

□

Definition 1.37. A set $S \subset \mathbb{R}$ is called *Lebesgue measurable* if it belongs to the λ-completion of $\mathcal{B}_\mathbb{R}$. □

The most versatile method of constructing measures is *Carathéodory Extension Theorem*. We need to introduce the appropriate concepts.

Definition 1.38. Fix a set Ω and an algebra $\mathcal{F} \subset 2^\Omega$.

(i) A function $\mu : \mathcal{F} \to [0, \infty]$ is called a *premeasure* if it satisfies the following conditions.

(a) $\mu[\emptyset] = 0$
(b) μ is *finitely additive*, i.e., for any finite collection of disjoint sets $A_1, \ldots, A_n \in \mathcal{F}$ we have
$$\mu\left[\bigcup_{k=1}^n A_k\right] = \sum_{k=1}^n \mu[A_k].$$
(c) μ is *countably additive*, i.e., for any sequence $(A_n)_{n \in \mathbb{N}}$ of disjoint sets in $\mathcal{F}$ whose union is a set $A \in \mathcal{F}$ we have
$$\mu[A] = \sum_{n \geq 1} \mu[A_n].$$

(ii) The premeasure μ is called *σ-finite* if there exists a sequence of sets $(\Omega_n)_{n \in \mathbb{N}}$ in $\mathcal{F}$ such that
$$\Omega = \bigcup_{n \in \mathbb{N}} \Omega_n,\ \mu[\Omega_n] < \infty,\ \forall n \in \mathbb{N}.$$

For a proof of the next central result we refer to [4, Sec. 1.3], [50, Chap. 3] or [92, Thm. 1.53, 1.65].

Theorem 1.39 (Carathéodory Extension Theorem). *Suppose that $\mathcal{F}$ is an algebra of subsets of Ω and $\mu : \mathcal{F} \to [0, \infty]$ is a σ-finite premeasure on $\mathcal{F}$. Then the following hold.*

(i) *The premeasure μ admits a unique extension to a measure $\tilde{\mu} : \sigma(\mathcal{F}) \to [0, \infty]$.*

(ii) For any $A \in \sigma(\mathcal{F})$ and any $\varepsilon > 0$ there exist mutually disjoint sets $A_1, \ldots, A_m \in \mathcal{F}$ and $B_1, \ldots, B_n \in \mathcal{F}$ such that

$$A \subset \bigcup_{j=1}^{m} A_j, \quad \tilde{\mu}\left[\bigcup_{j=1}^{m} A_j \setminus A\right] < \varepsilon,$$

and

$$\tilde{\mu}\left[A \triangle \bigcup_{k=1}^{n} B_k\right] < \varepsilon.$$

□

Example 1.40. Let $\mathcal{F}$ denote the collection of subsets of $\mathbb{R}$ that are union of intervals of the type $(a, b]$, $-\infty \leq a < b < \infty$. This is an algebra of sets. Any F can be written in a (non)unique way as a union

$$F = \bigcup_{j=1}^{n} (a_i, b_i], \quad a_i < b_i \leq a_{i+1} < b_{i+1}, \quad \forall i = 1, \ldots, n-1.$$

While this decomposition is not unique the sum

$$\boldsymbol{\lambda}[F] = \sum_{i=1}^{n} (b_i - a_i)$$

depends only on F and not on the decomposition. It is not very hard to show that the correspondence

$$\mathcal{F} \ni F \mapsto \boldsymbol{\lambda}[F] \in [0, \infty]$$

is finitely additive. The fact that $\boldsymbol{\lambda}$ is a premeasure, i.e., it is (conditionally) sigma-additive, is much more subtle, and it is ultimately rooted in the compactness of the closed and bounded intervals of $\mathbb{R}$. For details we refer to [4, Sec. 1.4] or [50, Chap. 3]. The resulting measure on $\mathcal{B}_{\mathbb{R}}$ is called the *Lebesgue measure* on $\mathbb{R}$ and we continue to denote it by $\boldsymbol{\lambda}$. □

Definition 1.41. A set $S \subset \mathbb{R}$ is called *Lebesgue measurable* if it belongs to the $\boldsymbol{\lambda}$-completion of $\mathcal{B}_{\mathbb{R}}$. □

Definition 1.42. A *distribution function* is a right-continuous nondecreasing function

$$F : \bar{\mathbb{R}} \to [0, 1]$$

such that $F(-\infty) = 0$ and $F(\infty) = 1$. □

Example 1.43. Suppose that X is a random variable defined on the probability space $(\Omega, \mathcal{S}, \mathbb{P})$. The function

$$F_X : \mathbb{R} \to [0, 1], \quad F_X(x) = \mathbb{P}[X \leq x]$$

is a distribution function called the *cumulative distribution function* or *cdf* of the random variable X. □

Example 1.44 (Lebesgue-Stieltjes measures). Suppose that $F : \overline{\mathbb{R}} \to [0,1]$ is a *distribution function*. Then there exists a unique Borel probability measure $\mu = \mu_F$ on $\mathcal{B}_\mathbb{R}$ such that

$$\mu\big[(x,y]\big] = F(y) - F(x), \quad \forall x \leq y \in \mathbb{R}. \tag{1.2.4}$$

To the uniqueness follows from the fact the collection of intervals $(-\infty, x]$ is a π system generates the Borel algebra of $\mathbb{R}$. The existence follows from Caratheodory's extension theorem; see [4, Sec. 1.4] or [50, Chap. 3]. Below we will describe another existence proof that relies only the existence of the usual Lebesgue measure.

The above measure μ_F is called the *Stieltjes probability measure* associated to the distribution function F. Its extension to the completion $\mathcal{B}_\mathbb{R}^\mu$ is called the Lebesgue-Stieltjes measure associated to the distribution function F.

Conversely, if μ is a Borel probability, measure on $\mathbb{R}$ then μ is the Stieltjes measure associated to its *cumulative distribution function* (cdf) $F : \overline{\mathbb{R}} \to [0,1]$, $F(x) = \mu\big[(-\infty, x]\big]$. □

Example 1.45 (Quantiles). Here is an alternate description of this measure based on a construction frequently used in statistics. Suppose that $F : \overline{\mathbb{R}} \to [0,1]$ is a distribution function. The *quantile function* of F is a generalized inverse of the nondecreasing function F. Define

$$\begin{aligned} Q : [0,1] \to \overline{\mathbb{R}}, \quad Q(p) &:= \inf\{x : p \leq F(x)\} \\ &= \inf F^{-1}\big([p,1]\big). \end{aligned} \tag{1.2.5}$$

Since F is right-continuous the above definition is equivalent to

$$F^{-1}\big([p,1]\big) = [Q(p), \infty].$$

Suppose that x_0 is a point of discontinuity of F and we set

$$p_0^- := \lim_{x \nearrow x_0} F(x) < F(x_0) =: p_0.$$

Note that $Q(p_0) = x_0$ and if $p \in (p_0^-, p_0]$, then $Q(p) = x_0$.

Note that for any $x \in \mathbb{R}$ we have

$$0 \leq y \leq F(x) \iff Q(y) \leq x, \tag{1.2.6}$$

$$Q^{-1}\big([-\infty, x]\big) = [0, F(x)]. \tag{1.2.7}$$

Indeed, $\ell \in Q^{-1}\big([-\infty, x]\big)$ if and only if $Q(\ell) \leq x$, i.e., $\ell \leq F(x)$. In particular,

$$Q^{-1}\big((x,y]\big) = \big(F(x), F(y)\big], \quad \forall -\infty \leq x \leq y \leq \infty.$$

The quantile is *left* continuous. Indeed, let $p_n \nearrow p_0$. We will show that

$$\lim_n Q(p_n) = Q(p_0).$$

Note that $\lim_n Q(p_n) \leq Q(p_0)$ since Q is nondecreasing. To prove that we have equality we argue by contradiction. Set $x_n := Q(p_n)$, $x_0 = Q(p_0)$. Suppose

$$\lim_n x_n = x_\infty < x_0 = \inf\{x; \ F(x) \geq p_0\}.$$

From the definition of inf as the *greatest* lower bound we deduce that there exists $x_* \in (x_\infty, x_0)$ such that $F(x_*) < p_0$. Thus $F(x_n) \leq F(x_*)$ Since $p_n \nearrow p_0$ we deduce $p_n > F(x^*)$ for all n sufficiently large. This implies

$$x^* \notin \{ x;\ F(x) \geq p_n \} = [Q(p_n), \infty)$$

i.e., $x_n = Q(p_n) > x_*$, for all n sufficiently large. This contradicts the fact that $x_n \to x_\infty < x_*$.

If $\boldsymbol{\lambda}_{[0,1]}$ denotes the Lebesgue measure[2] on $[0,1]$, then

$$Q_\# \boldsymbol{\lambda}_{[0,1]} [\, (x,y) \,] = \boldsymbol{\lambda} [\, Q^{-1}(\, (x,y) \,) \,] = F(y) - F(x).$$

Hence the pushforward measure $Q_\# \boldsymbol{\lambda}_{[0,1]}$ satisfies (1.2.4) since it coincides with μ_F on the π-system consisting of the intervals of the form $(a, b]$ it coincides with μ_F on the sigma-algebra of Borel sets.

When F is the cumulative distribution function of a random variable, the associated quantile function is called the quantile of the random variable X. $\square$

1.2.2 Independence and conditional probability

The next concepts are purely probabilistic in nature. They have no natural counterpart in the traditional measure theory.

Definition 1.46. (a) The events $A_1, A_2, \ldots, A_n$ of a sample space $(\Omega, \mathcal{S}, \mathbb{P})$ are called *independent* if, for any nonempty subset $\{i_1, \ldots, i_k\} \subset \{1, \ldots, n\}$, we have

$$\mathbb{P}[\, A_{i_1} \cap \cdots \cap A_{i_k} \,] = \mathbb{P}[A_{i_1}] \cdots \mathbb{P}[A_{i_k}].$$

(b) The families of events $\mathcal{A}_1, \ldots, \mathcal{A}_n \subset \mathcal{S}$ are called *independent* if for any $A_i \in \mathcal{A}_i$, $i = 1, \ldots, n$, the events $A_1, \ldots, A_n$ are independent.

(c) The (possibly infinite) collection of families of events $(\mathcal{A}_i)_{i \in I}$ is called *independent* if for any $i_1, \ldots, i_n \in I$ the finite collection $\mathcal{A}_{i_1}, \ldots, \mathcal{A}_{i_n}$ is independent.

(d) An *independency* is an independent collection $(\mathcal{S}_i)_{i \in I}$ of sigma-subalgebras of $\mathcal{S}$.

(d) The collection of random variables $X_i \in \mathcal{L}^0(\Omega, \mathcal{S})$, $i \in I$, is called *independent* if the collection of σ-algebras $\big(\sigma(X_i) \big)_{i \in I}$ is independent. $\square$

☞ *We will use the notation $X \perp\!\!\!\perp Y$ to indicate that the random variables X, Y are independent.*

Remark 1.47. (a) We want to emphasize that the independence condition is sensitive to the choice of probability measure involved in this definition.

---
[2] The proof of the existence of the Lebesgue measure is based on Caratheodory's extension theorem.

(b) It is possible that $n+1$ events be dependent although any n of them are independent. Here is one such instance, [146, Ex. 3.5]. Suppose we flip a fair coin n times. In this case a natural sample space is

$$\Omega = 2^{\mathbb{I}_n} = \{0,1\}^n,$$

with the uniform probability measure. (Above 1 = Heads.) For $k = 1, \ldots, n$ we denote by k the event "Heads at the k-th flip", i.e.,

$$E_k = \{ \omega = (\omega_1, \ldots, \omega_n) \in \Omega;\ \omega_k = 1 \}.$$

Denote by E_0 the event "the number of heads in these n flips is even", i.e.,

$$E_0 = \{ \omega \in \Omega;\ \omega_1 + \cdots + \omega_n \in 2\mathbb{Z} \}.$$

Clearly

$$\mathbb{P}[E_k] = \frac{1}{2},\ \forall k = 1, \ldots, n.$$

Since the probability of flipping an even number of Heads is equal to the probability of flipping an odd number of Heads, we deduce that

$$\mathbb{P}[E_0] = \frac{1}{2}.$$

For any subset $I \subset \{0, 1, \ldots, n\}$ we set

$$E_I := \bigcap_{i \in I} E_i.$$

The events $E_1, \ldots, E_n$ are independent. Observe that for any subset $I \subset \mathbb{I}_n$, $|I| = k < n$, we have

$$\mathbb{P}[E_0 \cap E_I] = \mathbb{P}\left[\left\{ \omega \in \Omega;\ \omega_i = 1\ \forall i \in I,\ \sum_{j \notin I} \omega_j \equiv |I| \bmod 2 \right\}\right]$$

$$= \underbrace{\mathbb{P}\big[\{\omega \in \Omega;\ \omega_i = 1\ \forall i \in I\}\big]}_{\frac{1}{2^k}} \cdot \underbrace{\mathbb{P}\left[\left\{ \omega \in \Omega;\ \sum_{j \notin I} \omega_j \equiv |I| \bmod 2 \right\}\right]}_{\frac{1}{2}}$$

$$= \frac{1}{2^{k+1}} = \mathbb{P}[E_0] \cdot \prod_{i \in I} \mathbb{P}[E_i].$$

Thus, any n of the events $E_0, E_1, \ldots, E_n$ are independent. Finally, note that

$$\prod_{i=0}^{n} \mathbb{P}[E_i] = \frac{1}{2^{n+1}} \text{ and } \mathbb{P}[E_0 \cap E_1 \cap \cdots \cap E_n] = \begin{cases} 0, & n \text{ odd}, \\ \frac{1}{2^n}, & n \text{ even}. \end{cases}$$

This shows the events $E_0, E_1, \ldots, E_n$ are dependent.

(c) If Ω is contained in each of the families of events $\mathcal{A}_1, \ldots, \mathcal{A}_n$, then these families are independent if and only if

$$\mathbb{P}[A_1 \cap \cdots \cap A_n] = \mathbb{P}[A_1] \cdots \mathbb{P}[A_n],\ \forall A_k \in \mathcal{A}_k,\ k = 1, \ldots, n. \qquad \square$$

Proposition 1.48. *Let $(\Omega, \mathcal{S}, \mathbb{P})$ be a sample space and that $\mathcal{P}_1, \ldots, \mathcal{P}_n \subset \mathcal{S}$ are π-systems each containing Ω. The following statements are equivalent*

(i) The families $\mathcal{P}_1, \ldots, \mathcal{P}_n$ are independent.
(ii) The collection of σ-algebras $\sigma(\mathcal{P}_1), \ldots, \sigma(\mathcal{P}_n)$ is independent.

Proof. Clearly it suffices to prove only (i) $\Rightarrow$ (ii). Fix $S_i \in \mathcal{P}_i$, $i = 2, \ldots, n$. Let
$$\mathcal{I} := \{ S \in \mathcal{S} : \ \mathbb{P}[S \cap S_2 \cap \cdots \cap S_n] = \mathbb{P}[S] \mathbb{P}[S_2] \cdots \mathbb{P}[S_n] \}.$$
Note that $\mathcal{P}_1 \subset \mathcal{I}$. Next let us observe that $\mathcal{I}$ is a λ-system. Indeed if $A, B \in \mathcal{I}$ and $A \subset B$ then
$$\mathbb{P}[(B \setminus A) \cap S_2 \cap \cdots \cap S_n] = \mathbb{P}[(B \cap S_2 \cap \cdots \cap S_n) \setminus (A \cap S_2 \cap \cdots \cap S_n)]$$
$$= \mathbb{P}[B] \mathbb{P}[S_2] \cdots \mathbb{P}[S_n] - \mathbb{P}[A] \mathbb{P}[S_2] \cdots \mathbb{P}[S_n] = \mathbb{P}[B \setminus A] \mathbb{P}[S_2] \cdots \mathbb{P}[S_n].$$
If $A_1 \subset A_2 \subset \cdots \subset A_\nu \subset$ is an increasing sequence of events in $\mathcal{I}$ and
$$A = \lim_{\nu \infty} A_\nu = \bigcup_{\nu \geq 1} A_\nu$$
then
$$\mathbb{P}[A \cap S_2 \cap \cdots \cap S_n] = \lim_{\nu \to \infty} \mathbb{P}[A_\nu \cap S_2 \cap \cdots \cap S_n]$$
$$= \lim_{\nu \to \infty} \mathbb{P}[A_\nu] \mathbb{P}[S_2] \cdots \mathbb{P}[S_n] = \mathbb{P}[A] \mathbb{P}[S_2] \cdots \mathbb{P}[S_n].$$
The $\pi - \lambda$ theorem implies that $\sigma(\mathcal{P}_1) \subset \mathcal{I}$ so that
$$\mathbb{P}[A_1 \cap S_2 \cap \cdots \cap S_n] = \mathbb{P}[A_1] \mathbb{P}[S_2] \cdots \mathbb{P}[S_n],$$
for all $A_1 \in \sigma(\mathcal{P}_1)$, $S_i \in \mathcal{P}_i$, $i = 2, \ldots, n$. Repeating the above argument we deduce
$$\mathbb{P}[A_1 \cap A_2 \cap \cdots \cap A_n] = \mathbb{P}[A_1] \mathbb{P}[A_2] \cdots \mathbb{P}[A_n], \ \forall A_k \in \sigma(\mathcal{P}_k), \ k = 1, \ldots, n.$$
Remark 1.47 shows that the σ-algebras $\sigma(\mathcal{P}_1), \ldots, \sigma(\mathcal{P}_n)$ are independent. $\square$

Corollary 1.49. *Consider the random variables $X_1, \ldots, X_n : (\Omega, \mathcal{S}, \mathbb{P}) \to \mathbb{R}$. The following statements are equivalent.*

(i) The random variables $X_1, \ldots, X_n$ are independent.
(ii) For any $x_1, \ldots, x_n \in \mathbb{R}$
$$\mathbb{P}[X_1 \leq x_1, \ldots, X_n \leq x_n] = \mathbb{P}[X_1 \leq x_1] \cdots \mathbb{P}[X_n \leq x_n].$$

Proof. It follows from Proposition 1.48 applied to the π-systems
$$\mathcal{P}_k := \Big\{ \{X_k \leq x_k\} : \ x_k \in \mathbb{R} \Big\}, \ k = 1, \ldots, n.$$

$\square$

Corollary 1.50 (Partition of independencies). *Suppose that $(\mathcal{S}_i)_{i \in I}$ is an independency of $(\Omega, \mathcal{S}, \mathbb{P})$. For any partition $(I_\alpha)_{\alpha \in A}$ of I we set*

$$\mathcal{F}_\alpha := \bigvee_{i \in I_\alpha} \mathcal{S}_i, \ \alpha \in A.$$

Then the collection $(\mathcal{F}_\alpha)_{\alpha \in A}$ is also an independency.

Proof. Denote by $\mathcal{C}_\alpha$ the π-system obtained by taking intersections of finitely many events from $\bigcup_{i \in I_\alpha} \mathcal{S}_i$. Then

$$\mathcal{F}_\alpha = \sigma(\mathcal{C}_\alpha), \ \forall \alpha \in A$$

and the family $(\mathcal{C}_\alpha)_{\alpha \in A}$ is independent. The conclusion now follows from Proposition 1.48. □

Corollary 1.51. *Suppose that the random variables $X_1, \ldots, X_n \in \mathcal{L}^0(\Omega, \mathcal{S}, \mathbb{P})$ are independent. Then for any $1 < k < n$ and any Borel measurable functions $f : \bar{\mathbb{R}}^k \to \bar{\mathbb{R}}$, $g : \bar{\mathbb{R}}^{n-k} \to \bar{\mathbb{R}}$ the random variables*

$$f(X_1, \ldots, X_k), \ g(X_{k+1}, \ldots, X_n)$$

are independent. □

Definition 1.52 (Tail algebra). Consider a sequence $(\mathcal{S}_n)_{n \in \mathbb{N}}$ of sub-σ-algebras of $(\Omega, \mathcal{S}, \mathbb{P})$. The *tail algebra* of this sequence is σ-algebra

$$\mathcal{T} = \mathcal{T}(\mathcal{S}_n) := \bigcap_{m \in \mathbb{N}} \mathcal{T}_m, \ \mathcal{T}_m := \bigvee_{n > m} \mathcal{S}_n. \tag{1.2.8}$$

The events in $\mathcal{T}$ are called *tail events*. □

Remark 1.53. (a) An event S is a tail event of the sequence $(\mathcal{S}_n)_{n \in \mathbb{N}}$ if

$$\forall m \in \mathbb{N}; \ S \in \bigvee_{n > m} \mathcal{S}_n.$$

The sequence of σ-algebras $(\mathcal{S}_n)_{n \in \mathbb{N}}$ can be viewed as an information stream. The tail events are described by a stream of information and are characterized by the fact that their occurrence is unaffected by information at finitely moments of time in the stream.

(b) To a sequence of random variables $X_n : (\Omega, \mathcal{S}, \mathbb{P}) \to \mathbb{R}$ we associate the sequence of σ-algebras $\mathcal{S}_n = \sigma(X_n)$ and the event $C=$*the sequence $(X_n)_{n \geq 1}$ converges*". To see that this is a tail event note that $\mathcal{T}_m = \sigma(X_{m+1}, X_{m+2}, \ldots)$ and

$$C = \bigcap_{m \in \mathbb{N}} C_m,$$

where C_m is the event "*the sequence $(X_k)_{k \geq m}$ converges*". Clearly $C_m \in \mathcal{T}_m$. □

Theorem 1.54 (Kolmogorov's 0-1 law). *If A is a tail event of the <u>independency</u> $(\mathcal{S}_n)_{n \in \mathbb{N}}$, then $\mathbb{P}[A] = 0$ or $\mathbb{P}[A] = 1$.*

Proof. Let $\mathfrak{T}_m$ as in (1.2.8). According to the principle of partition of independencies the collection $\mathcal{S}_1, \ldots, \mathcal{S}_m, \mathfrak{T}_m$ is an independency and, since $\mathfrak{T} \subset \mathfrak{T}_m$, the collection $\mathcal{S}_1, \ldots, \mathcal{S}_m, \mathfrak{T}$ is also an independency, $\forall m \in \mathbb{N}$. We deduce that for any $m \in \mathbb{N}$ the σ-algebras

$$\bigvee_{k=1}^{m} \mathcal{S}_k, \ \mathfrak{T}$$

are independent so $\{\mathfrak{T}_0, \mathfrak{T}\}$ is an independency. Hence, for any $A \in \mathfrak{T}$, and any $B \in \mathfrak{T}_0$, we have

$$\mathbb{P}[A \cap B] = \mathbb{P}[A]\mathbb{P}[B].$$

If above we choose $B = A \in \mathfrak{T} \subset \mathfrak{T}_0$ we deduce

$$\mathbb{P}[A] = \mathbb{P}[A]^2, \ \forall A \in \mathfrak{T} \Rightarrow \mathbb{P}[A] \in \{0, 1\}, \ \forall A \in \mathfrak{T}.$$

□

Definition 1.55. Let $(\Omega, \mathcal{S}, \mathbb{P})$ be a probability space. A *zero-one event* is a an event $S \in \mathcal{S}$ such that $\mathbb{P}[S] \in \{0, 1\}$. A *zero-one algebra* is a sigma-subalgebra $\mathcal{F} \subset \mathcal{S}$ consisting of zero-one events. □

Corollary 1.56. *Suppose that $(X_n)_{n \in \mathbb{N}}$ is a sequence of independent random variables on the probability space $(\Omega, \mathcal{S}, \mathbb{P})$. Then the series*

$$\sum_{n \in \mathbb{N}} X_n$$

is either almost surely convergent, or almost surely divergent. In other words, the almost sure convergence is a zero-one event. □

Definition 1.57. Suppose that A, B are events in the sample space $(\Omega, \mathbb{P}, \mathcal{S})$ such that $\mathbb{P}[B] \neq 0$. The *conditional probability of A given B* is the number

$$\mathbb{P}[A|B] := \frac{\mathbb{P}[A \cap B]}{\mathbb{P}[B]}.$$

□

Note that we have the useful *product formula*

$$\mathbb{P}[A \cap B] = \mathbb{P}[A|B]\mathbb{P}[B]. \tag{1.2.9}$$

In particular, we deduce that A, B are independent if and only if $\mathbb{P}[A] = \mathbb{P}[A|B]$. Note that the map

$$\mathbb{P}[-|B] : \mathcal{S} \to [0, 1], \ S \mapsto \mathbb{P}[S|B]$$

is also a probability measure on $\mathcal{S}$. We say that it is *the probability measure obtained by conditioning on B*.

Remark 1.58. Observe that n events $A_1, \ldots, A_n$, $n \geq 2$, are independent if and only if, for any nonempty subset $I \subset \{1, \ldots n\}$ of cardinality $< n$, and any $j \notin I$ we have

$$\mathbb{P}[A_j \,|\, A_I] = A_j, \ \text{where} \ A_I := \bigcap_{i \in I} A_i.$$

□

Suppose we are given a finite or countable measurable partition of $(\Omega, \mathcal{S}, \mathbb{P})$

$$\Omega = \bigsqcup_{i \in I} A_i, \ I \subset \mathbb{N}, \ \mathbb{P}[A_i] \neq 0, \ \forall i.$$

The *law of total probability* states that

$$\mathbb{P}[S] = \sum_{i \in I} \mathbb{P}[S|A_i]\mathbb{P}[A_i], \ \forall S \in \mathcal{S}. \tag{1.2.10}$$

Indeed,

$$\mathbb{P}[S] = \sum_{i \in I} \mathbb{P}[S \cap A_i] \stackrel{(1.2.9)}{=} \sum_{i \in I} \mathbb{P}[S|A_i]\mathbb{P}[A_i].$$

Example 1.59. Suppose that we have an urn containing b black balls and r red balls. A ball is drawn from the urn and discarded. Without knowing its color, what is the probability that a second ball drawn is black?

For $k = 1, 2$ denote by B_k the event *"the k-th drawn ball is black"*. We are asked to find $\mathbb{P}(B_2)$. The first drawn ball is either black (B_1) or not black (B_1^c). From the law of total probability we deduce

$$\mathbb{P}[B_2] = \mathbb{P}[B_2|B_1]\mathbb{P}[B_1] + \mathbb{P}[B_2|B_1^c]\mathbb{P}[B_1^c].$$

Observing that

$$\mathbb{P}[B_1] = \frac{b}{b+r} \text{ and } \mathbb{P}[B_1^c] = \frac{r}{b+r},$$

we conclude

$$\mathbb{P}[B_2] = \frac{b-1}{b+r-1} \cdot \frac{b}{b+r} + \frac{b}{b+r-1} \cdot \frac{r}{b+r} = \frac{b(b-1) + br}{(b+r)(b+r-1)}$$

$$= \frac{b(b+r-1)}{(b+r)(b+r-1)} = \frac{b}{b+r} = \mathbb{P}[B_1].$$

Thus, the probability that the second extracted ball is black is equal to the probability that the first extracted ball is black. This seems to contradict our intuition because when we extract the second ball the composition of available balls at that time is different from the initial composition.

This is a special case of a more general result, due to S. Poisson, [31, Sec. 5.3].

> Suppose in an urn containing b black and r red balls, n balls have been drawn first and discarded without their colors being noted. If another ball is drawn next, the probability that it is black is the same as if we had drawn this ball at the outset, without having discarded the n balls previously drawn.

To quote John Maynard Keynes, [90, p. 394],

> This is an exceedingly good example of the failure to perceive that a probability cannot be influenced by the *occurrence* of a material event but only by such *knowledge* as we may have, respecting the occurrence of the event. □

Example 1.60 (The ballot problem). This is one of the oldest problems in probability. A person starts at $S_0 \in \mathbb{Z}$ and every second (or epoch) he flips a fair coin: Heads, he moves ahead, Tails he takes one step back. We denote by S_n its location after n coin flips. The sequence of random variables $(S_n)_{n \in \mathbb{N}}$ is called the standard (or unbiased) random walk on $\mathbb{Z}$.

Formally we have a sequence of independent random variables $(X_n)_{n \in \mathbb{N}}$ such that
$$\mathbb{P}[X_n = 1] = \mathbb{P}[X_n = -1] = \frac{1}{2}, \quad \forall n \in \mathbb{N}.$$

The random variables with this distribution are called *Rademacher random variables*. Then
$$S_n = S_0 + X_1 + \cdots + X_n.$$
$S_0 = 0$, $\mathbb{I}_n := \{1, \ldots, n\}$
$$H_n := \#\{k \in \mathbb{I}_n; \; X_k = 1\}, \quad T_n = \{k \in \mathbb{I}_n; \; X_k = -1\}.$$

Thus H_n is the number of Heads during the first n coin flips, while T_n denotes the number of Tails during the first n coin flips. Note that
$$n = H_n + T_n, \quad S_n = S_0 + H_n - T_n = S_0 + 2H_n - n.$$

We deduce that
$$S_n = m \Longleftrightarrow n + m - S_0 = 2H_n.$$

In particular this shows that $S_n \equiv n - S \mod 2$, $\forall n \in \mathbb{N}$. Moreover,
$$S_n = m \Longleftrightarrow H_n = \frac{n + m - S_0}{2},$$
and we deduce.
$$\mathbb{P}[S_n = m] = \begin{cases} \binom{n}{(n-m-S_0)/2} 2^{-n} & m \equiv n - S_0 \mod 2, \\ 0, & \text{otherwise.} \end{cases}$$

It is convenient to visualize the random walk as a zig-zag obtained by successively connecting by a line segment the point $(n-1, S_{n-1})$ to the point (n, S_n), $n \in \mathbb{N}$. The connecting line segment has slope X_n; see Figure 1.1.

Suppose that $y \in \mathbb{N}$ and $S_0 = 0$. The *ballot problem* asks what is the probability p_y that
$$S_k > 0, \quad \forall k = 1, \ldots, n-1 \quad \text{given that} \quad S_n = y.$$

One can think of a zigzag as describing a succession of votes in favor of one of the two candidates H or T. When the zigzag goes up, a vote for H is cast, and when it goes down, a vote in favor of T is cast. We know that at the end of the election H was declared winner with y votes over T. Thus p_y is the probability that H was always ahead during the voting process.

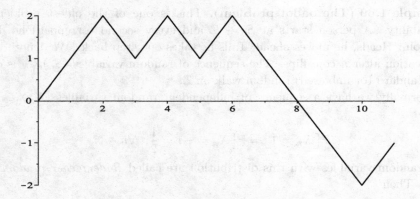

Fig. 1.1 A zig-zag describing a random walk started at $S_0 = 0$.

We set $H_n := a$, $T_n := b$ so $n = a+b$, $y = a-b$. The sample space in this problems is the space $\Omega_{n,y}$ of zigzags ω that start at the origin and end at (n, y). There are

$$|\Omega_{n,y}| = \binom{n}{a} = \binom{a+b}{b},$$

equally likely such zigzags. We seek the probability of the event

$$E := \{\omega \in \Omega_{n,y};\ \omega \text{ touches the horizontal axis}\}.$$

Then $p_y = 1 - \mathbb{P}[E]$.

We will compute $\mathbb{P}[E]$ by conditioning on S_1. There is a silent trap on our way. Since the first vote is equally likely to have been H or T, one might be tempted to think that $\mathbb{P}[S_1 = \pm 1] = \frac{1}{2}$. This is however not the case since the zig-zags in $\Omega_{n,y}$ are subject to an extra condition, namely the location (n, y) of their endpoints. We have

$$\mathbb{P}[E] = \mathbb{P}[E|\, S_1 = -1\,]\mathbb{P}[\,S_1 = -1\,] + \mathbb{P}[\,E|\, S_1 = 1\,]\mathbb{P}[\,S_1 = 1\,]$$

$$= \mathbb{P}[\,S_1 = -1\,] + \mathbb{P}[\,E|\, S_1 = 1\,]\mathbb{P}[\,S_1 = 1\,].$$

Note that there are

$$\binom{n-1}{a} = \binom{a+b-1}{a}$$

equally likely zigzags from $(1, -1)$ to $(n, a-b)$, so

$$\mathbb{P}[\,S_1 = -1\,] = \frac{\binom{n-1}{a}}{\binom{n}{a}} = \frac{b}{n} = \frac{b}{a+b}.$$

There are

$$\binom{n-1}{a-1} = \binom{a+b-1}{b}$$

equally likely zigzags form $(1,1)$ to $(n, a-b)$ so
$$\mathbb{P}[\,S_1 = 1\,] = \frac{\binom{n-1}{b}}{\binom{n}{a}} = \frac{a}{a+b}.$$
To count the number of zigzags from $(1,1)$ to $(n, a-b)$ that touch the horizontal axis we rely on a clever and versatile trick called *André's reflection trick*.

For each such zigzag Z denote by $k(Z)$ the first moment it touches the horizontal axis. Denote by Z^r the zigzag obtained from Z by reflecting in the horizontal axis the part of Z from $k(Z)$ to n; see Figure 1.2.

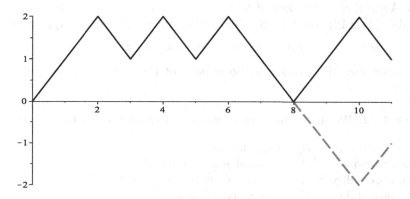

Fig. 1.2 *The zigzag Z^r traces Z until Z hits the horizontal axis. At this moment the zigzag Z^r follows the opposite motion of Z (dashed line).*

The end point of Z^r is $(n, -(a-b))$. The transformation $Z \to Z^r$ produces a bijection between the zigzags with origin $(1,1)$ and endpoint $(n, a-b)$ that touch the horizontal axis and the zigzags with origin $(1,1)$ and endpoint $(n, -(a-b))$. Indeed, any zigzag $Z' : (1,1) \to (n, b-a)$ must cross the horizontal axis. After the first touch we reflect it in this axis and obtain a zigzag $Z : (1,1) \to (n, a-b)$ such that $Z^r = Z'$. Clearly Z touches the horizontal axis.

The number of zigzags $(1,1) \to (n, b-a)$ is
$$\binom{n-1}{b-1} = \binom{a+b-1}{a}.$$
Hence
$$\mathbb{P}[\,E \,\|\, S_1 = 1\,] = \frac{\binom{a+b-1}{a}}{\binom{a+b-1}{a-1}} = \frac{(a-1)!b!}{a!(b-1)!} = \frac{b}{a}.$$
We deduce
$$\mathbb{P}[\,E\,] = \frac{b}{a} \cdot \frac{a}{a+b} + \frac{b}{a+b} = \frac{2b}{a+b}$$
and
$$p_y = 1 - \frac{2b}{a+b} = \frac{a-b}{a+b} = \frac{y}{n} = \frac{S_n}{n}. \qquad (1.2.11)$$

□

Proposition 1.61 (Bayes' formula). *Suppose we are given a finite or countable measurable partition of* $(\Omega, \mathcal{S}, \mathbb{P})$

$$\Omega = \bigsqcup_{i \in I} A_i, \quad I \subset \mathbb{N}, \quad \mathbb{P}[A_i] \neq 0, \quad \forall i.$$

Then, for any $S \in \mathcal{S}$ *such that* $\mathbb{P}[S] \neq 0$ *and* $i_0 \in I$ *we have*

$$\mathbb{P}[A_{i_0}|S] = \frac{\mathbb{P}[S|A_{i_0}]\mathbb{P}[A_{i_0}]}{\sum_{i \in I} \mathbb{P}[S|A_i]\mathbb{P}[A_i]}. \tag{1.2.12}$$

Proof. According to the law of total probability, the denominator in the right-hand-side of (1.2.12) equals $\mathbb{P}[S]$. Thus, the equality (1.2.12) is equivalent to

$$\mathbb{P}[A_{i_0}|S]\mathbb{P}[S] = \mathbb{P}[S|A_{i_0}]\mathbb{P}[A_{i_0}].$$

The product formula shows that both sides of the above equality are equal to $\mathbb{P}[A_{i_0} \cap S]$. □

Remark 1.62. We should mention here a terminology favored by statisticians.

- The events A_k are called *hypotheses*.
- The probability $\mathbb{P}[A_k]$ is called *prior* (probability).
- The probability $\mathbb{P}[A_k|S]$ is called *posterior* (probability).
- The probability $\mathbb{P}[S|A_k]$ is called *likelihood*.

Here is one frequent application of Bayes' principle. Suppose that we observed a random event S we know that it can be caused only by one of the random events A_i. To decide which of the events A_i is more likely to have caused S we need to find the largest of the posteriors $\mathbb{P}[A_i|S]$. Bayes' formula shows that the most likely cause maximizes the numerator $\mathbb{P}[S|A_i]\mathbb{P}[A_i]$. □

Example 1.63 (Biased coins). We say that a coin has bias $\theta \in (0,1)$ if the probability of showing Heads when flipped is θ. Suppose that we have an urn containing c_1 coins with bias θ_1 and c_2 coins with bias θ_2. Let $n := c_1 + c_2$ denote the total number of coins and set $p_i := \frac{c_i}{n}$, $i = 1, 2$. We assume that

$$c_1 < c_2 \quad \text{and} \quad \theta_1 > \theta_2, \tag{1.2.13}$$

i.e., there are fewer coins with higher bias. We draw a coin at random we flip it twice and we get Heads both times. What is the probability that the coin we have drawn has higher bias.

If θ denotes the (unknown) bias of the coin drawn at random, then we can think of θ as a random variable that takes two values θ_1, θ_2 with probabilities

$$\mathbb{P}[\theta_i] := \mathbb{P}[\theta = \theta_i] = p_i, \quad i = 1, 2.$$

Denote by E the event that two successive flips produce Heads. Then

$$\mathbb{P}[E|\theta_i] := \mathbb{P}[E|\theta = \theta_i] = \theta_i^2.$$

Bayes' formula shows that

$$\mathbb{P}[\theta_1|E] = \frac{\mathbb{P}[E|\theta_1]\mathbb{P}[\theta_1]}{\mathbb{P}[E|\theta_1]\mathbb{P}[\theta_1] + \mathbb{P}[E|\theta_2]\mathbb{P}[\theta_2]} = \frac{p_1\theta_1^2}{p_1\theta_1^2 + p_2\theta_2^2} = \frac{1}{1 + \frac{p_2}{p_1}\left(\frac{\theta_2}{\theta_1}\right)^2}.$$

Our assumption (1.2.13) shows that

$$\frac{c_2}{c_1} = \frac{p_2}{p_1} > 1 > \frac{\theta_2}{\theta_1}.$$

Observe that if $c_2\theta_2^2 > c_1\theta_1^2$, then

$$\mathbb{P}[\theta_1|E] < \frac{1}{2}.$$

Thus, in this case, if we observe two Heads, then the coin we randomly drew from the urn is less likely to be the one with bigger bias. For example if $\theta_1 = \frac{2}{3}$ and $\theta_2 = \frac{1}{3}$ and $c_2 > 8c_1$, then

$$\mathbb{P}[\theta_1|E] < \frac{1}{3},$$

so the randomly drawn coin is less likely to be the one heavily biased towards Heads. □

1.2.3 Integration of measurable functions

We outline below, mostly without proofs, the construction and the basic facts about integration of measurable functions. For details we refer to [50; 102; 148].

Fix a measured space $(\Omega, \mathcal{S}, \mu)$. Recall that $\mathrm{Elem}(\Omega, \mathcal{S})$ denotes the vector space of elementary $\mathcal{S}$-measurable functions (see Definition 1.19). We denote by $\mathrm{Elem}_+(\Omega, \mathcal{S})$ the convex cone of $\mathrm{Elem}(\Omega, \mathcal{S})$ consisting of nonnegative elementary functions. Define

$$\mu : \mathrm{Elem}_+(\Omega, \mathcal{S}) \to [0, \infty], \quad f \mapsto \mu[f] = \int_\Omega f(\omega)\mu[d\omega],$$

as follows. If

$$f = \sum_{k=1}^{M} a_i \boldsymbol{I}_{A_i}, \quad A_1, \ldots, A_M \text{ disjoint},$$

then

$$\mu[f] = \int_\Omega f(\omega)\mu[d\omega] := \sum_{i=1}^{M} a_i \mu[A_i].$$

Note that if

$$f = \sum_{j=1}^{N} b_j \boldsymbol{I}_{B_j}, \quad B_1, \ldots, B_n \text{ disjoint},$$

then $a_i = b_j$ if $A_i \cap B_j \neq \emptyset$. Hence

$$\sum_i a_i \mu[A_i] = \sum_i \sum_j a_i \mu[A_i \cap B_j] = \sum_j \sum_i b_j \mu[A_i \cap B_j] = \sum_j b_j \mu[B_j].$$

This shows that the value of $\int_\Omega f(\omega) \mu(d\omega)$ is independent of the decomposition of f as a linear combination of indicators of pairwise disjoint measurable sets.

The above integration map satisfies the following elementary properties.

$$\forall f, g \in \mathrm{Elem}_+(\Omega, \mathcal{S}) \ \ f \leq g \Rightarrow \mu[f] \leq \mu[g]. \tag{1.2.14a}$$

$$\forall a, b \geq 0, \ \ f, g \in \mathrm{Elem}_+(\Omega, \mathcal{S}) : \ \ \mu[af + bg] = a\mu[f] + b\mu[g]. \tag{1.2.14b}$$

For $f \in \mathcal{L}^0_+(\Omega, \mathcal{S})$ we set

$$\mathcal{E}^f_+ := \{ g \in \mathrm{Elem}_+(\Omega, \mathcal{S}); \ g \leq f.$$

The set $\mathcal{E}^f_+$ is nonempty since $0 \in \mathcal{E}^f_+$. Define

$$\boxed{\mu[f] = \int_\Omega f\, d\mu = \int_\Omega f(\omega) \mu[d\omega] := \sup_{g \in \mathcal{E}^f_+} \int_\Omega g(\omega) \mu[d\omega] \in [0, \infty)}. \tag{1.2.15}$$

Definition 1.64. A measurable function $f \in \mathcal{L}^0(\Omega, \mathcal{S})$ is called *μ-integrable* if

$$\mu[f^+], \ \mu[f^-] < \infty.$$

In this case we define its *Lebesgue integral* to be

$$\int_\Omega f\, d\mu = \int_\Omega f(\omega) \mu[d\omega] = \mu[f] := \mu[f_+] - \mu[f_-].$$

We denote by $\mathcal{L}^1(\Omega, \mathcal{S}, \mu)$ the set of μ-integrable functions and by $\mathcal{L}^1_+(\Omega, \mathcal{S}, \mu)$ the set of μ-integrable nonnegative functions. □

Note that

$$\forall f, g \in \mathcal{L}^0_+(\Omega, \mathcal{S}) \ \ f \leq g \Rightarrow \mu[f] \leq \mu[g]. \tag{1.2.16}$$

Moreover,

$$\forall f \in \mathcal{L}^0_+(\Omega, \mathcal{S}) : \ \mu[f > 0] = 0 \iff \int_\Omega f\, d\mu = 0. \tag{1.2.17}$$

The integral $\mathcal{L}^0_+ \ni f \mapsto \mu[f] \in [0, \infty]$ enjoys the following key continuity property which is the "workhorse" of the Lebesgue integration theory.

Theorem 1.65 (Monotone Convergence theorem). *Suppose that $(f_n)_{n \in \mathbb{N}}$ is a sequence in $\mathcal{L}^0_+(\Omega, \mathcal{S})$ that converges increasingly to $f \in \mathcal{L}^0_+(\Omega, \mathcal{S})$. Then*

$$\mu[f_n] \nearrow \mu[f] \ \ \text{as } n \to \infty.$$

Proof. The sequence $\mu[f_n]$ is nondecreasing and is bounded above by $\mu[f]$. Hence it has a, possibly infinite, limit and $\lim_{n\to\infty} \mu[f_n] \leq \mu[f]$. The proof of the opposite inequality

$$\lim_{n\to\infty} \mu[f_n] \geq \mu[f]$$

relies on a clever a clever trick. Fix $g \in \mathcal{E}_+^f$, $c \in (0,1)$, and set

$$S_n := \{\omega \in \Omega;\ f_n(\omega) \geq cg(\omega)\}.$$

Since $f = \lim f_n$ and (f_n) is a nondecreasing sequence of functions we deduce that S_n is a nondecreasing sequence of measurable sets whose union is Ω. For any elementary function h the product $\boldsymbol{I}_{S_n}h$ is also elementary. For any $n \in \mathbb{N}$ we have $f_n \geq f_n \boldsymbol{I}_{S_n} \geq cg\boldsymbol{I}_{S_n}$ so that

$$\mu[f_n] \geq \mu[\boldsymbol{I}_{S_n}f_n] \geq c\mu[g\boldsymbol{I}_{S_n}].$$

If we write g as a finite linear combination

$$g = \sum_j g_j \boldsymbol{I}_{A_j}$$

with A_j pairwise disjoint, then we deduce

$$\mu[f_n] \geq c\mu[g\boldsymbol{I}_{S_n}] = c\sum_j g_j \mu[A_j \cap S_n].$$

The sequence of sets $(A_j \cap S_n)_{n \in \mathbb{N}}$ is nondecreasing and its union is A_j so that

$$\lim_{n\to\infty} \mu[f_n] \geq c\sum_j g_j \lim_{n\to\infty} \mu[A_j \cap S_n] = c\sum_j g_j \mu[A_j] = c\mu[g].$$

Hence

$$\lim_{n\to\infty} \mu[f_n] \geq c\mu[g],\ \forall g \in \mathcal{E}_+^f,\ \forall c \in (0,1),$$

so that

$$\lim_{n\to\infty} \mu[f_n] \geq c\mu[f],\ \forall c \in (0,1).$$

Letting $c \nearrow 1$ we deduce $\lim_{n\to\infty} \mu[f_n] \geq \mu[f]$. $\square$

Corollary 1.66. *For any $f \in \mathcal{L}_+^0(\Omega, \mathcal{S})$ we have*

$$\mu[f] = \lim_{n\to\infty} \mu[D_n[f]]. \qquad \square$$

Corollary 1.67. *For any $f, g \in \mathcal{L}^1(\Omega, \mathcal{S}, \mu)$ and $a, b \in \mathbb{R}$ such that $af + bg$ is well defined we have $af + bg \in \mathcal{L}^1(\Omega, \mathcal{S}, \mu)$ and*

$$\int_\Omega (af + bg)d\mu = a\int_\Omega f d\mu + b\int_\Omega g d\mu. \qquad (1.2.18)$$

Moreover, if $f, g \in \mathcal{L}^1(\Omega, \mathcal{S}, \mu)$ and $f(\omega) \leq g(\omega)$, $\forall \omega \in \Omega$ then

$$\int_\Omega f d\mu \leq \int_\Omega g d\mu.$$

$\square$

Since $|f| = f^+ + f^-$ we deduce the following result.

Corollary 1.68. *Let $f \in \mathcal{L}^0(\Omega, \mathcal{S})$. Then*

$$f \in \mathcal{L}^1(\Omega, \mathcal{S}, \mu) \iff |f| \in \mathcal{L}^1(\Omega, \mathcal{S}, \mu). \qquad \square$$

Corollary 1.69 (Markov's Inequality). *Suppose that $f \in \mathcal{L}^1_+(\Omega, \mathcal{S}, \mu)$. Then, for any $C > 0$, we have*

$$\mu[\{f \geq C\}] \leq \frac{1}{C} \int_\Omega f d\mu. \qquad (1.2.19)$$

In particular, $f < \infty$, μ-a.e.

Proof. Note that

$$C\mathbf{I}_{\{f \geq C\}} \leq f \Rightarrow C\mu[\{f \geq C\}] = \int_\Omega C\mathbf{I}_{\{f \geq C\}} \leq \int_\Omega f d\mu.$$

□

Corollary 1.70. *If $f \in \mathcal{L}^1(\Omega, \mathcal{S}, \mu)$, then $\mu[\{|f| = \infty\}] = 0$.*

Proof. Note that

$$\mu[\{|f| = \infty\}] = \bigcap_{n \in \mathbb{N}} \mu[\{f > n\}].$$

On the other hand, Markov's inequality implies

$$\mu[\{f > n\}] \leq \frac{\mu[|f|]}{n} \to 0.$$

□

Proposition 1.71. *Suppose $f, g \in \mathcal{L}^0(\Omega, \mathcal{S})$ and $f = g$, μ-a.e. Then*

$$f \in \mathcal{L}^1(\Omega, \mathcal{S}, \mu) \iff g \in \mathcal{L}^1(\Omega, \mathcal{S}, \mu).$$

Moreover, if one of the above equivalent conditions hold, then $\mu[f] = \mu[g]$. □

Remark 1.72. The presentation so far had to tread carefully around a nagging problem: given f, g in $\mathcal{L}^1(\Omega, \mathcal{S}, \mu)$, then $f(\omega) + g(\omega)$ may not be well defined for some ω. For example, it could happen that $f(\omega) = \infty$, $g(\omega) = -\infty$. Fortunately, Corollary 1.70 shows that the set of such ω's is negligible. Moreover, if we redefine f and g to be equal to zero on the set where they had infinite values, then their integrals do not change. For this reason we alter the definition of $\mathcal{L}^1(\Omega, \mathcal{S}, \mu)$ as follows.

$$\mathcal{L}^1(\Omega, \mathcal{S}, \mu) := \left\{ f : (\Omega, \mathcal{S}) \to \mathbb{R}; \ f \text{ measurable } \int_\Omega |f| d\mu < \infty \right\}.$$

Thus, in the sequel the integrable functions will be assumed to be <u>everywhere</u> finite.

With this convention, the space $\mathcal{L}^1(\Omega, \mathcal{S}, \mu)$ is a vectors space and the Lebesgue integral is a linear functional

$$\mu : \mathcal{L}^1(\Omega, \mathcal{S}, \mu) \to \mathbb{R}, \ f \mapsto \mu[f].$$

□

Recall that for any sequence $(x_n)_{n\in\mathbb{N}}$ of real numbers we have

$$\liminf_{n\to\infty} x_n = \lim_{k\to\infty} x_k^* := \inf_{n\geq k} x_n.$$

The sequence (x_k^*) is nondecreasing. The Monotone Convergence Theorem has the following useful immediate consequence.

Theorem 1.73 (Fatou's Lemma). *Suppose that $(f_n)_{n\in\mathbb{N}}$ is a sequence in $\mathcal{L}_+^0(\Omega, \mathcal{S})$. Then*

$$\int_\Omega \liminf_{n\to\infty} f_n(\omega)\,\mu[\,d\omega\,] \leq \liminf_{n\to\infty} \int_\Omega f_n\,d\mu.$$

$\square$

Proof. Set

$$g_k := \inf_{n\geq k} f_n.$$

Proposition 1.17(iii) implies that $g_k \in \mathcal{L}_+^0(\Omega, \mathcal{S})$. The sequence (g_k) is nondecreasing and

$$\liminf_{n\to\infty} f_n = \lim_{k\to\infty} g_k.$$

The Monotone Convergence Theorem implies that

$$\int_\Omega \liminf_{n\to\infty} f_n(\omega)\,\mu[\,d\omega\,] = \lim_{k\to\infty} \int_\Omega g_k\,d\mu.$$

Note that $g_k \leq f_n$, $\forall n \geq k$, and thus

$$\int_\Omega g_k\,d\mu \leq \int_\Omega f_n\,d\mu,\ \ \forall n \geq k,$$

i.e.,

$$\int_\Omega g_k\,d\mu \leq \inf_{n\geq k} \int_\Omega f_n\,d\mu.$$

Letting $k \to \infty$ we deduce

$$\lim_{k\to\infty} \int_\Omega g_k\,d\mu \leq \lim_{k\to\infty} \inf_{n\geq k} \int_\Omega f_n\,d\mu = \liminf_{n\to\infty} \int_\Omega f_n\,d\mu.$$

$\square$

The next result illustrates one of the advantages of the Lebesgue integral over the Riemann integral: one needs less restrictive conditions to pass to the limit under the Lebesgue integral.

Theorem 1.74 (Dominated Convergence). *Suppose $(f_n)_{n\in\mathbb{N}}$ is a sequence in $\mathcal{L}^1(\Omega, \mathcal{S}, \mu)$ satisfying the following properties*

(i) *There exists $f \in \mathcal{L}^0(\Omega, \mathcal{S})$ such that*

$$\lim_{n\to\infty} f_n(\omega) = f(\omega),\ \ \forall \omega \in \Omega.$$

(ii) *There exists $g \in \mathcal{L}^1(\Omega, \mathcal{S}, \mu)$ such that*

$$|f_n(\omega)| \leq g(\omega),\ \ \forall \omega \in \Omega,\ n \in \mathbb{N}.$$

Then $f \in \mathcal{L}^1(\Omega, \mathcal{S}, \mu)$ and

$$\lim_{n\to\infty} \int f_n d\mu = \int_\Omega f d\mu, \qquad (1.2.20a)$$

$$\lim_{n\to\infty} \int_\Omega |f_n(\omega) - f(\omega)| d\mu = 0. \qquad (1.2.20b)$$

Proof. Set $g_n = |f| - f_n$. Then $g_n \geq 0$ and $\lim g_n = |f| - f$. Fatou's Lemma implies

$$\int_\Omega (|f| - f) d\mu \leq \liminf \int_\Omega (|f| - f_n) d\mu = \int_\Omega |f| d\mu - \limsup \int_\Omega f_n d\mu.$$

We deduce

$$\limsup \int_\Omega f_n d\mu \leq \int_\Omega f d\mu.$$

Arguing in the same fashion using the sequence $f_n - |f|$ we deduce

$$\int_\Omega f d\mu \leq \liminf \int_\Omega f_n d\mu.$$

Hence

$$\int_\Omega f d\mu \leq \liminf \int_\Omega f_n d\mu \leq \limsup \int_\Omega f_n d\mu \leq \int_\Omega f d\mu.$$

This proves (1.2.20a). The equality (1.2.20b) follows by applying (1.2.20a) to the sequence $g_n = |f_n - f|$. □

Theorem 1.75 (Change in variables). *Suppose that $(\Omega_0, \mathcal{S}_0)$, $(\Omega_1, \mathcal{S}_1)$ are measurable spaces and*

$$\Phi : (\Omega_0, \mathcal{S}_0) \to (\Omega_1, \mathcal{S}_1)$$

is a measurable map. Fix a measure $\mu_0 : \mathcal{S}_0 \to [0, \infty]$ and a measurable function $f \in \mathcal{L}^0(\Omega_1, \mathcal{S}_1)$. Then

$$f \in \mathcal{L}^1(\Omega_1, \mathcal{S}_1, \Phi_\# \mu_0) \Longleftrightarrow f \circ \Phi \in \mathcal{L}^1(\Omega_0, \mathcal{S}_0, \mu_0)$$

and

$$\int_{\Omega_0} f \circ \Phi \, d\mu_0 = \int_{\Omega_1} f d\Phi_\# \mu_0. \qquad (1.2.21)$$

Proof. Note that it suffices to prove the theorem in the case $f \geq 0$. The result is obviously true if $f \in \mathrm{Elem}_+(\Omega_1, \mathcal{S}_1)$. The general case follows from the Monotone Convergence Theorem using the increasing approximation $[f]_n \nearrow f$ of f by elementary functions; see (1.1.7). □

Remark 1.76. Unlike the well known change-in-variables formula, the map T in (1.2.21) need not be bijective, only measurable.

If T is bijective with measurable inverse, then for any measure μ_1 on $(\Omega_1, \mathcal{S}_1)$ then (1.2.21) applied to the map T^{-1} reads

$$\int_{\Omega_1} f(\omega_1) \mu_1[d\omega_1] = \int_{\Omega_0} f(T\omega_0) T^{-1}_\# \mu_1[d\omega_0], \qquad (1.2.22)$$

$\forall f \in \mathcal{L}^1(\Omega_1, \mathcal{S}_1, \mu_1)$.

In particular, if Ω_i are open subsets of $\mathbb{R}^n$, $T : \Omega_0 \to \Omega_1$ is a C^1-diffeomorphism onto, and μ_1 is the Lebesgue measure on Ω_1, then (1.2.22) reads

$$\int_{\Omega_1} f(y) \boldsymbol{\lambda}[dy] = \int_{\Omega_0} f(Tx) |\det J_T(x)| \boldsymbol{\lambda}[dx], \qquad (1.2.23)$$

where $J_T(x)$ is the Jacobian of the C^1 map $x \to Tx$. □

Proposition 1.77. *Let $f \in \mathcal{L}_+^0(\Omega, \mathcal{S})$. Suppose that $\mu : \mathcal{S} \to [0, \infty]$ is a sigma finite measure. Define*

$$\mu_f : \mathcal{S} \to [0, \infty], \quad \mu_f[S] = \int_S f \, d\mu := \int_\Omega \boldsymbol{I}_S f \, d\mu.$$

Then, μ_f is a measure. Moreover

$$\mu_{f_0} = \mu_{f_1} \iff f_0 = f_1, \quad \mu - \text{almost everywhere.} \qquad \square$$

The above result has an important converse. To state it we need to introduce the concept of *absolute continuity*.

Definition 1.78. Suppose that μ, ν are two measures on the measurable space. We say that ν is *absolutely continuous* with respect to μ, and we write this $\nu \ll \mu$ if

$$\forall S \in \mathcal{S}: \quad \mu[S] = 0 \Rightarrow \nu[S]. \qquad \square$$

For a proof of the next result we refer to [15; 33; 148].

Theorem 1.79 (Radon–Nikodym). *Suppose that μ, ν are two σ-finite measures on the measurable space $(\Omega, \mathcal{S})$. The following statements are equivalent.*

(i) $\nu \ll \mu$.
(ii) There exists $\rho \in \mathcal{L}_+^0(\Omega, \mathcal{S})$ such that $\nu = \mu_\rho$, i.e.,

$$\nu[S] = \int_S \rho \mu[d\omega], \quad \forall S \in \mathcal{S}.$$

The function ρ is not unique, but it defines a unique element in $L_+^0(\Omega, \mathcal{S}, \mu)$ which we denote by $\frac{d\nu}{d\mu}$ and we will refer to it as the density *of ν relative to μ.* $\square$

1.2.4 L^p spaces

We recall here an important class of Banach spaces. For proofs and many more details we refer to [50; 102; 148]. We define an equivalence relation $\sim_\mu$ on $\mathcal{L}^0(\Omega, \mathcal{S})$ by declaring $f \sim_\mu g$ iff $\mu[f \neq g] = 0$. Note that

$$f \in \mathcal{L}^1(\Omega, \mathcal{S}, \mu) \text{ and } g \sim_\mu f \Rightarrow g \in \mathcal{L}^1(\Omega, \mathcal{S}, \mu) \text{ and } \int_\Omega g \, d\mu = \int_\Omega f \, d\mu.$$

We set

$$L^0(\Omega, \mathcal{S}, \mu) := \mathcal{L}^0(\Omega, \mathcal{S}, \mu)/\sim_\mu, \quad L^1(\Omega, \mathcal{S}, \mu) := \mathcal{L}^1(\Omega, \mathcal{S}, \mu)/\sim_\mu.$$

For $p \in [1, \infty)$ we set

$$\mathcal{L}^p(\Omega, \mathcal{S}, \mu) := \left\{ f \in \mathcal{L}^0(\Omega, \mathcal{S}, \mu); \ |f|^p \in \mathcal{L}^1(\Omega, \mathcal{S}, \mu) \right\},$$

$$L^p(\Omega, \mathcal{S}, \mu) := \mathcal{L}^p(\Omega, \mathcal{S}, \mu)/\sim_\mu.$$

We will refer to the functions in $\mathcal{L}^p(\Omega, \mathcal{S}, \mu)$ as *p-integrable* functions. For $p \in [1, \infty)$ and $f \in \mathcal{L}^p(\Omega, \mathcal{S}, \mu)$ we set

$$\|f\|_p := \left(\int_\Omega |f|^p d\mu \right)^{\frac{1}{p}}.$$

Define

$$L^\infty(\Omega, \mathcal{S}, \mu) := \big\{ [f] \in L^0(\Omega, \mathcal{S}, \mu);\ \exists g \in \mathcal{L}^\infty(\Omega, \mathcal{S}),\ g \sim_\mu f \big\}.$$

For $f \in \mathcal{L}^0(\Omega, \mathcal{S})$ we define

$$\|f\|_\infty = \operatorname{ess\,sup} |f| := \inf \big\{ a \geq 0;\ \mu[\,|f| > a\,] = 0 \big\}.$$

Note that this quantity only depends on the $\sim_\mu$-equivalence class of f and

$$L^\infty(\Omega, \mathcal{S}, \mu) = \big\{ f \in L^1(\Omega, \mathcal{S}, \mu);\ \|f\|_\infty < \infty \big\}.$$

In this fashion we obtain for every $p \in [1, \infty]$ maps

$$\| - \|_p : L^p(\Omega, \mathcal{S}, \mu) \to [0, \infty).$$

Theorem 1.80 (Hölder inequality). *Let $p, q \in [1, \infty]$ such that*

$$\frac{1}{p} + \frac{1}{q} = 1.$$

Then for any $f \in \mathcal{L}^p(\Omega, \mathcal{S}, \mu)$ and $g \in \mathcal{L}^q(\Omega, \mathcal{S}, \mu)$ we have $fg \in \mathcal{L}^1(\Omega, \mathcal{S}, \mu)$ and

$$\int_\Omega |fg| d\mu \leq \|f\|_p \cdot \|g\|_q. \tag{1.2.24}$$

$\square$

Theorem 1.81 (Minkwoski's inequality). *Let $p \in [1, \infty]$, then,*

$$\forall f, g \in L^p(\Omega, \mathcal{S}, \mu): \quad \|f + g\|_p \leq \|f\|_p + \|g\|_p.$$

$\square$

Theorem 1.82. *Fix a sigma-finite measured space $(\Omega, \mathcal{S}, \mu)$.*

(i) For any $p \in [1, \infty]$, the pair $\big(L^p(\Omega, \mathcal{S}, \mu), \| - \|_p \big)$ is a Banach space.

(ii) If $p \in [1, \infty)$, the vector subspace of p-integrable elementary functions is dense in $L^p(\Omega, \mathcal{S}, \mathbb{P})$. In particular, if $\mathcal{S}$ is generated as a sigma-algebra by a countable collection of sets, then $L^p(\Omega, \mathcal{S}, \mu)$ is separable. $\square$

The above density result follows from a combined application of the Monotone Class Theorem and the Monotone Convergence Theorem; see Exercise 1.4.

Suppose that $(\Omega, \mathcal{S}, \mu)$ is a measured space and $p \in [1, \infty]$. Denote by q the exponent conjugate to p, i.e.,

$$\frac{1}{p} + \frac{1}{q} = 1 \Longleftrightarrow q = \frac{p}{p-1}.$$

If $g \in L^q(\Omega, \mathcal{S}, \mu)$, then Hölder's inequality shows that $fg \in L^1$, $\forall f \in L^p(\Omega, \mathcal{S}, \mu)$ and the resulting linear map

$$L^p(\Omega, \mathcal{S}, \mu) \ni f \mapsto \xi_g(f) := \int_\Omega gf d\mu \in \mathbb{R}$$

is continuous.

Theorem 1.83. *Suppose that $(\Omega, \mathcal{S}, \mu)$ is a sigma-finite measured space and $p \in (1, \infty)$. Then the map*

$L^q(\Omega, \mathcal{S}, \mu) \ni g \mapsto \xi_g \in L^p(\Omega, \mathcal{S}, \mu)^* =$ *the dual of the Banach space $L^p(\Omega, \mathcal{S}, \mu)$ is a bijective isometry of Banach spaces.* □

1.2.5 Measures on compact metric spaces

Up to this point we have indicated how one can use a measure to define an integral. The integral is a linear functional on an appropriate space of measurable spaces.

On certain measurable spaces one can invert this process. Suppose that X is a topological space and $\mathcal{B} = \mathcal{B}_X$ is the sigma algebra of Borel sets. We denote by $C_b(X)$ the vector space of bounded continuous functions on X. This is equipped with the sup-norm

$$\|f\|_\infty = \sup_{x \in X} |f(x)|.$$

Any finite Borel measure μ on $\mathcal{B}$ defines via integration a continuous linear functional

$$I_\mu : C_b(X) \to \mathbb{R}, \quad I_\mu[f] = \int_X f(x) \mu[dx].$$

This linear functional satisfies the positivity condition

$$I_\mu[f] \geq 0, \quad \forall f \in C_b(X), \ f \geq 0. \tag{Pos}$$

On metric spaces the measure μ is uniquely determined by the associated functional I_μ. More precisely we have the following fact.

Proposition 1.84. *If X is a metric space and μ, ν are two finite Borel measures such that*

$$I_\mu[f] = I_\mu[f], \quad \forall f \in C_b(X),$$

then $\mu[B] = \nu[B]$ for any subset $B \subset X$.

Proof. Since the Borel sigma-algebra of X is generated by the π-system $\mathcal{C}_X$ of closed subsets it suffices to show that

$$\mu[C] = \nu[C], \quad \forall C \in \mathcal{C}_X.$$

To see that this indeed the case fix $C \in \mathcal{C}_X$ and, for any $n \in \mathbb{N}$ denote by D_n the closed set

$$D_n := \{x \in X; \ \mathrm{dist}(x, C) \geq 1/n\}.$$

Define $f_n \in C_b(X)$

$$f_n(x) := \frac{\mathrm{dist}(x, D_n)}{\mathrm{dist}(x, D_n) + \mathrm{dist}(x, C)}.$$

The function f_n is identically 1 on C and identically 0 on D_n. Moreover

$$\lim_{n \to \infty} f_n(x) = \mathbf{I}_C(x), \quad \forall x \in X.$$

Using the Dominated Convergence Theorem we deduce

$$\mu[C] = \lim_{n \to \infty} I_\mu[f_n] = \lim_{n \to \infty} I_\nu[f_n] = \nu[C].$$

□

We want to include a useful consequence of the above proof.

Corollary 1.85. *Suppose that X is a metric space and μ is a finite Borel measure on X. Then the space $C_b(X)$ is dense in $L^1(X, \mathcal{B}_X, \mu)$.* □

We have the following remarkable result.

Theorem 1.86 (Riesz Representation). *Suppose that X is a compact metric space and L is a continuous linear functional on $C(X)$ satisfying the positivity condition (**Pos**). Then there exists a unique finite Borel measure μ on X such that*

$$L[f] = I_\mu[f], \quad \forall f \in C(X).$$

□

For a proof we refer to [52, Sec. IV.6, Thm. 3] or [148, Thm. 13.5].

Example 1.87. We can use the above result to construct probability measures on a smooth compact manifold M of dimension m. As shown in e.g. [122, Sec. 3.4.1] a Riemann metric g on M defines a continuous linear functional

$$C(M) \ni f \mapsto \int_M f dV_g \in \mathbb{R},$$

usually referred to as the integral with respect to the volume element determined by g. The Riesz Representation Theorem shows that this corresponds to the integral with respect to a finite Borel measure Vol_g on M called the *metric measure*. The metric volume of M is then

$$\mathrm{Vol}_g[M] = \int_M \boldsymbol{I}_M dV_g.$$

We can associate to it the metric probability measure $\mathbb{P}_g$

$$\mathbb{P}_g[B] := \frac{1}{\mathrm{Vol}_g[M]} \mathrm{Vol}_g[B],$$

for any Borel subset $B \subset M$.

In particular, if M is a compact submanifold of an Euclidean space $\mathbb{R}^N$, then it comes equipped with an induced metric and as such, with a finite metric measure μ_M and thus with a probability measure $\mathbb{P}_M$. We will refer to this probability measure as the *Euclidean probability measure*.

Suppose for example that $M = S^m$ is the unit sphere in $\mathbb{R}^m$

$$S^m := \{ (x_0, x_1 \ldots, x_m) \in \mathbb{R}^{m=1};\ x_0^2 + \cdots + x_m^2 = 1 \}.$$

The Euclidean volume of S^m is (see e.g. [122, Eq. (9.1.10)])

$$\boldsymbol{\sigma}_m := \frac{2\pi^{(m+1)/2}}{\Gamma(\frac{m+1}{2})}$$

and the Euclidean probability measure is
$$\mathbb{P}_{S^m} = \frac{1}{\sigma_m}\mu_{S^m}.$$
For example, if $m=1$, then μ_{S^1} is expressed traditionally as $d\theta$ where θ is the angular coordinate. Hence
$$\mathbb{P}_{S^1}[d\theta] = \frac{1}{2\pi}d\theta. \tag{1.2.25}$$
If we use spherical coordinates (φ,θ) on S^2, where φ denotes the Latitude and θ the Longitude, then
$$\mathbb{P}_{S^2}[d\varphi d\theta] = \frac{1}{4\pi}\sin\varphi d\varphi d\theta. \tag{1.2.26}$$
□

1.3 Invariants of random variables

We have defined the random variables as measurable functions on a probability space. In concrete examples this probability is not specifically mentioned. In fact there could be different looking random variables describing essentially the same random quantity.

Consider for example the simplest example of rolling a fair die and observing the number N that shows up. The possible values of N are $\{1,\ldots,6\}$. We equip $\mathbb{I}_6$ with the uniform probability measure and then we can view N as the map
$$N : \mathbb{I}_6 \to \mathbb{R}, \quad N(k) = k, \quad \forall k \in \mathbb{I}_6.$$
Consider now a different experiment. Pick a point x uniformly random in $(0,1]$. We receive a reward $R(x) = k \in \mathbb{I}_6$ if $\lceil 6x \rceil = k$. The functions N and R are obviously different but the random quantities they described are very similar and they should have many things in common.

This is analogous to the situation we encounter in geometry or physics when the same physical or geometric object can be given different descriptions using different coordinates. The laws of physics or geometry are however independent of coordinates. Technically, this means they are described in terms of tensors.

In this section we explain a few basic techniques for describing the behavior of random variables that capture the similarities we observe intuitively.

1.3.1 The distribution and the expectation of a random variable

Fix a probability space $(\Omega,\mathcal{S},\mathbb{P})$. For any random variable $X \in \mathcal{L}^0(\Omega,\mathcal{S})$ the most basic invariant is its *probability distribution* or the *law* of X, i.e., the pushforward
$$\mathbb{P}_X := X_\#\mathbb{P}. \tag{1.3.1}$$
Thus $\mathbb{P}_X$ is a Borel probability measure on $\bar{\mathbb{R}}$ and, as such, it is uniquely determined by the *cumulative distribution function* (cdf)
$$F(x) = F_X(x) := \mathbb{P}[X \le x].$$

More precisely, $\mathbb{P}_X$ can be identified with the associated Lebesgue-Stieltjes measure,

$$\mathbb{P}_X = dF_X.$$

When the random variable X is *discrete*, i.e., the range of X is a finite or countable discrete subset $\mathscr{X} \subset \mathbb{R}$, then $\mathbb{P}_X$ is completely determined by the "mass" of each $x \in \mathscr{X}$,

$$\mathbb{P}_X[\{x\}] = \mathbb{P}[X = x].$$

For this reason in this case the probability distribution of X is often referred as the *probability mass function* (or pmf) of X.

⚘ Given a Borel probability measure μ on $\bar{\mathbb{R}}$, we will use the notation $X \sim \mu$ to indicate that the probability distribution of X is μ, i.e., $\mathbb{P}_X = \mu$.

Any probability measure μ on $(\bar{\mathbb{R}}, \mathcal{B}_{\bar{\mathbb{R}}})$ tautologically defines a random variable with probability distribution μ. If we denote by $\mathbb{1}_{\bar{\mathbb{R}}}$ the identity map $\bar{\mathbb{R}} \to \bar{\mathbb{R}}$, then the random variable

$$X = \mathbb{1}_{\bar{\mathbb{R}}} : (\bar{\mathbb{R}}, \mathcal{B}_{\bar{\mathbb{R}}}, \mu) \to \bar{\mathbb{R}}$$

has probability distribution $\mathbb{P}_X = \mu$. Because of this fact random variables are often identified with their probability distributions. We will use the notations

$$X \stackrel{d}{=} Y \text{ or } X \sim Y$$

to indicate that X and Y have the same distribution.

Definition 1.88 (Expectation). The *expectation* or the *mean* of the integrable random variable $X \in \mathcal{L}^1(\Omega, \mathcal{S}, \mathbb{P})$ is the quantity

$$\mathbb{E}[X] = \mathbb{E}_{\mathbb{P}}[X] := \int_\Omega X(\omega) \mathbb{P}[d\omega]. \qquad \square$$

We deduce from the Change in Variables Theorem 1.75 that

$$\int_{\mathbb{R}} x \mathbb{P}_X[dx] = \int_{\mathbb{R}} \mathbb{1}_{\mathbb{R}}(x) X_\# \mathbb{P}[dx] = \int_\Omega \mathbb{1}_{\mathbb{R}}(X(\omega)) \mathbb{P}[d\omega] = \mathbb{E}[X]$$

so that obtain the useful formula

$$\mathbb{E}[X] = \int_{\mathbb{R}} x \mathbb{P}_X[dx]. \qquad (1.3.2)$$

If $F(x) = F_X(x)$ is the cdf of X, $F(x) = \mathbb{P}[X \leq x]$, then the distribution $\mathbb{P}_X$ is the Lebesgue-Stieltjes measure dF determined by F and (1.3.2) takes the classical form

$$\mathbb{E}[X] = \int_{\mathbb{R}} x dF(x). \qquad (1.3.3)$$

The above equality shows that

$$X \stackrel{d}{=} Y \Rightarrow \mathbb{E}[X] = \mathbb{E}[Y].$$

More generally, for any Borel measurable function $f : \mathbb{R} \to \mathbb{R}$ such that $f(X)$ is integrable or nonnegative we have[3]

$$\mathbb{E}[f(X)] = \int_{\mathbb{R}} f(x) \mathbb{P}_X[dx]. \tag{1.3.4}$$

In other words, the expectation of a random variable is determined by its probability distribution alone, and not on the precise nature of the sample space on which it is defined.

For example, the random variables N and R described at the beginning of this section have the same distribution and thus they have the same mean

$$\mathbb{E}[N] = \mathbb{E}[R] = \frac{1 + \cdots + 6}{6} = \frac{7}{2}.$$

Remark 1.89 (Bertrand's paradox). More often than not, in concrete problems the sample space where a random variable is defined is not explicitly mentioned. Sometimes this can create a problem. Consider the following classical example.

Pick a chord <u>at random</u> on a unit circle. What is the probability that its length is at most $\sqrt{3}$, the length of the edge of an equilateral triangle inscribed in that unit circle?

The answer depends on the concept of "*at random*" we utilize.

For example, we can think that a chord is determined by two points θ_1, θ_2 on the circle or, equivalently, by a pair of numbers in $[0, 2\pi]$. The corresponding chord has length $\leq \sqrt{3}$ if and only if $|\theta_1 - \theta_2| \leq \frac{2\pi}{3}$. The region in the square $[0, 2\pi]$ occupied by pairs (θ_1, θ_2) consists of two isosceles right triangles with legs of size $\frac{2\pi}{3}$ with vertices $(0, 2\pi)$ and $(2\pi, 0)$. By gluing these triangles along their hypothenuses we get a square one third the size of $[0, 2\pi]$. *Assuming* that the point (θ_1, θ_2) is chosen uniformly inside the square $[0, 2\pi]$ we deduce that the probability that the chord has length at most $\sqrt{3}$ is $\frac{1}{9}$.

On the other hand, a chord is uniquely determined by the location of its midpoint inside the unit circle. The chord has length at most $\sqrt{3}$ if and only if the midpoint is at distance at least $\frac{1}{2}$ from the center. *Assuming* that the midpoint is chosen uniformly inside this circle, we deduce that the probability that the chord is at most $\sqrt{3}$ is $\frac{3}{4}$ since the disk of radius $\frac{1}{2}$ occupies $\frac{1}{4}$ of the unit disk.

We can try to decide empirically which is correct answer, but any simulation/experiment must adopt a certain model of randomness. Things are even more complex. The set of chords has a natural symmetry given by the group of rotations about the origin. Any "reasonable" model of randomness ought to be compatible with this symmetry. In mathematical terms this means that the underlying probability measure ought to be invariant with respect to this symmetry.

As a set, we can identify the set of chords with the unit disk: we can describe a chord by indicating the location of its midpoint. The problem boils down to choosing a rotation invariant Borel measure on the unit disk. The quotient of the

---

[3] In undergraduate probability classes this formula is often referred as LOTUS: the **L**aw **O**f **T**he **U**nconscious **S**tatistician.

disk with respect to the group of rotation is a segment. In particular, any probability measure μ on the unit interval defines a rotation invariant probability measure $\mathbb{P}_\mu$ defined by on the unit disk, determined by the requirements

$$\mathbb{P}_\mu\big[\, 0 \leq r \leq r_1, \ \theta_0 \leq \theta \leq \theta_1 \,\big] = \frac{\theta_1 - \theta_0}{2\pi} \mu\big[\, [0, r_1]\,\big].$$

Hence, there are infinitely may geometric randomness models. In our first model of randomness, the measure μ is the distribution of $\cos\left(\frac{\Theta_1 + \Theta_2}{2}\right)$, where Θ_1, Θ_2 are independent uniformly distributed on $[0, 2\pi]$. In the second model of randomness the measure μ is $2r\,dr$. □

If $X, Y \in \mathcal{L}^1(\Omega, \mathcal{S}, \mathbb{P})$ and $a, b \in \mathbb{R}$, then $aX + bY \in \mathcal{L}^1(\Omega, \mathcal{S}, \mathbb{P})$ and

$$\mathbb{E}\big[\, aX + bY \,\big] = a\mathbb{E}\big[\, X \,\big] + b\mathbb{E}\big[\, Y \,\big]. \tag{1.3.5}$$

The above linearity of the expectation is a very powerful tool. Here is a simple illustration.

Example 1.90. Suppose that $n \geq 3$ birds are arranged along a circle looking towards the center. At a given moment each bird randomly and independently turns his head to the left or to the right, with equal probabilities. After they turn their heads, some birds will be visible by one of their neighbors, and some not. Denote by X_n the number of birds that are invisible to their neighbors. We want to compute $\mathbb{E}\big[\, X_n \,\big]$, the expected number of invisible birds. We leave the reader to convince herself/himself that X_n is indeed a well defined mathematical object.

For $k = 1, \ldots, n$ we denote by B_k the event that the k-th bird is invisible to its neighbors. Then

$$X_n = \sum_{k=1}^n \boldsymbol{I}_{B_k} \text{ and } \mathbb{E}\big[\, X_n \,\big] = \sum_{k=1}^n \mathbb{E}\big[\, \boldsymbol{I}_{B_k} \,\big] = \sum_{k=1}^n \mathbb{P}\big[\, B_k \,\big] = n\mathbb{P}\big[\, B_1 \,\big].$$

The probability that the first bird is invisible to is neighbors is computed by observing that this happens iff its right neighbor turn his head right and its left neighbor turn its head left. Since they do this independently with probabilities $\frac{1}{2}$ we deduce

$$\mathbb{P}\big[\, B_1 \,\big] = \frac{1}{2} \cdot \frac{1}{2} = \frac{1}{4}.$$

Hence

$$\mathbb{E}\big[\, X_n \,\big] = \frac{n}{4}.$$

To appreciate how efficient this computation is we present an alternate method.

We will determine the expectation by determining the probability distribution of X_n, or equivalently its *probability generating function* (pgf)

$$G_{X_n}(t) = \mathbb{E}\big[\, t^{X_n} \,\big] = \sum_{k \geq 0} \mathbb{P}\big[\, X_n = k \,\big] t^k.$$

I learned the argument below from *Luke Whitmer*, a student in one of my undergraduate probability courses.

Assume the birds sit on the edges of a convex n-gone $\mathcal{P}_n$. Orienting an edge corresponds to describing in which direction the corresponding bird is looking. We will refer to a choice of orientations of the edges of $\mathcal{P}_n$ as an *orientation* of $\mathcal{P}_n$. We denote by Ω_n the collection of orientations of $\mathcal{P}_n$. Note that $|\Omega_n| = 2^n$.

Fix a cyclic clockwise labelling of the vertices of n-gon, $v_1, v_2, \ldots, v_n$ and define v_m for $m \in \mathbb{N}$ by requiring $v_i = v_j$ if $i \equiv j \mod n$. The i-th bird sits on the edge $E_i := [v_i, v_{i+1}]$. The i-th bird, or equivalently the edge E_i is invisible to its neighbors if E_{i-1} is oriented from v_i to v_{i-1} and E_{i+1} is oriented from v_{i+1} to v_{i+2}. Given an orientation ω of $\mathcal{P}_n$ we denote by $x_n(\omega)$ the number of invisible edges in this orientation. Thus

$$\mathbb{P}[\,X_n = j\,] = \frac{\#\{\,\omega \in \Omega_n;\; x_n(\omega) = j\,\}}{2^n}.$$

We distinguish two cases.

1. $n = 2k$. Denote by $\mathcal{P}_n^+$ the polygon obtained from $\mathcal{P}_n$ by collapsing the edges $E_1, E_3, E_5, \ldots$. As vertices of the new polygon we can take the collapsed edges. The edges of the new polygon are

$$E_1^+ = E_2,\ E_2^+ = E_4, \ldots, E_k^+ = E_{2k}.$$

Similarly, we denote by $\mathcal{P}_n^-$ the polygon obtained from $\mathcal{P}_n$ by collapsing the edges $E_2, E_4, \ldots$. We can take the collapsed edges as vertices of the new polygon. Its edges are

$$E_1^- = E_1,\ E_2^- = E_3, \ldots, E_k^- = E_{2k-1}.$$

Note that an orientation of $\mathcal{P}_n$ induces orientations of $\mathcal{P}_n^\pm$ and conversely, orientations $\mathcal{P}_n^\pm$ determine an orientation of $\mathcal{P}_n$. We denote by $\Omega_n^\pm$ the set of orientations of $\mathcal{P}_n^\pm$. We thus have a bijection

$$\Omega_n \ni \omega \mapsto (\omega_+, \omega_-) \in \Omega_n^+ \times \Omega_n^-.$$

Suppose now that we have an oriented m-gon $\mathcal{Q}_m$. If $q_1, \ldots, q_m$ are the vertices $\mathcal{Q}_m$ we say that v_i is an *out-vertex* if both edges at v_i are oriented away from v_i and it is an *in-vertex*, if both edges at v_i are oriented towards v_i. A *neutral* vertex is a vertex with an incoming edge and one outgoing edge. For an orientation ω of $\mathcal{Q}_m$ we denote by $y_m(\omega)$ the number of out-vertices.

Fix an orientation on $\mathcal{P}_n$. An edge E_i is an invisible in this orientation if and only if the corresponding vertex in $\mathcal{P}_n^{(-1)^i}$ is an out vertex. More explicitly, if i is even/odd, then the corresponding vertex in $\mathcal{P}_n^\pm$ is an out-vertex. Note that,

$$x_{2k}(\omega) = y_k(\omega_+) + y_k(\omega_-). \tag{1.3.6}$$

We denote by $x_{n,j}$ the number of oriented n-gons with j invisible edges and we set

$$P_n(t) = \sum_{j \geq 0} x_{n,j} t^i = \sum_{\omega \in \Omega_n} t^{x_n(\omega)}.$$

Note that

$$G_{X_n}(t) = \frac{1}{2^n} P_n(t).$$

We denote by $y_{m,j}$ the number of oriented m-gons with j out-vertices and we set

$$Q_m(t) := \sum_{j \geq 0} y_{m,j} t^j = \sum_{\omega \in \Omega_m} t^{y_m(\omega)}.$$

From (1.3.6) we deduce

$$P_{2k}(t) = Q_k(t)^2. \tag{1.3.7}$$

2. $n = 2k + 1$. Fix an orientation of $\mathcal{P}_n$. Consider a new oriented n-gon $\mathcal{Q}_n$ with edges, in clockwise order

$$E_1', E_2', \ldots, E_n',$$

where E_i' carries the orientation of the edge $E_{(2i-1) \mod n}$ of $\mathcal{P}_n$. Denote the vertices of $\mathcal{Q}_n$ by $q_1, q_2, \ldots, q_n$, so the two edges that meet at q_i are E_{i-1}' and E_i'.

Imagine stepping in a clockwise fashion on the edges of $\mathcal{P}_n$ and skipping every other edge and labelling by E_i' the i-th edge we stepped on. Observe that the edge $E_{2i \mod n}$ of $\mathcal{P}_n$ is invisible iff the vertex q_{i+1} (where $E_i' \leftrightarrow E_{2i-1}$ and $E_{i+1}' \leftrightarrow E_{2i+1}$ meet) is an out-vertex. Thus, the number of invisible edges of $\mathcal{P}_n$ is equal to the number of out-vertices of $\mathcal{Q}_n$. Hence

$$P_{2k+1}(t) = Q_{2k+1}(t). \tag{1.3.8}$$

To determine $Q_m(t)$ fix an orientation ω of an m-gon $\mathcal{Q}_m$. As we travel clockwise from one vertex to the next, the out- and in-vertices alternate: once we leave an out-vertex, the first non-neutral vertex we meet is an in-vertex and similarly once we leave an in-vertex the first non-neutral vertex we encounter is an out-vertex. In particular this shows that there is an equal number of in and out-vertices. Fix a cyclic labelling $\{1, 2, \ldots, m\}$ of the vertices of $\mathfrak{Q}_m$. If $y_m(\omega) = j$ then $z_m(\omega) = j$ so the set S of locations of in-/out-vertices has cardinality $2j$,

$$S = \{1 \leq \ell_1 < \ell_2 < \cdots < \ell_{2j} \leq m, \}.$$

The above discussion shows that if ℓ_1 is an out/in-vertex, then all vertices $\ell_3, \ell_5, \ldots$ are out/in-vertices while the even vertices $\ell_2, \ell_4, \ldots$ are in/out-vertices. This shows that

$$y_{m,j} = 2\binom{n}{2j}, \quad Q_m(t) = \sum_{j \geq 0} \binom{n}{2j} t^j,$$

$$Q_m(t^2) = (1+t)^m + (1-t)^m.$$

Hence

$$P_{2k}(t^2) = \left((1+t)^k + (1-t)^k\right)^2 = (1+t)^{2k} + (1-t)^{2k} + 2(1-t^2)^k,$$

$$P_{2k+1}(t^2) = (1-t)^{2k+1} + (1+t)^{2k+1}.$$

We conclude that

$$G_{X_n}(t) = \frac{1}{2^n} \times \begin{cases} (1-\sqrt{t})^{2k+1} + (1+\sqrt{t})^{2k+1}, & n = 2k+1, \\ (1+\sqrt{t})^{2k} + (1-\sqrt{t})^{2k} + 2(1-t)^k, & n = 2k. \end{cases}$$

The mean of X_n is

$$\mathbb{E}[X_n] = G'_{X_n}(1).$$

$\square$

Theorem 1.91. *Suppose that $(\Omega, \mathcal{S}, \mathbb{P})$ and $\mathcal{F}, \mathcal{G} \subset \mathcal{S}$ are two independent sigma-subalgebras. If $X \in \mathcal{L}^1(\Omega, \mathcal{F}, \mathbb{P})$, $Y \in \mathcal{L}^1(\Omega, \mathcal{G}, \mathbb{P})$, then $XY \in \mathcal{L}^1(\Omega, \mathcal{S}, \mathbb{P})$ and*

$$\mathbb{E}[XY] = \mathbb{E}[X]\mathbb{E}[Y]. \tag{1.3.9}$$

Proof. Observe that the equality (1.3.9) is bilinear in X and Y. The equality holds for $X = \mathbf{I}_F$, $F \in \mathcal{F}$ and $Y = \mathbf{I}_G$, $G \in \mathcal{G}$ and thus it holds for $X \in \text{Elem}(\Omega, \mathcal{F})$ and $Y \in \text{Elem}(\Omega, \mathcal{G})$.

If X, Y are nonnegative, then $D_n[X]D_n[Y] \nearrow XY$ and the Monotone Convergence Theorem shows that (1.3.9) holds for $X, Y \geq 0$. The bilinearity of this equality implies that it holds in the claimed generality. $\square$

Corollary 1.92. *Suppose that $X, Y \in \mathcal{L}^1(\Omega, \mathcal{S}, \mathbb{P})$ are independent random variables such that $XY \in \mathcal{L}^1(\Omega, \mathcal{S}, \mathbb{P})$. Then*

$$\mathbb{E}[XY] = \mathbb{E}[X]\mathbb{E}[Y]. \tag{1.3.10}$$

Proof. Use Theorem 1.91 with $\mathcal{F} = \sigma(X)$ and $\mathcal{G} = \sigma(Y)$. $\square$

Corollary 1.93. *Suppose that the random variables $X_1, \ldots, X_n : (\Omega, \mathcal{S}, \mathbb{P}) \to \mathbb{R}$ are independent. Then, for any Borel measurable functions $f_1, \ldots, f_n : \mathbb{R} \to \mathbb{R}$ such that*

$$f_i(X_i) \in \mathcal{L}^1(\Omega, \mathcal{S}, \mathbb{P})$$

we have $f_1(X_1) \cdots f_n(X_n) \in \mathcal{L}^1(\Omega, \mathcal{S}, \mathbb{P})$ and

$$\mathbb{E}[f_1(X_1) \cdots f_n(X_n)] = \mathbb{E}[f_1(X_1)] \cdots \mathbb{E}[f_n(X_n)].$$

Proof. Follows inductively from Corollary 1.92 by observing that for any $k = 2, \ldots, n$ the random variables $f_1(X_1) \cdots f_{k-1}(X_{k-1})$ and $f_k(X_k)$ are independent. □

Corollary 1.94. *Let $X \in L^1(\Omega, \mathcal{S}, \mathbb{P})$ and suppose that $\mathcal{F} \subset \mathcal{S}$ is sigma-subalgebra. Then the following are equivalent.*

(i) *For any Borel measurable function $f : \mathbb{R} \to \mathbb{R}$ such that $f(X) \in L^1$ and any $F \in \mathcal{F}$*

$$\mathbb{E}[f(X)\mathbf{I}_F] = \mathbb{P}[F]\mathbb{E}[f(X)].$$

(ii) *The random variable X is independent of $\mathcal{F}$.*

Proof. The implication (i) ⇒ (ii) follows by using $f = \mathbf{I}_{(-\infty,x]}$, $x \in \mathbb{R}$. The converse follows from Theorem 1.92. □

The following is not the usual definition of a convex function (see Exercise 1.23) but it has the advantage that it is better suited for the applications we have in mind.

Definition 1.95. Let I be an interval of the real axis. A continuous function $\varphi : I \to \mathbb{R}$ is called *convex* if for any $x_0 \in I$ there exists a linear function $\ell(x)$ such that[4]

$$\ell(x_0) = \varphi(x_0), \quad \ell(x) \leq \varphi(x), \quad \forall x \in I.$$

The convex function is called *strictly convex* if for any $x_0 \in I$ there exists a linear function $\ell(x)$ such that

$$\ell(x_0) = \varphi(x_0), \quad \ell(x) < \varphi(x), \quad \forall x \in I \setminus \{x_0\}.$$
□

For example, if $\varphi : I \to \mathbb{R}$ is C^2, then φ is convex (resp. strictly convex) if $\varphi''(x) \geq 0$ (resp. $\varphi'(x) > 0$), $\forall x \in I$.

Theorem 1.96 (Jensen's Inequality). *Suppose that $(\Omega, \mathcal{S}, \mathbb{P})$ is a probability space, $X \in \mathcal{L}^1(\Omega, \mathcal{S}, \mathbb{P})$, and $\varphi : I \to \mathbb{R}$ is a convex function defined on an interval I that contains the range of X. Then $\mathbb{E}[\varphi(X)]$ is well defined (possibly infinite) and*

$$\varphi(\mathbb{E}[X]) \leq \mathbb{E}[\varphi(X)]. \tag{1.3.11}$$

Moreover, if φ is strictly convex, then $\varphi(\mathbb{E}[X]) = \mathbb{E}[\varphi(X)]$ iff X is a.s. constant.

Proof. Observe that the when φ is linear theorem is valid in the stronger form $\varphi(\mathbb{E}[X]) = \mathbb{E}[\varphi(X)]$. We can find a linear function $\ell : \mathbb{R} \to \mathbb{R}$ such that $\varphi(x) \geq \ell(x)$, $\forall x \in I$ and it is clear that if the theorem is valid for the nonnegative convex function $g := \varphi - \ell$, then it is also valid for φ. Note that $\mathbb{E}[g(X)] \in [0, \infty]$

---
[4] The graph of such an ℓ is tangent to the graph of φ at x_0.

and thus the addition $\mathbb{E}[g(X)] + \ell(\mathbb{E}[X])$ is well defined and yields a well defined $\mathbb{E}[\varphi(X)]$, when $\varphi(X)$ is integrable or nonnegative. Moreover $\varphi(X)$ is integrable if and only if $g(X)$ is so. Because of this, we set

$$\mathbb{E}[\varphi(X)] := \infty \text{ if } \varphi(X) \text{ is not integrable.}$$

Set $\mu := \mathbb{E}[X]$ and observe that $\mu \in I$ since $X \in I$ a.s. Choose a linear function $\ell : \mathbb{R} \to \mathbb{R}$ such that

$$\ell(x) \leq \varphi(x), \ \forall x \in I \text{ and } \ell(\mu) = \varphi(\mu).$$

Then

$$\varphi(\mathbb{E}[X]) = \varphi(\mu) = \ell(\mu) = \mathbb{E}[\ell(X)] \leq \mathbb{E}[\varphi(X)].$$

If φ is strictly convex, then we can choose $\ell(x)$ such that

$$\ell(x) < \varphi(x), \ \forall x \in I \setminus \{\mu\} \text{ and } \ell(\mu) = \varphi(\mu).$$

If X is not a.s. constant neither is the nonnegative random variable $\varphi(X) - \ell(X)$ so

$$\mathbb{E}[\varphi(X) - \varphi(\mu)] = \mathbb{E}[\varphi(X) - \ell(X)] > 0.$$

$\square$

For any convex function $\varphi : \mathbb{R} \to \mathbb{R}$ we define the φ-*entropy* of an integrable random variable X to be the quantity

$$\mathbb{H}_\varphi[X] := \mathbb{E}[\varphi(X)] - \varphi(\mathbb{E}[X]). \tag{1.3.12}$$

Jensen's inequality shows that $\mathbb{H}_\varphi[X] \geq 0$.

1.3.2 Higher order integral invariants of random variables

On a probability space $(\Omega, \mathcal{S}, \mathbb{P})$ we have the inclusions

$$L^{p_1}(\Omega, \mathcal{S}, \mathbb{P}) \subset L^{p_0}(\Omega, \mathcal{S}, \mathbb{P}), \ \forall 1 \leq p_0 < p_1 \leq \infty.$$

Indeed, let $X \in \mathcal{L}^{p_1}(\Omega, \mathcal{S}, \mathbb{P})$. Set

$$p := \frac{p_1}{p_0}, \ \varphi(x) = x^p, \ x \geq 0, \ Y = \|X\|^{p_0}.$$

Since $p_1 > p_0$ the function φ is convex and we have

$$\left(\|X\|_{p_0}\right)^{p_1} = \mathbb{E}\left[|X|_0^p\right]^p = h(\mathbb{E}[Y]) \overset{(1.3.11)}{\leq} \mathbb{E}[h(Y)] = \left(\|X\|_{p_1}\right)^{p_1}.$$

In particular, if $p_0 = 1 \leq p$ we deduce

$$\mathbb{E}[|X|]^p \leq \mathbb{E}[|X|^p]. \tag{1.3.13}$$

Given $k \in \mathbb{N}$ and $X \in L^k(\Omega, \mathcal{S}, \mathbb{P})$ we define the k-th momentum of X to be the quantity

$$\mu_k[X] := \mathbb{E}[X^k].$$

Note that $\mu_1[X] = \mathbb{E}[X]$.

Definition 1.97 (Variance). Let $(\Omega, \mathcal{S}, \mathbb{P})$ be a probability space. Suppose that $X \in \mathcal{L}^2(\Omega, \mathcal{S}, \mathbb{P})$ is a random variable with mean $\mu := \mathbb{E}[X]$. The *variance* of X is the real number
$$\mathrm{Var}[X] = \mathbb{E}[(X - \mu)^2].$$
The *standard deviation* of X is the quantity
$$\sigma[X] := \sqrt{\mathrm{Var}[X]}.$$
□

Note that
$$\mathrm{Var}[X] = 0 \Longleftrightarrow X = \mathbb{E}[X] \text{ a.s.}$$
The variance can be given the alternate description
$$\mathrm{Var}[X] = \mathbb{E}[X^2] - \mathbb{E}[X]^2 = \mu_2[X] - \mu_1[X]^2. \tag{1.3.14}$$
Indeed, if we set $\mu := \mathbb{E}[X]$, then
$$\mathrm{Var}[X] = \mathbb{E}[X^2 - 2\mu X + \mu^2] = \mathbb{E}[X^2] - 2\mu\mathbb{E}[X] + \mu^2 = \mathbb{E}[X^2] - \mathbb{E}[X]^2.$$
This shows that the variance is a special case of φ-entropy. More precisely,
$$\mathrm{Var}[X] = \mathbb{H}_\varphi[X], \quad \varphi(x) = x^2.$$
Note that
$$\mathrm{Var}[aX + b] = a^2 \mathrm{Var}[X], \quad \forall a, b \in \mathbb{R}. \tag{1.3.15}$$
Indeed, set $\bar{X} := X - \mu$ and $Z := aX + b$. Then
$$\mathrm{Var}[X] = \mathbb{E}[\bar{X}^2], \quad Z - \mathbb{E}[Z] = a(X - \mathbb{E}[X]) = a\bar{X},$$
$$\mathrm{Var}[Z] = \mathbb{E}[a^2 \bar{X}^2] = a^2 \mathrm{Var}[X].$$

Theorem 1.98 (Chebyshev's inequality). Let $X \in \mathcal{L}^2(\Omega,]\mathcal{S}, \mathbb{P})$. Set $\mu := \mathbb{E}[X]$ and $\sigma = \sigma[X]$. Then
$$\mathbb{P}[|X - \mu| \geq c\sigma] \leq \frac{1}{c^2}, \quad \forall c > 0. \tag{1.3.16}$$
Equivalently
$$\mathbb{P}[|X - \mu| \geq r] \leq \frac{\mathrm{Var}[X]}{r^2} = \frac{\sigma^2}{r^2}, \quad \forall r > 0. \tag{1.3.17}$$

Proof. Set $Y := |X - \mu|^2$. Then
$$\mathbb{P}[|X - \mu| > r] = \mathbb{P}[Y > r^2] \stackrel{(1.2.19)}{\leq} \frac{1}{r^2} \mathbb{E}[Y] = \frac{\mathrm{Var}[X]}{r^2}.$$
Chebyshev's inequality (1.3.16) now follows from (1.3.17) by setting $r = c\sigma$. □

Definition 1.99. Let $(\Omega, \mathcal{S}, \mathbb{P})$ be a probability space and $X, Y \in \mathcal{L}^2(\Omega, \mathcal{S}, \mathbb{P})$. We set

$$\mu_X := \mathbb{E}[X], \quad \mu_Y := \mathbb{E}[Y].$$

(i) The *covariance* of X, Y is the quantity

$$\mathrm{Cov}[X, Y] := \mathbb{E}[(X - \mu_X)(Y - \mu_Y)].$$

(ii) If X, Y are not deterministic we define the *correlation coefficient* of X and Y to be

$$\rho[X, Y] := \frac{\mathrm{Cov}[X, Y[}{\sigma[X]\sigma[Y]}.$$

$\square$

Proposition 1.100. *Let $X, Y \in \mathcal{L}^2(\Omega, \mathcal{S}, \mathbb{P})$. Then the following hold.*

(i) $\mathrm{Cov}[X, Y] = \mathbb{E}[XY] - \mathbb{E}[X]\mathbb{E}[Y]$.
(ii) If X, Y are independent, then $\mathrm{Cov}[X, Y] = 0$.
(iii) $\mathrm{Var}[X + Y] = \mathrm{Var}[X] + \mathrm{Var}[Y] + 2\mathrm{Cov}[X, Y]$.
(iv) If X, Y are independent, then $\mathrm{Var}[X + Y] = \mathrm{Var}[X] + \mathrm{Var}[Y]$.

Proof. Set

$$\mu_X := \mathbb{E}[X], \quad \bar{X} = X - \mu_X, \quad \mu_Y = \mathbb{E}[Y[, \quad \bar{Y} = Y - \mu_Y.$$

(i) We have

$$\mathrm{Cov}[X, Y] = \mathbb{E}[\bar{X}\bar{Y}] = \mathbb{E}[XY] - \underbrace{\mathbb{E}[\mu_X Y]}_{\mu_X \mu_Y} - \underbrace{\mathbb{E}[\mu_Y X]}_{\mu_X \mu_Y} + \mu_X \mu_Y$$

$$= \mathbb{E}[XY] - \mu_X \mu_Y.$$

(ii) Corollary 1.92 shows that if X, Y are independent, then $\mathbb{E}[XY] = \mu_X \mu_Y$, i.e., $\mathrm{Cov}[X, Y] = 0$.

(iii) Next

$$\mathrm{Var}[X + Y] = \mathbb{E}[(\bar{X} + \bar{Y})^2] = \mathbb{E}[\bar{X}^2] + \mathbb{E}[\bar{Y}^2] + 2\mathbb{E}[\bar{X}\bar{Y}]$$

$$= \mathrm{Var}[X] + \mathrm{Var}[Y] + 2\mathrm{Cov}[X, Y].$$

(iv) This follows from (ii) and (iii). $\square$

Corollary 1.101. *If $X_1, \ldots, X_n \in \mathcal{L}^2(\Omega, \mathcal{S}, \mathbb{P})$ are independent, then*

$$\mathrm{Var}[X_1 + \cdots + X_n] = \mathrm{Var}[X_1] + \cdots + \mathrm{Var}[X_n]. \tag{1.3.18}$$

$\square$

Example 1.102. Consider a probability space $(\Omega, \mathcal{S}, \mathbb{P})$ and two events $A, B \in \mathcal{S}$. We have

$$\operatorname{Cov}[\boldsymbol{I}_A, \boldsymbol{I}_B] = \mathbb{P}[A \cap B] - \mathbb{P}[A]\mathbb{P}[B].$$

Thus A, B are independent iff $\operatorname{Cov}[\boldsymbol{I}_A, \boldsymbol{I}_B] = 0$. □

Definition 1.103 (Moment generating function). Let X be a random variable defined on a probability space $(\Omega, \mathcal{S}, \mathbb{P})$ such that $e^{tX} \in \mathcal{L}^1(\Omega, \mathcal{S}, \mathbb{P})$ for all t in an open interval I containing 0. The *moment generating function* or *mgf* of X is the function

$$\operatorname{M}_X : I \to \mathbb{R}, \quad \operatorname{M}_X(t) = \mathbb{E}\bigl[e^{tX} \bigr].$$

□

The proof of following result is left to you as an exercise.

Proposition 1.104. *Let X such that $\operatorname{M}_X(t)$ is defined for all $t \in (-t_0, t_0)$.*

(i) The moment generating function determines the momenta of X. More precisely, the function

$$(-t_0, t_0) \ni t \mapsto \operatorname{M}_X(t)$$

is smooth and

$$\operatorname{M}_X^{(k)}(0) = \mu_k[X], \quad \forall k = 1, 2, \ldots. \tag{1.3.19}$$

(ii) The power series

$$\sum_{n=0}^\infty \mu_n[X] \frac{t^n}{n!},$$

converges to $\operatorname{M}_X(t)$, $\forall t \in (-t_0, t_0)$.

□

Corollary 1.105. *Suppose that $X_1, \ldots, X_n \in \mathcal{L}^0(\Omega, \mathcal{S}, \mathbb{P})$ are independent random variables such that $e^{tX_k} \in \mathcal{L}^1(\Omega, \mathcal{S}, \mathbb{P})$ for any $k = 1, \ldots, n$ and any t in an open interval $I \subset \mathbb{R}$ that contains the origin. Then*

$$\operatorname{M}_{X_1+\cdots+X_n}(t) = \operatorname{M}_{X_1}(t) \cdots \operatorname{M}_{X_n}(t), \quad \forall t \in I.$$

Proof. This is a special case of Corollary 1.93 corresponding to the choices

$$f_1(x) = \cdots = f_n(x) = e^{tx}, \quad t \in I.$$

□

Remark 1.106 (The moment problem). Denote by Prob the set of Borel probability measures on the real axis and by $\text{Prob}^{\infty-}$ the subset of Prob consisting of probability measures $\boldsymbol{p}$ such that

$$\int_{\mathbb{R}} |x|^k \boldsymbol{p}[dx] < \infty, \ \forall k \in \mathbb{N}.$$

For $\boldsymbol{p} \in \text{Prob}^{\infty-}$ and $k \in \mathbb{N}_0$ we set

$$\mu_k[\boldsymbol{p}] := \int_{\mathbb{R}} x^k \boldsymbol{p}[dx].$$

We denote by $\mathbb{R}^{\mathbb{N}_0}$ the set of sequences of real numbers $\underline{s} = (s_n)_{n \geq 0}$. We have a map

$$\boldsymbol{\mu} : \text{Prob}^{\infty-} \to \mathbb{R}^{\mathbb{N}_0}, \ \boldsymbol{\mu}[\boldsymbol{p}] = \big(\mu_n[\boldsymbol{p}]\big)_{n \geq 0}.$$

The *moment problem* asks the following.

(i) Describe the range of $\boldsymbol{\mu}$, i.e., given a sequence of real numbers $\underline{s} = s_0, s_1, \ldots$, decide if there exists $\boldsymbol{p} \in \text{Prob}^{\infty-}$ such that $\mu_n[\boldsymbol{p}] = s_n$, $\forall n \geq 0$.
(ii) Is it true that the moments uniquely determine a probability measure, i.e., given $\underline{s}$ in the range of $\boldsymbol{\mu}$ is it true that there exists a *unique* $\boldsymbol{p} \in \text{Prob}^{\infty-}$ such that $\boldsymbol{\mu}[\boldsymbol{p}] = \underline{s}$?

Party (i) of the moment problem is completely understood in the sense that there are known several necessary and sufficient conditions for a sequence $\underline{s}$ to be the sequence of momenta of a probability measure on $\mathbb{R}$. We refer to [137, Chap. 3] for more details.

As for part (ii), it is known that a sequence $\underline{s}$ can be the sequence of momenta of *several* probability measures; see Exercise 1.30. On the other hand, there are known sufficient conditions on $\underline{s}$ guaranteeing the uniqueness of measure with that sequence of momenta; see [137, Chap. 4] for more details. In particular, if X is a random variable such that e^{tX} is integrable for any t in an open interval containing 0, then $\mathbb{P}_X$ is uniquely determined by its moments, [137, Cor. 4.14]. $\square$

We formulate for the record the last uniqueness result mentioned above. In Exercise 2.45 we outline a proof of this special case.

Theorem 1.107. *Let $X, Y \in \mathcal{L}^0(\Omega, \mathcal{S}, \mathbb{P})$ such that there exist $r > 0$ with the property that*

$$\mathbb{E}\big[e^{tX}\big], \ \mathbb{E}\big[e^{tX}\big] < \infty, \ \forall |t| < r.$$

Then

$$X \stackrel{d}{=} Y \iff \mathbb{M}_X(t) = \mathbb{M}_Y(t), \ \forall |t| < r. \qquad \square$$

Corollary 1.108. *Suppose that* $\mathbb{P}_0, \mathbb{P}_1$ *are Borel probability measures on* $\mathbb{R}$ *supported on* $[0, 1]$, *i.e.*,

$$\mathbb{P}_0[\mathbb{R} \setminus [0,1]] = \mathbb{P}_1[\mathbb{R} \setminus [0,1]] = 0.$$

Then

$$\mathbb{P}_0 = \mathbb{P}_1 \iff \int_\mathbb{R} x^n \mu_0[dx] = \int_\mathbb{R} x^n \mu_1[dx], \quad \forall n \in \mathbb{N}.$$

□

Proof. Note that

$$\int_\mathbb{R} x^n \mathbb{P}_i[dx] \leq 1 \Rightarrow \int_\mathbb{R} e^{tx} \mathbb{P}_i[dx] < \infty, \quad \forall t \in \mathbb{R}$$

and

$$\int_\mathbb{R} e^{tx} \mathbb{P}_0[dx] = \int_\mathbb{R} e^{tx} \mathbb{P}_1[dx], \quad \forall t \in \mathbb{R}$$

$$\iff \int_\mathbb{R} x^n \mathbb{P}_0[dx] = \int_\mathbb{R} x^n \mathbb{P}_1[dx], \quad \forall n \in \mathbb{N}_0.$$

□

To a random variable X with range contained in $\mathbb{N}_0 = \{0, 1, 2, \dots\}$ we can associate its *probability generating function* (or pgf)

$$G_X(t) := \sum_{n \geq 0} \mathbb{P}[X = n] t^n = \mathbb{E}[t^X].$$

Note that

$$G_X(1) = 1, \quad G'_X(1) = \mathbb{E}[X], \quad G''_X(1) = \mathbb{E}[X(X-1)]. \tag{1.3.20}$$

Similarly, if X, Y are two independent $\mathbb{N}_0$-valued random variables, then

$$G_{X+Y}(t) = \mathbb{E}[t^{X+Y}] = \mathbb{E}[t^X] \mathbb{E}[t^Y] = G_X(t) G_Y(t).$$

1.3.3 Classical examples of discrete random variables

The theory of probability has grown mostly from concrete intriguing examples. In this process people encountered various frequently occurring patterns encoded by some ubiquitous random variables. We describe a few of them in the following subsections. These examples are part of the theory of probability and have many and varied uses. Their knowledge is absolutely necessary for a genuine understanding of probability.

Before Kolmogorov (and currently in most undergraduate probability courses), the world of random variables was divided into three categories: discrete, continuous and neither, or mixed. The discrete random variables are those whose ranges are discrete subsets of $\mathbb{R}$. A random variable X is called continuous if its probability distribution $\mathbb{P}_X$ is absolutely continuous with respect to the Lebesgue measure on

R. We throw in the third category the random variables that do not fit in these two categories. We want to describe a few classical example of discrete and continuous random variables that play an important role in probability. Throughout our presentation we will frequently assume that given a sequence $(\mu_n)_{n\in\mathbb{N}}$ of Borel probability measures on $\mathbb{R}$ there exists a probability space $(\Omega, \mathcal{S}, \mathbb{P})$ and independent random variables $X_n : (\Omega, \mathcal{S}, \mathbb{P}) \to \mathbb{R}$ such that $\mathbb{P}_{X_n} = \mu_n$, $\forall n \in \mathbb{N}$. The fact that such a thing is possible is a consequence of Kolmogorov's existence theorem, Theorem 1.195.

We begin by introducing some frequently occurring discrete random variables by describing the random experiments where they appear.

Example 1.109 (Bernoulli random variables). Suppose we perform a random experiment aiming to observe the occurrence of a certain event S, $p := \mathbb{P}[S]$. When S has occurred we say that we have registered a success. Traditionally such an experiment is called a *Bernoulli trial* with success probability p. When the event S is not observed we say that the experiment was a failure. The failure probability is $q := 1 - p$. The Bernoulli trial is encoded by the random variable $\boldsymbol{I}_S$ which takes the value 1 when we register a success, and the value 0 otherwise. We also say that $\boldsymbol{I}_S$ is a *Bernoulli random variable* observe that

$$\mathbb{E}[\boldsymbol{I}_S] = p, \ \operatorname{Var}[\boldsymbol{I}_S] = \mathbb{E}[\boldsymbol{I}_S^2] - (\mathbb{E}[\boldsymbol{I}_S])^2 = p - p^2 = pq.$$

Note that any random variable with range $\{0, 1\}$ is a Bernoulli random variable since $X = \boldsymbol{I}_{\{X=1\}}$. □

Example 1.110 (Binomial random variables). Suppose that we perform the experiment in the above example n times, and the results of these experiments are independent of each other. We denote by N the number of successes observed during these n trials.[5] We say that N is a *binomial random variable* corresponding to n trials with success probability p and we indicate this $N \sim \operatorname{Bin}(n, p)$.

For $k = 1, \ldots, n$ we denote by S_k the event *"the k-th trial was a success"*. Then

$$N = \sum_{k=1}^{n} \boldsymbol{I}_{S_k} \text{ and } \mathbb{E}[N] = \sum_{k=1}^{n} \mathbb{E}[\boldsymbol{I}_{S_k}] = np.$$

Since the events $(S_k)_{1 \leq k \leq n}$ are independent we deduce from Corollary 1.93 that

$$\operatorname{Var}[N] = \sum_{k=1}^{n} \operatorname{Var}[\boldsymbol{I}_{S_k}] = npq.$$

Next observe that

$$G_N(s) = q + ps, \ \mathrm{M}_N(t) = q + pe^t,$$

---

[5]Think for example that you roll a pair of dice 10 times and you aim to count how many times the sum of the numbers on the dice is 7. In this case success is when the sum is 7 and it is not hard to see that the probability of success is $\frac{1}{6}$.

so
$$G_N(s) = G_{I_{S_1}}(t)^n = (q+ps)^n, \quad \mathbb{M}_N(t) = \mathbb{M}_{I_{S_1}}(t)^n = (q+pe^t)^n. \qquad \square$$

This string of Bernoulli trials can be realized abstractly in the probability space
$$\left(\{0,1\}^n, 2^{\{0,1\}^n}, \beta_p^{\otimes n} \right)$$
described in Example 1.31(e). The events
$$S_k := \{ (\epsilon_1, \ldots, \epsilon_n) \in \{0,1\}^n; \; \epsilon_k = 1 \}, \quad k = 1, \ldots, n,$$
are independent and $\mathbb{P}[S_k] = p$, $\forall k = 1, \ldots, n$. Then
$$I_{S_k}(\epsilon) = \epsilon_k, \quad \forall \epsilon = (\epsilon_1, \ldots, \epsilon_n) \in \{0,1\}^n.$$
As explained in Example 1.31(e), the probability distribution of N is given by the equalities
$$\mathbb{P}[N = k] = \binom{n}{k} p^k q^{n-k}, \quad k = 0, 1, \ldots, n.$$
Equivalently,
$$\mathbb{P}_N = \sum_{k=0}^n \binom{n}{k} p^k q^{n-k} \delta_k.$$

$\square$

Example 1.111 (Waiting for successes). Suppose that we perform *independent* Bernoulli trials until we register the first success. We denote by T_1 the moment we observe the first success, $T_1 \in \{1, 2, \ldots, \infty\}$. The random variable T_1 is a *geometric random variable* with success probability p. We write this $T_1 \sim \text{Geom}(p)$.

Observe that $T_1 = n$ iff the first $n-1$ trials where failures and the n-th trial was a success. Thus
$$\mathbb{P}[T_1 = n] = q^{n-1} p.$$
In particular, $\mathbb{P}[N_1 = \infty] = 0$. We deduce that the probability distribution of T_1 is
$$\mathbb{P}_{T_1} = \sum_{n \geq 1} p q^{n-1} \delta_n.$$
Moreover
$$\mathbb{E}[T_1] = \sum_{n \geq 1} n p q^{n-1} = p \sum_{n \geq 1} n q^{n-1} = p \frac{d}{dq} \sum_{n \geq 0} q^n = \frac{p}{(1-q)^2} = \frac{1}{p}. \qquad (1.3.21)$$

Here is a simple plausibility test for this result. Suppose we role a die until we first roll a 1. The probability of rolling a 1 is $\frac{1}{6}$ so it is to be expected that we need 6 rolls until we roll our first 1.

We have

$$\mathbb{E}[T_1^2] - \mathbb{E}[T_1] = \sum_{n=1}^\infty n(n-1)pq^{n-1} = \sum_{n=2}^\infty n(n-1)pq^{n-1}$$

$$= pq \sum_{n=2}^\infty n(n-1)pq^{n-2} = pq \frac{d^2}{dq^2}\left(\frac{1}{1-q}\right) = \frac{2pq}{(1-q)^3} = \frac{2q}{p^2}.$$

We deduce that

$$\mathbb{E}[T_1^2] = \frac{2q}{p^2} + \frac{1}{p}, \ \ \operatorname{Var}[T_1] = \frac{2q}{p^2} + \frac{1}{p} - \frac{1}{p^2} = \frac{q}{p^2}.$$

Note that

$$\mathbb{M}_{T_1}(t) = \mathbb{E}[e^{tT_1}] = \sum_{n=1}^\infty pq^{n-1}e^{nt} = pe^t \sum_{m=0}^\infty (qe^t)^m = \frac{pe^t}{1-e^t}.$$

Consider now a more general situation. Fix $k \in \mathbb{N}$ and perform independent Bernoulli trials until we observe the k-th success. Denote by T_k the number trials until we record the k-th success. Note that

$$T_k = T_1 + (T_2 - T_1) + (T_3 - T_2) + \cdots + (T_k - T_{k-1}).$$

Due to the independence of the trials, once we observe the i-th success it is as if we start the experiment anew, so the waiting time $T_{i+1} - T_i$ until we observe the next success, the $(i+1)$-th, is a random variable with the same distribution as T_1

$$T_{i+1} - T_i \overset{d}{=} T_1, \ \forall i \in \mathbb{N}.$$

Hence $\mathbb{E}[T_{i+1} - T_i] = \mathbb{E}[T_1] = \frac{1}{p}$ so

$$\mathbb{E}[T_k] = k\mathbb{E}[T_1] = \frac{k}{p}. \tag{1.3.22}$$

The probability distribution of T_k is computed as follows. Note that $T_k = n$ if during the first $n-1$ trials we observed exactly $k-1$ successes, and at the n-th trial we observed another success. Hence

$$\mathbb{P}[T_k = n] = \binom{n-1}{k-1}p^{k-1}q^{n-k} \cdot p = \binom{n-1}{k-1}p^k q^{n-k}, \tag{1.3.23}$$

and

$$\mathbb{M}_{T_k}(t) = \left(\frac{pe^t}{1-e^t}\right)^k.$$

Since the waiting times between two consecutive successes are independent random variables we deduce

$$\operatorname{Var}[T_k] = k\operatorname{Var}[T_1] = \frac{kq}{p^2}.$$

The above probability measure on $\mathbb{R}$ is called the *negative binomial distribution* and T_k is called a *negative binomial random variable* corresponding to k successes with probability p. We write this $T_k \sim \operatorname{NegBin}(k,p)$. □

Let us describe a classical and less than obvious application of the geometric random variables.

Example 1.112 (The coupon collector problem). The coupon collector's problem arises from the following scenario.

Suppose that each box of cereal contains one of m different coupons. Once you obtain one of every type of coupons, you can send in for a prize. Ann wants that prize and, for that reason, she buys one box of cereals everyday. Assuming that the coupon in each box is chosen independently and uniformly at random from the m possibilities and that Ann does not collaborate with others to collect coupons, how many boxes of cereal is she expected to buy before she obtain at least one of every type of coupon?

Let N denote the number of boxes bought until Ann has at least one of every coupon. We want to determine $\mathbb{E}[N]$. For $i = 1, \ldots, n-1$ denote by N_i the number of boxes she bought while she had exactly i coupons. The first box she bought contained one coupon. Then she bought N_1 boxes containing the coupon you already had. After $1 + N_1$ boxes she has two coupons. Next, she bought N_2 boxes containing one of the two coupons you already had etc. Hence[6]

$$N = 1 + N_1 + \cdots + N_{m-1}.$$

Let us observe first that for $i = 1, \cdots, m-1$ we have

$$N_i \sim \text{Geom}(p_i), \quad p_i = \frac{m-i}{m}, \quad q_i = 1 - p_i = \frac{i}{m}.$$

Indeed, at the moment she has i coupons, a success occurs when she buys one of the remaining $m-i$ coupons. The probability of buying one such coupon is thus $\frac{m-i}{m}$. Think of buying a box at this time as a Bernoulli trial with success probability $\frac{m-i}{m}$. The number N_i is then equal to the number of trials until you register the first success. This argument also shows that the random variables N_i are independent. In particular,

$$\mathbb{E}[N_i] = \frac{1}{p_i} = \frac{m}{m-i}.$$

From the linearity of expectation we deduce

$$\mathbb{E}[N] = 1 + \mathbb{E}[N_1] + \mathbb{E}[N_2] + \cdots + \mathbb{E}[N_{m-1}]$$
$$= m \underbrace{\left(1 + \frac{1}{2} + \cdots + \frac{1}{m-1} + \frac{1}{m}\right)}_{=:H_m}.$$

Asymptotically H_m differs from $\log m$ by the mysterious Euler-Mascheroni constant $\gamma \approx 0.5772$, i.e.,

$$\lim_{m \to \infty}(H_m - \log m) = \gamma.$$

Thus the expected number of boxes needed to collect all the m coupons is about $m \log m + m\gamma$. □

---
[6]Here we tacitly assume that we can describe quantities N_i as measurable functions defined on the same probability space. In Exercise 1.7 we ask the reader to do this. It is more challenging than it looks.

Remark 1.113. We can ask a more general question. For $k \geq 1$ we at denote by $X_k = X_{k,m}$ the number of boxes Ann has to buy until she has at least k of these m coupons. We have seen that $\mathbb{E}[X_{1,m}] = mH_m$. One can show that as $m \to \infty$ we have

$$\mathbb{E}[X_{k,m}] = m\big(\log m + (k-1)\log\log m + \gamma - \log(k-1)! + o(1)\big),$$

where γ is the Euler-Mascheroni constant. For details we refer to [55; 120]. □

Example 1.114 (The hypergeometric distribution). Suppose that we have a bin containing w white balls and b black balls. We select n balls at random from the bin and we denote by X the number of white balls among the selected ones. This is a random variable with range $0, 1, \ldots, n$ called the *hypergeometric random variable* with parameters w, b, n. We will use the notation $X \sim \mathrm{HGeom}(w, b, n)$ to indicate this and we will refer to its pmf as the *hypergeometric distribution*. For example, if A is the number of aces in a random poker hand, then $A \sim \mathrm{HGeom}(4, 48, 5)$.

To compute $\mathbb{P}[X = k]$ when $X \sim \mathrm{HGeom}(w, b, n)$ note that a favorable outcome for the event $X = k$ is determined by a choice of k white balls (out of w) and another independent choice of $n - k$ black balls (out of b) so that the number of favorable outcomes is

$$\binom{w}{k}\binom{b}{n-k}.$$

The number of possible outcomes of a random draw of n balls $\binom{w+b}{n}$. Hence

$$\mathbb{P}[X = k] = \frac{\binom{w}{k}\binom{b}{n-k}}{\binom{w+b}{n}}.$$

Its probability generating function is

$$G_X(s) = \frac{1}{\binom{N}{n}} \sum_{k=0}^{w} \binom{w}{k}\binom{b}{n-k} s^k, \quad N := w + b.$$

We can identify $G_X(s)$ as the coefficient of x^n in the polynomial

$$Q(s,x) = \frac{1}{\binom{N}{n}}(1+sx)^w(1+x)^b.$$

We have

$$\frac{\partial Q}{\partial s}(s,x) = \frac{wx(1+x)^b}{\binom{N}{n}}(1+sx)^{w-1}.$$

The mean of X is $G'_X(1)$ and it is equal to the coefficient of x^n in

$$\frac{\partial Q}{\partial s}(1,x) = \frac{wx}{\binom{w+b}{n}}(1+x)^{N-1} = \frac{w\binom{N-1}{n-1}}{\binom{N}{n}} = \frac{wn}{N} = \frac{wn}{w+b}.$$

Hence

$$\mathbb{E}[\,\mathrm{HGeom}(w,b,n)\,] = \frac{w}{w+b} \cdot n. \tag{1.3.24}$$

□

Example 1.115 (Poisson random variables). These random variables count the number N of random rare events that occur in a given unit of time. E.g., N could mean the number of computers in a large organization that die during one fiscal year. They depend on a parameter λ and we indicate this using the notation $N \sim \text{Poi}(\lambda)$. If $N \sim \text{Poi}(\lambda)$, then

$$\mathbb{P}[N = n] = e^{-\lambda}\frac{\lambda^n}{n!}, \quad \text{i.e.,} \quad \mathbb{P}_N = \sum_{n=0}^{\infty} e^{-\lambda}\frac{\lambda^n}{n!}\delta_n.$$

Then

$$\mathbb{E}[N] = \sum_{n\geq 0} e^{-\lambda}\frac{n\lambda^n}{n!} = \lambda e^{-\lambda}\sum_{n\geq 1}\frac{\lambda^{n-1}}{(n-1)!} = \lambda.$$

The moment generating function of N is

$$\mathbb{M}_N(t) = \mathbb{E}[e^{tN}] = e^{-\lambda}\sum_{n\geq 0}\frac{(\lambda e^t)^n}{n!} = e^{\lambda(e^t-1)}.$$

We have

$$\mathbb{M}'_N(t) = \lambda e^t e^{\lambda(e^t-1)}, \quad \mathbb{M}''_N(t) = \lambda e^t e^{\lambda(e^t-1)} + (\lambda e^t)^2 e^{\lambda(e^t-1)}$$

so

$$\mathbb{E}[N^2] = \mathbb{M}''_N(0) = \lambda + \lambda^2, \quad \text{Var}[N] = \lambda.$$

$\square$

Example 1.116 (The inclusion-exclusion principle). Suppose that $(\Omega, \mathcal{S}, \mathbb{P})$ is a probability space and $A_1, \ldots, A_n \in \mathcal{S}$. Set

$$\mathbb{I}_n := \{1, \ldots, n\}.$$

For $m = 0, 1, \ldots, n$ we denote by Ω_m set of points $\omega \in \Omega$ that belong to *exactly* m of the sets $A_1, \ldots, A_n$. Note that

$$\Omega_0^c = A_1 \cup \cdots \cup A_n.$$

For $I \subset \mathbb{I}_n$ we set

$$A_I := \begin{cases} \bigcap_{k=1}^n A_i, & I \neq \emptyset, \\ \Omega, & I = \emptyset. \end{cases}$$

For $k \in \{0, 1, 2, \ldots n\}$ we define

$$s_k := \sum_{\substack{I \subset \mathbb{I}_n, \\ |I| = k}} \mathbb{P}[A_I].$$

The *inclusion-exclusion principle* states that

$$\mathbb{P}[\Omega_m] = \sum_{k=0}^{n-m}(-1)^k\binom{m+k}{m}s_{m+k}, \quad \forall m = 0, 1, \ldots, m. \tag{1.3.25}$$

Using the above equality with $m = 0$ we obtain the better known formula
$$\mathbb{P}[A_1 \cup \cdots \cup A_n] = 1 - \mathbb{P}[\Omega_0] = \sum_{k=1}^{n}(-1)^{k-1} \sum_{\substack{I \subset \mathbb{I}_n \\ |I|=k}} \mathbb{P}[A_I] = \sum_{k=1}^{n}(-1)^{k-1} s_k. \quad (1.3.26)$$

To prove (1.3.25) we set
$$T_k := \sum_{\substack{I \subset \mathbb{I}_n, \\ |I|=k}} \boldsymbol{I}_{A_I}.$$

Then
$$\boldsymbol{I}_{\Omega_m} = \sum_{k=0}^{n-m}(-1)^k \binom{m+k}{m} \boldsymbol{I}_{T_{m+k}}. \quad (1.3.27)$$

Indeed
$$\boldsymbol{I}_{\Omega_m} = \sum_{\substack{I \subset \mathbb{I}_n \\ |I|=m}} \left(\prod_{i \in I} \boldsymbol{I}_{A_i} \prod_{j \in \mathbb{I}_n \setminus I} \boldsymbol{I}_{A_j^c} \right) = \sum_{\substack{I \subset \mathbb{I}_n \\ |I|=m}} \left(\prod_{i \in I} \boldsymbol{I}_{A_i} \prod_{j \in \mathbb{I}_n \setminus I} (1 - \boldsymbol{I}_{A_j}) \right)$$
$$= \sum_{k=0}^{n-m}(-1)^k \sum_{|J|=m+k} c(J) \boldsymbol{I}_{A_J}.$$

Now observe that for any subset $J \subset \mathbb{I}_n$ of cardinality $m+k$ there are $\binom{m+k}{m}$ different way of writing $\boldsymbol{I}_{A_J}$ as a product
$$\boldsymbol{I}_{A_J} = \boldsymbol{I}_{A_I} \boldsymbol{I}_{A_{J \setminus I}}, \quad |I| = m.$$
Thus $c(J) = \binom{m+k}{m}$ for $|J| = m+k$. We deduce
$$\sum_{|J|=m+k} c(J) \boldsymbol{I}_{A_J} = \binom{m+k}{m} T_{m+k}.$$

Using the linearity of expectation we deduce from (1.3.27) that
$$\mathbb{P}[\Omega_m] = \mathbb{E}[\boldsymbol{I}_{\Omega_m}] = \sum_{k=0}^{n-m}(-1)^k \binom{m+k}{m} \mathbb{E}[\boldsymbol{I}_{T_{m+k}}],$$
where $\mathbb{E}[T_{m+k}] = s_{m+k}$.

Associated to the equality (1.3.25) there is a sequence of inequalities called the *Bonferroni inequalities*. For $\ell \in \mathbb{N}$ and $\frac{n-m}{2} \geq \ell$
$$\sum_{k=0}^{2\ell-1}(-1)^k \binom{m+k}{m} s_{m+k} \leq \mathbb{P}[\Omega_m] \leq \sum_{k=0}^{2\ell}(-1)^k \binom{m+k}{m} s_{m+k}. \quad (1.3.28)$$

The above inequalities follow from the *"motivic" Bonferroni inequalities*
$$\sum_{k=0}^{2\ell-1}(-1)^k \binom{m+k}{m} T_{m+k} \leq \boldsymbol{I}_{\Omega_m}$$
$$\leq \sum_{k=0}^{2\ell}(-1)^k \binom{m+k}{m} T_{m+k}, \quad 1 \leq \ell \leq \frac{n-m}{2}. \quad (1.3.29)$$

To prove this we fix $\omega \in \Omega$. We have to show that

$$\sum_{k=0}^{2\ell-1}(-1)^k\binom{m+k}{m}T_{m+k}(\omega) \leq \boldsymbol{I}_{\Omega_m}(\omega) \leq \sum_{k=0}^{2\ell}(-1)^k\binom{m+k}{m}T_{m+k}(\omega) \quad (1.3.30)$$

for $k \leq \frac{n-m}{2}$. Define

$$I_\omega := \{\, i \in \mathbb{I}_n;\ \omega \in A_i\,\}, \quad r(\omega) := |I_\omega| = \sum_{k=1}^n \boldsymbol{I}_{A_k}(\omega).$$

Note that $\boldsymbol{I}_{A_I}(\omega) = 0$ if $|I| > r(\omega)$. In particular, this shows that all the terms in the inequality (1.3.29) are equal to zero if $r(\omega) < m$.

Suppose that $r(\omega) \geq m$. Then, for any $k \leq r$, we have

$$T_k(\omega) = \sum_{\substack{I \subset I_\omega \\ |I|=k}} \boldsymbol{I}_{A_I}(\omega) = \binom{r}{k}.$$

Thus, the inequality (1.3.30) evaluated at ω is equivalent to

$$\sum_{k=0}^{\min(2\ell-1)}(-1)^k\binom{m+k}{m}\binom{r}{m+k} \leq \boldsymbol{I}_{\Omega_m}(\omega) \leq \sum_{k=0}^{2\ell}(-1)^k\binom{m+k}{m}\binom{r}{m+k}. \quad (1.3.31)$$

The inclusion exclusion-identity (1.3.25) shows that the inequalities become equalities for $2\ell > r - m$ so we assume $2\ell \leq r - m$.

For $r = m$ the inequality (1.3.31) is obvious since the sums in the left and right-hand sides consist of a single term equal to $1 = \boldsymbol{I}_{\Omega_m}(\omega)$. Assume $r > m$. In this case (1.3.31) is equivalent to

$$\sum_{k=0}^{2\ell-1}(-1)^k a_k \leq 0 \leq \sum_{k=0}^{2\ell}(-1)^k a_k, \quad a_k := \binom{m+k}{m}\binom{r}{m+k}. \quad (1.3.32)$$

Observe that

$$a_k = \binom{r}{m}\binom{p}{k}, \quad p = r - m.$$

The inequality (1.3.32) reduces to

$$\binom{p}{0} - \binom{p}{1} + \cdots + \binom{p}{2\ell-2} - \binom{p}{2\ell-1} \leq 0$$

$$0 \leq \binom{p}{0} - \binom{p}{1} + \binom{p}{2} + \cdots - \binom{p}{2\ell-1} + \binom{p}{2\ell},$$

where $2\ell \leq p$. These inequalities are immediate consequences of two well known properties of the binomial coefficients, namely their symmetry

$$\binom{p}{k} = \binom{p}{p-k},$$

and their *unimodality*

$$\binom{r}{0} \leq \binom{p}{1} \leq \cdots \leq \binom{p}{\lfloor p/2 \rfloor} = \binom{p}{\lfloor (p+1)/2 \rfloor} \geq \binom{p}{\lfloor p/2 \rfloor + 1} \geq \cdots \geq \binom{p}{p}.$$

For $m = 0$ we obtain the inequalities

$$\sum_{k=1}^{n} \mathbb{P}[A_k] - \sum_{1 \leq i < j \leq n} \mathbb{P}[A_i \cap A_j] \leq \mathbb{P}[A_1 \cup \cdots \cup A_n] \leq \sum_{k=1}^{n} \mathbb{P}[A_k].$$

The right-hand-side inequality is referred to as the *union bound*. □

Example 1.117 (Sieves and poissonization). Suppose now that we have an upper triangular array of measurable sets $(A_{n,i})_{i \in \mathbb{I}_n}$, $n \in \mathbb{N}$.

$$\begin{array}{l} A_{1,1} \\ A_{2,1}, \; A_{2,2} \\ \vdots \quad \vdots \\ A_{n,1}, \; A_{n,2}, \; A_{n,3} \cdots A_{n,n} \\ \vdots \quad \vdots \quad \vdots \quad \vdots \quad \vdots \end{array}$$

For $n \geq q$ we set

$$s_q^n := \sum_{\substack{I \subset \mathbb{I}_n, \\ |I|=q}} \mathbb{P}[A_{n,I}], \quad A_{n,I} = \bigcap_{i \in I} A_{n,i}.$$

Similarly, for $n \geq m$ we denote by Ω_m^n the set of points in Ω that belong to exactly m of the sets $A_{n,1}, \ldots, A_{n,n}$. Using Bonferroni inequalities we deduce that for fixed ℓ and $n > 2\ell + m$ we have

$$\sum_{k=0}^{2\ell-1} (-1)^k \binom{m+k}{m} s_{m+k}^n \leq \mathbb{P}[\Omega_m^n] \leq \sum_{k=0}^{2\ell} (-1)^k \binom{m+k}{m} s_{m+k}^n. \quad (1.3.33)$$

Suppose now that there exists $\lambda > 0$ such that, for any $q \in \mathbb{N}$ we have

$$\lim_{n \to \infty} s_q^n = \frac{\lambda^q}{q!}. \quad (1.3.34)$$

If we let $n \to \infty$ in (1.3.33) we obtain

$$\frac{1}{m!} \sum_{k=0}^{2\ell-1} (-1)^k \frac{\lambda^k}{k!} \leq \liminf_{n \to \infty} \mathbb{P}[\Omega_m^n] \leq \limsup_{n \to \infty} \mathbb{P}[\Omega^n] \leq \frac{1}{m!} \sum_{k=0}^{2\ell} (-1)^k \frac{\lambda^k}{k!}.$$

If we now let $\ell \to \infty$ we deduce

$$\lim_{n \to \infty} \mathbb{P}[\Omega_m^n] = \frac{e^{-\lambda} \lambda^m}{m!}.$$

We can rephrase this in an equivalent way. Set

$$X_n := \sum_{k=1}^{n} \boldsymbol{I}_{A_{n,k}}.$$

Then $\Omega_m^n = \{X_n = m\}$ and thus we showed that if (1.3.33) holds, then

$$\lim_{n\to\infty} \mathbb{P}[X_n = m] = \mathbb{P}[\mathrm{Poi}(\lambda) = m],$$

where we recall that $\mathrm{Poi}(\lambda)$ denotes a Poisson random variable with parameter λ.

The phenomenon depicted above is referred under the generic name of *poissonization* or *Poisson approximation*. Let us observe that if the events $A_{n,k}$ are independent and $\mathbb{P}[A_{n,i}] = \frac{\lambda}{n}$, then

$$s_k^n = \binom{n}{k}\left(\frac{\lambda}{n}\right)^k \sim \frac{\lambda^k}{k!} \text{ as } n \to \infty.$$

In this case $X_n = \mathrm{Bin}(n, \lambda/n)$. The success probability $\frac{\lambda}{n}$ is small for large n and for this reason the Poisson distribution is sometimes referred as the *law of rare events*:

The estimation techniques based on various versions of the inclusion-exclusion principle are called *sieves*. We refer to [143, Chaps. 2, 3] for a more detailed description of far reaching generalizations of the inclusion-exclusion principle and associated sieves. □

Example 1.118 (Fixed points of random permutations). Let us show how the above arguments work on the classical *derangements problem* Denote by $\mathfrak{S}_n$ the group of permutations of $\mathbb{I}_n$, we equip it with the uniform probability measure so each permutation σ has probability $\frac{1}{n!}$. For each $\sigma \in \mathfrak{S}_n$ we denote by $F(\sigma) = F_n(\sigma)$ its number of fixed points, i.e.,

$$F(\sigma) = \#\{k \in \mathbb{I}_n; \; \sigma(k) = k\}.$$

Thus $F : \mathfrak{S}_n \to \{0, 1, \ldots, n\}$ can be viewed as a random variable.

A *derangement* is a permutation σ with no fixed points, i.e., $F(\sigma) = 0$. A concrete occurrence of a derangement can be observed when a group of n, slightly inebriated, passengers board a plane and pick seats at random. A derangement occurs when none of them sits on his/her preassigned seat.

We want to compute the probability distribution of F, i.e., the probabilities

$$\mathbb{P}[F = m], \; k = 0, 1, \ldots, n.$$

For $j \in \mathbb{I}_n$ we denote by E_j the event $\sigma(j) = j$. The set of permutations that fix j can be identified with the set of permutations of $\mathbb{I}_n \setminus \{j\}$ so

$$\mathbb{P}[E_j] = \frac{(n-1)!}{n!} = \frac{1}{n}.$$

Observe that

$$F = \sum_{k=1}^n \boldsymbol{I}_{E_k},$$

so

$$\mathbb{E}[F] = \sum_{k=1}^n \mathbb{E}[\boldsymbol{I}_{E_k}] = \sum_{k=1}^n \mathbb{P}[E_k] = 1. \tag{1.3.35}$$

Thus the expected number of fixed points is rather low: a random permutation has, on average, one fixed point.

Let us compute the probability distribution of F. For each $I \subset \mathbb{I}_n$ we set
$$E_I = \bigcup_{i \in I} E_i.$$
Thus $\sigma \in E_I$ if and only if the permutation σ fixes all the points in I. We deduce that if $|I| = k$, then
$$\mathbb{P}[\,E_I\,] = \frac{(n-k)!}{n!} \text{ and } s_k := \sum_{|I|=k} \mathbb{P}[\,E_I\,] = \binom{n}{k}\frac{(n-k)!}{n!} = \frac{1}{k!}.$$
Note that $F(\sigma) = m$, then σ fixes exactly k points and (1.3.25) yields
$$\mathbb{P}[\,F=m\,] = \sum_{k=0}^{n-m} (-1)^k \binom{m+k}{m} s_{m+k} = \frac{1}{m!} \sum_{k=0}^{n-m} (-1)^k \frac{1}{k!}.$$
In particular, the number of derangements is
$$\mathbb{P}[\,F=0\,] = \sum_{k=0}^{n} (-1)^k \frac{1}{k!}.$$
The equality $\mathbb{E}[\,F\,] = 1$ yields an interesting identity
$$1 = \sum_{m=1}^{n} m\mathbb{P}[\,F=m\,] = \sum_{m=1}^{n} \frac{1}{(m-1)!}\left(\sum_{k=0}^{n-m} (-1)^k \frac{1}{k!}\right).$$
Note that
$$\lim_{n \to \infty} \mathbb{P}[\,F_n = m\,] = \frac{e^{-1}}{m!}. \tag{1.3.36}$$
The sequence $\frac{e^{-1}}{m!}$, $m \geq 0$ describes the Poisson distribution Poi(1). $\square$

1.3.4 Classical examples of continuous probability distributions

We want to describe a few example of random variables whose probability distributions are absolutely continuous with respect to the Lebesgue measure on the real. Such distributions are classically known as *continuous probability distributions*. Their probabilistic significance will gradually be revealed in the book.

Example 1.119 (Uniform distribution). A random variable X is said to be *uniformly distributed* or *uniform* in the interval $[a, b]$, and we write this $X \sim \text{Unif}(a, b)$, if
$$\mathbb{P}_X[\,dx\,] = \frac{1}{b-a} \boldsymbol{I}_{[a,b]} dx.$$
When $X \sim \text{Unif}(a, b)$ we have
$$\mathbb{M}_X(t) = \frac{1}{b-a} \int_a^b e^{tx} dx = \frac{e^{tb} - e^{ta}}{t(b-a)} = \sum_{n \geq 1} \frac{b^n - a^n}{n(b-a)} \frac{t^{n-1}}{(n-1)!}.$$
In particular we deduce
$$\mu_n[\,X\,] = \frac{1}{n+1} \frac{b^{n+1} - a^{n+1}}{b-a}. \tag{1.3.37}$$
$\square$

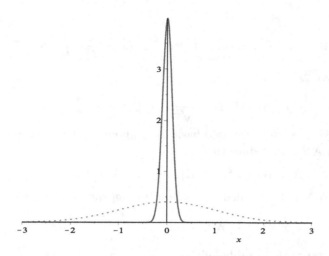

Fig. 1.3 The graph of $\gamma_{0,\sigma}$ for $\sigma=1$ (dotted curve) and $\sigma=0.1$ (continuous curve).

Example 1.120 (Gaussian random variables). The *Gaussian* or *normal* random variables form a 2-parameter family $N(\mu,\sigma^2)$, $\mu\in\mathbb{R}$, $\sigma>0$ where $X\sim N(\mu,\sigma^2)$ iff

$$\mathbb{P}_X[\,dx\,]=\gamma_{\mu,\sigma^2}(x)dx,\quad \gamma_{\mu,\sigma^2}(x):=\frac{1}{\sqrt{2\pi}\,\sigma}e^{-\frac{(x-\mu)^2}{2\sigma^2}}.$$

We will use the simpler notation $\gamma_{\sigma^2}(x):=\gamma_{0,\sigma^2}$. The measure

$$\mathbf{\Gamma}_{\mu,\sigma^2}[\,dx\,]:=\gamma_{\mu,\sigma^2}(x)dx$$

is called the *Gaussian measure* on $\mathbb{R}$ with mean μ and variance σ^2. Let us observe

$$X\sim N(\mu,\sigma^2)\iff \frac{1}{\sigma}(X-\mu)\sim N(0,1).$$

Indeed if we set

$$Y:=\frac{1}{\sigma}(X-\mu),$$

then

$$\mathbb{P}[\,Y\leq y\,]=\mathbb{P}[\,(x-\mu)/\sigma\leq y\,]=\mathbb{P}[\,x\leq\sigma y+\mu\,]=\int_{-\infty}^{\sigma y+\mu}\gamma_{\mu,\sigma^2}(x)dx$$

$$=\sigma\int_{-\infty}^{y}\gamma_{\mu,\sigma^2}(\sigma t+\mu)dt=\int_{-\infty}^{y}\gamma_{0,1}(t)dt.$$

Thus

$$\mathbb{E}[\,X\,]=\mathbb{E}[\,Y\,]+\mu,\quad \mathrm{Var}[\,X\,]=\sigma^2\mathrm{Var}[\,Y\,].$$

We have

$$\mathbb{E}[\,Y\,]=\frac{1}{\sqrt{2\pi}}\int_{\mathbb{R}}ye^{-y^2/2}dy=0,$$

and
$$\text{Var}[Y] = \mathbb{E}[Y^2] = \frac{1}{\sqrt{2\pi}} \int_{\mathbb{R}} y^2 e^{-y^2/2} dy = \frac{2}{\sqrt{2\pi}} \int_0^\infty y^2 e^{-y^2/2} dy$$

$(s = y^2/2, \ y = \sqrt{2s})$

$$= \frac{2}{\sqrt{\pi}} \int_0^\infty s^{1/2} e^{-s} ds = \frac{2}{\sqrt{\pi}} \Gamma(3/2) = \frac{2}{\sqrt{\pi}} \cdot \frac{1}{2} \Gamma(1/2) = 1,$$

where at the last two steps we used basic facts about the Gamma function recalled in Proposition A.2. We deduce that

$$X \sim N(\mu, \sigma^2) \ \Rightarrow \ \mathbb{E}[X] = \mu, \ \text{Var}[Y] = \sigma^2. \tag{1.3.38}$$

A variable $X \sim N(0,1)$ is called a *standard normal* random variable. Its cdf is

$$\Phi(x) := \mathbb{P}[X \leq x] = \frac{1}{\sqrt{2\pi}} \int_{-\infty}^x e^{-x^2/2} dx, \tag{1.3.39}$$

plays an important role in probability and statistics. The quantity

$$\frac{\mathbb{P}[X > x]}{\gamma_1(x)}$$

is called the *Mills ratio* of the standard normal random variable. It satisfies the inequalities

$$\frac{x}{x^2+1} \gamma_1(x) \leq \mathbb{P}[X > x] \leq \frac{1}{x} \gamma_1(x). \tag{1.3.40}$$

In Exercise 1.25 we outline a proof of this inequality.

Observe that if $X \sim N(0,1)$, and $\sigma \in \mathbb{R}$, then $\sigma X \in N(0, \sigma^2)$ and

$$\mathrm{M}_{\sigma X}(t) = \mathbb{E}[e^{t\sigma X}] = \mathrm{M}_X[\sigma t].$$

On the other hand, if $X \sim N(0,1)$, then

$$\mathrm{M}_X(t) = \frac{1}{\sqrt{2\pi}} \int_{\mathbb{R}} e^{tx - x^2/2} dx = \frac{1}{\sqrt{2\pi}} \int_{\mathbb{R}} e^{(2tx - x^2 - t^2)/2} e^{t^2/2} dx$$

$$= e^{t^2/2} \cdot \underbrace{\frac{1}{\sqrt{2\pi}} \int_{\mathbb{R}} e^{-(x-t)^2/2} dx}_{=1} = e^{t^2/2}.$$

Thus

$$\mu_{2m}[X] = \frac{(2m)!}{2^m m!} = (2m-1)!!, \ \mu_{2m-1}[X] = 0, \ \forall m \in \mathbb{N}. \qquad \square$$

Example 1.121 (Gamma distributions). The *Gamma distributions* with parameters ν, λ are defined by

$$\Gamma_{\nu, \lambda}[dx] = g_\nu(x; \lambda) dx$$

where $g_\nu(x; \lambda), \lambda, \nu > 0$ are given by

$$g_\nu(x; \lambda) = \frac{\lambda^\nu}{\Gamma(\nu)} x^{\nu-1} e^{-\lambda x} \boldsymbol{I}_{(0,\infty)}. \tag{1.3.41}$$

From the definition of the Gamma function we deduce that $g_\nu(x;\lambda)$ is indeed a probability density, i.e.,

$$\int_0^\infty g_\nu(x;\lambda)dx = 1.$$

We will use the notation $X \sim \text{Gamma}(\nu,\lambda)$ to indicate that $\mathbb{P}_X = \Gamma_{\nu,\lambda}$.

The Gamma$(1,\lambda)$-random variables play a special role in probability. They are called *exponential random variables* with parameter λ. We will use the notation $X \sim \text{Exp}(\lambda)$ to indicate that X is such a random variable. The distribution of $\text{Exp}(\lambda)$ is

$$\text{Exp}(\lambda) \sim \lambda e^{-\lambda x} \boldsymbol{I}_{(0,\infty)} dx.$$

We will have more to say about exponential variables in the next subsection.

The parameter ν is sometimes referred to as the *shape* parameter. Figure 1.4 may explain the reason for this terminology.

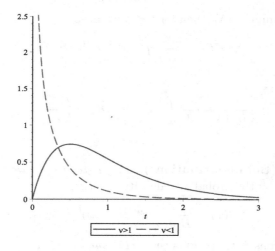

Fig. 1.4   The graphs of $g_\nu(x;\lambda)$ for $\nu > 1$ and $\nu < 1$.

For $n = 1, 2, 3, \ldots$ the distribution Gamma(n,λ) is also known as an *Erlang distribution* and has a simple probabilistic interpretation. If the waiting time T for a certain event is exponentially distributed with rate λ, e.g., the waiting time for a bus to arrive, then the waiting time for n of these events to occur independently and in succession is a Gamma(n,λ) random variable. We will prove this later.

The distribution $g_{n/2}(x;1/2)$, where $n = 1, 2, \ldots$, plays an important role in statistics it also known as the *chi-squared distribution with n degrees of freedom* and it is traditionally denoted by $\chi^2(n)$. One can show that if $X_1, \ldots, X_n$ are independent standard normal random variables, then the random variable

$$X_1^2 + \cdots + X_n^2$$

has a chi-squared distribution of degree n.

If $X \sim \text{Gamma}(\nu, \lambda)$ is a Gamma distributed random variable, then X is s-integrable for any $s \geq 1$. Moreover, for any $k \in \{1, 2, \dots\}$ we have

$$\mu_k[X] = \frac{\lambda^\nu}{\Gamma(\nu)} \int_0^\infty x^{k+\nu-1} e^{-\lambda x} dx$$

$(x = \lambda^{-1} t,\ dx = \lambda^{-1} dt,\ \lambda x = t,\ x^{k+\nu-1} = \lambda^{-(k+\nu-1)} t^{k+\nu-1})$

$$= \frac{1}{\lambda^k \Gamma(\nu)} \int_0^\infty t^{k+\nu-1} e^{-t} dt = \frac{\Gamma(k+\nu)}{\lambda^k \Gamma(\nu)}.$$

We deduce

$$\mathbb{E}[X] = \mu_1[X] = \frac{\Gamma(\nu+1)}{\lambda \Gamma(\nu)} = \frac{\nu}{\lambda},$$

$$\text{Var}[X] = \mu_2[X] - \mu_1[X]^2 = \frac{\Gamma(\nu+2)}{\lambda^2 \Gamma(\nu)} - \frac{\nu^2}{=} \frac{k(k+1) - k^2}{\lambda^2} = \frac{\nu}{\lambda^2}.$$

Finally, if $X \sim \text{Gamma}(\nu, \lambda)$, then for $t < \lambda$ we have

$$\mathbb{M}_X(t) = \frac{\lambda^\nu}{\Gamma(\nu)} \int_0^\infty x^{\nu-1} e^{-(\lambda-t)x} dx$$

$x = y/(\lambda - t)$

$$= \frac{\lambda^\nu}{\Gamma(\nu)(\lambda-t)^\nu} \int_0^\infty y^{\nu-1} e^{-y} dy = \left(\frac{\lambda}{\lambda-t}\right)^\nu.$$

$\square$

Example 1.122 (Beta distributions). The *Beta distribution* with parameters $a, b > 0$ is defined by the probability density function

$$\beta_{a,b}(x) = \frac{1}{B(a,b)} x^{a-1} (1-x)^{b-1} \boldsymbol{I}_{(0,1)}.$$

The normalizing constant $B(a, b)$ is the *Beta function* (A.1.2),

$$B(a, b) = \frac{\Gamma(a)\Gamma(b)}{\Gamma(a+b)}.$$

We will use the notation $X \sim \text{Beta}(a, b)$ to indicate that the pdf of X is a Beta distribution with parameters a, b.

Suppose that $X \sim \text{Beta}(a, b)$. Then

$$\mathbb{E}[X] = \frac{1}{B(a,b)} \int_0^1 x^a (1-x)^{b-1} dx = \frac{B(a+1, b)}{B(a, b)}$$

$$\stackrel{(A.1.4)}{=} \frac{\Gamma(a+1)\Gamma(a+b)}{\Gamma(a)\Gamma(a+b+1)} = \frac{a}{a+b},$$

$$\mathbb{E}[X^2] = \frac{1}{B(a,b)} \int_0^1 x^{a+1} (1-x)^{b-1} dx = \frac{\Gamma(a+2)\Gamma(a+b)}{\Gamma(a)\Gamma(a+b+2)} = \frac{a(a+1)}{(a+b)(a+b+1)}.$$

Hence
$$\operatorname{Var}[X] = \mathbb{E}[X^2] - \mathbb{E}[X]^2 = \frac{a}{a+b}\left(\frac{a+1}{a+b+1} - \frac{a}{a+b}\right)$$
$$= \frac{a}{a+b} \cdot \frac{(a+1)(a+b) - a(a+b+1)}{(a+b)(a+b+1)} = \frac{ab}{(a+b)^2(a+b+1)}.$$

The distribution Beta$(1/2, 1/2)$ is called the *arcsine distribution*. In this case
$$\beta_{1/2,1/2}(x) = \frac{1}{\pi}\frac{1}{\sqrt{x(1-x)}},$$
and
$$\int_0^x \beta_{1/2,1/2}(s)ds = \frac{1}{\pi}\arcsin\sqrt{x}.$$

We refer to Exercise 1.35 for an alternate interpretation of Beta$(1/2, 1/2)$. □

In Appendix A.2 we have listed the basic integral invariants of several frequently occurring probability distributions.

1.3.5 *Product probability spaces and independence*

Suppose $(\Omega_i, \mathcal{S}_i)$, $i = 0, 1$, are two measurable spaces. Recall that $\mathcal{S}_0 \otimes \mathcal{S}_1$ is the sigma-algebra of subsets of $\Omega_0 \times \Omega_1$ generated by the collection $\mathcal{R}$ of "rectangles" of the form $S_0 \times S_1$, $S_i \in \mathcal{S}_i$, $i = 0, 21$.

The goal of this subsection is to show that two sigma-finite measures μ_i on $\mathcal{S}_i$, $i = 0, 1$ induce in a canonical way a measure $\mu_0 \otimes \mu_1$ uniquely determined by the condition
$$\mu_0 \otimes \mu_1[S_0 \times S_1] = \mu_0[S_0]\mu_1[S_1], \quad \forall S_i \in \mathcal{S}_i, \ i = 0, 1.$$

The collection $\mathcal{A}$ of subsets of $\Omega_0 \times \Omega_1$ that are finite disjoint unions of rectangles is an algebra. This suggests using Carathéodory's existence theorem to prove this claim.

We choose a different route that bypasses Carathéodory's existence theorem. This alternate, more efficient approach, is driven by the Monotone Class Theorem and simultaneously proves a central result in integration theory, the Fubini-Tonelli Theorem. For every measurable space $(\Omega, \mathcal{S})$ we denote by $\mathcal{L}^0(\Omega, \mathcal{S})_*$ the space of $\mathcal{S}$ measurable functions $f : \Omega \to \mathbb{R}$.

Lemma 1.123. *Suppose that*
$$f \in \mathcal{L}^0(\Omega_0 \times \Omega_1, \mathcal{S}_0 \otimes \mathcal{S}_1)_* \cup \mathcal{L}^0_+(\Omega_0 \times \Omega_1, \mathcal{S}_0 \otimes \mathcal{S}_1).$$
Then, for any $\omega_1 \in \Omega_1$ the function $f^0_{\omega_1} : \Omega_0 \to \mathbb{R}$,
$$f^0_{\omega_1}(\omega_0) = f(\omega_0, \omega_1)$$
is $\mathcal{S}_0$-measurable and, for any $\omega_0 \in \Omega_0$, the function $f^1_{\omega_0} : (\Omega_1, \mathcal{S}_1) \to \mathbb{R}$,
$$f^1_{\omega_0}(\omega_1) = f(\omega_0, \omega_1)$$
is $\mathcal{S}_1$-measurable.

Proof. We prove only the statement concerning $f_{\omega_1}^0$. For simplicity will write f_{ω_1} instead of $f_{\omega_1}^0$. We will use the Monotone Class Theorem 1.21.

Denote by $\mathcal{M}$ the collection of functions $f \in \mathcal{L}^0(\Omega_0 \times \Omega_1, \mathcal{S}_0 \times \mathcal{S}_1)_*$ such that f_{ω_1} is $\mathcal{S}_0$-measurable, $\forall \omega_1 \in \Omega_1$. Clearly is $f, g \in \mathcal{M}$ are bounded then $af + bg \in \mathcal{M}$, $\forall a, b \in \mathbb{R}$

The collection $\mathcal{R}$ of rectangles is a π-system. Note that for any rectangle $R = S_0 \times S_1$ the function $f = \boldsymbol{I}_R$ belongs to $\mathcal{M}$. Indeed, for any $\omega_1 \in \Omega_1$ we have

$$f_{\omega_1} = \begin{cases} \boldsymbol{I}_{S_0}, & \omega_1 \in S_1, \\ 0, & \omega_1 \in \Omega_1 \setminus S_1. \end{cases}$$

If (f_n) is an increasing sequence of functions in $\mathcal{M}$ so is the sequence of slices f_{n,ω_1} so the limit f is also in $\mathcal{M}$. By the Monotone Class Theorem the collection $\mathcal{M}$ contains all the nonnegative measurable functions. Since $\mathcal{M}$ is a vector space, it must coincide with $\mathcal{L}^0(\Omega_0 \times \Omega_1, \mathcal{S}_0 \otimes \mathcal{S}_1)_\mathcal{B}$.

When $f \in \mathcal{L}_+^0$, but f is allowed to have infinite values, the function f is the increasing limit of a sequence in $\mathcal{M}$. Hence this situation is also included in the conclusions of the lemma. □

Theorem 1.124 (Fubini-Tonelli). *Let $(\Omega_i, \mathcal{S}_i, \mu_i)$, $i = 0, 1$ be two sigma-finite measured spaces.*

(i) There exists a measure μ on $\mathcal{S}_0 \otimes \mathcal{S}_1$ uniquely determined by the equalities

$$\mu[\, S_0 \times S_1 \,] = \mu_0[\, S_0 \,] \mu_1[\, S_1 \,], \quad \forall S_0 \in \mathcal{S}_0, \ S_1 \in \mathcal{S}_1.$$

We will denote this measure by $\mu_0 \otimes \mu_1$.

(ii) For each nonnegative function $f \in \mathcal{L}_+^0(\Omega_0 \times \Omega_1, \mathcal{S}_0 \otimes \mathcal{S}_1)$ the functions

$$\omega_0 \mapsto \boldsymbol{I}_1[\, f \,](\omega_0) := \int_{\Omega_1} f(\omega_0, \omega_1) \mu_1[\, d\omega_1 \,] \in [0, \infty],$$

$$\omega_1 \mapsto \boldsymbol{I}_0[\, f \,](\omega_1) := \int_{\Omega_0} f(\omega_0, \omega_1) \mu_0[\, d\omega_0 \,] \in [0, \infty]$$

are measurable and

$$\int_{\Omega_0} \left(\int_{\Omega_1} f(\omega_0, \omega_1) \mu_1[\, d\omega_1 \,] \right) \mu_0[\, d\omega_0 \,]$$
$$= \int_{\Omega_0 \times \Omega_1} f(\omega_0, \omega_1) \mu_0 \otimes \mu_1[\, d\omega_0 d\omega_1 \,] \quad (1.3.42)$$
$$= \int_{\Omega_1} \left(\int_{\Omega_0} f(\omega_0, \omega_1) \mu_0[\, d\omega_0 \,] \right) \mu_1[\, d\omega_1 \,].$$

In particular, if only one of the three terms above is finite, then all three are finite and equal.

(iii) Let $f \in \mathcal{L}^0(\Omega_0 \times \Omega_1, \mathcal{S}_0 \otimes \mathcal{S}_1, \mu_0 \otimes \mu_1)$. Then $f \in \mathcal{L}^1(\Omega_0 \times \Omega_1, \mathcal{S}_0 \otimes \mathcal{S}_1, \mu_0 \otimes \mu_1)$ if and only if at least one of the terms in (1.3.42) is well defined and finite. In this case all these terms are equal.

Proof. We will carry the proof in several steps.

Step 1. We will prove that for every positive function $f \in \mathcal{L}^0(\Omega_0 \times \Omega_1, \mathcal{S}_0 \times \mathcal{S}_1)$ the nonnegative function

$$\omega_0 \mapsto \boldsymbol{I}_1[\, f \,](\omega_0) = \int_{\Omega_1} f(\omega_0, \omega_1) \mu_1[\, d\omega_1 \,]$$

is measurable so that the integral

$$I_{1,0}[\, f \,] := \int_{\Omega_0} \left(\int_{\Omega_1} f(\omega_0, \omega_1) \mu_1[\, d\omega_1 \,] \right) \mu_0[\, d\omega_0 \,] \in [0, \infty]$$

is well defined.

This follows from the Monotone Class Theorem arguing exactly as in the proof of Lemma 1.123. For $S \in \mathcal{S}_0 \otimes \mathcal{S}_1$ we set
$$\mu_{1,0}[\,S\,] = I_{1,0}[\,\boldsymbol{I}_S\,].$$
Note that
$$\boldsymbol{I}_1[\,\boldsymbol{I}_{S_0 \times S_1}\,] = \int_{\Omega_1} \boldsymbol{I}_{\Omega_0 \times \Omega_1}(\omega_0, \omega_1) \mu_1[\,d\omega_1\,].$$
If $\omega_0 \in \Omega_0 \setminus S_0$ the integral is 0. If $\omega_0 \in S_0$ the integral is
$$\int_{\Omega_1} \boldsymbol{I}_{S_1} d\mu_1 = \mu_1[\,S_1\,].$$
Hence
$$\boldsymbol{I}_1[\,\boldsymbol{I}_{S_0 \times S_1}\,] = \mu_1[\,S_1\,]\boldsymbol{I}_{S_0}.$$
We deduce
$$I_{1,0}[\,S_0 \times S_1\,] = \mu_1[\,S_1\,] \int_{\Omega_0} \boldsymbol{I}_{S_0} d\mu_0 = \mu_0[\,S_0\,] \cdot \mu_1[\,S_1\,].$$
Clearly if $A, A' \in \mathcal{S}$ are disjoint, then $\boldsymbol{I}_{A \cup A'} = \boldsymbol{I}_A + \boldsymbol{I}_{A'}$ so that
$$I_{1,0}[\,\boldsymbol{I}_{A \cup A'}\,] = I_{1,0}[\,\boldsymbol{I}_A\,] + I_{1,0}[\,\boldsymbol{I}'_A\,]$$
and
$$\mu_{1,0}[\,A \cup A'\,] = \mu_{1,0}[\,A\,] + \mu_{1,0}[\,A'\,].$$
If
$$A_1 \subset A_2 \subset \cdots$$
is an increasing sequence of sets in $\mathcal{S}$ and
$$A = \bigcup_{n \geq 1} A_n,$$
then invoking the Monotone Convergence Theorem we first deduce that $I_{1,0}[\,\boldsymbol{I}_{A_n}\,]$ is a nondecreasing sequence of measurable functions converging to $I_{1,0}[\,\boldsymbol{I}_A\,]$ and then we conclude that $\mu_{1,0}[\,A_n\,]$ converges to $\mu_{1,0}[\,A\,]$. Hence $\mu_{1,0}$ is a measure on $\mathcal{S} = \mathcal{S}_0 \otimes \mathcal{S}_1$.

Step 2. A similar argument shows that
$$\mu_{0,1}[S] = \int_{\Omega_1} \left(\int_{\Omega_0} \boldsymbol{I}_S(\omega_0, \omega_1) \mu_0[\,d\omega_0\,] \right) \mu_1[\,d\omega_1\,]$$
is also a sigma-finite measure on $\mathcal{S} = \mathcal{S}_0 \otimes \mathcal{S}_1$. Note that
$$\mu_{1,0}[\,S_0 \times S_1\,] = \mu_{0,1}[\,S_0 \times S_1\,], \quad \forall S_0 \in \mathcal{S}_0, \ S_1 \in \mathcal{S}_1.$$
Thus $\mu_{1,0}[\,R\,] = \mu_{0,1}[\,R\,], \forall R \in \mathcal{R}$.

We want to show that if ν is another measure on $\mathcal{S}$ such that $\nu[\,R\,] = \mu_{1,0}[\,R\,]$ for any $R \in \mathcal{R}$, then $\nu[\,A\,] = \mu_{1,0}[\,A\,], \forall A \in \mathcal{S}$.

To see this assume first that μ_0 and μ_1 are finite measures. Then $\Omega_0 \times \Omega_1 \in \mathcal{R}$
$$\mu_{1,0}[\,\Omega_0 \times \Omega_1\,] = \nu[\,\Omega_0 \times \Omega_1\,] < \infty$$
and since $\mathcal{R}$ is a π-system we deduce from Proposition 1.29 that $\mu_{1,0} = \nu$ on $\mathcal{S}$.

To deal with the general case choose two increasing sequences $E_n^i \in \mathcal{S}_i$, $i = 0, 1$ such that
$$\mu_i[\,S_n^i\,] < \infty, \ \forall n \text{ and } \Omega_i = \bigcup_{n \geq 1} E_n^i, \ i = 0, 1.$$
Define
$$E_n := E_n^0 \times E_n^1, \mu_i^n[\,S_i\,] := \mu_i[\,S_i \cap E_n^i\,], \ S_i \in \mathcal{S}_i, \ i = 0, 1,$$
$$\nu^n[\,A\,] := \nu[\,A \cap E_n\,], \ \forall A \in \mathcal{S}.$$
Using the measures μ_i^n we form as above the measures $\mu_{1,0}^n$ and we observe that
$$\mu_{1,0}^n[\,A\,] = \mu_{0,1}[\,A \cap E_n\,], \ \forall n, \ \forall A \in \mathcal{S}.$$
For any rectangle R, the intersection $R \cap E_n$ is a rectangle and
$$\mu_{1,0}^n[\,R\,] = \nu^n[\,R\,], \ \forall n.$$
Thus
$$\mu_{1,0}^n[\,A\,] = \nu^n[\,A\,], \ \forall n \in \mathbb{N}, \ A \in \mathcal{S}.$$
If we let $n \to \infty$ in the above equality we deduce that $\mu_{1,0} = \nu$ on $\mathcal{S}$.

We deduce that $\mu_{0,1} = \mu_{1,0}$. Thus the measures $\mu_{0,1}$ and $\mu_{1,0}$ coincide on the algebra of sets generated by the rectangles and thus they must coincide on the $\mathcal{S}_0 \otimes \mathcal{S}_1$. This common measure is denoted by $\mu_0 \otimes \mu_1$ and it clearly satisfies statement (i) in the theorem.

Step 3. From **Step 2** we deduce that (1.3.42) is true for $f = \boldsymbol{I}_S, \forall S \in \mathcal{S}_0 \otimes \mathcal{S}_1$. From this, using the Monotone Class Theorem exactly as in the proof of Lemma 1.123 we deduce (1.3.42) in its entire generality. The claim in (iii) follows from the fact that any integrable function f is the difference of two nonnegative integrable functions $f = f^+ - f^-$ and the claim is true for $f^{\pm}$. $\square$

The above construction can be iterated. More precisely, given sigma-finite measured spaces $(\Omega_k, \mathcal{S}_k, \mu_k)$, $k = 1, \ldots, n$, we have a measure $\mu = \mu_1 \otimes \cdots \otimes \mu_n$ uniquely determined by the condition

$$\mu[S_1 \times S_2 \times \cdots \times S_n] = \mu_1[S_1]\mu_2[S_2]\cdots\mu_n[S_n], \quad \forall S_k \in \mathcal{S}_k, \quad k = 1, \ldots, n.$$

Remark 1.125. Recall that λ denotes the Lebesgue measure on $\mathbb{R}$. The measure $\lambda^{\otimes n}$ on $\mathcal{B}_{\mathbb{R}^n}$ is called the n-dimensional Lebesgue measure and will denoted by λ_n or simply λ, when no confusion is possible. A subset of $\mathbb{R}^n$ is called *Lebesgue measurable* if it belongs to the completion of the Borel sigma-algebra with respect to the Lebesgue measure.

One can prove that if a function $f : \mathbb{R}^n \to \mathbb{R}$ is absolutely Riemann integrable (see [121, Chap. 15]), then it is also Lebesgue integrable with respect to the Lebesgue measure on $\mathbb{R}^n$ and, moreover

$$\int_{\mathbb{R}^n} f(x)\,|dx| = \int_{\mathbb{R}^n} f(x)\,\boldsymbol{\lambda}[\,dx\,],$$

where the left-hand-side integral is the (improper) Riemann integral.

We recommend the reader to try to prove this fact or at least to try to understand why a Riemann integrable function defined on a cube is Lebesgue measurable. This is not obvious because there exist Riemann integrable functions that are not Borel measurable.

For example, if $C \subset [0,1]$ is the Cantor set, then there exists a subset A of C that are not Borel because the cardinality of the set 2^C is bigger than the cardinality of the family of Borel subsets of C. The subset A is Lebesgue measurable since C is Lebesgue negligible. The indicator function $\boldsymbol{I}_A$ is Riemann integrable but not Borel measurable.

The change in variables for the Riemann integral shows that if U, V are open subsets of $\mathbb{R}^n$ and $F : U \to V$ is a C^1-diffeomorphism onto V, then

$$F_{\#}^{-1}\boldsymbol{\lambda}_V[\,dx\,] = |\det J_F(x)|\boldsymbol{\lambda}_U[\,dx\,].$$

$\square$

Let us present a few useful consequences of Fubini's theorem.

Proposition 1.126. *Suppose that X is a nonnegative random variable defined on the probability space $(\Omega, \mathcal{S}, \mathbb{P})$. For any $p \in [1, \infty)$ we have*

$$\mathbb{E}[X^p] = p\int_0^\infty x^{p-1}\mathbb{P}[X > x]dx. \tag{1.3.43}$$

Proof. We have

$$p\int_0^\infty x^{p-1}\mathbb{P}[X > x]dx = \int_0^\infty \left(\int_\Omega \boldsymbol{I}_{\{X>x\}}(\omega)\mathbb{P}[\,d\omega\,]\right)px^{p-1}dx$$

$$= \int_{\substack{(\omega,x)\in\Omega\times[0,\infty)\\0\le x<X(\omega)}} px^{p-1}\mathbb{P}\otimes\boldsymbol{\lambda}[\,d\omega dx\,]$$

(use Fubini-Tonelli)

$$= \int_\Omega \left(\int_0^{X(\omega)} px^{p-1}dx \right) \mathbb{P}[\,d\omega\,] = \int_\Omega X^p(\omega)\mathbb{P}[\,d\omega\,] = \mathbb{E}[\,X^p\,].$$

□

We want to point out that when $p=1$ the equality

$$\int_{\substack{(\omega,x)\in\Omega\times[0,\infty)\\ 0\le x<X(\omega)}} \mathbb{P}\otimes\boldsymbol{\lambda}[\,d\omega dx\,] = \mathbb{E}[\,X\,]$$

simply says that $\mathbb{E}[\,X\,]$ is equal to the "area" below the graph of the function $X:\Omega\to[0,\infty)$.

Example 1.127. Suppose that X is a random variable that takes only nonnegative *integral* values. Then

$$\mathbb{P}_X = \sum_{n\ge 0} \mathbb{P}[\,X=n\,]\delta_n,$$

and

$$\mathbb{E}[\,X\,] \stackrel{(1.3.43)}{=} \int_0^\infty \mathbb{P}[\,X>x\,]dx$$
$$= \sum_{n\ge 0}\int_n^{n+1}\mathbb{P}[\,X>x\,]dx = \sum_{n\ge 0}\mathbb{P}[\,X>n\,]. \tag{1.3.44}$$

Let us apply this identity to a geometric random variable with success probability p, $T\sim\mathrm{Geom}(p)$. Note that $\mathbb{P}[\,T>n\,]$ is the probability that the waiting time for a success is $>n$ or, equivalently, the probability that the first n trials are failures. Hence

$$\mathbb{P}[\,T>n\,]=q^n \text{ so } \mathbb{E}[\,T\,] = \sum_{n\ge 0} q^n = \frac{1}{1-q} = \frac{1}{p}.$$

Similarly

$$\mu_2[\,T\,] = \mathbb{E}[\,T^2\,] = 2\sum_{n\ge 0} n\mathbb{P}[\,T>n\,]$$

$$= 2\sum_{n\ge 1} nq^n = 2q\sum_{n\ge 1} nq^{n-1} = \frac{2q}{(1-q)^2} = \frac{2q}{p^2}.$$

In particular

$$\mathrm{Var}[T] = \mathbb{E}[\,T^2\,] - \mathbb{E}[\,T\,]^2 = \frac{q}{p^2}.$$

□

Example 1.128. Suppose that T is an *exponential random variable with parameter* λ, i.e., a random variable with the exponential probability distribution
$$\mathbb{P}_T[\,dt\,] = \lambda e^{-\lambda t} \boldsymbol{I}_{(0,\infty)} dt.$$
This random variable describes the waiting time for an event to happen, e.g., the waiting time for a laptop to crash, or the waiting time for a bus to arrive at a bus station. The quantity $\lambda e^{-\lambda t} dt$ is the probability that the waiting time is in the interval $(t, t+dt]$. Then
$$\mathbb{P}[\,T > t\,] = \int_t^\infty \lambda e^{-\lambda \tau} d\tau = e^{-\lambda}, \;\; \mathbb{E}[\,T\,] = \int_0^\infty e^{-\lambda t} dt = \frac{1}{\lambda}.$$
We see that $\frac{1}{\lambda}$ is measured in units of time. For this reason λ is called the *rate* and describes how many rare events take place per unit of time.

Similarly
$$\mu_2[\,T\,] = \mathbb{E}[\,T^2\,] = 2\int_0^\infty t\mathbb{P}[\,T > t\,]dt = 2\int_0^\infty t e^{-\lambda t} dt = \frac{2}{\lambda^2}\int_0^\infty s e^{-s} ds$$
$$= \frac{2}{\lambda^2}\Gamma(2) = \frac{2}{\lambda^2}.$$
The function $S(t) := \mathbb{P}[\,T > t\,]$ is called the *survival function*. For example, if T denotes the life span of a laptop, then $S(t)$ is the probability that a laptop survives more than g units of time.

The exponential distribution enjoys the so called *memoryless property*
$$\mathbb{P}[\,T > t+s \,|\, T > s\,] = \mathbb{P}[\,T > t\,]. \tag{1.3.45}$$
For example, if T is the waiting time for a bus to arrive then, given that you've waited more that s units of time, the probability that you will have to wait at least t extra is the same as if you have not waited at all. The proof of (1.3.45) is immediate.
$$\mathbb{P}[\,T > t+s \,|\, T > s\,] = \frac{\mathbb{P}[\,T > t+s\,]}{\mathbb{P}[\,T > s\,]} = \frac{e^{-\lambda(t+s)}}{e^{-\lambda s}} = e^{-\lambda t} = \mathbb{P}[\,T > t\,]. \quad \square$$

Example 1.129 (Integration by parts). Suppose that μ_0, μ_1 are two Borel probability measures on $\mathbb{R}$ supported on $[0, \infty)$, i.e.
$$\mu_k[\,(-\infty, 0)\,] = 0, \;\; k = 0, 1.$$
We set
$$F_k(x) = \mu_k[\,(-\infty, x]\,], \;\; k = 0, 1,$$
so that μ_k is the Lebesgue-Stieltjes measure determined by F_k. Note that
$$F_k(0) = \mu_k[\,\{0\}\,].$$
Classically, the integral
$$\int_{[0,a]} u(x)\mu_k[\,dx\,]$$

was denoted by
$$\int_0^a u(x)dF_k(x).$$
This classical notation is a bit ambiguous due to the following simple fact
$$\int_{[0,a]} u(x)\mu_k[\,dx\,] = u(0)F_k(0) + \int_{(0,a]} u(x)\mu_k[\,dx\,].$$
We want to prove a version of the integration by parts formula. Namely, we will show that if one of the functions F_0, F_1 is continuous, then
$$\int_0^a F_0(x)dF_1(x) = F_0(a)F_1(a) - F_0(0)F_1(0) - \int_0^a F_1(x)dF_0(x). \quad (1.3.46)$$
Assume for simplicity that F_1 is continuous so $F_1(0) = 0$. Set $\mu := \mu_0 \otimes \mu_1$. Observe that since
$$F_0(a)F_1(a) - F_0(0)F_1(0) = F_0(a)F_1(a) = \mu\big[\underbrace{[0,a]\times[0,a]}_{S_a}\big].$$
Using the Fubini-Tonelli theorem we deduce
$$\int_0^a F_1(x)dF_1(x) = \int_{[0,a]} \left(\int_{\mathbb{R}} \boxed{I_{(-\infty,x]}(y)}\mu_1[\,dy\,]\right)\mu_0[\,dx\,]$$
(F_1 is continuous)
$$= \int_{[0,a]} \left(\int_{[0,a]} \boxed{I_{[0,x)}(y)}\mu_1[\,dy\,]\right)\mu_0[\,dx\,] = \mu[\,R_0\,],$$
where
$$R_0 := \{\,(x,y) \in \mathbb{R}^2;\ 0 \le y < x \le a,\ y < x\,\}.$$
Similarly
$$\int_0^a F_0(y)dF_1(y) = \int_{[0,a]} \left(\int_{[0,a]} I_{[0,y]}\mu_0[\,dx\,]\right)\mu_1[\,dy\,] = \mu[\,R_1\,],$$
$$R_1 := \{\,(x,y) \in \mathbb{R}^2;\ 0 \le x \le y \le a\,\}.$$
Observe that the regions R_0, R_1 are disjoint.

The region R_0 is the part of the square $S_a = [0,a] \times [0,a]$ strictly below the diagonal $y = x$, while R_1 is the part of this square above or this diagonal. Hence $S_a = R_0 \cup R_1$ and thus
$$\mu[\,R_0\,] + \mu[\,R_1\,] = \mu[\,S_a\,].$$
Let us observe that the integration by parts formula is not true if both F_0, F_1 are discontinuous. Take for example the case $\mu_0 = \mu_1 = \tfrac{1}{2}(\delta_1 + \delta_3)$. Then
$$F_0(x) = F_1(x) = F(x) = \begin{cases} 0, & x < 1, \\ \tfrac{1}{2}, & 1 \le x < 3, \\ 1, & x \ge 3. \end{cases}$$

In this case we have

$$\int_0^2 F(x)dF(x) = \int_{[0,2]} F(x)\mu_0[\,dx\,] = \frac{1}{2}F(1) = \frac{1}{4}, \quad F(2)^2 = \frac{1}{4}$$

so

$$2\int_0^2 F(x)dF(x) \neq F(2)^2.$$

The reason for this failure has a simple geometric origin: the diagonal $\{y = x\}$ may not be $\mu_0 \otimes \mu_1$-negligible. The continuity assumption allowed us to discard the diagonal of the square because in this case it is indeed negligible. □

Definition 1.130. Fix a probability space $(\Omega, \mathcal{S}, \mathbb{P})$.

(i) Suppose that V is a finite dimensional vector space. We denote by $\mathcal{B}_V$ the sigma-algebra of Borel subsets of V. A V-valued *random vector* is a measurable map

$$\boldsymbol{X} : (\Omega, \mathcal{S}, \mathbb{P}) \to (V, \mathcal{B}_V).$$

Its *probability distribution* is the pushforward measure $\mathbb{P}_{\boldsymbol{X}} := \boldsymbol{X}_\# \mathbb{P}$. By definition, $\mathbb{P}_{\boldsymbol{X}}$ is a Borel probability measure on V.

(ii) The *joint probability distribution* of the random variables

$$X_1, \ldots, X_n : (\Omega, \mathcal{S}, \mathbb{P}) \to \mathbb{R}$$

is the probability distribution of the random vector

$$\boldsymbol{X} := (X_1, \ldots, X_n) : (\Omega, \mathcal{S}, \mathbb{P}) \to \mathbb{R}^n.$$

We will denote by $\mathbb{P}_{X_1, \ldots, X_n}$ the joint distribution.

□

Observe that joint probability distribution $\mathbb{P}_{X_1, \ldots, X_n}$ is uniquely determined by the probabilities

$$\mathbb{P}[\,X_1 \leq x_1, \ldots, X_n \leq x_n\,], \quad x_1, \ldots, x_n \in \mathbb{R}.$$

Proposition 1.131. *Suppose that $(\Omega, \mathcal{S}, \mathbb{P})$ is a probability space and*

$$X_1, \ldots, X_n \in \mathcal{L}^0(\Omega, \mathcal{S}, \mathbb{P})$$

are random variables with probability distributions $\mathbb{P}_{X_1}, \ldots, \mathbb{P}_{X_n}$. The following statements are equivalent.

(i) *The random variables $X_1, \ldots, X_n$ are independent.*
(ii) $\mathbb{P}_{X_1, \ldots, X_n} = \mathbb{P}_{X_1} \otimes \cdots \otimes \mathbb{P}_{X_n}.$

Proof. The random variables $X_1, \ldots, X_n$ are independent iff for any Borel sets $B_1, \ldots, B_n \subset \mathbb{R}$ we have

$$\mathbb{P}\big[\, X_1 \in B_1, \ldots, X_n \in B_n \,\big] = \mathbb{P}[X_1 \in B_1] \cdots \mathbb{P}[X_n \in B_n]$$

$$\Longleftrightarrow \mathbb{P}_{X_1,\ldots,X_n}\big[\, B_1 \times \cdots \times B_n \,\big] = \mathbb{P}_{X_1} \otimes \cdots \otimes \mathbb{P}_{X_n}\big[\, B_1 \times \cdots \times B_n \,\big].$$

Thus the random variables $X_1, \ldots, X_n$ are independent iff the measures $\mathbb{P}_{X_1,\ldots,X_n}$ and $\mathbb{P}_{X_1} \otimes \cdots \otimes \mathbb{P}_{X_n}$ coincide on the set of rectangles $B_1 \times \cdots \times B_n$, i.e.,

$$\mathbb{P}_{X_1,\ldots,X_n} = \mathbb{P}_{X_1} \otimes \cdots \otimes \mathbb{P}_{X_n}.$$

$\square$

1.3.6 Convolutions of Borel measures on the real axis

Definition 1.132. Let μ, ν be two probability measures on $(\mathbb{R}, \mathcal{B}_\mathbb{R})$. The *convolution* of μ with ν is the probability measure $\mu * \nu$ on $(\mathbb{R}, \mathcal{B}_\mathbb{R})$ defined by

$$\mu * \nu [\, B \,] = \int_\mathbb{R} \mu [\, B - y \,] \nu [\, dy \,], \quad \forall B \in \mathcal{B}_\mathbb{R}. \tag{1.3.47}$$

$\square$

Denote by S_y the shift $S_y : \mathbb{R} \to \mathbb{R}$, $S_y(x) = x + y$ and set $\mu_y := (S_y)_\# \mu$. Note that for any Borel set $B \subset \mathbb{R}$ we have

$$\mu_y [\, B \,] = \mu \big[\, S^{-1}(B) \,\big] = \mu [\, B - y \,],$$

so we can rewrite (1.3.47) in the form

$$\mu * \nu [\, - \,] = \int_\mathbb{R} \mu_y [\, - \,] \nu [\, dy \,].$$

A simple argument based on the Monotone Convergence Theorem shows that $\mu * \nu$ is indeed a Borel measure on $\mathbb{R}$. By letting $B = \mathbb{R}$ in (1.3.47) we see that $\mu * \nu$ is indeed a probability measure.

Note that $\mu * \nu$ is a *mixture* in the sense that it is obtained by averaging of the family of probability measures $(\mu_y)_{y \in \mathbb{R}}$ with respect to the probability measure $\nu[dy]$. For example, if

$$\nu = \sum_{i=1}^n \frac{1}{n} \delta_{x_i},$$

then

$$\mu * \nu = \frac{1}{n} \sum_{i=1}^n \mu_{x_i}.$$

Proposition 1.133. *Let μ, ν be probability measures on $(\mathbb{R}, \mathcal{B}_\mathbb{R})$ and*

$$\Phi : \mathbb{R}^2 \to \mathbb{R}, \quad \Phi(x, y) = x + y.$$

Then $\mu * \nu = \Phi_\#(\mu \otimes \nu) = \nu * \mu$.

Proof. Let $B \in \mathcal{B}_{\mathbb{R}}$ and set $\hat{B} = \Phi^{-1}(B)$. Set
$$\hat{B}_y := \{\, x;\ (x,y) \in \hat{B} \,\} = B - y.$$
Then
$$\Phi_\#(\mu \otimes \nu)[B] = \int_{\mathbb{R}^2} \boldsymbol{I}_{\hat{B}} \mu \otimes \nu[dxdy]$$
(use Fubini-Tonelli)
$$= \int_{\mathbb{R}} \left(\int_{\mathbb{R}} \boldsymbol{I}_{\hat{B}_y} \mu[dx] \right) \nu[dy] = \int_{\mathbb{R}} \mu[B - y]\nu[dy] = \mu * \nu[B].$$
The equality $\mu * \nu = \nu * \mu$ follows by changing the order of integration in the Fubini-Tonelli theorem. $\square$

Corollary 1.134. *Let $X, Y \in \mathcal{L}^0(\Omega, \mathcal{S}, \mathbb{P})$ be two independent random variables with distributions $\mathbb{P}_X$ and $\mathbb{P}_Y$. Then*
$$\mathbb{P}_{X+Y} = \mathbb{P}_X * \mathbb{P}_Y.$$

Proof. Since X, Y are independent we have $\mathbb{P}_{X,Y} = \mathbb{P}_X \otimes \mathbb{P}_Y$. Note that $\mathbb{P}_{X+Y} = \Phi_\# \mathbb{P}_{X,Y}$. The conclusion now follows from Proposition 1.133. $\square$

Remark 1.135. (a) Suppose that F_μ is the cdf of the probability measure μ, i.e., $F_\mu(c) = \mu\big[(-\infty, c]\big]$, $\forall c \in \mathbb{R}$. Then the cdf $F_{\mu*\nu}$ of $\mu * \nu$ satisfies
$$F_{\mu*\nu}(c) = \int_{\mathbb{R}} F_\mu(c - x)\nu[dx], \quad \forall c \in \mathbb{R}.$$
We write this equality as
$$F_{\mu*\nu} = F_\mu * \nu. \tag{1.3.48}$$
If μ and ν are absolutely continuous with respect to the Lebesgue measure $\boldsymbol{\lambda}$ on $\mathbb{R}$ so
$$\mu[dx] = \rho_\mu(x)dx, \quad \nu[dx] = \rho_\nu(x)dx, \quad \rho_\mu, \rho_\nu \in L^1(\mathbb{R}, \boldsymbol{\lambda}),$$
then $\mu * \nu \ll \boldsymbol{\lambda}$ and
$$\mu * \nu[dx] = \rho_{\mu*\nu}(x)dx, \quad \rho_{\mu*\nu}(x) = \rho_\mu * \rho_\nu(x) := \int_{\mathbb{R}} \rho_\mu(x - y)\nu[dy].$$
To see this it suffices to check that for any $c \in \mathbb{R}$ we have
$$\mu * \nu\big[(-\infty, c]\big] = \int_{-\infty}^c \rho_{\mu*\nu}(x)dx.$$
We have
$$\mu * \nu\big[(-\infty, c]\big] = \int_{\mathbb{R}} \mu\big[(-\infty, c - y]\big]\nu[dy] = \int_{\mathbb{R}} \left(\int_\infty^{c-y} \rho(x)dx \right) \nu[dy]$$
$$= \int_{\mathbb{R}} \left(\int_\infty^{c-y} \rho_\mu(x)dx \right) \nu[dy] = \int_{\mathbb{R}} \left(\int_\infty^{c} \rho(z - y)dz \right) \nu[dy]$$

(use Fubini)
$$= \int_{-\infty}^{c} \left(\int_{\mathbb{R}} \rho_\mu(z-y)\nu[dy] \right) dz = \int_{-\infty}^{c} \rho_{\mu * \nu}(z)[dz].$$

(b) Any Borel probability measure μ on $\mathbb{R}$ is the probability distribution of the random variable

$$\mathbb{1}_{\mathbb{R}} : (\mathbb{R}, \mathcal{B}_{\mathbb{R}}, \mu) \to \mathbb{R}, \quad \mathbb{1}_{\mathbb{R}}(x) = x.$$

If μ_1, μ_2, μ_3 are different Borel probability measures on $\mathbb{R}$, then we can define three independent random variables

$$X_1, X_2, X_3 : \left(\mathbb{R}^3, \mathcal{B}_{\mathbb{R}^3}, \mu_1 \otimes \mu_2 \otimes \mu_2 \right) \to \mathbb{R},$$

$$X_k(x_1, x_2, x_3) = x_k, \quad k = 1, 2, 3.$$

Note that $\mathbb{P}_{X_k} = \mu_k$, $\forall k = 1, 2, 3$. Since $(X_1 + X_2) \perp\!\!\!\perp X_3$ and $X_1 \perp\!\!\!\perp (X_2 + X_3)$ we deduce

$$(\mu_1 * \mu_2) * \mu_3 = \mathbb{P}_{(X_1+X_2)+X_3} = \mathbb{P}_{X_1+(X_2+X_3)} = \mu_1 * (\mu_2 * \mu_3).$$

Similarly

$$\mu_1 * \mu_2 = \mathbb{P}_{X_1+X_2} = \mathbb{P}_{X_2+X_1} = \mu_2 * \mu_1.$$

(c) The operation of convolution makes sense for any finite Borel measures μ, ν on $\mathbb{R}$ and satisfies the same commutativity and associativity properties we encountered in the case of probability measures. Note that $\mu * \nu[\mathbb{R}] = \mu[\mathbb{R}] \cdot \nu[\mathbb{R}]$. □

Example 1.136 (Poisson processes). Suppose that we have a stream of events occurring in succession at random times $S_1 \leq S_2 \leq S_3 \leq \cdots$ such that the waiting times between two successive occurrences

$$T_1 = S_1, \quad T_2 = S_2 - S_1, \ldots, T_n = S_n - S_{n-1}, \ldots$$

are i.i.d. exponential random variables $T_n \sim \text{Exp}(\lambda)$, $n = 1, 2, \ldots$. We set $S_0 := 0$.

It may help to think of the sequence (T_n) as inter-arrival times for a bus. The first bus arrives at the station at time $S_1 = T_1$. Once the n-th bus has left the station, the waiting time for the next bus to arrive is an exponential random variable T_{n+1} independent of the preceding waiting times. From this point of view, S_n is the arrival time of the n-th bus.

For $t > 0$ we denote by $N(t)$ the number events that of occurred during the time interval $[0, t]$. In terms of streams of busses, $N(t)$ would count the number of buses that have arrived at the station in the interval $[0, t]$. In other words

$$N(t) = \max \{ n \geq 1; \; S_n \leq t \} = \#\{ n \geq 1; \; S_n \leq t \}.$$

This is a discrete random variable with range $\{0,1,2,3,\ldots\}$. The collection of random variables $\{N(t),\ t\geq 0\}$ is called the *Poisson process* with intensity λ. Note that

$$N(t) = \sum_{n=1}^{\infty} \mathbf{I}_{[0,t]}(S_n).$$

Let us find the distribution (pmf) of $N(t)$. We have

$$\mathbb{P}[N(t)=0] = \mathbb{P}[T_1 > t] = e^{-\lambda t} = \text{the survival function of Exp}(\lambda).$$

If $n > 0$, then $N(t) = n$ if and only if the n-th bus arrived sometime during the interval $[0,t]$, i.e., $S_n \leq t$, but the $(n+1)$-th bus has not arrived in this time interval. We deduce

$$\mathbb{P}[N(t)=n] = \mathbb{P}\Big[\,\{S_n \leq t\} \setminus \{S_{n+1} \leq t\}\,\Big] = \mathbb{P}[S_n \leq t] - \mathbb{P}[S_{n+1} \leq t].$$

If we denote by $F_n(t)$ the cdf of S_n, then we can rewrite the above equality in the form

$$\mathbb{P}[N(t)=n] = F_n(t) - F_{n+1}(t).$$

We have

$$\mathbb{P}_{S_n} = \underbrace{\mathrm{Exp}(\lambda) * \cdots * \mathrm{Exp}(\lambda)}_{n}$$

$$= \underbrace{\mathrm{Gamma}(\lambda,1) * \cdots * \mathrm{Gamma}(\lambda,1)}_{n} \stackrel{(1.6.6a)}{=} \mathrm{Gamma}(\lambda,n).$$

Hence, for $n > 0$

$$F_{n+1}(t) = \frac{\lambda^{n+1}}{\Gamma(n+1)}\int_0^t s^n e^{-\lambda s}ds = \frac{\lambda^{n+1}}{n!}\int_0^t s^n e^{-\lambda s}ds.$$

For $n > 0$, we integrate by parts to obtain

$$F_{n+1}(t) = -\left(\frac{\lambda^n}{n!}s^n e^{-\lambda s}\right)\bigg|_{s=0}^{s=t} + \frac{\lambda^n}{(n-1)!}\int_0^t s^{n-1}e^{-\lambda s}ds = -\frac{(t\lambda)^n}{n!}e^{-\lambda t} + F_n(t).$$

Hence

$$\mathbb{P}[N(t)=n] = F_n(t) - F_{n+1}(t) = \frac{(t\lambda)^n}{n!}e^{-\lambda t},\quad n > 0. \tag{1.3.49}$$

This shows that $N(t)$ is a Poisson random variable, $N(t) \sim \mathrm{Poi}(\lambda t)$.

The Poisson process plays an important role in probability since it appears in many situations and displays many surprising phenomena. One such interesting phenomenon is the *waiting time paradox*, [59, I.4]. To better appreciate this paradox we consider two separate situations.

Suppose first that buses arrive at a bus station following a Poisson stream with frequency λ. Bob arrives at the bus station at a time $t \geq 0$, the bus is not there and he is waiting for the next one. His *waiting time* is

$$W_t := S_{N(t)+1} - t.$$

We want to compute its expectation $w_t := \mathbb{E}[W_t]$. There are two possible heuristic arguments.

(i) The memoryless property of the exponential distribution shows that w_t should be independent of t so $w_t = w_0 = \frac{1}{\lambda}$.
(ii) Bob's arrival time t is uniformly distributed in the inter-arrivals interval $\left(S_{N(t)}, S_{N(t)+1}\right)$ of expected length $\frac{1}{\lambda}$ and, as in the earlier deterministic computation, the expectation should be half its length, $\frac{1}{2\lambda}$.

We will show that (i) provides the correct answer. However, even the reasoning (ii) holds a bit of truth. To see what is happening we compute the expectations of $S_{N(t)}$ and $S_{N(t)+1}$. We have

$$\mathbb{E}\left[S_{N(t)}\right] = \int_0^t \mathbb{P}\left[S_{N(t)} > x\right] dx.$$

Note that

$$\mathbb{P}\left[S_{N(t)} > x\right] = \sum_{n \geq 0} \mathbb{P}\left[S_{N(t)} > x, \ N(t) = n\right].$$

On the other hand,

$$\mathbb{P}\left[S_{N(t)} > x, \ N(t) = n\right] = \mathbb{P}\left[x < S_n \leq t, \ S_n + T_{n+1} > t\right].$$

The random variables S_n and T_{n+1} are independent and the joint distribution of (S_n, T_{n+1}) is

$$\mathbb{P}_{S_n, T_{n+1}}[dsdt] = \underbrace{\frac{\lambda^n}{(n-1)!} s^{n-1} e^{-\lambda s} \lambda e^{-\lambda \tau}}_{\rho(s,\tau)} ds d\tau$$

so

$$\mathbb{P}\left[x < S_n < t, S_n + T_{n+1} > t\right] = \int_{\substack{x < s \leq t \\ s+\tau > t}} \rho(s,\tau) ds d\tau$$

$$= \int_x^t \left(\int_{t-s}^\infty \rho(s,\tau) d\tau\right) ds = \int_x^t \mathbb{P}\left[T_{n+1} > t - s\right] \frac{\lambda^n}{(n-1)!} s^{n-1} e^{-\lambda s} ds$$

$$= \int_x^t e^{-\lambda(t-s)} \frac{\lambda^n}{(n-1)!} s^{n-1} e^{-\lambda s} ds = \frac{e^{-\lambda t} \lambda^n}{(n-1)!} \int_x^t s^{n-1} ds = \frac{e^{-\lambda t} \lambda^n}{n!} \left(t^n - x^n\right).$$

We deduce

$$\mathbb{P}\left[S_{N(t)} > x\right] = \sum_{n \geq 0} \frac{e^{-\lambda t} \lambda^n}{n!} \left(t^n - x^n\right) = 1 - e^{\lambda(-\lambda(t-x))},$$

$$\mathbb{E}\left[S_{N(t)}\right] = \int_0^t \left(1 - e^{-\lambda(t-x)}\right) dx = t - e^{-\lambda t} \int_0^t e^{\lambda x} dt = t - \frac{e^{-t}}{\lambda}(e^{\lambda t} - 1).$$

Hence

$$\mathbb{E}\left[S_{N(t)}\right] = t - \frac{1}{\lambda} + \frac{e^{-\lambda t}}{\lambda} = \frac{1}{\lambda} \mathbb{E}\left[N(t) - 1 + e^{-\lambda t}\right]. \qquad (1.3.50)$$

Let us compute $\mathbb{E}\big[S_{N(t)+1}\big]$. Again, we have
$$\mathbb{P}\big[S_{N(t)+1} > x\big] = \sum_{n\geq 0} \mathbb{P}\big[S_{N(t)+1} > x,\ N(t) = n\big],$$
and
$$\mathbb{P}\big[S_{N(t)+1} > x,\ N(t) = n\big] = \mathbb{P}\big[S_n \leq t\ S_{n+1} \geq \max(t,x)\big]$$
$$= \begin{cases} \mathbb{P}\big[S_n \leq t,\ S_n + T_{n+1} \geq t\big], & x \leq t, \\ \mathbb{P}\big[S_n \leq t,\ S_n + T_{n+1} \geq x\big], & x > t. \end{cases}$$
For any $c \geq t$ we have
$$\mathbb{P}\big[S_n \leq t,\ S_n + T_{n+1} \geq c\big] = \int_{\substack{s\leq t,\\ s+\tau\geq c}} \rho(s,t)\,ds\,d\tau$$
$$= \int_0^t \left(\int_{c-s}^\infty \rho(s,\tau)d\tau\right)ds = \frac{\lambda^n}{(n-1)!}\int_0^t e^{-\lambda(c-s)}s^{n-1}e^{-\lambda s}ds = \frac{e^{-\lambda c}(\lambda t)^n}{n!}.$$
Observing that
$$\sum_{n\geq 0} \frac{e^{-\lambda c}(\lambda t)^n}{n!} = e^{-\lambda(c-t)}$$
we deduce that
$$\mathbb{P}\big[S_{N(t)} > x\big] = \begin{cases} 1, & x \leq t, \\ e^{-\lambda(x-t)}, & x > t. \end{cases}$$
Hence
$$\mathbb{E}\big[S_{N(t)+1}\big] = \int_0^t dx + e^{\lambda t}\int_t^\infty e^{-\lambda x}dx = t + \frac{1}{\lambda} = \frac{1}{\lambda}\mathbb{E}\big[N(t)+1\big], \qquad (1.3.51)$$
and
$$w_t = \mathbb{E}\big[S_{N)t)+1}\big] - t = \frac{1}{\lambda}.$$
In fact much more is true. One can show (see [132, Sec. 3.6]) that the waiting time W_t is an exponential random variable, $W_t \sim \text{Exp}(\lambda)$, in agreement with the conclusion of the argument (i).

The above computations are a bit counterintuitive. The number of busses arriving during a time interval $[0,t]$ is $N(t)$. The busses arrive with a frequency of $\frac{1}{\lambda}$ per unit of time, so we should expect to wait $t = \frac{1}{\lambda}\mathbb{E}\big[N(t)\big]$ units of time for $N(t)$ busses to arrive. Formula (1.3.50) shows that we should expect less. On the other hand, formula (1.3.51) shows that we should expect $\frac{1}{\lambda}\mathbb{E}\big[N(t)+1\big]$ units of time for $N(t)+1$ busses to arrive! We refer to Remark 3.71 for an explanation for this paradoxical divergence of conclusions.

The above computation show that the expectation of $L_t = S_{N(t)+1} - S_{N(t)}$ is
$$\mathbb{E}\big[L_t\big] = \frac{2}{\lambda} - \frac{e^{-\lambda t}}{\lambda} \approx \frac{2}{\lambda} \text{ for } t \text{ large.}$$
This shows that even the argument (ii) captures a bit of what is going on since w_t is close to half the expected length of the inter-arrival interval $\big(S_{N(t)}, S_{N(t)+1}\big)$.

The Poisson processes are special cases of *renewal processes*. For an enjoyable and highly readable introduction to renewal processes we refer to [59] or [132, Chap. 3]. For a more in-depth presentation of these processes and some of their practical applications we refer to [5]. □

1.3.7 Modes of convergence of random variables

Fix a probability space $(\Omega, \mathcal{S}, \mathbb{P})$.

Definition 1.137 (Almost sure convergence). We say that the sequence of random variables

$$X_n \in L^0(\Omega, \mathcal{S}, \mathbb{P}), \quad n \in \mathbb{N},$$

converges *almost surely (or a.s.)* to $X \in L^0(\Omega, \mathcal{S}, \mathbb{P})$ if there exist $\Omega_0 \in \mathcal{S}$ such that

$$\mathbb{P}[\Omega_0] = 1, \quad \lim_{n \to \infty} X_n(\omega) = X(\omega), \quad \forall \omega \in \Omega_0.$$

We will use the notation $X_n \xrightarrow{\text{a.s.}} X$ to indicate the a.s. convergence. $\square$

To describe a useful criterion for a.s. convergence we need to rely on a very versatile classical result.

Definition 1.138. For any sequence of events $(A_n)_{n \in \mathbb{N}} \subset \mathcal{S}$ we denote by A_n i.o. the event "A_n occurs infinitely often",

$$A_n \text{ i.o.} := \bigcap_{m \geq 1} \bigcup_{n \geq m} A_n.$$

Thus

$$\omega \in A_n \text{ i.o.} \iff \forall m \in \mathbb{N} \ \exists n \geq m : \ \omega \in A_n. \qquad \square$$

Theorem 1.139 (Borel-Cantelli Lemma). *Consider a sequence of events* $(A_n)_{n \in \mathbb{N}} \subset \mathcal{S}$.

(i) If

$$\sum_{n \geq 1} \mathbb{P}[A_n] < \infty.$$

Then $\mathbb{P}[A_n \text{ i.o.}] = 0$.

(ii) Conversely, if the events $(A_n)_{n \in \mathbb{N}}$ are <u>independent</u> then $\mathbb{P}[A_n \text{ i.o.}] \in \{0, 1\}$, and

$$\mathbb{P}[A_n \text{ i.o.}] = 0 \iff \sum_{n \geq 1} \mathbb{P}[A_n] < \infty. \tag{1.3.52}$$

Proof. (i) We set

$$N := \sum_{n \geq 1} \boldsymbol{I}_{A_n}.$$

Note that $\{A_n \text{ i.o.}\} = \{N = \infty\}$. From the Monotone Convergence Theorem we deduce

$$\mathbb{E}[N] = \sum_{n \geq q} \mathbb{E}[\boldsymbol{I}_{A_n}] = \sum_{n \geq 1} \mathbb{P}[A_n] < \infty$$

so $\mathbb{P}[N = \infty] = 0$.

(ii) Kolmogorov's 0-1 theorem shows that when the events $(A_n)_{n\geq 1}$ are independent we have $\mathbb{P}[\,A_n \text{ i.o.}\,] \in \{0,1\}$.

To prove (1.3.52) we have to show that if

$$\sum_{n\geq 1} \mathbb{P}[\,A_n\,] = \infty,$$

then $\mathbb{P}[\,A_n \text{ i.o.}\,] = 1$. We have

$$\mathbb{P}\left[\bigcup_{n\geq m} A_n\right] = 1 - \mathbb{P}\left[\bigcap_{n\geq m} A_n^c\right]$$

(use the independence of A_n)

$$= 1 - \prod_{n\geq m}\left(1 - \mathbb{P}[\,A_n\,]\right)$$

$(1 - x \leq e^{-x}, \forall x \in \mathbb{R})$

$$\geq 1 - e^{-\sum_{n\geq m} \mathbb{P}[A_n]} = 1.$$

Hence

$$\mathbb{P}[\,A_n \text{ i.o.}\,] = \lim_{m\to\infty} \mathbb{P}\left[\bigcup_{n\geq m} A_n\right] = 1.$$

$\square$

Remark 1.140. Statement (i) in Theorem 1.139 is usually referred to as the *First Borel-Cantelli Lemma* while statement (ii) is usually referred to as the *Second Borel-Cantelli Lemma*. Exercises 3.12 and 3.18 present refinements of the Borel-Cantelli lemmas. $\square$

Observe that $X_n \to X$ a.s. if and only if, for any $\nu \in \mathbb{N}$

$$\mathbb{P}[\,\{\,|X_n - X| > 1/\nu\,\} \text{ i.o.}\,] = 0.$$

The Borel-Cantelli Lemma now implies the following result.

Corollary 1.141. *Suppose that there exists $X \in L^0(\Omega, \mathcal{S}, \mathbb{P})$ such that the sequence $X_n \in L^0(\Omega, \mathcal{S}, \mathbb{P})$ satisfies*

$$\sum_{n\geq 1} \mathbb{P}[\,|X_n - X| > \varepsilon\,] < \infty, \quad \forall \varepsilon > 0.$$

Then $X_n \xrightarrow{\text{a.s.}} X$. $\square$

Proof. The Borel-Cantelli Lemma implies that
$$\mathbb{P}\big[\,|X_n - X| > \varepsilon \text{ i.o.}\,\big] = 0, \quad \forall \varepsilon > 0.$$
Hence, for any $\varepsilon > 0$ there exists a negligible set $S_\varepsilon \in \mathcal{S}$ such that, for any $\omega \in \Omega \setminus S_\varepsilon$ we have
$$\limsup_{n \to \infty} |X_n(\omega) - X(\omega)| \leq \varepsilon.$$
Set
$$S_\infty = \bigcup_{k \in \mathbb{N}} S_{1/k}.$$
We deduce that for any $\omega \in \Omega \setminus S_\infty$ we have
$$\limsup_{n \to \infty} |X_n(\omega) - X(\omega)| \leq 1/k, \quad \forall k \in \mathbb{N}.$$
$\square$

Definition 1.142. We say that the sequence $X_n \in L^0(\Omega, \mathcal{S}, \mathbb{P})$ converges *in probability* to the random variable $X \in L^0(\Omega, \mathcal{S}, \mathbb{P})$ if, $\forall \varepsilon > 0$, we have
$$\lim_{n \to \infty} \mathbb{P}\big[\,|X_n - X| > \varepsilon\,\big] = 0.$$
We will use the notation $X_n \xrightarrow{P} X$ to indicate convergence in probability. $\square$

Observe that if $X_n \to X$ in probability and, for any $n \in \mathbb{N}$, we have $X_n = X_n'$ a.s., then $X_n' \to X$ in probability. Thus the convergence in probability is correctly defined in $L^0(\Omega, \mathcal{S}, \mathbb{P})$.

The convergence in probability is equivalent to the convergence defined by a metric on $L^0(\Omega, \mathcal{S}, \mathbb{P})$. For $X, Y \in \mathcal{L}^0(\Omega, \mathcal{S}, \mathbb{P})$ we set
$$\operatorname{dist}(X, Y) := \mathbb{E}\big[\min(|X - Y|, 1)\big]. \tag{1.3.53}$$
Clearly $\operatorname{dist}(X, Y) = \operatorname{dist}(Y, X)$ and
$$\operatorname{dist}(X, Z) \leq \operatorname{dist}(X, Y) + \operatorname{dist}(Y, Z).$$
Note that $\operatorname{dist}(X, Y) = 0$ iff $X = Y$ a.s. so "dist" is a metric on $L^0(\Omega, \mathcal{S}, \mathbb{P})$.

Proposition 1.143. *Let $X, X_n \in \mathcal{L}^0(\Omega, \mathcal{S}, \mathbb{P})$. Then the following statements are equivalent.*

(i) $X_n \to X$ in probability as $n \to \infty$.
(ii) $\operatorname{dist}(X_n, X) \to 0$ as $n \to \infty$.

Proof. Set
$$\rho(x) := \min(|x|, 1), \quad Y_n := X_n - X.$$
Using Markov's inequality we deduce that for any $n \geq 1$ and any $\varepsilon \in (0, 1)$ we have
$$\varepsilon \mathbb{P}\big[\,|Y_n| > \varepsilon\,\big] = \varepsilon \mathbb{P}\big[\rho(Y_n) > \varepsilon\big] \leq \mathbb{E}\big[\rho(Y_n)\big] = \operatorname{dist}(Y_n, 0).$$
This shows that (ii) $\Rightarrow$ (i).

Conversely, observe that, for any $\varepsilon > 0$, we have
$$\mathbb{E}\big[\rho(Y_n)\big] = \int_{|Y_n| \leq \varepsilon} \rho(Y_n) d\mathbb{P} + \int_{|Y_n| > \varepsilon} \rho(Y_n) d\mathbb{P} \leq \varepsilon + \mathbb{P}\big[\,|Y_n| > \varepsilon\,\big].$$
This proves that $0 \leq \liminf \operatorname{dist}(Y_n, 0) \leq \limsup \operatorname{dist}(Y_n, 0) \leq \varepsilon, \forall \varepsilon > 0$. $\square$

The next result describes the relationships between a.s. convergence and convergence in probability.

Theorem 1.144. *Let $X, X_n \in \mathcal{L}^0(\Omega, \mathcal{S}, \mathbb{P})$. Then the following hold.*

(i) If $X_n \to X$ a.s., then $X_n \to X$ in probability.
(ii) If $X_n \to X$ in probability, then (X_n) contains a subsequence that converges a.s. to X.
(iii) The sequence X_n converges in probability to X if and only if any subsequence contains a further subsequence that is a.s. convergent to X.

Proof. (i) Set $Y_n := X_n - X$. Since $Y_n \to 0$ a.s. we have $\min(|Y_n|, 1) \to 0$ a.s. From the Dominated Convergence Theorem we deduce
$$\mathrm{dist}(X_n, X) = \mathbb{E}\big[\,|Y_n|\,\big] \to 0,$$
so that $Y_n \xrightarrow{p} 0$.

(ii) Suppose that $Y_n \to 0$ in probability. We deduce that for any $k \in \mathbb{N}$ there exists $n_k \in \mathbb{N}$ such that
$$\forall n \geq n_k: \quad \mathbb{P}\big[\,|Y_n| > 1/k\,\big] < \frac{1}{2^k}.$$
Now observe that for any $m > 0$, the series
$$\sum_{k \geq 1} \mathbb{P}\big[\,|Y_{n_k}| > 1/m\,\big]$$
is convergent since, for $k > m$ we have
$$\mathbb{P}\big[\,|Y_{n_k}| > 1/m\,\big] \leq \mathbb{P}\big[\,|Y_{n_k}| > 1/k\,\big] < \frac{1}{2^k}.$$
The desired conclusion now follows from Corollary 1.141.

(iii) Recall that a sequence in a metric space converges to a given point if and only if any subsequence contains a sub-subsequence converging to that point. The properties (i) and (ii) show that the sequence (X_n) satisfies this condition with respect to the metric dist defined by ρ. □

Corollary 1.145. *If the sequence (X_n) in $L^0(\Omega, \mathcal{S}, \mathbb{P})$ converges in probability to X, then for any continuous function $f : \mathbb{R} \to \mathbb{R}$ the sequence $f(X_n)$ converges in probability to $f(X)$.*

Proof. The sequence (X_n) satisfies the necessary and sufficient conditions (iii) in Theorem 1.144. Since f is continuous, the sequence $f(X_n)$ satisfies these necessary and sufficient conditions as well. □

Definition 1.146. *Let $p \in [1, \infty)$. We say that the sequence $(X_n)_{n \in \mathbb{N}} \subset L^0(\Omega, \mathcal{S}, \mathbb{P})$ converges in p-mean or in L^p to $X \in L^0(\Omega, \mathcal{S}, \mathbb{P})$ if*
$$X, X_n \in L^p(\Omega, \mathcal{S}, \mathbb{P}), \quad \forall n \in \mathbb{N},$$
and
$$\lim_{n \to \infty} \mathbb{E}\big[\,|X_n - X|^p\,\big] = 0.$$
□

Proposition 1.147. *If $X_n \to X$ in p-mean, then $X_n \to X$ in probability. In particular, X_n admits a subsequence that converges* a.s. *to X.*

Proof. Set $Y_n := X_n - X$. Then
$$\mathbb{P}\big[\,|Y_n| > \varepsilon\,\big] = \mathbb{P}\big[\,|Y_n|^p > \varepsilon^p\,\big] \overset{(1.2.19)}{\leq} \frac{1}{\varepsilon^p}\mathbb{E}\big[\,|Y_n|^p\,\big] \to 0 \text{ as } n \to \infty.$$
□

Example 1.148. For each $n \in \mathbb{N}$ and each $1 \leq k \leq n$ we set
$$A_{k,n} = [(k-1)/n, k/n], \quad X_{k,n} = \boldsymbol{I}_{A_{k,n}} : [0,1] \to \mathbb{R}.$$
Then the sequence of random variables
$$X_{1,1}, X_{1,2}, X_{2,2}, X_{1,3}, X_{2,3}, X_{3,3}, \ldots$$
converges in mean and in probability to 0. It does not converge a.s. to 0 because for any $x \in [0,1]$ infinitely many of these random variables are equal to 1 at x.

The related sequence $Y_{k,n} = nX_{k,n}$ converges in probability to 0 but not in mean since $\|Y_{k,n}\|_{L^1} = 1$. □

Example 1.149 (Bernoulli). Suppose that $(X_n)_{n \geq 1}$ is a sequence of i.i.d. Bernoulli random variables with wining probability $p = \frac{1}{2}$. Set
$$S_n = X_1 + \cdots + X_n \sim \text{Bin}(n, 1/2), \quad M_n = \frac{1}{n}S_n.$$
Then
$$\text{Var}\big[\,M_n\,\big] = \frac{1}{n^2}\text{Var}\big[\,S_n\,\big] = \frac{1}{n}\text{Var}\big[\,\text{Ber}(1/2)\,\big] = \frac{1}{4}n.$$
Hence
$$\|M_n - 1/2\|_{L^2} = \frac{1}{2\sqrt{n}} \to 0 \text{ as } n \to \infty,$$
so that M_n converges in 2-mean to $\frac{1}{2}$ and thus, in probability to $\frac{1}{2}$. Intuitively, M_n is the fraction of Heads in a string on n independent fair con flips. From Chebyshev's inequality we deduce that
$$\mathbb{P}\big[\,|M_n - 1/2| > \varepsilon\,\big]\big[\leq \frac{\varepsilon^2}{4n}.$$
It turns out that this probability is much smaller. In (2.3.11a) we will show that
$$\mathbb{P}\big[\,|M_n - 1/2| > \varepsilon\,\big] \leq 2e^{-2n\varepsilon^2}.$$
For example if $n = 0.1$, then
$$\mathbb{P}\big[\,|M_{1000} - 1/2| > \varepsilon\,\big] \leq 2e^{-20} \approx 4.2 \times 10^{-9}.$$
□

Example 1.150 (Longest common subsequence). Consider a finite set $\mathcal{A}$, $|\mathcal{A}| = k$, called *alphabet*. A word of length n in the alphabet $\mathcal{A}$ is a finite sequence of the form

$$\underline{x} := (x_1, \ldots, x_n) \in \mathcal{A}^n.$$

A *subsequence* of such a word is a word of the form

$$(x_{f(1)}, \ldots x_{f(\ell)}) \in \mathcal{A}^\ell,$$

where f an increasing function $f : \{1, \ldots, \ell\} \to \{1, \ldots, n\}$. The natural number ℓ is called the length of the subsequence.

A common subsequence of two words $\underline{x}, \underline{y} \in \mathcal{A}^n$ is a word $w \in \mathcal{A}^\ell$ that is a subsequence of both. For example, if $\mathcal{A} = \{H, T\}$, then H, T, H, T, T is a subsequence of both words

$$\underline{H, T}, T, H, \underline{H, T, T} \quad \text{and} \quad T, \underline{H, T}, H, \underline{T, T}, H$$

We are interested in the length of the longest common subsequence of two *random* words of length n on the alphabet $\mathcal{A}$. Such a problem arises in genetics. In that case the alphabet is $\{A, C, T, G\}$. The DNA molecules are described by (very long) words in this alphabet. The existence a long common subsequence of two such words is an indication of a common ancestor of two living organisms with those DNAs.

From a mathematical point of view, we fix a probability measure π on an alphabet $\mathcal{A}$ and we choose independent random variables

$$\{X_n, Y_n; \ n \in \mathbb{N}\}$$

where X_n, Y_n are $\mathcal{A}$-valued and have common distribution π.

One can think that these random variables are obtained as follows. Two individuals independently roll identical "dice" with faces labeled by $\mathcal{A}$ and whose occurrences are governed by π. The first individual generates the sequence (X_n) while the second individual generates the sequence Y_n. We denote by L_n the length of the longest common subsequence of the words

$$(X_1, \ldots, X_n) \quad \text{and} \quad (Y_1, \ldots, Y_n).$$

We want to prove at a.s. and L^1 we have

$$\lim_{n \to \infty} \frac{L_n}{n} = R(\pi) := \sup_{n \geq 1} \frac{L_n}{n}. \tag{1.3.54}$$

In particular, this shows that

$$\lim_{n \to \infty} \frac{L_n}{n} > L_1 > 0.$$

Note that L_1 is a Bernoulli random variable with success probability

$$p = \sum_{a \in \mathcal{A}} \pi[a]^2.$$

The equality (1.3.54) is due to Chvátal and Sankoff [32], but we will follow the presentation in [144, Chap. 1].

The key observation is that the sequence $(\ell_n)_{n \in \mathbb{N}}$ is *superadditive*, i.e.,

$$\ell_n + \ell_m \leq \ell_{m+n}, \quad \forall m, n \in \mathbb{N}. \tag{1.3.55}$$

The proof is very simple. We set $Z_n = (X_n, Y_n)$ and we observe that the random variable L_n is an invariant of the sequence of pairs $(Z_1, \ldots, Z_n)$, $L_n = L(Z_1, \ldots, Z_n)$. Clearly

$$L_m = L(Z_{n+1}, \ldots, Z_{n+m}), \quad \forall m, n \in \mathbb{N}.$$

If we concatenate the longest common subsequence of $(Z_1, \ldots, Z_n)$ with the longest common subsequence of $(Z_{n+1}, \ldots, Z_{n+m})$ we obtain a common subsequence of $(Z_1, \ldots, Z_n, Z_{n+1}, \ldots, Z_{n+m})$ of length

$$L(Z_1, \ldots, Z_n) + L(Z_{n+1}, \ldots, Z_{n+m})$$

showing that

$$L(Z_1, \ldots, Z_n) + L(Z_{n+1}, \ldots, Z_{n+m}) \leq L(Z_1, \ldots, Z_n, Z_{n+1}, \ldots, Z_{n+m}),$$

i.e.,

$$L_m + L_n \leq L_{m+n}, \quad \forall m, n \in \mathbb{N}. \tag{1.3.56}$$

Taking the expectations of both sides in the above inequality we obtain (1.3.55).

The conclusion (1.3.54) is now an immediate consequence of the following elementary result.

Lemma 1.151 (Fekete). *Suppose that $(x_n)_{n \geq 1}$ is a* subadditive *sequence of real numbers, i.e.,*

$$x_{m+n} \leq x_m + x_n, \quad \forall m, n \in \mathbb{N}.$$

Then

$$\lim_{n \to \infty} \frac{x_n}{n} = \mu := \inf_{n \geq 1} \frac{x_n}{n}.$$

Proof. Then, for any $c > \mu$ we can find $k = k(c)$ such that $x_k \leq kc$. The subadditivity condition implies $x_{kn} \leq n x_k$, $\forall n \in \mathbb{N}$, so that

$$\mu \leq \frac{x_{nk}}{nk} < c, \quad \forall n \in \mathbb{N}.$$

Hence

$$\mu \leq \liminf_{n \to \infty} \frac{x_n}{n} \leq c, \quad \forall c > \mu,$$

i.e.,

$$\mu = \liminf_{n \to \infty} \frac{x_n}{n}.$$

Now observe that for any $n \geq k$, there exist $m \in \mathbb{N}$ and $r \in \{0, 1, \ldots, k-1\}$ such that $n = mk + r$. Hence

$$x_n \leq ma_k + a_r < mk(\mu + \varepsilon) + x_r$$

so that

$$\frac{x_n}{n} < \frac{(n-r)c}{n} + \frac{M}{n}, \quad M = \sup\{|a_1| + \cdots + |a_{k_1}|\}.$$

Hence

$$\limsup_{n \to \infty} \frac{x_n}{n} \leq \limsup_{n \to \infty} \frac{(n-r)c}{n} = c, \quad \forall c > \mu.$$

This completes the proof of the lemma. $\square$

The conclusion (1.3.54) follows from Fekete's Lemma applied to the sequence $x_n = -L_n$. The inequality (1.3.56) show that

$$\frac{L_n}{n} \to R := \sup_n \frac{L_n}{n}.$$

Set $r = r(\pi) := \mathbb{E}[R]$. The Cauchy-Schwartz inequality

$$r \geq \mathbb{E}[L_1] = \sum_{a \in \mathcal{A}} \pi[a]^2 \geq \frac{1}{k}\left(\sum_{a \in \mathcal{A}} \pi(a)\right)^2 = \frac{1}{k} > 0.$$

The Dominated Convergence Theorem implies that

$$r = \lim_{n \to \infty} \frac{1}{n}\mathbb{E}[L_n].$$

The exact value of $r(\pi)$ is not known in general. In Example 3.34, using more sophisticated techniques, we will show that the limit $R(\pi)$ is constant, $R(\pi) = r$ and $\frac{L_n}{n}$ is highly concentrated around its mean r_n. $\square$

The concept of convergence in probability is weaker than the concepts of convergence a.s. or in p-mean. In many applications it is useful to know sufficient additional assumptions that will guarantee that a sequence convergent in probability is also convergent in p-mean. The a.s. convergence does not guarantee convergence in mean. The next elementary example is typical of what can go wrong.

Example 1.152. Consider the interval $[-1, 1]$ equipped with the uniform probability measure $\frac{1}{2}dx$. Consider the sequence of bounded, nonnegative, random variables

$$X_n = 2^n \boldsymbol{I}_{[-2^{-n}, 2^{-n}]}.$$

Note that $X_n \to 0$ a.s. but

$$\mathbb{E}[X_n] = \frac{2^n}{2} \int_{-2^{-n}}^{2^{-n}} dx = 1, \quad \forall n.$$

As we will see later in Chapter 3, the reason why the convergence in mean fails is the high concentration of X_n on sets of smaller and smaller measures. $\square$

Our next result is an example of a sufficient condition for a sequence converging in probability to also converge in the mean. It is a stepping stone towards the more refined results that we will discuss in Chapter 3.

Theorem 1.153 (Bounded Convergence Theorem). *Suppose that (X_n) is a sequence in $L^1(\Omega, \mathcal{S}, \mathbb{P})$ that converges in probability to $X \in L^1(\Omega, \mathcal{S}, \mathbb{P})$. If the sequence (X_n) is bounded in $L^\infty(\Omega, \mathcal{S}, \mathbb{P})$, i.e.,*

$$M := \sup_{n \in \mathbb{N}} \|X_n\|_\infty < \infty,$$

then $X_n \to X$ in L^1 and

$$\lim_{n \to \infty} \mathbb{E}[X_n] = \mathbb{E}[X]. \tag{1.3.57}$$

Proof. We follow the approach in [153, Thm. 1.4]. Since

$$|\mathbb{E}[X_n] - \mathbb{E}[X]| \leq \mathbb{E}[|X_n - X|],$$

and $|X_n - X| \to 0$ in probability, it suffices to consider only the special case $X = 0$, $X_n \geq 0$, and $X_n \geq 0$ a.s.. In such an instance the claimed L^1-convergence follows from (1.3.57).

For any $\varepsilon > 0$ we have

$$\mathbb{E}[X_n] = \mathbb{E}[X_n \boldsymbol{I}_{\{X_n \leq \varepsilon\}}] + \mathbb{E}[X_n \boldsymbol{I}_{\{X_n > \varepsilon\}}] \leq M(\varepsilon + \mathbb{P}[X_n > \varepsilon]).$$

Letting $n \to \infty$ taking to account that $X_n \geq 0$ and $X_n \to 0$ in probability we deduce

$$0 \leq \liminf_{n \to \infty} \mathbb{E}[X_n] \leq \limsup_{n \to \infty} \mathbb{E}[X_n] \leq M\varepsilon, \quad \forall \varepsilon > 0.$$

$\square$

Remark 1.154. The Bounded Convergence theorem does not follow immediately from the Dominated Convergence Theorem which involves a.s. convergence. However, using Theorem 1.144(iii) we can use the Dominated Convergence Theorem to provide an alternate proof of the Bounded Convergence Theorem. $\square$

1.4 Conditional expectation

The concept of *conditioning* is a central pillar of the theory of probability. It has a genuinely probabilistic origin and very rich and subtle ramifications. Also, it takes some time getting used to it. This concept is one important reason why in probability sigma-algebras play a much more important role than in analysis.

Fix a probability space $(\Omega, \mathcal{S}, \mathbb{P})$.

1.4.1 Conditioning on a sigma sub-algebra

The main formal constructions of this section are best understood if we first consider a special but very useful example.

Example 1.155 (Conditioning on a partition). Suppose that $(\Omega, \mathcal{S}, \mathbb{P})$ and $(F_\alpha)_{\alpha \in A}$, $A \subset \mathbb{N}$, is a finite or countable partition of Ω with measurable and non-negligible chambers, i.e.,

$$F_\alpha \in \mathcal{S}, \ \mathbb{P}[F_\alpha] > 0, \ \forall \alpha \in A.$$

We denote by $\mathcal{F}$ the sigma-algebra generated by this partition. In other words, $F \subset \mathcal{F}$ if and only if it is a union of chambers F_α. This means that $\exists B \subset A$ such that

$$F = \bigcup_{\beta \in B} F_\beta.$$

Observe that a function $Y : \Omega \to \mathbb{R}$ is $\mathcal{F}$-measurable if and only there exist real numbers $(y_\alpha)_{\alpha \in A}$ such that

$$Y = \sum_{\alpha \in A} y_\alpha \boldsymbol{I}_\alpha, \ \boldsymbol{I}_\alpha := \boldsymbol{I}_{F_\alpha}.$$

Moreover

$$Y \in L^1 \iff \sum_\alpha |y_\alpha| \mathbb{P}[F_\alpha] < \infty.$$

Suppose now that $X \in L^1(\Omega, \mathcal{S}, \mathbb{P})$. We define the *expectation of X given the event F_α* to be the expectation of X with respect to the conditional probability $\mathbb{P}[-|F_\alpha]$, i.e., the *number*

$$\bar{x}_\alpha = \mathbb{E}[X|F_\alpha] := \frac{1}{\mathbb{P}[F_\alpha]} \mathbb{E}[X\boldsymbol{I}_\alpha] = \frac{1}{\mathbb{P}[F_\alpha]} \int_{F_\alpha} X(\omega) \mathbb{P}[d\omega]. \quad (1.4.1)$$

We obtain an $\mathcal{F}$-measurable *random variable*

$$\bar{X} = \sum_\alpha \bar{x}_\alpha \boldsymbol{I}_\alpha.$$

Note that

$$|\bar{x}_\alpha| \leq \frac{1}{\mathbb{P}[F_\alpha]} \mathbb{E}[|X|\boldsymbol{I}_\alpha]$$

so

$$\mathbb{E}[|\bar{X}|] \leq \sum_\alpha \mathbb{E}[|X|\boldsymbol{I}_\alpha] = \mathbb{E}[|X|] < \infty.$$

Since

$$\mathbb{E}[X\boldsymbol{I}_\alpha] = \mathbb{E}[\bar{X}\boldsymbol{I}_\alpha], \ \forall \alpha \in A,$$

we deduce

$$\mathbb{E}[X\boldsymbol{I}_F] = \mathbb{E}[\bar{X}\boldsymbol{I}_F], \ \forall F \in \mathcal{F}. \quad (1.4.2)$$

Note that if
$$Y = \sum_\alpha y_\alpha \boldsymbol{I}_\alpha$$
is another $\mathcal{F}$-measurable, integrable random variable that satisfies (1.4.2), then
$$\mathbb{P}[F_\alpha] x_\alpha = \mathbb{E}[X \boldsymbol{I}_\alpha] = \mathbb{E}[Y \boldsymbol{I}_\alpha] = \mathbb{P}[F_\alpha] y_\alpha, \quad \forall \alpha \in A,$$
so that $y_\alpha = x_\alpha$, $\forall \alpha$, i.e., $\overline{X}$ is uniquely determined by (1.4.2).

If in (1.4.2) we set $F = \Omega$ we deduce
$$\mathbb{E}[X] = \mathbb{E}[\overline{X}] = \sum_\alpha \bar{x}_\alpha \mathbb{P}[F_\alpha] = \sum_\alpha \mathbb{E}[X| F_\alpha] \mathbb{P}[F_\alpha]. \tag{1.4.3}$$

When $X = \boldsymbol{I}_S$, then
$$\mathbb{E}[\boldsymbol{I}_S| F_\alpha] = \frac{\mathbb{P}[S \cap F_\alpha]}{\mathbb{P}[F_\alpha]} = \mathbb{P}[S| F_\alpha].$$

In this special case the equality (1.4.3) becomes the *law of total probability*
$$\mathbb{P}[S] = \sum_\alpha \mathbb{P}[S| F_\alpha] \mathbb{P}[F_\alpha]. \tag{1.4.4}$$
$\square$

The next result explains why the condition (1.4.2) is key to our further developments.

Proposition 1.156. *If $\mathcal{F} \subset \mathcal{S}$ is a sigma sub-algebra and $Y_0, Y_1 \in \mathcal{L}^1(\Omega, \mathcal{F}, \mathbb{P})$ are two $\mathcal{F}$-measurable random variables such that*
$$\mathbb{E}[Y_0 \boldsymbol{I}_F] = \mathbb{E}[Y_1 \boldsymbol{I}_F], \quad \forall F \in \mathcal{F}, \tag{1.4.5}$$
then $Y_0 = Y_1$ a.s.

Proof. Set $Z = Y_0 - Y_1$. Then Z is $\mathcal{F}$-measurable, integrable and satisfies
$$\mathbb{E}[Z \boldsymbol{I}_F] = 0, \quad \forall F \in \mathcal{F}. \tag{1.4.6}$$
If we let $F = \{Z > 1/n\}$, $n \in \mathbb{N}$, we deduce that
$$\frac{1}{n} \mathbb{P}[Z > 1/n] \leq \mathbb{E}[Z \boldsymbol{I}_{\{Z > 1/n\}}] = 0, \quad \forall n \in \mathbb{N}.$$
Thus
$$\mathbb{P}[Z > 1/n] = 0, \quad \forall n \in \mathbb{N} \Rightarrow \mathbb{P}[Z > 0] = 0.$$
A similar argument shows that $\mathbb{P}[Z < 0] = 0$. $\square$

Definition 1.157. Let $\mathcal{F} \subset \mathcal{S}$ be a sigma sub-algebra and $X \in L^1(\Omega, \mathcal{F}, \mathbb{P})$. A *version of the conditional expectation of X given $\mathcal{F}$* is an $\mathcal{F}$-measurable random variable $\overline{X} \in \mathcal{L}^1(\Omega, \mathcal{F}, \mathbb{P})$ such that
$$\mathbb{E}[X \boldsymbol{I}_F] = \mathbb{E}[\overline{X} \boldsymbol{I}_F], \quad \forall F \in \mathcal{F}. \tag{1.4.7}$$
$\square$

According to Proposition 1.156, any two random variables $\overline{X}_0, \overline{X}_1 \in \mathcal{L}^1(\Omega, \mathcal{F}, \mathbb{P})$ satisfying (1.4.7) are a.s. equal. Their equivalence class in $L^1(\Omega, \mathcal{F}, \mathbb{P})$ is denoted by $\mathbb{E}[X \| \mathcal{F}]$ and it is called *the conditional expectation* of X given $\mathcal{F}$.

I am using different notations for the conditional expectation given and event, $\mathbb{E}[X|F]$, and the conditional expectation given a sigma-subalgebra, $\mathbb{E}[X \| \mathcal{F}]$, for a simple reason: I want to emphasize visually that the first is a number and the latter is a function.

Remark 1.158. Using the Monotone Convergence Theorem and the Monotone Class Theorem we deduce that the following are equivalent.

(i) The random variable $\overline{X} \in \mathcal{L}^1(\Omega, \mathcal{F}, \mathbb{P})$ is a representative of $\mathbb{E}[X \| \mathcal{F}]$.
(ii) For any $Y \in L^\infty(\Omega, \mathcal{F}, \mathbb{P})$

$$\mathbb{E}[XY] = \mathbb{E}[\overline{X}Y]. \tag{1.4.8}$$

(iii) There exists a π-system $\mathcal{A} \subset \mathcal{F}$ that generates $\mathcal{F}$ such that

$$\mathbb{E}[X\boldsymbol{I}_A] = \mathbb{E}[\overline{X}\boldsymbol{I}_A], \quad \forall A \in \mathcal{A}. \tag{1.4.9}$$

From Corollary 1.94 we deduce that

$$\overline{X} = \mathbb{E}[X \| \mathcal{F}] \iff \overline{X} \text{ is } \mathcal{F} \text{ measurable and } (X - \overline{X}) \perp\!\!\!\perp \mathcal{F}. \tag{1.4.10}$$

$\square$

Definition 1.159. Given random variables $X \in L^0(\Omega, \mathcal{S}, \mathbb{P})$, $Y \in L^1(\Omega, \mathcal{S}, \mathbb{P})$ we write

$$\boxed{\mathbb{E}[Y \| X] := \mathbb{E}[Y \| \sigma(X)]},$$

where $\sigma(X)$ denotes the sigma-subalgebra generated by X. This random variable is called the *conditional expectation of Y given X*. $\square$

Remark 1.160. A function $\overline{Y} \in L^1(\Omega, \sigma(X), \mathbb{P})$ represents $\mathbb{E}[Y \| X]$ if, for any $x \in \mathbb{R}$ we have

$$\int_{\{X \leq x\}} Y(\omega)\mathbb{P}[d\omega] = \int_{\{X \leq x\}} \overline{Y}(\omega)\mathbb{P}[d\omega].$$

Since $\mathbb{E}[Y \| X]$ is $\sigma(X)$-measurable we deduce from Dynkin's Theorem 1.23 that there exists a Borel measurable function $f : \mathbb{R} \to \mathbb{R}$ such that

$$f(X) = \mathbb{E}[Y \| X] \quad \text{a.s.}$$

This is equivalent to the statement

$$\mathbb{E}[Y\boldsymbol{I}_{\{X \leq x\}}] = \mathbb{E}[f(X)\boldsymbol{I}_{\{X \leq x\}}], \quad \forall x \in \mathbb{R}. \tag{1.4.11}$$

The function $f(x)$ is called the *conditional expectation of Y given $X = x$* and it is denoted by $\mathbb{E}[Y|X = x]$. Think of it as the conditional expectation of Y given the possible negligible event $\{X = x\}$.

Note that
$$\mathbb{E}[Y] = \mathbb{E}[\overline{Y}] = \mathbb{E}\big[\mathbb{E}[Y \,\|\, X]\big] = \mathbb{E}[f(X)].$$
Thus
$$\mathbb{E}[Y] = \mathbb{E}[f(X)] = \int_{\mathbb{R}} f(x)\mathbb{P}_X[dx].$$
We can rewrite the last equality as
$$\mathbb{E}[Y] = \int_{\mathbb{R}} \mathbb{E}[Y|X=x]\mathbb{P}_X[dx]. \qquad (1.4.12)$$

This approach to computing the expectation of Y by relying on the above identity is referred to *computing the expectation of Y by conditioning* on X. This generalizes the elementary situation in Exercise 1.12. □

Example 1.161. Suppose that $X, Y : (\Omega, \mathcal{S}, \mathbb{P}) \to \mathbb{R}$ are two random variables such that their joint probability distribution $\mathbb{P}_{X,Y} \in \mathrm{Prob}(\mathbb{R}^2)$ is absolutely continuous with respect to the Lebesgue measure on $\mathbb{R}^2$. This means that there exists a Lebesgue integrable function
$$p_{X,Y} : \mathbb{R}^2 \to [0, \infty)$$
such that
$$\mathbb{P}[(X,Y) \in B] = \int_B p_{X,Y}(x,y)dxdy, \quad \forall B \in \mathcal{B}_{\mathbb{R}^2}.$$
We denote by $\mathbb{P}_X$ and respectively $\mathbb{P}_Y$ the probability distributions of X and respectively Y. Note that the cumulative distribution function F_X of X is
$$F_X(c) = \mathbb{P}[X \le c] = \int_{-\infty}^{c} \underbrace{\left(\int_{\mathbb{R}} p_{X,Y}(x,y)dy\right)}_{=:p_X(x)} dx = \int_{-\infty}^{c} p_X(x)dx.$$
This shows that $\mathbb{P}_X$ is absolutely continuous with respect to the Lebesgue measure on $\mathbb{R}$ and
$$\mathbb{P}_X[dx] = p_X(x)dx.$$
Similarly
$$\mathbb{P}_Y[dy] = p_Y(y)dy = \int_{\mathbb{R}} p_{X,Y}(x,y)dx.$$
Classically, the probability distributions $\mathbb{P}_X$ and $\mathbb{P}_Y$ are called the *marginal distributions* of the random vector (X, Y). We define
$$p_{Y|X=x}(y) = \begin{cases} \frac{p_{X,Y}(x,y)}{p_X(x)}, & p_X(x) \ne 0, \\ 0, & p_X(x) := 0. \end{cases}$$

Assume that Y is integrable. Define

$$f:\mathbb{R}\to\mathbb{R},\quad f(x)=\int_{\mathbb{R}}yp_{Y|X=x}(y)dy=\begin{cases}\frac{1}{p_X(x)}\int_{\mathbb{R}}yp_{X,Y}(x,y)dy,&p_X(x)\neq 0,\\ 0,&p_X(x)=0.\end{cases}$$

Using Fubini and the integrability of Y we deduce that the above integrals are well defined and the resulting function f is Borel measurable. Note that

$$f(x)p_X(x)=\int_{\mathbb{R}}yp_{X,Y}(x,y)dy,\quad\forall x\in\mathbb{R}.$$

We want to show that $f(x)=\mathbb{E}\big[Y\,\big|\,X=x\big]$, i.e., $f(X)$ is a version of $\mathbb{E}\big[Y\,\|\,X\big]$. We will show that it satisfies (1.4.11).

Let $c\in\mathbb{R}$. We have

$$\begin{aligned}\mathbb{E}\big[f(X)\boldsymbol{I}_{X\leq c}\big]&=\int_{-\infty}^{c}f(x)p_X(x)dx=\int_{\mathbb{R}}\left(\int_{\mathbb{R}}yp_{X,y}dy\right)\boldsymbol{I}_{(-\infty,c]}(x)dx\\ &=\int_{\mathbb{R}^2}y\boldsymbol{I}_{(-\infty,c]}(x)p_{X,Y}(x,y)dxdy=\mathbb{E}\big[Y\boldsymbol{I}_{X\leq c}\big].\end{aligned}\quad(1.4.13)$$

The function $f(x)$ is the conditional expectation $\mathbb{E}\big[Y|X=x\big]$ discussed in Remark 1.158.

Note that the event $\{X=x\}$ has probability zero so this nomenclature should be taken with a grain of sand since we cannot apply (1.4.1). Intuitively

$$\mathbb{E}\big[Y|X=x\big]=\lim_{\varepsilon\searrow 0}\mathbb{E}\big[Y\,\big|\,\{|X-x|<\varepsilon\}\big]\stackrel{(1.4.1)}{=}\lim_{\varepsilon\searrow 0}\frac{\mathbb{E}\big[Y\boldsymbol{I}_{|X-x|<\varepsilon}\big]}{\mathbb{P}\big[|X-x|<\varepsilon\big]}.$$

$\square$

One issue we need to address is the existence of the conditional expectation. There is a fast proof based on the Radon–Nikodym theorem. We will use a more roundabout approach that sheds additional light on the nature of conditional expectation. As an aside, let us mention that this approach leads to an alternate proof of the Radon–Nikodym theorem that does not rely on the concept of signed-measure.

Theorem 1.162. *For any $X\in L^1(\Omega,\mathcal{S},\mathbb{P})$ and any sigma sub-algebra $\mathcal{F}\subset\mathcal{S}$ there exists a conditional expectation $\mathbb{E}\big[X\,\|\,\mathcal{F}\big]\in L^1(\Omega,\mathcal{F},\mathbb{P})$.*

Proof. We follow the approach in [160]. We establish the existence gradually, first under more restrictive assumptions.

Step 1. Assume $X\in L^2(\Omega,\mathcal{S},\mathbb{P})$. Then $L^2(\Omega,\mathcal{F},\mathbb{P})$ is a closed subspace of $L^2(\Omega,\mathcal{S},\mathbb{P})$. Denote by $P_{\mathcal{F}}X$ the orthogonal projection of X on this closed subspace. We claim that

$$P_{\mathcal{F}}X=\mathbb{E}\big[X\,\|\,\mathcal{F}\big],\quad(1.4.14a)$$

$$X\geq 0\Rightarrow\mathbb{E}\big[X\,\|\,\mathcal{F}\big]\geq 0.\quad(1.4.14b)$$

Set $Y = P_{\mathcal{F}} X$. Since $X - Y \perp L^2(\Omega, \mathcal{F}, \mathbb{P})$ we deduce
$$\mathbb{E}\big[(X - Y)Z\big] = 0, \quad \forall Z \in L^2(\Omega, \mathcal{F}, \mathbb{P}).$$
In particular,
$$\mathbb{E}\big[(X - Y)\boldsymbol{I}_F\big] = 0, \quad \forall F \in \mathcal{F}.$$
This proves (1.4.14a). Now suppose that $X \geq 0$. For any $n \in \mathbb{N}$ we have
$$0 \leq \mathbb{E}\big[X\boldsymbol{I}_{\{Y \leq -1/n\}}\big] = \mathbb{E}\big[Y\boldsymbol{I}_{\{Y \leq -1/n\}}\big] \leq -\frac{1}{n}\mathbb{P}\big[Y \leq -1/n\big],$$
so
$$\mathbb{P}\big[Y \leq -1/n\big] = 0, \quad \forall n \in \mathbb{N}.$$
This proves (1.4.14b). Clearly, the resulting map
$$L^2(\Omega, \mathcal{S}, \mathbb{P}) \ni X \mapsto \mathbb{E}\big[X \,\|\, \mathcal{F}\big] \in L^2(\Omega, \mathcal{F}, \mathbb{P})$$
is linear.

Step 2. Assume $X \in L^1(\Omega, \mathcal{S}, \mathbb{P})$. Decompose $X = X^+ - X^-$ and, for $n \in \mathbb{N}$, set
$$X_n^{\pm} = \min(X^{\pm}, n).$$
Note that $X_n^{\pm} \in L^{\infty}(\Omega, \mathcal{S}, \mathbb{P})$ and, as $n \to \infty$, $X_n^{\pm} \nearrow X^{\pm}$ a.s.. From Step 1 we deduce that the random variables $X_n^{\pm}$ have conditional expectations given $\mathcal{F}$. Choose versions
$$Y_n^{\pm} := \mathbb{E}\big[X_n^{\pm} \,\|\, \mathcal{F}\big].$$
Since $X_n^{\pm} - X_m^{\pm} \geq 0$ a.s. if $m \leq n$ we deduce from (1.4.14b) that
$$0 \leq Y_m^{\pm} \leq Y_n^{\pm}, \quad \text{a.s.}, \quad \forall m \leq n.$$
We set
$$Y^{\pm} := \lim_{n \to \infty} Y_n^{\pm}.$$
From the Monotone Convergence Theorem we deduce that
$$\infty > \mathbb{E}\big[X^{\pm}\big] = \lim_{n \to \infty} \mathbb{E}\big[X_n^{\pm}\big] = \lim_{n \to \infty} \mathbb{E}\big[Y_n^{\pm}\big] = \mathbb{E}\big[Y^{\pm}\big].$$
This shows that the random variables $Y_{\pm}$ are integrable and in particular a.s. finite. We set
$$Y := Y^+ - Y^-.$$
We will show that Y is a version of the conditional expectation of X given $\mathcal{F}$. Let $F \in \mathcal{F}$. Then
$$\mathbb{E}\big[X\boldsymbol{I}_F\big] = \mathbb{E}\big[X_+\boldsymbol{I}_F\big] - \mathbb{E}\big[X_-\boldsymbol{I}_F\big] = \lim_{n \to \infty}\mathbb{E}\big[X_n^+\boldsymbol{I}_F\big] - \lim_{n \to \infty}\mathbb{E}\big[X_n^-\boldsymbol{I}_F\big]$$
$$\lim_{n \to \infty}\mathbb{E}\big[Y_n^+\boldsymbol{I}_F\big] - \lim_{n \to \infty}\mathbb{E}\big[Y_n^-\boldsymbol{I}_F\big] = \mathbb{E}\big[Y^+\boldsymbol{I}_F\big] - \mathbb{E}\big[Y^-\boldsymbol{I}_F\big] = \mathbb{E}\big[Y\boldsymbol{I}_F\big].$$
This proves that Y is a version of $\mathbb{E}\big[X \,\|\, \mathcal{F}\big]$. $\square$

Remark 1.163. (a) The sigma sub-algebra $\mathcal{F}$ should be viewed as encoding partial information that we have about a random experiment. Following a terminology frequently used in statistics, we refer to the $\mathcal{F}$-measurable random variables as *predictors* determined by the information contained in $\mathcal{F}$.

Step 1 in the above proof shows that the conditional expectation $\overline{X}$ of a random variable X, given the partial information $\mathcal{F}$, should be viewed as the predictor that best approximates X given the information $\mathcal{F}$. The missing part $X - \overline{X}$ is independent of $\mathcal{F}$ so it is unknowable given only the information encoded by $\mathcal{F}$.

Note that when $\mathcal{F} = \{\emptyset, \Omega\}$, then

$$\mathbb{E}[X] = \mathbb{E}[X]\boldsymbol{I}_\Omega.$$

To put it differently, if the only information we have about a random experiment is that there will be an outcome, then the most/best we can predict about a numerical characteristic of that outcome is its expectation.

(b) A random variable $X \in L^1(\Omega, \mathcal{S}, \mathbb{P})$ defines a signed measure

$$\mu_X : \mathcal{F} \to [0, \infty), \;\; \mu_X[F] = \int_F X(\omega)\mathbb{P}[d\omega], \;\; \forall F \in \mathcal{F}.$$

This measure is absolutely continuous with $\mathbb{P}$ (restricted to $\mathcal{F}$). The Radon–Nicodym theorem implies that there exists an $\mathcal{F}$-measurable integrable function $\rho_X \in L^1(\Omega, \mathcal{F}, \mathbb{P})$ such that $\mu_X[d\omega] = \rho_X(\omega)\mathbb{P}[d\omega]$, i.e.,

$$\int_F X(\omega)\mathbb{P}[d\omega] = \int_F \rho_X(\omega)\mathbb{P}[d\omega], \;\; \forall F \in \mathcal{F}.$$

This shows that $\rho_X = \mathbb{E}[X \,\|\, \mathcal{F}]$ a.s. $\square$

Definition 1.164. Given a sigma subalgebra $\mathcal{F} \subset \mathcal{S}$, and an event $S \in \mathcal{S}$, we define the *conditional probability* of S given $\mathcal{F}$ to be the <u>random variable</u>

$$\mathbb{P}[S \,\|\, \mathcal{F}] := \mathbb{E}[\boldsymbol{I}_S \,\|\, \mathcal{F}]. \qquad \square$$

Example 1.165 (Conditioning on an event). Suppose that $S \in \mathcal{S}$ is an event such $0 < \mathbb{P}[S] < 1$. Let $Y \in L^1(\Omega, \mathcal{S}, \mathbb{P})$. Then

$$\mathbb{E}[Y \,\|\, \boldsymbol{I}_S] = \mathbb{E}[Y \,|\, S]\boldsymbol{I}_S + \mathbb{E}[Y \,|\, S^c]\boldsymbol{I}_{S^c},$$

where we recall that (see (1.4.1))

$$\mathbb{E}[Y \,|\, S] = \frac{1}{\mathbb{P}[S]}\mathbb{E}[Y\boldsymbol{I}_S]. \qquad \square$$

Our next result lists the main properties of the conditional expectation.

Theorem 1.166. *Suppose that $\mathcal{F} \subset \mathcal{S}$ is a sigma sub-algebra. Then the following hold.*

(i) *Let $X \in L^1(\Omega, \mathcal{S}, \mathbb{P})$. If Y is any version of $\mathbb{E}[X \,\|\, \mathcal{F}]$, then $\mathbb{E}[Y] = \mathbb{E}[X]$. In other words*

$$\mathbb{E}\Big[\mathbb{E}[X \,\|\, \mathcal{F}]\Big] = \mathbb{E}[X]. \qquad (1.4.15)$$

(ii) If $X, Y \in L^1(\Omega, \mathcal{S}, \mathbb{P})$ and $X \leq Y$ a.s., then $\mathbb{E}[X \| \mathcal{F}] \leq \mathbb{E}[Y \| \mathcal{F}]$ a.s.

(iii) The map
$$L^1(\Omega, \mathcal{S}, \mathbb{P}) \ni X \mapsto \mathbb{E}[X \| \mathcal{F}] \in L^1(\Omega, \mathcal{F}, \mathbb{P})$$
is a linear contraction, i.e., it is linear and satisfies
$$\| \mathbb{E}[X \| \mathcal{F}] \|_{L^1} \leq \|X\|_{L^1}, \quad \forall X \in L^1(\Omega, \mathcal{S}, \mathbb{P}).$$

(iv) If $X \in L^1(\Omega, \mathcal{S}, \mathbb{P})$ and $Y \in L^\infty(\Omega, \mathcal{F}, \mathbb{P})$, then
$$\mathbb{E}[XY \| \mathcal{F}] = Y \mathbb{E}[X \| \mathcal{F}].$$

(v) If $\mathcal{G} \subset \mathcal{F}$ is another sigma sub-algebra, then for any $X \in L^1(\Omega, \mathcal{S}, \mathbb{P})$ we have
$$\mathbb{E}[X \| \mathcal{G}] = \mathbb{E}\Big[\mathbb{E}[X \| \mathcal{F}] \| \mathcal{G}\Big].$$

(vi) If $0 \leq X_n \nearrow X$ a.s., $X \in L^1(\Omega, \mathcal{S}, \mathbb{P})$, then
$$\mathbb{E}[X_n \| \mathcal{F}] \nearrow \mathbb{E}[X \| \mathcal{F}], \text{ a.s.}$$

(vii) If $X_n \in L^1(\Omega, \mathcal{S}, \mathbb{P})$, $n \in \mathbb{N}$, $X_n \geq 0$ a.s., $\liminf X_n \in L^1$ a.s., then
$$\mathbb{E}[\liminf X_n \| \mathcal{F}] \leq \liminf \mathbb{E}[X_n \| \mathcal{F}] \text{ a.s.}$$

(viii) If $X_n \to X$ a.s. and there exists $Y \in L^1(\Omega, \mathcal{S}, \mathbb{P})$ such that $|X_n| \leq Y$ a.s., then
$$\mathbb{E}[X_n \| \mathcal{F}] \to \mathbb{E}[X \| \mathcal{F}] \text{ a.s.}$$

(ix) If $X \in L^1(\Omega, \mathcal{S}, \mathbb{P})$ and $\varphi : \mathbb{R} \to \mathbb{R}$ is a convex function such that $\varphi(X)$ is integrable, then
$$\varphi\Big(\mathbb{E}[X \| \mathcal{F}]\Big) \leq \mathbb{E}[\varphi(X) \| \mathcal{F}] \text{ a.s.}$$

In particular, if we choose $\varphi(x) = |x|^p$, $p \geq 1$ we deduce that the conditional expectation defines a linear map
$$\mathbb{E}[- \| \mathcal{F}] : L^p(\Omega, \mathcal{S}, \mathbb{P}) \to L^p(\Omega, \mathcal{F}, \mathbb{P})$$
that is linear contraction, i.e.,
$$\| \mathbb{E}[X \| \mathcal{F}] \|_{L^p} \leq \|X\|_{L^p}.$$

(x) If $\mathcal{G}$ is another sigma-algebra that is independent of $\sigma(X) \vee \mathcal{F}$, then
$$\mathbb{E}[X \| \mathcal{F} \vee \mathcal{G}] = \mathbb{E}[X \| \mathcal{F}].$$

In particular, if $X \in L^1(\Omega, \mathcal{S}, \mathbb{P})$ is independent of $\mathcal{G}$, then
$$\mathbb{E}[X \| \mathcal{G}] = \mathbb{E}[X].$$

Proof. (i) Follows by choosing $F = \Omega$ in (1.4.7). (ii) Follows from the proof of Theorem 1.162.

(iii) The linearity follows from the fact that the defining condition (1.4.7) is linear in X. Now let $X \in L^1(\Omega, \mathcal{S}, \mathbb{P})$. We have $X = X^+ - X^-$. Choose versions $Y^\pm$ of $\mathbb{E}[X^\pm \,\|\, \mathcal{F}]$. Then $Y_\pm \geq 0$ and

$$\left|\mathbb{E}[X \,\|\, \mathcal{F}]\right| = |Y^+ - Y^-| \leq Y^+ + Y^- = \mathbb{E}[X^+ + X^- \,\|\, \mathcal{F}] = \mathbb{E}[|X| \,\|\, \mathcal{F}].$$

Hence

$$\left\|\mathbb{E}[X \,\|\, \mathcal{F}]\right\|_{L^1} \leq \mathbb{E}\Big[\mathbb{E}[|X| \,\|\, \mathcal{F}]\Big] = \mathbb{E}[|X|] = \|X\|_{L^1}.$$

(iv) Choose a version Z of $\mathbb{E}[X \,\|\, \mathcal{F}]$. Let $Y \in L^\infty(\Omega, \mathcal{F}, \mathbb{P})$. We have to show that YZ is a version of $\mathbb{E}[XY \,\|\, \mathcal{F}]$, i.e.,

$$\mathbb{E}[XY\mathbf{I}_F] = \mathbb{E}[ZY\mathbf{I}_F], \quad \forall F \in \mathcal{F}. \qquad (1.4.16)$$

Let $F \in \mathcal{F}$. Since Z is a version of $\mathbb{E}[X \,\|\, \mathcal{F}]$ we deduce from (1.4.8) that

$$\mathbb{E}[XU] = \mathbb{E}[ZU], \quad \forall U \in L^\infty(\Omega, \mathcal{F}, \mathbb{P}).$$

In particular, $\forall F \in \mathcal{F}$ we have

$$\mathbb{E}[X\underbrace{Y\mathbf{I}_F}_{U}] = \mathbb{E}[ZU] = \mathbb{E}[ZY\mathbf{I}_F].$$

Thus ZY satisfies (1.4.16).

(v) Choose a version Y of $\mathbb{E}[X \,\|\, \mathcal{F}]$, and a version Z of $\mathbb{E}[Y \,\|\, \mathcal{G}]$. We have to show that Z is also a version of $\mathbb{E}[X \,\|\, \mathcal{G}]$. Let $G \in \mathcal{G}$. We have

$$\mathbb{E}[Y\mathbf{I}_B] = \mathbb{E}[Z\mathbf{I}_B],$$

$$\mathbb{E}[X\mathbf{I}_G] \stackrel{G \in \mathcal{F}}{=} \mathbb{E}[Y\mathbf{I}_G] = \mathbb{E}[Z\mathbf{I}_B].$$

(vi) Choose versions Y_n of $\mathbb{E}[X_n \,\|\, \mathcal{F}]$ and Y of $\mathbb{E}[X \,\|\, \mathcal{F}]$. Note that Y_n is increasing. The Monotone Convergence theorem implies that $\|X - X_n\|_{L^1} \to 0$. From (iii) we deduce

$$\|Y_n - Y\|_{L^1} \leq \|X_n - Y\|_{L^1}.$$

Proposition 1.147 implies that Y_n admits a subsequence that converges a.s. to Y. Since the sequence Y_n is increasing we deduce that the whole sequence converges a.s. to Y.

(vii) Set

$$Y_k = \inf_{n \geq k} X_n.$$

The sequence of random variables (Y_k) is increasing and converges a.s. to $X := \liminf X_n$. We deduce from (vi) that

$$\mathbb{E}[Y_k \,\|\, \mathcal{F}] \nearrow \mathbb{E}[X \,\|\, \mathcal{F}].$$

Note that since $Y_k \leq X_n$, $\forall n \geq k$, we have
$$\mathbb{E}\big[\,Y_k \,\|\, \mathcal{F}\,\big] \leq Z_k := \inf_{n \geq k} \mathbb{E}\big[\,X_n \,\|\, \mathcal{F}\,\big]$$
so
$$\mathbb{E}\big[\,X \,\|\, \mathcal{F}\,\big] = \lim_k \mathbb{E}\big[\,Y_k \,\|\, \mathcal{F}\,\big] \leq \lim_k Z_k = \liminf \mathbb{E}\big[\,X_n \,\|\, \mathcal{F}\,\big].$$

(viii) Set $Y_n := X_n + Y$. Then $Y_n \geq 0$ and $Y_n \to X + Y$ a.s. We deduce from (vii) that
$$\mathbb{E}\big[\,X \,\|\, \mathcal{F}\,\big] + \mathbb{E}\big[\,Y \,\|\, \mathcal{F}\,\big] \leq \liminf \mathbb{E}\big[\,X_n \,\|\, \mathcal{F}\,\big] + \mathbb{E}\big[\,Y \,\|\, \mathcal{F}\,\big]$$
i.e.,
$$\mathbb{E}\big[\,X \,\|\, \mathcal{F}\,\big] \leq \liminf \mathbb{E}\big[\,X_n \,\|\, \mathcal{F}\,\big].$$

Similarly, we set $Z_n = Y - X_n$. Then $Z_n \geq 0$ and $Z_n \to Y - X$ a.s. Applying (vii) to Z_n we deduce
$$\limsup \mathbb{E}\big[\,X_n \,\|\, \mathcal{F}\,\big] \leq \mathbb{E}\big[\,X \,\|\, \mathcal{F}\,\big].$$

(ix) We need to use a less familiar property of convex functions, [4, Thm. 6.3.4]. More precisely, there exist *sequences* of real numbers $(a_n)_{n \in \mathbb{N}}$ and $(b_n)_{n \in \mathbb{N}}$ such that
$$\varphi(x) = \sup_{n \in \mathbb{N}}(a_n x + b_n), \quad \forall x \in \mathbb{R}.$$

Set $\ell_n(x) = a_n x + b_n$.[7] Clearly
$$\ell_n\Big(\mathbb{E}\big[\,X \,\|\, \mathcal{F}\,\big] \Big) = \mathbb{E}\big[\,\ell_n(X) \,\|\, \mathcal{F}\,\big] \leq \mathbb{E}\big[\,\varphi(X) \,\|\, \mathcal{F}\,\big].$$
Hence
$$\varphi\Big(\mathbb{E}\big[\,X \,\|\, \mathcal{F}\,\big] \Big) = \sup_{n \in \mathbb{N}} \ell_n\Big(\mathbb{E}\big[\,X \,\|\, \mathcal{F}\,\big] \Big) = \sup_{n \in \mathbb{N}} \mathbb{E}\big[\,\ell_n(X) \,\|\, \mathcal{F}\,\big] \leq \mathbb{E}\big[\,\varphi(X) \,\|\, \mathcal{F}\,\big].$$

(x) Let $G \in \mathcal{G}$ and F in $\mathcal{F}$. Then, the random variables $\boldsymbol{I}_G$ and $X\boldsymbol{I}_F$ are independent so
$$\mathbb{E}\big[\,X\boldsymbol{I}_{F \cap G}\,\big] = \mathbb{E}\big[\,X F_F \boldsymbol{I}_G\,\big] = \mathbb{E}\big[\,X\boldsymbol{I}_F\,\big]\mathbb{P}\big[\,G\,\big].$$

If Y is a version of $\mathbb{E}\big[\,X \,\|\, \mathcal{F}\,\big]$, then Y is $\mathcal{F}$-measurable and thus independent of G, so
$$\mathbb{E}\big[\,Y\boldsymbol{I}_{F \cap G}\,\big] = \mathbb{E}\big[\,Y\boldsymbol{I}_F \boldsymbol{I}_G\,\big] = \mathbb{E}\big[\,Y\boldsymbol{I}_F\,\big]\mathbb{P}\big[\,G\,\big]$$
$$= \mathbb{E}\big[\,X\boldsymbol{I}_F\,\big]\mathbb{P}\big[\,G\,\big] = \mathbb{E}\big[\,X\boldsymbol{I}_{F \cap G}\,\big], \quad \forall F \in \mathcal{F},\ G \in \mathcal{G}.$$

Since the collection
$$\big\{ F \cap G;\ F \in \mathcal{F},\ G \in \mathcal{G} \big\}$$
is a π-system generating $\mathcal{F} \vee \mathcal{G}$, we deduce from Dynkin's $(\pi - \lambda)$ theorem that
$$\mathbb{E}\big[\,Y\boldsymbol{I}_S\,\big] = \mathbb{E}\big[\,X\boldsymbol{I}_S\,\big], \quad \forall S \in \mathcal{F} \vee \mathcal{G},$$
so that $\mathbb{E}\big[\,X \,\|\, \mathcal{F} \vee \mathcal{G}\,\big] = Y$, i.e., $\mathbb{E}\big[\,X \,\|\, \mathcal{F}\,\big] = \mathbb{E}\big[\,X \,\|\, \mathcal{F} \vee \mathcal{G}\,\big]$. $\square$

---
[7]When φ is C^1 the family ℓ_n coincides with the family $(\ell_q)_{q \in \mathbb{Q}}$, $\ell_q(x) = \varphi'(q)(x - q) + \varphi(q)$.

1.4.2 Some applications of conditioning

To give the reader a taste of the power and uses of conditional expectation we describe some nontrivial and less advertised uses of conditional expectation.

Example 1.167. Suppose that a player rolls a die an indefinite amount of times. More formally, we are given a sequence independent random variables $(X_n)_{n\in\mathbb{N}}$, uniformly distributed on $\mathbb{I}_6 := \{1, 2, \ldots, 6\}$.

For $k \in \mathbb{N}$, we say that a k-run of length k occurred at time n if $n \geq k$ and

$$X_n = X_{n-1} = \cdots = X_{n-k+1} = 6.$$

We set

$$R = R_k := \{ n; \text{ a } k\text{-run occurred at time } n \} \subset \mathbb{N} \cup \{\infty\}, \quad T = T_k = \inf R_k,$$

where $\inf \emptyset := \infty$. Thus T is the moment when the first k-run occurs. We want to show that $\mathbb{E}[T] < \infty$.

Note that for each $n \in \mathbb{N}$ the event $\{T \leq n\}$ belongs to the sigma algebra $\mathcal{F}_n$ generated by $X_1, \ldots, X_n$. The explanation is simple: if we know the results of the first n rolls of the die we can decide if a k-run was occurred. Consider the conditional probability

$$\mathbb{P}\big[\{T \leq n+k\} \,\|\, \mathcal{F}_n\big] = \mathbb{E}\big[\, \boldsymbol{I}_{\{T \leq n+k\}} \,\|\, \mathcal{F}_n\big].$$

This conditional probability is a *random variable*. Since the sigma-algebra $\mathcal{F}_n$ is defined by the partition

$$S_{i_1,\ldots,i_n} := \{X_1 = i_1, \ldots, X_n = i_n\}, \quad i_j \in \{1, \ldots, 6\},$$

we see that $\mathbb{P}\big[T \leq n+k \,\|\, \mathcal{F}_n\big]$ has the form

$$\mathbb{P}\big[T \leq n+k \,\|\, \mathcal{F}_n\big] = \sum_{i_1,\ldots,i_n=1}^{6} p_{i_1,\ldots,i_n|k} \boldsymbol{I}_{S_{i_1,\ldots,i_n}},$$

where

$$p_{i_1,\ldots,i_n|k} = \mathbb{P}\big[T \leq n+k \,\big|\, X_1 = i_1, \ldots, X_n = i_n \big].$$

Note that, irrespective of the i_j-s, we have

$$p_{i_1,\ldots,i_n|k} \geq \frac{1}{6^k} =: r.$$

Hence

$$\mathbb{P}\big[T \leq n+k \,\|\, \mathcal{F}_n\big] \geq r, \quad \forall n.$$

In particular

$$\mathbb{P}\big[T > n+k \,\|\, \mathcal{F}_n\big] \leq (1-r) < 1, \quad \forall n \in \mathbb{N}.$$

Now observe that for any $n \in \mathbb{N}$, $\ell \in \mathbb{N}_0$ we have $\{T > n + \ell k\} \in \mathcal{F}_{n+\ell k}$. Hence

$$\mathbb{P}\big[T > n + (\ell+1)k\big] = \mathbb{E}\big[\, \boldsymbol{I}_{\{T > n+(\ell+1)k\}} \boldsymbol{I}_{\{T > n+\ell k\}} \,\big]$$

$$= \mathbb{E}\Big[\, \boldsymbol{I}_{\{T>\ell k\}} \mathbb{E}\big[\, T > n+(\ell+1)k \,\|\, \mathcal{F}_{n+\ell k}\,\big]\,\Big]$$

$$\leq (1-r)\mathbb{E}\big[\, \boldsymbol{I}_{\{T>\ell k\}}\,\big] = (1-r)\mathbb{P}\big[\, T > n+\ell k\,\big].$$

Iterating, we deduce that for any $i \in \{1,\dots,k\}$ and any $\ell \in \mathbb{N}$ we have

$$\mathbb{P}\big[\,T > i+\ell k\,\big] < (1-r)^\ell \mathbb{P}\big[\,T>i\,\big] \leq (1-r)^\ell.$$

Now observe that

$$\mathbb{E}\big[\,T\,\big] = \sum_{n\in\mathbb{N}_0} \mathbb{P}\big[\,T>n\,\big] = \sum_{i=1}^k \sum_{\ell\in\mathbb{N}_0} \mathbb{P}\big[\,T>i+\ell k\,\big] < \sum_{i=1}^k \sum_{\ell\in\mathbb{N}_0} (1-r)^\ell = \frac{k}{r} < \infty.$$

This proves that $\mathbb{E}\big[\,T\,\big]$ is finite. In Example 3.31 we will use martingale techniques to show that

$$\mathbb{E}\big[\,T\,\big] = \frac{6^{k+1}-6}{5}.$$

$\square$

Example 1.168 (Optimal stopping with finite horizon). Let us consider the following abstract situation. Suppose we are given N random variables

$$X_1,\dots,X_N \in \mathcal{L}^0(\Omega,\mathcal{S},\mathbb{P}).$$

For $n \in \mathbb{I}_N := \{1,2,\dots,N\}$ we denote by $\mathcal{F}_n$ the sigma-algebra generated by $X_1,\dots,X_n$. Suppose that we are also given a sequence of rewards

$$R_n \in \mathcal{L}^1(\Omega,\mathcal{F}_n,\mathbb{P}),\ \ n\in\mathbb{I}_N.$$

A *stopping time* is a random variable $T:(\Omega,\mathcal{S},\mathbb{P})\to\mathbb{I}_N$ such that $\{T\leq n\}\in\mathcal{F}_n$, $\forall n\in\mathbb{I}_N$. Equivalently, T is a stopping time if and only if $\{T=n\}\in\mathcal{F}_n$, $\forall n$. Note that if T is a stopping time, then $\{T\geq n\} = \Omega\setminus\{T\leq n-1\}\in\mathcal{F}_{n-1}$.

One should think of the collection $X_1,\dots,X_N$ as a finite stream of random quantities flowing in time, one quantity per unit of time. The reward R_n depends only on the observed values $X_1,\dots,X_n$, i.e., $R_n = R_n(X_1,\dots,X_n)$. A stopping time describes a decision when to stop the stream based only on the information accumulated up to the decision moment. After we observe the first quantity X_1, we can decide if $T=1$. If this not the case, we observe a second quantity and, using the information about X_1, and X_2 we can decide to stop, i.e., if $T=2$ or not. We continue until we either observe all the random quantities or at the first n such that $T=n$.

We set

$$R_T = \sum_{n\in\mathbb{I}_N} R_n \boldsymbol{I}_{\{T=n\}}.$$

In other words R_T is the reward at the random stopping time T. We denote by $\mathcal{T}$ the collection of all possible stopping times. Note that

$$\mathbb{E}\big[\,|R_T|\,\big] \leq \sum_{n=1}^N \mathbb{E}\big[\,|R_n|\,\big] < \infty,\ \ \forall T\in\mathcal{T}.$$

We want to show that there exists $T_* \in \mathcal{T}$ such that

$$\mathbb{E}[R_{T_*}] = r := \sup_{T \in \mathcal{T}} \mathbb{E}[R_T].$$

Such a T_* is called an *optimal stopping time*. To prove the existence of an optimal time we establish a Fermat-like optimality condition that the optimal stopping times satisfy. We follow [28, Chap. 3].

For $n \in \mathbb{I}_N$ we set

$$\mathcal{T}_n := \{T \in \mathcal{T};\ T \geq n\}.$$

Note that

$$\mathcal{T} = \mathcal{T}_1 \supset \mathcal{T}_2 \supset \cdots \supset \mathcal{T}_N.$$

A stopping time T belongs to $\mathcal{T}_n$ if and only if the decision to stop comes only after we have observed the first n random variables in the stream, $X_1, \ldots, X_n$.

We will detect an optimal stopping strategy using a process of "successive approximations". The first approximation is the simplest strategy: pick the reward only at the end, after we have observed all the N variables in the stream. In this case the reward is $Y_N = R_N$. This may not give us the largest expected reward because some of the up-stream rewards could have been higher. We tweak this strategy a bit to produce a better outcome.

We wait to observe the first $N-1$ variables in the stream, and then decide what to do. At this moment our reward is R_{N-1}. To decide what to do next we compare this reward with the expected reward R_N given that we observed $X_1, \ldots, X_{N-1}$, i.e., with the conditional expectation $\mathbb{E}[Y_N \| \mathcal{F}_{N-1}] = \mathbb{E}[R_N \| \mathcal{F}_{N-1}]$. This is an $\mathcal{F}_{N-1}$-measurable quantity, i.e., a quantity that is computable from the knowledge of $X_1, \ldots, X_{N-1}$.

If the reward R_{N-1} that what we have in our hands is bigger than we expect to gain given our current information, we choose it and we stop. If not, we wait one more step to stop. More formally, we stop after $N-1$ steps if $R_{N-1} \geq \mathbb{E}[R_N \| \mathcal{F}_{N-1}]$ and we continue one more step otherwise. The decision is thus based on the random variable $Y_{N-1} = \max(R_{N-1}, \mathbb{E}[Y_N \| \mathcal{F}_{N-1}])$.

This heuristic suggests the following backwards induction.

$$\begin{aligned}Y_N &:= R_N,\ Y_n := \max\{R_n,\ \mathbb{E}[Y_{n+1} \| \mathcal{F}_n]\},\\ T_n &:= \min\{i \geq n;\ R_i \geq Y_i\} = \min\{i \geq n;\ R_i = Y_i\}.\end{aligned} \quad (1.4.17)$$

Note that $T_n \geq n$ and, for any $k \geq n$,

$$\{T_n > k\} = \{R_k < \mathbb{E}[Y_{k+1} \| \mathcal{F}_k]\} \in \mathcal{F}_k.$$

Hence $T_n \in \mathcal{T}_n$. We claim that for any $n = 1, \ldots, N$ we have

$$Y_n \geq \mathbb{E}[R_T \| \mathcal{F}_n],\ \forall T \in \mathcal{T}_n. \qquad (1.4.18a)$$

$$\mathbb{E}[R_{T_n} \| \mathcal{F}_n] = Y_n. \qquad (1.4.18b)$$

Hence
$$\mathbb{E}[R_{T_n} \| \mathcal{F}_n] \geq \mathbb{E}[Y_n] = \mathbb{E}[R_T \| \mathcal{F}_n], \quad \forall T \in \mathcal{T}_n.$$
By taking expectations we deduce
$$\mathbb{E}[R_{T_n}] = \sup_{T \in \mathcal{T}_n} \mathbb{E}[R_T]. \qquad (1.4.19)$$
This shows that the stopping time T_1 is optimal.

The optimal stopping strategy T_1 has a natural description: stop at the first moment when the reward at hand is bigger that the expected future reward, given the information we have at that moment. The stopping strategy T_n is similar, but delayed for n units of times.

We will prove (1.4.18a) and (1.4.18b) by backwards induction on n.

The inequality (1.4.18a) is clearly true for $n = N$. Assume it is true for n. Let $T \in \mathcal{T}_{n-1}$ and set $T' = \max\{T, n\}$. Then $T' \in \mathcal{T}_n$. For $A \in \mathcal{F}_{n-1}$ we have
$$\int_A R_T = \int_{A \cap \{T=n-1\}} R_{n-1} + \int_{A \cap \{T \geq n\}} R_{T'}$$
($\{T \geq n\} \in \mathcal{F}_{n-1}$)
$$= \int_{A \cap \{T=n-1\}} R_{n-1} + \int_{A \cap \{T \geq n\}} \mathbb{E}[R_{T'} \| \mathcal{F}_{n-1}]$$
$$= \int_{A \cap \{T=n-1\}} R_{n-1} + \int_{A \cap \{T \geq n\}} \mathbb{E}\big[\mathbb{E}[R_{T'} \| \mathcal{F}_n] \,\|\, \mathcal{F}_{n-1}\big]$$
(use the induction assumption $\mathbb{E}[R_{T'} \| \mathcal{F}_n] \leq Y_n$)
$$\leq \int_{A \cap \{T=n-1\}} \underbrace{R_{n-1}}_{\leq Y_{n-1}} + \int_{A \cap \{T \geq n\}} \underbrace{\mathbb{E}[Y_n \| \mathcal{F}_{n-1}]}_{\leq Y_{n-1}} \leq \int_A Y_{n-1}.$$
This proves the inequality (1.4.18a).

To prove the equality (1.4.18b), we run the above argument with $T = T_{n-1}$. Observe that in this case
$$\mathcal{U}_n := \{T = n-1\} = \{R_{n-1} \geq \mathbb{E}[Y_n \| \mathcal{F}_{n-1}]\} = \{Y_{n-1} = R_{n-1}\}, \qquad (1.4.20a)$$
$$\mathcal{V}_n := \{T_{n-1} > n-1\} = \{R_{n-1} < \mathbb{E}[Y_n \| \mathcal{F}_{n-1}]\}$$
$$= \{Y_{n-1} = \mathbb{E}[Y_n \| \mathcal{F}_{n-1}]\}. \qquad (1.4.20b)$$
We have $T_{n-1} = n-1$ on $\mathcal{U}_n$ and $T_{n-1} = T_n$ on $\mathcal{V}_n$ so that
$$\int_A R_{T_{n-1}} = \int_{A \cap \mathcal{U}_n} R_{n-1} + \int_{A \cap \mathcal{V}_n} R_{T_n}$$
($\mathcal{V}_n \in \mathcal{F}_{n-1}$)
$$= \int_{A \cap \mathcal{U}_n} R_{n-1} + \int_{A \cap \mathcal{V}_n} \mathbb{E}\big[\mathbb{E}[R_{T_n} \| \mathcal{F}_n] \,\|\, \mathcal{F}_{n-1}\big]$$
($Y_n = \mathbb{E}[R_{T_n} \| \mathcal{F}_n]$ by induction)
$$\int_{A \cap \mathcal{U}_n} R_{n-1} + \int_{A \cap \mathcal{V}_n} \mathbb{E}[Y_n \| \mathcal{F}_{n-1}]$$
(use (1.4.20a) and (1.4.20b))
$$= \int_A \max\{R_{n-1}, \mathbb{E}[Y_n \| \mathcal{F}_{n-1}]\} = \int_A Y_{n-1}.$$

□

Remark 1.169. The procedure for determining the optimal time T_1 outlined in the above example is a bit counterintuitive. The maximal expected reward is $\mathbb{E}[Y_1]$. By construction, the random variable Y_1 is $\mathcal{F}_1$-measurable, by construction, and thus has the form $f(X_1)$ for some Borel measurable function $f : \mathbb{R} \to \mathbb{R}$. Thus we can determine Y_1 knowing only the initial input X_1. On the other hand the definition of Y_1 by descending induction used the knowledge of the entire stream $X_1, \ldots, X_N$, not just the initial input X_1.

What it is true is that we can compute the maximal expected reward without running the stream. On the other hand, the moment we stop and the actual reward when stopped are *random quantities*. It is conceivable that if we do not stop when T_1 tells us to stop we could get a higher reward later on. However, on average, we cannot beat the stopping strategy T_1.

We will illustrate this process on the classical *secretary problem*. □

Example 1.170 (The secretary problem). Suppose we have a box with N prizes with values $v_1 < \cdots < v_N$. Bob would like to pick the most valuable item but he does not know the actual values v_n. He is allowed to sample them successively without replacement. At the j-th draw he is told the value V_j of the j-th prize. He can either accept the j-th prize or he can decline it and ask to sample another one. A prize once declined cannot be accepted later on. We are interested in a strategy that maximizes the probability that Bob picks the most valuable prize.[8]

Consider the relative rankings

$$X_n := \#\{ j \leq n; \; V_j \geq V_n \}. \tag{1.4.21}$$

Thus, X_n counts how may gifts unveiled up to the moment n are at least as valuable as the n-gift revealed. In particular, if $X_n = 1$, then V_n is the largest of the observed values $V_1, \ldots, V_n$.

We might be tempted to set the reward $R_n = \boldsymbol{I}_{\{V_n = v_N\}}$, but this is not $\mathcal{F}_n$-measurable. We can fix this issue by setting

$$R_n := \mathbb{E}\big[\boldsymbol{I}_{\{V_n = v_N\}} \,\|\, X_n\big].$$

Observe that for any stopping time T we have

$$\mathbb{E}[R_T] = \sum_{n=1}^{N} \int_{T=n} R_n = \sum_{n=1}^{N} \int_{T=n} \boldsymbol{I}_{\{V_n = v_N\}}$$

$$= \sum_{n=1}^{N} \mathbb{P}[V_n = V_N, \, T = n] = \mathbb{P}[V_T = v_N].$$

We want to find a stopping time T that maximizes $\mathbb{E}[R_T]$, i.e., the probability that Bob pick the biggest prize. Let us make a few remarks.

---
[8]Think of N secretaries interviewing for a single job and the values $v_1, \ldots, v_N$ rank their job suitability, the higher the value the more suitable. The interviewer learns the value v_k only at the time of the interview.

1. Observe that rankings $(X_n)_{n \in \mathbb{N}}$ defined in (1.4.21) are *independent* and
$$\mathbb{P}[X_n = j] = \frac{1}{n}, \quad \forall 1 \leq j \leq n \leq N. \tag{1.4.22}$$
Indeed, the random vector $(V_1, \ldots, V_N)$ can be identified with a random permutation $\varphi \in \mathfrak{S}_N$ of $\mathbb{I}_N$
$$(V_1, \ldots, V_N) = (v_{\varphi(1)}, \ldots, v_{\varphi(N)}).$$
The rank X_n is then a function of φ
$$X_n(\varphi) := \#\{\, j \leq n;\ \varphi(j) \geq \varphi(n)\,\}.$$
To reach the desired conclusion observe that the map
$$\vec{X} : \mathfrak{S}_N \to \mathbb{I}_1 \times \mathbb{I}_2 \times \cdots \times \mathbb{I}_N, \quad \varphi \mapsto \bigl(X_1(\varphi), \ldots, X_N(\varphi)\bigr)$$
is a bijection.[9]

2. We have
$$R_n = \frac{n}{N} \boldsymbol{I}_{\{X_n = 1\}} = \frac{n}{N} \boldsymbol{I}_{\{V_n = v_N\}}.$$
Indeed, the conditional expectation $R_n = \mathbb{E}\bigl[\,\boldsymbol{I}_{\{V_n = v_N\}} \,\|\, X_n\,\bigr]$ is a function of $x_n \in \mathbb{I}_n$ and we have
$$R_n(x_n) = \mathbb{E}\bigl[\,\boldsymbol{I}_{\{V_n = v_N\}} \,\bigl|\, X_n = x_n\,\bigr] = \mathbb{P}\bigl[\,V_n = v_N \,\bigl|\, X_n = x_n\,\bigr].$$
This probability is zero if $X_n > 1$. Now observe that
$$\mathbb{P}\bigl[\,V_n = v_N \,\bigl|\, X_n = 1\,\bigr] = \frac{\mathbb{P}[V_n = v_N]}{\mathbb{P}[X_n = 1]} = \frac{(N-1)!}{\binom{N}{n}(n-1)!(N-n)!} = \frac{n}{N}.$$
Following (1.4.17) and (1.4.18a) we set $y_n = \mathbb{E}[Y_n]$. The quantity y_n is the probability of Bob obtaining the largest prize among the strategies that discard the first $(n-1)$ selected prizes. We have
$$Y_N = R_N = \boldsymbol{I}_{\{V_N = v_N\}}, \quad y_N = \frac{1}{N}.$$
Since $\{V_N = v_N\} = \{X_N = 1\}$ is *independent* of $\mathcal{F}_{N-1}$ we deduce
$$\mathbb{E}\bigl[\,\boldsymbol{I}_{\{V_N = v_N\}} \,\|\, \mathcal{F}_{N-1}\,\bigr] = \mathbb{E}\bigl[\,\boldsymbol{I}_{\{V_N = v_N\}}\,\bigr] \stackrel{(1.4.22)}{=} \frac{1}{N} = y_N,$$
$$Y_{N-1} = \max\Bigl\{R_{N-1},\ \mathbb{E}\bigl[\,\boldsymbol{I}_{\{V_N = v_N\}} \,\|\, \mathcal{F}_{N-1}\,\bigr]\Bigr\}$$
$$= \max\bigl\{R_{N-1},\ y_N\bigr\} = \frac{N-1}{N}\boldsymbol{I}_{\{X_{N-1} = 1\}} + \frac{1}{N}\boldsymbol{I}_{\{X_{N-1} > 1\}},$$
$$y_{N-1} = \frac{1}{N} + \frac{(N-2)}{(N-1)}y_N.$$

---
[9]From the equality $\varphi^{-1}(N) = \max\{j,\ X_j(\varphi) = 1\}$ we deduce inductively that $\vec{X}$ is injective. It is also surjective since $\mathfrak{S}_N$ and $\prod_{n=1}^{N} \mathbb{I}_n$ have the same cardinality.

Similarly

$$\mathbb{E}[Y_{N-1} \| \mathcal{F}_{N-2}] = \mathbb{E}[Y_{N-1}] = y_{N-1}$$

$$Y_{N-2} = \max\{R_{N-2}, y_{N-1}\}$$

$$= \max\{(N-2)/N, y_{N-1}\} I_{\{X_{N-2}=1\}} + y_{N-1} I_{\{X_{N-2}>1\}},$$

$$y_{N-2} \stackrel{(1.4.22)}{=} \max\{(N-2)/N, y_{N-1}\} \frac{1}{N-2} + \frac{N-3}{N-2} y_{N-1}.$$

Iterating we deduce

$$Y_n = \max\{R_n, y_{n+1}\} = \max\{n/N, y_{n+1}\} I_{\{X_n=1\}} + y_{n+1} I_{\{X_n>1\}},$$

$$y_n = \max\{n/N, y_{n+1}\} \frac{1}{n} + \frac{n-1}{n} y_{n+1}.$$

While it is difficult to find an explicit formula for y_n, the above equalities can be easily implemented on a computer. The optimal probability is $p_N = y_1$. Here is a less than optimal but simple R code that computes y_1 given N.

```
optimal<-function(N){
  p<-1/N
  m<-N-1
  for (i in 1:m){
    p<-max((N-i)/N,p)/(N-i)+((N-i-1)/(N-i))*p
  }
p
}
```

Here are some results. Below, p_N denotes the optimal probability of choosing the largest among N prizes.

N	3	4	5	6	8	100	200
p_N	0.5	0.458	0.433	0.4277	0.4098	0.3710	0.3694

Note that $y_{n+1} < y_n$ with equality when $y_{n+1} > \frac{n}{N}$. We deduce that

$$y_{n+1} \geq \frac{n}{N} \Rightarrow y_{n+1} = y_n = \cdots = y_1.$$

We set

$$N_* := \max\{n; \ y_n \geq (n-1)/N\}$$

so $y_{N_*+1} < y_{N_*} = y_{N_*-1} = \cdots = y_1$. The optimal strategy is given by the stopping time T_{N_*}: reject the first $N_* - 1$ selected gifts and then pick the first gift that is more valuable than any of the preceding ones.

N	3	4	8	10	50	100	1000
N_*	3	3	5	5	20	39	370

For example, for $N = 10$ we have

n	1	2	3	4	5	6	7	8	9	10
y_n	0.398	0.398	0.398	0.398	0.398	0.372	0.32	0.26	0.18	0.1

In this case $N_* = 5$ and the optimal strategy corresponds to the stopping time T_5: reject the first four gifts and then accept the first gift more valuable then any of the previously chosen. In this case the probability of choosing the most valuable gifts is $p_{10} \approx 0.398$.

Let us sketch what happens as $N \to \infty$. Consider the sequence $z^N := (z_n)_{1 \leq n \leq N+1}$ defined by backwards induction

$$z_{N+1} = 0, \quad z_n = \frac{n-1}{n} z_{n+1} + \frac{1}{N}, \quad 1 \leq n \leq N.$$

One can show by backwards induction that $z_n \leq y_n, \forall n \leq N$ and $z_n = y_n, \forall n \geq N_*$.

Denote by $f_N : [0,1] \to \mathbb{R}$ the continuous function $[0,1] \to \mathbb{R}$ that is linear on each on the intervals $[(i-1)/N, i/N]$ and such that

$$f_N(i/N) = z_{N+1-i}, \quad i = 0, 1, \ldots, N.$$

Note that

$$f_N\big((i+1)/N\big) - f(i/N) = z_{N-i} - z_{N-i+1} = \frac{1}{N} - \frac{1}{N-i} z_{N-i+1}$$

$$= \frac{1}{N}\left(1 - \frac{1}{1-i/N} f_N(i/N)\right).$$

We recognize here the Euler scheme for the initial value problem

$$f' = 1 - \frac{1}{1-t} f, \quad f(0) = 0 \tag{1.4.23}$$

corresponding to the subdivision i/N of $[0,1]$.

The unique solution of this equation is $f(t) = -(1-t)\log(1-t)$ and $f_N(t)$ converge to $f(t)$ uniformly on the compacts of $[0,1)$. In fact, (see [25, Sec. 212]) for every $T \in (0,1)$, there exists $C = C_T > 0$ such that

$$\sup_{t \in [0,T]} \big|f_N(t) - f(t)\big| \leq \frac{C_T}{N}.$$

Set $g_N(t) = f_N(1-t)$; see Figure 1.5.

Note that $z_n = z_n^N = g_N\big((n-1)/N\big)$, $n = 1, \ldots, N+1$. We deduce that if $n/N \to \tau \in (0,1]$ as $N \to \infty$ we have

$$z_n^N \to g(\tau) = -\tau \log \tau, \quad \frac{N}{n} z_n \to -\log \tau.$$

From the equality

$$N(z_n - z_{n+1}) = 1 - \frac{N}{n} z_{n+1}, \quad \forall 1 \leq n \leq N$$

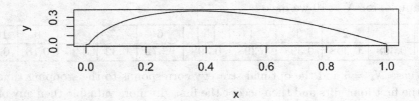

Fig. 1.5 The graph of g_{100}.

we deduce that
$$\lim_{N/n \to \tau} N(z_n - z_{n+1}) = 1 + \log \tau = \begin{cases} < 0, & \tau > 1/e, \\ > 0, & \tau < 1/e. \end{cases}$$
This implies that as $N \to \infty$ we have
$$\frac{N_*}{N} \to \frac{1}{e} \approx 0.368, \quad y_{N_*} = z_{N_*} \to \frac{1}{e}$$
as $N \to \infty$. For details we refer to [28, Sec. 3.3] or [69].

As explained in [69] a (nearly) optimal strategy is as follows. Denote by m the largest integer satisfying
$$\frac{N - 1/2}{e} + \frac{1}{2} \leq m \leq \frac{N - 1/2}{e} + \frac{3}{2}.$$
Reject the first m prizes and accept the next prize more valuable than any of the preceding ones. □

1.4.3 Conditional independence

Suppose that $(\Omega, \mathcal{S}, \mathbb{P})$ is a probability space.

Definition 1.171. Fix a sigma-subalgebra $\mathcal{G}$ of $\mathcal{S}$. The family $(\mathcal{F}_i)_{i \in I}$ of sigma-subalgebras of $\mathcal{S}$ is said to be *conditionally independent given* $\mathcal{G}$ if, for any finite subset $J \subset I$ and any events $F_j \in \mathcal{F}_j$, $j \in J$, we have
$$\mathbb{E}\left[\prod_{j \in J} I_{F_j} \,\|\, \mathcal{G} \right] = \prod_{j \in J} \mathbb{E}\left[I_{F_j} \,\|\, \mathcal{G} \right] \text{ a.s.}$$
Given sigma algebras $\mathcal{F}, \mathcal{G}, \mathcal{H} \subset \mathcal{S}$ we use the notation $\mathcal{F} \perp\!\!\!\perp_\mathcal{G} \mathcal{H}$ to indicated that $\mathcal{F}$ is independent of $\mathcal{H}$ given $\mathcal{G}$. □

The next proposition generalizes the result in Exercise 1.8.

Proposition 1.172 (Doob-Markov). *Given sigma algebras $\mathcal{F}_\pm, \mathcal{F}_0, \subset \mathcal{S}$ the following are equivalent.*

(i) $\mathbb{E}[X_+ \| \mathcal{F}_- \vee \mathcal{F}_0] = \mathbb{E}[X_+ \| \mathcal{F}_0]$ a.s., $\forall X_+ \in L^1(\Omega, \mathcal{F}_+, \mathbb{P})$.
(ii) $\mathcal{F}_+ \perp\!\!\!\perp_{\mathcal{F}_0} \mathcal{F}_-$.

Proof. The condition (i) is equivalent to

$$\mathbb{E}[XX_+] = \mathbb{E}\Big[X\mathbb{E}[X_+ \| \mathcal{F}_0]\Big], \quad \forall X \in L^\infty(\Omega, \mathcal{F}_0 \vee \mathcal{F}_-, \mathbb{P}). \tag{1.4.24}$$

The condition (ii) equivalent to

$$\mathbb{E}[X_+ X_- \| \mathcal{F}_0] = \mathbb{E}[X_+ \| \mathcal{F}_0]\mathbb{E}[X_- \| \mathcal{F}_0], \quad \forall X_\pm \in L^\infty(\Omega, \mathcal{F}_\pm, \mathbb{P}).$$

Note that since $\mathbb{E}[X_+ \| \mathcal{F}_0]$ is an $\mathcal{F}_0$-measurable random variable we have

$$\mathbb{E}[X_+ \| \mathcal{F}_0]\mathbb{E}[X_- \| \mathcal{F}_0] = \mathbb{E}\Big[X_- \mathbb{E}[X_+ \| \mathcal{F}_0] \, \Big\| \, \mathcal{F}_0\Big].$$

Thus, (ii) is equivalent to

$$\mathbb{E}[X_+ X_- \| \mathcal{F}_0] = \mathbb{E}\Big[X_- \mathbb{E}[X_+ \| \mathcal{F}_0] \, \Big\| \, \mathcal{F}_0\Big],$$

i.e., for any nonnegative, bounded, $\mathcal{F}_0$-measurable random variable X_0 we have

$$\mathbb{E}[X_0 X_- X_+] = \mathbb{E}\Big[X_0 X_- \mathbb{E}[X_+ \| \mathcal{F}_0]\Big].$$

Since $\mathcal{F}_0 \vee \mathcal{F}_+$ coincides with the sigma-algebra generated collection of random variables $X_0 X_+$, $X_0 \in L^\infty(\Omega, \mathcal{F}_0, \mathbb{P})$, $X_- \in L^\infty(\Omega, \mathcal{F}_-, \mathbb{P})$ we deduce that the last equality is equivalent to (1.4.24), i.e., (i) is equivalent to (ii). □

Remark 1.173. You should think of a system evolving in time. Then $\mathcal{F}_0$ collects the present information about the system, $\mathcal{F}_-$ collects the past information and $\mathcal{F}_+$ collects the future information. Roughly speaking, the above proposition shows that the information about an event given the present and the past coincides with the information given the present if and only if the future is independent of the past given the present. □

1.4.4 Kernels and regular conditional distributions

Suppose that $(\Omega_0, \mathcal{F}_0)$ and $(\Omega_1, \mathcal{F}_1)$ are two measurable spaces. A *kernel* from $(\Omega_0, \mathcal{F}_0)$ to $(\Omega_1, \mathcal{F}_1)$ is a function

$$K : \Omega_0 \times \mathcal{F}_1 \to [0, \infty], \quad (\omega_0, F_1) \mapsto K_\omega[F_1]$$

with the following properties.

(**K**$_1$) For each $\omega_0 \in \Omega_0$, the map

$$\mathcal{F}_1 \ni F_1 \mapsto K_{\omega_0}[F_1] \in [0, \infty]$$

is a measure. We will denote this measure by $K_{\omega_0}[d\omega_1]$.

(**K**$_2$) For each $F_1 \in \mathcal{F}$ the function

$$\Omega_0 \ni \omega_0 \mapsto K_{\omega_0}[F_1] \in [0, \infty]$$

is $\mathcal{F}_0$-measurable. We will denote this random variable by $K_\square[F_1]$.

The kernel K is called a *probability kernel* or a *Markovian kernel* if $K_{\omega_0}[-]$ is a probability measure on $(\Omega_1, \mathcal{F}_1)$, for any $\omega_0 \in \Omega_0$.

The condition $(\mathbf{K_1})$ above shows that a kernel is a family $(K_{\omega_0}[-])_{\omega_0 \in \Omega_0}$ of measures on $(\Omega_1, \mathcal{F}_1)$ parametrized by Ω_0. Condition $(\mathbf{K_2})$ is a measurability condition on this family. For this reason kernels are also know as *random measures*.

Example 1.174. Consider the Bernoulli measure

$$\beta_p := q\delta_0 + p\delta_1 \in \mathrm{Prob}(\mathbb{R}), \ \ p \in [0,1], \ \ q = 1 - p.$$

To obtain a random measure we let p be a random quantity. More precisely, if $f : (\Omega, \mathcal{S}) \to [0,1]$ is a measurable function, then

$$\beta_{X(\omega)} = \big(1 - f(\omega)\big)\delta_0 + f(\omega)\delta_1$$

defines a Markov kernel $K : \Omega \times \mathcal{B}_\mathbb{R} \to [0,1]$,

$$K_\omega[B] = \big(1 - X(\omega)\big)\delta_0[B] + X(\omega)\delta_1[B]. \qquad \square$$

Given a measure μ on the measurable space $(\Omega, \mathcal{F})$ and a nonnegative measurable function $f \in \mathcal{L}^0_+(\Omega, \mathcal{F})$ we set

$$\langle \mu, f \rangle := \mu[f] = \int_\Omega f(\omega)\mu[d\omega] \in [0, \infty].$$

Theorem 1.175. *Suppose that $K : \Omega_0 \times \mathcal{F}_1 \to [0,\infty]$ is a kernel from $(\Omega_0, \mathcal{F}_0)$ to $(\Omega_1, \mathcal{F}_1)$.*

(i) For any $f \in \mathcal{L}^0_+(\Omega_1, \mathcal{F}_1)$ we define its pullback *by K to be the function*

$$K^*f : \Omega_0 \to [0, \infty], \ \ K^*f(\omega_0) = \int_{\Omega_1} f(\omega_1) K_{\omega_0}[d\omega_1].$$

Then $K^*f \in \mathcal{L}^0_+(\Omega_0, \mathcal{F}_0)$.

(ii) For any measure $\mu : \mathcal{F}_0 \to [0,\infty]$ we define its push-forward *by K to be the measure $K_*\mu : \mathcal{F}_1 \to [0,\infty]$*

$$K_*\mu[F_1] := \int_{\Omega_0} K_{\omega_0}[F_1]\mu[d\omega_0] \in [0,\infty], \ \ F_1 \in \mathcal{F}_1.$$

Then $K_*\mu$ is a σ-additive measure on $(\Omega_1, \mathcal{F}_1)$.

(iii) The pullback and push-forward by K are adjoints of each other. More precisely, for any measure μ on $(\Omega_0, \mathcal{F}_0)$ and any measurable function $f \in \mathcal{L}^0_+(\Omega_1, \mathcal{F}_1)$ we have

$$\langle \mu, K^*f \rangle = \langle K_*\mu, f \rangle. \tag{1.4.25}$$

Proof. (i) For any $F \in \mathcal{F}_1$ we have $K^*\boldsymbol{I}_F(\omega_0) = K_{\omega_0}[F]$ so $K^*\boldsymbol{I}_F \in \mathcal{L}^0(\Omega_0, \mathcal{F}_0)$. Clearly the correspondence $f \mapsto K^*f$ is monotone and the conclusion follows from the fact that a nonnegative function is measurable iff it is the limit of an increasing sequence of simple functions.

The statement (ii) follows from the Monotone Convergence theorem and $(\mathbf{K_1})$. For part (iii), fix the measure μ. Observe that for $F \in \mathcal{F}_1$ we have

$$\langle \mu, K^* I_F \rangle = \int_{\Omega_0} K^* I_F(\omega_0) \mu[d\omega_0] = \int_{\Omega_0} \int_{\Omega_1} K_{\omega_0}[F][d\omega_1]$$
$$= K_* \mu[F] = \langle K_* \mu, I_F \rangle.$$

Thus (1.4.25) holds for $f = I_H$, $F \in \mathcal{F}_1$. The general case follows by invoking the Monotone Class Theorem. $\square$

When K is a Markovian kernel and μ is a probability measure then the pushforward $K_* \mu$ is also a probability measure. For any $F_1 \in \mathcal{F}_1$ the measure $K_* \mu[F_1]$ is the expectation of the random variable $\omega_0 \mapsto K_{\omega_0}[F_1]$ with respect to μ. The measure $K_* \mu$ is said to be a *mixture* of the random measure $\omega_0 \mapsto K_{\omega_0}[-]$ driven by μ.

Example 1.176. (a) Suppose that $(\Omega_0, \mathcal{F}_0)$, $(\Omega_1, \mathcal{F}_1)$ are two measurable spaces and

$$T : (\Omega_0, \mathcal{F}_0) \to (\Omega_1, \mathcal{F}_1)$$

is a measurable map. Then T defines a kernel

$$K^T : \Omega_0 \times \mathcal{F}_1, \quad K^T_{\omega_0}[F_1] = \delta_{T(\omega_0)}[F_1],$$

where δ_{ω_1} denotes the Dirac measure on $(\Omega_1, \mathcal{F}_1)$ concentrated at ω_1; see Example 1.31(a). Observe that for any measure μ on $\mathcal{F}_0$ and any $f \in \mathcal{L}^0_+(\Omega_1, \mathcal{F}_1)$ we have

$$K^T_* \mu = T_\# \mu, \quad (K^T)^* f = T^* f := f \circ T.$$

Thus, (1.4.25 contains as a special case the change in variables formula (1.2.21).

(b) Consider the random measure $K : \Omega \times \mathcal{B}_\mathbb{R} \to [0,1]$ in Example 1.174. Given a probability measure μ on $(\Omega, \mathcal{S})$ we have

$$K_* \mu = \text{Ber}(\bar{f}) = (1 - \bar{f})\delta_0 + \bar{f} \delta_1, \quad \bar{f} := \mathbb{E}_\mu[f].$$

(c) Suppose that $\mathscr{X}$ is a finite or countable set. A kernel $(\mathscr{X}, 2^\mathscr{X}) \to (X, 2^\mathscr{X})$ is defined by a function (matrix) $K : \mathscr{X} \times \mathscr{X} \to [0, \infty]$, via the equality

$$K_x[S] = \sum_{s \in S} K(x, s), \quad \forall x \in \mathscr{X}, \ S \subset \mathscr{X}.$$

The kernel is Markovian if

$$\sum_{x' \in \mathscr{X}} K(x, x') = 1, \quad \forall x \in \mathscr{X}.$$

(d) Suppose that $f : \mathbb{R}^2 \to [0, \infty)$ is an integrable function such that

$$\int_\mathbb{R} f(x, y) dy = 1, \quad \forall x \in \mathbb{R}.$$

It defines a Markovian kernel

$$K : \mathbb{R} \times \mathcal{B}_\mathbb{R} \to [0,1], \quad K_x[B] = \int_B f(x,y) dy,$$

$\forall x \in \mathbb{R}$ and any Borel subset $B \subset \mathbb{R}$. We can rewrite this as $K_x[dy] = f(x,y) dy$.
$\square$

Suppose that $(\Omega, \mathcal{F}, \mathbb{P})$ is a probability space and $\mathcal{S} \subset \mathcal{F}$ is a sigma sub-algebra. For every event $F \in \mathcal{F}$ the *random variable*

$$\mathbb{P}[F \,\|\, \mathcal{S}] := \mathbb{E}[\boldsymbol{I}_F \,\|\, \mathcal{S}]$$

is called the *conditional probability* of F given $\mathcal{S}$. The random variable $\mathbb{P}[F \,\|\, \mathcal{S}]$ is unique only up to equality off a negligible set.

Note that for any increasing family $(F_n)_{n \geq 1} \subset \mathcal{F}$ there exists a negligible set $\mathcal{N} \subset \Omega$ such that

$$\lim_n \mathbb{P}[F_n \,\|\, \mathcal{S}](\omega) = \mathbb{P}[\lim_n F_n \,\|\, \mathcal{S}](\omega), \quad \forall \omega \in \Omega \setminus \mathcal{N}.$$

A priori, the negligible set $\mathcal{N}$ depends on the family $(F_n)_{n \geq 1}$, and there might not exist one negligible set that works for all such increasing families. When such a thing is possible we say that the conditional probability $\mathbb{P}[-\,\|\, \mathcal{S}]$ admits a *regular version*. Here is the precise definition.

Definition 1.177. Let $(\omega, \mathcal{S}, \mathbb{P})$ be a probability space. A *regular version* of $\mathbb{P}[-\,\|\, \mathcal{S}]$ is a probability or Markovian kernel Q from $(\Omega, \mathcal{S})$ to $(\Omega, \mathcal{F})$

$$Q : \Omega \times \mathcal{F} \to [0, 1]$$

such that, for any $F \in \mathcal{F}$, the random variable $\Omega \ni \omega \mapsto Q_\omega[F]$ is a version of $\mathbb{P}[F \,\|\, \mathcal{S}]$. In other words,

- the map $\omega \mapsto Q_\omega[F]$ is $\mathcal{S}$-measurable and
- for any $S \in \mathcal{S}$, $F \in \mathcal{F}$ we have

$$\mathbb{P}[S \cap F] = \int_S Q_\omega[F] \mathbb{P}[d\omega].$$

$\square$

Proposition 1.178. *If*

$$Q : \Omega \times \mathcal{F} \to [0,1], \quad (\omega, F) \mapsto Q_\omega[F]$$

is a regular version of $\mathbb{P}[-\,\|\, \mathcal{S}]$, *then*

$$\forall X \in L^1(\Omega, \mathcal{F}, \mathbb{P}), \quad \mathbb{E}[X \,\|\, \mathcal{S}]_\omega = \int_\Omega X(\eta) Q_\omega[d\eta] = Q^* X(\omega) \quad \text{a.s.} \tag{1.4.26}$$

Proof. Note that (1.4.26) holds in the special case $X = \boldsymbol{I}_F$ because

$$Q^* \boldsymbol{I}_F(\omega) = Q_\omega[F] = \mathbb{P}[F \,\|\, \mathcal{S}](\omega) = \mathbb{E}[\boldsymbol{I}_F \,\|\, \mathcal{S}](\omega).$$

The general case follows from the Monotone Class theorem. $\square$

The equality (1.4.26) can be written in the less precise but more intuitive way

$$\mathbb{E}[X \,\|\, \mathcal{S}] = \int_\Omega X(\eta) \mathbb{P}[d\eta \,\|\, \mathcal{S}]. \tag{1.4.27}$$

More generally, consider a measurable map $T : (\widetilde{\Omega}, \widetilde{\mathcal{F}}) \to (\Omega, \mathcal{F})$. Let $\widetilde{\mathbb{P}}$ be a probability measure on $(\widetilde{\Omega}, \widetilde{\mathcal{F}})$ and suppose that $\widetilde{\mathcal{S}} \subset \widetilde{\mathcal{F}}$ is a sigma subalgebra. For every $F \in \mathcal{F}$ we set

$$\mathbb{P}_T[\,F \,\|\, \widetilde{\mathcal{S}}\,] := \widetilde{\mathbb{P}}[\,T \in F \,\|\, \widetilde{\mathcal{S}}\,] = \mathbb{E}_{\widetilde{\mathbb{P}}}[\,T^* \boldsymbol{I}_F \,\|\, \widetilde{\mathcal{S}}\,] = \mathbb{E}_{\widetilde{\mathbb{P}}}[\,\boldsymbol{I}_{T^{-1}(F)} \,\|\, \widetilde{\mathcal{S}}\,]. \qquad (1.4.28)$$

We will refer to $\mathbb{P}_T[\,-\,\|\, \widetilde{\mathcal{S}}\,]$ as the *conditional distribution* of T given $\widetilde{\mathcal{S}}$. Observe that when

$$(\widetilde{\Omega}, \widetilde{\mathcal{F}}) = (\Omega, \mathcal{F})\ \widetilde{\mathbb{P}} = \mathbb{P} \text{ and } T = \mathbb{1}_\Omega,$$

then

$$\mathbb{P}_{\mathbb{1}_\Omega}[\,-\,\|\, \widetilde{\mathcal{S}}\,] = \mathbb{P}[\,-\,\|\, \widetilde{\mathcal{S}}\,].$$

Note that for any increasing family $(F_n)_{n \geq 1} \subset \mathcal{F}$ we have

$$\lim_{n \to \infty} \mathbb{P}_T[\,F_n \,\|\, \widetilde{\mathcal{S}}\,] = \mathbb{P}_T[\,\lim_n F_n \,\|\, \widetilde{\mathcal{S}}\,] \text{ a.s.}$$

We say that $\mathbb{P}_T[\,-\,\|\, \widetilde{\mathcal{S}}\,]$ admits a regular version if we can choose representatives for each $\mathbb{P}_T[\,F \,\|\, \widetilde{\mathcal{S}}\,]$, $F \in \mathcal{F}$ so that the above equality holds for any increasing sequence (F_n). Here is a more precise definition.

Definition 1.179. Let $(\widetilde{\Omega}, \widetilde{\mathcal{F}}, \widetilde{\mathbb{P}})$ be a probability space and $T : (\widetilde{\Omega}, \widetilde{\mathcal{F}}) \to (\Omega, \mathcal{F})$ be a measurable map. Fix a sigma-subalgebra $\widetilde{\mathcal{S}} \subset \widetilde{\mathcal{F}}$. A *regular version* of the conditional probability distribution $\mathbb{P}_T[\,-\,\|\, \widetilde{\mathcal{S}}\,]$ of the map T conditioned on $\widetilde{\mathcal{S}}$ is a probability kernel Q from $(\widetilde{\Omega}, \widetilde{\mathcal{F}})$ to $(\Omega, \mathcal{F})$ such that, for any $F \in \mathcal{F}$, the random variable $Q_\square[\,F\,]$ is a version of $\mathbb{P}_T[\,F \,\|\, \widetilde{\mathcal{S}}\,]$. In other words,

- the random variable $Q_\square[\,F\,]$ is $\widetilde{\mathcal{S}}$-measurable and
- for any $\widetilde{S} \in \widetilde{\mathcal{S}}$, $F \in \mathcal{F}$ we have

$$\widetilde{\mathbb{P}}[\,\widetilde{S} \cap T^{-1}(F)\,] = \int_{\widetilde{\Omega}} Q_{\widetilde{\omega}}[\,F\,] \widetilde{\mathbb{P}}[\,d\widetilde{\omega}\,]. \qquad (1.4.29)$$

$\square$

A conditional probability distribution need not admit a regular version. For that to happen we need to impose conditions on $\mathcal{F}$, the sigma algebra on the target space. We need to make a brief topological digression.

Definition 1.180. A *Lusin space* is a topological space homeomorphic to a Borel subset of a compact metric space. $\square$

Remark 1.181. (a) The above is not the usual definition of a Lusin space but it has the advantage that emphasizes the compactness feature we need in the proof of Kolmogorov's existence theorem.

There are plenty of Lusin spaces. In fact, a topological space that is not Lusin is rather unusual. We refer to [15; 39] for a more in depth presentation of these spaces and their applications in measure theory and probability. To give the reader a taste of the fauna of Lusin spaces we list a few examples.

- The Euclidean spaces $\mathbb{R}^n$ are Lusin spaces.
- A Borel subset of a Lusin space is also Lusin space.
- The Cartesian product of two Lusin spaces is a Lusin space.
- A less obvious example is that of *Polish spaces*, i.e., complete separable metric spaces. More precisely every Polish space is homeomorphic to a countable intersection of open subsets of $[0,1]^{\mathbb{N}}$; see [18, Chap. IX, Sec. 6.1, Corollary 1].
- A space is Lusin if and only if it is homeomorphic to a Borel subset of a Polish space.

(b) From a measure theoretic point of view the Lusin spaces are indistinguishable from the Polish spaces. More precisely, for any Lusin space X, there exists a Polish space Y and a Borel measurable bijection $\Phi : X \to Y$ such that the inverse is also Borel measurable; see [35, Prop. 8.6.13].

The Polish spaces have another important property. More precisely, a Polish space equipped with the σ-algebra of Borel subsets is *isomorphic as a measurable space* to a Borel subset E of $[0,1]$ equipped with the σ-algebra of Borel subsets. For a proof we refer to [126, Sec. I.2]. Moreover, any two Borel subsets of $\mathbb{R}$ are measurably isomorphic if and only if they have the same cardinality, [126, Ch. I, Thm. 2.12].

On the other hand, it is known that the continuum hypothesis holds for the Borel subsets of a Polish space; see [39, Appendix III.80] or [98, XII.6]. In particular, any Borel subset of $\mathbb{R}$ is either finite, countable or has the continuum cardinality. In particular, this shows that any Borel subset of $[0,1]$ is measurably isomorphic to a compact subset of $[0,1]$. Hence *any Lusin space is Borel isomorphic to a compact metric space*!

The measurable spaces isomorphic to a Borel subset E of $[0,1]$ equipped with the σ-algebra of Borel subsets are called *standard measurable* spaces and play an important role in probability. Hence the Lusin spaces are standard measurable spaces. □

We have the following general existence result.

Theorem 1.182 (Existence of regular conditional probabilities). *Suppose that*

- $(\Omega, \mathcal{F}, \mathbb{P})$ *is a probability space,*
- $\mathcal{Y}$ *is a Lusin space and*
- $\mathcal{B}_\mathcal{Y}$ *is the sigma algebra of Borel subsets of* $\mathcal{Y}$.

Then, for every measurable map $Y : (\Omega, \mathcal{F}) \to (\mathcal{Y}, \mathcal{B}_\mathcal{Y})$, *and every σ-subalgebra* $\mathcal{S} \subset \mathcal{F}$ *there exists a regular version*
$$Q : \Omega \times \mathcal{B}_\mathcal{Y} \to [0,1], \quad (\omega, B) \mapsto Q_\omega[B]$$
of the conditional distribution $\mathbb{P}_Y[\,-\,\|\,\mathcal{S}\,]$. *This means that*
$$Q_\square[B] = \mathbb{P}[Y \in B \,\|\, \mathcal{S}] \quad \text{a.s.,} \quad \forall B \subset \mathcal{B}_\mathcal{Y}.$$

Moreover, for any measurable function $f : (\mathcal{Y}, \mathcal{B}_Y) \to \mathbb{R}$, we have

$$\mathbb{E}[f \circ Y \| \mathcal{S}](\omega) = \int_{\mathcal{Y}} f(y) Q_\omega[dy], \quad \forall \omega \in \Omega. \tag{1.4.30}$$

Idea of proof. For a complete proof we refer to [33, Th. IV2.10], [39, III.71], [40, IX.11] or [135, II.89].

We can assume that $\mathcal{Y}$ is a compact metric space. Fix a dense countable subset $\mathcal{U} \subset C(Y)$ such that $1 \in \mathcal{U}$ and $\mathcal{U}$ is a vector space over $\mathbb{Q}$. we can find representatives $\Phi(u)$ of $\mathbb{E}[u(Y) \| \ ]$ such that the map

$$\mathcal{U} \ni u \mapsto \Phi(u) \in \mathcal{L}^1(\Omega, \mathcal{S}, \mathbb{P})$$

is $\mathbb{Q}$-linear, $\Phi(1) = 1$ and $\Phi(u) \geq 0$ if $u \geq 0$. For every nonnegative $f \in C(U)$ we set

$$\Phi^*(f) := \sup \{ \Phi(u); \ u \in \mathcal{U}, \ 0 \leq u \leq f \}.$$

One can show that

$$\Phi^*(f) := \inf \{ \Phi(u); \ u \in \mathcal{U}, \ u \geq f \}.$$

For arbitrary $f \in C(\mathcal{Y})$ we set

$$\Phi^*(f) = \Phi^*(f^+) - \Phi^*(f^-).$$

One can show that the resulting map

$$C(\mathcal{Y}) \ni f \mapsto \Phi^*(f) \in \mathcal{L}^1(\Omega, \mathcal{S}, \mathbb{P})$$

is $\mathbb{R}$-linear, $\Phi^*(1) = 1$ and $\Phi^*(f) \geq 0$ if $f \geq 0$. The Riesz Representation Theorem 1.86 implies that for ay $\omega \in \Omega$ there exists a probability measure $\mu_\omega : \mathcal{B}_\mathcal{Y} \to [0,1]$ such that

$$\Phi^*(f)(\omega) = \int_Y f(y) \mu_\omega[dy].$$

One then shows that for any $B \in \mathcal{B}_\mathcal{Y}$ the map $\Omega \ni \omega \mapsto \mu_\omega[B] \in [0,1]$ is $\mathcal{S}$-measurable and thus it is a regular version of the conditional distribution of Y given $\mathcal{S}$. □

In the special case when $\mathcal{S}$ is the σ-algebra generated by a measurable map $X : \Omega \to \mathcal{X}$, $\mathcal{X}$ some measurable space, we use the notation

$$\mathbb{P}_Y[dy \| X] := \mathbb{P}_Y[dy \| \sigma(X)]$$

to denote a regular version for the conditional distribution of Y given X. This is a random Borel measure on $\mathcal{Y}$.

Example 1.183. Consider the special case of Theorem 1.182 where $\mathcal{Y} = \mathbb{R}$ and $Y \in L^1(\Omega, \mathcal{F}, \mathbb{P})$. For any sigma subalgebra $\mathcal{S} \subset \mathcal{F}$ there exists a kernel

$$Q : \Omega \times \mathcal{B}_\mathbb{R} \to [0,1]$$

such that

$$\mathbb{P}[Y \leq y \| \mathcal{S}] = Q_\square[(-\infty, y]].$$

Moreover

$$\mathbb{E}[Y \| \mathcal{S}] = \int_\mathbb{R} y Q_\square[dy], \quad \mathbb{P}\text{-a.s. on } \Omega.$$

□

Example 1.184. Suppose that X_0, Y_0, X_1, Y_1 are random variables and $T : \mathbb{R}^2 \to \mathbb{R}^k$ is a Borel measurable map. Denote by $\mathbb{P}^0$ the joint probability distribution of (X_0, Y_0). Suppose that the joint distribution of (X_1, Y_1) has the form

$$\mathbb{P}^1[\,dxdy\,] = g(T(x,y))\mathbb{P}^0[\,dxdy\,]$$

for some nonnegative measurable function $g : \mathbb{R}^k \to [0, \infty)$.

We denote by $\mathbb{P}^i[\,-\,\|\,T\,]$ the regular conditional probability $\mathbb{P}^i[\,-\,\|\,\sigma(T)\,]$. In other words, for any bounded nonnegative measurable function $f : \mathbb{R}^k \to [0, \infty)$ and any Borel set $B \subset \mathbb{R}$ we have $\mathbb{P}^i[\,B\,\|\,T\,] \in \mathcal{L}^0_+(\mathbb{R}^2, \sigma(T))$ and

$$\int_{\mathbb{R}^2} \boldsymbol{I}_B f(T(x,y)) \mathbb{P}^i[\,dxdy\,] = \int_{\mathbb{R}^2} \mathbb{P}^i[\,B\,\|\,T\,] f(T(x,y)) \mathbb{P}^i[\,dxdy\,], \quad i=0,1.$$

Note that

$$\int_{\mathbb{R}^2} \boldsymbol{I}_B f(T(x,y)) \mathbb{P}^1[\,dxdy\,] = \int_{\mathbb{R}^2} \boldsymbol{I}_B f(T(x,y)) g(T(x,y)) \mathbb{P}^0[\,dxdy\,]$$

$$= \int_{\mathbb{R}^2} \mathbb{P}^0[\,B\,\|\,T\,] f(T(x,y)) g(T(x,y)) \mathbb{P}^0[\,dxdy\,]$$

$$= \int_{\mathbb{R}^2} \mathbb{P}^0[\,B\,\|\,T\,] f(T(x,y)) \mathbb{P}^1[\,dxdy\,].$$

Hence

$$\mathbb{P}^1[\,B\,\|\,T\,] = \mathbb{P}^0[\,A\,\|\,T\,], \quad \forall B \in \mathcal{B}_{\mathbb{R}^2}.$$

Suppose that the distribution $\mathbb{P}^0$ is known and would like to get information about the distribution of (X_1, Y_1) by investigating $T(X_0, Y_0)$. The above equality shows that knowledge of T adds nothing to our understanding of the density $g(T(x,y))$ beyond what we know from (X_0, Y_0). $\square$

Example 1.185. Suppose that $X_1, \ldots, X_n$ are independent and uniformly distributed in the interval $[0, L]$. Set

$$X_{(n)} := \max_{1 \le k \le n} X_k, \quad X_{(1)} := \min_{1 \le k \le n} X_k.$$

Note that

$$\mathbb{P}[\,X_{(n)} \le x\,] = \mathbb{P}[\,X_k \le x, \; \forall k=1,\ldots,n\,] = \left(\frac{x}{L}\right)^n,$$

so that the probability distribution of $X_{(n)}$ is

$$\mathbb{P}_n[\,dx\,] = n\frac{x^{n-1}}{L^n} \boldsymbol{I}_{[0,L]}(x) dx.$$

Similarly,

$$\mathbb{P}[\,X_{(1)} > x\,] = \mathbb{P}[\,X_k > x, \; \forall k=1,\ldots,n\,] = \left(\frac{(L-x)}{L}\right)^n,$$

so the probability distribution of $X_{(1)}$ is

$$\mathbb{P}_1[\,dx\,] = n\underbrace{\frac{(L-x)^{n-1}}{L^n}}_{=:\rho_1(x)}\,\boldsymbol{I}_{[0,L]}(x)dx.$$

Let us compute the conditional distribution of $\mathbb{P}_{X_{(n)}}[\,dx_n\,\|\,X_{(1)}\,]$. We begin by computing the random variables.

$$\mathbb{P}[\,X_{(n)} \leq x_n \,\|\, X_{(1)}\,],\quad 0 \leq x_n \leq L.$$

Observe first that ; $\forall 0 \leq x_1, x_n \leq L$,

$$\mathbb{E}\big[\,\boldsymbol{I}_{X_{(n)}\leq x_n}\boldsymbol{I}_{X_{(1)}\geq x_1}\,\big] = \mathbb{P}\big[\,x_1 \leq X_1, \ldots X_n \leq x_n\,\big] = \frac{(x_n - x_1)_+^n}{L^n}.$$

We need to find a function $f(X_{(1)}) = f_{x_n}(X_{(1)})$ such that

$$\mathbb{E}\big[\,f(X_{(1)})\boldsymbol{I}_{X_{(1)}\geq x_1}\,\big] = \frac{(x_n - x_1)_+^n}{L^n},\quad \forall x_1,$$

i.e.,

$$\int_{[x_1, L]} f(x)\rho_1(x)dx = \frac{(x_n - x_1)_+^n}{L^n},\quad \forall x_1.$$

Derivating with respect to x_1 we deduce

$$f(x_1)\rho_1(x_1) = n\frac{(x_n - x_1)_+^{n-1}}{L^n}.$$

Hence

$$\mathbb{P}\big[\,X_{(n)} \leq y\,\big|\,X_{(1)} = x_1\,\big] = n\frac{(x_n - x_1)_+^{n-1}}{L^n \rho_1(x_1)} = \frac{(y - x_1)_+^{n-1}}{(L - x_1)^{n-1}}.$$

Thus, the conditional probability of $X_{(n)}$ given that $X_{(1)} = x_1$ is

$$\mathbb{P}_{X_{(n)}}[\,dx_n\,|\,X_{(1)} = x_1\,] = \frac{(n-1)(x_n - x_1)_+^{n-2}}{(L - x_1)^{n-1}}.$$

We define the *empirical gap* or *sample range* to be the random variable $G = X_{(n)} - X_{(1)}$. To find the distribution of G we condition on $X_{(1)}$ and we have

$$\mathbb{P}[\,G \leq g\,] = \int_{[0,L]} \mathbb{P}\big[\,X_{(n)} \leq x_1 + g\,\big|\,X_{(1)} = x_1\,\big]\mathbb{P}_{X_{(1)}}[\,dx_1\,]$$

$$= \int_{[0,L]} \mathbb{P}\big[\,X_{(n)} \leq X_{(1)} + g\,\big|\,X_{(1)} = x_1\,\big]\rho_1(x_1)dx_1.$$

Now observe that

$$\mathbb{P}\big[\,X_{(n)} \leq X_{(1)} + g\,\big|\,X_{(1)} = x_1\,\big] = \int_{[0,\min(L, x_1 + g)]} \frac{(n-1)(x_n - x_1)_+^{n-2}}{(L - x_1)^n}dx_n$$

$$= \frac{g^{n-1}}{(L-x_1)^{n-1}} \boldsymbol{I}_{[0,L-g]}(x_1) + \boldsymbol{I}_{[L-g,L]}(x_1).$$

Thus

$$\mathbb{P}[G \le g] = \frac{ng^{n-1}}{L^n} \int_0^{L-g} dx_1 + \int_{[L-g,L]} \rho_1(x_1) dx_1 = \frac{ng^{n-1}(L-g)}{L^n} + \frac{g^n}{L^n}.$$

We deduce

$$\frac{d}{dg}\mathbb{P}[G \le g] = \frac{n(n-1)g^{n-2}}{L^{n-1}} + \frac{n^2 g^{n-1}}{L^n} - \frac{ng^{n-1}}{L^n} = \frac{n(n-1)g^{n-2}}{L^{n-1}}\left(1 - \frac{g}{L}\right).$$

Thus, the probability distribution of G is

$$P_G[dg] = \frac{n(n-1)g^{n-2}}{L^{n-1}}\left(1 - \frac{g}{L}\right) \boldsymbol{I}_{[0,L]}(g)\, dg.$$

If $L = 1$ then the above distribution is the Beta distribution Beta$(n-1, 2)$. □

1.4.5 *Disintegration of measures*

Suppose that $(\Omega_i, \mathcal{F}_i)$, $i = 0, 1$ are two measurable spaces and

$$K : \Omega_0 \times \mathcal{F}_1 \to [0, \infty)$$

is a kernel from $(\Omega_0, \mathcal{F}_0)$ to $(\Omega_1, \mathcal{F}_1)$. Then any measure μ_0 on (Ω_0, μ_0) defines a measure $\mu = \mu_{K,\mu_0}$ on $(\Omega, \mathcal{F}) := (\Omega_0 \times \Omega_1, \mathcal{F}_0 \otimes \mathcal{F}_1)$ via the equality

$$\mathbb{P}[S] = \int_{\Omega_0} \left(\int_{\Omega_1} \boldsymbol{I}_S(\omega_0, \omega_1) K_{\omega_0}[d\omega_1] \right) \mu_0[d\omega_0]. \tag{1.4.31}$$

We say that a measure μ on $(\Omega_0 \times \Omega_1, \mathcal{F}_0 \otimes \mathcal{F}_1)$ *is disintegrated by* μ_0 or μ_0 *disintegrates* μ if μ is of the form μ_{K,μ_0} defined above. In this case K is called the *disintegration kernel*. Often we will use the notation

$$\mu[d\omega_0 d\omega_1] = \mu_0[d\omega_0] K_{\omega_0}[d\omega_1]. \tag{1.4.32}$$

Observe that if K is a Markovian kernel and μ_0 is a probability measure, then μ_{K,μ_0} is a probability measure. In this case, for emphasis, we use the notation $\mathbb{P}_{K,\mu_0}$.

Example 1.186. For any probability measures μ_i on $(\Omega_i, \mathcal{F}_i)$, $i = 0, 1$, the product measure $\mu = \mu_0 \otimes \mu_1$ is disintegrated by μ_0 since

$$\mu = \mathbb{P}_{K,\mu_0}, \quad K_{\omega_0}[-] = \mu_1[-].$$

□

Suppose that the measure μ on $(\Omega, \mathcal{F}) := (\Omega_0 \times \Omega_1, \mathcal{F}_0 \otimes \mathcal{F}_1)$ is disintegrated by μ_0, $\mu = \mathbb{P}_{K,\mu_0}$. Consider the natural projections

$$\pi_i : \Omega \to \Omega_i, \quad \pi_i(\omega_0, \omega_1) = \omega_i, \quad i = 0, 1,$$

and set $\widetilde{\mathcal{F}}_0 := \pi_0^{-1}(\mathcal{F}_0) \subset \mathcal{F} := \mathcal{F}_0 \otimes \mathcal{F}_1$.

For $F_0 \in \mathcal{F}_0$ we have
$$\mu[F_0 \times \Omega_1] = \int_{F_0} \underbrace{\left(\int_{\Omega_1} K_{\omega_0}[d\omega_1]\right)}_{K_{\omega_0}[\Omega_1]=1} \mu_0[d\omega_0] = \mu_0[F_0],$$
so that $\mu_0 = (\pi_0)_\# \mu$. This shows that the measure μ_0 is uniquely and a priori determined by μ. We can rewrite (1.4.32) as
$$\mu[d\omega_0 d\omega_1] = (\pi_0)_\# \mu[d\omega_0] K_{\omega_0}[d\omega_1]. \tag{1.4.33}$$
Next, for any $S = F_0 \times \Omega_1 \in \widetilde{\mathcal{F}}_0$, and any $F_1 \in \mathcal{F}_1$, we have
$$\mu[\pi_1^{-1}(F_1) \cap S] = \mu[F_0 \times F_1] \stackrel{(1.4.31)}{=} \int_{F_0} K_{\omega_0}[F_1] \mu_0[d\omega_0].$$
Thus, the disintegration kernel K is a regular version of the conditional distribution of the measurable map π_1 conditioned on $\widetilde{\mathcal{F}}_0$; see (1.4.28). Note also that if $\mu_1 := (\pi_1)_\# \mu$, then, for any $F_1 \in \mathcal{F}_1$, we have
$$\mu_1[F_1] = \mu[\Omega_0 \times F_1] = \int_{F_0} K_{\omega_0}[F_1] \mu_0[d\omega_0].$$
In other words, $\mu_1 = K_* \mu_0$. Thus, μ_1 is a mixture of the measures $\left(K_{\omega_0}[-]\right)_{\omega_0 \in \Omega_0}$ driven by μ_0.

Conversely, any regular version of the conditional distribution $\mu[-\|\widetilde{\mathcal{F}}_0]$ produces a disintegration kernel of the measure μ. Theorem 1.182 implies the next result.

Corollary 1.187. *If $(\Omega_1, \mathcal{F}_1)$ is isomorphic as a measurable space with a Lusin space equipped with the Borel sigma algebra then, for any measurable space $(\Omega_0, \mathcal{F}_0)$, any probability measure on $(\Omega_0 \times \Omega_1, \mathcal{F}_0 \otimes \mathcal{F}_1)$ is disintegrated by $\mu_0 = (\pi_0)_\# \mu$.* □

Example 1.188. Consider a random 2-dimensional vector (X, Y) with joint distribution
$$\mathbb{P}_{X,Y} \in \text{Prob}(\mathbb{R}^2).$$
According to Corollary 1.187, the distribution $\mathbb{P}_X$ of X disintegrates the joint distribution $\mathbb{P}_{X,Y}$. Suppose that $K_x[dy]$ is the associated disintegration kernel, i.e.,
$$\mathbb{P}_{X,Y}[dxdy] = K_x[dy] \mathbb{P}_X[dx].$$
Then, for any measurable function $f : \mathbb{R} \to \mathbb{R}$ such that $f(Y) \in L^1$ there exists a measurable function $g : \mathbb{R} \to \mathbb{R}$ such that $g(X) = \mathbb{E}[f(Y) \| X]$. As in Remark 1.160 we denote $g(x)$ by $\mathbb{E}[f(Y) \| X = x]$. We can give a more explicit description of $g(x)$ using the disintegration kernel. More precisely we will show that
$$g(x) = \int_{\mathbb{R}} f(y) K_x[dy].$$

A Monotone Class argument shows that g is Borel measurable. For any $x_0 \in \mathbb{R}$ we have

$$\mathbb{E}\big[f(Y)\boldsymbol{I}_{\{X\leq x_0\}}\big] = \int_{\mathbb{R}^2} f(y)\boldsymbol{I}_{(-\infty,x_0]}(x)\mathbb{P}_{X,Y}\big[dxdy\big]$$

$$= \int_{\mathbb{R}}\left(\int_{\mathbb{R}} f(y)K_x\big[dy\big]\right)\boldsymbol{I}_{(-\infty,x_0]}(x)\mathbb{P}_X\big[dx\big] = \int_{\mathbb{R}} g(x)\boldsymbol{I}_{(-\infty,x_0]}(x)\mathbb{P}_X\big[dx\big]$$

$$= \mathbb{E}\big[g(X)\boldsymbol{I}_{\{X\leq x_0\}}\big].$$

Since the sets $\{X \leq x_0\}$ form a π-system that generate $\sigma(X)$ we deduce that

$$\mathbb{E}\big[f(Y)\boldsymbol{I}_S\big] = \mathbb{E}\big[g(X)\boldsymbol{I}_S\big], \quad \forall S \in \sigma(X).$$

Thus

$$g(X) = \mathbb{E}\big[f(Y)\,\|\,X\big].$$

We write this as

$$\mathbb{E}\big[f(Y)\,\|\,X\big] = \int_{\mathbb{R}} f(y)K_X\big[dy\big].$$

Hence the conditional expectations $\mathbb{E}\big[f(Y)\,\|\,X\big]$ are determined by the kernel K that disintegrates the joint probability distribution $\mathbb{P}_{X,Y}$.

In particular, if $B \subset \mathbb{R}$ is a Borel set, and $f = \boldsymbol{I}_B$ we have the *law of total probability*

$$\mathbb{P}\big[Y \in B\big] = \mathbb{E}\big[\boldsymbol{I}_B(Y)\big] = \int_{\mathbb{R}} \mathbb{E}\big[\boldsymbol{I}_B(Y)\big|\,X=x\big]\mathbb{P}_X\big[dx\big],$$

where

$$\mathbb{E}\big[\boldsymbol{I}_B(Y)\big|\,X=x\big] = \mathbb{P}\big[Y \in B\big|\,X=x\big] = \int_B K_x\big[dy\big].$$

This proves that the disintegration kernel $K_x\big[dy\big]$ is a regular conditional distribution of Y given X, i.e.,

$$K_x\big[dy\big] = \mathbb{P}\big[Y \in [y, y+dy]\big|\,X=x\big] \text{``}=\text{''} \frac{\mathbb{P}\big[X \in [x,x+dx], Y \in [y,y+dy]\big]}{\mathbb{P}\big[X \in [x,x+dx]\big]}.$$

Observe that if $\mathbb{P}_{X,Y}$ is absolutely continuous with respect to the Lebesgue measure on $\mathbb{R}^2$ so that

$$\mathbb{P}_{X,Y}\big[dxdy\big] = p(x,y)dxdy,$$

then

$$K_x\big[dy\big] = \frac{p(x,y)}{p_0(x)}dy, \quad p_0(x) = \int_{\mathbb{R}} p(x,y)dy,$$

where we set $\frac{p(x,y)}{p_0(x)} = 0$ if $p_0(x) = 0$. Then

$$\mathbb{E}\big[f(Y)\big|\,X=x\big] = \int_{\mathbb{R}} f(y)\frac{p(x,y)}{p_0(x)}dy.$$

$\square$

Example 1.189. Suppose that $f : [0,1] \to \mathbb{R}$ is a C^1-function of length L. Define a random measure

$$K : \big([0,1], \mathcal{B} \big) \to \big(\mathbb{R}, \mathcal{B} \big), \quad K_x = \delta_{f(x)}.$$

Let

$$\mu_0[\, dx \,] = \frac{\sqrt{1 + |f'(x)|^2}}{L} \cdot \boldsymbol{\lambda}[\, dx \,] \in \mathrm{Prob}\big([0,1]\big).$$

Then the Borel probability measure $\mathbb{P}_{K,\mu_0}$ on $[0,1] \times \mathbb{R}$ corresponds to the integration with respect to the normalized arclength along the graph of f. $\square$

Example 1.190. Suppose that $X_1, \ldots, X_n$ are independent random variables with common distribution $p(x)\boldsymbol{\lambda}[\, dx \,]$. Denote by $\boldsymbol{X}$ the random vector $(X_1, \ldots, X_n)$. Suppose that $f : \mathbb{R}^n \to \mathbb{R}$ is a Borel measurable function. Denote by $\mathbb{P}$ the distribution of the random vector $\big(\boldsymbol{X}, f(\boldsymbol{X}) \big)$. This is disintegrated by the distribution $\mu_0 := \mathbb{P}_{\boldsymbol{X}}$ of the random vector $\boldsymbol{X}$. The disintegration kernel K is the conditional distribution of $f(\boldsymbol{X})$ given $\boldsymbol{X}$. We deduce that

$$K_{x_1, \ldots, x_n}[\, - \,] = \delta_{f(x_1, \ldots, x_n)}.$$

If B_0 is a Borel subset of $\mathbb{R}^n$ and B_1 is a Borel subset of $\mathbb{R}$, then

$$\mathbb{P}[\, B_0 \times B_1 \,] = \int_{B_0} \boldsymbol{I}_{B_1}\big(f(x_1, \ldots, x_n) \big) p(x_1) \cdots p(x_n) dx_1 \cdots dx_n.$$

Using a notation dear to theoretical physicists we can rewrite the above equality as

$$\mathbb{P}[\, dx_1 \cdots dx_n dy \,] = \delta\big(y - f(x_1, \ldots, x_n) \big) p(x_1) \cdots p(x_n) \big) dx_1 \cdots dx_n dy,$$

where $\delta(z)$ denotes the Dirac "function" on the real axis. $\square$

1.5 What are stochastic processes?

We have already met stochastic processes though we have not called them so. This section is meant to be a first encounter with this vast subject. We have a rather restricted goal namely, to explain what they are, describe a few basic features and more importantly, show that stochastic processes with prescribed statistics do exist as mathematical objects.

1.5.1 *Definition and examples*

A *stochastic process* is simply a family $(X_t)_{t \in T}$ of random variables parametrized by a set T. They are all defined on the same probability space $(\Omega, \mathcal{S}, \mathbb{P})$. The variables could be real valued, vector valued or we can allow them to be valued in a measurable space $(\mathbb{X}, \mathcal{F})$, where $\mathcal{F}$ is a sigma-algebra of subsets of $\mathbb{X}$. Frequently $\mathbb{X} = \mathbb{R}^n$ for some n but, as we will see below, it is very easy to produce more complicated examples.

Obviously stochastic processes exist, but once we impose some restriction on their behavior, the existence of such stochastic processes is less obvious. A classical situation very investigated in probability is that of families $(X_t)_{t \in T}$ of real valued random variables that are *independent, identically distributed* (or i.i.d. for brevity). We denote by $\mathbb{P}_X$ common distribution.

A basic question arises. Given a Borel probability measure μ on $\mathbb{R}$ and a set T can we find a probability space $(\Omega, \mathcal{S}, \mathbb{R})$ and independent random variables

$$X_t : (\Omega, \mathcal{S}, \mathbb{R}) \to \mathbb{R}, \ t \in T,$$

such that $\mathbb{P}_{X_t} = \mu, \forall t \in \mathbb{R}$?

When T is finite, say $T := \{1, 2, \ldots, n\}$ the answer is positive. As probability space we can take

$$(\Omega, \mathcal{S}, \mathbb{P}) := \big(\mathbb{R}^n, \mathcal{B}_{\mathbb{R}^n}, \mu^{\otimes n} \big).$$

The random variables are then the coordinate functions

$$X_k : \mathbb{R}^n \to \mathbb{R}, \ X_k(x_1, \ldots, x_n) = x_k, \ k = 1, \ldots, n.$$

Using the notation $\mathbb{R}^T$ instead of $\mathbb{R}^n$ we see that we have defined a probability measure on the space of functions $T \to \mathbb{R}$.

What happens if T is infinite, say $T = \mathbb{N}$, in which case we seek a sequence $(X_n)_{n \in \mathbb{N}}$ of i.i.d. random variables with common probability distribution μ. A substantial portion of probability is devoted to investigating such sequences and it would be embarrassing, to say the least, if it turned out they do not exist. We will see that this is not the case.

It is also very easy to stumble into situations in which the random variables are not independent, or take value in some infinite dimensional space. Here is such a situation.

Suppose that $A_0, A_1, \ldots, A_n$ is a family of i.i.d. (real valued) random variables defined on the probability space $(\Omega, \mathcal{S}, \mathbb{P})$. For every $t \in [0, 1]$ we set

$$A_t := A_0 + A_1 t + \cdots + A_n t^n.$$

We now have on our hands a family of random variables $(A_t)_{t \in [0,1]}$. These are dependent. To understand why, suppose for simplicity that the variables A_k have mean zero and variance 1. Then A_t has mean zero and for any $s, t \in [0, 1]$

$$\mathrm{Cov}\big[\, A_s, A_t \,\big] = \mathbb{E}\big[\, A_s A_t \,\big] = 1 + (st) + \cdots + (st)^n > 1.$$

Thus the random variables (A_t) are dependent.

Let $\mathbb{X}$ denote the Banach space $C\big([0, 1]\big)$ equipped with the sup norm. The family (A_t) defines a map

$$A : \Omega \to \mathbb{X}, \ \Omega \ni \omega \mapsto A_t(\omega) = \sum_{k=0}^{n} A_k(\omega) t^k \in \mathbb{X}$$

and one can show that this is measurable with respect to the Borel sigma-algebra on $\mathbb{X} = C\big([0,1]\big)$. The push-forward of $\mathbb{P}$ via the map A defines a Borel probability $\mathbb{P}_A$ measure on $\mathbb{X}$ so $(\mathbb{X}, \mathcal{B}_\mathbb{X}, \mathbb{P}_A)$ is a probability space.

It comes with a natural family of random variables $X_t : C([0,1]) \to \mathbb{R}$, $t \in [0,1]$. More precisely, the random variable X_t associates to a function $f \in C([0,1])$ its value at t, $X_t(f) = f(t)$. Note that $A_t = X_t \circ A$. Thus we can view $A_\bullet$ as a random continuous function.

Suppose now that $(X_t)_{t \in T}$ is a general family of random variables

$$X_t : (\Omega, \mathcal{S}, \mathbb{P}) \to (\mathbb{X}, \mathcal{F}).$$

This defines a map

$$X : T \times \Omega \to \mathbb{X}, \quad T \times \Omega \ni (t, \omega) \mapsto X(t, \omega) := X_t(\omega) \in \mathbb{X},$$

such that X_t is measurable for any t.

Equivalently, we can view this as a map

$$X : \Omega \to \mathbb{X}^T = \text{the space of functions } f : T \to \mathbb{X}, \tag{1.5.1}$$

where for each $\omega \in \Omega$ we have a function $X(\omega) : T \to \mathbb{X}$, $t \mapsto X_t(\omega)$.

It is convenient to regard $\mathbb{X}^T$ as a product of copies $\mathbb{X}_t$ of $\mathbb{X}$

$$\mathbb{X}^T = \prod_{t \in T} \mathbb{X}_t.$$

Each copy $\mathbb{X}_t$ is equipped with a copy $\mathcal{F}_t$ of the sigma-algebra $\mathcal{F}$.

The map (1.5.1) is measurable with respect to the sigma algebra $\mathcal{F}^T$ in $\mathbb{X}^T$, the smallest sigma-algebra $\mathcal{S}$ in $\mathbb{X}^T$ such that all the evaluation maps

$$\mathbf{Ev}_t : (\mathbb{X}^T, \mathcal{S}) \to (\mathbb{X}, \mathcal{F}), \quad \mathbf{Ev}_t(f) := f(t),$$

are measurable. Equivalently,

$$\mathcal{F}^T = \bigvee_{t \in T} \mathbf{Ev}_t^{-1}(\mathcal{F}).$$

The push-forward by X defines a probability measure $\mathbb{P}_X$ on $(\mathbb{X}^T, \mathcal{F}^T)$ called the *distribution* of the stochastic process $(X_t)_{t \in T}$. In this way we can view the process as defining a random function $T \to \mathbb{X}$.

For any finite set $I = \{t_1, \ldots, t_m\} \subset T$ we have a sigma-algebra $\mathcal{F}^I$ in $\mathbb{X}^I$,

$$\mathcal{F}^I = \mathcal{F}_{t_1} \otimes \cdots \otimes \mathcal{F}_{t_m},$$

and we obtain a random "vector"

$$X^I : (\Omega, \mathcal{S}) \to (\mathbb{X}^I, \mathcal{F}^I), \quad \omega \mapsto (X_{t_1}(\omega), \ldots, X_{t_m}(\omega)) \in \mathbb{X}^I.$$

We denote by $\mathbb{P}_I$ its probability distribution $\mathbb{P}_I := (X^I)_\# \mathbb{P}$. Note that we have a tautological measurable projection $\pi_I : \mathbb{X}^T \to \mathbb{X}^I$, and

$$\mathbb{P}_I = (\pi_I)_\# (\mathbb{P}_X).$$

Suppose now that $J \subset T$ is another finite set containing I

$$J = \{t_1, \ldots, t_m, t_{m+1}, \ldots, t_n\}, \quad n > m.$$

We get in a similar fashion a probability measure on $\mathbb{X}^J$. Now observe that we have a canonical projection

$$\mathcal{P}_{IJ} : \mathbb{X}^J \to \mathbb{X}^I, \ \ (x_{t_1}, \ldots, x_{t_m}, x_{t_{m+1}}, \ldots, x_{t_n}) \mapsto (x_{t_1}, \ldots, x_{t_m})$$

and, tautologically, we have[10]

$$(\mathcal{P}_{IJ})_\# \mathbb{P}_J = \mathbb{P}_I \tag{1.5.2}$$

since $X^I = \mathcal{P}_{IJ}(X^J)$. Proposition 1.29 shows that $\mathbb{P}_X$ is the unique probability measure $\mathbb{P}$ on $\mathbb{X}^T$ such that for any finite subset $I \subset T$ we have

$$\mathbb{P}_I = (\pi_I)_\#(\mathbb{P}).$$

A family of measures $\mathbb{P}_I$ on $\mathbb{X}^I$, I finite subset of T constrained by the compatibility condition (1.5.2) for any finite subsets $I \subset J \subset T$ is said to be a *projective family*. Note that to any probability measure $\bar{\mathbb{P}}$ on $(\mathbb{X}^T, \mathcal{F}^T)$ there is an associated projective the family of probability measures

$$\mathbb{P}_I := (\pi_I)_\#(\bar{\mathbb{P}}).$$

There are other ways of constructing projective families.

Example 1.191. Suppose that we are given a sequence of measurable spaces $(\mathscr{X}_n, \mathcal{F}_n)_{n \geq 0}$ is a measurable space. For $n \in \mathbb{N}_0 := \{0, 1, \ldots\}$ we set

$$\mathscr{X}^{\widehat{\mathbb{I}}_n} := \prod_{k=0}^n \mathscr{X}_k, \ \ \mathcal{F}^{\widehat{\mathbb{I}}_n} := \bigotimes_{k=0}^n \mathcal{F}_k.$$

Consider a family of Markovian kernels $K_n : (\mathscr{X}^{\widehat{\mathbb{I}}_n}, \mathcal{F}^{\widehat{\mathbb{I}}_n}) \to (\mathscr{X}_{n+1}, \mathcal{F}_{n+1})$, $n \in \mathbb{N}_0$. In other words we have random probability measure

$$\mathscr{X}^{\widehat{\mathbb{I}}_n} \ni (x_0, \ldots, x_n) \to K_{x_0, x_1, \ldots, x_n}[dx_{n+1}]$$

on $(\mathscr{X}_{n+1}, \mathcal{F}_{n+1})$. Then, starting with a probability measure μ_0 on $(\mathscr{X}, \mathcal{F})$, we obtain inductively using the prescription (1.4.32) a projective family of probability measures,

$$\mathbb{P}_0 = \mu_0, \ \ \mathbb{P}_{n+1} = \mathbb{P}_{K_n, \mathbb{P}_n}. \tag{1.5.3}$$

This means that for any $S \in \mathcal{F}^{\widehat{\mathbb{I}}_{n+1}}$ we have

$$\mathbb{P}_{n+1}[S] = \int_{\mathscr{X}} \int_{\mathscr{X}^{\widehat{\mathbb{I}}_n}} K_{\vec{x}}[dx_{n+1}] \boldsymbol{I}_S(\vec{x}, x_{n+1}) \mathbb{P}_n[d\vec{x}], \ \ \vec{x} = (x_0, \ldots, x_n).$$

Equivalently, $\mathbb{P}_n$ disintegrates $\mathbb{P}_{n+1}$ and K_n is the disintegration kernel.

Denote by $\mathcal{P}_{n,n+1}$ the natural projection $\mathscr{X}^{\widehat{\mathbb{I}}_{n+1}} \to \mathscr{X}^{\widehat{\mathbb{I}}_n}$,

$$(x_0, x_1, \ldots, x_n, x_{n+1}) \mapsto (x_0, x_1, \ldots, x_n).$$

Since K is a *Markovian* kernel, i.e.,

$$\int_{\mathscr{X}} K_{\vec{x}}[dx'] = 1, \ \ \forall \vec{x} \in \mathscr{X}^{\widehat{\mathbb{I}}_n},$$

---
[10] Take a few seconds to convince yourself of the validity of (1.5.2).

we deduce that $\mathbb{P}_n = (\mathcal{P}_{n,n+1})_\# \mathbb{P}_{n+1}$, $\forall n \in \mathbb{N}_0$. This shows that the collection $(\widehat{\mathbb{P}}_n)_{n \in \mathbb{N}_0}$ is a projective family of probability measures.

Note that if K_n is deterministic, i.e., $K_{x_0,\ldots,x_n}[-]$ is independent of $x_0,\ldots,x_n$, then we can think of K_n as a probability measure μ_n on $\mathscr{X}$. In this case

$$\mathbb{P}_n = \mu_0 \otimes \cdots \otimes \mu_n.$$

If $(\mathscr{X}_n, \mathcal{F}_n) = (\mathscr{X}, \mathcal{F})$ for all $n \geq 0$ can obtain kernels K_n as above starting from a single Markovian kernel $K = (\mathscr{X}, \mathcal{F}) \to (\mathscr{X}, \mathcal{F})$

$$K : \mathscr{X} \times \mathcal{F} \to [0,1], \quad (x, F) \mapsto K_x[F].$$

More precisely, we set $K_{x_0,\ldots,x_n}[dx] = K_{x_n}[dx]$.

In this case the measures $\mathbb{P}_n$ on $\mathcal{F}^{\widehat{1}_n}$ are defined by

$$\mathbb{P}_n[dx_0 dx_1 \cdots dx_n] = \mu_0[dx_0] K_{x_0}[dx_1] \cdots K_{x_{n-1}}[dx_n].$$

More precisely, for any $S \in \mathcal{F}^{\widehat{1}_n}$ we have

$$\mathbb{P}_n[S] = \int_{\mathscr{X}^{\widehat{1}_n}} \boldsymbol{I}_S(\vec{x}) \mu_0[dx_0] K_{x_0}[dx_1] \cdots K_{x_{n-2}}[dx_{n-1}] K_{x_{n-1}}[dx_n]. \quad (1.5.4)$$

The above is an iterated integral, going from right to left, i.e., we first integrate with respect to x_n, next with respect to x_{n-1} etc.

Such a situation occurs in the context of Markov chains. □

1.5.2 Kolmogorov's existence theorem

Fix a topological space $\mathbb{X}$ and a parameter set T. We denote by 2_0^T the collection of *finite* subsets of T. For $I \in 2_0^T$ we denote by $\mathcal{B}_I$ the Borel σ-algebra in $\mathbb{X}^I$ equipped with the product topology. For any finite subsets $I \subset J \subset T$ we denote by $\mathcal{P}_{IJ}$ the natural projection $\mathbb{X}^J \to \mathbb{X}^I$.

For $t \in T$ we denote by π_t the natural projection

$$\pi_t : \mathbb{X}^T \to \mathbb{X}, \quad \pi_t(\underline{x}) = x_t.$$

More generally, for any $I \in 2_0^T$ we define $\pi_I : \mathbb{X}^T \to \mathbb{X}^I$ by setting

$$\mathbb{X}^T \ni \underline{x} \mapsto \pi_I(\underline{x}) = (x_i)_{i \in I} \in \mathbb{X}^I.$$

Definition 1.192. The *natural σ-algebra* $\mathcal{E}_T$ in $\mathbb{X}^T$ is the smallest σ-algebra $\mathcal{E} \subset 2^{\mathbb{X}^T}$ such that all the maps π_t, $t \in T$, are $(\mathcal{E}, \mathcal{B}_{\mathbb{X}})$-measurable, i.e., the σ-algebra generated by the family of σ-algebras $\pi_t^{-1}(\mathcal{B}_{\mathbb{X}})$. □

Remark 1.193. The sigma-algebra $\mathcal{E}_T$ can also be identified with the σ-algebra of the Borel subsets of $\mathbb{X}^T$ equipped with the product topology. □

A *cylinder* is a subset of $\mathbb{X}^T$ of the form
$$\pi_I^{-1}(S) = S \times \mathbb{X}^{T\setminus I}, \quad I \in 2_0^T, \; S \in \mathcal{B}_I.$$
We denote by $\mathcal{C}_T$ the collection of cylinders. Clearly $\mathcal{C}_T$ is an algebra of sets that generates the natural σ-algebra $\mathcal{E}_T$.

Definition 1.194. A *projective family* of probability measures on $\mathbb{X}^T$ is a family $\mathbb{P}_I$ of probability measures on $(\mathbb{X}^I, \mathcal{B}_I)$, $I \in 2_0^T$, such that for any $I \subset J$ in 2_0^T we have
$$\mathbb{P}_I = (\mathcal{P}_{IJ})_\# \mathbb{P}_J. \tag{1.5.5}$$
$\square$

As discussed in the previous subsection any measure on $\mathbb{X}^T$ defines a canonical projective family. *Kolmogorov's existence (or consistency) theorem* states that all projective families are obtained in this fashion.

Theorem 1.195 (Kolmogorov existence theorem). *Suppose that $\mathbb{X}$ is a Lusin space, i.e., a Borel subset of a compact metric space; see Definition 1.180. For any projective family $(\mathbb{P}_I)_{I \in 2_0^T}$ of probability measures on $\mathbb{X}^T$ there exists a probability measure $\widehat{\mathbb{P}}$ on $\mathcal{E}_T$ uniquely determined by the requirement: $\forall I \in 2_0^T$ and $\mathbb{P}_I = (\mathcal{P}_I)_\#(\widehat{\mathbb{P}})$. This means that for any $B_I \in \mathcal{B}_I$,*
$$\widehat{\mathbb{P}}\bigl[\pi_I^{-1}(P_I)\bigr] = \mathbb{P}_I[B_I]. \tag{1.5.6}$$

Proof. The uniqueness follows from Proposition 1.29.

The existence is a rather deep result ultimately based on Tikhonov's compactness result. We follow the approach in [135, Secs. 30, 31].

Observe that C is a cylinder if and only if
$$\exists I \in 2_0^T \text{ and } B_I \in \mathcal{B}_I \text{ such that } C = \pi_I^{-1}(B_I).$$
For $I \in 2_0^T$ we set $\mathcal{C}_T^I := \pi^{-1}(\mathcal{B}_I) \subset \mathcal{E}_T$. Note that
$$C \in \mathcal{C}_T^I \Longleftrightarrow C = B_I \times \mathbb{X}^{T \setminus I}, \; B_I \in \mathcal{B}_I, \tag{1.5.7a}$$
$$C \in \mathcal{C}_T^I \cap \mathcal{C}_T^J \neq \emptyset \Rightarrow C \in \mathcal{C}_T^{I \cap J}. \tag{1.5.7b}$$
Define
$$\widehat{\mathbb{P}}_I : \mathcal{C}_T^I \to [0, \infty), \; \widehat{\mathbb{P}}_I[C] = \mathbb{P}_I[\pi_I(C)].$$
Note that if $C \in \mathcal{C}_T^I \cap \mathcal{C}_T^J$, then, according to (1.5.7b), $C \in \mathcal{C}_T^K$ for some $K \subset I \cap J$. Then
$$\pi_I(C) = \mathcal{P}_{KI}^{-1}\bigl(\pi_K(C)\bigr), \; \pi_J(C) = \mathcal{P}_{KJ}^{-1}\bigl(\pi_K(C)\bigr).$$
Thus
$$\mathbb{P}_I[\pi_I(C)] = \mathbb{P}_I\bigl[\mathcal{P}_{KI}^{-1}(\pi_K(C))\bigr] = (\mathcal{P}_{KI})_\# \mathbb{P}_I[\pi_K(C)] \stackrel{(1.5.5)}{=} \mathbb{P}_K[\pi_K(C)],$$

and, similarly,

$$\mathbb{P}_J[\pi_J(C)] = \mathbb{P}_I\left[\mathcal{P}_{KJ}^{-1}(\mathcal{P}_K(C))\right] = (\mathcal{P}_{KJ})_\# \mathbb{P}_J[\pi_K(C)] \stackrel{(1.5.5)}{=} \mathbb{P}_K[\pi_K(C)].$$

Hence, if $C \in \mathcal{C}_T^I \cap \mathcal{C}_T^J$, then $\widehat{\mathbb{P}}_I[C] = \widehat{\mathbb{P}}_J[C]$.

We have thus defined a *finitely additive* measure $\widehat{\mathbb{P}}$ on the *algebra*

$$\mathcal{C}_T = \bigcup_{I \in 2_0^T} \mathcal{C}_T^I.$$

To invoke Carathéodory's extension theorem (Theorem 1.39) it suffices to show that $\widehat{\mathbb{P}}$ is countably additive of $\mathcal{C}_T$.

Suppose that $(C_n)_{n \in \mathbb{N}}$ is a sequence of disjoint sets in $\mathcal{C}_T$ and

$$C_\infty = \bigcup_{n \geq 1} C_n \in \mathcal{C}_T.$$

We have to show that

$$\widehat{\mathbb{P}}[C_\infty] = \sum_{n \geq 1} \widehat{\mathbb{P}}[C_n].$$

Equivalently, if we set

$$B_n := C_\infty \setminus \bigcup_{k=1}^n C_k \in \mathcal{F},$$

we have to show that

$$\lim_{n \to \infty} \widehat{\mathbb{P}}[B_n] = 0.$$

More explicitly, we will show that if $(B_n)_{n \geq 1}$ is a decreasing sequence of sets in $\mathcal{C}_T$ with empty intersection, then $\widehat{\mathbb{P}}[B_n] \to 0$ as $n \to \infty$. To complete this step we need to make a brief foundational digression.

Digression 1.196 (Regularity of Borel measures). When dealing with measures on topological spaces there are several desirable compatibility conditions between the measure-theoretic objects and the topological ones.

Definition 1.197. Let X be a topological space and μ a Borel measure on X.

(i) The measure μ is called *outer regular* if for any Borel set $B \in \mathcal{B}_X$ we have

$$\mu[B] = \inf_{\substack{U \supset B, \\ U \text{ open}}} \mu[U].$$

(ii) The measure μ is called *inner regular* if for any Borel set $B \in \mathcal{B}_X$ we have

$$\mu[B] = \sup_{\substack{C \subset B, \\ C \text{ closed}}} \mu[C].$$

(iii) The measure μ is called *regular* if it is both inner and outer regular.

(iv) The measure μ is called *Radon* if it is outer regular, and for any Borel set $B \in \mathcal{B}_X$, we have

$$\mu[B] = \sup_{\substack{K \subset B, \\ K \text{ compact}}} \mu[K].$$

From the above definition it is clear that

$$\mu \text{ is Radon} \Rightarrow \mu \text{ is regular}.$$

A deep result in measure theory states that any Borel probability measure on a Lusin space is Radon, [15, Thm. 7.4.3]. For our immediate needs we can get away by with a lot less. We have the following useful result, [126, Chap. II, Thm. 1.2]. A proof is outlined in Exercise 1.53.

Theorem 1.198. *Any Borel probability measure on a metric space is regular.*

This concludes our digression. □

As mentioned in Remark 1.181(b), any Lusin space is Borel isomorphic to a compact metric space. Thus it suffices to prove Kolmogorov's theorem only in the special when $\mathscr{X}$ *is a compact metric space*. From Theorem 1.198 we deduce the following result.

Lemma 1.199. *Let Y be a compact metric space. Then any Borel probability measure on Y is Radon.* □

We can now complete the proof of Kolmogorov's theorem. Suppose there exists a decreasing sequence of sets $(B_n)_{n \geq 1}$ in $\mathcal{C}_T$ such that

$$\bigcap_{n \in \mathbb{N}} B_n = \emptyset.$$

We want to prove that

$$\lim_{n \to \infty} \widehat{\mathbb{P}}[B_n] = 0.$$

We argue by contradiction. Suppose that

$$\lim_{n \to \infty} \widehat{\mathbb{P}}[B_n] = \delta > 0.$$

We can find a strictly increasing sequence of finite subsets of T

$$I_1 \subset I_2 \subset \cdots,$$

and subsets $S_n \subset \mathbb{X}^{I_n}$, $n \in \mathbb{N}$, such that

$$B_n = S_n \times \mathbb{X}^{T \setminus I_n}, \quad S_n \subset \mathbb{X}^{I_n}, \quad \forall n \in \mathbb{N}.$$

We set

$$I_\infty := \bigcup_{n \geq 1} I_n \subset T.$$

For any $n \in \mathbb{N}$, the space $\mathscr{X}^{I_n}$ is a compact metric space and Lemma 1.199 implies that the Borel probability measure $\mathbb{P}_{I_n}$ on $\mathscr{X}^{I_n}$ is Radon. Hence, for any $n > 0$, there exists a compact set $K_n \subset S_n$ such that

$$\mathbb{P}_{I_n}[\, S_n \setminus K_n \,] < \frac{\delta}{2^{n+1}}.$$

Set

$$C_n := K_n \times \mathbb{X}^{T \setminus I_n} \subset S_n \times \mathbb{X}^{T \setminus I_n} = B_n.$$

Tikhonov's compactness theorem shows that all products $\mathbb{X}^{T \setminus I_n}$ are compact with respect to the product topology. Hence the sets $C_n = K_n \times \mathbb{X}^{T \setminus I_n}$ are also compact. Note that

$$\widehat{\mathbb{P}}[\, B_n \setminus C_n \,] = \mathbb{P}_{I_n}[\, S_n \setminus K_n \,] < \frac{\delta}{2^{n+1}}, \quad \forall n \in \mathbb{N}. \tag{1.5.8}$$

Set

$$D_n := \bigcap_{j=1}^{n} C_j \in \mathcal{C}_T.$$

Observe that $(D_n)_{n \in \mathbb{N}}$ is a decreasing sequence of compact subsets of $\mathbb{X}^T$ and

$$\widehat{\mathbb{P}}[\, B_n \setminus D_n \,] = \widehat{\mathbb{P}}\left[\bigcup_{j=1}^{n}(B_n \setminus C_j)\right] \leq \sum_{j=1}^{n} \widehat{\mathbb{P}}[\, B_n \setminus C_j \,]$$

$$\stackrel{(1.5.8)}{\leq} \sum_{j=1}^{n} \widehat{\mathbb{P}}[\, B_j \setminus C_j \,] \leq \sum_{j=1}^{n} \frac{\delta}{2^{j+1}} < \frac{\delta}{2}.$$

Hence

$$\widehat{\mathbb{P}}[\, D_n \,] > \frac{\delta}{2}, \quad \forall n \geq 1.$$

This shows that D_n is nonempty $\forall n \in \mathbb{N}$ so the decreasing sequence of nonempty and *compact* sets D_n has a *nonempty* intersection. We have reached a contradiction since

$$\emptyset = \bigcap_{n \geq 1} B_n \supset \bigcap_{n \geq 1} D_n \neq \emptyset.$$

□

The real axis $\mathbb{R}$ is a Lusin space. Given a Borel probability measure $\mathbb{P}$ on $\mathbb{R}$ we can construct trivially a projective family $\mathbb{P}_I$, $I \in 2_0^{\mathbb{N}}$. More precisely $\mathbb{P}_I = \mathbb{P}^{\otimes |I|}$ on $\mathbb{R}^I$. We deduce that we have a natural Borel probability measure $\mathbb{R}^{\mathbb{N}}$. We have natural random variables on this probability space

$$X_n : \mathbb{R}^{\mathbb{N}} \to \mathbb{R}, \quad X_n(\underline{x}) = x_n, \quad \forall \underline{x} = (x_1, x_2, \ldots,) \in \mathbb{R}^{\mathbb{N}}.$$

Note that $\mathbb{P}_{X_n} = \mathbb{P}$, $\forall n$ and the joint distribution of $X_1, \ldots, X_n$ is $\mathbb{P}^{\otimes n}$. Thus, the random variables (X_n) are independent and have identical distributions. We have thus proved the following fact.

Corollary 1.200. *For any probability measure $\mathbb{P} \in \mathrm{Prob}(\mathbb{R}, \mathcal{B}_\mathbb{R})$, there exists a probability space $(\Omega, \mathcal{S}, \bar{\mathbb{P}})$ and a sequence of independent identically distributed (or i.i.d. for brevity) random variables $X_n : (\Omega, \mathcal{S}, \bar{\mathbb{P}}) \to \mathbb{R}$, $n \in \mathbb{N}$, with common distribution $\mathbb{P}$.* □

Remark 1.201. The proof of Theorem 1.195 uses in an essential fashion the topological nature of the projective family of measures $(\mathbb{P}_I)_{I \in 2_0^T}$. We want to emphasize that in this theorem the set of parameters T is arbitrary.

If the set of parameters T is countable, say $T = \mathbb{N}_0$, then one can avoid the topological assumptions.

Consider for example the projective family of measures $\mathbb{P}_n$ constructed in Example 1.191. Recall briefly its construction that we are given a sequence of measurable spaces $(\mathcal{X}_n, \mathcal{F}_n)_{n \geq 0}$ and measures $\mathbb{P}_n$ on

$$(\mathcal{X}_0 \times \cdots \times \mathcal{X}_n, \mathcal{F}_0 \otimes \cdots \otimes \mathcal{F}_n)$$

such that $\mathbb{P}_n$ disintegrates $\mathbb{P}_{n+1}$, $\forall n \geq 0$. (Observe that this condition is automatically satisfied if each $\mathcal{X}_n$ is a Lusin space.) Set

$$\mathcal{X}^\infty := \prod_{n=0}^\infty \mathcal{X}_n,$$

denote by π_n the natural projection $\mathcal{X}^\infty \to \mathcal{X}_n$ and by $\mathcal{F}^{\otimes \infty}$ the sigma-algebra

$$\mathcal{F}^{\otimes \infty} := \bigvee_{n \geq 0} \pi_n^{-1}(\mathcal{F}_n).$$

A theorem of C. *Ionescu-Tulcea* (see [85, Thm. 8.24]) states that there exists a unique probability measure $\mathbb{P}_\infty$ on $\mathcal{F}^{\otimes \infty}$ such that

$$(\mathcal{P}_n)_\# \mathbb{P}_\infty = \mathbb{P}_n, \quad \forall n \geq 0,$$

where $\mathcal{P}_n$ denotes the natural projection $\mathcal{X}^\infty \to \mathcal{X}_0 \times \cdots \times \mathcal{X}_n$.

As a special case of this result let us mention an infinite-dimensional version of Fubini-Tonelli: given measures μ_n on $\mathcal{F}_n$, there exists a unique measure μ_∞ on $\mathcal{F}^{\otimes \infty}$ such that

$$(\mathcal{P}_n)_\# \mu_\infty = \bigotimes_{k=0}^n \mu_k.$$

For this reason we will denote the measure μ_∞ by

$$\bigotimes_{n=0}^\infty \mu_n.$$

□

1.6 Exercises

Exercise 1.1. Let S_0, S_1 be two sigma-algebras of a set Ω. Prove that the following are equivalent.

(i) The union $S_0 \cup S_1$ is a sigma-algebra.
(ii) Either $S_0 \subset S_1$ or $S_1 \subset S_0$.

$\square$

Exercise 1.2 (Alexandrov). Suppose that K is a compact metric space, $\mathcal{F}$ is an algebra of subsets of K and $\mu : \mathcal{F} \to [0,1]$ is a finitely additive function satisfying the following property. For any $F \in \mathcal{F}$ and any $\varepsilon > 0$ there exist sets $F_\pm \in \mathcal{F}$ such that
$$cl(F_-) \subset F \subset int(F_+), \quad \mu\big[F_+ \setminus F_-\big] < \varepsilon.$$
Prove that μ is a premeasure.

$\square$

Exercise 1.3. Fix a set Ω of finite cardinality m and a probability measure π on Ω. Set
$$\Omega^\infty := \Omega^\mathbb{N}$$
so the elements of Ω^∞ are functions $\underline{\omega} : \mathbb{N} \to \Omega$. For every $n \in \mathbb{N}$ define
$$\pi_n : \Omega^\infty \to \Omega^n, \quad \pi_n(\underline{\omega}) = (\omega_1, \ldots, \omega_n),$$
and denote by $\mathcal{C}_n$ the collection of sets of the form
$$C = \pi_n^{-1}(S), \quad S \subset \Omega^n, \quad n \in \mathbb{N}.$$
Note that $\mathcal{C}_1 \subset \mathcal{C}_2 \subset \cdots$. Set
$$\mathcal{C} := \bigcup_{n \in \mathbb{N}} \mathcal{C}_n.$$

(i) Show that $\mathcal{C}_n$ is a σ-algebra of subsets of Ω^∞, $\forall n \in \mathbb{N}$.
(ii) For any $n \in \mathbb{N}$ define $\beta_n = \beta_{n,\pi} : \mathcal{C}_n \to [0,1]$,
$$\beta_n\big[\pi_n^{-1}(S)\big] := \pi^{\otimes n}[S] = \sum_{(\omega_1, \ldots, \omega_n) \in S} \prod_{j=1}^n \pi[\{\omega_j\}].$$
Show that β_n is a well defined measure on $\mathcal{C}_n$ and
$$\beta_{n+1}\big|_{\mathcal{C}_n} = \beta_n.$$

(iii) Equip Ω^∞ with the metric
$$d(\underline{\omega}, \underline{\eta}) = \sum_{n \in \mathbb{N}} \frac{1}{2^n} h(\omega_m, \eta_n), \quad h(\omega, \eta) = \begin{cases} 0, & \omega = \eta, \\ 1, & \omega \neq \eta. \end{cases}$$

Prove that (Ω^∞, d) is a compact metric space. **Hint.** Use the diagonal procedure to show that any sequence if Ω admits a convergent subsequence.

(iv) Define $\beta = \beta_\pi : \mathcal{C} \to [0,1]$,
$$\beta\big|_{\mathcal{C}_n} = \beta_n.$$
Show that β is a well defined <u>premeasure</u> on $\mathcal{C}$. **Hint.** Use Exercise 1.2.

(v) Denote by $\bar\beta = \bar\beta_\pi$ the extension of β as measure to the σ-algebra $\sigma(\mathcal{C})$. (Its existence is guaranteed by the Caratheodory extension theorem.) For $omega_0 \in \Omega$ we set
$$\Delta_\omega := \{\, \underline{\omega} \in \Omega : \ \exists m \in \mathbb{N} \text{ such that } \omega_n = \omega,\ \forall n > m\,\}.$$
Show that $\Delta_\Omega \in \sigma(\mathcal{C})$ and $\bar\beta[\Delta_\omega] = 0$.

(vi) Define $X_n : \Omega \to \Omega$, $X_n(\underline{\omega}) = \omega_n$. Show that the collection of random variables $(X_n)_{n \in \mathbb{N}}$ is independent and have the same distribution π.

(vii) Let $\Omega = \{0,1\}$, $\pi=$the uniform measure on $\{0,1\}$, and consider $\Omega^\infty = \{0,1\}^\mathbb{N}$ equipped with the measure $\bar\beta = \bar\beta_\pi$ constructed as above. Show that the map
$$B : (\Omega^\infty, \sigma(\mathcal{C})) \to \big([0,1], \mathcal{B}_{[0,1]}\big), \quad B = \sum_{n \in \mathbb{N}} \frac{1}{2^n} X_n$$
is measurable and find $B_\# \bar\beta$. □

Exercise 1.4. Suppose that $(\Omega, \mathcal{S}, \mu)$ is a *finite*[11] measured space and $\mathcal{A} \subset \mathcal{S}$ a countable family of measurable subsets that generates $\mathcal{S}$, $\sigma(\mathcal{A}) = \mathcal{S}$. Assume $\Omega \in \mathcal{A}$. Denote by $\mathbb{R}[\mathcal{A}]$ the vector space spanned by I_A, $A \in \mathcal{A}$. Fix $p \in [1, \infty)$ and denote by $\mathcal{M}_p$ the intersection of $L^\infty(\Omega, \mathcal{S}, \mu)$ with the L^p-closure of $\mathbb{R}[\mathcal{A}]$.

(i) Prove $\mathcal{M}_p = L^\infty(\Omega, \mathcal{S}, \mathbb{P})$.
(ii) Prove that $\mathbb{R}[\mathcal{A}]$ is dense in $L^p(\Omega, \mathcal{S}, \mu)$.
(iii) Prove that $L^p(\Omega, \mathcal{S}, \mu)$ is separable. □

Exercise 1.5. Suppose that $(\Omega, \mathcal{F}, \mu)$ is a measured space and (S, d) a metric space. Consider a function
$$F : S \times \Omega \to \mathbb{R}, \quad (s, \omega) \mapsto F_s(\omega)$$
satisfying the following properties.

(i) For any $s \in S$ the function $\Omega \ni \omega \mapsto F_s(\omega) \in \mathbb{R}$ is measurable.
(ii) For any $\omega \in \Omega$ the function $S \ni s \mapsto F_s(\omega) \in \mathbb{R}$ is continuous.
(iii) There exists $h \in \mathcal{L}^1(\Omega, \mathcal{S}, \mu)$ such that $|F_s(\omega)| \leq h(\omega),\ \forall (s, \omega) \in S \times \Omega$.

Prove that $F_s \in \mathcal{L}^1(\Omega, \mathcal{S}, \mu)$, $\forall s \in S$, and the resulting function
$$S \ni s \mapsto \int_\Omega F_s(\omega) \mu[\,d\omega\,] \in \mathbb{R}$$
is <u>continuous</u>. **Hint.** Use the Dominated Convergence Theorem. □

---
[11]The sigma-finite situation follows from the finite situation in a standard fashion.

Exercise 1.6. Suppose that $(\Omega, \mathcal{F}, \mu)$ is a measured space and $I \subset \mathbb{R}$ is an open interval. Consider a function
$$F : I \times \Omega \to \mathbb{R}, \quad (t, \omega) \mapsto F(t, \omega)$$
satisfying the following properties.

(i) For any $t \in I$ the function $F(t, -) : \Omega \to \mathbb{R}$ is integrable,
$$\int_\Omega |F(t, \omega)| \, \mu[d\omega] < \infty.$$
(ii) For any $\omega \in \Omega$ the function $I \ni t \mapsto F(t, \omega) \in \mathbb{R}$ is differentiable at $t_0 \in I$. We denote by $F'(t_0, \omega)$ its derivative.
(iii) There exists $h \in \mathcal{L}^1(\Omega, \mathcal{S}, \mu)$ and $c > 0$ such that
$$|F(t, \omega) - F(t_0, \omega)| \leq h(\omega)|t - t_0|, \quad \forall (t, \omega) \in I \times \Omega.$$

Prove that the function
$$I \ni t \mapsto \int_\Omega F(t, \omega) \mu[d\omega] \in \mathbb{R}$$
is differentiable at t_0 and
$$\frac{d}{dt}\bigg|_{t=t_0} \left(\int_\Omega F(t, \omega) \mu[d\omega] \right) = \int_\Omega F'(t_0, \omega) \mu[d\omega]. \qquad \square$$

Exercise 1.7. Prove that the random variables $N_1, \ldots, N_m$ that appear in Example 1.112 on the coupon collector problem can be realized as measurable functions defined on the same probability space. **Hint.** Use Exercise 1.3. $\qquad \square$

Exercise 1.8 (Markov). Let $(\Omega, \mathcal{S}, \mathbb{P})$ be a sample space and A_-, A_0, A_+, $\mathbb{P}[A_0] \neq 0$. We say that A_+ is *independent of* A_- *given* A_0 if
$$\mathbb{P}[A_+ \cap A_- | A_0] = \mathbb{P}[A_+ | A_0]\mathbb{P}[A_- | A_0].$$
Show that A_+ is independent of A_- given A_0 if and only if
$$\mathbb{P}[A_+ | A_0 \cap A_-] = \mathbb{P}[A_+ | A_0]. \qquad \square$$

Exercise 1.9 (M. Gardner). A family has two children. Find the conditional probability that both children are boys in each of the following situations.

(i) One of the children is a boy.
(ii) One of the children is a boy born on a Thursday. $\qquad \square$

Exercise 1.10. A random experiment is performed repeatedly and the outcome of an experiment is independent of the outcomes of the previous experiments. While performing these experiments we keep track of the occurrence of the mutually exclusive events A and B, i.e., $A \cap B = \emptyset$. We assume that A and B have positive probabilities.[12] What is the probability that A occurs before B? **Hint.** Consider the event $C = (A \cup B)^c$ = neither A, nor B. Condition of the result of the first experiment which can be A, B or C. □

Exercise 1.11. Consider the standard random walk on $\mathbb{Z}$ started at 0. More precisely are given a sequence of i.i.d. random variables $(X_n)_{n \in \mathbb{N}}$ such that $\mathbb{P}[X_n = 1] = \mathbb{P}[X_n = -1] = \frac{1}{2}$, $\forall n$ and we set
$$S_n := X_1 + \cdots + X_n.$$
Let T denote the time of the first return to 0,
$$T := \min\{n \in \mathbb{N};\ S_n = 0\}.$$
Set $f_n = \mathbb{P}[T = n]$, $u_n := \mathbb{P}[S_n = 0]$.

(i) Prove that $u_{2n} = \mathbb{P}[S_1 \neq 0,\ S_2 \neq 0, \ldots, S_{2n} \neq 0]$. Deduce that $f_{2n} = u_{2n-2} - u_{2n}$. **Hint.** Use André's reflection principle in Example 1.60.

(ii) Visualize the random walk as a zig-zag of the kind depicted in Figure 1.1. For such a zigzag we denote by $L_n(z)$ the number of its first n segments that are above the x axis. Equivalently,
$$L_n(z) := \#\{k;\ 1 \leq k \leq n;\ S_k > -\varepsilon\}.$$
For example, for the zig-zag z in Figure 1.1 we have
$$L_8(z) = L_9(z) = L_{10}(z) = 8.$$
Show that
$$\mathbb{P}[L_{2n} = m] = \begin{cases} u_{2k} u_{2n-2k}, & m = 2k \leq 2n, \\ 0, & m \equiv 1 \bmod 2. \end{cases}$$

(iii) Prove that $\mathbb{P}[L_{2n} = 2k \mid S_{2n} = 0] = \frac{1}{n+1}$. □

Exercise 1.12. Suppose that $X, Y : (\Omega, \mathcal{S}, \mathbb{P}) \to \mathbb{R}$ are two random variables whose ranges $\mathcal{X}$ and $\mathcal{Y}$ are countable subsets of $\mathbb{R}$. We set
$$\mathbb{E}[X \| Y] = \sum_{y \in \mathcal{Y}} \mathbb{E}[X \| Y = y] \mathbf{I}_{\{Y=y\}} \in \mathcal{L}^0(\Omega, \sigma(Y), \mathbb{P}),$$
where
$$\mathbb{E}[X \| Y = y] := \sum_{x \in \mathcal{X}} x \mathbb{P}[X = x \| Y = y].$$
The random variable $\mathbb{E}[X \| Y]$ is called the *conditional expectation of X given Y*. Prove that
$$\mathbb{E}[X] = \mathbb{E}[\mathbb{E}[X \| Y]]. \qquad \square$$

---
[12] For example if we roll a pair of dice, A could be the event "*the sum is 4*" and B could be the event "*the sum is 7*". In this case
$$\mathbb{P}[A] = \frac{3}{36} = \frac{1}{12},\quad \mathbb{P}[B] = \frac{6}{36} = \frac{1}{6}.$$

Exercise 1.13 (Polya's urn). An urn U contains r_0 red balls and g_0 red balls. At each stage a ball is selected at random from the urn, we observe its color, we return it to the urn and then we add another ball of the same color. We denote by R_n the number of red balls and by G_n the number of green balls at stage n. Finally, we denote by C_n the "concentration" of red balls at stage n,
$$C_n = \frac{R_n}{R_n + G_n}.$$
(i) Show that $\mathbb{E}[\,C_{n+1} \,\|\, R_n\,] = C_n$, where the conditional expectation $\mathbb{E}[\,C_{n+1} \,\|\, R_n\,]$ is defined in Exercise 1.12.
(ii) Show that $\mathbb{E}[C_n] = \frac{r_0}{r_0+g_0}$, $\forall n \in \mathbb{N}$. □

Exercise 1.14. Prove the claim about the events S_k at the end of Example 1.110. □

Exercise 1.15 (Banach's matchbox problem). An eminent mathematician fuels a smoking habit by keeping matches in both trouser pockets. When impelled by need, he reaches a hand into a randomly selected pocket and grabs about for a match. Suppose he starts with n matches in each pocket. What is the probability that when he first discovers a pocket to be empty of matches the other pocket contains exactly m matches? □

Exercise 1.16. Suppose that $X_n : \mathcal{L}^1(\Omega, \mathcal{S}, \mathbb{P})$, $n \in \mathbb{N}$, is a sequence of independent and identically distributed (i.i.d.) random variables and $T \in \mathcal{L}^1(\Omega, \mathcal{S}, \mathbb{P})$ is a random variable with range contained in $\mathbb{N}$ and independent of the variables X_n. Define $S_T : \Omega \to \mathbb{R}$
$$S_T(\omega) = \sum_{n=1}^{T(\omega)} X_n(\omega).$$
Prove *Wald's formula*
$$\mathbb{E}[\,S_T\,] = \mathbb{E}[\,T\,]\mathbb{E}[\,X_1\,]. \tag{1.6.1}$$
□

Exercise 1.17. A box contains n identical balls labelled $1, \ldots, n$. Draw one ball, uniformly random, and record its label N. Next flip a fair coin N times. What is the expected number of heads you roll? **Hint.** Use Wald's formula. □

Exercise 1.18. Suppose that $X \in L^0(\Omega, \mathcal{S}, \mathbb{P})$ is a *nonnegative* random variable. Prove that if the range of $\mathscr{X}$ is contained in $\mathbb{N}_0$, then
$$\mathbb{E}[\,X\,] - 1 \leq \sum_{n \geq 0} \mathbb{P}[X > n] \leq \mathbb{E}[X].$$
In particular, conclude that
$$X \in L^1(\Omega, \mathcal{S}, \mathbb{P}) \iff \sum_{n \geq 0} \mathbb{P}[X > n] < \infty.$$
Hint. Use (1.3.44). □

Exercise 1.19. There are n unstable molecules $m_1, \ldots, m_n$ in a row. One of the $n-1$ pairs of neighbors, chosen uniformly at random, combine to form a stable dimer. This process continues until there remain U_n isolated molecules, no two of which are adjacent.

(i) Show that the probability p_n that m_1 remains uncombined satisfies
$$(n-1)p_n = p_1 + p_2 + \cdots + p_{n-2}.$$
Deduce that
$$p_n = \sum_{k=0}^{n-1} \frac{(-1)^k}{k!} \to e^{-1} \text{ as } n \to \infty.$$
Hint. Condition on the first pair of molecules (m_r, m_{r+1}) that gets combined.

(ii) Show that the probability $q_{r,n}$ that the molecule m_r remains uncombined is $p_r p_{n-r+1}$.

(iii) Show that
$$\mathbb{E}[U_n] = \sum_{r=1}^{n} q_{r,n}.$$

(iv) Show that
$$\lim_{n \to \infty} \frac{1}{n} \mathbb{E}[U_n] = e^{-2}.$$
□

Exercise 1.20. Let $N = N_m$ be the random variable defined in the coupon collector problem described in Example 1.112. Show that
$$\operatorname{Var}[N_m] = m \sum_{k=1}^{m} \frac{m-k}{k^2}.$$
□

Exercise 1.21 (The Birthday Problem). Let $N \in \mathbb{N}$. Consider a sequence $(X_n)_{n \in \mathbb{N}}$ of independent random variables uniformly distributed on the finite set $\{1, \ldots, N\}$. Define B_N to be the *birthday random variable*[13]
$$B_N(\omega) = \min\{j \in \mathbb{N}: \ \exists 1 \leq i < j \text{ such that } X_j(\omega) = X_i(\omega)\}.$$
Compute the probabilities
$$\mathbb{P}[B_N \leq k], \quad k = 1, \ldots, N.$$
□

Exercise 1.22 (Buffon's Problem). A needle of length ℓ is thrown at random on a plane ruled by parallel lines distance d apart. Denote by N_ℓ the number of lines that intersect the needle.

---
[13]You should think of B_N as follows. Suppose that you have an urn with N balls labelled $1, \ldots, N$. Suppose we perform the following experiment: draw a ball at random, record its label, put it back in the box, and then repeat until you notice that the label you've drawn has appeared before. The random variable B_N is the first moment when you've noticed a label that was drawn before. The classical birthday problem is the special case $N = 365$.

(i) Prove that $\mathbb{P}[N_\ell = 1]$ when $\ell < d$.
(ii) Prove that $\mathbb{E}[N_{\ell_0+\ell_1}] = \mathbb{E}[N_{\ell_0}] + \mathbb{E}[N_{\ell_1}]$, $\forall \ell_0, \ell_1 > 0$.
(iii) Compute $\mathbb{E}[N_\ell]$, $\ell > 0$.

□

Exercise 1.23. Suppose that I is an interval of the real axis and $f : I \to \mathbb{R}$ is a continuous function. Prove that the following are equivalent.

(i) For any $x, y \in I$, and any $t \in (0,1)$ we have $f((1-t)x+ty) \leq (1-t)f(x)+tf(y)$.
(ii) For any $x_0 \in I$ there exists a linear function $\ell : \mathbb{R} \to \mathbb{R}$ such that

$$\ell(x_0) = f(x_0), \quad \ell(x) \leq f(x), \quad \forall x \in I.$$

□

Exercise 1.24 (Hermite polynomials). Suppose that $X \sim N(0,1)$ so

$$\mathbb{P}_X[dx] = \mathbf{\Gamma}_1[dx] := \gamma_1(x)\boldsymbol{\lambda}[dx], \quad \gamma_1(x) = \frac{1}{\sqrt{2\pi}} e^{-\frac{x^2}{2}}.$$

For $k \in \mathbb{N}_0$ we denote by $\mathbb{R}[x]$ the space of polynomial with real coefficients. Define the linear operators

$$P, Q : \mathbb{R}[x] \to \mathbb{R}[x],$$

$$(Pf)(x) = f'(x), \quad (Qf)(x) = -f'(x) + xf(x). \tag{1.6.2}$$

(i) Prove that for any $f \in \mathbb{R}[x]$ we have

$$(PQ - QP)f = f.$$

(ii) Denote by $H_0 \in \mathbb{R}[x]$ the constant polynomial identically equal to 1. Show that for any $n \in \mathbb{N}$ the function

$$H_n := Q^n H_0$$

is a degree n polynomial satisfying

$$PH_n = nH_{n-1}, \quad QPH_n = nH_n, \quad \forall n \in \mathbb{N},$$

and

$$H_n = xH_{n-1} - (n-1)H_{n-2}, \quad \forall n \geq 2.$$

The polynomials $H_n(x)$ are called the *Hermite polynomials*.

(iii) Show that for any $f, g \in \mathbb{R}[x]$

$$\int_\mathbb{R} Pf(x)g(x)\,\mathbf{\Gamma}_1[dx] = \int_\mathbb{R} f(x)Qg(x)\,\mathbf{\Gamma}_1[dx]. \tag{1.6.3}$$

(iv) Show that

$$H_n(x) = (-1)^n e^{\frac{x^2}{2}} P^n\bigl(e^{-\frac{x^2}{2}}\bigr), \quad \forall n \in \mathbb{N}. \tag{1.6.4}$$

(v) Show that for any $m, n \in \mathbb{N}_0$ we have
$$\int_{\mathbb{R}} H_n(x) H_m(x) \mathbf{\Gamma}_1[dx] = n! \delta_{mn}.$$

(vi) Show that
$$\sum_{n \geq 0} H_n(x) \frac{\lambda^n}{n!} = e^{\lambda x - \lambda^2/2}. \tag{1.6.5}$$

(vii) Suppose that $f \in \mathbb{R}[x]$, $\deg f \leq n$. Prove that
$$f(x) = \sum_{k=1}^{n} \frac{1}{k!} \mathbb{E}\big[f^{(k)}(X) \big] H_k(x).$$

$\square$

Exercise 1.25. Suppose that $X \sim N(0,1)$, i.e.,
$$\mathbb{P}_X[dx] = \gamma_1(x) dx, \quad \gamma_1(x) = \frac{1}{\sqrt{2\pi}} e^{-\frac{x^2}{2}}.$$

Set $\overline{\Phi}(x) := \mathbb{P}[X > x]$. Prove the Mills ratio inequalities (1.3.40), i.e.,
$$\frac{x}{x^2+1} \gamma_1(x) \leq \overline{\Phi}(x) \leq \frac{1}{x} \gamma_1(x), \quad \forall x > 0.$$

Hint. For the upper bound observe that
$$-Q\overline{\Phi} = \int_x^{\infty} \overline{\Phi}(x) dx > 0,$$
where Q is the operator defined in (1.6.2). Next express
$$\int_x^{\infty} Q\overline{\Phi}(t) dt \leq 0$$
in terms of $\overline{\Phi}$ and γ_1.

$\square$

Exercise 1.26. We denote by $\mathrm{Dens}(\mathbb{R})$ the space of probability densities on $\mathbb{R}$, i.e., functions $p \in L^1(\mathbb{R}, \boldsymbol{\lambda})$ such that
$$\int_{\mathbb{R}} p(x) dx = 1 \text{ and } p(x) \geq 0 \text{ almost everywhere.}$$

For $p \in \mathrm{Dens}(\mathbb{R})$ we set
$$\mathbb{E}[p] := \int_{\mathbb{R}} x p(x) dx, \quad \mathrm{Var}[p] := \int_{\mathbb{R}} x^2 p(x) dx - \mathbb{E}[p]^2.$$

The *entropy*[14] of $p \in \mathrm{Dens}(\mathbb{R})$ is the quantity
$$\mathrm{Ent}[p] := -\int_{\mathbb{R}} p(x) \log p(x) dx \in [0, \infty],$$

where we set $0 \cdot \log 0 = 0$.

---
[14] The entropy is a measure of disorder or randomness of the probability density: the higher the entropy the less predictable is the associated random variable.

(i) Show that if
$$\gamma_1(x) := \frac{1}{\sqrt{2\pi}} e^{-x^2/2},$$
then
$$\text{Ent}[\gamma_1] = \frac{1 + \log 2\pi}{2}.$$

(ii) Show that if $p, q \in \text{Dens}(\mathbb{R})$ and $q(x) > 0$, $\forall x \in \mathbb{R}$, then
$$\text{Ent}[p] \leq -\int_{\mathbb{R}} p(x) \log q(x) dx$$
if the integral on the right hand side is finite. Moreover equality holds iff $p = q$.

Hint. Show that $p(x) - p(x) \log p(x) \leq q(x) - p(x) \log q(x)$, $\forall x \in \mathbb{R}$.

(iii) Show that if $p \in \text{Dens}(\mathbb{R})$ satisfies
$$\mathbb{E}[p] = 0 = \mathbb{E}[\gamma_1], \quad \text{Var}[p] = 1 = \text{Var}[\gamma_1],$$
then $\text{Ent}[p] \leq \text{Ent}[\gamma_1]$ with equality iff $p = \gamma_1$.

□

Exercise 1.27. Let $X : (\Omega, \mathcal{S}, \mathbb{P} \to \mathbb{N}_0)$ be a random variable and $\lambda > 0$. Prove that the following are equivalent.

(i) $X \sim \text{Poi}(\lambda)$.
(ii) $\mathbb{E}[\lambda f(X+1) - X f(X)] = 0$, for any bounded function $f : \mathbb{N}_0 \to \mathbb{R}$.

□

Exercise 1.28. Prove Proposition 1.104. □

Exercise 1.29. Show that
$$\mathbb{M}_N(t) = \frac{pe^t}{1 - qe^t} \quad \text{if } N \sim \text{Geom}(p),$$

$$\mathbb{M}_N(t) = e^{\lambda(e^t - 1)} \quad \text{if } N \sim \text{Poi}(\lambda),$$

and
$$\mathbb{M}_X(t) = \frac{\lambda}{\lambda - t} \quad \text{if } X \sim \text{Exp}(\lambda).$$

□

Exercise 1.30. Let $Y \sim N(0, 1)$ be a standard normal random variable and set $X := \exp(Y)$.

(i) Show that
$$\mathbb{E}[X^n] = e^{n^2/2}, \quad \forall n \in \mathbb{N}.$$

(ii) Prove that the probability distribution $\mathbb{P}_X$ of X is given by the *log-normal* law
$$\mathbb{P}_X[dx] = p(x)dx, \quad p(x) = \begin{cases} \frac{1}{x\sqrt{2\pi}} e^{-\frac{1}{2}(\log x)^2}, & x > 0, \\ 0, & x \leq 0, \end{cases}$$
where log denotes the natural logarithm.

(iii) For $\alpha \in [-1, 1]$ we set
$$p_\alpha(x) = \begin{cases} p(x)\big(1 + \alpha \sin(2\pi \log x)\big), & x > 0, \\ 0, & x \leq 0. \end{cases}$$
Prove that for any $\alpha \in [-1, 1]$ and any $n \in \mathbb{N}_0$ we have
$$\int_\mathbb{R} x^n p_\alpha(x) dx = e^{n^2/2}.$$

Thus, for any $\alpha \in [-1, 1]$, the function $p_\alpha(x)dx$ is a probability density on $\mathbb{R}$ and the probability measure $p_\alpha(x)$ has the same moments as X.

□

Exercise 1.31. Let $X : (\Omega, \mathcal{S}, \mathbb{P}) \to \mathbb{R}$ be a random variable with range contained in
$$\mathbb{N}_0 = \{0, 1, 2, \dots\}.$$
Its *probability generating function* (or *pgf* for brevity) is the formal power series
$$PG_X(s) = \sum_{n \geq 0} \mathbb{P}[X = n] s^n.$$

(i) Show that the power series defining PG_X is convergent for any $|s| < 1$. Moreover, $\forall t \leq 0$ we have
$$\mathbb{M}_X(t) = PG_X(e^t).$$

(ii) Compute PG_X when $X \sim \text{Bin}(n, p)$, $X \sim \text{Geom}(p)$, $X \sim \text{Poi}(\lambda)$.

□

Exercise 1.32. Show that
$$\text{Gamma}(\nu_0, \lambda) * \text{Gamma}(\nu_1, \lambda) = \text{Gamma}(\nu_0 + \nu_1, \lambda), \quad \forall \nu_0, \nu_1 > 0, \quad (1.6.6a)$$
$$N(0, v_0) * N(0, v_1) = N(0, v_0 + v_1), \quad \forall v_0, v_1 > 0, \quad (1.6.6b)$$
$$\text{Poi}(\lambda_0) * \text{Poi}(\lambda_1) = \text{Poi}(\lambda_0 + \lambda_1), \quad \forall \lambda_0, \lambda_1 > 0. \quad (1.6.6c)$$

Hint. Use Theorem 1.107, Corollary 1.105 and Corollary 1.134. □

Exercise 1.33. Let $\mu_0, \mu_1 \in \text{Prob}([0,1])$ be two Borel probability measures. Prove that the following statements are equivalent.

(i)
$$\int_0^1 x^n \, \mu_0[\,dx\,] = \int_0^1 x^n \, \mu_1[\,dx\,], \quad \forall n \in \mathbb{N}.$$

(ii) For any Borel subset $B \subset [0,1]$
$$\mu_0[\,B\,] = \mu_1[\,B\,].$$

□

Exercise 1.34. Denote by $\text{Prob} = \text{Prob}(\mathbb{R}, \mathcal{B}_\mathbb{R})$ the space of probability measures on $(\mathbb{R}, \mathcal{B}_\mathbb{R})$. Show that (Prob, μ) is a commutative semigroup with unit δ_0, the Dirac measure concentrated at 0. □

Exercise 1.35. Consider the interval $[-\pi/2, \pi/2]$ equipped with the probability measure
$$\mathbb{P}[\,dx\,] = \frac{1}{\pi} \lambda[\,dx\,],$$
$\lambda =$ the Lebesgue measure. We regard the function
$$X : [-\pi/2, \pi/2] \to \mathbb{R}, \quad X(t) = \sin^2 t$$
as a random variable on this probability spaces. Prove that $X \sim \text{Beta}(1/2, 1/2)$. □

Exercise 1.36. For any $a, b > 0$ we define the *incomplete Beta function*
$$B_{a,b} : (0,1) \to \mathbb{R}, \quad B_{a,b}(x) = \frac{1}{B(a,b)} \int_0^x t^{a-1}(1-t)^{b-1} dt,$$
where $B(a,b)$ is the Beta function (A.1.2).

(i) Prove that
$$\frac{x^a(1-x)^b}{aB(a,b)} = B_{a,b}(x) - B_{a+1,b}(x). \tag{1.6.7a}$$

$$\frac{x^a(1-x)^b}{bB(a,b)} = B_{a,b+1}(x) - B_{a,b}(x). \tag{1.6.7b}$$

(ii) Show that if $k, n \in \mathbb{N}$, $k < n$ we have
$$B_{k,n+1-k}(x) = \sum_{a=k}^n \binom{n}{a} x^a (1-x)^{n-a}. \tag{1.6.8}$$

Exercise 1.37. Suppose that $X_1, \ldots, X_n : (\Omega, \mathcal{S}, \mu) \to \mathbb{R}$ are random variables with joint probability distribution

$$\mathbb{P}_{X_1,\ldots,X_n}[\, dx_1 \cdots dx_n \,] = p(x_1, \ldots, x_n) dx_1 \cdots dx_n,$$

$$p \geq 0, \quad \int_{\mathbb{R}^n} p(x_1, \ldots, x_n) dx_1 \cdots dx_n = 1.$$

Consider the new random variables

$$Y_i = \sum_{k=1}^n a_{ij} X_j, \quad a_{ij} \in \mathbb{R}$$

where the matrix $A = (a_{ij})_{1 \leq i,j \leq n}$ is invertible with inverse $A^{-1} = (a^{ij})_{1 \leq i,j \leq n}$. Prove that the joint distribution of $Y_1, \ldots, Y_n$ is given by the density

$$q(y_1, \ldots, y_n) = \frac{1}{|\det A|} p(a^{11} y_1 + \cdots + a^{1,n} y_n, \ldots, a^{n1} y_1 + \cdots + a^{nn} y_n). \quad \square$$

Exercise 1.38. Suppose that $X_1, \ldots, X_N$ are independent standard normal random variables. For $n = 1, \ldots$, we denote by R_n^2 the random variable $X_1^2 + \cdots + X_n^2$.

(i) Prove that

$$R_n^2 \sim \chi^2(n) := \mathrm{Gamma}(\nu, \lambda), \quad \nu = \frac{n}{2}, \quad \lambda = \frac{1}{2}.$$

(ii) Prove that

$$\frac{R_n^2}{R_N^2} \sim \mathrm{Beta}(a, b), \quad \text{where} \quad a = \frac{n}{2}, \quad b = \frac{N-n}{2}.$$

(iii) Set

$$\overline{X} := \frac{1}{n}(X_1 + \cdots + X_n), \quad S^2 := \frac{1}{n-1} \sum_{i=1}^n (X_i - \overline{X})^2.$$

Prove that $(n-1) S^2 \sim \chi^2(n-1)$.

(iv) Set

$$T_n := \frac{\overline{X}}{S/\sqrt{n}}.$$

Prove that $T_n \sim \mathrm{Stud}_{n-1}$, where Stud_p denotes the *Student t-distribution* with p degrees of freedom

$$\mathrm{Stud}_p = \frac{1}{\sqrt{p\pi}} \frac{\Gamma(\frac{p+1}{2})}{\Gamma(\frac{p}{2})} \frac{1}{(1 + t^2/p)^{(p+1)/2}} dt, \quad t \in \mathbb{R}, \quad p > 0.$$

$\square$

Exercise 1.39. Fix a probability space $(\Omega, \mathcal{S}, \mathbb{P})$. Show that $L^0(\Omega, \mathcal{S}, \mathbb{P})$ equipped with the metric dist defined in (1.3.53) is a complete metric space. More precisely, show that if a sequence of random variables $X_n \in L^0(\Omega, \mathcal{S}, \mathbb{P})$ is Cauchy in probability, i.e.,

$$\lim_{m,n\to\infty} \mathbb{P}\big[\, |X_m - X_n| > r \,\big] = 0, \quad \forall r > 0,$$

then there exists a random variable $X \in L^0(\Omega, \mathcal{S}, \mathbb{P})$ such that $X_n \to X$ in probability. □

Exercise 1.40. Prove the claim in Remark 1.158. □

Exercise 1.41. Suppose that X, Y are independent random variables with distributions $\mathbb{P}_X$ and respectively $\mathbb{P}_Y$. Let $f : \mathbb{R}^2 \to \mathbb{R}$ be a Borel measurable function such that $f(X, Y)$ is integrable. Show that

$$\mathbb{E}\big[\, f(X, Y) \,\|\, X \,\big] = h(X),$$

where

$$h(x) = \int_{\mathbb{R}} f(x, y) \mathbb{P}_Y \big[\, dy \,\big].$$

□

Exercise 1.42. Suppose that $(\Omega, \mathcal{S}, \mathbb{P})$ is a probability space, $\mathcal{F} \subset \mathcal{S}$ a sigma-subalgebra and $X \in \mathcal{L}^0(\Omega, \mathcal{S})$, $Y \in \mathcal{L}^0(\Omega, \mathcal{F})$. Prove that the following are equivalent.

(i) $X = Y$ a.s.
(ii) For any bounded Borel measurable function $f : \mathbb{R} \to \mathbb{R}$, $\mathbb{E}\big[\, f(X) \,\|\, \mathcal{F} \,\big] = f(Y)$ a.s.

□

Exercise 1.43. Suppose that the sequence of independent random variables $(X_n)_{n\in\mathbb{N}}$ converges in probability. Prove that it is a.s. constant. □

Exercise 1.44. For $n \in \mathbb{N}$ we denote by C_n the cone in $\mathbb{R}^n$ defined by

$$C_n := \big\{\, (x_1, \ldots, x_n) \in \mathbb{R}^n : \, x_1 \leq x_2 \leq \cdots \leq x_m \,\big\}.$$

Define ord $: \mathbb{R}^n \to C_n$

$$(x_1, \ldots, x_n) \mapsto \operatorname{ord}(x_1, \ldots, x_n) = (x_{(1)}, x_{(2)}, \ldots, x_{(n)}),$$

where

$$x_{(1)} = \min\{x_1, \ldots, x_n\}, \quad x_{(2)} = \min\big(\{x_1, \ldots, x_n\} \setminus \{x_{(1)}\}\big), \ldots.$$

In other words, $x_{(1)}, \ldots, x_{(n)}$ are the numbers $x_1, \ldots, x_n$ rearranged in increasing order.

Suppose $X_1, \ldots, X_n$ are n i.i.d. random variables with common cdf
$$F(x) = \int_{-\infty}^x p(s)ds, \ p \in L^1(\mathbb{R}, \boldsymbol{\lambda}).$$
The *order statistics* of the random variables $X_1, \ldots, X_n$ is the random vector
$$\text{ord}(\boldsymbol{X}) := (X_{(1)}, \ldots, X_{(n)}),$$
where $\boldsymbol{X} = (X_1, \ldots, X_n)$.

(i) Show that the distribution of $\text{ord}(\boldsymbol{X})$ is
$$\mathbb{P}_{\text{ord}(\boldsymbol{X})}[dx_1 \cdots dx_n] = n! p(x_1) \cdots p(x_n) \boldsymbol{I}_{C_n}(x_1, \ldots, x_n) dx_1 \cdots dx_n.$$

(ii) Denote by $F_{(j)}$ the cdf of the component $X_{(j)}$, $F_{(j)}(x) = \mathbb{P}[X_{(j)} \leq x]$. Prove that
$$F_{(j)}(x) = \sum_{k=j}^n \binom{n}{k} F(x)^k \big(1 - F(x)\big)^{n-k}.$$

(iii) Suppose that $X_1, \ldots, X_n \sim \text{Unif}(0,1)$. Show that
$$X_{(j)} \sim \text{Beta}(j, n+1-j), \ \mathbb{E}[X_{(j)}] = \frac{j}{n+1}.$$

(iv) Suppose that $X_1, \ldots, X_n \sim \text{Unif}(0,1)$ and consider the random vector
$$Y = (X_{(2)}, \ldots, X_{(n)}).$$
Compute the conditional distribution of Y given $X_{(1)}$
$$\mathbb{P}_Y[dy_2 \cdots dy_n \| X_{(1)} = x].$$

(v) Suppose that $X_1, \ldots, X_n \sim \text{Exp}(\lambda)$. Show that[15]
$$X_{(1)} \sim \text{Exp}(n\lambda), \ \mathbb{E}[X_{(1)}] = \frac{1}{n\lambda}.$$

(vi) Suppose that $X_1, \ldots, X_n \sim \text{Exp}(\lambda)$. Show that
$$nX_{(1)}, \ (n-1)(X_{(2)} - X_{(1)}), \ldots, 2(X_{(n-1)} - X_{(n-2)}), \ X_{(n)} - X_{(n-1)}$$
are independent $\text{Exp}(\lambda)$ random variables. **Hint.** Use (i) and Exercise 1.37.

□

Exercise 1.45. Suppose that $X_1, \ldots, X_{n-1}$ are independent and uniformly distributed in $[0,1]$. Consider their order statistics
$$X_{(1)} \leq \cdots \leq X_{(n-1)}$$
and the corresponding spacings[16]
$$S_1 = X_{(1)}, \ S_2 = X_{(2)} - X_{(1)}, \ldots, S_n = 1 - X_{(n-1)}.$$
Denote by L_n the largest spacing, $L_n = \max(S_1, \ldots, S_n)$.

---

[15] To appreciate how surprising then conclusion (v) think that an institution buys a large number n of computers, all of the same brand, and $X_1, \ldots, X_n$ denote the lifetimes of these machines. Each is expected to last $1/\lambda$ years. The random variable $X_{(1)}$ is the lifetime of the first computer that breaks down. The result in (v) show that we should expect the first break down pretty soon, in $\frac{1}{n\lambda}$ years!

[16] The $n-1$ points $X_1, \ldots, X_{n-1}$ divide the interval $[0,1]$ into n subintervals and the spacings are the lengths of these subintervals.

(i) Prove that $(S_1,\ldots,S_n)$ is uniformly distributed in the simplex
$$\Delta_n := \left\{ (s_1,\ldots,s_n) \in [0,1]^n;\ \sum_{k=1}^n s_k = 1 \right\}.$$
Deduce that $\mathbb{E}[S_k] = \frac{1}{n}$, $\forall k = 1,\ldots,n$.

(ii) Use (i) to show that
$$\mathbb{E}[L_n] = \frac{1}{n}\sum_{k=1}^n (-1)^{k+1} \frac{1}{k}\binom{n}{k}.$$

(iii) Let $Y_1,\ldots,Y_n$ be independent $\mathrm{Exp}(1)$ random variables. Set $T_n = Y_1 + \cdots + Y_n$. Find the joint distribution of $(Y_1,\ldots,Y_m,T_n)$ and show that the random variables
$$\frac{Y_1}{T_n},\ldots,\frac{Y_n}{T_n}$$
has the same joint distribution as $S_1,\ldots,S_n$. Deduce that L_n has the same distribution as
$$\frac{\max_{1\le k\le n} Y_k}{T_n}.$$

(iv) Prove that L_n and
$$\frac{1}{T_n}\sum_{k=1}^n \frac{Y_k}{k}$$
have the same distribution. **Hint.** Use (iii) and Exercise 1.44(vi). Deduce that[17]
$$\mathbb{E}[L_n] := \frac{1}{n}\sum_{k=1}^n \frac{1}{k}.$$

□

Remark 1.202. Observe that the above exercise produces a strange identity,
$$\sum_{k=1}^n \frac{1}{k} = \sum_{k=1}^n (-1)^{k+1} \frac{1}{k}\binom{n}{k}.$$

□

Exercise 1.46. Consider the Poisson process $(N(t))_{t\ge 0}$ with intensity λ described in Example 1.136.

(i) Find the distribution of $W_t = N(t) + 1 - t$.
(ii) Show that $N(t+h) - N(t) \sim \mathrm{Poi}\,\lambda h$, $t \ge 0$, $h > 0$.

□

---

[17] This equality shows that $\mathbb{E}[L_n] \sim \frac{\log n}{n}$, which substantially higher than the mean of each individual spacing, $\mathbb{E}[S_k] = \frac{1}{n}$, $\forall k$.

Exercise 1.47. Consider the Poisson process $(N(t))_{t\geq 0}$ with intensity λ described in Example 1.136. Let S be a nonnegative random variable independent of the arrival times $(T_n)_{n\geq 0}$ of the Poisson process. For any arrival time T_n we denote by $Z_{T_n,S}$ the number of arrival times located in the interval $(T_n, T_n + S]$

$$Z_{T_n,S} := \#\{k > n;\ T_n < T_k \leq T_n + S\}.$$

Prove that

$$\mathbb{P}\big[\,Z_{T_n,S} = k\,\big] = \int_0^\infty e^{-k\lambda s}\frac{(\lambda s)^k}{k!}\mathbb{P}_S\big[\,ds\,\big]. \qquad \square$$

Exercise 1.48. Suppose that $N(t)$ is a Poisson process (see Example 1.136) with intensity λ and arrival times

$$T_1 \leq T_2 \leq \cdots.$$

Fix $t > 0$ and let $(X_n)_{n\geq 1}$ be i.i.d. random variables uniformly distributed in $[0, t]$. Prove that, conditional on $N(t) = n$, the random vectors

$$(T_1, \ldots, T_n)\ \text{and}\ (X_{(1)}, \ldots, X_{(n)})$$

have the same distribution. $\qquad \square$

Exercise 1.49. Suppose that the 20 contestants at a quiz show are each given the same question, and that each answers it correctly, independently of the others, with probability P. But the difficulty of the question is that P itself is a random variable.[18] Suppose, for the sake of illustration, that P is uniformly distributed over the interval $(0, 1]$.

(i) What is the probability that exactly two of the contestants answer the question correctly?
(ii) What is the expected number of contestants that answer a question correctly?

$\qquad \square$

Exercise 1.50 (Skhorohod). Denote by $\mathrm{Prob}^0(\mathbb{R})$ the set of Borel probability measures on $\mathbb{R}$ such that

$$\int_{\mathbb{R}} x\mu[\,dx\,] = 0.$$

Clearly $\mathrm{Prob}^0(\mathbb{R})$ is a convex subset of the set $\mathrm{Prob}(\mathbb{R})$ of Borel probability measures on $\mathbb{R}$.

For $u, v \geq 0$ such that $u + v > 0$ we define the bipolar measure

$$\beta_{u,v} := \frac{v}{u+v}\delta_{-u} + \frac{u}{u+v}\delta_v \in \mathrm{Prob}^0(\mathbb{R}).$$

Let $Q := \{\,(u,v) \in \mathbb{R}^2;\ u, v \geq 0,\ u+v \geq 0\,\}$. We regard $\beta_{u,v}$ as a random measure (or Markov kernel) $\beta : Q \times \mathcal{B}_{\mathbb{R}} \to \mathbb{R}$

$$\beta\big((u,v), B\big) = \beta_{u,v}\big[B\big].$$

---
[18]Think of P as a random Bernoulli measure of the kind discussed in Example 1.174.

Prove that for any $\mu \in \mathrm{Prob}^0(\mathbb{R})$ there exists a Borel probability measure ν on Q such that $\mu := \beta_* \nu$. In other words, any measure $\mu \in \mathrm{Prob}^0(\mathbb{R})$ is a mixture of bipolar measures. □

Exercise 1.51. Given sigma algebras $\mathcal{F}_\pm, \mathcal{F}_0, \subset \mathcal{S}$, prove that the following are equivalent.

(i) $\mathcal{F}_+ \perp\!\!\!\perp_{\mathcal{F}_0} \mathcal{F}_-$.
(ii) $\mathcal{F}_+ \perp\!\!\!\perp_{\mathcal{F}_0} \mathcal{F}_0 \vee \mathcal{F}_-$.

□

Exercise 1.52. Given sigma algebras $\mathcal{F}_\pm, \mathcal{F}_0, \subset \mathcal{S}$, prove that the following are equivalent.

(i) $\mathcal{F}_+ \perp\!\!\!\perp \mathcal{F}_0 \vee \mathcal{F}_-$
(ii) $\mathcal{F}_+ \perp\!\!\!\perp \mathcal{F}_0$ and $\mathcal{F}_+ \perp\!\!\!\perp_{\mathcal{F}_0} \mathcal{F}_-$.

□

Exercise 1.53. Suppose that μ is a Borel probability measure on the metric space (X, d). Denote by $\mathcal{C}$ the collection of Borel subsets S of X satisfying the regularity property: for any ε_0 there exists a closed subset $C_\varepsilon \subset S$ and an open subset $\mathcal{O}_\varepsilon \supset S$ such that
$$\mu[\mathcal{O}_\varepsilon \setminus C_\varepsilon] < \varepsilon.$$

(i) Show that $S \in \mathcal{C} \Rightarrow S^c := X \setminus C \in \mathcal{C}$.
(ii) Show that any closed set belongs to C.[19]
(iii) Show that $\mathcal{C}$ is a π-system.
(iv) Show that $\mathcal{C}$ is a λ-system.
(v) Show that $\mathcal{C}$ coincides with the family of Borel subsets.

□

Exercise 1.54. Suppose that (X, d) is a compact metric space and μ is a finite Borel measure on X. Prove that for any $p \in [1, \infty)$ the space $C(X)$ of continuous functions on X is dense in $L^p(X, \mu)$. **Hint.** Use Exercise 1.53 to show that for any Borel subset $B \subset X$ the indicator function $\boldsymbol{I}_B$ can be approximated in L^p by continuous functions. □

---

[19]This is where the fact is a X metric space plays an important role.

Chapter 2

Limit theorems

The limit theorems have preoccupied mathematicians from the dawns of probability. The first law of large numbers goes back to Jacob Bernoulli at the end of the seventeenth century. The Golden Theorem in his *Ars Conjectandi* is what we call today a weak law of large numbers. Bernoulli considers an urn that contains a large number of black and white balls. If $p \in (0,1)$ is the proportion of white balls in the urn and we draw with replacement a large number n of balls, then the proportion p_n of white balls among the extracted ones is with high confidence within a given open interval containing p.

His result lacked foundations since the concept of probability lacked a proper definition. The situation improved at the beginning of the twentieth century when E. Borel proved a strong form of Bernoulli's law. Borel too lacked a good definition of a probability space, but he worked rigorously. In modern terms, he used the interval $[0,1]$ with the Lebesgue measure as probability space. He then proceeded to construct explicitly a sequence of functions $X_n : [0,1] \to \mathbb{R}$ which, viewed as random variables are i.i.d. with common distribution $\text{Bin}(1/2)$.

It took the efforts of Hinchin and Kolmogorov to settle the general case. The strong law of large numbers states that if $(X_n)_{n \in \mathbb{N}}$ are i.i.d. random variables with finite mean μ, then the empirical mean

$$M_n = \frac{1}{n} \sum_{k=1}^{n} X_n$$

converges a.s. to the theoretical mean μ.

This chapter is devoted to these limit theorems. In the first section we investigate the SLLN = Strong Law of Large Numbers. The approach we use is due to Kolmogorov. It reduces this law to the convergence of random series of independent random variables.

The second section is devoted to the Central Limit Theorem stating that the distribution of M_n is very close to the distribution of a Gaussian random variable with the same mean and variance as M_n. The third section, is more modern, and it is devoted to concentration inequalities. These state in a quantitative fashion that the probability that M_n deviates from the mean μ by a certain amount is extremely small under certain conditions. The fourth section is devoted to uniform limit theorem of the Glivenko-Cantelli type. We have included this section due to its applications in machine learning. In particular, we show how such results coupled with the concentration inequalities lead to **P**robably **A**pproximatively **C**orrect, or PAC, learning.

The last section of this chapter is devoted to a brief introduction to the Brownian motion. This is such a fundamental object that we thought that any student of probability ought to make its acquaintance as soon as possible. As always, along the way we present many, we hope, interesting examples.

2.1 The Law of Large Numbers

This section is devoted to the (Strong) Law of Large Numbers. We follow Kolmogorov's approach based on random series, a subject of independent interest.

2.1.1 *Random series*

Fix a probability space $(\Omega, \mathcal{S}, \mathbb{P})$ and consider a sequence of *independent* random variables

$$X_n : (\Omega, \mathcal{S}, \mathbb{P}) \to \mathbb{R}, \quad n \in \mathbb{N}.$$

The independence of the random variables (X_n) allows us to invoke Kolmogorov's 0-1 theorem and conclude that the random series

$$\sum_{n \in \mathbb{N}} X_n \qquad (2.1.1)$$

either converges almost surely, or diverges almost surely. We want to describe by describing one simple sufficient condition for convergence.

Theorem 2.1 (Kolmogorov's one series). *Suppose that*

$$\mathbb{E}[X_n] = 0, \quad \forall n \in \mathbb{N}, \qquad (2.1.2a)$$

$$\sum_{n \geq 1} \mathrm{Var}[X_n] < \infty. \qquad (2.1.2b)$$

Then the series (2.1.1) converges almost surely and in L^2.

Proof. For $n \in \mathbb{N}$ we denote by S_n the n-th partial sum of the series (2.1.1),

$$S_n := \sum_{k=1}^{n} X_k.$$

The L^2-convergence follows immediately from (2.1.2b) which, coupled with the independence of the random variables (X_n) implies that the sequence (S_n) is Cauchy in L^2 since

$$\|S_{n+k} - S_n\|_{L^2}^2 = \sum_{j=1}^{k} \mathrm{Var}[X_{n+j}], \quad \forall k, n \in \mathbb{N}.$$

The proof of the a.s. convergence is more difficult. It relies on a fundamental inequality which we will further generalize in the next chapter. The independence of the random variables (X_n) is used crucially in its proof.

Lemma 2.2 (Kolmogorov's maximal inequality). *Set*
$$M_n := \max_{1 \leq k \leq n} |S_k|.$$
Then, for all $a > 0$, we have
$$\mathbb{P}[\, M_n > a \,] \leq \frac{1}{a^2} \operatorname{Var}[S_n] = \frac{1}{a^2} \sum_{k=1}^n \operatorname{Var}[X_k]. \tag{2.1.3}$$

Proof of Kolmogorov's maximal inequality. Define
$$N : \Omega \to \mathbb{N} \cup \{\infty\}, \quad N(\omega) := \inf\{\, n \geq 1;\ |S_n(\omega)| > a \,\}.$$
Notice that $N(\omega)$ is the first $n \in \mathbb{N} \cup \{\infty\}$ such that $S_n(\omega) > a$, i.e.,
$$N(\omega) = k \iff S_1(\omega), \ldots, S_{k-1}(\omega) \leq a \text{ and } S_k(\omega) > a.$$
This shows that the event $A_k = \{N = k\}$ is in the σ-algebra generated by $X_1, \ldots, X_k$. Since $S_n - S_k = X_{k+1} + \cdots + X_n$ we deduce that $\boldsymbol{I}_{A_k}$, $\boldsymbol{I}_{A_k} S_k$ are independent of $S_n - S_k$. We have
$$\operatorname{Var}[\, S_n \,] = \mathbb{E}[\, S_n^2 \,] \geq \mathbb{E}[\, S_n^2 \boldsymbol{I}_{\{M_n \geq a\}} \,]$$
$$= \sum_{k=1}^n \mathbb{E}[\, \boldsymbol{I}_{A_k} S_n^2 \,] = \sum_{k=1}^n \mathbb{E}\Big[\, \boldsymbol{I}_{A_k} \big(S_k^2 + 2S_k(S_n - S_k) + (S_n - S_k)^2 \big) \,\Big]$$
$(\boldsymbol{I}_{A_k}, \boldsymbol{I}_{A_k} S_k \perp\!\!\!\perp S_n - S_k)$
$$= \sum_{k=1}^n \Big(\mathbb{E}[\, \boldsymbol{I}_{A_k} S_k^2 \,] + 2\mathbb{E}[\, \boldsymbol{I}_{A_k} S_k \,] \underbrace{\mathbb{E}[\, S_n - S_k \,]}_{=0} + \underbrace{\mathbb{E}[\, \boldsymbol{I}_{A_k} \,] \mathbb{E}[\, (S_n - S_k)^2 \,]}_{\geq 0} \Big)$$
$$\geq \sum_{k=1}^n \underbrace{\mathbb{E}[\, \boldsymbol{I}_{A_k} S_k^2 \,]}_{S_k^2 \geq a^2 \text{ on } A_k} \geq a^2 \sum_{k=1}^n \mathbb{P}[A_k] = a^2 \mathbb{P}[M_n \geq a \,].$$
$\square$

We can now complete the proof of Theorem 2.1. Using Kolmogorov's maximal inequality for the sequence $(X_{m+n})_{n \in \mathbb{N}}$ we deduce that for any $n \in \mathbb{N}$ we have
$$\mathbb{P}\left[\max_{1 \leq k \leq n} |S_{m+k} - S_m| > \varepsilon \right] \leq \frac{1}{\varepsilon^2} \operatorname{Var}[\, S_{m+n} - S_m \,] = \frac{1}{\varepsilon^2} \sum_{k=1}^n \operatorname{Var}[X_{m+k}]$$
$$\leq \frac{1}{\varepsilon^2} \underbrace{\sum_{k \geq 1} \operatorname{Var}[X_{m+k}]}_{=: r_m}.$$

Thus
$$\mathbb{P}\left[\sup_{n\geq 1}|S_{m+n}-S_m|>\varepsilon\right]\leq \frac{r_m}{\varepsilon^2}. \qquad (2.1.4)$$

We set
$$Y_m := \sup_{i,j\geq m}|S_i-S_j|, \quad Z_m := \sup_{n\geq 1}|S_{m+n}-S_m|.$$

Now observe that S_m converges a.s. iff $Y_m \to 0$ a.s. The sequence Y_m is nonincreasing and thus it converges a.s. to a random variable $Y \geq 0$. We will show that $Y = 0$ a.s.

Note that, for $i, j > m$ we have
$$|S_i-S_j| \leq |S_i-S_m|+|S_j-S_m| \leq 2Z_m,$$
so $Y_m \leq 2Z_m$, $\forall m$ so
$$Y_m > 2\varepsilon \Rightarrow Z_m > \varepsilon \Rightarrow \mathbb{P}[Y_m > 2\varepsilon] \leq \mathbb{P}[Z_m > \varepsilon].$$

The equality (2.1.4) reads
$$\mathbb{P}[Z_m > \varepsilon] \leq \frac{r_m}{\varepsilon^2} \quad \forall m \geq 1, \forall \varepsilon > 0.$$

Hence
$$\lim_{m\to\infty}\mathbb{P}[Y_m>\varepsilon] = \lim_{m\to\infty}\mathbb{P}[|Z_m|>\varepsilon] = 0.$$

On the other hand, for any $\varepsilon > 0$ we have
$$0 \leq Y \leq Y_m, \ \forall m \Rightarrow \mathbb{P}[Y>\varepsilon] \leq \mathbb{P}[Y_m>\varepsilon], \ \forall m.$$

We conclude that
$$\mathbb{P}[Y>\varepsilon] = 0, \ \forall \varepsilon > 0 \Rightarrow Y = 0 \text{ a.s.}$$

$\square$

Example 2.3. Consider a sequence of i.i.d. Bernoulli random variables $(N_k)_{n\in\mathbb{N}}$ with success probability $\frac{1}{2}$. The resulting random variables $R_k = (-1)^{N_k}$ are called *Rademacher random variables* and take only the values ± 1 with equal probabilities.

We obtain the random series
$$\sum_{n\geq 1}\frac{(-1)^{N_k}}{k} = \sum_{n\geq 1}\frac{R_n}{n}.$$

Loosely speaking, this is a version of the harmonic series with random signs
$$\pm 1 \pm \frac{1}{2} \pm \frac{1}{3} \pm \cdots, \qquad (2.1.5)$$

where the $\pm$-choices at any term are equally likely and also independent of the choices at the other terms of the series. We set
$$X_k = \frac{(-1)^{N_k}}{k}.$$

We know that if all the terms are positive, a probability zero event, then we obtain the harmonic series which is divergent. On the other hand,

$$\mathbb{E}[X_k] = 0, \quad \mathrm{Var}[X_k] = \frac{1}{k^2}.$$

Since

$$\sum_{k \geq 1} \frac{1}{k^2} < \infty,$$

we deduce from Kolmogorov's one series theorem that the series

$$\sum_{k \geq 1} X_k$$

is a.s. convergent. Thus, if we flip a fair coin with two sides, a + side and a − side and we assign the signs in (2.1.5) according to the coin flips, the resulting series is convergent with probability 1! □

Remark 2.4. Kolmogorov also established necessary and sufficient conditions for convergence in his *three series theorem*. Before we state it let us introduce a convenient notation. For any random variable X and any positive constant C we denote by X^C the truncation

$$X^C := X\boldsymbol{I}_{\{|X| \leq C\}} = \begin{cases} X, & |X| \leq C, \\ 0, & |X| > C. \end{cases} \tag{2.1.6}$$

Theorem 2.5 (Kolmogorov's three series theorem). *Consider a sequence of independent random variables $X_n \in \mathcal{L}^0(\Omega, \mathcal{S}, \mathbb{P})$. The following statements are equivalent.*

(i) The series

$$\sum_{n \geq 1} X_n \tag{2.1.7}$$

converges almost surely.

(ii) For any $C > 0$ the following three series are convergent.

$$\sum_{n \geq 1} \mathbb{P}[|X_n| > C] = \sum_{n \geq 1} \mathbb{P}[X_n \neq X_n^C].$$

$$\sum_{n \geq 1} \mathbb{E}[X_n^C].$$

$$\sum_{n \geq 1} \mathrm{Var}[X_n^C].$$

For a proof we refer to [33, Sec. 3.7] or [140, IV.§2]. □

2.1.2 The Law of Large Numbers

The frequentist interpretation of probability asserts that the probability of an event is roughly the frequency of the occurrence of that event in a very large number of independent trials. The Law of Large Numbers formalizes this intuition. The surprising thing, at least to this author, is that reality respects the theory so closely: the Law of Large Numbers adds a surprising level of predictability to uncertainty!

Throughout this section $(X_n)_{n\geq 1}$ is a sequence of iid random variables $X_n \in L^1(\Omega, \mathcal{S}, \mathbb{P})$. Set

$$\mu := \mathbb{E}[X_n], \quad S_n := X_1 + \cdots + X_n.$$

The various versions of the Law of Large Numbers state that the empirical means S_n/n converge in an appropriate sense to the theoretical mean μ. The convergence in probability is usually referred to as the *Weak Law of Large Numbers* (or WLLN) while the a.s. convergence is known as the *Strong Law of Large Numbers* (or SLLN). We begin by presenting a few special, but historically important, cases.

Theorem 2.6 (Markov). *If $X_n \in L^2(\Omega, \mathcal{S}, \mathbb{P})$, then $\frac{1}{n}S_n \to \mu$ in probability.*

Proof. We use the same strategy as in Example 1.149. Denote by σ^2 the common variance of the random variables X_n. Since they are independent we have $\mathrm{Var}[S_n] = n\sigma^2$, so

$$\mathrm{Var}[S_n/n] = \frac{1}{n^2}\mathrm{Var}[S_n] = \frac{1}{n\sigma^2}.$$

Let $\varepsilon > 0$. Note that $\mathbb{E}[S_n/n] = \mu$. Chebyshev's inequality (1.3.17) implies

$$\mathbb{P}[|S_n/n - \mu| > \varepsilon] \leq \frac{1}{n\sigma^2\varepsilon^2} \to 0 \text{ as } n \to \infty.$$

Thus $S_n/n \to \mu$ in probability. $\square$

Theorem 2.7 (Cantelli). *If $X_n \in L^4(\Omega, \mathcal{S}, \mathbb{P})$, then $\frac{1}{n}S_n \to \mu$ almost surely.*

Proof. By replacing X_n with $Y_n := X_n - \mu$ we can assume $\mu = 0$. We set

$$\sigma^2 := \mu_2[X_k], \quad r^4 := \mu_4[X_k], \quad M_n := S_n/n.$$

Note that

$$\mathbb{P}[|M_n| > \varepsilon] = \mathbb{P}[|M_n|^4 > \varepsilon^4] \leq \frac{1}{\varepsilon^4}\mathbb{E}[M_n^4] = \frac{1}{n^4\varepsilon^4}\mathbb{E}[S_n^4].$$

Observe that

$$\mathbb{E}[S_n^4] = \sum_{i,j,k,\ell=1}^{n} \mathbb{E}[X_i X_j X_k X_\ell]. \tag{2.1.8}$$

Let $i \neq j$. Due to the independence of the random variables $(X_n)_{n\in\mathbb{N}}$ we have

$$\mathbb{E}[X_i^2 X_j^2] = \mathbb{E}[X_i^2]\mathbb{E}[X_j^2] = \sigma^4, \quad \mathbb{E}[X_i X_j^3] = \mathbb{E}[X_i]\mathbb{E}[X_j^3] = 0.$$

Similarly, for distinct i, j, k, ℓ, we have
$$\mathbb{E}\left[X_i X_j X_k X_\ell \right] = 0.$$
Thus
$$\mathbb{E}\left[S_n^4 \right] = nr^4 + 2\binom{4}{2}\sum_{j<k}\sigma^4 = nr^4 + 6\binom{n}{2}\sigma^4 = O(n^2) \text{ as } n \to \infty.$$
Hence
$$\mathbb{P}\left[|M_n| > \varepsilon \right] = O\left(\frac{1}{n^2 \varepsilon^4}\right) \text{ as } n \to \infty$$
so that, for any $\varepsilon > 0$,
$$\sum_{n \geq 1} \mathbb{P}\left[|M_n| > \varepsilon \right] < \infty.$$
Corollary 1.141 implies that $M_n \to 0$ a.s. $\square$

Remark 2.8. The above Strong Law of Large Numbers is not the most general, but its proof makes the role of independence much more visible. More precisely the independence, or the small correlations force the fourth moment of S_n to be "unnaturally" small and thus the large fluctuations around the mean are highly unlike, i.e. the $\mathbb{P}\left[|M_n| > \varepsilon \right]$ is very small for large n. $\square$

The next result, due to Kolmogorov, generalizes both results above.

Theorem 2.9 (The Strong Law of Large Numbers). *Suppose that $(X_n)_{n \geq 1}$ is a sequence of iid random variables $X_n \in L^1(\Omega, \mathcal{S}, \mathbb{P})$. Then*
$$\lim_{n \to \infty} \frac{1}{n} S_n = \mu \text{ a.s.}$$

Proof. We accomplish this in several steps.

Step 1. Truncate. Set
$$Y_n := X_n \mathbf{I}_{\{|X_n| < n\}}, \quad T_n := Y_1 + \cdots + Y_n.$$
We claim that
$$\mathbb{P}\left[X_n \neq Y_n \text{ i.o.} \right] = 0. \tag{2.1.9}$$
Indeed, since the random variables (X_n) are identically distributed we have
$$\sum_{k \geq 1} \mathbb{P}\left[|X_k| > k \right] = \sum_{k \geq 1} \mathbb{P}\left[|X_1| > k \right] \leq \int_0^\infty \mathbb{P}\left[|X_1| > t \right] dt \stackrel{(1.3.43)}{=} \mathbb{E}\left[|X_1| \right] < \infty$$
and Borel-Cantelli's Lemma implies that
$$\mathbb{P}\left[|X_k| > k \text{ i.o.} \right] = 0.$$
This is equivalent to (2.1.9). We deduce from (2.1.9) that
$$\lim_{n \to \infty} \frac{1}{n} |S_n - T_n| = 0 \text{ a.s.}$$

Thus, it suffices to show that

$$\lim_{n\to\infty} \frac{1}{n} T_n = \mu \quad \text{a.s.} \qquad (2.1.10)$$

Step 2. Centering. The sequence $\left(\mathbb{E}[Y_k]\right)_{k\geq 1}$ converges to $\mu = \mathbb{E}[X]$ as $k \to \infty$. Indeed, since the random variables are identically distributed we have

$$\mathbb{E}[Y_k] = \mathbb{E}[X_k I_{\{|X_k|\leq k\}}] = \mathbb{E}[X_1 I_{\{|X_1|\leq k\}}] \to \mathbb{E}[X_1],$$

where at the last step we used the Dominated Convergence theorem. It follows that the sequence $\mathbb{E}[Y_n]$ is also *Cèsaro convergent*[1] to the same limit, i.e.,

$$\lim_{n\to\infty} \frac{1}{n} \mathbb{E}[T_n] = \lim_{n\to\infty} \frac{1}{n} \sum_{k=1}^{n} \mathbb{E}[Y_k] = \mu.$$

Thus, it suffices to prove that

$$\lim_{n\to\infty} \left(\frac{1}{n} \sum_{k=1}^{n} Y_k - \frac{1}{n} \sum_{k=1}^{n} \mathbb{E}[Y_k] \right) = 0, \quad \text{a.s.}$$

$$Z_n := Y_n - \mathbb{E}[Y_n].$$

We have to prove that the *Cèsaro means* of Z_n converge to 0 a.s., i.e.,

$$\lim_{n\to\infty} \frac{1}{n} \sum_{k=1}^{n} Z_k = 0 \quad \text{a.s.} \qquad (2.1.11)$$

Step 3. Conclusion. We will rely on the following elementary result.

Lemma 2.10 (Kronecker's Lemma). *Suppose that $(a_n)_{n\in\mathbb{N}}$ and $(x_n)_{n\in\mathbb{N}}$ are sequences of real numbers satisfying the following conditions.*

(i) The sequence (a_n) is increasing, positive and unbounded.
(ii) The series $\sum_{n\geq 1} \frac{x_n}{a_n}$ is convergent.

Then

$$\lim_{n\to\infty} \frac{1}{a_n} \sum_{k=1}^{n} x_k = 0.$$

Assume temporarily the validity of Kronecker's lemma. Thus, to prove (2.1.11) it suffices to show that the random series

$$\sum_{n\geq 1} \frac{Z_n}{n}$$

is a.s. convergent. The independence assumption will finally play a role because we will invoke the one-series theorem. Clearly the random variables $\frac{Z_n}{n}$ are independent. We claim that

$$\sum_{k\geq 1} \frac{\text{Var}[Z_k]}{k^2} < \infty. \qquad (2.1.12)$$

---
[1] Use Exercise 2.3 with $p_{k,n} = 1/n$.

We have
$$\operatorname{Var}[Z_k] = \operatorname{Var}[Y_k] = \mathbb{E}[Y_k^2] - \mathbb{E}[Y_k]^2 \leq \mathbb{E}[Y_k^2]$$
$$\stackrel{(1.3.43)}{=} \int_0^\infty 2y\mathbb{P}[|Y_k| > y]\,dy = \int_0^\infty 2y\mathbb{P}[k \geq |X_k| > y]\boldsymbol{I}_{\{y<k\}}\,dy$$
$$\leq \int_0^\infty 2y\mathbb{P}[|X_k| > y]\boldsymbol{I}_{\{y<k\}}\,dy.$$

Thus
$$\sum_{k\geq 1} \frac{\operatorname{Var}[Z_k]}{k^2} \leq \sum_{k\geq 1} \frac{1}{k^2}\int_0^\infty 2y\mathbb{P}[|X_k| > y]\boldsymbol{I}_{\{y<k\}}\,dy$$

$$= \int_0^\infty \left(\sum_{k\geq 1}\frac{1}{k^2}\boldsymbol{I}_{\{y\leq k\}}\right) 2y\mathbb{P}[|X_1| > y]\,dy = \int_0^\infty \underbrace{\left(\sum_{k\geq y}\frac{1}{k^2}\right) 2y}_{=:w(y)}\,\mathbb{P}[|X_1| > y]\,dy.$$

We claim that
$$w(y) < 6, \quad \forall y \geq 0. \tag{2.1.13}$$

Indeed, for $y \leq 1$ we have
$$w(y) = 2y\sum_{k\geq 1}\frac{1}{k^2} \leq 4y < 4.$$

For $y \in (1, 2]$ we have
$$w(y) = 2y\sum_{k\geq 2}\frac{1}{k^2} < 2y \leq 4.$$

For $y > 2$ we have
$$\sum_{k\geq y}\frac{1}{k^2} \leq \int_{\lfloor y\rfloor - 1}^\infty \frac{1}{t^2}\,dt = \frac{1}{\lfloor y\rfloor - 1}$$

so
$$w(y) \leq \frac{2y}{\lfloor y\rfloor - 1} \leq \frac{2\lfloor y\rfloor + 2}{\lfloor y\rfloor - 1} = 2 + \frac{4}{\lfloor y\rfloor - 1} < 6.$$

Using (2.1.13) we deduce
$$\sum_{k\geq 1}\frac{\operatorname{Var}[Z_k]}{k^2} < 6\int_0^\infty \mathbb{P}[|X_1| > y]\,dy = 6\mathbb{E}[|X_1|] < \infty.$$

This proves (2.1.12) and completes the proof of the SLLN, assuming Lemma 2.10. $\square$

Proof of Lemma 2.10. Set

$$y_n := \frac{x_n}{a_n}, \;\; s_0 = a_0 := 0, \;\; s_n := \sum_{k=1}^n y_k, \;\; n \geq 1,$$

so that the sequence $(s_n)_{n\geq 1}$ is convergent. We have to show that

$$\lim_{n\to\infty} \frac{1}{a_n} \sum_{k=1}^n a_k y_k = 0.$$

We have[2]

$$\sum_{k=1}^n a_k y_k = \sum_{k=1}^n a_k (s_k - s_{k-1}) = a_n s_n - \sum_{k=1}^n s_{k-1}(a_k - a_{k-1}).$$

Now set

$$w_k := a_k - a_{k-1}, \;\; p_{n,k} := \frac{w_k}{a_n}.$$

Since $(a_n)_{n\in\mathbb{N}}$ is increasing, positive and unbounded we deduce

$$\sum_{k=1}^n p_{n,k} = 1, \;\; \forall n \geq 1, \;\; \lim_{n\to\infty} p_{n,k} = 0, \;\; \forall k. \tag{2.1.14}$$

Observe that

$$\frac{1}{a_n} \sum_{k=1}^n a_k y_k = s_n - \sum_{k=1}^n p_{k,n} s_{k-1}.$$

The conditions (2.1.14) imply that (see Exercise 2.3)

$$\lim_{n\to\infty} \sum_{k=1}^n p_{k,n} s_{k-1} = \lim_{n\to\infty} s_n.$$

□

Since a.s. convergence implies convergence in probability we deduce from the SLLN the *Weak Law of Large Numbers* (or WLLN).

Corollary 2.11. *Suppose that* $X_n \in L^1(\Omega, \mathcal{S}, \mathbb{P})$, $n \in \mathbb{N}$, *is a sequence of i.i.d. random variables with common mean* μ. *We set*

$$S_n = \sum_{k=1}^n X_k.$$

Then the empirical mean $\frac{1}{n} S_n$ *converges in probability to* μ. □

---
[2]This is classically known as Abel's trick. It is a discrete version of the integration by parts trick.

Remark 2.12. Let us observe that in Theorem 2.6 the random variables X_n need not be independent or identically distributed. Assuming all have mean 0, all we need for that for the Weak Law of Large Numbers to hold is that the random variables are pairwise uncorrelated,

$$\mathbb{E}[X_m X_n] = \mathbb{E}[X_m]\mathbb{E}[X_n], \quad \forall m \neq n, \tag{2.1.15}$$

and the only constraint on their distribution is

$$\sup_n \mathbb{E}[X_n^2] < \infty.$$

In Exercise 2.6 we ask the reader to show that the WLLN holds even if we assume something weaker than (2.1.15) namely that if $|m - n| \gg 1$, the random variables X_m and X_n are weakly correlated. More precisely

$$\lim_{k \to \infty} \sup_{m \in \mathbb{N}} |\mathbb{E}[X_m X_{m+k}]| = 0.$$

Similarly, for the strong Strong Law of Large Numbers to hold we do not need the variables to be independent. The theorem continues to hold if the variables are identically distributed, integrable and only pairwise independent. For a proof we refer to [53, Sec. 2.4].

The arguments in the proof Theorem 2.7 show that the SLLN holds even when the variables X_n are neither independent, nor identically distributed. Assuming that all the variables have mean zero, the SLLN holds if any four of them are independent, and the only assumptions about their distributions is

$$\sup_n \mathbb{E}[X_n^4] < \infty.$$

A natural philosophical question arises. What makes the Law of Large Numbers possible? The above discussion suggests that it is a consequence of a mysterious interplay between some form of independence and some "asynchronicity": their fluctuations around the mean cannot be in resonance. These features can be observed in the other Laws of Large Numbers we will discuss in this text.

If the random variables are independent, but not necessarily identically distributed, there are known necessary and sufficient conditions for the WLLN to hold. We refer to [59, IX], [70, §22.], or [127, Chap. 4] for details. □

Remark 2.13. Suppose that $(X_n)_{n \geq 1}$ is a sequence of i.i.d. variables. This Strong Law of Large Numbers shows that if they have finite mean μ, then the empirical means

$$M_n = \frac{1}{n}(X_1 + \cdots + X_n)$$

converge a.s. to μ. If $\mu = \infty$ and M_n converge a.s. to a random variable M_∞, then M_∞ is a.s. constant. Exercise 2.9 outlines a proof of this fact. □

Example 2.14. Suppose we roll a fair die a large number n of times and we denote by S_n the number of times we roll a 1. Intuition tells us that if the die is fair, then for large n, the fraction of times we get a 1 should be close to $\frac{1}{6}$, i.e.,

$$\frac{S_n}{n} \approx \frac{1}{6} \text{ for } n \gg 0.$$

This follows from the SLLN. Indeed, the above experiment is encoded by a sequence $(X_n)_{n \in \mathbb{N}}$ of i.i.d. Bernoulli random variables with success probability $p = \frac{1}{6}$. Then

$$S_n = \sum_{k=1}^{n} X_k,$$

and the SLLN

$$\frac{S_n}{n} \to \mathbb{E}[X_1] = \frac{1}{6} \text{ a.s. as } n \to \infty.$$

It helps to visualize a computer simulation of such an experiment. Suppose we roll a die a large number N of times. For $i = 1, \ldots, N$ we denote by f_i the frequency of 1-s during the first i trials, i.e.,

$$f_i = \frac{S_i}{i}.$$

The resulting vector $(f_i)_{1 \leq i \leq N} \in \mathbb{R}^N$ is called relative or cumulative frequency.

The R-code below simulates one such experiment when we roll the die 12,000 times.

```
N<-12000
x<-sample(1:6, N, replace=TRUE)
rolls<-x==1
rel_freq<-cumsum(rolls)/(1:N)

plot(1:N,rel_freq,type="l", xlab="Number of rolls",

ylab="The frequency of occurrence of 1",
    main="Average number 1-s during random rolls of die")
abline(h=1/6,col="red")
```

The output is a plot of the collection of points (i, f_i) depicted in Figure 2.1. □

Example 2.15 (The Monte-Carlo method). Consider a box (parallelepiped)

$$B_k := I_1 \times \cdots \times I_k \subset \mathbb{R}^k$$

where $I_1, \ldots, I_k \subset \mathbb{R}$ are nontrivial bounded intervals. Consider independent random variables $X_1, \ldots, X_k$, where X_j is uniformly distributed on I_j. The the probability distribution of the random vector $\boldsymbol{X} = (X_1, \ldots, X_k)$ is

$$\frac{1}{\lambda_k[B_k]} I_{B_k} \lambda_k,$$

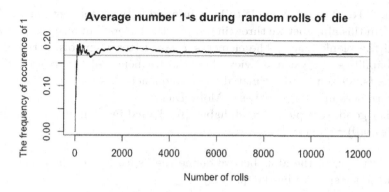

Fig. 2.1 The frequencies f_i fluctuates wildly initially and then stabilizes around the horizontal line $y = 1/6$ in perfect agreement with SLLN.

where we recall that $\boldsymbol{\lambda}_k$ denotes the Lebesgue measure on $\mathbb{R}^k$. If $f : B_k \to \mathbb{R}$ is integrable, then

$$\frac{1}{\boldsymbol{\lambda}_k[B_k]} \int_{B_k} f(\boldsymbol{x}) \boldsymbol{\lambda}_k(d\boldsymbol{x}) = \mathbb{E}\big[\, f(\boldsymbol{X})\,\big].$$

Suppose that $\boldsymbol{X}_n = (X_{n,1}, \ldots, X_{n,k})$, $n \in \mathbb{N}$, is a sequence of i.i.d. random vectors uniformly distributed in B_k, then the sequence of random variables $(f(\boldsymbol{X}_n))_{n \in \mathbb{N}}$ is i.i.d., with the same distribution as $f(\boldsymbol{X})$. The SLLN implies that the sequence random variables

$$Z_n = \frac{1}{n}\big(f(\boldsymbol{X}_1) + \cdots + f(\boldsymbol{X}_n) \big)$$

converges a.s. to

$$\frac{1}{\boldsymbol{\lambda}_k[B_k]} \int_{B_k} f(\boldsymbol{x}) \boldsymbol{\lambda}_k(d\boldsymbol{x}).$$

This fact can be used to produce approximations to integrals using probabilistic methods. When the dimension k is large these methods are, to this day, the only viable methods for approximating integrals of functions of many variables.

In Example A.21 we describe a computer implementation of this strategy using the programming language R. □

2.1.3 Entropy and compression

Let us describe a surprising application of the law of large numbers. Suppose that we are given a finite set $\mathscr{X}$ equipped with a probability measure $\mathbb{P}$ defined by the function $p : \mathscr{X} \to [0,1]$

$$p(x) := \mathbb{P}\big[\,\{x\}\,\big].$$

We will refer to the pair $(\mathscr{X}, p)$ as *alphabet*.

Example 2.16. A good example to have in mind is the "alphabet" of the English language. In this alphabet we throw in not just the letters, but also the punctuation signs and the blank space. The elements x_i are letters/symbols of the alphabet. The probabilities $p(x_i)$ can be viewed as the frequency of the symbol x_i in the written texts. One way to estimated these frequencies[3] is to count the number of their occurrences in a large text, say Moby Dick.

Another good example is the alphabet $\{0,1\}$ used in computer languages. The frequencies $p(0) = p(1) = \frac{1}{2}$. □

For a letter x_i of the alphabet we define the "surprise" or "information" contained in the letter x_i to be the quantity

$$S(x_i) := -\log_2 p(x_i).$$

The base 2 of the logarithm is the convention used in information theory and we will stick with it. The unit of measure of surprise/information is the *bit*. Note that $S(x_i) \in [0, \infty]$. Observe that the less likely the letter x_i, the bigger the surprise. The *Shanon entropy* or the *information entropy* of the alphabet is the quantity

$$\operatorname{Ent}_2[p] := \mathbb{E}_p[S] := -\sum_{x \in \mathscr{X}} p(x) \log_2 p(x), \tag{2.1.16}$$

where we adhere to the convention $0 \cdot \log 0 = 0$. Thus the entropy is the expected "surprise" of the alphabet. For example, if an urn contains 99 black balls and only one white ball. We would be extremely surprised if when we randomly draw a ball from the urn it urns out to be the white one. The average amount of surprise in this case is

$$-0.99 \log_2(0.99) - 0.01 \log_2(0.01) \approx 0.08.$$

If p_0 is the uniform probability measure on $\mathscr{X}$, then

$$\operatorname{Ent}_2[p_0] = \log_2 |\mathscr{X}|.$$

Let $m := |\mathscr{X}|$. Note that $\operatorname{Prob}(\mathscr{X})$ can be identified with the $(m-1)$-dimensional simplex

$$\Delta_m = \{ p = (p_1, \ldots, p_m) \in [0, \infty)^m; \ p_1 + \cdots + p_m = 1 \}.$$

We can view the entropy as a function $\operatorname{Ent}_2 : \Delta_{m-1} \to [0, \infty)$. One can check that it is concave since the function $[0, \infty) \ni x \mapsto f(x) = -x \log_2 x$ is strictly concave. We have

$$\operatorname{Ent}_2[p] = \sum_{i=1}^{m} f(p_i).$$

Jensen's inequality shows that

$$\frac{1}{m} \sum_{i=1}^{m} f(p_i) \leq f\left(\frac{1}{m} \sum_{i=1}^{m} p_i \right) = f(1/m) = \frac{\log_2 m}{m},$$

---

[3]As a curiosity, the letter "e" is the most frequent letter of he English language; it appears 13% of the time in large texts. It is for this reason that it has the simplest Morse code, a dot.

with equality if and only if $p_1 = \cdots = p_m = \frac{1}{m}$. We deduce
$$\mathrm{Ent}_2[p] \leq \log_2 |\mathscr{X}|, \quad \forall p \in \mathrm{Prob}(\mathscr{X}), \tag{2.1.17}$$
with equality if and only if p is the uniform probability measure. We will see later that the above is a special case of the Gibbs' inequality (2.3.8). Intuitively, this inequality says that among all the probability measures on a finite set, the uniform one is the most "chaotic", the least "predictable".

We will refer to the elements of $\mathscr{X}^n$ as *words* of length n. The term "word" is a bit misleading. For example, when $\mathscr{X}$ is the English alphabet as above, an element of $\mathscr{X}^n$ with large n can be thought of as the sequence of symbols appearing in a large text. On the other hand, we can think of $\mathscr{X}^n$ itself as a new alphabet with frequencies
$$p_n(x_1, \ldots, x_n) = p(x_1) \cdots p(x_n).$$
The amount of "surprise" of a word $(x_1, \ldots, x_n)$ is
$$S(x_1, \ldots, x_n) = \sum_{k=1}^{n} S(x_k).$$
The entropy of $(\mathscr{X}^n, p_n)$ is
$$\mathrm{Ent}_2[p_n] = n \, \mathrm{Ent}_2[p].$$
We denote by $\mathscr{X}^*$ the disjoint union of the sets $\mathscr{X}^n$,
$$\mathscr{X}^* = \bigsqcup_{n \in \mathbb{N}} \mathscr{X}^n,$$
and we will refer to it as the *vocabulary* of the alphabet $\mathscr{X}$.

Fix and alphabet $(\mathscr{X}, p)$. We want to describe an efficient way of encoding the words in $\mathscr{X}^n$ by words in the vocabulary of the binary alphabet $\mathcal{B} := \{0, 1\}$. Thus, we want to construct a code map $\mathcal{C} : \mathscr{X}^n \to \mathcal{B}^*$ such that the words $x \in \mathscr{X}^n$ with high frequency are encoded by words in $\mathcal{B}^*$ of short length. Normally we would require that $\mathcal{C}$ be injective but we are willing to sacrifice precision a bit for the sake of efficiency. We would be happy if the probability that two different words have the same code is very small, i.e., the event
$$x, x' \in \mathscr{X}^n, \quad x \neq x' \text{ and } \mathcal{C}(x) = \mathcal{C}(x)$$
has a very small probability.

Definition 2.17. Let $\varepsilon > 0$. The ε-*typical set* $A_\varepsilon^{(n)}$ with respect to $p(x)$ is the set $A_\varepsilon^{(n)} \subset \mathscr{X}^n$ consisting of words $(x_1, x_2, \ldots, x_n)$ with the property
$$2^{-n(\mathrm{Ent}_2[p]+\varepsilon)} \leq p(x_1, x_2, \ldots, x_n) \leq 2^{-n(\mathrm{Ent}_2[p]-\varepsilon)}. \tag{2.1.18}$$
$\square$

Theorem 2.18 (Asymptotic Equipartition Property). *For any $\varepsilon > 0$ there exists $N = N(\varepsilon)$ such that for any $n > N(\varepsilon)$, the following hold.*

(i) $p_n[A_\varepsilon^{(n)}] > 1 - \varepsilon$.
(ii) $|A_\varepsilon^{(n)}| \leq 2^{n(\text{Ent}_2[p]+\varepsilon)}$.
(iii) $|A_\varepsilon^{(n)}| \geq (1-\varepsilon)2^{n(\text{Ent}_2[p]-\varepsilon)}$.

Proof. We sample $(\mathscr{X}, p)$ it according to the frequencies $p(x_k)$ and we obtain a sequence $(X_n)_{n \in \mathbb{N}}$ of i.i.d. $\mathscr{X}$-valued random variables distributed according to p. We obtain random words $(X_1, \ldots, X_n)$, $n \in \mathbb{N}$. The average amount of surprise per letter in this word is

$$\frac{1}{n}S(X_1, \ldots, X_n) = \frac{1}{n}\sum_{k=1}^{n} S(X_k).$$

The law of large numbers shows that the random variables $\frac{1}{n}S(X_1, \ldots, X_n)$ converge in probability to $\text{Ent}_2[p]$. Now observe that

$$(X_1, \ldots, X_n) \in A_\varepsilon^{(n)} \iff \text{Ent}_2[p] - \varepsilon \leq \frac{1}{n}\sum_{k=1}^{n} S(X_k) \leq \text{Ent}_2[p] + \varepsilon$$

so

$$\mathbb{P}_n[A_\varepsilon^{(n)}] = \mathbb{P}\Big[\text{Ent}_2[p] - \varepsilon \leq \frac{1}{n}\sum_{k=1}^{n} S(X_k) \leq \text{Ent}_2[p] + \varepsilon\Big] \to 1$$

as $n \to \infty$. Fix $N = N(\varepsilon)$ such that

$$p_n[A_\varepsilon^{(n)}] > 1 - \varepsilon, \quad \forall n > N(\varepsilon).$$

Note that for $n > N(\varepsilon)$

$$1 = \sum_{x \in \mathscr{X}^n} p(x) \geq \sum_{x \in A_\varepsilon^{(n)}} p_n(x) \geq 2^{-n(\text{Ent}_2[p]+\varepsilon)}|A_\varepsilon^{(n)}|,$$

and thus we have

$$|A_\varepsilon^{(n)}| \leq 2^{n(\text{Ent}_2[p]+\varepsilon)}.$$

Finally, for $n > N(\varepsilon)$ we have

$$1 - \varepsilon < \mathbb{P}_n[A_\varepsilon^{(n)}] \leq \sum_{x \in A_\varepsilon^{(n)}} 2^{-n(\text{Ent}_2[p]-\varepsilon)} = 2^{-n(\text{Ent}_2[p]-\varepsilon)}|A_\varepsilon^{(n)}|,$$

and conclude that $|A_\varepsilon^{(n)}| \geq (1-\varepsilon)2^{n(\text{Ent}_2[p]-\varepsilon)}$. □

The *Asymptotic Equipartion Property* (or AEP) shows that a typical set has probability nearly 1, all its elements are nearly equiprobable, and its cardinality is nearly $2^{n\,\text{Ent}_2[p]}$. The inequality (2.1.17) shows that if p is not he uniform probability measure on $\mathscr{X}$, then

$$2^{\text{Ent}_2[p]} \ll |\mathscr{X}|.$$

Hence, if $\varepsilon > 0$ is sufficiently small, then

$$\frac{|A_\varepsilon^{(n)}|}{|\mathscr{X}^n|} \to 0$$

exponentially fast as $n \to \infty$. That is, the typical sets have high probability and are "extremely small" if the entropy is small.

This suggests the following coding procedure. Fix $\varepsilon > 0$ so that $1 - \varepsilon$ will be our confidence level. For $n > N(\varepsilon)$ the set $A_\varepsilon^{(n)}$ has about 2^L elements where $L = \lceil n \operatorname{Ent}_2[p] \rceil$ elements and thus we can find an injection

$$\mathcal{I} : A_\varepsilon^{(n)} \to \mathcal{B}^L.$$

For $x \in A_\varepsilon^{(n)}$ we attach the symbol 1 at the beginning of the word $\mathcal{I}(x) \in \mathcal{B}^L$ and the resulting word in $\mathcal{B}^{L+1}$ will encode x. It uses $L + 1$ bits. The first bit is 1 and indicates that the word x is typical.

We are less careful with the atypical words. Chose *any* map

$$\mathcal{J} : \mathcal{X}^n \setminus A_\varepsilon^{(n)} \to \mathcal{B}^L$$

and we encode an atypical word x using the binary word $\mathcal{J}(x)$ with a prefix 0 attached to indicate that it is atypical. The resulting map $\mathcal{C} : \mathcal{X}^n \to \mathcal{B}^{L+1}$ is not injective, but if two words have the same code, they must be atypical and thus occur with very small frequency. This is an example of *compression*.

Take for example the English language. There are various estimates for its entropy, starting with the pioneering word of Claude Shannon. Most recent ones[4] vary from 1 to 1.5 bits. How do we encode efficiently texts consisting of $n = 10^6$ symbols say? For example, "*Moby Dick*" has $206,052$ words and the average length of an English word is 5 letters so "*Moby Dick*" consists of about 1.03 million symbols.

Forgetting capitalization and punctuation there are 26^n such texts and a brute encoding would require 26^n codewords to cover all the possibilities. The above result however says that roughly $2^{1.5n}$ texts suffice to capture nearly surely almost everything. The term compression is fully justified since this is a much smaller fraction of the total number of possible texts. Also we only need codewords of lengths 1.5 million. Thus we need is roughly 1.5 gigabits to encode such a text. If the letters of the alphabet where uniformly distributed in human texts[5] then the entropy would be $\log_2(26) \approx 4.70 > 3 \times 1.5$ and we would need more than three times amount of memory to store it.

Remark 2.19. The story does not end here and much more precise results are available. To describe some of them note first that for any alphabet $\mathcal{X}$ there is an obvious operation of concatenation

$$* : \mathcal{X}^m \times \mathcal{X}^n \to \mathcal{X}^{m+n}, \quad (x, x') \mapsto x * x'$$

where the word $x * x'$ is obtained by writing in succession the word x followed by x'. Note that this code uses on average $\frac{L+1}{n} \approx \operatorname{Ent}_2[p]$ bits per symbol in a word. This is an example of compression.

---

[4] A Google search with the keywords "entropy of the English language" will provide many more details on this subject.

[5] The famous monkey on a typewriter produces texts where the letters are uniformly distributed, but we can safely call the resulting texts highly atypical of the English texts humans are used to.

A *binary code* for the alphabet $(\mathcal{X}, p)$ is an injection

$$C : \mathcal{C} \to \mathcal{B}^*.$$

For each $x \in \mathcal{X}$ we denote by $L_C(x)$ the length of the code word $C(x)$. The expected length of a codeword is

$$\ell_C := \mathbb{E}[L_C] = \sum_{x \in \mathcal{X}} L_C(x) p(x).$$

Note that C extends to a map

$$C^* : \mathcal{X}^* \to \mathcal{B}^*, \quad C^*(x_1, \ldots, x_n) = C(x_1) * \cdots * C(x_n).$$

The code C is called *uniquely decodable* if its extension $C^* : \mathcal{X}^* \to \mathcal{B}^*$ is also injective.

An important subclass of uniquely decodable codes are *instantaneous codes*. A code C is called *instantaneous* if no codeword is a prefix of some other code word. E.g., if one of the codewords is 10, then no other codeword can begin with 10.

Here is a very revealing example. Consider an alphabet $\mathcal{A}$ consisting of four letters $\mathcal{A} := \{a, b, c, d\}$ with frequencies

$$p_a = 1/2, \quad p_b = 1/3, \quad p_c = p_d = 1/12.$$

Consider the following instantaneous code

$$a \to 1, \quad b \to 01, \quad c \to 001 \quad d \to 000.$$

The expected code length is

$$\frac{1}{2} + \frac{2}{3} + \frac{3}{12} + \frac{3}{12} = \frac{5}{3} \approx 1.666.$$

The entropy of alphabet is

$$\mathrm{Ent}_2[\mathcal{A}] = \frac{\log 2}{2} + \frac{\log 3}{3} + \frac{\log 12}{6} \approx 1.625.$$

Kraft's inequality shows that for any uniquely decodable code C we have

$$\ell_C \geq \mathrm{Ent}_2[\mathcal{A}].$$

Moreover, there exist optimal codes C such that

$$\ell_C \leq \mathrm{Ent}_2[\mathcal{A}] + 1.$$

Such codes are called *Shannon codes*. The above code is a Shannon code. In fact it is a special example of the famous *Huffman code*, [37].

Let us discuss a particularly suggestive experiment that highlights a defining feature of Huffman codes and reveals one interpretation of the entropy of an alphabet.

Suppose we have an urn containing the letters a, b, c, d, in proportions p_a, p_b, p_c, p_d. A person randomly draws a letter from the urn and you are supposed to guess what it is by asking YES/NO question. Think YES = 1, NO = 0. The above code describes an optimal guessing strategy. Here it is.

(1) Ask first if the letter is $a \to 1$. If the answer is YES ($= 1$), the game is over. The game has length 1 with probability $1/2$.

(01) If the answer is NO ($= 0$) the letter can only be b, c or d. Ask if the letter is $b \to 01$. If the answer is YES ($= 1$) the game is over. The game has length 2 with probability $1/3$.

(001) If the answer is NO ($= 0$) ask if the letter is $c \to 001$. The game has length 3 with probability $1/6$.

For more details about information theory and its application we refer to [37; 112]. For a more informal introduction to information theory we refer to [60]. The eminently readable [71] contains historical perspective on the evolution of information theory. Kolmogorov's brief but very rich in intuition survey [95] is a good place to start learning about the mathematical theory of information. □

2.2 The Central Limit Theorem

The goal of this section is to prove a striking classical result that adds additional information to the Law of Large Numbers.

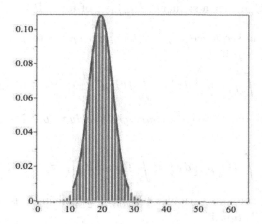

Fig. 2.2 *Visualizing the Central Limit Theorem.*

Suppose that $(X_n)_{n \in \mathbb{N}}$ is a sequence of i.i.d. random variables with mean μ and *finite* variance σ^2. Note that the sum $S_n := X_1 + \cdots + X_n$ has mean $n\mu$ and variance $n\sigma^2$. Loosely speaking, the central limit theorem states that for large n the probability distribution of S_n "resembles" very much a Gaussian with the same mean and variance.

For example, if the X_n-s are Bernoulli random variables with success probability p, then $\mu = p$, $\sigma^2 = pq$ and $S_n \sim \text{Bin}(n, p)$. In Figure 2.2 we have illustrated what happens in the case $p = 0.3$ and $n = 65$.

The vertical lines depict the probability mass function of the binomial distribution while the curve wrapping them is the Gaussian with the same mean and variance. They obviously do "resemble". However, we need to define precisely what we mean by "resemble".

2.2.1 Weak and vague convergence

Let (X, d) be a metric space. Denote by $\text{Meas}(X)$ the set of *finite* Borel measures on X and by $\text{Prob}(X) \subset \text{Meas}(X)$ the space of Borel probability measures on X.

We denote by $C_0(X)$ the space of continuous functions $X \to \mathbb{R}$ with compact support and by $C_b(X)$ the space of bounded continuous functions $X \to \mathbb{R}$. This is a Banach space with respect to the sup-norm

$$\|f\|_\infty := \sup_{x \in X} |f(x)|.$$

For any $f \in C_b(X)$ and $\mu \in \text{Meas}(X)$ we set

$$\mu[f] := \int_X f(x) \mu[dx] < \infty.$$

Definition 2.20. Consider a sequence $(\mu_n)_{n \in \mathbb{N}}$ of finite Borel measures on X.

(i) We say that the sequence (μ_n) *converges vaguely* to $\mu \in \text{Meas}(X)$, and we write this $\mu_n \dashrightarrow \mu$ if

$$\lim_{n \to \infty} \int_\mathbb{R} f(x) \mu_n[dx] = \int_\mathbb{R} f(x) \mu[dx], \quad \forall f \in C_0(X). \tag{2.2.1}$$

(ii) We say that the sequence (μ_n) *converges weakly* to $\mu \in \text{Meas}(X)$, and we write this $\mu_n \Rightarrow \mu$ if

$$\lim_{n \to \infty} \int_\mathbb{R} f(x) \mu_n[dx] = \int_\mathbb{R} f(x) \mu[dx], \quad \forall f \in C_b(X). \tag{2.2.2}$$

(iii) A sequence of random variables $(X_n)_{n \in \mathbb{N}}$ valued in X is said to *converge in law* or *in distribution* if

$$\mathbb{P}_{X_n} \Rightarrow \mathbb{P}_X \text{ in } \text{Prob}(X),$$

i.e.,

$$\lim_{n \to \infty} \mathbb{E}[f(X_n)] = \mathbb{E}[f(X)], \quad \forall f \in C_b(X). \tag{2.2.3}$$

We will use the notation $X_n \xrightarrow{d} X$ to indicate that X_n converges to X in distribution. □

Definition 2.21. A collection $F \subset C_b(X)$ is called *separating* if given $\mu_0, \mu_1 \in \text{Meas}(\mathbb{R})$ such that $\mu_0[f] = \mu_1[f], \forall f \in F$, then $\mu_0 = \mu_1$. □

As shown in Proposition 1.84, the collection $C_b(X)$ is separating so the above definition is not vacuous for any metric space.

In the remainder of the subsection we will focus exclusively on the special case when $X = \mathbb{R}^k$ equipped with its natural metric.

Lemma 2.22. *The collection $C_0(\mathbb{R}^k)$ is separating. More precisely, let μ_0, $\mu_1 \in \mathrm{Meas}(\mathbb{R}^k)$. If*

$$\mu_0[f] = \int_{\mathbb{R}} \mu_1[f], \ \forall f \in C_0(\mathbb{R}^k),$$

then $\mu_0 = \mu_1$.

Proof. According to Proposition 1.29 it suffices so show that

$$\mu_0[\overline{B}_r(x_0)] = \mu_1[\overline{B}_r(x)], \ \forall r > 0, \ x_0 \in \mathbb{R}^k,$$

where $\overline{B}_r(x_0)$ denotes the closed ball of radius r centered at x_0. Set

$$S_n := \{ x \in \mathbb{R}^k; \ \mathrm{dist}(x, x_0) \geq r + 1/n \}.$$

Fix $r > 0$, $x_0 \in \mathbb{R}^k$. For $n \in \mathbb{N}$ define $f_n : \mathbb{R} \to [0,1]$

$$f_n(x) = \frac{\mathrm{dist}(x, S_n)}{\mathrm{dist}(x, \overline{B}_r(x_0)) + \mathrm{dist}(x, S_n)}.$$

Observe that f_n is continuous, $\mathrm{supp}\, f_n = \overline{B}_{r+1/n}(x_0)$ and

$$\lim_{n \to \infty} f_n(x) = \boldsymbol{I}_{\overline{B}_r(x_0)} \quad \text{a.s.}$$

The Dominated Convergence Theorem implies that

$$\int_{\mathbb{R}} \boldsymbol{I}_{\overline{B}_r(x_0)}(x) \mu_0[dx] = \lim_{n \to \infty} \int_{\mathbb{R}} f_n(x) \mu_0[dx]$$

$$= \lim_{n \to \infty} \int_{\mathbb{R}} f_n(x) \mu_1[dx] = \int_{\mathbb{R}} \boldsymbol{I}_{\overline{B}_r(x_0)} \mu_1[dx].$$

□

Lemma 2.22 shows a sequence of Borel probability measures on $\mathbb{R}^k$ has at most one vague limit, i.e., if $\mu_n \dashrightarrow$ and $\mu_n \dashrightarrow \mu'$, then $\mu = \mu'$.

Example 2.23. Let

$$\mu_n = \frac{1}{n} \sum_{k=1}^{n} \delta_{k/n}.$$

Then

$$\mu_n \Rightarrow \mu = \boldsymbol{I}_{[0,1]}(x) dx \sim \mathrm{Unif}(0,1).$$

Indeed, if $f \in C_b(\mathbb{R})$, then

$$\int_{\mathbb{R}} f(x) \mu_n[dx] := \frac{1}{n} \sum_{k=1}^{n} f(k/n).$$

The sum in the right-hand-side of the above equality is a Riemann sum for f corresponding to the uniform partition

$$0 < \frac{1}{n} < \frac{2}{n} < \cdots < \frac{n-1}{n} < 1.$$

Since f is Riemann integrable we deduce

$$\lim_{n\to\infty} \frac{1}{n} \sum_{k=1}^{n} f(k/n) = \int_0^1 f(x)dx = \int_{\mathbb{R}} f(x)\mu[dx].$$

$\square$

Example 2.24. There exist vaguely convergent sequences of Borel probability measures on $\mathbb{R}$ that are not weakly convergent. Take for example $\mu_n = \delta_n$, $n \in \mathbb{N}$. Then $\mu_n \dashrightarrow 0$ yet μ_n does not converge weakly to 0 since $\mu_n[\mathbb{R}] = 1$, $\forall n$. $\square$

Theorem 2.25 (Mapping theorem). *Suppose that $F : \mathbb{R}^k \to \mathbb{R}^m$ is a continuous function and $X_n : (\Omega, \mathcal{S}, \mathbb{P}) \to \mathbb{R}^k$, $n \in \mathbb{N}$ is a sequence of random vectors converging in distribution to the random vector X. Then the sequence of random vectors $Y_n = F(X_n)$ converges in distribution to $Y = F(X)$.*

Proof. Let $f \in C_b(\mathbb{R}^m)$. Then $f \circ F \in C_b(\mathbb{R}^n)$ and

$$\mathbb{E}[f(Y_n)] = \mathbb{E}[f \circ F(X_n)] \to \mathbb{E}[f \circ F(X)] = \mathbb{E}[f(Y_n)].$$

$\square$

Proposition 2.26. *If the random variables X_n converge in probability to X, then they also converge in law to X. In particular, if X_n converge in p-mean to X, then they also converge in law to X.*

Proof. We deduce from Corollary 1.145 that for any $f \in C_b(\mathbb{R})$ the random variables $f(X_n)$ converge in probability to $f(X)$. The bounded convergence theorem implies

$$\lim_{n\to\infty} \mathbb{E}[f(X_n)] = \mathbb{E}[f(X)], \quad \forall f \in C_b(\mathbb{R}).$$

$\square$

Example 2.27. Fix a standard normal random variable X. Then $\mathbb{P}_X = \mathbb{P}_{-X}$ so $-X$ is a standard normal random variable as well. Consider the constant sequence

$$X_n = X, \quad n \in \mathbb{N}.$$

Then $\mathbb{P}_{X_n} \Rightarrow \mathbb{P}_{-X}$, but X_n does not converge to $-X$ in probability. $\square$

Theorem 2.28 (Portmanteau theorem). *Let $\mu_n \in \mathrm{Prob}(\mathbb{R}^k)$, $n \in \mathbb{N}$, be a sequence of Borel probability measures on $\mathbb{R}^k$. The following statements are equivalent.*

(i) *The sequence $(\mu_n)_{n\in\mathbb{N}}$ converges weakly to $\mu \in \mathrm{Meas}(\mathbb{R}^k)$.*

(ii) For any open set $U \subset \mathbb{R}^k$ we have
$$\mu[U] \leq \liminf \mu_n[U].$$
(iii) For any closet set $C \subset \mathbb{R}^k$ we have
$$\mu[C] \geq \limsup \mu_n[C].$$
(iv) For any Borel set $B \subset \mathbb{R}^k$ such that $\mu[\partial B] = 0$ we have
$$\mu[B] = \lim_{n \to \infty} \mu_n[B].$$

Proof. (i) $\Rightarrow$ (ii) According to Theorem 1.198 the measure μ is regular, i.e., for any $\varepsilon > 0$ there exists a closed set $C_\varepsilon \subset U$ such that
$$\mu[U] > \mu[C_\varepsilon] > \mu[U] - \varepsilon.$$
Consider the continuous function
$$f : \mathbb{R} \to [0,1], \quad f(x) = \frac{\operatorname{dist}(x, U^c)}{\operatorname{dist}(x, U^c) + \operatorname{dist}(x, C_\varepsilon)}.$$
Note that $f = 1$ on C_ε and $f = 0$ outside U we have
$$\mu_n[f] \leq \mu_n[U], \quad \forall n \in \mathbb{N} >.$$
In particular, we deduce that, $\forall \varepsilon > 0$, we have
$$\mu[U] - \varepsilon < \mu[C_\varepsilon] \leq \mu[f] = \lim_{n \to \infty} \mu_n[f] \leq \liminf_n \mu_n[U].$$
This proves (ii).

(ii) $\iff$ (iii) Follows from the following facts

- The set U is open iff U^c is closed.
- For any Borel set $B \subset \mathbb{R}$, $\mu[B^c] = 1 - \mu[B]$.

(ii) + (iii) $\Rightarrow$ (iv). Let $B \subset \mathbb{R}^k$ be a Borel set such that $\mu[\partial B] = 0$. Denote by U the interior of B and by C its closure so that $\partial B = C \setminus U$. We deduce
$$\mu[B] = \mu[C] = \mu[U].$$
Thus
$$\limsup \mu_n[C] \leq \mu[C] = \mu[B] = \mu[U] \leq \liminf \mu_n[U].$$
Since ∂B is closed we deduce
$$\limsup \mu_n[\partial B] \leq \mu[\partial B] = 0.$$
Hence
$$\mu_n[U] = \mu_n[C] + \mu_n[\partial B], \quad \lim_n \mu_n[\partial B] = 0,$$
so
$$\liminf \mu_n[U] = \liminf \mu_n[C].$$

Hence
$$\lim_n \mu_n[C] = \mu[C], \ \lim_n \mu_n[B] = \lim_n \mu_n[C] + \lim_n \mu_n[\partial B] = \mu[B].$$

(iv) $\Rightarrow$ (i). Clearly it suffices to show that $\mu_n[f] \to \mu[f]$, for any nonnegative, bounded, continuous function f on $\mathbb{R}^k$.

Suppose that f be such a function. Set $K = \sup f$. For any $\nu \in \operatorname{Prob}(\mathbb{R}^k)$ we can regard f as a random variable $(\mathbb{R}^k, \mathcal{B}_{\mathbb{R}^k}, \nu) \to \mathbb{R}$. The integral $\nu[f]$ is then the expectation of this random variable. Using Proposition 1.126 with $p=1$ we deduce that
$$\mathbb{E}_\nu[f] = \int_\mathbb{R} f(x)\nu[dx] = \int_\mathbb{R} \nu[f>t] = \int_0^K \nu[f>t]\, dt.$$

Note that
$$\nu[f=t] = 0 \Rightarrow \nu[\partial\{f>t\}] = 0.$$

Observe next that for any $n \in \mathbb{N}$ we have
$$\#\{t \in \mathbb{R};\ \nu[f=t] \geq 1/n\} \leq n,$$
so, for any $\nu \in \operatorname{Prob}(\mathbb{R}^k)$ the set
$$\{t \in \mathbb{R};\ \mu[f=t] > 0\}$$
is at most a countable set. We deduce from (iv) that
$$\lim_{n\to\infty} \mu_n[f>t] = \mu[f>t] \ \text{ for almost any } t.$$

From the Dominated Convergence Theorem we deduce
$$\lim_{n\to\infty} \mu_n[f] = \lim_{n\to\infty} \int_0^K \mu_n[f>t]\, dt = \int_0^K \mu[f>t]\, dt = \mu[f].$$

$\square$

Corollary 2.29. *Let X_n, $n \in \mathbb{N}$, be a sequence of random variables. Denote by $F_n(x)$ the cdf of X_n,*
$$F_n(x) = \mathbb{P}[X_n \leq x], \ x \in \mathbb{R}.$$
The following statements are equivalent.

(i) The random variables X_n converge in law to the random variable X.
(ii) If $F(x)$ is the cdf of X, then
$$\lim_{n\to\infty} F_n(x) = F(x),$$
for any point of continuity x of F.

Proof. Set $\mu_n := \mathbb{P}_{X_n}$, $\mu := \mathbb{P}_X$. The condition (ii) is a special case of condition (iv) of the Portmanteau Theorem so (i) $\Rightarrow$ (ii).

(ii) $\Rightarrow$ (i) Denote by $\mathscr{X} \subset \mathbb{R}$ the set of points continuity of F. Note that its complement $\mathbb{R} \setminus \mathscr{X}$ is at most countable so $\mathscr{X}$ is dense. Note that for $a, b \in \mathscr{X}$, $a < b$ we have
$$\mathbb{P}[\, a < X < b\,] = F(b) - F(a).$$
For any $a, b \in \mathbb{R}$, $a < b$ and any $\varepsilon > 0$ there exist $a_\varepsilon, b_\varepsilon \in \mathscr{X}$, $a < a_\varepsilon < b_\varepsilon < b$ such that
$$F(b_\varepsilon) - F(a_\varepsilon) = \mathbb{P}[\, a_\varepsilon < X < b_\varepsilon\,] > \mathbb{P}[\, a < X < b\,] - \varepsilon.$$
Hence
$$\lim_{n \to \infty} \bigl(F_n(b_\varepsilon) - F_n(a_\varepsilon) \bigr) = F(b_\varepsilon) - F(a_\varepsilon) > \mathbb{P}[\, a < X < b\,] - \varepsilon.$$
On the other hand
$$\mathbb{P}[\, a < X_n < b\,] \geq \mathbb{P}[\, a_\varepsilon < X_n < b_\varepsilon\,], \quad \forall n,$$
so that
$$\liminf_{n \to \infty} \mathbb{P}[\, a < X_n < b\,] \geq \mathbb{P}[\, a < X < b\,] - \varepsilon, \quad \forall \varepsilon > 0,$$
i.e.,
$$\liminf_{n \to \infty} \mathbb{P}[\, a < X_n < b\,] \geq \mathbb{P}[\, a < X < b\,], \quad \forall a < b \in \mathbb{R}.$$
Thus, the sequence μ_n satisfies the condition (ii) in the portmanteau theorem Theorem 2.28, where U is any open interval of the real axis. Since any open set of the real axis is a disjoint union of countably many open intervals, we deduce that condition (ii) in the Portmanteau Theorem is satisfied for *all* the open sets $U \subset \mathbb{R}$.
$\square$

Theorem 2.30 (Slutsky). *Suppose that $(X_n)_{n \in \mathbb{N}}$ and $(Y_n)_{n \in \mathbb{N}}$ are sequences of random variables such that (X_n) converges in distribution to X and Y_n converges in probability to $c \in \mathbb{R}$. Then the sum $X_n + Y_n$ converges in distribution to $X + c$.*

Proof. Without loss of generality we can assume $c = 0$. We follow the argument in [12, Chap. 1, Sec. 3]. Fix a closed subset $C \subset \mathbb{R}$. For $\varepsilon > 0$ set
$$C_\varepsilon := \bigl\{\, x \in \mathbb{R}; \ \mathrm{dist}(x, C) \leq \varepsilon \,\bigr\}.$$
The set C_ε is closed and we have
$$\mathbb{P}[\, X_n + Y_n \in C\,] \leq \mathbb{P}[\, |Y_n| > \varepsilon\,] + \mathbb{P}[\, X_n \in C_\varepsilon\,].$$
Letting $n \to \infty$ we deduce from the assumptions and the Portmanteau Theorem that
$$\limsup_{n \to \infty} \mathbb{P}[\, X_n + Y_n \in C\,] \leq \limsup_{n \to \infty} \mathbb{P}[\, X_n \in C_\varepsilon\,] \leq \mathbb{P}[\, X \in C_\varepsilon\,].$$
Now let $\varepsilon \searrow 0$ observing that $C_\varepsilon \searrow C$. $\square$

We can now formulate and prove the main convergence criterion of this subsection.

Theorem 2.31. *Suppose that $(\mu_n)_{n\in\mathbb{N}}$ is a sequence of finite Borel measures on $\mathbb{R}$. Fix a subset $\mathcal{F} \subset C_b(\mathbb{R})$ whose closure in $C_b(\mathbb{R})$ contains $C_0(\mathbb{R})$. The following statements are equivalent.*

(i) The sequence (μ_n) converges weakly to $\mu \in \mathrm{Meas}(\mathbb{R})$.

(ii) The sequence (μ_n) converges vaguely to $\mu \in \mathrm{Meas}(\mathbb{R})$ and

$$\mu[\mathbb{R}] = \lim_{n\to\infty} \mu_n[\mathbb{R}].$$

(iii) $\mu \in \mathrm{Meas}(\mathbb{R})$ and

$$\lim_{n\to\infty} \int_\mathbb{R} f(x)\mu_n[\,dx\,] = \int_\mathbb{R} f(x)\mu[\,dx\,], \quad \forall f \in \mathcal{F}, \tag{2.2.4}$$
$$\mu[\mathbb{R}] = \lim_{n\to\infty} \mu_n[\mathbb{R}].$$

Proof. Since

$$\mu[\mathbb{R}] = \int_\mathbb{R} \boldsymbol{I}_\mathbb{R} \mu[\,dx\,] = \lim_{n\to\infty} \int_\mathbb{R} 1 \mu_n[\,dx\,], \quad \forall \nu \in \mathrm{Meas}(\mathbb{R})$$

we can replace the measures μ_n by $\frac{1}{\mu_n[\mathbb{R}]}\mu_n$ and thus we can assume that all the measures μ_n are probability measures. In this case (ii) reads

μ_n *converges vaguely to μ and μ is a probability measure,*

while (iii) reads

μ *is a probability measure and $\mu_n[f] \to \mu[f]$, $\forall f \in \mathcal{F}$.*

Obviously (i) $\Rightarrow$ (ii) and (ii) $\Rightarrow$ (iii). It suffices to prove that (ii) $\Rightarrow$ (i) and (iii) $\Rightarrow$ (ii). We will need the following result.

Lemma 2.32. *Any finite Borel probability measure $\mu \in \mathrm{Meas}(\mathbb{R}^k)$ is on $\mathbb{R}^k$ is Radon, i.e., for any Borel set $B \subset \mathbb{R}$ and any $\varepsilon > 0$, there exists a compact set $K \subset B$ such that $\mu[B \setminus K] < \varepsilon$.*

Proof. Let $B \subset \mathbb{R}^k$ be a Borel set and $\varepsilon > 0$. According to Theorem 1.198, the measure μ is regular. Hence, there exists a closed set $C \subset B$ such that

$$\mu[B \setminus C] < \frac{\varepsilon}{2}.$$

On the other hand, we can find $R > 0$ sufficiently large such that

$$\mu[\,\overline{B}_R(0)\,] > \mu[\mathbb{R}^k] - \frac{\varepsilon}{2}.$$

We set $K := \overline{B}_R \cap C$. The set K is clearly compact and

$$\mu[C \setminus K] \leq \mu[\mathbb{R}^k \setminus \overline{B}_R(0)] < \frac{\varepsilon}{2}.$$

Thus $\mu[B \setminus K] = \mu[B \setminus K] + \mu[C \setminus K] < \varepsilon$. $\square$

(ii) ⇒ (i) We will show that the sequence (μ_n) satisfies the condition (ii) in the portmanteau theorem. Now let $U \subset \mathbb{R}^k$ be an open set and $\varepsilon > 0$. Lemma 2.32 shows that for any $\varepsilon > 0$ there exists a compact set $K \subset U$ such that

$$\mu[K] > \mu[U] - \varepsilon.$$

Now choose $r < \frac{1}{2} \operatorname{dist}(K, U^c)$ and set

$$C_r := \{ x \in \mathbb{R}^k;\ \operatorname{dist}(x, K) \geq r \}.$$

The set C_r is closed and its complement

$$V_r := \{ x \in \mathbb{R}^k;\ \operatorname{dist}(x, K) < r \} \subset U$$

is precompact. Consider the continuous function

$$\varphi : \mathbb{R}^k \to [0, \infty),\ \varphi(x) = \frac{\operatorname{dist}(x, C_r)}{\operatorname{dist}(x, K) + \operatorname{dist}(x, C_r)}.$$

Observe that it vanishes on C_r and thus it has compact support contained in U. Moreover, $\varphi = 1$ on K. Thus

$$\mu_n[K] \leq \mu_n[\varphi] \leq \mu_n[U].$$

Letting $n \to \infty$ we deduce

$$\mu[U] - \varepsilon < \mu[K] \leq \mu[\varphi] = \lim_n \mu_n[\varphi] \leq \liminf_n \mu_n[U],\ \forall \varepsilon > 0.$$

This establishes condition (ii) of the Portmanteau Theorem.

(iii) ⇒ (ii) Let $\varphi \in C_0(\mathbb{R}^k)$. For any $\varepsilon > 0$ choose $f_\varepsilon \in \mathcal{F}$ such that $\|\varphi - f_\varepsilon\|_\infty < \frac{\varepsilon}{2}$, i.e.,

$$f_\varepsilon - \frac{\varepsilon}{2} \leq \varphi \leq f_\varepsilon + \frac{\varepsilon}{2}.$$

Then

$$\mu_n[f_\varepsilon] - \frac{\varepsilon}{2} \leq \mu_n[\varphi] \leq \mu_n[f_\varepsilon] + \frac{\varepsilon}{2}.$$

Letting $n \to \infty$ we deduce

$$\mu[\varphi] - \varepsilon < \mu[f_\varepsilon] - \frac{\varepsilon}{2} = \lim \mu_n[f_\varepsilon] - \frac{\varepsilon}{2} \leq \liminf_n \mu_n[\varphi] \leq \limsup_n \mu_n[\varphi]$$

$$\leq \lim \mu_n[f_\varepsilon] + \frac{\varepsilon}{2} = \mu[f_\varepsilon] + \frac{\varepsilon}{2} < \mu[\varphi] + \varepsilon.$$

The above inequalities hold for any $\varepsilon > 0$ so

$$\liminf_n \mu_n[\varphi] = \limsup_n \mu_n[\varphi] = \mu[\varphi].$$

$\square$

Corollary 2.33. *Consider a sequence $\mu_n \in \operatorname{Prob}(\mathbb{R}^k)$ and $\mu \in \operatorname{Meas}(\mathbb{R})$. Then the following are equivalent.*

(i) The sequence (μ_n) converges weakly to μ.

(ii) For any bounded Lipschitz function $f : \mathbb{R} \to \mathbb{R}$ we have

$$\mu_n[f] = \mu[f].$$

Proof. The implication (i) $\Rightarrow$ (ii) is obvious. To prove that (ii) $\Rightarrow$ (i) observe first that any compactly supported continuous function can be uniformly approximated by compactly supported smooth functions[6] so the closure in $C_b(\mathbb{R}^k)$ of the set of bounded Lipschitz functions contains $C_0(\mathbb{R}^k)$. The measure μ is a probability measure since the constant function $\boldsymbol{I}_{\mathbb{R}^k}$ is bounded and Lipschitz and thus

$$\mu[\boldsymbol{I}_{\mathbb{R}^k}] = \lim_{n \to \infty} \mu_n[\boldsymbol{I}_{\mathbb{R}^k}] = 1.$$

The conclusion now follows from Theorem 2.31. □

Corollary 2.34. *If a sequence $\mu_n \in \mathrm{Prob}(\mathbb{R})$ converges vaguely to a probability measure, then it also converges weakly.* □

Corollary 2.35. *Suppose that $(X_n)_{n \in \mathbb{N}}$ and X are random variables with ranges contained in $\mathbb{Z}$. Then $X_n \Rightarrow X$ if and only if*

$$\lim_{n \to \infty} \mathbb{P}[X_n = k] = \mathbb{P}[X = k], \quad \forall k \in \mathbb{Z}. \tag{2.2.5}$$

Proof. The condition (2.2.5) is clearly satisfied if $X_n \Rightarrow X$ since

$$\mathbb{P}[X = k] = \mathbb{P}[k - 1/2 < X \leq k + 1/2]$$

$$= \lim_{n \to \infty} \mathbb{P}[k - 1/2 < X_n \leq k + 1/2] = \lim_{n \to \infty} \mathbb{P}[X_n = k].$$

Conversely, if (2.2.5) is satisfied, then

$$\mathbb{E}[\varphi(X_n)] \to \mathbb{E}[f(X)], \quad \forall \varphi \in C_0(\mathbb{R}).$$

The conclusion now follows from Theorem 2.31. □

The next result generalizes Fatou's Lemma. However, our proof relies on Fatou's Lemma.

Proposition 2.36. *Suppose that the sequence of random variables $(X_n)_{n \in b\mathbb{N}}$ converges in distribution to X. Then*

$$\mathbb{E}[|X|] \leq \liminf_{n \to \infty} \mathbb{E}[|X_n|].$$

In particular, X is integrable if the sequence $(X_n)_{n \in \mathbb{N}}$ is bounded in L^1, i.e.,

$$\sup_n \mathbb{E}[|X_n|] < \infty.$$

---
[6]On simple way to see this is to use Weierstrasss approximation theorem.

Proof. The Mapping Theorem 2.25 implies that the sequence $(|X_n|)_{n\in\mathbb{N}}$ converges in distribution to $|X|$. Thus

$$\lim_{n\in\mathbb{N}} \mathbb{P}\bigl[\,|X_n|>t\,\bigr] = \mathbb{P}\bigl[\,|X|>t\,\bigr],$$

for all t outside a countable subset of $[0,\infty)$. Using (1.3.43) we deduce

$$\mathbb{E}\bigl[\,|X|\,\bigr] = \int_0^\infty \mathbb{P}\bigl[\,|X|>t\,\bigr]dt, \quad \mathbb{E}\bigl[\,|X_n|\,\bigr] = \int_0^\infty \mathbb{P}\bigl[\,|X_n|>t\,\bigr]dt, \quad \forall n.$$

Fatou's Lemma implies

$$\int_0^\infty \mathbb{P}\bigl[\,|X|>t\,\bigr]dt \leq \liminf_{n\to\infty} \int_0^\infty \mathbb{P}\bigl[\,|X_n|>t\,\bigr]dt.$$

□

2.2.2 The characteristic function

The key ingredient in the proof of the CLT is that of *Fourier transform* or *characteristic function* of a finite Borel measure $\mu \in \mathrm{Meas}(\mathbb{R})$. This is the complex valued function

$$\widehat{\mu}:\mathbb{R}\to\mathbb{C}, \quad \widehat{\mu}(\xi) = \int_\mathbb{R} e^{i\xi x}\mu[dx].$$

Note that μ is a *probability* measure if and only if $\widehat{\mu}(0)=1$.

The characteristic function of a random variable X is the Fourier transform $\Phi_X(\xi)$ of its probability distribution,

$$\Phi_X(\xi) = \widehat{\mathbb{P}}_X(\xi) = \mathbb{E}\bigl[\,e^{i\xi X}\,\bigr].$$

Note that

$$|\Phi_X(\xi)| \leq 1, \quad \forall \xi \in \mathbb{C}.$$

Moreover, $\Phi_X(0) = \mathbb{E}[\,1\,] = 1$.

From the Dominated Convergence theorem we deduce that Φ_X is a continuous function $\mathbb{R}\to\mathbb{C}$. Thus, Fourier transform is a map

$$\mathrm{Prob}(\mathbb{R}) \ni \mu \mapsto \widehat{\mu} \in C_b(\mathbb{R},\mathbb{C}).$$

Proposition 2.37. *Let $X \in L^2(\Omega, \mathcal{S}, \mathbb{P})$. Then $\Phi_X \in C^2(\mathbb{R})$ and*

$$\Phi_X'(0) = i\mathbb{E}\bigl[\,X\,\bigr], \quad \Phi_X''(0) = -\mathbb{E}\bigl[\,X^2\,\bigr].$$

Proof. Denote by $\mathbb{P}_X$ the probability distribution of X so $\mathbb{P}_X \in \mathrm{Prob}(\mathbb{R})$. Then

$$\Phi_X(\xi) = \int_\mathbb{R} e^{ix\xi}\mathbb{P}_X[dx].$$

Note that since $X \in L^2$ we have

$$\int_\mathbb{R} |x|\,\mathbb{P}_X[dx], \quad \int_\mathbb{R} x^2\,\mathbb{P}_X[\,dx\,] < \infty$$

so
$$\partial_\xi e^{ix\xi} = ixe^{ix\xi} \in L^1(\mathbb{R}, \mathbb{P}_X),$$
$$\partial_\xi^2 e^{ix\xi} = -x^2 e^{ix\xi} \in L^1(\mathbb{R}, \mathbb{P}_X).$$

This shows (see Exercise 1.6) that the integral
$$\int_\mathbb{R} e^{ix\xi} \mathbb{P}_X[dx]$$
is twice differentiable with respect to the parameter ξ and we have
$$\Phi'_X(\xi) = -i \int_\mathbb{R} xe^{i\xi x} \mathbb{P}_X[dx], \quad \Phi''_X(\xi) = -\int_\mathbb{R} x^2 e^{i\xi x} \mathbb{P}_X[dx].$$
Using the Dominated Convergence theorem we deduce that the function
$$\xi \mapsto -\int_\mathbb{R} x^2 e^{i\xi x} \mathbb{P}_X[dx]$$
is continuous so $\Phi_X \in C^2(\mathbb{R})$. □

Let $\Gamma_v \in \mathrm{Prob}(\mathbb{R})$ be the Gaussian measure with mean 0 and variance $v > 0$
$$\Gamma_v[dx] = \gamma_v(x)dx, \quad \gamma_v(x) = \frac{1}{\sqrt{2\pi v}} e^{-\frac{x^2}{2v}}.$$

Proposition 2.38.
$$\widehat{\Gamma}_v(\xi) = e^{-\frac{v\xi^2}{2}} = \sqrt{\frac{2\pi}{v}} \gamma_{1/v}(\xi), \quad \forall v > 0. \qquad (2.2.6)$$

Proof. We have
$$\widehat{\Gamma}_v(\xi) = \frac{1}{\sqrt{2\pi v}} \int_\mathbb{R} e^{-\frac{x^2}{2v}} e^{i\xi x} dx = \frac{1}{\sqrt{2\pi}} \int_\mathbb{R} e^{-\frac{y^2}{2}} e^{i\sqrt{v}\xi y} dy = \widehat{\Gamma}_1(\sqrt{v}\,\eta).$$
Thus it suffices to determine
$$f(\xi) = \widehat{\Gamma}_1(\xi) = \frac{1}{\sqrt{2\pi}} \int_\mathbb{R} e^{-\frac{x^2}{2}} e^{i\xi x} dx.$$
The imaginary part of the above integrand is odd function (in x) so $f(\xi)$ is real, $\forall \xi$, i.e.,
$$f(\xi) = \frac{1}{\sqrt{2\pi}} \int_\mathbb{R} e^{-\frac{x^2}{2}} \cos(\xi x) dx.$$
The function
$$\frac{d}{d\xi}\left(e^{-\frac{x^2}{2}} \cos(\xi x) \right) = -xe^{-\frac{x^2}{2}} \sin(\xi x)$$
is integrable (in the x variable). This shows that $f(\xi)$ is differentiable (see Exercise 1.6) and
$$f'(\xi) = -\frac{1}{\sqrt{2\pi}} \int_\mathbb{R} xe^{-\frac{x^2}{2}} \sin(\xi x) dx = \frac{1}{\sqrt{2\pi}} \int_\mathbb{R} \frac{d}{dx}\left(e^{-\frac{x^2}{2}} \right) \sin(\xi x) dx$$

(integrate by parts)
$$= -\frac{\xi}{\sqrt{2\pi}} \int_{\mathbb{R}} e^{-\frac{x^2}{2}} \cos(\xi x) dx = -\xi f(\xi).$$
Thus
$$f'(\xi) + \xi f(\xi) = 0$$
so that
$$\frac{d}{d\xi}\left(e^{\xi^2/2} f(\xi)\right) = 0 \Longleftrightarrow f(\xi) = Ce^{-\frac{\xi^2}{2}}.$$
Since $f(0) = 1$ we deduce $C = 1$ and thus $\widehat{\mathbf{\Gamma}}_1(\xi) = e^{-\frac{\xi^2}{2}}$. $\square$

Theorem 2.39. *A probability measure $\mu \in \mathrm{Prob}(\mathbb{R})$ is uniquely determined by its characteristic function, i.e., the map*
$$\mathrm{Prob}(\mathbb{R}) \ni \mu \mapsto \widehat{\mu} \in C_b(\mathbb{R}, \mathbb{C})$$
is injective.

Proof. For any $v > 0$ and $\mu \in \mathrm{Prob}(\mathbb{R})$ we set $\mu_v := \mathbf{\Gamma}_v * \mu$. According to Remark 1.135 we have
$$\mu_v[dx] := \rho_v(x)dx, \quad \rho_v(x) = \int_{\mathbb{R}} \boldsymbol{\gamma}_v(x-y) \mu[dy].$$
The theorem follows from the following two facts.

Fact 1. The family $(\mu_v)_{v>0}$ is completely determined by $\widehat{\mu}$.

Fact 2. The family $(\mu_v)_{v>0}$ converges weakly to μ as $v \searrow 0$, i.e.,
$$\lim_{v \searrow 0} \mu_v[f] = \mu[f], \quad \forall f \in C_b(\mathbb{R}).$$

Proof of Fact 1. The idea behind this fact is that the Fourier transform and the convolution interact in a nice way. More precisely we will show that
$$\rho_v(x) = \frac{1}{\sqrt{2\pi v}} \int_{\mathbb{R}} e^{ix\xi} \boldsymbol{\gamma}_{1/v}(\xi) \widehat{\mu}(-\xi) d\xi. \tag{2.2.7}$$
Using (2.2.6) we deduce
$$\sqrt{2\pi v} \boldsymbol{\gamma}_v(x) = e^{-\frac{x^2}{2v}} = \int_{\mathbb{R}} e^{ix\xi} \boldsymbol{\gamma}_{1/v}(\xi) d\xi.$$
We deduce
$$\rho_v(x) = \frac{1}{\sqrt{2\pi v}} \int_{\mathbb{R}} \left(\int_{\mathbb{R}} e^{i(x-y)\xi} \boldsymbol{\gamma}_{1/v}(\xi) d\xi \right) \mu[dy]$$
(use Fubini)
$$= \frac{1}{\sqrt{2\pi v}} \int_{\mathbb{R}} e^{ix\xi} \boldsymbol{\gamma}_{1/v}(\xi) \left(\int_{\mathbb{R}} e^{-iy\xi} \mu[dy] \right) d\xi = \frac{1}{\sqrt{2\pi v}} \int_{\mathbb{R}} e^{ix\xi} \boldsymbol{\gamma}_{1/v}(\xi) \widehat{\mu}(-\xi) d\xi.$$

Proof of Fact 2. Let $f \in C_b(\mathbb{R})$. A simple application of Fubini shows that

$$\int_{\mathbb{R}} f(x)\mu_v[dx] = \int_{\mathbb{R}} f(x)\left(\int_{\mathbb{R}} \gamma_v(x-y)\mu[dy]\right)dx$$

$$= \int_{\mathbb{R}} \underbrace{\left(\int_{\mathbb{R}} \gamma_v(x-y)f(x)dx\right)}_{=:f_v(y)} \mu[dy].$$

The function f_v is obviously continuous. If $M := \sup_{x \in \mathbb{R}} |f(x)|$, then

$$|f_v(y)| \leq M \int_{\mathbb{R}} \gamma_v(x-y)dx \stackrel{x \to z+y}{=} M \int_{\mathbb{R}} \gamma_v(z)dz = M.$$

On the other hand

$$f_v(y) = \int_{\mathbb{R}} \gamma_v(x-y)f(x)dx = \int_{\mathbb{R}} \gamma_v(z)f(z+y)dz = \mathbf{\Gamma}_v[T_y f],$$

where $T_y f(z) := f(z+y)$. Fix y, $\varepsilon > 0$ and a $\delta = \delta(\varepsilon) > 0$ such that

$$\sup_{|z|<\delta} |f(z+y) - f(y)| < \frac{\varepsilon}{2}.$$

Then

$$|f_v(y) - f(y)| = |\mathbf{\Gamma}_v[T_y f] - f(y)| = \left|\int_{\mathbb{R}} (f(z+y) - f(y))\mathbf{\Gamma}_v[dz]\right|$$

$$\leq \int_{|z|<\delta} |f(z+y) - f(y)|\mathbf{\Gamma}_v[dz] + \int_{|z|\geq\delta} |f(z+y) - f(y)|\mathbf{\Gamma}_v[dz]$$

$$\leq \sup_{|z|<\delta} |f(z+y) - f(y)| + 2M\mathbf{\Gamma}_v[|z|>\delta]$$

$$\stackrel{(1.3.17)}{\leq} \sup_{|z|<\delta} |f(z+y) - f(y)| + \frac{2Mv}{\delta^2} < \frac{\varepsilon}{2} + \frac{2Mv}{\delta^2}.$$

Hence

$$\limsup_{v \searrow 0} |f_v(y) - f(y)| \leq \frac{\varepsilon}{2}, \quad \forall \varepsilon > 0,$$

so that

$$\lim_{v \searrow 0} f_v(y) = f(y), \quad \forall y \in \mathbb{R}.$$

The Dominated Convergence theorem now implies

$$\lim_{v \searrow 0} \mu_v[f] = \lim_{v \searrow 0} \mu[f_v] = \mu\left[\lim_{v \searrow 0} f_v\right] = \mu[f].$$

$\square$

Remark 2.40. (a) The above theorem can be rephrased as stating that the collection of trigonometric functions

$$\{\, \mathbb{R} \ni x \mapsto \cos(\xi x), \sin(\xi x);\ \xi \in \mathbb{R} \,\}$$

is separating. However, the smaller family

$$\{\, \mathbb{R} \ni x \mapsto \cos(\xi x), \sin(\xi x);\ |\xi| < 1 \,\},$$

is *not separating*! More precisely, there exists two *distinct* probability measures μ_0, μ_1 such that

$$\widehat{\mu}_0(\xi) = \widehat{\mu}_1(\xi),\ \forall |\xi| < 1.$$

We refer to [111, Chap. IV, Sec. 15, p. 231] for more details.

(b) The range of the Fourier transform

$$\mathrm{Prob}(\mathbb{R}) \ni \mu \mapsto \hat{\mu} \in C_b(\mathbb{R})$$

can also be characterized. Note first that $\forall \mu \in \mathrm{Prob}(\mathbb{R})$

$$\hat{\mu}(0) = \mu[\,\mathbb{R}\,] = 1,\ \widehat{\mu}(-\xi) = \overline{\widehat{\mu}(\xi)},\ \forall \xi \in \mathbb{R}.$$

Additionally, the function $\widehat{\mu}$ is *positive definite*. This means that, for any $n \in \mathbb{N}$ and any $\xi_1, \ldots, \xi_n \in \mathbb{R}$, the hermitian matrix

$$\bigl(\widehat{\mu}(\xi_i - \xi_j) \bigr)_{1 \leq i,j \leq n}$$

is positive semidefinite, i.e., for any $z_1, \ldots, z_n$ we have

$$\sum_{1 \leq i,j \leq n} \widehat{\mu}(\xi_i - \xi_j) z_i \bar{z}_j \geq 0.$$

This follows by observing that

$$\sum_{1 \leq i,j \leq n} \widehat{\mu}(\xi_i - \xi_j) z_i \bar{z}_j = \int_{\mathbb{R}} \Bigl| \sum_{i=1}^n z_k e^{i\xi_k x} \Bigr|^2 \mu[dx].$$

It turns out that these above necessary conditions characterize the range of the Fourier transform. This is the content of the celebrated Bochner theorem. For a proof we refer to [59, p. 622], [68, §II.3], or [135, I.24]. □

Theorem 2.41 (Lévy's Continuity Theorem). *Let $(\mu_n)_{n \in \mathbb{N}}$ be a sequence in* $\mathrm{Prob}(\mathbb{R})$ *and* $\mu \in \mathrm{Prob}(\mathbb{R})$. *The following statements are equivalent.*

(i) The sequence $(\mu_n)_{n \in \mathbb{N}}$ converges weakly to μ.
(ii) For any $\xi \in \mathbb{R}$

$$\lim_{n \to \infty} \widehat{\mu}_n(\xi) = \widehat{\mu}(\xi).$$

Proof. Our presentation is influenced by François Le Gall's lecture notes [102].

(i) $\Rightarrow$ (ii) Since $\mu_n \Rightarrow \mu$ we deduce that for any $\xi \in \mathbb{R}$ we have

$$\lim_{n\to\infty} \int_{\mathbb{R}} \cos(\xi x)\mu_n[dx] = \int_{\mathbb{R}} \cos(\xi x)\mu[dx],$$

$$\lim_{n\to\infty} \int_{\mathbb{R}} \sin(\xi x)\mu_n[dx] = \int_{\mathbb{R}} \sin(\xi x)\mu[dx].$$

(ii) $\Rightarrow$ (i) We carry the proof in two steps. For any $v > 0$ and any $f \in C_b(\mathbb{R})$ we define $f_v : \mathbb{R} \to \mathbb{R}$

$$f_v(x) = \int_{\mathbb{R}} f(x-y)\mathbf{\Gamma}_v[dy].$$

It is easy to see that $f_v \in C_b(\mathbb{R})$.

Step 1. We will show that

$$\lim_{n\to\infty} \mu_n[f_v] = \mu[f_v], \quad \forall v > 0, \quad \forall f \in C_0(\mathbb{R}). \tag{2.2.8}$$

Observe that

$$f_v = \int_{\mathbb{R}} f(x-y)\gamma_v(y)dy = \int_{\mathbb{R}} f(z)\gamma_v(x-z)dz.$$

Let $\nu \in \mathrm{Prob}(\mathbb{R})$. Then

$$\mu[f_v] = \int_{\mathbb{R}} \left(\int_{\mathbb{R}} f(z)\gamma_v(z-x)dz \right) \nu[dx] = \int_{\mathbb{R}} f(z) \underbrace{\left(\int_{\mathbb{R}} \gamma_v(z-x))\nu[dx] \right)}_{\rho_v(z)} dz$$

$$\stackrel{(2.2.7)}{=} \frac{1}{\sqrt{2\pi v}} \int_{\mathbb{R}} \left(\int_{\mathbb{R}} e^{ix\xi}\gamma_{1/v}(\xi)\widehat{\nu}(-\xi)d\xi \right) f(x)dx$$

$$= \frac{1}{\sqrt{2\pi v}} \int_{\mathbb{R}} \underbrace{\left(\int_{\mathbb{R}} e^{ix\xi} f(x)d\xi \right)}_{=:\widehat{f}(\xi)} \gamma_{1/v}(\xi)\widehat{\nu}(-\xi)d\xi = \frac{1}{\sqrt{2\pi v}} \int_{\mathbb{R}} \widehat{f}(\xi)\gamma_{1/v}(\xi)\widehat{\nu}(-\xi)d\xi.$$

The function $\widehat{f}(\xi)$ is well defined since $f \in C_0(\mathbb{R})$. The Dominated Convergence theorem shows that $\widehat{f}$ is continuous. Moreover

$$|\widehat{f}(\xi)| \leq \int_{\mathbb{R}} |f(x)|\, dx.$$

We deduce that, $\forall n \in \mathbb{N}$,

$$\mu_n[f_v] = \frac{1}{\sqrt{2\pi v}} \int_{\mathbb{R}} \widehat{f}(\xi)\gamma_{1/v}(\xi)\widehat{\mu}_n(-\xi)d\xi.$$

The Dominated Convergence theorem shows that

$$\lim_{n\to\infty} \int_{\mathbb{R}} \widehat{f}(\xi)\gamma_{1/v}(\xi)\widehat{\mu}_n(-\xi)d\xi = \int_{\mathbb{R}} \widehat{f}(\xi)\gamma_{1/v}(\xi)\widehat{\mu}(-\xi)d\xi = \mu[f_v].$$

Step 2. If $f \in C_b(\mathbb{R})$ is *uniformly* continuous, then f_v converges to f uniformly as $v \searrow 0$.

Set
$$M := \sup_{x \in \mathbb{R}} |f(x)|, \quad \omega(r) := \sup_{\substack{x,y \in \mathbb{R} \\ |x-y| \leq r}} |f(x) - f(y)|, \quad r > 0.$$

Since f is uniformly continuous we have
$$\lim_{r \searrow 0} \omega(r) = 0.$$

For $x \in \mathbb{R}$
$$\left| f_v(x) - f(x) \right| = \left| \int_{\mathbb{R}} f(x-y) \mathbf{\Gamma}_v[dy] - f(x) \right|$$
$$= \left| \int_{\mathbb{R}} f(x-y) \mathbf{\Gamma}_v[dy] - \int_{\mathbb{R}} f(x) \mathbf{\Gamma}_v[dy] \right| \leq \int_{\mathbb{R}} |f(x-y) - f(x)| \mathbf{\Gamma}_v[dy]$$
$$= \int_{|y| \leq r} |f(x-y) - f(x)| \mathbf{\Gamma}_v[dy] + \int_{|y| > r} |f(x-y) - f(x)| \mathbf{\Gamma}_v[dy]$$
$$\leq \omega(r) \int_{|y| \leq r} \mathbf{\Gamma}_v[dy] + 2M \int_{|y| > r} \mathbf{\Gamma}_v[dy]$$

(use Chebyshev's inequality to estimate the second integral)
$$\leq \omega(r) + 2M \frac{v}{r^2}, \quad \forall v, r > 0.$$

Now choose $\alpha \in (0, 1/2)$, $r = v^\alpha$. We deduce that
$$\sup_{x \in \mathbb{R}} \left| f_v(x) - f(x) \right| \leq \omega(v^\alpha) + 2M v^{1-2\alpha} \to 0 \text{ as } v \searrow 0.$$

This completes Step 2.

We deduce that the family
$$\mathcal{F} := \{ \varphi_v; \ v > 0, \ \varphi \in C_0(\mathbb{R}) \}$$
contains $C_0(\mathbb{R})$ in its closure and $\mu_n[f] \to \mu[f]$ for any $f \in \mathcal{F}$. The conclusion follows from Theorem 2.31. $\square$

Remark 2.42. (a) One can show that if a sequence $\mu_n \in \mathrm{Prob}(\mathbb{R})$ converges weakly to a probability measure μ, then $\widehat{\mu}_n(\xi)$ converges ro $\widehat{\mu}(\xi)$ uniformly on compacts; see Exercise 2.37.

(b) In Theorem 2.41 we assumed that the limit of the sequence of characteristic functions $(\widehat{\mu}_n)_{n \in \mathbb{N}}$ is the characteristic function of a probability measure μ. This assumption is not necessary.

One can show that if the characteristic functions $\widehat{\mu}_n(\xi)$ converge pointwisely to a continuous function f, then f is the characteristic function of a probability measure. In Exercise 2.36 we describe the main steps of the rather involved proof of this more general result. $\square$

Remark 2.43. P. Lévy, [105, §17, p. 47], introduced a metric d_L on $\mathrm{Prob}(\mathbb{R})$. More precisely, given $\mu_0, \mu_i \in \mathrm{Prob}(\mathbb{R})$ with cumulative distribution functions

$$F_i(x) = \mu_i\big[(-\infty, x]\big], \quad x \in \mathbb{R}, \quad i = 0, 1,$$

then the Lévy metric is the length of the largest segment cut-out by the graphs Γ_0, Γ_1 of F_0, F_1 along a line of the form $x + y = a$. The graphs are made continuous by adding vertical segments connecting $F_i(x-0)$ to $F_i(x)$ at the points of discontinuity. Intuitively, the distance is the diagonal if the largest square with sides parallel to the axes that can be squeezed between the curves Γ_0 and Γ_1.

More precisely

$$d_L(\mu_0, \mu_1) = \sup_{a \in \mathbb{R}} \mathrm{dist}_{\mathbb{R}^2}\big(p_0(a), p_1(a)\big),$$

where $p_i(a)$ is the intersection of the graph Γ_i with the line $x + y = a$. Note that if we write $p_i(a) = (x_i, y_i)$, then $y_i = F(x_i)$,[7] then

$$d_L(\mu_0, \mu_1) = \sup\big\{\sqrt{2}|x_0 - x_1|; \; x_0 + F_0(x_0) = x_1 + F_1(x_1)\big\}.$$

Lévy refers to the convergence with respect to the metric d_L as "*convergence from the point of view of Bernoulli*". He shows (see [105, §17]) that a sequence of probability measures μ_n converges in the metric d_L to a probability measure μ if and only if the characteristic functions $\widehat{\mu}_n$ converge to the characteristic function μ. Hence, the convergence in the metric d_L is the weak convergence so that d_L metrizes the weak convergence. □

2.2.3 The Central Limit Theorem

We can now state and prove the main result of this section.

Theorem 2.44 (Central Limit Theorem). *Suppose that $X_n \in L^2(\Omega, \mathcal{S}, \mathbb{P})$ is a sequence of i.i.d. with common mean μ and common variance v. Set*

$$\bar{X}_n = X_n - \mu, \quad \bar{S}_n = \sum_{k=1}^{n}(X_k - \mu), \quad Z_n = \frac{1}{\sqrt{nv}}\bar{S}_n = \frac{1}{\sqrt{nv}}\left(\sum_{k=1}^{n} X_k - n\mu\right).$$

Then $Z_n \Rightarrow N(0,1)$.

Proof. According to Lévy's continuity theorem it suffices to show that

$$\lim_{n \to \infty} \Phi_{Z_n}(\xi) = \Phi_{\Gamma_1}(\xi) = e^{-\frac{\xi^2}{2}}.$$

Observe that $\overline{X}_n$ are i.i.d. with mean 0 and variance v, while Z_n has mean 0 and variance 1. Denote by $\Phi(\xi)$ their common characteristic function, $\Phi(\xi) = \mathbb{E}\big[e^{i\bar{X}_1}\big]$. We have

$$\Phi_{Z_n}(\xi) = \Phi_{\bar{S}_n/\sqrt{nv}}(\xi) = \Phi_{\bar{S}_n}(\xi/\sqrt{nv}) = \mathbb{E}\left[\prod_{k=1}^{n} \exp\left(i\frac{\xi}{\sqrt{nv}}\overline{X}_k\right)\right]$$

---
[7]At a point of discontinuity this reads $y_i \in \big(F_i(x_i - 0), F_i(x_i)\big)$.

(the variables $\exp\left(i\frac{\xi}{\sqrt{nv}}\overline{X}_k\right)$, $1 \leq k \leq n$ are independent)

$$= \prod_{k=1}^{n} \mathbb{E}\left[\exp\left(i\frac{\xi}{\sqrt{nv}}\overline{X}_k\right)\right] = \Phi\left(\xi/\sqrt{nv}\right)^n.$$

Proposition 2.37 shows that the function $\Phi(\eta)$ is C^2, so as $\eta \to 0$ we have

$$\Phi(\eta) = \Phi(0) + \Phi'(0)\eta + \frac{1}{2}\Phi''(0)\eta^2 + o(\eta^2) = 1 + i\mathbb{E}[\overline{X}_1]\eta - \frac{1}{2}\mathbb{E}[\overline{X}_1^2]\eta^2 + o(\eta^2)$$

$(\mathbb{E}[\overline{X}_1] = 0$, $\mathbb{E}[\overline{X}_1^2] = \text{Var}[\overline{X}_1] = v)$

$$= 1 - \frac{v}{2}\eta^2 + o(\eta^2).$$

Now let $\eta = \xi/\sqrt{nv}$, $n \gg 0$. We deduce

$$\Phi\left(\xi/\sqrt{nv}\right)^n = \left(1 - \frac{\xi^2}{2n} + o(1/n)\right)^n.$$

At this point we want to invoke the following result.

Lemma 2.45. *Suppose that $(c_n)_{n \geq 1}$ is a convergent sequence of <u>complex</u> numbers and*

$$c = \lim_{n \to \infty} c_n.$$

Then

$$\lim_{n \to \infty}\left(1 + \frac{c_n}{n}\right)^n = e^c.$$

Assuming Lemma 2.45 we deduce that, for any $\xi \in \mathbb{R}$ we have

$$\lim_{n \to \infty} \Phi_{Z_n}(\xi) = \lim_{n \to \infty}\left(1 - \frac{\xi^2}{2n} + o(1/n)\right)^n = e^{-\frac{\xi^2}{2}} = \Phi_{\Gamma_1}(\xi).$$

Proof of Lemma 2.45. Set $c = a + bi$, $c_n = a_n + b_n i$, so that $a_n \to a$, $b_n \to b$. We set

$$z_n = 1 + \frac{c_n}{n} = 1 + \frac{a_n}{n} + \frac{b_n}{n}i.$$

For large n $z_n = r_n e^{i\theta_n}$, where

$$r_n = \sqrt{(1 + a_n/n)^2 + b_n^2/n^2} = \left(1 + 2a/n + o(1/n)\right)^{1/2},$$

$$|\theta_n| < \frac{\pi}{2}, \quad \tan\theta_n = \frac{1}{n}\frac{b_n}{1 + a_n/n}.$$

Thus

$$\theta_n = \arctan\left(\frac{1}{n}\frac{b_n}{1 + a_n/n}\right) = \frac{b}{n} + o(1/n) \text{ as } n \to \infty.$$

We deduce that as $n \to \infty$ we have

$$z_n^n = \left(1 + 2a/n + o(1/n)\right)^{n/2} \cdot e^{i(b + o(1))} \to e^a \cdot e^{ib} = e^c.$$

$\square$

Remark 2.46. There is a more refined version of the Central Limit Theorem that does not require that the random variables be identically distributed, only independent. More precisely, we have the following result of Lindeberg.

Suppose that $(X_n)_{n\geq 0}$ is a sequence of independent random variables with zero means and finite variances. We set $S_n := X_1 + \cdots + X_n$ and $s_n^2 := \mathrm{Var}[S_n]$. Suppose that, $\forall t > 0$

$$\lim_{n\to\infty} \frac{1}{s_n^2} \sum_{k=1}^{n} \mathbb{E}\big[\boldsymbol{I}_{\{|X_k|>ts_n\}} X_k^2\big] = 0. \qquad (2.2.9)$$

Then $\frac{1}{s_n} S_n$ converges in distribution to a standard normal random variable.

Note that if the random variables X_n are also identically distributed with common variances σ^2, then $\sigma_n^2 = n\sigma^2$. Then

$$\frac{1}{s_n^2} \sum_{k=1}^{n} \mathbb{E}\big[\boldsymbol{I}_{\{|X_k|>ts_n\}} X_k^2\big] = \frac{1}{\sigma^2} \mathbb{E}\big[\boldsymbol{I}_{\{|X_1|>t\sigma\sqrt{n}\}} X_1^2\big] \to 0$$

as $n \to \infty$. Hence condition (2.2.9) is satisfied when the random variables are i.i.d. For a proof of Lindeberg's theorem we refer to [59, Sec. VIII.4].

For even more general versions of the CLT we refer to [72; 127]. □

2.3 Concentration inequalities

Suppose that $(X_n)_{n\in\mathbb{N}}$ is a sequence of i.i.d. random variables with mean 0. Let

$$S_n := X_1 + \cdots + X_n.$$

The Strong Law of Large of Numbers shows that $\frac{1}{n} S_n \to 0$ a.s. A concentration inequality offers a quantitative information on the probability that $\frac{1}{n} S_n$ deviates from 0 by a given amount ε. More concretely, it gives an upper bound for the probability that $\frac{1}{n}|S_n| > \varepsilon$. If the random variables X_n have finite second moments, $\sigma^2 = \mathrm{Var}[X_1]$, then we have seen that Chebyshev's inequality yields the estimate

$$\mathbb{P}\big[|S_n| > n\varepsilon\big] = \mathbb{P}\big[S_n^2 > n^2\varepsilon^2\big] < \frac{\mathrm{Var}[S_n]}{n^2\varepsilon^2} = \frac{\sigma^2}{n\varepsilon^2}.$$

In the proof of Theorem 2.7 we have shown that if the variables X_n have a stronger integrability property namely $\mathbb{E}[X_n^4] < \infty$, then there exists a constant $C > 0$ such that for any $\varepsilon > 0$ and any $\varepsilon > 0$ we have

$$\mathbb{P}\big[|S_n| > n\varepsilon\big] \leq \frac{C}{n^2\varepsilon^4},$$

showing that $\frac{1}{n} S_n$ is even more concentrated around its mean. Loosely speaking, we expect higher concentration around if X_n have lighter tails, i.e., the probabilities

$$\mathbb{P}\big[|X_n| > x\big]$$

decay fast as $x \to \infty$.

In this section we want to describe some quantitative results stating that, under appropriate light-tail assumptions, for any $\varepsilon > 0$ the probability $\mathbb{P}\big[|S_n| > n\varepsilon\big]$ decays exponentially fast to 0 as $n \to \infty$. The subject of concentration inequalities has witnessed and explosive growth in the last three decades so we will only be able to scratch the surface. For more on this subject we refer to [17].

2.3.1 The Chernoff bound

Many useful concentration inequalities are based on the *Chernoff method*. Let us describe its basics.

Suppose that X is a centered, i.e., mean zero, random variable such that

$$\mathbb{M}_X(\lambda) := \mathbb{E}\big[e^{\lambda X}\big] < \infty, \quad \forall \lambda \in J,$$

where J is an open interval containing the origin. We set

$$J_\pm := \big\{\lambda \in I;\ \pm\lambda > 0\,\big\}.$$

Note that this implies that X has moments of any order and thus it imposes severe restrictions on the tail of X. We define the *cumulant* of X to be the function,

$$\Psi_X : J \to \mathbb{R}, \quad \Psi_X(\lambda) := \log \mathbb{M}_X(\lambda)\big].$$

The function $x \mapsto e^{tx}$ is convex and Jensen's inequality shows

$$\mathbb{E}\big[e^{\lambda X}\big] \geq e^{\lambda \mathbb{E}[X]} = 1$$

so $\Psi_X(\lambda) \geq 0$.

Here is the key idea of Chernoff's method. For $x > 0$ we have

$$\mathbb{P}\big[X > x\big] = \mathbb{P}\big[e^{\lambda X} > e^{\lambda x}\big] \leq \frac{1}{e^{\lambda x}} \mathbb{E}\big[e^{\lambda X}\big], \quad \forall \lambda \in J_+,$$

where at the last step we used Markov's inequality. Hence

$$\mathbb{P}\big[X > x\big] \leq e^{-(x\lambda - \Psi_X(\lambda))}, \quad \forall \lambda \in J_+.$$

Set

$$I_+(x) := \sup_{\lambda \in J_+} \big(x\lambda - \Psi_X(\lambda)\big).$$

We obtain in this fashion the *Chernoff bound*

$$\mathbb{P}\big[X > x\big] \leq e^{-I_+(x)}, \quad I_+(x) := \sup_{\lambda \in (0,r)} \big(x\lambda - \Psi_X(\lambda)\big), \quad \forall x > 0. \tag{2.3.1}$$

Note that $I_+(x) \geq 0$ since $\Psi_X(\lambda) \geq 0$. Arguing in a similar fashion we deduce

$$\mathbb{P}\big[X < x\big] \leq e^{-I_-(x)}, \quad I_-(x) := \sup_{\lambda \in J_-} \big(x\lambda - \Psi_X(\lambda)\big), \quad \forall x < 0. \tag{2.3.2}$$

More generally, if X has a nonzero mean μ, then $\overline{X} = X - \mu$ is centered. If $\mathbb{E}\big[e^{\lambda X}\big]$ exists for $\lambda \in J$, then

$$\Psi_X(\lambda) = \Psi_{\bar X} = \Psi_X(\lambda) - \lambda\mu,$$

and we deduce

$$\mathbb{P}\big[X > x + \mu\big] \leq e^{-I_+(x)}, \quad I_+(x) := \sup_{\lambda \in J_+} \big((x+\mu)\lambda - \Psi_X(\lambda)\big), \quad \forall x > 0 \tag{2.3.3}$$

and

$$\mathbb{P}\big[X > x + \mu\big] \leq e^{-I_-(x)}, \quad I_-(x) := \sup_{\lambda \in J_-} \big((x+\mu)\lambda - \Psi_X(\lambda)\big), \quad \forall x < 0. \tag{2.3.4}$$

Suppose that $(X_n)_{n\in\mathbb{N}}$ is a sequence of i.i.d. random variables such that
$$\mathbb{M}(\lambda) = \mathbb{M}_{X_k}(\lambda) < \infty,$$
for any λ in an open interval J containing 0. Set
$$\mu := \mathbb{E}[X_k], \quad S_n := X_1 + \cdots + X_n.$$
Then
$$\mathbb{E}[S_n] = n\mu, \quad \mathbb{M}_{S_n}(\lambda) = \mathbb{M}(\lambda)^n, \quad \Psi_{S_n}(\lambda) = n\Psi(\Lambda).$$
We deduce that
$$\sup_{\lambda \in J_+} \big((nx + n\mu)\lambda - \Psi_{S_n}(\lambda)\big) = nI_+(x), \quad \forall x > 0,$$
and
$$\sup_{\lambda \in J_-} \big((nx + n\mu)\lambda - \Psi_{S_n}(\lambda)\big) = nI_-(x), \quad \forall x < 0.$$
We deduce
$$\mathbb{P}\left[\frac{1}{n}S_n - \mu > x\right] = \mathbb{P}[S_n - n\mu > nx] \leq e^{-nI_+(x)}, \quad \forall x > 0, \tag{2.3.5a}$$

$$\mathbb{P}\left[\frac{1}{n}S_n - \mu < x\right] = \mathbb{P}[S_n - n\mu < nx] \leq e^{-nI_-(x)}, \quad \forall x < 0. \tag{2.3.5b}$$

We have reached a remarkable conclusion. The assumption $\mathbb{M}(\lambda) < \infty$ for λ in an open neighborhood of the origin implies that the probability that the empirical mean $\frac{1}{n}S_n$ deviates from the theoretical mean μ by a fixed amount x decays exponentially to 0 as $n \to \infty$. In other words, $\frac{1}{n}S_n$ is highly concentrated around its mean and the above inequalities quantify this fact.

To gain some more insight on the above estimates it is useful to list a few properties of the function $I_+(x)$.

Proposition 2.47. *Suppose that the centered random variable X satisfies*
$$\mathbb{M}_X(\lambda) = \mathbb{E}[e^{\lambda X}] < \infty, \quad \forall \lambda \in J,$$
where $J \subset \mathbb{R}$ is an open interval containing 0. Then the following hold.

(i) $\mathbb{M}_X(0) = 1$, $\mathbb{M}'_X(0) = 0$, $\mathbb{M}''_X(0) = \mathrm{Var}[X]$.
(ii) *The function $J \ni \lambda \mapsto \Psi_X(\lambda) \in \mathbb{R}$ is convex and nonnegative. Moreover $\Psi''_X(0) > 0$.*
(iii) *The function*
$$I : \mathbb{R} \to [0, \infty], \quad I(x) = \sup_{\lambda \in J}\big(\lambda x - \Psi_X(\lambda)\big)$$
is convex. Moreover $I(x) = I_X(x) = I_\pm(x)$ for $\pm x > 0$.
(iv) $I(x) > 0$ *if $x \neq 0$.*

Proof. (i) is an elementary computation.

(ii) To prove that $\Psi_X(\lambda)$ is convex let $t_1, t_2 \in (0,1)$ such that $t_1 + t_2 = 1$. Then, using Hölder's inequality with $p = \frac{1}{t_1}$ and $q = \frac{1}{t_2}$ we deduce that for any $\lambda_1, \lambda_2 \in \mathbb{R}$ we have

$$\mathbb{E}\big[e^{t_1\lambda_1 X + t_2\lambda_2 X}\big] \leq \mathbb{E}\big[\big(e^{t_1\lambda_1 X}\big)^{1/t_1}\big]^{t_1} \mathbb{E}\big[\big(e^{t_2\lambda_2 X}\big)^{1/t_2}\big]^{t_2} = \mathbb{E}\big[e^{\lambda_1 X}\big]^{t_1} \mathbb{E}\big[e^{\lambda_2 X}\big]^{t_2}.$$

Taking the logarithm of both sides of the above inequality we obtain the convexity of $\Psi_X(\lambda)$. Next observe that

$$\Psi_X'(0) = \frac{\mathbb{M}_X'(0)}{\mathbb{M}_X(0)} = 0.$$

Since $\Psi_X(\lambda)$ is convex is graph sits above the tangent at $\lambda = 0$ so $\Psi_X(\lambda) \geq 0$, $\forall \lambda \in J$.

(iii) For $t_1, t_2 \in (0,1)$ such that $t_1 + t_2 = 1$ and for $x_1, x_2 > 0$ we have

$$I_+(t_1 x_1 + t_1 x_2) = \sup_{\lambda \in (0,r)} \big((t_1 x_2 + t_2 x_2) - \Psi_X(\lambda)\big)$$

$$= \sup_{\lambda \in (0,r)} \big(t_1(x_1 - -\Psi_X(\lambda)) - (t_2 x_2 - \Psi_X(\lambda))\big) \leq t_1 I_+(x_1) + t_2 I_+(x_2).$$

Observe that for $x > 0$ we have

$$\lambda x - \Psi_X(\lambda) \leq 0, \ \forall \lambda \leq 0$$

proving that

$$I(x) = \sup_{\lambda \in J}\big(\lambda x - \Psi_X(\lambda)\big) = \sup_{\lambda \in J_+}\big(\lambda x - \Psi_X(\lambda)\big).$$

(iv) Observe that

$$\Psi_X''(\lambda) = \frac{\mathbb{M}_X''(\lambda)\mathbb{M}_X(\lambda) - \mathbb{M}_X'(\lambda)^2}{\mathbb{M}_X(\lambda)^2} \tag{2.3.6}$$

so $\Psi_X''(0) = \mathbb{M}_X''(0) = \operatorname{Var}[X] > 0$. This proves that $\lambda x - \Psi_X(\lambda) > 0$ for $|\lambda|$ small and $x \neq 0$ so $I(x) > 0$ if $x \neq 0$. $\square$

Remark 2.48. As explained in [134, §12], to any convex lower semicontinuous function $f : \mathbb{R}^n \to (0, \infty]$ we can associate a *conjugate*

$$f^* : \mathbb{R}^n \to (-\infty, \infty], \ f^*(p) = \sup_{x \in \mathbb{R}^n}\big(\langle p, x \rangle - f(x)\big),$$

where $\langle -, - \rangle$ denotes the canonical inner product in $\mathbb{R}^n$. One can show that f^* is also convex and lower semicontinuous and $f = (f^*)^*$. The conjugate f^* is sometimes called the *Fenchel-Legendre conjugate* of f. Observe that $I(x)$ is the conjugate of the convex function $\Psi_X(\lambda)$. $\square$

Example 2.49. Suppose that $X \sim \mathrm{Bin}(p)$. Then $\mathbb{E}[X] = p$, $\mathbb{M}_X(\lambda) = (q + pe^\lambda)$. For $x \in \mathbb{R}$ we have

$$f_x(\lambda) := (x+p)\lambda - \Psi_X(\lambda) = (x+p)\lambda - \log(q + pe^\lambda)$$

$$f'_x(\lambda) = x + p - \frac{pe^\lambda}{q + pe^\lambda}$$

and $f'_x(\lambda) = 0$ if

$$p(x+p-1)e^\lambda = -q(x+p), \quad \text{i.e.,} \quad pe^\lambda = q\frac{x+p}{q-x}.$$

This forces $x \in (-p, q)$. In this case

$$\lambda = \log q - \log p + \log(x+p) - \log(q-x) = \log\frac{x+p}{p} - \log\frac{q-x}{q}$$

$$I(x) = (x+p)\log\frac{x+p}{p} - (x+p)\log\frac{q-x}{q} + \log\frac{q-x}{q}$$

$$= (x+p)\log\frac{x+p}{p} + (q-x)\log\frac{q-x}{q}.$$

□

Remark 2.50. Suppose that $\mathbb{P}, \mathbb{Q}$ are two Borel probability measures on $\mathbb{R}$ that are mutually absolutely continuous,

$$\mathbb{P} \ll \mathbb{Q} \quad \text{and} \quad \mathbb{Q} \ll \mathbb{P}.$$

We denote by $\rho_{\mathbb{P}|\mathbb{Q}} := \frac{d\mathbb{P}}{d\mathbb{Q}}$ the density of $\mathbb{P}$ with respect to $\mathbb{Q}$. We define the *Kullback-Leibler divergence*

$$\mathbb{D}_{KL}[\mathbb{P} \| \mathbb{Q}] := \int_\mathbb{R} \log\frac{d\mathbb{P}}{d\mathbb{Q}} \mathbb{P}[dx]. \qquad (2.3.7)$$

(a) Suppose that $\mathbb{P}$ is the probability distribution $\mathrm{Bin}(p)$,

$$\mathbb{P} = q\delta_0 + p\delta_1.$$

For $x \in (-p, q)$ consider the probability distribution

$$\mathbb{Q}_x = (q-x)\delta_0 + (p+x)\delta_1.$$

Then

$$\mathbb{D}_{KL}[\mathbb{Q}_x \| \mathbb{P}] = (x+p)\log\frac{x+p}{p} + (q-x)\log\frac{q-x}{q}.$$

This is the rate $I(x)$ we found in Example 2.49.

(b) Let X be a random variable with probability distribution $\mathbb{Q}$ and set $Z := \rho_{\mathbb{P}|\mathbb{Q}}(X)$. Then

$$\mathbb{E}[Z] = \int_\mathbb{R} \frac{d\mathbb{P}}{d\mathbb{Q}} d\mathbb{Q} = \int_\mathbb{R} d\mathbb{P} = 1,$$

$$\mathbb{E}\big[\, Z \log Z\,\big] = \int_{\mathbb{R}} \frac{d\mathbb{P}}{d\mathbb{Q}} \log \frac{d\mathbb{P}}{d\mathbb{Q}} \, d\mathbb{Q} = \int_{\mathbb{R}} \log \frac{d\mathbb{P}}{d\mathbb{Q}} \, d\mathbb{P} = \mathbb{D}_{KL}\big[\,\mathbb{P} \,\|\, \mathbb{Q}\,\big].$$

Thus
$$\mathbb{E}\big[\, Z \log Z \,\big] - \mathbb{E}\big[\, Z \,\big] \log \mathbb{E}\big[\, Z \,\big] = \mathbb{D}_{KL}\big[\,\mathbb{P} \,\|\, \mathbb{Q}\,\big]$$
showing that Kullback-Leibler divergence is a special case of φ-entropy (1.3.12). More precisely, the above equality shows that
$$\mathbb{D}_{KL}\big[\,\mathbb{P} \,\|\, \mathbb{Q}\,\big] = \mathbb{H}_\varphi\big[\, Z \,\big], \quad \varphi(z) = z \log z, \quad z > 0.$$

In particular this yields *Gibbs' inequality*
$$\mathbb{D}_{KL}\big[\, P \,\|\, \mathbb{Q}\,\big] \geq 0. \tag{2.3.8}$$

Above, we could have used instead of the natural logarithm any logarithm in a base > 1 and reach the same conclusion. In particular, if we work with $\log_2$ and we set
$$\mathbb{D}_2\big[\,\mathbb{P} \,\|\, \mathbb{Q}\,\big] = \int_{\mathbb{R}} \log_2 \frac{d\mathbb{Q}}{d\mathbb{P}} \mathbb{P}\big[\, dx \,\big].$$

Then Gibbs' inequality continues to hold in this case as well
$$\mathbb{D}_2\big[\,\mathbb{P} \,\|\, \mathbb{Q}\,\big] \geq 0. \tag{2.3.9}$$

Let $\mathscr{X}$ be a finite subset of $\mathbb{R}$. Assume that we are given a function $p: \mathscr{X} \to (0, 1]$ such that
$$\sum_{x \in \mathscr{X}} p(x) = 1$$
so p defines the probability measure
$$\mathbb{P}_p = \sum_{x \in \mathscr{X}} p(x) \delta_x \in \mathrm{Prob}(\mathbb{R}).$$

Recall that its *Shannon entropy* is (see (2.1.16) is the quantity
$$\mathrm{Ent}_2\big[\, p \,\big] = - \sum_{x \in \mathscr{X}} p(x) \log_2 p(x).$$

The uniform probability measure on $\mathscr{X}$ is
$$\mathbb{P}_0 = \sum_{x \in \mathscr{X}} p_0(x) \delta_x = \frac{1}{|\mathscr{X}|} \sum_{x \in \mathscr{X}} \delta_x.$$

Note that $\mathbb{P}_p$ and $\mathbb{P}_0$ are mutually absolutely continuous. Gibbs' inequality shows that
$$\mathbb{D}_2\big[\,\mathbb{P} \,\|\, \mathbb{P}_0\,\big] \geq 0.$$

On the other hand
$$\mathbb{D}_2\big[\,\mathbb{P} \,\|\, \mathbb{P}_0\,\big] = \sum_{x \in \mathscr{X}} \log_2\big(|\mathscr{X}| \cdot p(x)\big) p(x) = \log_2 |\mathscr{X}| + \sum_{x \in \mathscr{X}} p(x) \log_2 p(x) \geq 0.$$

We obtained again the inequality (2.1.17).
$$\mathrm{Ent}_2\big[\, p \,\big] \leq \log_2 |\mathscr{X}| = \mathrm{Ent}_2\big[\, p_0 \,\big]. \tag{2.3.10}$$

$\square$

Example 2.51. Suppose that $X \sim N(0,1)$. Then, for any $\lambda \in \mathbb{R}$,

$$\mathbb{M}_X(\lambda) = \frac{1}{\sqrt{2\pi}} \int_{\mathbb{R}} e^{\lambda x} e^{-x^2/2} dx = \frac{1}{\sqrt{2\pi}} \int_{\mathbb{R}} e^{-(x^2 - 2\lambda x + \lambda^2)/2} e^{\lambda^2/2} dx = e^{\lambda^2/2}.$$

Note that $Y = \sigma X \sim N(0, \sigma)$ and

$$\mathbb{M}_Y(\lambda) = \mathbb{M}_X(\sigma \lambda) = e^{\sigma^2 \lambda^2/2}, \quad \Psi_Y(\lambda) = \frac{\sigma^2 \lambda^2}{2}.$$

The supremum

$$I(x) := \sup_{\lambda \in \mathbb{R}} \left(x\lambda - \frac{\sigma^2 \lambda^2}{2} \right)$$

is achieved for $\lambda = \lambda_x = \frac{x}{\sigma^2}$ and it is equal to

$$I(x) = \frac{x^2}{2\sigma^2}.$$

In other words, if $X \sim N(0, \sigma^2)$, then

$$\mathbb{P}\big[|X| > \varepsilon\big] \leq 2 \max\big(\mathbb{P}\big[X < -x\big], \mathbb{P}\big[X > x\big]\big) \leq 2 e^{-\frac{x^2}{2\sigma^2}}.$$

□

2.3.2 Some applications

Often an explicit description of $\Psi_X(\lambda)$ may either not be possible, or it could be too complicated to be useful. That is why it is more practical to have simple ways of producing upper bounds for the moment generating function.

Definition 2.52. A random variable X with mean μ said to be *subgaussian* of type σ^2, and we write this $X \in \mathbb{G}(\sigma^2)$, if

$$\Psi_{X-\mu}(\lambda) \leq \frac{\sigma^2 \lambda^2}{2}, \quad \forall \lambda \in \mathbb{R} \iff \mathbb{E}\big[e^{\lambda(X-\mu)}\big] \leq e^{\frac{\lambda^2 \sigma^2}{2}}, \quad \forall \lambda \in \mathbb{R}. \quad □$$

Note that if $X \in \mathbb{G}(\sigma^2)$ and $\pm x > 0$

$$\sup_{\pm \lambda \geq 0} \big(x\lambda - \Psi_X(\lambda) \big) \geq \frac{x^2}{2\sigma^2},$$

and thus

$$\max\big(\mathbb{P}\big[X - \mu < -x\big], \mathbb{P}\big[X - \mu > x\big] \leq e^{-\frac{x^2}{2\sigma^2}}\big), \quad \forall x > 0, \quad (2.3.11a)$$

$$\mathbb{P}\big[|X - \mu| > x\big] \leq 2 e^{-\frac{x^2}{2\sigma^2}}, \quad \forall x > 0. \quad (2.3.11b)$$

Observe that if X_1, X_2 are independent random variables and $X_k \in \mathbb{G}(\sigma_k^2)$, $k = 1, 2$, then

$$a_1 X_1 + a_1 X_2 \in \mathbb{G}(a_1^2 \sigma_1^2 + a_2^2 \sigma_2^2), \quad \forall a_1, a_2 \in \mathbb{R}.$$

In particular, if $X_1, \ldots, X_n$ are centered, independent random variables in $\mathbb{G}(\sigma^2)$, then we have
$$\frac{1}{n}(X_1 + \cdots + X_n) \in \mathbb{G}(\sigma^2/n),$$
and thus we obtain *Hoeffding's inequality*
$$\mathbb{P}\left[\left|\frac{1}{n}(X_1 + \cdots + X_n)\right| > x\right] \leq 2e^{-\frac{nx^2}{2\sigma^2}}, \quad \forall x > 0. \tag{2.3.12}$$

Example 2.53. Suppose that R is a Rademacher random variable, i.e., it takes only the values ± 1 with equal probabilities. Then
$$\mathbb{E}[e^{\lambda R}] = \cosh \lambda \leq e^{\lambda^2/2},$$
where the last inequality is obtained by inspecting the Taylor series of the two terms and using the inequality $2^n n! \leq (2n)!$. Hence $R \in \mathbb{G}(1)$. Similarly, $cR \in \mathbb{G}(1)$, $\forall c \in [0,1]$. $\square$

For these estimates to be useful we need to have some simple ways of recognizing subgaussian random variables.

Proposition 2.54. *Suppose that X is a centered random variable, i.e., $\mathbb{E}[X] = 0$. If there exists $C > 0$ such that*
$$\mathbb{E}[X^{2k}] \leq k! C^k, \quad \forall k \in \mathbb{N},$$
then $X \in \mathbb{G}(4C)$.

Proof. We rely on a very useful symmetrization trick. Choose a random variable X' independent of X but with the same distribution as X. Then the random variable $Y = X - X'$ is symmetric, i.e., Y and $-Y$ have the same probability distributions. Observe next that since $-X'$ is centered we have
$$\mathbb{E}[e^{-\lambda X'}] \geq e^{-\lambda \mathbb{E}[X']} = 1, \quad \forall \lambda \in \mathbb{R}.$$
We deduce
$$\mathbb{E}[e^{\lambda X}] \leq \mathbb{E}[e^{\lambda X}] \cdot \mathbb{E}[e^{-\lambda X'}] = \mathbb{E}[e^{\lambda(X-X')}] = \sum_{k=0}^{\infty} \frac{\lambda^{2k}}{(2k)!} \mathbb{E}[(X - X')^{2k}].$$
Since the function x^{2k} is convex we have
$$(x+y)^{2k} \leq 2^{2k-1}(x^{2k} + y^{2k}), \quad \forall x, y \in \mathbb{R}$$
so
$$\mathbb{E}[(X - X')^{2k}] \leq 2^{2k}\mathbb{E}[X^{2k}] \leq 2^{2k} k! C^k = \frac{(2k)!}{(2k-1)!!}(2C)^k \leq \frac{(2k)!}{k!}(2C)^k.$$
Hence
$$\mathbb{E}[e^{\lambda X}] \leq \sum_{k=0}^{\infty} \frac{(2C\lambda^2)^k}{k!} = e^{2C\lambda^2}.$$
Hence $X \in \mathbb{G}(4C)$. $\square$

Example 2.55. Suppose that R is a Rademacher random variable. Clearly

$$\mathbb{E}[R^{2k}] = 1 \leq k! 1^k, \quad \forall k \in \mathbb{N}$$

so that $R \in \mathbb{G}(4)$. We see that this estimate is not as good as the one in Example 2.53. □

The next result offers a sharper estimate under certain conditions.

Proposition 2.56 (Hoeffding's lemma). *Suppose that X is a random variable such that $X \in [a,b]$ a.s. Then $X \in \mathbb{G}((b-a)^2/4)$, i.e.,*

$$\mathbb{E}[e^{\lambda(X-\mu)}] \leq e^{\frac{\lambda^2(b-a)^2}{8}}, \quad \forall \lambda \in \mathbb{R}. \tag{2.3.13}$$

Proof. Let us first observe that any random variable Y such that $Y \in [a,b]$ a.s. satisfies

$$\mathrm{Var}[Y] \leq \frac{(b-a)^2}{4}.$$

Indeed, if $\mu = \mathbb{E}[Y]$, then $Y - \mu \in [a-\mu, b-\mu]$. If

$$m = \frac{(a-\mu) + (b-\mu)}{2}$$

is the midpoint of $[a-\mu, b-\mu]$, then

$$|(Y-\mu) - m| \leq \frac{b-a}{2}$$

and

$$\mathrm{Var}[Y] \leq \mathbb{E}[(Y-\mu)^2] + m^2 = \mathbb{E}[((Y-\mu) - m)^2] \leq \frac{(b-a)^2}{4}.$$

Observe next that we can assume that X is centered. Indeed, if $\mu = \mathbb{E}[X]$, then the centered variable $X - \mu$ satisfies $X - \mu \in [a-\mu, b-\mu]$ and $(b-a) = (b-\mu) - (a-\mu)$.

Denote by $\mathbb{P}$ the probability distribution of X. For any $\lambda \in \mathbb{R}$ we denote by $\mathbb{P}_\lambda$ the probability measure on $\mathbb{R}$ given by

$$\mathbb{P}_\lambda[dx] = \frac{e^{\lambda x}}{\mathbb{E}[e^{\lambda X}]} \mathbb{P}[dx]. \tag{2.3.14}$$

Note that $\mathbb{P}_\lambda$ is also supported on $[a,b]$. Since $\mathbb{E}[X] = 0$ we have $\Psi'_X(0) = 0$. We deduce from (2.3.6) that

$$\Psi''_X(\lambda) = \frac{1}{\mathbb{E}[e^{\lambda X}]} \mathbb{E}[X^2 e^{\lambda X}] - \left(\frac{\mathbb{E}[X e^{\lambda X}]}{\mathbb{E}[e^{\lambda X}]}\right)^2$$

$$= \int_\mathbb{R} x^2 \mathbb{P}_\lambda[dx] - \left(\int_\mathbb{R} x \mathbb{P}_\lambda[dx]\right)^2.$$

The last term is the variance of a random variable Z with probability distribution $\mathbb{P}_\lambda$. Since $\mathbb{P}_\lambda$ is supported in $[a,b]$ we have $Z \in [a,b]$ and we deduce

$$\Psi_X''(\lambda) = \mathrm{Var}[Z] \leq \frac{(b-a)^2}{4}. \tag{2.3.15}$$

Using the Taylor approximation with Lagrange remainder we deduce that for some $\xi \in [0, \lambda]$ we have

$$\Psi_X(\lambda) = \underbrace{\Psi_X(0) + \lambda \Psi_X'(0)}_{=0} + \frac{1}{2} \Psi_X''(\xi) \leq \frac{\lambda^2 (b-a)^2}{8}.$$

Hence $X \in \mathbb{G}((b-a)^2/4)$. $\square$

Hoeffding's Lemma shows that if R is a Rademacher random variable, then $R \in \mathbb{G}(1)$ as in Example 2.53, which is an improvement over Proposition 2.54.

If $R_1, \ldots, R_n$ are independent Rademacher random variables, then for any $c_1, \ldots, c_n \in [-1, 1]$ we have $c_k R_k \in \mathbb{G}(1)$ and we deduce from Hoeffding's inequality that

$$\mathbb{P}\left[\frac{1}{n} \big| c_1 R_1 + \cdots + c_n R_n \big| > r \right] \leq 2 e^{-\frac{nr^2}{2}}. \tag{2.3.16}$$

Example 2.57 (The Poincaré phenomenon). Suppose that X is a standard normal random variable and $Y = X^2$

$$\mathbb{M}_Y(\lambda) = \mathbb{E}\left[e^{\lambda X^2}\right] = \frac{1}{\sqrt{2\pi}} \int_\mathbb{R} e^{\frac{(2\lambda-1)x^2}{2}} dt.$$

This integral converges only for $\lambda < \frac{1}{2}$ and in this case it is equal to

$$\mathbb{M}_Y(\lambda) = \frac{1}{\sqrt{1-2\lambda}}.$$

In particular, X^2 is not subgaussian. Note that $\mathbb{E}[Y] = \mathbb{E}[X^2] = 1$. Hence

$$\mathbb{M}_{Y-1}(\lambda) = \frac{e^{-\lambda}}{\sqrt{1-2\lambda}}, \quad \Psi_{Y-1}(\lambda) = -\lambda - \frac{1}{2} \log(1-2\lambda).$$

Since $Y \geq 0$ we have $\mathbb{P}[Y - 1 < y] = 0$ for $y \leq -1$. For $y \in (-1, \infty)$ the supremum

$$I(y) := \sup_{\lambda < 1/2} \left(\lambda y - \Psi_{Y-1}(\lambda) \right)$$

is achieved when

$$\frac{d}{d\lambda}\left(\lambda y - \Psi_{Y-1}(\lambda)\right) = y + 1 - \frac{1}{1-2\lambda} = 0.$$

Solving this equation for λ we get

$$1 - 2\lambda = \frac{1}{y+1} \iff \lambda = \frac{y}{2(y+1)}$$

and

$$I(y) = \frac{y^2}{2(y+1)} + \frac{y}{2(y+1)} - \frac{1}{2}\log(1+y) = \frac{y}{2} - \frac{1}{2}\log(y+1) \geq \frac{y^2}{4}.$$

Hence
$$\mathbb{P}[Y-1<-y] \vee \mathbb{P}[Y-1>y] \leq e^{-\frac{y^2}{4}}, \ \forall y>0.$$
Suppose now that
$$\vec{X} = (X_1, \ldots, X_n)$$
is a Gaussian random vector, where X_k are independent standard normal random variables. The square of its Euclidean norm is the chi-squared random variable
$$Z_n = \|\vec{X}\|^2 = \sum_{k=1}^n X_k^2.$$
We deduce that
$$\mathbb{P}\left[\frac{1}{n}Z_n - 1 > y\right] < e^{-\frac{ny^2}{4}}, \ \forall y > 0$$
$$\mathbb{P}\left[\frac{1}{n}Z_n - 1 < y\right] < e^{-\frac{ny^2}{4}}, \ \forall y \in (-1, 0).$$
Thus, for large n the random vector $\frac{1}{\sqrt{n}}\vec{X}$ is highly concentrated around the unit sphere in $\mathbb{R}^n$. This is sometimes referred to as the *Poincaré phenomenon*. In Exercise 2.52 we describe another proof of this result. □

We conclude this section with a remarkable application of the Poincaré phenomenon. Consider a Gaussian random vector in $\mathbb{R}^n$
$$\vec{X} = (X_1, \ldots, X_n),$$
where the components X_k are independent standard normal random variables. Note that for any unit vector $\vec{u} = (u_1, \ldots, u_n)$ the inner product
$$\langle \vec{u}, \vec{X} \rangle = u_1 X_1 + \cdots + u_n X_n$$
is a mean zero Gaussian random variable. Moreover
$$\operatorname{Var}[\langle \vec{u}, \vec{X} \rangle] = \mathbb{E}[|\langle \vec{u}, \vec{X} \rangle|^2] = 1 = \|\vec{u}\|^2.$$
Suppose that we are now given d such independent[8] random vectors
$$\vec{X}_j = (X_{1,j}, \ldots, X_{n,j}), \ 1 \leq j \leq d.$$
We obtain a random map
$$A : \mathbb{R}^n \to \mathbb{R}^d, \ \mathbb{R}^n \ni \vec{u} \mapsto (Y_1, \ldots, Y_d) := (\langle \vec{u}, \vec{X}_1 \rangle, \ldots, \langle \vec{u}, \vec{X}_d \rangle). \quad (2.3.17)$$
If $\|\vec{u}\| = 1$ components of $A\vec{u}$ are independent standard normal random variables so that $\|A\vec{u}\|^2$ is a chi-squared random variable. We set $B := \frac{1}{\sqrt{d}}A$ and we deduce from Example 2.57 that for any $\varepsilon \in (0,1)$ and any unit vector $\vec{u}$ we have
$$\mathbb{P}\big[\,\big|\|B\vec{u}\|^2 - 1\big| > \varepsilon\,\big] \leq 2e^{-\frac{d\varepsilon^2}{4}}.$$

---
[8]Independence is meant in probabilistic sense, not linear independence.

Suppose now that we have a large cloud of points
$$C = \{x_1, \ldots, x_N\} \subset \mathbb{R}^n.$$
For $1 \leq i < j \leq N$ we write $v_{ij} = x_j - x_i$. We deduce that
$$\mathbb{P}\left[1 - \varepsilon \leq \frac{\|Bv_{ij}\|}{\|v_{ij}\|} \leq 1 + \varepsilon, \; \forall 1 \leq i < j \leq N\right] \leq 2\binom{N}{2} e^{-\frac{d\varepsilon^2}{4}} \leq N^2 e^{-\frac{d\varepsilon^2}{2}}.$$
Now fix a confidence level $0 < p_0 < 1$ and observe that
$$N^2 e^{-\frac{d\varepsilon^2}{2}} < p_0 \iff d\varepsilon^2 > 4\log\frac{N}{p_0} \iff d > \frac{4}{\varepsilon^2}\log\frac{N}{p_0}.$$
We have thus proved the following remarkable result.

Theorem 2.58 (Lindenstrauss-Johnson). *Fix $\varepsilon > 0$ and $p_0 \in (0,1)$ and a cloud of C of N points in $\mathbb{R}^n$. If*
$$d = d(N, \varepsilon, p_0) := \left\lceil \frac{4}{\varepsilon^2} \log \frac{N}{p_0} \right\rceil, \tag{2.3.18}$$
then, with probability at least $1 - p_0$, the random Gaussian map $B = \frac{1}{\sqrt{d}}A$, where A is described by (2.3.17), distorts very little the relative distances between the points in C, i.e.,
$$(1-\varepsilon)\|Bx - By\| \leq \|x - y\| \leq (1+\varepsilon)\|Bx - By\|.$$
$\square$

Remark 2.59. Let us highlight some remarkable features of the above result. Note first that the dimension $d(N, \varepsilon, p_0)$ is *independent* of the dimension of the ambient space $\mathbb{R}^n$ where the cloud C resides. Moreover, $d(N, \varepsilon, p_0)$ is substantially smaller than the size N of the cloud.

For example, if we choose the confidence level $p_0 = 10^{-3}$, the distortion factor $\varepsilon = 10^{-1}$ and the size of the cloud $N = 10^{12}$, then
$$\frac{4}{\varepsilon^2}\log\frac{N}{p_0} = 60 \cdot 10^2 \log 10 < 14 \cdot 10^3 \ll 10^{12}.$$
The cloud C could be chosen in a Hilbert space and we can choose as ambient space the subspace $\mathrm{span}(C)$ that has dimension $n \leq N$. In this case the vectors $Y_k := \frac{1}{\sqrt{N}} \vec{X}_k$, $k = 1, \ldots, d$, have with high confidence norm 1.
$$\mathbb{P}\big[\,\big|\|Y_k\| - 1\big| > \delta, \; \forall 1 \leq k \leq d\,\big] \leq 2de^{-\frac{N\delta^2}{4}}.$$
They are also, with high confidence, mutually orthogonal. Indeed, Exercise 2.49 shows that for $|r| < \frac{1}{2}$
$$\mathbb{P}\big[\,|\langle Y_i, Y_j\rangle| > r, \; \forall i < j\,\big] \leq 2\binom{d}{2} e^{-\frac{Nr^2}{12}}.$$

This shows that the operator $\frac{1}{\sqrt{N}}A$ is with high confidence very close to the orthogonal projection $P_{\vec{X}_1,\ldots,\vec{X}_d}$ onto the random d-dimensional[9] subspace span$\{\vec{X}_1,\ldots,\vec{X}_d\}$. This shows that, with high confidence, the operator

$$\sqrt{\frac{N}{d}}P_{\vec{X}_1,\ldots,\vec{X}_d}$$

distorts very little the distances between the points in C. The projected cloud has identical size, similar geometry but lives in a subspace of much smaller dimension. □

2.4 Uniform laws of large numbers

Fix a Borel probability measure μ on $\mathbb{R}$. Suppose that

$$X_n : (\Omega, \mathcal{S}, \mathbb{P}) \to \mathbb{R}, \ n \in \mathbb{N}$$

is a sequence of i.i.d. random variables with common probability distribution μ. For any Borel set $B \subset \mathbb{R}$ the random variables $\boldsymbol{I}_B(X_n)$ are i.i.d. and have finite means

$$m_B := \mathbb{P}[X_1 \in B] = \mu[B].$$

The Strong Law of Large Numbers shows that the empirical means

$$M_n[B] := \frac{1}{n}\bigl(\boldsymbol{I}_B(X_1) + \cdots + \boldsymbol{I}_B(X_n)\bigr) = \frac{\#\{1 \leq k \leq n;\ X_k \in B\}}{n}$$

converge a.s. to $\mu[B]$. In particular, this provides an asymptotic confirmation of the "frequentist" interpretation of probability as the ratio of favorable cases to the number of possible cases.

If we choose B of the form $(-\infty, x]$, then we obtain the *empirical cdf*

$$F_n(x) = M_n\bigl[(-\infty, x]\bigr] = \frac{1}{n}\frac{\#\{1 \leq k \leq n;\ X_k \leq x\}}{n}.$$

This is a *random* quantity (variable), $F_n(x) = F_n(x,\omega)$, $\omega \in \Omega$. For each $n \in \mathbb{N}$, the collection $\bigl(F_n(x)\bigr)_{x\in\mathbb{R}}$ is an example of *empirical process*.

For any $x \in \mathbb{R}$, the random variable $F_n(x)$ converges a.s. to $F(x)$, where F is the cdf of μ

$$F(x) = \mu\bigl[(-\infty, x]\bigr].$$

For $x \in \Omega$ the set $N_x \subset \Omega$ such that $F_n(x,\omega)$ does not converge to $F(x)$ is negligible but, since $\mathbb{R}$ is not countable, the union

$$N = \bigcup_{x \in \mathbb{R}} N_x$$

need not be negligible. In other words, the set of ω's such that the *functions* $F_n(-,\omega)$ do not converge pointwisely to the function $F(-)$ need not by negligible. We will show that this is not the case.

---

[9] It is not hard to see that dim span$\{\vec{X}_1,\ldots,\vec{X}_d\} = d$ a.s.

2.4.1 The Glivenko-Cantelli theorem

Define
$$D_n = D_n^F : \Omega \to [0, \infty), \quad D_n(\omega) := \sup_{x \in \mathbb{R}} \big| F_n(x, \omega) - F(x) \big|. \qquad (2.4.1)$$

For a fixed $\omega \in \Omega$ the sequence of functions $\big(F_n(-, \omega)\big)_{n \in \Omega}$ converges uniformly to $F(-)$ if and only if $D_n(\omega) \to 0$. We will show that this is the case for almost all ω.

Denote by $U(y)$ the cdf of the uniform distribution on $[0, 1]$,
$$U(y) = \begin{cases} 0, & y < 0, \\ y, & y \in [0, 1], \\ 1, & y > 1, \end{cases}$$

and by Q the quantile of F defined in (1.2.5), $Q : [0, 1] \to \overline{\mathbb{R}}$
$$Q(\ell) := \inf \big\{ x : \; \ell \leq F(x) \big\} = \inf F^{-1}\big([\ell, \infty]\big) = \inf F^{-1}\big([\ell, 1]\big).$$

Lemma 2.60. *The function D_n^F is measurable and $D_n^F \leq D_n^U$, with equality if F is continuous.*

Proof. Let us first show that D_n is indeed measurable. We will show that
$$D_n = \sup_{x \in \mathbb{Q}} \big| F_n(x) - F(x) \big|. \qquad (2.4.2)$$

According to Proposition 1.17(iii) the quantity in the right-hand-side is measurable.

Fix $\omega \in \Omega$. There exists then a sequence of real numbers $(x_n)_{n \in \mathbb{N}}$ such that
$$\lim_{n \to \infty} \big| F_n(x_n, \omega) - F(x) \big| = D_n(\omega).$$

Now observe that the functions $x \mapsto F_n(x, \omega)$, $F(x)$ are right-continuous so there exists a sequence of rational numbers $(q_n)_{n \in \mathbb{N}}$ such that $q_n > x_n$ and
$$\Big| \big| F_n(x_n, \omega) - F(x_n) \big| - \big| F_n(q_n, \omega) - F(q_n) \big| \Big| < \frac{1}{n}.$$

Hence
$$\lim_{n \to \infty} \big| F_n(q_n, \omega) - F(q_n) \big| = \lim_{n \to \infty} \big| F_n(x_n, \omega) - F(x_n) \big|$$

thus proving that the functions (2.4.2) are measurable.

Consider now a sequence of i.i.d. random variables $(Y_n)_{n \in \mathbb{N}}$ uniformly distributed on $[0, 1]$. Denote by U_n the associated empirical c.d.f.-s,
$$U_n(x) = \frac{1}{n} \sum_{k=1}^{n} \boldsymbol{I}_{(-\infty, x]}(Y_k).$$

Then $X_n = Q(Y_n)$ are i.i.d. with common cdf F. Note that
$$U_n\big(F(x)\big) - F(x) = \frac{1}{n} \sum_{k=1}^{n} \boldsymbol{I}_{Y_k \leq F(x)} - F(x)$$

$$\stackrel{(1.2.6)}{=} \frac{1}{n} \sum_{k=1}^{n} \boldsymbol{I}_{Q(Y_k) \leq x} - F(x) = F_n(x) - F(x).$$

Thus

$$D_n^F = \sup_{x \in \mathbb{R}} \big| F_n(x) - F(x) \big| = \sup_{x \in \mathbb{R}} \big| U_n(F(x)) - U(F(x)) \big|$$

$$\leq \sup_{y \in \mathbb{R}} \big| U_n(y) - U(y) \big| = D_n^U.$$

Observe that if F is continuous, then $\forall y \in (0,1), \exists x \in \mathbb{R}$, such that $F(x) = y$ so

$$\sum_{x \in \mathbb{R}} \big| U(F_n(x)) - U(F(x)) \big| = \sup_{y \in \mathbb{R}} \big| U_n(y) - U(y) \big|.$$

$\square$

Theorem 2.61 (Glivenko-Cantelli). *Suppose that $(X_n)_{n \in \mathbb{N}}$ is a sequence of i.i.d. random variables with common distribution μ and cdf F. Denote by $F_n(x)$ the empirical cdf-s*

$$F_n(x) = \frac{1}{n} \sum_{k=1}^{n} \boldsymbol{I}_{(-\infty, x]}(X_k).$$

Then, almost surely, $F_n(x)$ converges uniformly to $F(x)$, i.e.,

$$D_n^F \to 0 \text{ a.s. as } n \to \infty,$$

where D_n is defined by (2.4.1).

Proof. Lemma 2.60 shows that it suffices to prove the theorem only in the special case when that random variables are uniformly distributed. Thus we assume $F = U$.

Fix a partition $\mathcal{P}$ of $[0,1]$, $\mathcal{P} = \{0 = x_0 < x_1 < x_2 < \cdots < x_m = 1\}$. Set

$$\|\mathcal{P}\| := \max_{1 \leq k \leq m} (x_k - x_{k-1}).$$

For $x \in [x_{k-1}, x_k)$ and $n \in \mathbb{N}$ we have

$$U_n(x_{k-1}) \leq U_n(x) \leq U_n(x_k), \ |U_n(x) - x| = x - U_n(x),$$

$$x_k - U_n(x_k) \leq (x_k - x) + x - U(x) \leq \|\mathcal{P}\| + (x - U_n(x)),$$

$$x - U_n(x) \leq x - x_{k-1}+ \leq x_k - U_n(x_{k-1}) \leq \|\mathcal{P}\| + x_{k-1} - U_n(x_{k-1}),$$

$$x_k - U_n(x_k) - \|\mathcal{P}\| \leq x - U_n(x) \leq \|\mathcal{P}\| + x_{k-1} - U_n(x_{k-1}).$$

If we set

$$D_n(\mathcal{P}) := \max_{1 \leq k \leq m} x_k - U_n(x_k),$$

we deduce that for any partition $\mathcal{P}$ of $[0,1]$ we have

$$D_n(\mathcal{P}) \geq 0, \ D_n(\mathcal{P}) - \|\mathcal{P}\| \leq D_n^U \leq D_n(\mathcal{P}) + \|\mathcal{P}\|.$$

Now consider a sequence $\mathcal{P}_k$ of partitions such that

$$\|\mathcal{P}_k\| < \frac{1}{k}, \quad \forall k \in \mathbb{N}.$$

As we mentioned at the beginning of this subsection, the Strong Law of Large Numbers implies that

$$U(x) - U_n(x) = x - U_n(x) \to 0 \text{ a.s. as } n \to \infty.$$

Thus

$$\forall k \in \mathbb{N} \quad \lim_{n \to \infty} D_n(\mathcal{P}_k) = 0 \text{ a.s.}$$

Hence

$$-\frac{1}{k} < -\|\mathcal{P}_k\| \leq \liminf_{n \to \infty} D_n \leq \limsup_{n \to \infty} D_n \leq \|\mathcal{P}_k\| < \frac{1}{k}, \quad \forall k \in \mathbb{N}.$$

Letting $k \to \infty$ we deduce the desired conclusion. □

Remark 2.62. Suppose that $(X_n)_{n \in \mathbb{N}}$ is a sequence of i.i.d. random variables with common cdf $F(x)$. Form the empirical (cumulative) distribution function

$$F_n(x) = \frac{1}{n} \sum_{k=1}^{n} \boldsymbol{I}_{(-\infty,x]}(X_k),$$

and the corresponding deviation $D_n := \sup_{x \in \mathbb{R}} |F_n(x) - F(x)|$. The Glivenko-Cantelli theorem shows that $D_n \to 0$ a.s.

On the other hand, observe that for each $x \in \mathbb{R}$ the random variables $\boldsymbol{I}_{(-\infty,x]}(X_n)$ are i.i.d. random Bernoulli random variables with success probability $F(x)$. The central limit theorem shows

$$\sqrt{n}\big(F_n(x) - F(x)\big) \Rightarrow N\big(0, F(x)(1 - F(x))\big).$$

The *Kolmogorov-Smirnov theorem* states that

$$\sqrt{n} D_n \Rightarrow D_\infty, \quad \mathbb{P}\big[D_\infty > c\big] = 2 \sum_{m \geq 1} (-1)^{m-1} e^{-2c^2 m^2}.$$

For an "elementary" proof of this fact we refer to [57]. For a more sophisticated proof that reveals the significance of the strange series above we refer to [12] or [51]. □

2.4.2 VC-theory

We want to present a generalization of the Glivenko-Cantelli theorem based on ideas pioneered by V. N. Vapnik and A. Ja. Cervonenkis [151] that turned out to be very useful in machine learning. Our presentation follows [129, Chap. II]. For more recent developments we refer to [51; 70; 150; 156].

Fix a Borel probability measure μ on $\mathscr{X} := \mathbb{R}^N$. Any sequence of i.i.d. random vectors

$$X_n : (\Omega, \mathcal{S}, \mathbb{P}) \to \mathscr{X} = \mathbb{R}^N$$

with common distribution μ defines empirical probabilities

$$P_n := \frac{1}{n}\sum_{k=1}^{n}\delta_{X_k}.$$

The empirical probabilities are random measures on $(\mathscr{X}, \mathcal{B}_{\mathscr{X}})$. More precisely, for any Borel subset $B \subset \mathscr{X}$, $P_n[B]$ is the random variable

$$P_n[B] = \frac{1}{n}\sum_{k=1}^{n} I_B(X_k).$$

Suppose we are given a family $\mathcal{F} := (B_t)_{t \in T}$ of Borel subsets of $\mathscr{X} = \mathbb{R}^N$, $N \geq 1$, parametrized by a set T. We assume T is a Borel subset of another Euclidean space $\mathbb{R}^p$ and we denote by $\mathcal{B}_T$ its Borel algebra. For example, we can choose $\mathscr{X} = \mathbb{R}$,

$$B_t = (-\infty, t], \quad t \in T = \mathbb{R}.$$

For each $n \in \mathbb{N}$ we obtain a stochastic process parametrized by T,

$$P_n : T \times \Omega \to [0,1], \quad P_n(t, \omega) = P_n[B_t](\omega) = \frac{1}{n}\sum_{k=1}^{n} I_{B_t}(X_k(\omega)).$$

For each $n \in \mathbb{N}$ we obtain a random variable

$$P_n(t) : \Omega \to [0,1], \quad \omega \mapsto P_n(t, \omega).$$

The collection of random variables $(P_n(-))_{t \in T}$ is an example of *empirical process*. Note that

$$\mathbb{E}[P_n(t)] = \mu[B_t], \quad \mathrm{Var}[P_n(t)] = \frac{1}{n}\mathrm{Var}[P_1(t)] = \frac{v_t}{n},$$

where

$$v_t := \mu[B_t]\bigl(1 - \mu[B_t]\bigr) \leq \frac{1}{4}.$$

The Strong Law of Large Numbers implies that

$$Z_n(t) := P_n(t) - \mu[B_t] = \frac{1}{n}\sum_{k=1}^{n}\bigl(Y_k(t) - \mathbb{E}[Y_k(t)]\bigr) \to 0 \text{ a.s. as } n \to \infty.$$

Moreover, Chebyshev's inequality shows that

$$\mathbb{P}\bigl[\,|Z_n(t)| > \varepsilon\,\bigr] \leq \frac{v_t}{n\varepsilon} \leq \frac{1}{4n\varepsilon^2}. \tag{2.4.3}$$

Can we conclude that $Z_n(t) \to 0$ uniformly a.s. in the precise sense described in Glivenko-Cantelli's theorem?

To proceed further we will need to make some further assumptions on the family $(B_t)_{t \in T}$. Later we will have a few things to say about their feasibility. Set

$$D_n := \sup_{t \in T}|Z_n(t)| : \Omega \to [0,1].$$

Here is our first measure theoretic assumption.

M₁. *The function D_n is measurable*

To prove that $D_n \to 0$ a.s. we will employ a different strategy than before. More precisely we intend to show that, under certain assumptions on the family $(B_t)_{t \in T}$, the probability $\mathbb{P}[D_n > \varepsilon]$ decays very fast as $n \to \infty$, for any $\varepsilon > 0$. This will guarantee that the series

$$\sum_{n \in \mathbb{N}} \mathbb{P}[D_n > \varepsilon]$$

is convergent for any $\varepsilon > 0$ and thus, according to Corollary 1.141, the sequence D_n converges a.s. to 0. To obtain these tail estimates we will rely on some clever symmetrization tricks.

To state the first symmetrization result choose another sequence $X'_n : \Omega \to \mathscr{X}$, $n \in \mathbb{N}$, of i.i.d. random variables, independent of $(X_n)_{n \in \mathbb{N}}$, but with the same distribution. Set

$$Y'_k(t) := \boldsymbol{I}_{B_t}(X'_k), \quad Z'_n(t) := \frac{1}{n} \sum_{k=1}^n \bigl(Y'_k(t) - \mu[Y'_k(t)]\bigr), \quad \forall n \in \mathbb{N},\ t \in T,$$

$$D_{n,n} := \sup_{t \in T} \bigl| Z'_n(t) - Z_n(t) \bigr|. \qquad (2.4.4)$$

Equivalently,

$$D_{n,n} = \sup_{t \in T} \frac{1}{n} \Bigl| \sum_{k=1}^n \bigl(Y_{n+k}(t) - Y_k(t)\bigr) \Bigr|.$$

Here are our next measure theoretic assumption.

M$'_1$. *The function $D_{n,n}$ is measurable*

M$_2$. *For any $n > 0$ and any $\varepsilon > 0$ there exists a measurable map*

$$\tau : \bigl(\Omega, \sigma(X_1, \ldots, X_n)\bigr) \to (T, \mathcal{B}_T)$$

such that $|Z_n(\tau)| > \varepsilon$ on $\{D_n > \varepsilon\}$, i.e.,

$$D_n(\omega) > \varepsilon \Rightarrow \bigl| Z_n(\tau(\omega)) \bigr| > \varepsilon. \qquad (2.4.5)$$

Lemma 2.63 (First symmetrization lemma).

$$\mathbb{P}[D_n > \varepsilon] \leq 2\mathbb{P}[D_{n,n} > \varepsilon/2], \quad \forall \varepsilon > 0,\ \forall n > \frac{1}{2\varepsilon^2}. \qquad (2.4.6)$$

Proof. Choose a measurable map $\tau : \bigl(\Omega, \sigma(X_1, \ldots, X_n)\bigr) \to (T, \mathcal{B}_T)$ satisfying **M$_2$**. Then τ is independent of Z'_n and we deduce

$$\mathbb{E}\bigl[\boldsymbol{I}_{\{|Z'_n(\tau)| \leq \varepsilon/2\}} \,\|\, \sigma(X_1, \ldots, X_n)\bigr] = \mathbb{E}\bigl[\boldsymbol{I}_{\{|Z'_n(\tau(x_1,\ldots,x_n))| \leq \varepsilon/2\}}\bigr] \stackrel{(2.4.3)}{\geq} 1 - \frac{1}{n\varepsilon^2},$$

$$\mathbb{P}\bigl[|Z'_n(\tau)| \leq \varepsilon/2 \,\|\, D_n\bigr] = \mathbb{E}\Bigl[\mathbb{E}\bigl[\boldsymbol{I}_{\{|Z'_n(\tau)| \leq \varepsilon/2\}} \,\|\, \sigma(X_1, \ldots, X_n)\bigr] \,\|\, D_n\Bigr]$$

$$\geq 1 - \frac{1}{n\varepsilon^2}.$$

Integrating over $\{D_n > \varepsilon\}$ we deduce

$$\left(1 - \frac{1}{n\varepsilon^2}\right)\mathbb{P}[\,D_n > \varepsilon\,] \leq \mathbb{P}[\,|Z_n'(\tau)| \leq \varepsilon/2,\ D_n > \varepsilon\,]$$

$$\stackrel{(2.4.5)}{\leq} \mathbb{P}[\,|Z_n'(\tau)| \leq \varepsilon/2,\ |Z_n(\tau)| > \varepsilon\,] \leq \mathbb{P}[\,|Z_n'(\tau) - Z_n(\tau)| > \varepsilon/2\,]$$

$$\leq \mathbb{P}\big[\sup_{t \in T} |Z_n'(t) - Z_n(t)| > \varepsilon/2\big].$$

The inequality (2.4.6) follows by observing that for $n > \frac{1}{2\varepsilon^2}$ we have $1 - \frac{1}{n\varepsilon^2} > \frac{1}{2}$.
$\square$

Note that the variables $(Y_n(t))_{n\in\mathbb{N}}$ are independent Bernoulli random variables with success probability $p_t = \mu[\,B_t\,]$. The random variables $(Y_n'(t))$ are also of the same kind and also independent of the Y's. The key gain is that the random variables

$$\Xi_n = Y_k'(t) - Y_k(t)$$

are symmetric, i.e., Ξ_n and $-\Xi_n$ have the same distributions. They take only the values $-1, 0, 1$ with distributions

$$\mathbb{P}[\,\Xi_t = \pm 1\,] = p_1(1 - p_t),\ \ \mathbb{P}[\,\Xi_t = 0\,] = 1 - 2p_t(1 - p_t).$$

The advantage of working with symmetric random variables will become apparent after describe our second symmetrization trick known as *Rademacher symmetrization*.

Recall that a Rademacher random variable is a random variable that takes the only the values ± 1, with equal probabilities. Suppose that $(R_n)_{n\in\mathbb{N}}$ is sequence of independent Rademacher random variables[10] that are also independent of the variables X_n and X_n'.

Observe that the random variables $\overline{Y}_n := R_n Y_n$ are also symmetric.

Lemma 2.64 (Rademacher symmetrization). *For any $n \in \mathbb{N}$ we have*

$$\mathbb{P}\left[\sup_{t\in\mathbb{R}} \frac{1}{n}\left|\sum_{k=1}^{n}\big(Y_k'(t) - Y_k(t)\big)\right| > \frac{\varepsilon}{2}\right] \leq 2\mathbb{P}\left[\sup_{t\in\mathbb{R}} \frac{1}{n}\left|\sum_{k=1}^{n}\overline{Y}_k(t)\right| > \frac{\varepsilon}{4}\right]. \quad (2.4.7)$$

Proof. The key observation is that, because $\Xi_k(t) = Y_k'(t) - Y_k(t)$ is symmetric, it has the same distribution as $R_k \Xi_k(t)$. Set

$$S_n(t) := \frac{1}{n}\sum_{k=1}^{n} R_k Y_k(t),\ \ S_n'(t) := \frac{1}{n}\sum_{k=1}^{n} R_k Y_k'(t),\ \ \overline{S}_n(t) := \sum_{k=1}^{n} \overline{Y}_k(t)$$

---

[10] Here we are making a tacit assumption that there exists such a sequence random variables R_n defined on Ω. For example if we can choose Ω to be the probability space $(\mathscr{X}, \mu^{\otimes\mathbb{N}}) \otimes (\mathscr{X}, \mu^{\otimes\mathbb{N}}) \otimes \{-1,1\}^{\otimes\mathbb{N}}$ all the above choices are possible. The choice of Ω is irrelevant because the Glivenko-Cantelli theorem is a result about $(\mathscr{X}, \mu^{\otimes\mathbb{N}})$.

$$\mathbb{P}\left[\sup_{t\in\mathbb{R}}\frac{1}{n}\left|\sum_{k=1}^{n}(Y_k'(t)-Y_k(t))\right|>\frac{\varepsilon}{2}\right]=\mathbb{P}\left[\sup_{t\in\mathbb{R}}\frac{1}{n}|S_n(t)-S_n'(t)|>\frac{\varepsilon}{2}\right]$$
$$\leq \mathbb{P}\left[\sup_{t\in\mathbb{R}}\frac{1}{n}|S_n(t)|>\frac{\varepsilon}{4}\right]+\mathbb{P}\left[\sup_{t\in\mathbb{R}}\frac{1}{n}|S_n'(t)|>\frac{\varepsilon}{4}\right]=2\mathbb{P}\left[\sup_{t\in\mathbb{R}}\frac{1}{n}|\bar{S}_n(t)|>\frac{\varepsilon}{4}\right],$$

where we used the fact that $R_k Y_k'(t)$ and $R_k Y_k(t)$ have the same distributions. □

Putting together all of the above we deduce

$$\mathbb{P}[\,D_n>\varepsilon\,]\leq 4\mathbb{P}\left[\sup_{t\in\mathbb{R}}\frac{1}{n}\left|\sum_{k=1}^{n}R_k Y_k\right|>\frac{\varepsilon}{4}\right],\quad \forall \varepsilon>0,\ n>\frac{1}{2\varepsilon^2}. \qquad (2.4.8)$$

To make further progress we condition on the variables (X_n) and we deduce

$$\mathbb{P}\left[\sup_{t\in\mathbb{R}}\frac{1}{n}\left|\sum_{k=1}^{n}R_k Y_k(t)\right|>\frac{\varepsilon}{4}\right]$$

$$=\int_{\mathscr{X}^n}\mathbb{P}\left[\sup_{t\in\mathbb{R}}\underbrace{\frac{1}{n}\left|\sum_{k=1}^{n}R_k y_k(t,\vec{x})\right|}_{=:S_t(\vec{x})}>\frac{\varepsilon}{4}\right]\mu^{\otimes n}[\,dx_1\cdots dx_n\,],$$

where $\vec{x}:=(x_1,\ldots,x_n)\in\mathscr{X}^n$ and
$$y_k(t,\vec{x})=\boldsymbol{I}_{B_t}(x_k)\in\{0,1\},\quad \forall k=1,\ldots,n,\ t\in T.$$

Hence

$$\mathbb{P}[\,D_n>\varepsilon\,]\leq 4\int_{\mathscr{X}^n}\mathbb{P}\left[\sup_{t\in\mathbb{R}}S_t(\vec{x})>\frac{\varepsilon}{4}\right]\mu^{\otimes n}[\,dx_1\cdots dx_n\,]. \qquad (2.4.9)$$

For each $n\in\mathbb{N}$, $t\in T$ and $\vec{x}\in\mathscr{X}^n$ we set $\mathbb{I}_n:=\{1,\ldots,n\}$,
$$C_t(\vec{x}):=\{\,k\in\mathbb{I}_n;\ y_k(t,\vec{x})=1\,\}=\{\,k\in\mathbb{I}_n;\ x_k\in B_t\,\}.$$
Roughly speaking, $C_t(\vec{x})=B_t\cap\{x_1,\ldots,x_n\}$.
$$\mathcal{C}_n(\vec{x}):=\{\,C\subset\mathbb{I}_n;\ \exists t\in T,\ C=C_t(\vec{x})\,\}.$$

For every $C\subset\mathbb{I}_n$ we set
$$S_C:=\frac{1}{n}\left|\sum_{k\in C}R_k\right|,$$

so that $S_t(\vec{x})=S_{C_t(\vec{x})}$. Hence

$$\mathbb{P}\left[\sup_{t\in T}S_t(\vec{x})>\varepsilon/4\right]=\mathbb{P}\left[\sup_{C\in\mathcal{C}_n(\vec{x})}S_C>\varepsilon/4\right]\leq\sum_{C\in\mathcal{C}_n(x)}\mathbb{P}[\,S_C>\varepsilon/4\,].$$

We can now finally understand the role of the Rademacher symmetrization. The sums

$$\sum_{k=1}^{n}R_k y_k(t,\vec{x})$$

are of the type appearing in Hoeffding's inequality (2.3.12), where $R_k y_k(t, \vec{x}) \in \mathbb{G}(1)$ by the computation in Example 2.53. We deduce

$$\mathbb{P}\big[S_C > \varepsilon/4 \big] \leq 2e^{-n^2\varepsilon^2/32}, \ \forall C \subset \mathbb{I}_n.$$

We deduce
$$\mathbb{P}\big[\sup_{t \in T} S_t(\vec{x}) > \varepsilon/4 \big] \leq 2|\mathcal{C}_n(\vec{x})| e^{-n\varepsilon^2/32}. \qquad (2.4.10)$$

Using this in (2.4.9) we deduce

$$\mathbb{P}\big[D_n > \varepsilon \big] \leq 8 e^{-n\varepsilon^2/32} \int_{\mathscr{X}^n} |\mathcal{C}_n(\vec{x})| \, \mu^{\otimes n}\big[dx_1 \cdots dx_n \big]. \qquad (2.4.11)$$

We have a rough bound $|\mathcal{C}_n(\vec{x})| \leq 2^n$ but it is not helpful. At this point we add our last and crucial assumption.

VC. *The family $\mathcal{F} = (B_t)_{t \in T}$ satisfies VC-condition.*[11] *This means that there exists $d \in \mathbb{N}$ such that*

$$\sup_{\vec{x} \in \mathscr{X}^n} |\mathcal{C}_n(\vec{x})| = O(n^d) \ \text{as } n \to \infty.$$

With this assumption in place we deduce that there exists $K > 0$ such that

$$2|\mathcal{C}_n(\vec{x})| \leq K(n^d + 1), \ \forall n \in \mathbb{N}, \ \forall \vec{x} \in \mathscr{X}^n$$

so that
$$\mathbb{P}\big[D_n > \varepsilon \big] \leq 8K e^{-n\varepsilon^2/32}(n^d + 1). \qquad (2.4.12)$$

In the above estimate the constant K is *independent* of the distribution μ. Since the series

$$\sum_{n \in \mathbb{N}} e^{-n\varepsilon^2/32}(n^d + 1) < \infty, \ \forall \varepsilon > 0,$$

we deduce that $D_n \to 0$ a.s. We have thus proved the following wide ranging generalization of the Glivenko-Cantelli theorem.

Theorem 2.65 (Vapnik-Chervonenkis). *Suppose that $\mathcal{F} = (B_t)_{t \in T}$ is a family of Borel subsets of $\mathscr{X} = \mathbb{R}^N$ parametrized by a Borel subset T of some Euclidean space, and μ is a Borel probability measure on $\mathscr{X}$. Assume that $\mu, \mathcal{F}$ satisfy the conditions $\mathbf{M_1}, \mathbf{M_1}, \mathbf{M_2}$.*

Fix a sequence of independent random vectors $X_n : \Omega \to \mathscr{X}$ with common distribution μ. Form the empirical measures

$$\mu_n : \Omega \times \mathcal{B}_X \to [0, \infty], \ \mu_n^\omega[B] = \frac{1}{n} \sum_{k=1}^n \mathbf{I}_B\big[X_k(\omega) \big].$$

If $\mathcal{F}$ satisfies the VC-condition, then, almost surely,

$$\mu_n[B] \to \mu[B] \ \text{as } n \to \infty$$

uniformly in $B \in \mathcal{F}$, i.e.,

$$\lim_{n \to \infty} \sup_{B \in \mathcal{F}} \big| \mu_n[B] - \mu[B] \big| = 0 \ \text{a.s.}$$

$\square$

---

[11]VC = Vapnik-Chervonenkis.

Remark 2.66. (a) The technical assumptions $\mathbf{M_1}$, $\mathbf{M'_1}$, $\mathbf{M_2}$ are measure-theoretic in nature and are automatically satisfied if the space of parameters T is countable. There are quite general (and very technical) results that guarantee that these results hold in a rather broad range of situations, [129, Appendix C].

There are more sophisticated ways of bypassing $\mathbf{M_1}$ and $\mathbf{M'_1}$ and we refer to [51], [70] or [150] for details. Section 1.1 in [150] does a particularly clear and efficient job of describing these measurability issues and the methods that were proposed over the years to circumvent them.

If one assumes the condition $\mathbf{VC}$, one can bypass assumption $\mathbf{M_2}$ by using a weaker form of the first symmetrization trick. Observe first that

$$\mathbb{E}[D_n] \leq \mathbb{E}[D_{n,n}]. \tag{2.4.13}$$

Indeed

$$\frac{1}{n}\left|\sum_{k=1}^{n}(Y_k(t) - \mathbb{E}[Y_k(t)])\right| = \frac{1}{n}\left|\mathbb{E}\left[\sum_{k=1}^{n} Y_k(t) - \mathbb{E}[Y'_k(t)] \,\|\, Y_k, 1 \leq k \leq n\right]\right|$$

$$= \frac{1}{n}\left|\mathbb{E}\left[\sum_{k=1}^{n}(Y_k(t) - Y'_k(t)) \,\|\, Y_k, 1 \leq k \leq n\right]\right|$$

$$\leq \mathbb{E}\left[\left|\sum_{k=1}^{n}(Y_k(t) - Y'_k(t))\right| \,\|\, Y_k, 1 \leq k \leq n\right] \leq \mathbb{E}[D_{n,n} \,\|\, Y_k, 1 \leq k \leq n].$$

Hence

$$D_n(t) = \sup \frac{1}{n}\left|\sum_{k=1}^{n}(Y_k(t) - \mathbb{E}[Y_k(t)])\right| \leq \mathbb{E}[D_{n,n} \,\|\, Y_k, 1 \leq k \leq n].$$

By taking the expectations of both sides of the above inequality we obtain (2.4.13). A similar argument as in the proof of the Rademacher symmetrization lemma yields

$$\mathbb{E}[D_{n,n}] \leq 2\,\mathbb{E}\underbrace{\left[\sup_{t\in T}\frac{1}{n}\left|\sum_{k=1}^{n}\overline{Y}_k(t)\right|\right]}_{=:\mathcal{R}_n(T)}.$$

The sequence $\mathcal{R}_n(T)$ is called the *Rademacher complexity* of the family $(B_t)_{t\in T}$.

Azuma's inequality (3.1.14), a refined concentration inequality, shows that D_n is highly concentrated around its mean. The $\mathbf{VC}$ condition can be used to show that the Rademacher complexity goes to 0 as $n \to \infty$. Thus the mean of D_n goes to 0 as $n \to \infty$. Combining these facts one can obtain an inequality very similar to (2.4.11). For details we refer to [156, Sec. 4.2] or Exercise 3.20.

(b) One can obtain bounds for the tails of D_n by a Chernoff-like technique, by obtaining bounds for $\mathbb{E}[\Phi(D_n)]$, where $\Phi : [0, \infty) \to \mathbb{R}$ is a convex increasing function; see Exercise 2.53. We refer to [130] or [150] for details. □

The key assumption is **VC** and we want to discuss it in some detail and describe several nontrivial examples of families of sets satisfying this condition.

Fix an ambient space $\mathscr{X}$ and $\mathcal{F} \subset 2^{\mathscr{X}}$ a family of subsets of $\mathscr{X}$. The *shadow* of $\mathcal{F}$ on a subset A is the family

$$\mathcal{F}_A := \{ F \cap A;\ F \in \mathcal{F} \} \subset 2^A.$$

Note that for a finite set A we have

$$|\mathcal{F}_A| \leq 2^{|A|}.$$

When we have equality above we say that A is *shattered* by $\mathcal{F}$. Thus, A is shattered by $\mathcal{F}$ if *any* subset of A is in the shadow of $\mathcal{F}$. We set

$$s_{\mathcal{F}}(n) := \max\{\, |\mathcal{F}_A|;\ |A| = n \,\}.$$

Thus $s_{\mathcal{F}}(n)$ is the size of the largest shadow on a subset of $\mathscr{X}$ of cardinality n. Note that $s_{\mathcal{F}}(n) \leq 2^n$.

For a nonempty $\mathcal{F}$ we define its *VC-dimension* to be

$$\dim_{VC}(\mathcal{F}) := \max\{\, n \in \mathbb{N};\ s_{\mathcal{F}}(n) = 2^n \,\}.$$

Thus, any subset A such that $|A| \leq \dim_{VC}(\mathcal{F})$ is shattered by $\mathcal{F}$. In other words, if $k = \dim_{VC}(\mathcal{F})$, then for any $n \leq k$ we have

$$s_{\mathcal{F}}(n) = 2^n = \sum_{j=1}^{\min(n,k)} \binom{n}{j}.$$

We have the following remarkable dichotomy. For proof we refer to [51, Thm. 4.1.2] or [70, Thm. 3.6.3].

Theorem 2.67 (Sauer Lemma). *If* $\dim_{VC}(\mathcal{F}) = k < \infty$, *then*

$$\forall n > k:\ s_{\mathcal{F}}(n) \leq P_k(n) := \sum_{j=0}^{\min(n,k)} \binom{n}{j}.$$

Note that $P_k(n)$ is a polynomial of degree k in n. □

Define the *density* of $\mathcal{F}$ to be

$$\operatorname{dens}(\mathcal{F}) = \inf\{r > 0;\ s_{\mathcal{F}}(n) = O(n^r),\ \text{as } n \to \infty\,\}.$$

We see that the family $\mathcal{F}$ satisfies the condition **VC** if and only if $\operatorname{dens}(\mathcal{F}) < \infty$. Sauer's lemma implies that $\operatorname{dens}(\mathcal{F}) = \dim_{VC}(\mathcal{F})$ so that

$$\operatorname{dens}(\mathcal{F}) < \infty \iff \dim_{VC}(\mathcal{F}) < \infty.$$

We see that a family $\mathcal{F}$ satisfies the condition **VC** if and only if its VC-dimension is finite. A family with finite VC-dimension is called a *VC-family*.

Note that $\dim_{VC}(\mathcal{F}) < k$ if and only if any set $A \subset \mathscr{X}$ of cardinality k contains a subset A_0 with the property that any set in $\mathcal{F}$ that contains A_0 also contains an element in $A \setminus A_0$. Intuitively, the sets in $\mathcal{F}$ cannot separate A_0 from its complement in A. Let us give some examples of VC families.

(i) Suppose that $\mathcal{F}$ consists of all the lower half-lines $(-\infty, t] \subset \mathbb{R}$, $t \in \mathbb{R}$. Note that if $A = \{a_1, a_2\}$, $a_1 < a_2$, then any half-line that contains a_2 must also contain a_1 so that $\dim_{VC}(\mathcal{F}) \leq 1$.

(ii) Suppose that $\mathcal{F}$ consists of all the open-half spaces of the vector space $\mathbb{R}^n$. A classical theorem of Radon [114, Thm. 1.3.1] shows that any subset $A \subset \mathbb{R}^n$ of cardinality $n+2$ contains a subset A_0 that cannot be separated from its complement $A \setminus A_0$ by a hyperplane. Thus $\dim_{VC}(\mathcal{F}) \leq n+1$. With a bit more work one can show that in fact we have equality.

(iii) The above example is a special case of the following general result, [51, Thm. 4.2.1].

Theorem 2.68. *Let $\mathscr{X}$ be a set. Suppose that V is a finite dimensional vector space of functions $f : \mathscr{X} \to \mathbb{R}$. The space V defines two families of subsets of $\mathscr{X}$,*

$$\mathcal{F}_V^{>0} = \big\{ \{f > 0\},\ f \in V \big\}, \quad \mathcal{F}_V^{\geq 0} = \big\{ \{f \geq 0\},\ f \in V \big\}.$$

Then

$$\dim_{VC}\big(\mathcal{F}_V^{>0}\big) = \dim_{VC}\big(\mathcal{F}_V^{\geq 0}\big) = \dim V.$$

(iv) If $\mathcal{F}_0, \mathcal{F}_1$ are two VC-families of subsets of a set $\mathscr{X}$, then $\mathcal{F}_0 \cup \mathcal{F}_1$ is also a VC family. Moreover (see [51, Thm. 4.5.1])

$$\mathrm{dens}(\mathcal{F}_0 \cup \mathcal{F}_1) = \max\big(\mathrm{dens}(\mathcal{F}_0), \mathrm{dens}(\mathcal{F}_1)\big),$$

and (see [51, Prop. 4.5.2])

$$\dim_{VC}\big(\mathcal{F}_0 \cup \mathcal{F}_1\big) \leq \dim \mathcal{F}_0 + \dim \mathcal{F}_1 + 1.$$

The above equality is optimal.

(v) If $\mathcal{F}_0, \mathcal{F}_1$ are two VC-families of subsets of a set $\mathscr{X}$ and we set

$$\mathcal{F}_0 \sqcap \mathcal{F}_1 := \big\{ F_0 \cap F_1;\ F_k \in \mathcal{F}_k,\ k = 0, 1 \big\},$$

then (see [51, Thm. 4.5.3])

$$\mathrm{dens}\big(\mathcal{F}_0 \sqcap \mathcal{F}_1\big) \leq \mathrm{dens}(\mathcal{F}_0) + \mathrm{dens}(\mathcal{F}_1).$$

(vi) If $\mathcal{F}_k$ is a VC family of subsets of $\mathscr{X}_k$, $k = 0, 1$, and we define

$$\mathcal{F}_0 \otimes \mathcal{F}_1 := \big\{ F_0 \times F_1;\ F_k \in \mathcal{F}_k,\ k = 0, 1 \big\},$$

then $\mathcal{F}_0 \otimes \mathcal{F}_1$ is a VC family of $\mathscr{X}_0 \times \mathscr{X}_1$; see [51, Thm. 4.5.3]. Moreover

$$\mathrm{dens}(\mathcal{F}_0 \otimes \mathcal{F}_1) \leq \mathrm{dens}(\mathcal{F}_0) + \mathrm{dens}(\mathcal{F}_1).$$

2.4.3 PAC learning

Let us explain why the above results are relevant in machine learning. Suppose that we are dealing with a 0-1 good/bad decision problem.

More precisely we want to determine when a parameter $x \in \mathbb{R}^N$ is "good", i.e., determine the set G of "good" parameters. For example we know from other considerations that a parameter $x \in \mathbb{R}$ is good if and only if $x \leq t_0$, but we do not know the precise value of t_0. However, we have some information about the "good" set: it is of the form $(-\infty, t]$, $t \in \mathbb{R}$.

More generally, for one reason or another we are lead to believe that the set G belongs to a family $(B_t)_{t \in T}$, where $T \subset \mathbb{R}^p$ and B_t is a Borel subset of $\mathbb{R}^N$. The family is $(B_t)_{t \in T}$ called a *hypothesis class*. Thus we seek $t_0 \in T$ such that $B_{t_0} = G$.

Consider a silly but suggestive example. Suppose that we want to decide when a banana is good. The goodness of a banana is decided by, say, three parameters: Color, Flavor, Softness, or CFS. Hence the good bananas are defined by some measurable subset in the CFS space. Suppose we have a collection $\mathcal{F}$ of categories of bananas, each category being defined by constraints in the CFS.

We are allowed to ask an Oracle to pick banana at random and answer then following yes/no questions. Does the chosen banana belong to a given category B_t? Is the chosen banana a good banana? However, the Oracle won't tell us which of the categories of bananas is the good category. Saying that a banana is good and it belongs to a category B_t only says that the banana belongs to $B_t \cap G$. We are suppose to learn the good category G by repeating the above experiment many times and recording the answers.

Technically, the Oracle puts at our disposal a sequence of i.i.d. $\mathbb{R}^N$-valued random vectors ($\mathbb{R}^N$ plays the role of the CFS space)

$$X_n : (\Omega, \mathcal{S}, \mathbb{P}) \to \mathbb{R}^N, \ n \in \mathbb{N},$$

and the values $Y_n = I_G(X_n)$, $n \in \mathbb{N}$. However, we do not know the common probability distribution μ of these random vectors.

If we knew this probability distribution, then we could find $G = B_{t_0}$ as a minimizer of the *deterministic* functional $L_\mu : T \to [0, 1]$

$$L_\mu(t) = \frac{1}{n} \sum_{k=1}^{n} \mathbb{P}\big[\, I_{B_t}(X_k) \neq Y_k \,\big] = \frac{1}{n} \sum_{k=1}^{n} \mathbb{P}\big[\, I_{B_t}(X_k) \neq I_G(X_k) \,\big]$$

$$= \frac{1}{n} \sum_{k=1}^{n} \mathbb{P}\big[\, I_{B_t \Delta G}(X_k) = 1 \,\big] = \mu\big[\, B_t \Delta G \,\big].$$

In fact $L_\mu(t_0) = 0$. Note that

$$\mu\big[\, B_t \Delta G \,\big] = \mathbb{E}\big[\, I_{B_t \Delta G} \,\big] = \mathbb{E}\big[\, I_{B_t} + I_G - 2 I_{B_t \cap G} \,\big].$$

The law of large numbers shows that $\mathbb{P}$-a.s. we have

$$\lim_{n \to \infty} \frac{1}{n} \sum_{k=1}^{n} \big(I_{B_t}(X_k) + I_G(X_k) - 2 I_{B_t \cap G}(X_k) \big)$$

$$= \mathbb{E}\bigl[\, \boldsymbol{I}_{B_t} + \boldsymbol{I}_G - 2\boldsymbol{I}_{B_t}\boldsymbol{I}_G \,\bigr] = L_\mu(t).$$

Thus, even if we do not know μ we can estimate $L_\mu(t)$ using the *random* functionals

$$L_n(t) = \frac{1}{n}\sum_{k=1}^n \bigl(\, \boldsymbol{I}_{B_t}(X_k) + \boldsymbol{I}_G(X_k) - 2\boldsymbol{I}_{B_t \cap G}(X_k) \,\bigr)$$

$$= \frac{1}{n}\sum_{k=1}^n \bigl(\, \boldsymbol{I}_{B_t}(X_k) + Y_k - 2Y_k \boldsymbol{I}_{B_t}(X_k) \,\bigr).$$

If $(B_t)_{t \in B_t}$ is a VC-family, then so is the family $(B_t \cap G)_{t \in T}$ and (2.4.11) shows that there exists constants $K, c > 0$, *independent of the mysterious* μ, such that

$$\mathbb{P}\bigl[\, \sup_{t \in T} |L_n(t) - L_\mu(t)| > \varepsilon \,\bigr] \le K e^{-cn\varepsilon^2}, \quad \forall n.$$

Thus, if we ask the oracle to give us a large sample $(x_1, y_1), \ldots, (x_n, y_n)$ of $(X_1, Y_1), \ldots, (X_n, Y_n)$ we obtain a *deterministic* functional

$$L_n(t; x_1, \ldots, x_n) = \frac{1}{n}\sum_{k=1}^n \bigl(\, \boldsymbol{I}_{B_t}(x_k) + y_k - 2\boldsymbol{I}_{B_t}(x_k) y_k \,\bigr).$$

If we find t_n such that $L_n(t_n; x_1, \ldots, x_n) < \frac{\varepsilon}{2}$, then

$$\mathbb{P}\bigl[\, L_\mu(t_n) > \varepsilon \,\bigr] \le \mathbb{P}\bigl[\, |L_n(t_n) - L_\mu(t_n)| > \varepsilon/2 \,\bigr] \le K e^{-cn\varepsilon^2/4}.$$

Thus, for large n, $L_n(t_n)$ is, with high confidence, within ε of the absolute minimum $L_\mathbb{P}(t_0) = 0$. Hopefully, this signifies that t_n is close to t_0. In the language of machine learning we say that the hypothesis class $(B_t)_{t \in T}$ is PAC learnable, where PAC stands for **P**robably **A**pproximatively **C**orrect. For more details we refer to [139; 152].

Remark 2.69. The results in this section only scratch the surface of the vast subject concerned with the limits of empirical processes. We have limited our presentation to 0-1-functions. The theory is more general than that.

Suppose that $(\mathbb{U}, \mathcal{U})$ is a measurable space and

$$X_n : \bigl(\Omega, \mathcal{S}, \mu\bigr) \to (\mathbb{U}, \mathcal{U})$$

is a sequence of i.i.d. measurable maps with common distribution $\mathbb{P} = (X_n)_\# \mu$, $\forall n$. Fix a family $\mathcal{F}$ of bounded measurable functions $\mathbb{U} \to \mathbb{R}$. We obtain a random measure

$$\mathbb{P}_n := \frac{1}{n}\sum_{k=1}^n \delta_{X_n}.$$

We obtain a stochastic process parametrized by $f \in \mathcal{F}$

$$(\mathbb{P}_n - \mathbb{P})[f] := \frac{1}{n}\sum_{k=1}^n \bigl(\, f(X_n) - \mathbb{E}[f(X_n)] \,\bigr) \in L^\infty(\Omega, \mathcal{S}, \mu), \quad f \in \mathcal{F}.$$

When $\mathcal{F}$ consists of indicator functions of measurable sets we obtain the situation described in this section.

For each f the SLLN shows that
$$(\mathbb{P}_n - \mathbb{P})[f] \to 0 \text{ a.s.}$$
while the CLT shows that
$$\sqrt{n}(\mathbb{P}_n - \mathbb{P})[f] \Rightarrow N(0, v(f)), \quad v(f) := \mathrm{Var}[f(X_n)], \quad \forall n.$$
What can be said about the limit of the *process* $\mathbb{P}_n - \mathbb{P}$?

Just like there are different flavors of convergence of random variables, there are many ways in which stochastic processes can converge. Various measurability issues make empirical processes trickier to handle. We refer to [51; 70; 129; 150; 156] for more details about this problem. □

2.5 The Brownian motion

The Brownian motion bears the name of its discoverer, the botanist R. Brown who observed in 1827 the chaotic motion of a particle of pollen in a fluid. Its study took off at the beginning of the 20th century and has since witnessed dramatic growth. It popped up in many branches of sciences and has lead to the development of many new branches of mathematics. In the theory of stochastic processes it plays a role similar to the role of Gaussian random variables in classical probability. It is such a fundamental and rich object that I believe any student learning the basic principles of probability needs to have a minimal introduction to it.

I drew my inspiration from many sources and I want to mention a few that we used more extensively, [12; 53; 103; 106; 136; 145]. My approach is not the most "efficient" one since I wanted to use the discussion of the Brownian motion as an opportunity to introduce the reader to other several important concepts concerning stochastic processes.

2.5.1 *Heuristics*

To get a grasp on the Brownian motion on a line, we consider first a discretization. We assume that the pollen particle performs a random walk along the line starting at the origin. Every unit of time τ it moves to the right or to the left, with equal probabilities, a distance δ. We denote by $S_n^{\delta,\tau}$ its location after n steps, or equivalently, its location at time $n\tau$, assuming we start the clock when the motion begins.

When $\delta = \tau = 1$ we obtain the standard random walk on $\mathbb{Z}$
$$S_n^{1,1} = S_n := \sum_{k=1}^{n} X_k,$$
where $(X_n)_{n \geq 1}$ is a sequence of independent Rademacher variables, i.e., random variables taking the values ± 1 with equal probabilities.

We assume that during the $(n+1)$-th jump the particle travels with constant speed 1 so we can assume that its location at time $t \in [n, n+1)$ is

$$W^1(t) = S_n + (t-n)X_{n+1} = S_{\lfloor t \rfloor} + (t - \lfloor t \rfloor)X_{\lfloor t \rfloor + 1}.$$

If we sample the random variables (X_n), then of $W^1(t)$ is a piecewise linear function with linear pieces of slopes ± 1. Its graph is a zig-zag of the type depicted in Figure 2.3.

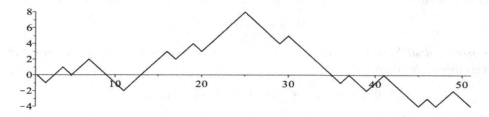

Fig. 2.3 *The zig-zag depicting a random walk.*

Suppose now that the pollen particle performs these random jumps at a much faster rate say ν-jumps per second and the size (in absolute value) of the jump is δ meters. We choose δ to depend on the frequency ν and we intend to let $\nu \to \infty$. Assuming that during a jump its speed is constant we deduce that this speed is $\delta \nu$ meters per second and its location at time t will be

$$W^{\nu,\delta}(t) = \delta S_{\lfloor \nu t \rfloor} + \underbrace{\delta(\nu t - \lfloor \nu t \rfloor)X_{\lfloor \nu t \rfloor + 1}}_{=:R_{\nu,\delta}(t)}.$$

To understand this formula observe that in the time interval $[0,t]$ the particle performed $\lfloor \nu t \rfloor$ complete jumps of size δ. It completed the last one at time $\frac{\lfloor \nu t \rfloor}{\nu}$. From this moment to t it travels in the direction $X_{\lfloor \nu t \rfloor + 1}$ with speed $\delta \nu$ for a duration of time $t - \frac{\lfloor \nu t \rfloor}{\nu}$.

Assuming that in finite time the particle will stay within a bounded region it is reasonable to assume that

$$\forall t, \ \sup_\nu \mathbb{E}\big[W^{\nu,\delta}(t)^2 \big] < \infty. \tag{2.5.1}$$

Now observe that $\delta S_{\lfloor \nu t \rfloor}$ and $R_{\nu,\delta}$ are mean zero independent random variables so that

$$\mathbb{E}\big[W^{\nu,\delta}(t)^2 \big] = \delta^2 \mathbb{E}\big[S_{\lfloor \nu t \rfloor}^2 \big] + \mathbb{E}\big[R_{\nu,\delta}(t)^2 \big] = \delta^2 \lfloor \nu t \rfloor + \mathbb{E}\big[R_{\nu,\delta}(t)^2 \big].$$

Clearly $\mathbb{E}\big[R_{\nu,\delta}(t)^2 \big] \in [0, \delta]$ so for (2.5.1) to hold we need

$$\sup_\nu \delta^2 \nu < \infty.$$

We achieve this by setting $\delta = \nu^{-1/2}$ and we set

$$\begin{aligned} W^\nu(t) &:= W^{\nu, \nu^{-1/2}}(t) = \nu^{-1/2} S_{\lfloor \nu t \rfloor} + R_\nu(t), \\ R_\nu(t) &:= \nu^{-1/2}(\nu t - \lfloor \nu t \rfloor) X_{\lfloor \nu t \rfloor + 1}. \end{aligned} \tag{2.5.2}$$

For each ν, the collection $(W^\nu(t))_{t\geq 0}$ is a real valued random process parametrized by $[0,\infty)$. Think of it as a random real valued function defined on $[0,\infty)$. It turns out that the random processes $(W^\nu(t))_{t\geq 0}$ have a sort of limit as $\nu \to \infty$. The next result states this in a more precise form.

Proposition 2.70. *Let $0 \leq s < t$. Then as $\nu \to \infty$ the random variable $W^\nu(t) - W^\nu(s)$ converges in distribution to a Gaussian random variable with mean zero and variance $t-s$. In particular, since $W^\nu(0) = 0$ we deduce that the limit*

$$W(t) = \lim_\nu W^\nu(t)$$

exists in distribution and it is a Gaussian random variable with mean zero and variance t. Moreover, if

$$0 \leq s_0 < t_0 \leq s_1 < t_1 \leq \cdots \leq s_k < t_k, \ \ k \geq 1,$$

then the increments

$$W(t_0) - W(s_0), \ W(t_1) - W(s_1), \ \ldots, \ W(t_k) - W(s_k)$$

are independent.

Proof. Fix $0 \leq s < t$. For ν sufficiently large we have $\lfloor \nu s \rfloor < \lfloor \nu t \rfloor$ and

$$W^\nu(t) - W^\nu(s) = \underbrace{\nu^{-1}\bigl(S_{\lfloor \nu t\rfloor} - S_{\lfloor \nu s\rfloor}\bigr)}_{Y_\nu} + \underbrace{\bigl((R_\nu(t) - R_\nu(s))\bigr)}_{Z_\nu}.$$

Observe first that

$$\lim_{n\to\infty} \mathbb{E}\bigl[\, Z_\nu^2 \,\bigr] = 0.$$

In particular, this shows that Z_ν converges in probability to 0. On the other hand

$$Y_\nu = \frac{\sqrt{\lfloor \nu t\rfloor - \lfloor \nu s\rfloor}}{\sqrt{\nu}} \cdot \underbrace{\frac{1}{\sqrt{\lfloor \nu t\rfloor - \lfloor \nu s\rfloor}} \sum_{k=\lfloor \nu s\rfloor+1}^{\lfloor \nu t\rfloor} X_k}_{\overline{Y}_\nu}.$$

The Central Limit Theorem shows that $\overline{Y}_\nu$ converges in distribution to a standard normal random variable. Since

$$\lim_{\nu\to\infty} \frac{\sqrt{\lfloor \nu t\rfloor - \lfloor \nu s\rfloor}}{\sqrt{\nu}} = \sqrt{t-s}$$

we deduce that Y_ν converges in distribution to a Gaussian random variable with mean zero and variance $t-s$. Invoking Slutsky's theorem (Theorem 2.30) we deduce that $Y_\nu + Z_\nu$ converges in distribution to a Gaussian random variable with mean zero and variance $t-s$.

Now let

$$0 \leq s_0 < t_0 \leq s_1 < t_1 \leq \cdots \leq s_k < t_k, \ \ k \geq 1.$$

For large ν the random variables
$$\nu^{-1/2}\big(S_{\lfloor t_j\rfloor}-S_{\lfloor s_j\rfloor}\big),\ \ j=0,1,\ldots,k$$
are independent and the above argument shows that they converge in law to the Gaussian
$$W(t_j)-W(s_j),\ \ j=0,1,\ldots,k.$$
Corollary 2.29 implies that these increments are also independent. □

Definition 2.71 (Pre-Brownian motion). A *pre-Brownian motion* on $[0,\infty)$ is a collection of real valued random variables $\big(W(t)\big)_{t\geq 0}$ with the following properties.

(i) $W(0)=0$.
(ii) For any $0\leq s<t$ the increment $W(t)-W(s)$ is a Gaussian random variable with mean zero and variance $t-s$.
(iii) For any
$$0\leq s_0<t_0\leq s_1<t_1\leq\cdots\leq s_k<t_k,\ \ k\geq 1,$$
increments
$$W(t_0)-W(s_0),\ W(t_1)-W(s_1),\ \ldots,\ W(t_k)-W(s_k)$$
are independent.

A pre-Brownian motion on $[0,1]$ is a collection of real valued random variables $\big(W(t)\big)_{t\in[0,1]}$ satisfying (i)–(iii) above with the s's and t's in $[0,1]$. □

We have thus proved that a suitable rescaling of the standard random walk on $\mathbb{Z}$ converges to a pre-Browning motion. In Figure 2.4 we have depicted the graph of a sample of $W^\nu(t)$ for $\nu=100$. Its graph is also a piecewise linear curve, but its linear pieces are much steeper, of slopes $\pm\nu^{1/2}$.

Suppose that $\big(W(t)\big)_{t\geq 0}$ is a pre-Brownian motion on $[0,\infty)$. As explained in Subsection 1.5.1, this process defines a probability measure on $\mathbb{R}^{[0,\infty)}$ equipped with the product sigma-algebra $\mathcal{B}_{\mathbb{T}}^{[0,\infty)}$ called the distribution of the process. We want to show that any two pre-Brownian motions have the same distributions. This requires a small digression in the world of Gaussian measures and processes. In the next subsection we survey some basic facts concerning these concepts. In Exercise 2.54 we ask the reader to fill in some of the details of this digression.

2.5.2 *Gaussian measures and processes*

Let V be an n-dimensional real vector space. We denote by V^* its dual, $V^*=\mathrm{Hom}(V,\mathbb{R})$. We have a natural pairing
$$\langle -,-\rangle:V^*\times V\to\mathbb{R},\ \ \langle\xi,x\rangle:=\xi(x),\ \ \forall\xi\in V^*,\ x\in V.$$

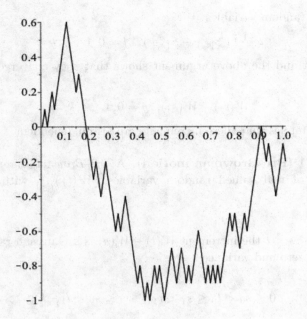

Fig. 2.4   *Approximating the Brownian motion.*

A Borel probability measure $\mu \in \mathrm{Prob}(V)$ is called *Gaussian* if for every linear functional $\xi \in V^*$, the resulting random variable $\xi : (V, \mathcal{B}_V, \mu) \to \mathbb{R}$ is Gaussian with mean $m[\xi]$ and variance $v[\xi]$, i.e., (see Example 1.120)

$$\mathbb{P}_\xi[dx] = \mathbf{\Gamma}_{m[\xi], v[\xi]}[dx] = \begin{cases} \frac{1}{(2\pi)^{n/2}} \cdot e^{-\frac{(x-m[\xi])^2}{2v[\xi]}} \, dx, & v[\xi] \neq 0, \\ \delta_{m[\xi]}, & v[\xi] = 0. \end{cases}$$

Equivalently, this means that the characteristic function of $\mathbb{P}_\xi$ is

$$\widehat{\mathbb{P}_\xi}(t) = \mathbb{E}\left[e^{it\xi}\right] = e^{-\frac{v[\xi]t^2}{2} + itm[\xi]}.$$

A random vector $\boldsymbol{X} : (\Omega, \mathcal{S}, \mathbb{P}) \to V$ is called *Gaussian* if its probability distribution is a Gaussian measure on V. The random variables $X_1, \ldots, X_n : (\Omega, \mathcal{S}, \mathbb{P}) \to \mathbb{R}$ are called *jointly Gaussian* if the random vector

$$\vec{X} : \Omega \to \mathbb{R}^n, \quad \vec{X}(\omega) = (X_1(\omega), \ldots, X_n(\omega)),$$

is Gaussian. This means that for any real constants $\xi_1, \ldots, \xi_n$, the linear combination

$$\xi_1 X_1 + \cdots + \xi_n X_n$$

is a Gaussian random variable.

For any Gaussian measure μ on the finite dimensional vector space V with mean $m_\mu[\xi]$ and variance $v_\mu[\xi]$ we define its *covariance form* to be

$$C = C_\mu : V^* \times V^* \to \mathbb{R},$$

$$C(\xi,\eta) = \frac{1}{4}\big(v_\mu[\xi+\eta] - v_\mu[\xi-\eta]\big) = \mathbb{E}_\mu\big[\,(\xi - m_\mu[\xi])(\eta - m_\mu[\eta])\,\big].$$

Then (see Exercises 2.54(ii) + (iii)) the mean m_μ is a linear functional $m_\mu : V^* \to \mathbb{R}$ and the covariance C_μ is a symmetric and positive semidefinite bilinear form on V^*.

Proposition 2.72. *A Gaussian measure on a vector space is uniquely determined by its mean and covariance form.* □

The proof of the above result is based on the Fourier transform and its main steps are described in Exercise 2.54. In the sequel we will refer to the mean zero Gaussian measures as *centered*.

Example 2.73. (a) If $X_1, \ldots, X_n$ are independent Gaussian random variables, then any linear combination

$$\xi_1 X_1 + \cdots + \xi_n X_n$$

is also Gaussian. In particular, if $X_1, \ldots, X_n$ are independent standard normal random variables, then the random vector $\vec{X} = (X_1, \ldots, X_n)$ is Gaussian and its distribution is the *standard Gaussian measure* on $\mathbb{R}^n$

$$\mathbf{\Gamma}_1[\,dx\,] = \frac{1}{(2\pi)^{n/2}} e^{-\frac{1}{2}\|x\|^2} dx.$$

(b) If $\vec{X} = (X_1, \ldots, X_n)$ is a Gaussian random vector, then the mean of its distribution is the vector

$$m[\vec{X}] := (\mathbb{E}[X_1], \ldots, \mathbb{E}[X_n])$$

and the covariance form of its distribution is the $n \times n$ matrix C with entries the covariances of the components, i.e.,

$$C_{ij} = \mathrm{Cov}[X_i, X_j] = \mathbb{E}\big[(X_i - \mathbb{E}[X_i])(X_j - \mathbb{E}[X_j])\big], \quad 1 \leq i, j \leq n.$$

(c) If μ is Gaussian measure on a finite dimensional vector space and $A : U \to V$ is a linear map to another vector space then the pushforward $A_\#\mu$ is also a Gaussian measure on V. In particular if

$$\vec{X} = (X_1, \ldots, X_n)$$

is a Gaussian vector and A is an $m \times m$ matrix then the vector $\vec{Y} = A\vec{X}$ is also Gaussian. Note that

$$\vec{Y} = (Y_1, \ldots, Y_m), \quad Y_i = \sum_{j=1}^n a_{ij} X_j, \quad i = 1, \ldots, m.$$

(d) Suppose $(-,-)$ is an inner product on the vector space V with associated norm $\|-\|$. We can then identify V^* with V and the symmetric bilinear forms on V^* with symmetric operators. The centered Gaussian measure on V whose covariance form is given by the inner product is

$$\mathbf{\Gamma}_1[\,dx\,] = \frac{1}{(2\pi)^{\dim V/2}} e^{-\frac{1}{2}\|x\|^2} dx.$$

If $A : V \to V$ is a symmetric linear operator, then pushforward $A_\# \mathbf{\Gamma}_1$ is the Gaussian measure with covariance form $C = A^2$. More precisely,

$$C(v_1, v_2) = (Av_1, Av_2) = (A^2 v_1, v_2).$$

If, additionally A is invertible, then

$$A_\# \mathbf{\Gamma}_1 \bigl[\, dx \,\bigr] = \frac{1}{\sqrt{\det(2\pi A^2)}} e^{-\frac{1}{2} \|A^{-2} x\|^2} dx.$$

We deduce that for any bilinear, symmetric positive semidefinite form

$$C : V^* \times V^* \to \mathbb{R}$$

there exists a centered Gaussian measure admitting C as covariance form. Indeed, if we fix a metric on V then we can identify C with a symmetric, positive semidefinite operator $C \to V$. If $A = \sqrt{C}$, then the Gaussian measure $A_\# \mathbf{\Gamma}_1$ is centered and has covariance form C. $\square$

Definition 2.74 (Gaussian processes). A *Gaussian process* parametrized by a set T is a collection of random variables $\bigl(X(t) \bigr)_{t \in T}$ defined on the same probability space $(\Omega, \mathcal{S}, \mathbb{P})$ such that, for any finite subset $I = \{t_1, \ldots, t_n\} \subset T$, the random vector $X_I := \bigl(X(t_1), \ldots, X(t_n) \bigr)$ is Gaussian. We denote by $\mathbf{\Gamma}_I$ its distribution. The process is called *centered* if $\mathbb{E}\bigl[X(t) \bigr] = 0$, $\forall t \in T$. $\square$

Suppose that $\bigl(X(t) \bigr)_{t \in T}$ is a Gaussian process. Its distribution is a probability measure on $\mathbb{R}^T$ uniquely determined by the Gaussian measures $\mathbf{\Gamma}_I$, I finite subset of T. In turn, these probability measures are *uniquely determined* by the *mean function*

$$m : T \to \mathbb{R}, \quad m(t) = \mathbb{E}\bigl[X(t) \bigr]$$

and the *covariance kernel*

$$K : T \times T \to \mathbb{R}, \quad K(s, t) = \mathrm{Cov}\bigl[X(s), X(t) \bigr].$$

Example 2.75. Suppose that $\bigl(W(t) \bigr)_{t \geq 0}$ is a pre-Brownian motion. For any $0 \leq t_1 < \cdots < t_n$ the random vector

$$\bigl(X_1, \ldots, X_n \bigr) = \bigl(W(t_1), W(t_2) - W(t_1), \ldots, W(t_n) - W(t_{n-1}) \bigr)$$

is Gaussian since its components are independent Gaussian random variables; see Example 2.73(a). Observing that

$$\bigl((W(t_1), \ldots, W(t_n)) \bigr) = (X_1, X_1 + X_2, \ldots, X_1 + \cdots + X_n)$$

we deduce from Example 2.73(c) that the vector $\bigl((W(t_1), \ldots, W(t_n)) \bigr)$ is also Gaussian as linear image of a Gaussian vector. Thus, any pre-Brownian motion is a Gaussian process. It is centered since all the random variables $W(t)$ have mean zero. Its distribution is a probability measure on then path space $\mathbb{R}^{[0,\infty)}$ uniquely determined by the covariance kernel

$$K : [0,\infty) \times [0,\infty) \to \mathbb{R}, \quad K(s,t) = \mathbb{E}\big[\, W(s) W(t) \,\big].$$

We claim that

$$K(s,t) = \min(s,t), \quad \forall s, t \geq 0. \tag{2.5.3}$$

Indeed, assume without any loss of generality that $s \leq t$. Then

$$\mathbb{E}\big[\, W(s) W(t) \,\big] = \mathbb{E}\big[\, W(s)^2 \,\big] + \mathbb{E}\big[\, W(s) \big(W(t) - W(s) \big) \,\big].$$

The first summand is equal to s according to property (ii) of a pre-Brownian motion. Property (iii) implies

$$\mathbb{E}\big[\, W(s) \big(W(t) - W(s) \big) \,\big] = \mathbb{E}\big[\, W(s) \,\big] \cdot \mathbb{E}\big[\, W(t) - W(s) \,\big] = 0.$$

Hence

$$\mathbb{E}\big[\, W(s) W(t) \,\big] = s = \min(s,t).$$

We see that *all pre-Brownian motions have the same covariance form and thus they all have the same distribution.*

Conversely, suppose that $\big(X(t) \big)_{t \geq 0}$ is a centered Gaussian process whose covariance form is given by (2.5.3). Then this process is a pre-Brownian motion. Indeed,

$$\mathbb{E}\big[\, X(0)^2 \,\big] = K(0,0) = 0$$

so $X(0) = 0$ a.s. Next, observe that

$$\mathbb{E}\big[\, X(t)^2 \,\big] = K(t,t) = t.$$

Each increment $X(t) - X(s)$, $s < t$, is Gaussian and

$$\mathrm{Var}\big[\, X(t) - X(s) \,\big] = K(t,t) - 2K(s,t) + K(s,s) = t - s.$$

Finally suppose that $0 \leq s_1 < t_1 \leq \cdots \leq s_n < t_n$. Then the n-dimensional random vector of increments

$$\vec{Y} := \big(X(t_1) - X(s_1), \ldots, X(t_n) - X(s_n) \big)$$

is centered Gaussian. The equality (2.5.3) implies that

$$\mathrm{Cov}\big[\, Y_i, Y_j \,\big] = 0, \quad \forall i \neq j$$

and we deduce from Exercise 2.55 that the components of $\vec{Y}$ are independent. This proves that $\big(X(t) \big)_{t \geq 0}$ is a pre-Brownian motion. □

Remark 2.76 (Brownian events). Consider an *arbitrary* pre-Brownian motion

$$B_t : (\Omega, \mathcal{F}, \mathbb{P}) \to \mathbb{R}, \ t \geq 0.$$

We define the σ-algebra of *Brownian events* to be the σ-subalgebra of $\mathcal{F}$ generated by the family of random variables B_t, $t \geq 0$. Concretely, any Brownian event E has the form

$$(B_{\tau(n)})_{n \in \mathbb{N}} \in S,$$

where $S \subset [0, \infty)^{\mathbb{N}}$ is a measurable subset and $\tau : \mathbb{N} \to [0, \infty)$ is an injection.

The restriction of $\mathbb{P}$ to the σ-algebra of Brownian events is *uniquely determined* by the distributions of the *Gaussian* random vectors

$$(B_{t_1}, \ldots, B_{t_n}), \ n \in \mathbb{N}, \ t_1, \ldots, t_n.$$

In turn, the distribution of such a vector is uniquely determined by the covariances

$$\mathbb{E}[B_s B_t] = \mathbb{E}[B_s(B_s + B_t - B_s)] = \mathbb{E}[B_s^2] = s = \min(s, t).$$

We see that these distributions *are independent* of the choice of pre-Brownian motion B. This shows that if

$$B^i : (\Omega^i, \mathcal{F}^i, \mathbb{P}^i) \to \mathbb{R}, \ i = 1, 2,$$

are two pre-Brownian motions, then for any measurable set $S \subset [0, \infty)$ and any injection $\tau : \mathbb{N} \to [0, \infty)$ we have

$$\mathbb{P}^1\big[(B^1_{\tau(n)})_{n \in \mathbb{N}} \in S\big] = \mathbb{P}^2\big[(B^2_{\tau(n)})_{n \in \mathbb{N}} \in S\big]. \qquad \square$$

Example 2.77 (Gaussian random functions). Suppose that $f_n : T \to \mathbb{R}$, $n \in \mathbb{N}$, is a sequence of functions defined on a set T and $(X_n)_{n \in \mathbb{N}}$ is a sequence of independent standard normal random variables defined on a probability space $(\Omega, \mathcal{S}, \mathbb{P})$. For each $t \in T$ we have a series of random variables

$$F(t) = \sum_{n \in \mathbb{N}} X_n f_n(t).$$

We want to emphasize that $F(t)$ also depends on the random parameter $\omega \in \Omega$,

$$F(t) = F(t, \omega) = \sum_{n \in \mathbb{N}} X_n(\omega) f_n(t). \tag{2.5.4}$$

The above is a series of *real numbers*.

Observe that if the sequence of functions f_n satisfies the condition

$$\sum_{n \in \mathbb{N}} f_n(t)^2 < \infty, \ \forall t \in T, \tag{2.5.5}$$

then the series defining $F(t)$ converges in $L^2(\Omega, \mathcal{S}, \mathbb{P})$, for any $t \in T$. To see this, consider the partial sums

$$F_n(t) = \sum_{k=1}^{n} X_k f_k(t).$$

Then, for $m < n$, we have

$$\mathbb{E}\big[\,\big(F_n(t) - F_m(t)\big)^2\,\big] = \sum_{k=m+1}^{n} f_k(t)^2 \mathbb{E}\big[\,X_k^2\,\big] = \sum_{k=m+1}^{n} f_k(t)^2.$$

This proves that the sequence $\big(F_n(t)\big)_{n\in\mathbb{N}}$ is Cauchy in $L^2(\Omega, \mathcal{S}, \mathbb{P})$. The family $F = \big(F(t)\big)_{t\in T}$ is a centered Gaussian random process. It is convenient to think of F as a random function. Its value $F(t)$ at t is not a deterministic quantity, it is random.

The covariance kernel is

$$K(s,t) = K_F(s,t) = \mathbb{E}\big[\,F(s)F(t)\,\big] = \sum_{n\in\mathbb{N}} f_n(s)f_n(t).$$

The above series is absolutely convergent since

$$2|f_n(s)f(t)| \leq f_n(s)^2 + f_n(t)^2, \quad \forall n, s, t.$$

Note that since the random vector $(F(s), F(t))$ is Gaussian, the random variables are independent iff they are not correlated, i.e., $\mathbb{E}\big[\,F(s)F(t)\,\big] = 0$. Thus the covariance kernel can be viewed as a measure of dependency between the values of F at different points $s, t \geq 0$.

Using Kolmogorov's one series theorem we deduce from the L^2 convergence that for any $t \in T$ there exists a measurable subset $\mathcal{N}_t \subset \Omega$ such that $\mathbb{P}\big[\,\mathcal{N}_t\,\big] = 0$ and, for any $\omega \in \Omega \setminus \mathcal{N}_t$ the series $F(t, \omega)$ in (2.5.4) converges. We will denote by $F(t, \omega)$ its sum. Set

$$\mathcal{N} := \bigcup_{t \in T} \mathcal{N}_t.$$

For $\omega \in \Omega \setminus \mathcal{N}$ we obtain a genuine function

$$F_\omega : T \to \mathbb{R}, \quad F_\omega(t) = F(t, \omega).$$

The function F_ω is referred to as a *path* of the stochastic process. We encounter here one of the recurring headaches in the theory of stochastic processes. Namely, if T is not countable, the set $\mathcal{N}$ may not negligible so the paths may not exists a.s.

If the parameter space T has additional structure, one could ask if the paths are compatible in some fashion with that structure. For example, if T is an interval of the real axis, we could ask if the paths are continuous functions of t. □

Example 2.78. A *Gaussian white noise* is a triplet $\big(H, (\Omega, \mathcal{S}, \mathbb{P}), W\big)$, where

- H is a separable real Hilbert space,
- $(\Omega, \mathcal{S}, \mathbb{P})$ is a probability space and,
- $W : H \to L^2(\Omega, \mathcal{S}, \mathbb{P})$, $h \mapsto W(h)$ is an isometry of H into $L^2(\Omega, \mathcal{S}, \mathbb{P})$ such that, for any $h \in H$, the random variable W_h is centered Gaussian.

Since X is an isometry we deduce that
$$\operatorname{Var}[W(h)] = \mathbb{E}[W(h)^2] = \|h\|_H^2.$$
In particular, this also shows that the image $W(H)$ of X is a *closed* subspace of $L^2(\Omega, \mathcal{S}, \mathbb{P})$ consisting of centered Gaussian random variables. Such a subspace is called a *Gaussian Hilbert space*. Obviously there is a natural bijection between Gaussian white noises and Gaussian Hilbert spaces.

Here is how one can construct Gaussian white noises. Fix a separable Hilbert space H with inner product $(-,-)$. Next, fix a Hilbert basis of $(e_n)_{n \in \mathbb{N}}$. Every element in H can then be decomposed along this basis
$$h = \sum_{n \in \mathbb{N}} a_n(h) e_n, \quad a_n(h) := (h, e_n).$$
Choose a sequence of independent standard normal random variables $(X_n)_{n \in \mathbb{N}}$ defined on a probability space $(\Omega, \mathcal{S}, \mathbb{P})$. For $h \in H$ we set
$$W(h) = \sum_{n \in \mathbb{N}} a_n(h) X_n.$$
From Parseval's identity we deduce that
$$\sum_{n \in \mathbb{N}} a_n(h)^2 = \|h\|_H^2$$
proving that the series defining $W(h)$ converges in L^2. The collection $(W(h))_{h \in H}$ is a Gaussian process and its covariance is
$$K(h_0, h_1) = \mathbb{E}[W(h_0) W(h_1)] = \sum_{n \in \mathbb{N}} a_n(h_0) a_n(h_1) = (h_0, h_1).$$
In particular, this proves that the correspondence $h \mapsto W(h)$ is an isometry, and thus we have produced a Gaussian white noise.

As a special example, suppose that $H = L^2([0, \infty), \boldsymbol{\lambda})$. Fix a Hilbert basis $(f_n)_{n \in \mathbb{N}}$ and construct the Gaussian noise as above
$$L^2([0, \infty), \boldsymbol{\lambda}) \ni f \mapsto W_f = \sum a_n(h) W_n \quad a_n(f) = \int_0^\infty f(t) f_n(t) dt.$$
For each $t \in [0, \infty)$ we set
$$B(t) := W(\boldsymbol{I}_{[0,t]}) = \sum_{n \in \mathbb{N}} \left(\int_0^t f_n(s) ds \right) X_n. \tag{2.5.6}$$
Note that
$$\boldsymbol{E}[B(s) B(t)] = \int_0^\infty \boldsymbol{I}_{[0,s]}(x) \boldsymbol{I}_{[0,t]}(x) dx = \min(s, t).$$
This shows that $W(t)$ is a pre-Brownian motion.

Observe that if $s \neq t$ and $|u| < |t - s|/2$, then the random variables
$$\frac{1}{u}(B(s+u) - B(s)) \quad \text{and} \quad \frac{1}{u}(B(t+u) - B(t))$$
are independent.

Now we need to make a leap of faith and pretend we can derivate with respect to t. (We really cannot.) Letting $u \to 0$ we deduce that $F'(t)$ and $F'(s)$ are independent Gaussian random variables. Derivating with the same abandon the equality (2.5.6) we deduce

$$B'(t) = \sum_{n \in \mathbb{N}} f_n(t) X_n. \qquad (2.5.7)$$

Thus, the elusive $B'(t)$ is a random "function" of the kind described in Example 2.77 with one big difference: in this case the condition (2.5.5) is not satisfied. Observe that the "value" of F' at a point t is independent of its value at a point s. Thus, the value F' at a point carries no information about its value at a different point so $F'(t)$ is a completely chaotic random "function" and it is what is commonly referred to as *white noise*.

As we will see in the next subsection the function $B(t)$ cannot be derivated at any point. Moreover, the series (2.5.7) does not converge in a classical sense. However it can be shown to converge in the sense of distributions. For an excellent discussion of this aspect we refer to [68, Sec. III.4].

For any function $f \in L^2([0, \infty))$ we define its *Wiener integral*

$$\int_0^t f(s) dB(s) := W(\boldsymbol{I}_{[0,t]} f). \qquad (2.5.8)$$

In Exercise 2.60 we give an alternate definition of the this object that justifies this choice of notation. In particular we deduce that

$$B(t) = \int_0^t dB(s).$$

Even though $B'(t)$ does not exist in any meaningful way, the above intuition is nevertheless very important since it is what lead to the very important concepts of Ito integral and stochastic differential equations. $\square$

2.5.3 The Brownian motion

We have almost everything we need to define the concept of Brownian motion and prove its existence.

Definition 2.79. A stochastic process $(B(t))_{t \geq 0}$ defined on a probability space $(\Omega, \mathcal{S}, \mathbb{P})$ is called a *standard Brownian motion* or *Wiener process* if the following hold.

(i) $B(t)$ is a pre-Brownian motion.
(ii) For any $\omega \in \Omega$ the path $B_\omega : [0, \infty) \to \mathbb{R}$, $t \mapsto B(t, \omega)$ is continuous. $\square$

To prove the existence of a standard Brownian motions we need a bit more terminology and another fundamental result of Kolmogorov.

Definition 2.80. Let $(\Omega, \mathcal{S}, \mathbb{P})$ be a probability space, T a set, and $(\mathbb{X}, \mathcal{F})$ a measurable set. Consider stochastic processes
$$X, Y : T \times \Omega \to \mathbb{X}, \quad (t, \omega) \mapsto X_t(\omega),\ Y_t(\omega).$$

(i) The process Y is said to be a *modification* or *version* X, and we denote this $X \sim Y$ if for any $t \in T$ there exists a negligible subset $\mathcal{N}_t$ such that
$$X_t(\omega) = Y_t(\omega), \quad \forall \omega \in \Omega \setminus \mathcal{N}_t.$$

(ii) The processes X, Y are said to be *indistinguishable* and we denote this $X \approx Y$, if there exists a negligible subset $\mathcal{N}$ such that
$$X_t(\omega) = Y_t(\omega), \quad \forall t \in T, \ \forall \omega \in \Omega \setminus \mathcal{N}.$$

(iii) The processes X, Y are said to be *stochastically equivalent*, and we denote this $X \sim_s Y$, if for any finite subset $I \subset T$ the random vectors X^I and Y^I have the same distribution.

$\square$

Note that $\approx, \sim, \sim_s$ are equivalence relations and
$$X \approx Y \implies X \sim Y \implies X \sim_s Y.$$

We have shown that any two pre-Brownian motions are stochastically equivalent. We want to prove something stronger namely, that any pre-Brownian motion admits a version whose paths are almost surely continuous maps $[0, \infty) \to \mathbb{R}$. We begin by proving a more general result.

Theorem 2.81 (Kolmogorov's Continuity Theorem). *Suppose that T is a compact interval of the real axis, $(\Omega, \mathcal{F}, \mathbb{P})$ is a probability space and*
$$X : T \times \Omega \to \mathbb{R}, \quad (t, \omega) \mapsto X_t(\omega)$$
is a stochastic process such that, there exist constant $q, r, K > 0$ with the property that
$$\mathbb{E}\big[\,|X_s - X_t|^q\,\big] \leq K|s - t|^{1+r}, \quad \forall s, t \in T. \tag{2.5.9}$$
Then, for any $\alpha \in (0, r/q)$, the process X admits a modification Y whose paths are almost surely Hölder continuous with exponent α. This means, that for any $\alpha \in (0, r/q)$ there exists a stochastic process $(Y_t)_{t \in T}$, a negligible subset $\mathcal{N}_\alpha \subset \Omega$ and a measurable function
$$C = C_\alpha : \Omega \to [0, \infty),$$
such that
- *$\forall t \in T,\ X_t = Y_t$ a.s. and,*
- *for any $\omega \in \Omega \setminus \mathcal{N}_\alpha$, and any $s, t \in T$ we have*
$$\big|Y_s(\omega) - Y_t(\omega)\big| \leq C(\omega)|s - t|^\alpha.$$

Proof. We follow the presentation in [103, Thm. 29]. Without loss of generality we can assume that $T = [0,1]$. We denote by $\mathcal{D}$ the set of diadic numbers in $[0,1]$

$$\mathcal{D} = \bigcup_{n \geq 0} D_n, \quad D_n = \left\{ \frac{k}{2^n}; \ 0 \leq k \leq 2^n \right\}.$$

Fix $\alpha \in (0, r/q)$. We carry the proof in two steps.

Step 1. We will show that there exists a measurable negligible subset $\mathcal{N} \subset \Omega$ and a measurable function $C : \Omega \to [0, \infty)$ such that

$$\big| X_t(\omega) - X_s(\omega) \big| \leq C(\omega)|t-s|^\alpha, \quad \forall s, t \in \mathcal{D}. \tag{2.5.10}$$

From the assumption (2.5.9) and Markov's inequality we deduce that for any $s, t \in T$ and any $a > 0$ we have

$$\mathbb{P}\big[\,|X_s - X_t| > a\,\big] \leq \frac{1}{a^q} \mathbb{E}\big[\,|X_s - X_t|^q\,\big] \leq \frac{K}{a^q}|t-s|^{1+r}.$$

Applying this inequality to $s = \frac{j-1}{2^n}$, $t = j/2^n$ and $a = \frac{1}{2^{n\alpha}}$ we deduce

$$\mathbb{P}\big[\,|X_{(k-1)/2^n} - X_{k/2^n}| > 2^{-n\alpha}\,\big] \leq K 2^{nq\alpha} 2^{-nq-nr} = K(\rho/2)^n, \quad \rho := 2^{(\alpha q - r)}.$$

Hence

$$\mathbb{P}\left[\underbrace{\bigcup_{k=1}^{2^n} \{|X_{(k-1)/2^n} - X_{k/2^n}| > 2^{-n\alpha}\}}_{H_n}\right] \leq K\rho^n.$$

Note that since $\alpha < r/q$ we have $\rho \in (0,1)$. From the Borel-Cantelli Lemma we deduce that

$$\mathbb{P}\big[\,H_n \text{ i.o.}\,\big] = 0.$$

Thus, there exists a negligible set $\mathcal{N}$ with the following property: for any $\omega \in \Omega \setminus \mathcal{N}$ there exists $n_0(\omega) \in \mathbb{N}$ so that for any $n \geq n_0(\omega)$, and any $k = 1, \ldots, 2^n$ we have

$$|X_{(k-1)/2^n}(\omega) - X_{k/2^n}(\omega)| \leq K 2^{-n\alpha}.$$

At this point we invoke an elementary result whose proof we postpone.

Lemma 2.82. *Let $f : \mathcal{D} \to \mathbb{R}$ be a function. Suppose that there exist $\alpha \in (0,1)$, $n_0 \in \mathbb{N}$ and $K > 0$ such that, $\forall n \geq n_0$ and any $k = 1, \ldots, 2^n$*

$$\big| f\big((k-1)/2^n\big) - f\big(k/2^n\big)\big| \leq K 2^{-n\alpha}. \tag{2.5.11}$$

Then there exists $C = C(n_0, \alpha, K)$ such that

$$\big|f(s) - f(t)\big| \leq C|s-t|^\alpha, \quad \forall t, s \in \mathcal{D}.$$

□

This shows that for every $\omega \in \Omega \setminus \mathcal{N}$ the map $\mathcal{D} \ni t \mapsto X_t(\omega)$ is Hölder continuous with exponent α. This completes the proof of (2.5.10).

Step 2. We can now produce the claimed modification. For every $\omega \in \Omega \setminus \mathcal{N}$ the map

$$\mathcal{D} \ni t \mapsto X_t(\omega)$$

admits a unique α-Hölder extension $T \ni t \mapsto \overline{X}_t(\omega) \in \mathbb{R}$. For $t_0 \in T$ we have

$$\lim_{\substack{t \to t_0 \\ t \in \mathcal{D}}} X_t = \overline{X}_{t_0}.$$

Since

$$\lim_{\substack{t \to t_0 \\ t \in \mathcal{D}}} \mathbb{E}\big[\,|X_t - X_{t_0}|^q\,\big] = 0$$

we deduce that $X_{t_0} = \bar{X}_{t_0}$ a.s. Hence the process $(\overline{X}_t)_{t \in T}$ is a modification of $(X_t)_{t \in \mathbb{T}}$ whose paths are a.s. α-Hölder continuous. $\square$

Proof of Lemma 2.82. Let $0 \leq m < n_0$ $s = \frac{i-1}{2^m}$ and $t = \frac{i}{2^m}$. For $j = 0, 1, \ldots, 2^{n_0 - m}$ we set $s_j = s + \frac{j}{2^{n_0}}$. Then

$$\big|f(t) - f(s)\big| \leq K \sum_{j=1}^{2^{n_0 - m}} 2^{-n_0 \alpha} = K 2^{n_0 - m} 2^{-n_0 \alpha} = \underbrace{K 2^{n_0 - m + \alpha(m - n_0)}}_{=: K_1} 2^{-m\alpha}.$$

Thus, f satisfies (2.5.11) *for any* $n \geq 1$, possibly with a different constant $K' = \max(K_1, K)$.

Let $s, t \in \mathcal{D}$. Assume $s < t$. Let p be the largest positive integer such that $t - s < 2^{-p}$. Note that $2^{-(p+1)} \leq t - s \leq 2^{-p}$. Set $\ell := \lfloor 2^p s \rfloor$. Then

$$\frac{\ell}{2^p} \leq s \leq \underbrace{\frac{\ell}{2^p} + \frac{1}{2^{p+1}}}_{=: u} \leq t \leq \frac{\ell + 1}{2^p}.$$

Then for some $m, n \in \mathbb{N}$ we have

$$s = u - \frac{\epsilon_1}{2^{p+1}} - \cdots - \frac{\epsilon_i}{2^{p+m}}, \quad t = u + \frac{\eta_1}{2^{p+1}} + \cdots + \frac{\eta_i}{2^{p+n}},$$

where $\epsilon_i, \eta_j \in \{0, 1\}$. For $i = 1, \ldots, m$, and $j = 1, \ldots, n$ we set

$$s_i = u - \frac{\epsilon_1}{2^{p+1}} - \cdots - \frac{\epsilon_i}{2^{p+i}}, \quad t_j := u + \frac{\eta_1}{2^{p+1}} + \cdots + \frac{\eta_i}{2^{p+j}}.$$

Using (2.5.11) we deduce

$$\big|f(u) - f(s)\big| \leq \big|f(u) - f(s_1)\big| + \cdots + \big[f(s_{m-1}) - f(s_m)\big] \leq \sum_{i=1}^{m} K_1 2^{-(p+i)\alpha}$$

$$\leq K_1 2^{-p\alpha} \sum_{i=1}^{\infty} 2^{-i\alpha} = K_1 \underbrace{\frac{2^\alpha}{2^\alpha - 1}}_{=: K_2} 2^{-p\alpha} \leq 2^\alpha K_2 |s - t|^\alpha,$$

where at the last step we used the fact that $2^{-p} < 2|s-t|$. Similarly
$$|f(u) - f(t)| \leq 2^\alpha K_2 |s-t|^\alpha.$$
Hence
$$|f(s) - f(t)| \leq |f(s) - f(u)| + |f(u) - f(t)| \leq 2^{1+\alpha} K_2 |t-s|^\alpha.$$
This proves the lemma with $C(n_0, \alpha, K) = 2^{1+\alpha} K_2$. □

Remark 2.83. (a) Using Exercise 2.59 one can modify the modification in Theorem 2.81 to be α-Hölder continuous for any $\alpha \in (0, q/r)$, not just for a fixed α in this range.

(b) The argument in the proof of Lemma 2.82 is an elementary incarnation of the *chaining technique*. For a wide ranging generalization of the continuity Theorem 2.81 and the chaining technique we refer to [101, Chap. 11]. □

Corollary 2.84. *Suppose that $(W_t)_{t \geq 0}$ is a pre-Brownian motion. Then for any $\alpha \in (0, 1/2)$ the process (W_t) admits a modification whose paths are a.s. α-Hölder continuous. In particular, Brownian motions exist.*

Proof. Set $\delta = \frac{1}{2} - \alpha$. Note that since $W_t - W_s$ is Gaussian with mean 0 and variance $|t-s|$. Then $D := \frac{1}{\sqrt{|t-s|}}(W_t - W_s) \sim N(0,1)$ so that, $\forall q \geq 1$, we have
$$\mathbb{E}[|W_t - W_s|^q] = |t-s|^{q/2} \mathbb{E}[|D|^q].$$
If we choose $q > \frac{1}{\delta}$, then we deduce that
$$\frac{q/2 - 1}{q} = \frac{1}{2} - \frac{1}{q} > \alpha$$
and Theorem 2.81 implies that (W_t) admits a modification $(\overline{W}_t)_{t \geq 0}$ whose paths are a.s. α-Hölder continuous.

Recall that this means that there exists a measurable negligible set $\mathcal{N} \subset \Omega$ such that $\forall \omega \in \Omega \setminus \mathcal{N}$ the path $t \mapsto \overline{W}_t(\omega)$ is continuous. Now define
$$B : [0, \infty) \times \Omega, \quad (t, \omega) \mapsto B_t(\omega) = \begin{cases} \overline{W}_t(\omega), & \omega \in \Omega \setminus \mathcal{N}, \\ 0, & \omega \in \mathcal{N}. \end{cases}$$
Clearly $(B_t)_{t \geq 0}$ is a (standard) Brownian motion. □

Remark 2.85. I want to say a few words about Paul Lévy's elegant construction of the Brownian motion, [106, Sec. 1].

He produces the Brownian motion on $[0,1]$ as a limit of random piecewise linear functions L_n with nodes on the dyadic sets
$$D_n := \left\{ \frac{k}{2^n}; \ 0 \leq k \leq 2^n \right\}, \ n \geq 0.$$
They are successively better approximations of the Brownian motion. The 0-th order approximation is the random linear function $L_0(t)$ such that $L_0(0) = 0$ and $L_0(1)$ is a standard normal random variable.

The n-th order approximation L_n satisfies the following conditions.

- It is linear on each of the intervals $\big((k-1)/2^n, k/2^n\big)$, $L_n(0)=0$.
- The increments

$$L_n(k/2^n) - L_n\big((k-1)/2^b\big), \quad k=1,\ldots,2^n$$

are normal random variables with mean zero and variance $1/2^n$.
- $L_n(t) = L_{n-1}(t), \forall t \in D_{n-1}$.

To explain how to produce $L_n(t)$ given $L_{n-1}(t)$ we only need to explain how to produce $L_n\big((2k-1)/2^n\big)$ given that

$$L_n(j/2^{n-1}) = L_{n-1}(j/2^{n-1}), \quad j = k-1, k.$$

To "guess" what $L_n\big((2k-1)/2^n\big)$ should be, we take our inspiration from the Brownian motion that we want to approximate.

Consider two moments of time $t_0 < t_1$ in $[0,1]$. Then $B(t_0) \sim N(0,t_0)$, $B(t_1) \sim N(0,t_1)$ and $B(t_1) - B(t_0)$ is a normal random variable with mean 0, variance $t_1 - t_0$, independent of $B(t_0)$. Denote by t_* the midpoint of $[t_0, t_1]$, $t_* = (t_0 + t_1)/2$.

Consider the linear interpolation

$$Z = \frac{1}{2}\big(B(t_0) + B(t_1)\big).$$

The difference

$$\Delta := B(t_*) - Z = \frac{1}{2}\big(B(t_*) - B(t_0)\big) + \frac{1}{2}\big(B(t_*) - B(t_1)\big)$$

is a sum of two independent normal random variables, that are also independent of $B(t_0)$. Thus Δ is a normal random variable with mean 0, variance $(t_1-t_0)/4$, independent of $B(t_0)$. We write

$$B(t_*) = Z + \Delta = \frac{1}{2}\big(B(t_0) + B(t_1)\big) + \frac{\sqrt{t_1-t_0}}{2}X. \qquad (2.5.12)$$

We can now describe Lévy's prescription. We set

$$\mathcal{D} = \bigcup_{n \geq 0} D_n$$

and consider a family $(X_t)_{t \in \mathcal{D}}$ of independent standard normal random variables. Then

$$L_0(t) := tX_1.$$

The approximation L_{n+1} is obtained from L_n as follows. If $t_0 < t_1$ are two consecutive points in D_n and $t_* \in D_{n+1}$ is the midpoint of $[t_0, t_1]$, then $L_{n+1}(t_*)$ is obtained by mimicking (2.5.12), i.e.,

$$L_{n+1}(t_*) = \frac{1}{2}\big(L_n(t_0) + L_n(t_1)\big) + \frac{\sqrt{t_1-t_0}}{2}X_{t_*}$$

$$= \frac{1}{2}\big(L_n(t_0) + L_n(t_1)\big) + \frac{1}{2^{1+n/2}}X_{t_*}.$$

To prove that the sequence $L_n(t)$ converges uniformly a.s. it suffices to show that the series of random variables

$$\sum_{n\geq 0} \underbrace{\sup_{t\in[0,1]} \big| L_{n+1}(t) - L_n(t) \big|}_{=:U_n}$$

converges a.s.

Denote by M_n the set of midpoints of the 2^n intervals determined by D_n, $M_n = D_{n+1} \setminus D_n$. From the construction of L_n we deduce that

$$U_n = \frac{1}{2^{1+n/2}} \max_{\tau \in M_n} |X_\tau|.$$

We deduce that for any $c > 0$ we have

$$\mathbb{P}\big[U_n > c \big] \leq 2^n \mathbb{P}\big[Y > 2^{1+n/2} c \big], \quad Y \sim N(0,1).$$

The Mills ratio inequalities (1.3.40) coupled with the Borel-Cantelli lemma lead to the claimed convergence. □

Let us observe that if $(B(t))$ is a standard Brownian motion, then $B(0) = 0$ a.s. For this reason, the standard Brownian motion is also referred to as the *Brownian motion started at 0*. For $x \in \mathbb{R}$ we set $B^x(t) = x + B(t)$. We will refer to $B^x(t)$ as the *Brownian motion started at x*.

Remark 2.86 (The Wiener measure). The space $\mathcal{C} := C\big([0,\infty)\big)$ of continuous functions $[0,\infty) \to \mathbb{R}$ is equipped with a natural metric d,

$$d(f,g) = \sum_{n \in \mathbb{N}} \frac{1}{2^n} \min\big(1, d_n(f,g)\big), \quad d_n(f,g) := \sup_{t \in [n-1,n]} |f(t) - g(t)|.$$

The topology induced by this metric is the topology of uniform convergence on the compact subsets of $[0,\infty)$. One can prove (see Exercise 2.61) that the Borel algebra of this metric space coincides with the sigma algebra generated by the functions

$$\mathbf{Ev}_t : \mathcal{C} \to \mathbb{R}, \quad \mathbf{Ev}_t(f) = f(t).$$

More generally, for any finite subset $I \subset [0,\infty)$ we have a measurable evaluation maps

$$\mathbf{Ev}_I : \mathcal{C} \to \mathbb{R}^I, \quad f \mapsto f|_I.$$

Proposition 1.29 shows that if μ_0, μ_1 are two probability measures on $\mathcal{C}$ such that

$$(\mathbf{Ev}_I)_\# \mu_0 = (\mathbf{Ev}_I)_\# \mu_1$$

for any finite subset $I \subset [0,\infty)$, then $\mu_0 = \mu_1$.

Note that if $(X_t)_{t \geq 0}$ is a stochastic process defined on a probability space $(\Omega, \mathcal{S}, \mathbb{P})$ whose paths are continuous, then it defines a map

$$X : \Omega \to \mathcal{C}, \quad \Omega \ni \omega \mapsto X(\omega) \in \mathcal{C}, \quad X(\omega)(t) = X_t(\omega).$$

The map X is measurable since its composition with all the evaluation maps $\mathbf{Ev}_I$ are measurable. Thus the stochastic process defines a probability measure
$$\mathbb{P}_X := X_\# \mathbb{P} \in \mathrm{Prob}(\mathcal{C}, \mathcal{B}_\mathcal{C})$$
called the *distribution* of the process.

Suppose that B^0, B^1 are two Brownian motions defined on possibly different probability spaces. They have distributions
$$\mathbb{W}_0, \mathbb{W}_1 \in \mathrm{Prob}(\mathcal{C}, \mathcal{B}_\mathcal{C}).$$
These distributions coincide since the finite dimensional distributions $\pi_I \mathbb{W}_j$, $i = 0, 1$ are centered Gaussian with identical covariances
$$\mathbb{E}[B^i_{t_1} B^i_{t_1}] = \min(t_1, t_2), \quad \forall t_1, t_2 \in I, \; i = 0, 1.$$
Thus, the Brownian motions determine a probability measure $\mathbb{W}$ on $\mathcal{C}$ uniquely determined by the requirement that for any finite subset $\{t_1, \ldots, t_n\} \subset [0, \infty)$ the random vector
$$(\mathbf{Ev}_{t_1}, \ldots, \mathbf{Ev}_{t_n})$$
is centered Gaussian with covariances $\mathbb{E}[\mathbf{Ev}_{t_i} \mathbf{Ev}_{t_j}] = \min(t_i, t_j)$. This measure is known as the *Wiener measure*. We denote it by $\mathbb{W}$.

Note that $\mathbb{W}$ is unique probability measure on $\mathcal{C}$ such that the canonical process
$$B_t : (\mathcal{C}, \mathcal{B}_\mathcal{C}, \mathbb{W}) \to \mathbb{R}, \quad \mathcal{C} \ni f \mapsto \mathbf{Ev}_t(f) = f(t)$$
is itself a Brownian motion, i.e.,
$$\mathbb{E}_\mathbb{W}[B_s B_t] = \min(s, t), \quad \forall s, t \geq 0. \tag{2.5.13}$$

We have proved the existence of Wiener's measure by relying on the existence of Brownian motion. Conversely, if by some other method we can construct the Wiener measure on $\mathcal{C}$, then as a bonus we deduce the existence of Brownian motions. Here is one such alternate method.

Consider a sequence of i.i.d. random variables $(X_n)_{n \in \mathbb{N}}$ with mean 0 and variance 1. We set
$$S_0 = 0, \quad S_n = X_1 + \cdots + X_n, \quad n \in \mathbb{N}.$$
Imitating (2.5.2), for $\nu \in \mathbb{N}$ and $t \geq 0$ we set
$$W^\nu(t) := \nu^{-1/2} S_{\lfloor \nu t \rfloor} + R_\nu(t), \quad R_\nu(t) := \nu^{-1/2}(\nu t - \lfloor \nu t \rfloor) X_{\lfloor \nu t \rfloor + 1}. \tag{2.5.14}$$
For each ν, the paths of the random process are continuous and piecewise linear. The above discussion shows that it defines a Borel probability measure $\mathbb{P}_\nu = \mathbb{P}_{W^\nu}$ on $\mathcal{C}$.

Donsker's Invariance Principle shows that the sequence $\mathbb{P}_\nu$ converges weakly to a probability measure on $\mathbb{P}_\infty$ satisfying (2.5.13). In other words, $\mathbb{P}_\infty$ is the Wiener measure. We can view the Invariance Principle as a functional version of the Central Limit Theorem. Its proof requires an in depth investigation of the space of probability measures on Polish spaces[12] and is beyond the scope of this text. For a most readable presentation of Donsker's theorem and some of its consequences we refer to [12], [19, Chap. 13]. □

---

[12] Recall that a Polish space is a complete separable metric space.

The next result suggests that the paths of a Brownian motion are very rough, i.e., they have poor differentiability properties.

Proposition 2.87 (The quadratic variation of Brownian paths). *Consider a Brownian motion $B_t)_{t\geq 0}$ defined on the probability space $(\Omega, \mathcal{S}, \mathbb{P})$. Fix $c > 0$ and let*

$$0 = t_0^n < t_1^n < \cdots < t_{p_n}^n = c, \quad n \in \mathbb{N}$$

be a sequence of subdivisions of $[0, t]$ with mesh

$$\mu_n := \sup_{1 \leq k \leq p_n} (t_k^n - t_{k-1}^n)$$

tending to 0 as $n \to \infty$. Define the quadratic variations

$$Q_n(c) := \sum_{k=1}^{p_n} \left(B_{t_k^n} - B_{t_{k-1}^n} \right)^2.$$

Then $\mathbb{E}[Q_n(c)] = c$, $\forall n$ and $Q_n(c) \to c$ in $L^2(\Omega, \mathcal{S}, \mathbb{P})$ as $n \to \infty$.

Proof. The Gaussian random variables $X_k^n = B_{t_k^n} - B_{t_{k-1}^n}$, $1 \leq k \leq p_n$, are independent, have mean zero and momenta

$$\mathbb{E}[(X_k^n)^2] = t_k^n - t_{k-1}^n, \quad \mathbb{E}[(X_k^n)^4] = 3(t_k^n - t_{k-1}^n)^2.$$

From the first equality we deduce $\mathbb{E}[Q_n(c)] = c$. Moreover

$$\sum_{k=1}^{p_n} (X_k^n)^2 - c = \sum_{k=1}^{p_n} \underbrace{\left((X_k^n)^2 - (t_k^n - t_{k-1}^n) \right)}_{=:Y_k^n}.$$

The random variables Y_k^n are independent and have mean zero so

$$\left\| \sum_{k=1}^{p_n} (X_k^n)^2 - c \right\|_{L^2}^2 = \sum_{k=1}^{n} \|Y_k^n\|_{L^2}^2.$$

Now observe that

$$\|Y_k^n\|_{L^2}^2 = \mathbb{E}[(X_k^n)^4] - 2(t_k^n - t_{k-1}^n)\mathbb{E}[(X_k^n)^2] + (t_k^n - t_{k-1}^n)^2 = 2(t_k^n - t_{k-1}^n)^2.$$

Hence

$$\left\| \sum_{k=1}^{p_n} (X_k^n)^2 - c \right\|_{L^2}^2 = 2 \sum_{k=1}^{p_n} (t_k^n - t_{k-1}^n)^2$$

$$\leq 2\mu_n \sum_{k=1}^{p_n} (t_k^n - t_{k-1}^n) = 2\mu_n c \to 0 \quad \text{as } n \to \infty.$$

$\square$

On a subsequence n_j we have $Q_{n_j}(c) \to c > 0$ a.s. On the other hand, if for some $\omega \in \Omega$ the function $t \to B_t(\omega)$ where Hölder with exponent $\alpha > 1/2$ on $[0, c]$, then for some constant $C = C_\omega > 0$ independent of n we would have

$$0 \leq Q_n(t)(\omega) \leq C_\omega^2 \sum_k \left| t_k^n - t_{k-1}^n \right|^{2\alpha} \leq C_\omega^2 \mu_n^{2\alpha-1} c \to 0.$$

This prove that B_t is a.s. not α-Hölder on $[0, c]$, $\alpha > 1/2$.

On the other hand, we know that the paths of the Brownian motion are Hölder continuous for any exponent $< 1/2$. A 1933 result of Paley, Wiener, Zygmund [124] shows that they have very poor differentiability properties. First some historical context.

One question raised in the 19th century was whether there exist continuous functions on an interval that are nowhere differentiable. Apparently Gauss believed that there are no such functions. K. Weierstrass explicitly produced in 1872 such examples defined by lacunary (or sparse) Fourier series. In 1931 S. Banach [7] and S. Mazurkewicz [115] independently showed that the complement of the set of nowhere differentiable functions in the metric space of continuous functions on a compact interval is very small, meagre in the Baire category sense.

The 1933 result of Paley, Wiener, Zygmund that we want discuss is similar in nature. They prove that the complement set of continuous nowhere differentiable functions $f \in \mathcal{C}$ is negligible with respect to the Wiener measure.

Theorem 2.88 (Paley, Wiener, Zygmund). *The paths of a Brownian motion $(B_t)_{t \geq 0}$ are a.s. nowhere differentiable.*

Proof. We follow the very elegant argument of Dvoretzky, Erdös, Kakutani [54]. We will show that for any interval $I = [a, b) \subset [0, \infty)$ the paths of (B_t) are a.s. nowhere differentiable on I. Assume the Brownian motion is defined on a probability space $(\Omega, \mathcal{S}, \mathbb{P})$. This probability space could be the space $\mathcal{C}$ equipped with the Wiener measure. For ease of presentation we assume that $I = [0, 1)$. Consider the set

$$S := \big\{ \omega \in \Omega; \text{ the path } B_t(\omega) \text{ is nowhere differentiable on } [0, 1) \big\}.$$

The set S may not be measurable[13] but we will show that its complement is contained in a measurable subset of Ω of measure zero.

Let us observe that if $\omega \in \Omega \setminus S$, i.e., the path $t \mapsto B_t(\omega)$ is differentiable at a point $t_0 \in [0, 1]$, then there exist $M, N \in \mathbb{N}$ such that for any $n \geq N$ there exists $k \in \{1, \ldots, n-2\}$ with the property that

$$\left| B_{(k-1+i)/n}(\omega) - B_{(k+i)/n}(\omega) \right| \leq \frac{M}{n}, \quad \forall i = 0, 1, 2.$$

To see this set $f(t) = B_t(\omega)$, $m = |f'(t_0)|$, $M = \lfloor m \rfloor + 2$. Then there exists $\varepsilon > 0$ so that if $s, t \in (t_0 - \varepsilon, t_0 + \varepsilon)$, $s < t$ we have

$$|f(s) - f(t)| \leq M(t - s).$$

---

[13]In 1936 S. Mazurkewicz proved that the set S is *not* a Borel subset of $\mathcal{C}$.

Now choose N such that $\frac{1}{N} < \frac{\varepsilon}{6}$ and, for $n \geq N$ choose $k \in \{1, 2, \ldots, n\}$ such that

$$t_0 - \varepsilon < \frac{k-1}{n}, \frac{k}{n}, \frac{k+1}{n}, \frac{k+2}{n} < t_0 + \varepsilon. \qquad (2.5.15)$$

We deduce that

$$\Omega \setminus S \subset \bigcup_{M \in \mathbb{N}} \bigcup_{N \in \mathbb{N}} \underbrace{\left(\bigcap_{n \geq N} \bigcup_{k=1}^{n} \bigcap_{i=0}^{2} \{\,|B_{(k-1+i)/n} - B_{(k+i)/n}| \leq M/n\,\} \right)}_{=:X_{M,N}}.$$

Clearly, the set $X_{M,N}$ is measurable and it suffices to show it is negligible. We have

$$\mathbb{P}[X_{M,N}] \leq \inf_{n \geq N} \sum_{k=1}^{n-2} \mathbb{P}\left[\max_{0 \leq i \leq 2} |B_{(k-1+i)/n} - B_{(k+i)/n}| \leq M/n\right]. \qquad (2.5.16)$$

Now observe that the increments $B_{(k-1)/n} - B_{k/n}$ are independent Gaussians with mean zero and variance $1/n$. We deduce

$$\mathbb{P}[X_{M,N}] \leq \inf_{n \geq N} \sum_{k=1}^{n-2} \mathbb{P}\bigl[\,|B_{(k-1)/n} - B_{k/n}| \leq M/n\,\bigr]^3.$$

The exponent 3 above will make all the difference. It appears because of the constraint (2.5.15) on N. Since $\sqrt{n}\,|B_{(k-1)/n} - B_{k/n}|$ is standard normal, the random variable $|B_{(k-1)/n} - B_{k/n}|$ is normal with variance $\frac{1}{n}$ and we have

$$\mathbb{P}\bigl[\,|B_{(k-1)/n} - B_{k/n}| \leq M/n\,\bigr] = 2\sqrt{\frac{n}{2\pi}} \int_0^{M/n} e^{-x^2 n/2} dx$$

$(x = My/n)$

$$2\sqrt{\frac{n}{2\pi}} \frac{M}{n} \int_0^1 e^{-\frac{My^2}{2n}} dy \leq \underbrace{\frac{2}{\sqrt{2\pi}}}_{=:C} M n^{-1/2}.$$

Hence

$$\sum_{k=1}^{n-2} \mathbb{P}\bigl[\,|B_{(k-1)/n} - B_{k/n}| \leq M/n\,\bigr]^3 \leq n C^3 M^3 n^{-3/2} = C^3 M^3 n^{-1/2}, \quad \forall n \geq N,$$

and (2.5.16) implies that $\mathbb{P}[X_{M,N}] = 0$. $\hfill\square$

2.6 Exercises

Exercise 2.1 (Skorokhod). Suppose that $X_1,\ldots,X_n$ are independent random variables. We set $S_k = X_1 + \cdots + X_k$, $k = 1,\ldots n$. Let $\alpha > 0$ and set
$$c := \sup_{1 \leq j \leq n} \mathbb{P}\big[|S_n - S_j| > \alpha\big], \quad M_n := \sup_{1 \leq j \leq n} |S_j|.$$
Prove that if $c < 1$, then
$$\mathbb{P}\big[M_n > 2\alpha\big] \leq \frac{1}{1-c}\mathbb{P}\big[|S_n| > \alpha\big].$$

Hint. Denote by J the first j such that $|S_j| > 2\alpha$. Note that $\mathbb{P}[M_n > 2\alpha] = \mathbb{P}[J \leq n]$
$$\mathbb{P}[|S_n|>\alpha] \geq \mathbb{P}[|S_n|>\alpha,\ M_n>2\alpha] = \sum_{j=1}^n \mathbb{P}[|S_n|>\alpha,\ J=j] \geq \sum_{j=1}^n \mathbb{P}[|S_n-S_j|\leq\alpha,\ J=j].$$
Observe that the event $\{J = j\}$ is independent of $S_n - S_j$. □

Exercise 2.2. Suppose that $(X_n)_{n\geq 1}$ is a sequence of independent random variables. Prove that the following statements are equivalent.

(i) The series $\sum_{n\geq 1} X_n$ converges in probability.
(ii) The series $\sum_{n\geq 1} X_n$ converges a.s.

Hint. Use Exercise 2.1. □

Remark 2.89. The so called *Lévy equivalence theorem*, [47, §III.2], [50, §9.7] or [105, §43] states that a series with independent terms converges a.s. iff converges in probability, iff converges in distribution. □

Exercise 2.3. Consider an infinite array of *nonnegative* numbers $P = (p_{n,k})_{k,n\geq 1}$ satisfying the following conditions.

(i) The array is lower triangular, i.e., $p_{n,k} = 0$, $\forall k > n$.
(ii) For every n, the n-th row of P defines a probability distribution on $\mathbb{I}_n = \{1, 2, \ldots, n\}$, i.e.,
$$\sum_{k=1}^n p_{n,k} = 1, \quad \forall n \geq 1.$$
(iii) The sequence determined by each column of P converges to 0, i.e.,
$$\lim_{n\to\infty} p_{n,k} = 0, \quad \forall k \geq 1.$$

Show that if (x_n) is a sequence of real numbers that converges to a number x, then the sequence of weighted averages
$$y_n := \sum_{k=1}^n p_{n,k} x_k$$
converges to the same number x. □

Exercise 2.4. In this exercise we describe the *acceptance-rejection method* frequently used in Monte-Carlo simulations. For any nonnegative function $f : \mathbb{R} \to [0, \infty)$ we denote by G_f the region bellow its graph

$$G_f := \{(x,y) \in \mathbb{R}^2;\ 0 \leq y \leq f(x((x)\}.$$

(i) Suppose that we are given a probability density $p : \mathbb{R} \to [0, \infty)$

$$\int_{\mathbb{R}} p(x)dx = 1.$$

For any positive constant c we set

$$\mu_p^c = \frac{1}{c} \boldsymbol{I}_{G_{cp}}(x,y)dxdy.$$

Since area$(G_{cp}) = c$ we deduce that μ_p^c defines a Borel probability measure on $\mathbb{R}^2$. The natural projection $\mathbb{R}^2 \ni (x,y) \mapsto x \in \mathbb{R}$ is a random variable X defined on the probability space $(\mathbb{R}^2, \mathcal{B}_{\mathbb{R}^2}, \mu_p^c)$. Prove that the probability distribution of X is $p(x)dx$.

(ii) Suppose that X is a random variable with probability distribution $p(x)dx$. Let U be a random variable independent of X and uniformly distributed over $[0,1]$. Prove that the probability distribution of the random vector $(X, cp(X)U)$ is μ_p^c.

(iii) Let $q : \mathbb{R} \to [0, \infty)$ be another probability density such that, there exists $c > 0$ with the property that

$$q(x) \leq cp(x), \quad \forall x \in \mathbb{R}.$$

Suppose that $(U_n)_{n \in \mathbb{N}}$: is a sequence of i.i.d. random variables uniformly distributed on $[0,1]$ and $(X_n)_{n \in \mathbb{N}}$ is a sequence of i.i.d., independent of the U_n's and with common distribution $p(x)dx$. Denote by N the random variable

$$N = \inf\{n \in \mathbb{N} :\ cp(X_n)U_n \leq q(X_n)\}.$$

Prove that

$$\mathbb{E}[N] = c.$$

Hint. Consider the random vector $V_n = (X_n, cp(X_n)U_n)$, observe that

$$N = \inf\{n \in \mathbb{N};\ V_n \in G_q\},$$

and use part (ii) to show that N is a geometric random variable.

(iv) Define $Y = X_N$, i.e.,

$$Y(\omega) = X_{N(\omega)}(\omega).$$

From (iii) we know that $\mathbb{P}[N < \infty] = 1$ so Y is defined outside a probability zero set. Prove that the probability distribution of the random variable Y is $q(y)dy$.

□

Remark 2.90 (Acceptance-Rejection method). Suppose that a computer can sample the distribution $\text{Unif}(0,1)$ and it can sample the distribution $p(x)dx$. We can then sample the distribution $q(y)dy$ as follows. Sample successively and independently $\text{Unif}(0,1)$ and $p(x)dx$ and denote by U_n and respectively X_n the samples obtained at the n-th trial. Stop at the first trial N when the inequality $cU_n \leq \frac{q(X_n)}{p(X_n)}$ is observed. Set $Y = X_N$. The results in the above exercise show that the expected waiting time to observe this inequality is c and the random number Y samples the distribution $q(y)dy$. $\square$

Exercise 2.5 (Bernstein). For each $x \in [0,1]$ we consider a sequence $(B_k^x)_{k \in \mathbb{N}}$ of i.i.d. Bernoulli random variables with probability of success x. We set

$$S_n^x = \sum_{k=1}^n B_k^x.$$

Note that $S_n^x/n \in [0,1]$ and the *SLLN* shows that

$$S_n^x/n \to x \text{ a.s. as } n \to \infty.$$

The dominated converges theorem implies that for any continuous function $f : [0,1] \to \mathbb{R}$ we have

$$\lim_{n \to \infty} \mathbb{E}\big[\, f(S_n^x/n)\,\big] = f(x).$$

Set

$$B_n^f(x) := \mathbb{E}\big[\, f(S_n^x/n)\,\big].$$

(i) Show that

$$B_n^f(x) = \sum_{k=0}^n \binom{n}{k} x^k (1-x)^k f(k/n).$$

(ii) Prove that as $n \to \infty$ the polynomials $B_n^f(x)$ converge *uniformly on* $[0,1]$ to $f(x)$.

Hint. For (ii) imitate the argument in Step 2 of the proof of Theorem 2.41. $\square$

Exercise 2.6. Suppose that $X_n \in L^2(\Omega, \mathcal{S}, \mathbb{P})$ is a sequence of random variables with mean zero and variance one such that

$$\lim_{k \to \infty} \mathbb{E}\big[\, X_m X_{m+k}\,\big] = 0, \text{ uniformly in } m.$$

Prove that

$$\frac{1}{n}\big(X_1 + \cdots + X_n\big) \xrightarrow{p} 0 \text{ as } n \to \infty. \qquad \square$$

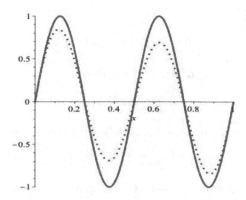

Fig. 2.5 The graph of $f(x) = \sin(4\pi x)$ (the continuous curve) and of the degree 50 Bernstein polynomial $B_{50}^f(x)$ (the dotted curve).

Exercise 2.7. Suppose that a player rolls a die an indefinite amount of times. More formally, we are given a sequence independent random variables $(X_n)_{n \in \mathbb{N}}$, uniformly distributed on $\mathbb{I}_6 := \{1, 2, \ldots, 6\}$. For $k \in \mathbb{N}$, we say that *a k-run occurred at time n* if $n \geq k$ and

$$X_n = X_{n-1} = \cdots = X_{n-k+1} = 6.$$

For $n \in \mathbb{N}$ we set

$$R_n = R_n^k := \#\{\, m \leq n;\ \text{a } k\text{-run occurred at time } m\,\},$$

$$T = T_k = \min\{\, n \geq k;\ R_n > 0\,\}.$$

Thus T is the moment when the first k-run occurs. As shown in Example 1.167, $\mathbb{E}[T] < \infty$.

(i) Compute $\mathbb{E}[T]$.
(ii) Prove that $\frac{R_n}{n}$ converges in probability to $\frac{1}{6^k}$. **Hint.** For $n \geq k$ set

$$Y_n := \boldsymbol{I}_{\{X_n = 6\}} \cdots \boldsymbol{I}_{X_{\{n-k+1 = 6\}}}.$$

Observe that $R_n = Y_k + \cdots + Y_n$.

$\square$

Exercise 2.8 (A. Renyi). Suppose that $(A_n)_{n \geq 0}$ is a sequence of events in the sample space $(\Omega, \mathcal{S}, \mathbb{P})$ with the following properties.

- $A_0 = \Omega$.
- $\mathbb{P}[A_n] \neq 0, \forall n \geq 0$.
- There exists $\rho \in (0, 1]$ satisfying

$$\lim_{n \to \infty} \mathbb{P}[A_n | A_k] = \rho, \ \ \forall k \geq 0. \tag{2.6.1}$$

Set $X_n := \boldsymbol{I}_{A_n} - \rho$.

(i) Prove that
$$\lim_{n\to\infty} \mathbb{E}[X_n X_k] = 0, \quad \forall k \in \mathbb{N}.$$

(ii) Prove that for any $X \in L^2(\Omega, \mathcal{S}, \mathbb{P})$ we have
$$\lim_{n\to\infty} \mathbb{E}[X_n X] = 0.$$

(iii) Conclude that the sequence (A_n) satisfies the *mixing condition with density ρ*
$$\lim_{n\to\infty} \mathbb{P}[A_n \cap A] = \rho \mathbb{P}[A], \quad \forall A \in \mathcal{S}. \tag{2.6.2}$$

Thus, in the long run, the set A_n occupies the same proportion ρ of any measurable set A.

□

Exercise 2.9 (A. Renyi). Suppose that $(X_n)_{n \in \mathbb{N}}$ is a sequence of i.i.d., almost surely finite random variables. Set
$$M_n := \frac{X_1 + \cdots + X_n}{n}.$$
Assume that the empirical means M_n converge in probability to a random variable M. The goal of the exercise is to prove that M is a.s. constant. We argue by contradiction. *Assume M is not a.s. constant.* Let $F : \mathbb{R} \to [0,1]$ the cdf of M, $F(m) = \mathbb{P}[M \leq m]$.

(i) Prove that there exist two continuity points $a < b$ of $F(x)$ such that
$$p_0 := F(b) - F(a) = \mathbb{P}[a < M \leq b] \in (0,1).$$

(ii) Prove that there exists $\nu_0 \in \mathbb{N}$ such that
$$\mathbb{P}[a < M_n \leq b] > 0, \quad \forall n \geq \nu_0.$$

(iii) Set $A_0 = \Omega$ and
$$A_n = \{a < M_{\nu_0 + n} \leq b\}, \quad n \geq 1.$$
Prove that the sequence (A_n) satisfies the condition (2.6.1) with $\rho = p_0$.

(iv) Set $B := \{a < M \leq b\}$. Prove that the restriction of M_n to $(B, \mathcal{S}|_B, \mathbb{P}[-|B])$ converges in probability to $M|_B$. Here
$$\mathcal{S}|_B = \{S \cap B; \ S \in \mathcal{S}\}.$$

(v) Deduce that $p_0 = 1$, thus contradicting (i).

□

Exercise 2.10 (Vitali-Hahn-Saks). Suppose that $(\Omega, \mathcal{S}, \mu)$ is a probability space. Define an equivalence relation on $\mathcal{S}$ by setting $S \sim S'$ if $\mu[S \Delta S'] = 0$, where Δ denotes the symmetric difference

$$S \Delta S' = (S \setminus S') \cup (S' \cup S).$$

Define $d : \mathcal{S} \times \mathcal{S} \to [0, \infty)$

$$d(S_0, S_1) = \mu[S_0 \Delta S_1].$$

(i) Prove that $\forall S_0, S_1, S_2 \in \mathcal{S}$ we have

$$d(S_0, S_1) = d(S_1, S_0), \quad d(S_0, S_2) \leq d(S_0, S_1) + d(S_1, S_2)$$

and $d(S_0, S_1) = 0$ iff $S_0 \sim S_1$.

(ii) Prove that d defines a *complete* metric d on $\bar{\mathcal{S}} := \mathcal{S}/\sim$.

(iii) Suppose that $\lambda : \mathcal{S} \to \mathbb{R}$ is a probability measure that is absolutely continuous with respect to μ. Hence $\lambda[S_0] = \lambda[S_1] = 0$ if $S_0 \sim S_1$. Prove that the induced function

$$\bar{\lambda} : \bar{\mathcal{S}} \to \mathbb{R}$$

is continuous with respect to the metric d.

(iv) Suppose that (λ_n) is a sequence of probability measures on $\mathcal{S}$ such that $\lambda_n \ll \mu$ for any $n \in \mathbb{N}$ and, $\forall S \in \mathcal{S}$, the sequence $\lambda_n[S]$ has a finite limit $\lambda[S]$. Prove that $\lambda : \mathcal{S} \to \mathbb{R}$ is finitely additive and $\lambda[S] = 0$ if $\mu[S] = 0$.

(v) For any $\varepsilon > 0$ and $k \in \mathbb{N}$ we set

$$\bar{\mathcal{S}}_{k,\varepsilon} := \left\{ S \in \bar{\mathcal{S}}; \ \sup_{m \in \mathbb{N}} |\lambda_l[S] - \lambda_{k+n}[S]| \leq \varepsilon \right\}.$$

Prove that the sets $\bar{\mathcal{S}}_{k,\varepsilon} \subset \bar{\mathcal{S}}$ are closed with respect to the metric d and

$$\bar{\mathcal{S}} = \bigcup_{k \in \mathbb{N}} \bar{\mathcal{S}}_{k,\varepsilon}, \quad \forall \varepsilon > 0.$$

(vi) Prove that $\bar{\lambda} : \mathcal{S} \to [0,1]$ is continuous and deduce that $\bar{\lambda}$ is a probability measure. **Hint.** It suffice to show that for any decreasing sequence (S_n) in $\mathcal{S}$ with empty intersection we have $\lim \lambda[S_n] = 0$. Deduce this from (v) and Baire's theorem.

□

Exercise 2.11 (A. Renyi). Let $(\Omega, \mathcal{S}, \mathbb{P})$ be a probability space and suppose that (A_n) is a *stable sequence of events*, i.e., for any $B \in \mathcal{S}$ the sequence $\mathbb{P}[A_n \cap B]$ has a finite limit $\lambda[B]$ and $\lambda[\Omega] \in (0,1)$. Prove that $\lambda : \mathcal{S} \to [0,1]$ is a finite measure absolutely continuous with respect to $\mathbb{P}$, $\lambda \ll \mathbb{P}$. Denote by ρ the density of λ with respect to $\mathbb{P}$, $\rho = \frac{d\lambda}{d\mathbb{P}}$. The function ρ is called the *density of the stable sequence of events*. **Hint.** Use Exercise 2.10.

□

Exercise 2.12 (A. Renyi). Let $(\Omega, \mathcal{S}, \mathbb{P})$ be a probability space and suppose that $(A_n)_{n \in \mathbb{N}}$ is a sequence of events such that the limits

$$\lambda_0 = \lim_{n \to \infty} \mathbb{P}[A_n], \quad \lambda_k := \lim_{n \to \infty} \mathbb{P}[A_k \cap A_n], \quad k \in \mathbb{N}$$

exist and $\lambda_0 \in (0,1)$. Denote by X linear span of the indicators $\boldsymbol{I}_{A_n}$ and by $\overline{X}$ its closure in L^2.

(i) Prove that $\forall \xi \in \overline{X}$ there exists a limit

$$L(\xi) := \lim_{n \to \infty} \mathbb{E}[\xi \boldsymbol{I}_{A_n}] = \mathbb{E}[\rho \xi].$$

(ii) Prove that $\forall \xi \in L^2(\Omega, \mathcal{S}, \mathbb{P})$ there exists a limit

$$L(\xi) = \lim_{n \to \infty} \mathbb{E}[\xi \boldsymbol{I}_{A_n}] = \mathbb{E}[\rho \xi].$$

(iii) Show that $\overline{X} = L^2(\Omega, \mathcal{S}, \mathbb{P})$ and $\exists \rho \in L^2(\Omega, \mathcal{S}, \mathbb{P})$ such that $L(\xi) = \mathbb{E}[\rho \xi]$, $\forall \xi \in L^2(\Omega, \mathcal{S}, \mathbb{P})$.

(iv) Show that $(A_n)_{n \in \mathbb{N}}$ is a stable sequence with density ρ. (Note that when ρ is constant the sequence satisfies the mixing condition (2.6.1) with density $\rho = \lambda_0$.)

□

Exercise 2.13. Suppose that $f : [0,1] \to [0,1]$ is a continuous function that is not identically 0 or 1. For $n \in \mathbb{N}$ we set

$$A_n = \bigcup_{k=0}^{n-1} [k/n, k/n + f(k/n)].$$

Show that $(A_n)_{n \geq 1}$ is a stable sequence of events and compute its density. □

Exercise 2.14. Suppose that π is a probability measure on $\mathbb{I}_n = \{1, 2, \ldots, n\}$, $p_i = \pi[\{i\}]$. Consider a sequence $(X_n)_{n \in \mathbb{N}}$ of i.i.d. random variables uniformly distributed on $[0,1]$. For $j \in \mathbb{I}_n$ and $m \in \mathbb{N}$ we set

$$Z_{m,j} := \#\left\{ 1 \leq k \leq m; \; \sum_{i=0}^{j-1} p_i \leq X_k < \sum_{i=0}^{j} p_i \right\}, \quad H_m := \frac{1}{m} \sum_{j=1}^{n} Z_{m,j} \log_2 p_j.$$

Prove that

$$\lim_{m \to \infty} H_m = -\operatorname{Ent}_2[\pi] = \sum_{j=1}^{n} p_j \log_2 p_j, \quad \text{a.s.}$$

□

Exercise 2.15. Let $(X_n)_{n \in \mathbb{N}}$ be a sequence of i.i.d. Bernoulli random variables with success probability $\frac{1}{2}$ and $(Y_n)_{n \in \mathbb{N}}$ a sequence of i.i.d. Bernoulli random variables with success probability $\frac{1}{3}$. (The sequences (X_n) and (Y_n) may not be independent

of each other.) Set $\mathcal{B} = \{0,1\}$ and denote by $\mathcal{F}_n$ the sigma-algebra of $\mathcal{B}^{\mathbb{N}}$ generated by the cylinders
$$C_\epsilon^k := \{\underline{x} = (x_1, x_2, \dots) \in \mathcal{B}^{\mathbb{N}}; \; x_k = \epsilon\}, \quad k = 1, 2, \dots, n, \quad \epsilon = 0, 1.$$
We set
$$\mathcal{F} := \bigcup_{n \in \mathbb{N}} \mathcal{F}_n.$$
The sequence (X_n) (resp. Y_n) define a probability measures $\mathbb{P} = \text{Ber}(1/2)^{\otimes \mathbb{N}}$ (resp. $\mathbb{Q} = \text{Ber}(1/3)^{\otimes \mathbb{N}}$) on $\mathcal{B}^{\mathbb{N}}$; see Subsection 1.5.1. Denote by $\mathbb{P}_n$ (resp. $\mathbb{Q}_n$) the restrictions of $\mathbb{P}$ (resp. $\mathbb{Q}$) to $\mathcal{F}_n$.

(i) Prove that for any $n \in \mathbb{N}$ the measure $\mathbb{Q}_n$ is absolutely continuous with respect to $\mathbb{P}_n$. Compute the density $\frac{d\mathbb{Q}_n}{d\mathbb{P}_n}$ of $\mathbb{Q}_n$ with respect to $\mathbb{P}_n$.
(ii) Prove that $\mathbb{Q}$ is not absolutely continuous with respect to $\mathbb{P}$. **Hint.** Use the Law of Large Numbers.

□

Exercise 2.16. Show that the Gaussian measures $\mathbf{\Gamma}_v[dx] = \gamma_{0,v}(x)[dx]$,
$$\gamma_{0,v}(x) := \frac{1}{\sqrt{2\pi v}} e^{-\frac{x^2}{2v}},$$
converge weakly to the Dirac measure δ_0 on $\mathbb{R}$ as their variances v converge to 0.
Hint. Use the Chebyshev's inequality (1.3.17) $\mathbf{\Gamma}_v[|x| > c] \leq \frac{v}{c^2}$. □

Exercise 2.17. Let (X_n) be a sequence of geometric random variables $X_n \sim \text{Geom}(1/n)$. Prove that
$$Y_n := \frac{1}{n} X_n \Rightarrow X \sim \text{Exp}(1).$$
Hint. Show that $\mathbb{P}[Y_n > y] \to e^{-y}$ as $n \to \infty$, $\forall y \in \mathbb{R}$. □

Exercise 2.18. Fix $\lambda > 0$. Show that as $n \to \infty$ we have $\text{Bin}(n, \lambda/n) \Rightarrow \text{Poi}(\lambda)$, where $\text{Bin}(n, \lambda/n)$ denotes the binomial probability distribution corresponding to n independent trials with success probability λ/n and $\text{Poi}(\lambda)$ denotes the Poisson distribution with parameter λ. □

Exercise 2.19 (Occupancy Problem). Suppose that n balls are successively and randomly placed in r boxes, i.e., for all boxes are equally likely to be the destination of a given ball. Let $N_r = N_{r,n}$ the number of empty boxes.

(i) Compute the expectation and variance of N_r. **Hint.** $N_r = \sum_{k=1}^r I_{B_{k,n}}$, $B_{k,n} = $ box k is empty after all the n balls have been randomly placed in the r boxes.
(ii) Show that if $n/r \to c > 0$ as $r \to \infty$, then $\frac{N_r}{r} \to e^{-c}$ in probability.
(iii) Compute $\mathbb{P}[N_{r,n} = k]$. **Hint.** Use the inclusion-exclusion equality (1.3.25).

(iv) Show that if $re^{-n/r} \to \lambda$ as $r \to \infty$, then $N_{r,n}$ converges in distribution to Poi(λ). □

Remark 2.91. Let me comment why the result in Exercise 2.19 is surprising. Consider the following concrete situation.

Assume $n = 2r$ and suppose that we want to distribute $2r$ gifts to r children. We want to do this in the "fairest" possible way since the gifts, of equal value, are different, and several kids may desire the same gift. To remove any bias, "common sense" suggests that each gift should be given to a child chosen uniformly at random. There are twice as many gifts as children so what can go wrong? Part (ii) of this exercise shows that for n large nearly surely $e^{-2}r \approx 0.13r$ children will receive no gifts! □

Exercise 2.20 (Coupon collector problem). For $n \in \mathbb{N}$ denote by N_n the number of boxes of cereals one has to purchase in order to obtain all the n coupons of a collection; see Example 1.112. Recall that $\mathbb{E}[N_n] \sim n \log n$ as $n \to \infty$. Prove that

$$\lim_{n \to \infty} \mathbb{P}[N_n - n \log n \le nx] = \exp(e^{-x}).$$

Hint. Reduce to Exercise 2.19(iv). □

Exercise 2.21. For $N \in \mathbb{N}$ denote by B_N the birthday random variable defined in Exercise 1.21.

(i) Show that as $N \to \infty$, the sequence of random variables

$$X_N := \frac{1}{\sqrt{N}} B_N$$

converges in law to a *Raleigh random variable*, i.e., a random variable X with probability distribution

$$\mathbb{P}_X[dx] = xe^{-\frac{x^2}{2}} \boldsymbol{I}_{[0,\infty)}(x) dx.$$

(ii) Prove that

$$\lim_{N \to \infty} \mathbb{E}[X_N] = \mathbb{E}[X] = \sqrt{\frac{\pi}{2}}.$$

Hint. Observe that $\mathbb{P}[X > x] = e^{-\frac{x^2}{2}}$. Using the Taylor expansion of $\log(1-t)$ at $t = 0$ show that

$$\log \mathbb{P}[X_N > x] \le -\frac{x^2}{2} \quad \text{and} \quad \lim_{N \to \infty} \log \mathbb{P}[X_N > x] = -\frac{x^2}{2}, \quad \forall x > 0.$$

□

Exercise 2.22. [P. Lévy] Consider the random variables L_n defined in Exercise 1.11. Prove that as $n \to \infty$ the random variables $\frac{L_n}{n}$ converge in distribution to the arcsine distribution Beta$(1/2, 1/2)$; see Example 1.122. **Hint.** You need to use Stirling formula (A.1.7) with error estimate (A.1.8). □

Exercise 2.23. Suppose that $(X_n)_{n\in\mathbb{N}}$ and $(Y_n)_{n\in\mathbb{N}}$ are two sequences of random variables such that $X_n \to X$ in distribution and
$$\lim_{n\to\infty} \mathbb{P}\big[\, X_n \neq Y_n\,\big] = 0.$$
Then $Y_n \to X$ in distribution. □

Exercise 2.24. Suppose that $(X_n)_{n\in\mathbb{N}}$ and $(Y_n)_{n\in\mathbb{N}}$ are two sequences of random variables such that

- X_n converges in distribution X.
- Y_n converges in distribution to Y.
- X_n is independent of Y_n for every n and X is independent of Y.

Prove the following.

(i) The random vector (X_n, Y_n) converges in distribution to (X, Y).
(ii) The sum $X_n + Y_n$ converges in distribution to $X + Y$.

□

Exercise 2.25. Suppose that $(X_n)_{n\in\mathbb{N}}$ and $(Y_n)_{n\in\mathbb{N}}$ are sequences of random variables with the following properties.

(i) The random variables $(X_n)_{n\in\mathbb{N}}$ are identically distributed.
(ii) The sequence of random vectors (X_n, Y_n) converges in distribution to the random vector (X, Y).

Prove that for any Borel measurable function $f : \mathbb{R} \to \mathbb{R}$ the sequence of random vectors $(\,f(X_n), Y_n)$ converges in distribution to $(\,f(X), Y)$. **Hint.** Fix a Borel measurable function f. It suffices to show that for any continuous and bounded functions $u, v : \mathbb{R} \to \mathbb{R}$ we have
$$\lim_{n\to\infty} \mathbb{E}[\,u(\,f(X_n)\,)v(Y_n)\,] = \mathbb{E}[\,u(\,f(X)\,)v(Y)\,].$$
Consider the Borel measurable functions v_n defined by $v_n(X_n) = \mathbb{E}[\,v(Y_n)\,\|\,X_n\,] = v_n(X_n)$. □

Exercise 2.26. Suppose that $(X_n)_{n\in\mathbb{N}}$ and $(Y_n)_{n\in\mathbb{N}}$ are two sequences of random variables such that X_n converges in distribution to X and Y converges in probability to the constant c. Prove that the random vector (X_n, Y_n) converges in distribution to (X, c). **Hint.** Prove that (X_n, c) converges in probability to (X, c) and then use Exercise 2.23. □

Exercise 2.27. Suppose that $(X_n)_{n\in\mathbb{N}}$ is a sequence of i.i.d. L^2 random variables with $\mu = \mathbb{E}[\,X_n\,]$, $\sigma^2 = \mathrm{Var}[\,X_n\,]$. Set
$$\overline{X}_n = \frac{1}{n}(X_1 + \cdots + X_n), \quad Y_n = \frac{1}{n-1}\sum_{k=1}^{n-1}(X_k - \bar{X}_n)^2.$$
Prove that $\mathbb{E}[\,Y_n\,] = \sigma^2$ and $Y_n \to \sigma^2$ in probability. □

Exercise 2.28 (Trotter). We outline a proof of the CLT that does not rely on the characteristic function.

For any random variable X and any $f \in C_b(\mathbb{R})$ we denote by $T_X f$ the function $\mathbb{R} \to \mathbb{R}$ given by
$$T_X f(y) = \mathbb{E}[f(X+y)], \ y \in \mathbb{R}.$$

(i) Prove the correspondence $f \mapsto T_X f$ induces a bounded linear map $C_b(\mathbb{R}) \to C_b(\mathbb{R})$ satisfying $\|T_X f\| \leq \|f\|$, where
$$\|g\| = \sup_{x \in \mathbb{R}} |g(x)|, \ \forall g \in C_b(\mathbb{R}).$$

(ii) Show that if $X \perp\!\!\!\perp Y$, then $T_X \circ T_Y = T_{X+Y} = T_Y \circ T_X$.

(iii) Let $(X_n)_{n \in \mathbb{N}}$ be a sequence i.i.d. square integrable random variables such that
$$\mathbb{E}[X_n] = 0, \ \operatorname{Var}[X_n] = 1, \ \forall n \in \mathbb{N}.$$
Additionally, fix a sequence $(Y_n)_{n \in \mathbb{N}}$ of independent standard normal random variables that are also independent of the X_n's. Set
$$U_n = \frac{1}{\sqrt{n}} \sum_{k=1}^n X_k, \ V_n = \frac{1}{\sqrt{n}} \sum_{k=1}^n Y_k,$$
Prove that for any $f \in C_b(\mathbb{R})$ we have
$$\|T_{U_n} f - T_{V_n} f\| \leq n \|T_{\frac{X_1}{\sqrt{n}}} f - T_{\frac{Y_1}{\sqrt{n}}} f\|.$$

(iv) Let $f \in C_0^2(\mathbb{R})$. Show that
$$\lim_{n \to \infty} n \|T_{\frac{X_1}{\sqrt{n}}} f - T_{\frac{Y_1}{\sqrt{n}}} f\| = 0.$$

(v) Prove that $T_{V_n} = T_Z$, where Z is a standard normal random variable.

(vi) Show that for any $f \in C_0^2(\mathbb{R})$ we have
$$\lim_{n \to \infty} \mathbb{E}[f(U_n)] = \mathbb{E}[f(Z)].$$

□

Exercise 2.29. Suppose that $(X_n)_{n \in \mathbb{N}}$ is a sequence of i.i.d. Bernoulli random variables with success probability $p = \frac{1}{2}$. For each $n \in \mathbb{N}$ we set
$$S_n := \sum_{k=1}^b \frac{1}{2^k} X_k.$$

(i) Find the probability distribution of S_n.

(ii) Prove that for any $p \in [1, \infty]$ the sequence S_n converges a.s. and L^p to a random variable S uniformly distributed on $[0, 1]$.

(iii) Compute the characteristic functions $F_n(\xi) = \mathbb{E}[\exp(i\xi S_n)]$ and deduce Viète's formula
$$\frac{\sin \xi}{\xi} = \prod_{n=1}^{\infty} \cos(\xi/2^n).$$

(iv) Suppose that μ is a Borel probability measure on $\mathbb{R}$ with quantile $Q : [0,1] \to \mathbb{R}$,
$$Q(p) = \inf\{\, x \in \mathbb{R};\ \mu\big[(-\infty, x]\big) \geq p \,\}.$$
Prove that the sequence $Q(S_n)$ converges a.s. to a random variable with distribution μ. **Have a look at Example 1.45.**

□

Remark 2.92. Part (iv) of the above exercise is essentially a universality property of the simplest random experiment: tossing a fair coin. If we are able to perform this experiment repeatedly and independently, then we can approximate any probability distribution. In other words, we can approximatively sample any probability distribution by flipping fair coins. □

Exercise 2.30. Suppose that $(X_n)_{n \in \mathbb{N}}$ is a sequence of i.i.d. random variables uniformly distributed in $[0, L]$, $L > 0$. For $n \in \mathbb{N}$ we set
$$X_{(n)} := \max(\, X_1, X_2, \ldots, X_n \,).$$
Prove that $\lim_{n \to \infty} \mathbb{E}[\, X_{(n)} \,] = L$ and $X_{(n)} \to L$ in probability. **Hint.** Have a look at Exercise 1.44. □

Exercise 2.31. Suppose that $(X_n)_{n \in \mathbb{N}}$ is a sequence of i.i.d. random variables uniformly distributed in $[0, 1]$. Denote by $X^n_{(1)}, X^n_{(2)}, \ldots, X^{(n)}_n$ the order statistics of the first n of them; see Exercise 1.44. Prove that for any $k \in \mathbb{N}$ the random variable $nX^n_{(k)}$ converges in distribution to $\mathrm{Gamma}(1, k)$. □

Exercise 2.32. Suppose that $(\mu_n)_{n \geq 0}$ is a sequence of finite Borel measures. Denote their
$$F_n(x) = \mu_n\big[(-\infty, x]\big], \quad \forall n \in \mathbb{N},\ x \in \mathbb{R}.$$
Let $\mu \in \mathrm{Meas}(\mathbb{R})$ be a finite Borel measure with potential
$$F(x) = \mu\big[(-\infty, x]\big], \quad \forall x \in \mathbb{R}.$$
Prove that the following are equivalent.

(i) The measures (μ_n) converge vaguely to μ
(ii) For every point x of continuity of F
$$\lim_{n \to \infty} F_n(x) = F(x).$$

□

Exercise 2.33. Suppose that $(\mu_n)_{n \in \mathbb{R}}$ is a sequence of Borel measures on $\mathbb{R}$ such that
$$\sup_n \mu_n[\mathbb{R}] < \infty.$$
Set $F_n(x) = \mu_n\big[(-\infty, x]\big]$, $x \in \mathbb{R}$. Prove that (μ_n) contains a subsequence converging *vaguely* to a finite probability measure. **Hint.** Construct a subsequence (n_k) such that, the sequence $F_{n_k}(q)$ is convergent for any $q \in \mathbb{Q}$. □

Exercise 2.34. A subset $\mathcal{M} \subset \mathrm{Prob}(\mathbb{R})$ is called *tight* if, for any $\varepsilon > 0$ there exists a compact set $K \subset \mathbb{R}$ such that $\mu[\mathbb{R} \setminus K] < \varepsilon$, $\forall \mu \in \mathbb{R}$. Prove that a tight sequence of probability measures admits a *weakly* convergent subsequence. **Hint.** Use Exercise 2.33.

□

Exercise 2.35. Let $\mu \in \mathrm{Prob}(\mathbb{R})$ be a Borel probability measure with characteristic function $\widehat{\mu}$. Prove that for any $r > 0$ we have

$$\mu[\{|x| > r\}] \leq \frac{1}{C} \int_0^1 |1 - \mathrm{Re}\, \widehat{\mu}(t/r)|\, dt, \quad C := \inf_{x \in [0,1]} \left(1 - \frac{\sin x}{x}\right).$$

□

Exercise 2.36. This exercise describes a strengthening of Levy's continuity theorem. Suppose that (μ_n) is a sequence of Borel probability measures on $\mathbb{R}$ with characteristic functions $\widehat{\mu}_n(\xi)$. Assume that the functions $\widehat{\mu}_n(\xi)$ converge pointwisely to a function $f : \mathbb{R} \to \mathbb{C}$ that is continuous at 0.

(i) Prove that the sequence $(\mu_n)_{n \in \mathbb{N}}$ is tight. **Hint.** Use Exercise 2.35.
(ii) Show that f is the characteristic function of a Borel *probability* measure μ. **Hint.** Use Exercise 2.34.
(iii) Prove that μ_n converges weakly to μ.

□

Exercise 2.37. Suppose that (μ_n) is a sequence of Borel probability measures on $\mathbb{R}$ that converges weakly to a probability measure μ. Prove that the characteristic functions $\widehat{\mu}_n$ converge to $\widehat{\mu}$ *uniformly on the compacts of* $\mathbb{R}$.

□

Exercise 2.38. Suppose that X is a random variable and $\varphi(\xi)$ is its characteristic function

$$\varphi(\xi) = \mathbb{E}\left[e^{i\xi X}\right].$$

Prove that the following are equivalent.

(i) X is a.s. constant.
(ii) There exists $r > 0$ such that $|\varphi(\xi)| = 1$, $\forall \xi \in [-r, r]$.

Hint. Use an independent copy X' of X.

□

Exercise 2.39 (Lévy). The concentration function of a random variable X is the function

$$C_X : [0, \infty) \to [0, 1], \quad C_X(r) := \inf_{x \in \mathbb{R}} \mathbb{P}\big[|X - x| > r\big].$$

(i) Prove that for any $r > 0$ there exists $x_r \in \mathbb{R}$ such that
$$C_X(r) = \mathbb{P}\big[\,|X - x_r| > r\big].$$

(ii) Prove that if $\mathrm{Var}[X] < \infty$ is integrable, then
$$C_X(r) \leq \frac{1}{r^2}\mathrm{Var}[X].$$

(iii) Prove that if X, Y are independent random variables, then
$$C_{X+Y}(r) \geq \max\bigl(C_X(r), C_Y(r)\bigr), \quad \forall r > 0.$$

(iv) Suppose that (X_n) is a sequence of independent random variables. We set
$$S_n := X_1 + \cdots + X_n, \quad C_{n,N} = C_{S_N - S_n}.$$

Show that the limits
$$C_n(r) = \lim_{N \to \infty} C_{n,N}(r)$$
exists for every n, and the resulting sequence $C_n(r)$ is nondecreasing.

(v) Show that
$$\lim_{n \to \infty} C_n(r)$$
is independent of r and it is either 0 or 1.

$\square$

Exercise 2.40. A probability measure $\mu \in \mathrm{Prob}(\mathbb{R})$ is said to be an *infinitely divisible distribution* if for any $n \in \mathbb{N}$, there exists $\mu_n \in \mathbb{N}$ such that
$$\mu = \mu_n^{*n} := \underbrace{\mu * \cdots * \mu}_{n}.$$

A random variable is called *infinitely divisible* if its distribution is such.

(i) Prove that the Poisson distributions and the Gaussian distributions are infinitely divisible.
(ii) Prove that any linear combination of independent infinitely divisible random variables is an infinitely divisible random variable.
(iii) Suppose that $(X_n)_{n \in \mathbb{N}}$ is a sequence of i.i.d. random variables with common distribution $\nu \in \mathrm{Prob}(\mathbb{R})$. Denote by $N(t)$, $t \geq 0$ a Poisson process with intensity $\lambda > 0$; see Example 1.136. For $t \geq 0$ we set
$$Y(t) = \sum_{k=1}^{N(t)} X_k.$$

The distribution of $Y(t)$ denoted by Q_t, is called a *compound Poisson distribution*. The distribution ν is called the *compounding distribution*. Show that
$$Q_t = e^{-\lambda t} \sum_{n=0}^{\infty} \frac{(\lambda t)^n}{n!} \nu^{*n}$$
and deduce that $Q_t * Q_s = Q_{t+s}$, $\forall t, s \geq 0$. In particular Q_t is infinitely divisible.

(iv) Compute the characteristic function of Q_t.
(v) Prove that any weak limit of infinitely divisible distribution is also infinitely divisible.

□

Exercise 2.41. Give an example of a sequence of random variables $X_n \in L^1(\Omega, \mathcal{S}, \mathbb{P})$ such that X_n converge in distribution to 0 but

$$\lim_{n \to \infty} \mathbb{E}[X_n] = \infty.$$

□

Exercise 2.42 (Skhorohod). Suppose that $\mu_n \in \mathrm{Prob}(\mathbb{R})$, $n \in \mathbb{N}$, is a sequence converging weakly to μ. Denote by $F_n : \mathbb{R} \to [0, 1]$ the distribution function of μ_n,

$$F_n(x) = \mu_n[(-\infty, x]],$$

and by Q_n the associated quantile function (see (1.2.5))

$$Q_n : [0, 1] \to \mathbb{R}, \quad Q_n(t) = \inf\{x; \ t \leq F_n(x)\}.$$

We can regard Q_n as random variables defined on the probability space

$$([0, 1], \mathcal{B}_{[0,1]}, \boldsymbol{\lambda}_{[0,1]}),$$

where $\boldsymbol{\lambda}_{[0,1]}$ denotes the Lebesgue measure on $[0, 1]$. As shown in Example 1.44, $\mu_n = (Q_n)_\# \boldsymbol{\lambda}_{[0,1]}$, so that μ_n is the probability distribution of Q_n. Prove that the sequence Q_n converges a.s. on $[0, 1]$ to a random variable with probability distribution μ.

In other words, given any sequence $\mu_n \in \mathrm{Prob}(\mathbb{R})$ that converges weakly to $\mu \in \mathrm{Prob}(\mathbb{R})$, we can find a sequence of random variables X_n, defined on the same probability space with $\mathbb{P}_{X_n} = \mu_n$ and such that X_n converges a.s. to a random variable X with distribution μ.

□

Exercise 2.43. Suppose that the sequence of random variables $X_n : (\Omega, \mathcal{S}, \mathbb{P}) \to \mathbb{R}$, $n \in \mathbb{N}$, converges in distribution to the random variable X. Prove that for any continuous function $f : \mathbb{R} \to \mathbb{R}$ the random variables $f(X_n)$ converge in distribution to $f(X)$. **Hint.** Use Exercise 2.42.

□

Exercise 2.44. Let μ be a Borel probability measure on $\mathbb{R}$ satisfying

$$\exists r_0 > 0 : \int_\mathbb{R} e^{tx} \mu[dx] < \infty, \ \forall |t| < r_0.$$

(i) Let $p \in [1, \infty)$. Prove that the map

$$L^p(\mathbb{R}, \mu) \ni f \mapsto Tf \in C_b(\mathbb{R}, \mathbb{C}), \quad (Tf)(\xi) = \int_\mathbb{R} e^{i\xi x} f(x) \mu[dx]$$

is injective. **Hint.** Reduce to Theorem 2.39 by writing $f = f_+ - f_-$.

(ii) If $f \in L^2(\mathbb{R}, \mu)$. Prove that there exists $r_1 > 0$ such that for any complex number such that $|\operatorname{Re} z| < r_1$ the complex valued function $\mathbb{R} \ni x \mapsto e^{izx} f(x) \in \mathbb{C}$ is μ integrable and the resulting function

$$z \mapsto F(z) = \int_{\mathbb{R}} e^{izx} f(x) \mu[\, dx\,]$$

is holomorphic in the strip $\{\,|\operatorname{Re} z| < r_1\,\}$.

(iii) Prove that $\mathbb{R}[\,x\,]$, the space of polynomials with real coefficients, is dense in $L^2(\mathbb{R}, \mu)$. **Hint.** You have to show that if $f \in L^2(\mathbb{R}, \nu)$ satisfies

$$\int_{\mathbb{R}} f(x) x^n \mu[\, dx\,] = 0, \quad \forall n \geq 0,$$

the $f = 0$ μ-a.s. Use (i) and (ii) to achieve this.

(iv) Consider the Hermite polynomials $\bigl(H_n(x)\bigr)_{n \geq 0}$ described in Exercise 1.24. Prove that the collection

$$\frac{1}{\sqrt{n!}} H_n, \quad n \geq 0$$

is a complete orthonormal basis of the Hilbert space $L^2(\mathbb{R}, \gamma_1)$, where γ_1 is the standard Gaussian measure on $\mathbb{R}$.

□

Exercise 2.45. Suppose that μ_0, μ_1 are two Borel probability measures such that $\exists t_0 > 0$

$$\int_{\mathbb{R}} e^{tx} \mu_0[\, dx\,] = \int_{\mathbb{R}} e^{tx} \mu_1[\, dx\,], \quad \forall |t| < t_0.$$

Fix $r > 0$ as in Exercise 2.44(ii) such that for any complex number the functions

$$z \mapsto F_k(z) = \int_{\mathbb{R}} e^{izx} f(x) \mu_k[\, dx\,], \quad k = 0, 1,$$

are well defined and holomorphic in the strip $\{\,|\operatorname{Re} z| < r\,\}$. Show that $F_0 = F_1$ and deduce that $\mu_0 = \mu_1$. **Hint.** Set $F = F_1 - F_0$. Use the Cauchy-Riemann equations to prove that $\frac{d^n F}{dz^n}\bigr|_{z=t} = 0, \forall n \in \mathbb{N}, \forall t \in (-r, r)$.

□

Exercise 2.46 (De Moivre). Let $X_n \sim \operatorname{Bin}(n, 1/2)$ and $Y \sim N(0, 1)$. Prove that

$$\lim_{n \to \infty} \frac{\mathbb{P}\bigl[\,|X_n - n/2| \leq \frac{r}{2}\sqrt{n}\,\bigr]}{\mathbb{P}\bigl[\,|Y| < r\,\bigr]} = 1, \quad \forall r > 0.$$

□

Exercise 2.47 (t-statistic). Suppose that $(X_n)_{n \in \mathbb{N}}$ is a sequence of i.i.d. random variables such that $\mathbb{E}[\,X_n\,] = 0$, $\mathbb{E}[\,X_n^2\,] = \sigma^2 < \infty$, $\forall n$. We set

$$M_n = \frac{1}{n} \sum_{k=1}^{n} X_n, \quad V_n = \frac{1}{n-1} \sum_{k=1}^{n} (X_k - M_n)^2, \quad T_n = \sqrt{n} \frac{M_n}{\sqrt{V_n}}.$$

(i) Prove that V_n converges in probability to σ^2.

(ii) Prove that T_n converges in distribution to a standard normal random variable.
Hint. Use CLT and Slutsky's theorem. □

Exercise 2.48. Suppose that $X = X_\lambda$ is a Gamma$(1,\lambda)$ random variable (see Example 1.121) and $Y = Y_\lambda$ is a random variable such that
$$\mathbb{P}[Y = n \,\|\, X] = \frac{X^n}{n!} e^{-nX}, \quad \forall n = 0, 1, 2, \dots.$$
In other words, conditioned on $X = x$ the random variable Y is Poi(x).

(i) Compute the characteristic function of Y.
(ii) Show that the random variable
$$\frac{1}{\sqrt{\mathrm{Var}[Y_\lambda]}}(Y_\lambda - \mathbb{E}[Y_\lambda])$$
converges in distribution to $N(0,1)$ as $\lambda \to \infty$.

□

Exercise 2.49. Suppose that X, Y are independent random normal variables. Set $Z = XY$.

(i) Show that
$$M_Z(\lambda) = \frac{1}{\sqrt{1-\lambda^2}}, \quad |\lambda| < 1,$$
and deduce that
$$\Psi_Z(\lambda) \leq \frac{\lambda^2}{2(1-\lambda^2)}, \quad \forall \lambda \in (-1,1).$$

(ii) Prove that I_Z (see Proposition 2.47) satisfies
$$I_Z(z) \geq \frac{1}{3}(\sin z)^2 \geq \frac{1}{12} z^2, \quad \forall |z| < \frac{\pi}{6}.$$

□

Exercise 2.50. Let $\mathscr{X}$ be a finite set. The *entropy* of a random variable $X : (\Omega, \mathcal{S}, \mathbb{P}) \to \mathscr{X}$ is
$$\mathrm{Ent}_2[X] := \mathrm{Ent}_2[\mathbb{P}_X] = -\sum_{x \in \mathscr{X}} p_X(x) \log_2 p(x), \quad p_X(x) = \mathbb{P}[\{X = x\}].$$
Given two random variables $X_i : (\Omega, \mathcal{S}, \mathbb{P}) \to \mathscr{X}_i$, $i = 1, 2$, we define their *relative entropy* to be
$$\mathrm{Ent}_2[X_2 | X_1] := \sum_{(x_1, x_2) \in \mathscr{X}_1 \times \mathscr{X}_2} p_{X_1, X_2}(x_1, x_2) \log_2 \left(\frac{p_{X_1 \cdot X_2}(x_1, x_2)}{p_{X_1}(x_1)} \right),$$
where $p_{X_1, X_2}(x_1, x_2) = \mathbb{P}[\{X_1 = x_1, X_2 = x_2\}]$.

(i) Show that if $X_i : (\Omega, \mathcal{S}, \mathbb{P}) \to \mathcal{X}_i$, $i = 1, 2$, are random variables, then
$$\mathrm{Ent}_2\big[\, X_2\,\big] - \mathrm{Ent}_2\big[\, X_2 \,\big|\, X_1\,\big] = \mathbb{D}_{KL}\big[\, \mathbb{P}_{(X_1,X_2)}, \mathbb{P}_{X_1} \otimes \mathbb{P}_{X_2}\,\big],$$
where $\mathbb{D}_{KL}$ is the Kullback-Leibler divergence defined in (2.3.7).

(ii) Suppose that we are given n finite sets $\mathcal{X}_i$, $i = 1, \dots, n$ and n maps
$$X_i : (\Omega, \mathcal{S}, \mathbb{P}) \to \mathcal{X}_i.$$
We denote by $\mathrm{Ent}_2(X_1, \dots, X_n)$ the entropy of the product random variable
$$(X_1, \dots, X_n) : \Omega \to \mathcal{X}_1 \times \cdots \times \mathcal{X}_n.$$
Prove that
$$\mathrm{Ent}_2\big[\, X_1, \dots, X_n\,\big] = \sum_{k=1}^n \mathrm{Ent}_2\big[\, X_k \,\big|\, (X_{k-1}, \dots, X_1)\,\big].$$
☐

Exercise 2.51 (Herbst). Let $\phi : [0, \infty) \to \mathbb{R}$, $\phi(x) = x \log x$, where $0 \cdot \log 0 := 0$. For any nonnegative random variable Z we set
$$\mathbb{H}\big[\, Z\,\big] = \mathrm{End}_\phi\big[\, Z\,\big] = \mathbb{E}\big[\, \phi(Z)\,\big] - \phi\big(\mathbb{E}\big[\, Z\,\big]\big).$$
Suppose that X is a random variable such that $\mathbb{M}_X(\lambda) = \mathbb{E}\big[\, e^{\lambda X}\,\big] < \infty$, for all λ is an open interval J containing 0. We set $\mathbb{H}_X(\lambda) := \mathbb{H}\big[\, e^{\lambda X}\,\big]$. Prove that if
$$\mathbb{H}_X(\lambda) \leq \frac{\lambda^2 \sigma^2}{2} \mathbb{M}_X(\lambda),$$
then $X \in \mathbb{G}(\sigma^2)$. ☐

Exercise 2.52 (Poincaré phenomenon). Denote by S^n the unit sphere in $\mathbb{R}^{n+1}$,
$$S^n := \left\{ (x_0, x_1, \dots, x_n);\ \sum_{k=0}^n x_k^2 = 1 \right\}.$$
Suppose that $(X_0, \dots, X_n)$ is a random point uniformly distributed on S^n with respect to the canonical Euclidean volume on S^n. Prove that there exists $c > 0$ such that
$$\forall n \in \mathbb{N},\ r \in [0, 1];\ \mathbb{P}\big[\, |X_0| > r\,\big] = C e^{-\frac{nr^2}{2}}.$$
Thus for spheres of large dimension n most of the volume is concentrated near the Equator $\{x_0 = 0\}$! **Hint.** Choose independent standard random variables $Y_0, \dots, Y_n$ set $Z = Y_0^2 + \cdots + Y_n^2$. Show that the random vector
$$(X_0, \dots, X_n) = \frac{1}{\sqrt{Z}} (Y_0, \dots Y_n)$$
is uniformly distributed on S^n. To conclude use Exercise 1.38. ☐

Exercise 2.53. Suppose that $\psi : [0, \infty) \to [0, \infty)$ is an *Orlicz function*, i.e., it is convex, increasing and $\psi(x) \to \infty$ as $x \to \infty$. Fix a probability space $(\Omega, \mathcal{S}, \mathbb{P})$. For any Orlicz function ψ and any random variable $X \in \mathcal{L}^0(\Omega, \mathcal{S}, \mathbb{P})$ we set
$$\|X\|_\psi = \inf\big\{\, t > 0;\ \mathbb{E}\big[\, \psi(X/t)\,\big] \leq 1\,\big\},$$
where $\infty \emptyset := \infty$. We set
$$\mathcal{L}_\psi(\Omega, \mathcal{S}, \mathbb{P}) = \big\{\, X \in \mathcal{L}^0(\Omega, \mathcal{S}, \mathbb{P});\ \|X\|_\psi < \infty\,\big\}$$
and we denote by L_ψ the quotient of $\mathcal{L}_\psi$ modulo the a.s. equality.

(i) Prove that $L_\psi(\Omega, \mathcal{S}, \mathbb{P})$ is a normed space.
(ii) Show that when $\psi(x) = x^p$, $p \in [1, \infty)$ we have $L_\psi = L^p$.
(iii) Let $\psi(x) = e^{x^2} - 1$. Prove that X is subgaussian if and only if $X \in L^\Psi$. □

Exercise 2.54. Let V be an n-dimensional real vector space. We denote by V^* its dual, $V^* = \mathrm{Hom}(V, \mathbb{R})$. We have a natural pairing

$$\langle -, - \rangle : V^* \times V \to \mathbb{R}, \quad \langle \xi, x \rangle := \xi(x), \quad \forall \xi \in V^*, \ x \in V.$$

A Borel probability measure $\mu \in \mathrm{Prob}(V)$ is called *Gaussian* if for every linear functional $\xi \in V^*$, the resulting random variable

$$\xi : (V, \mathcal{B}_V, \mu) \to \mathbb{R}$$

is Gaussian with mean $m[\xi]$ and variance $v[\xi]$, i.e., (see Example 1.120)

$$\mathbb{P}_\xi[dx] = \mathbf{\Gamma}_{m[\xi], v[\xi]}[dx] = \frac{1}{(2\pi)^{n/2}} \cdot e^{-\frac{(x - m[\xi])^2}{2v[\xi]}} dx.$$

A random vector $\boldsymbol{X} : (\Omega, \mathcal{S}, \mathbb{P}) \to V$ is called *Gaussian* if its probability distribution is a Gaussian measure on V

(i) Show that the map $V^* \ni \xi \mapsto m[\xi] \in \mathbb{R}$ is linear and thus defines an element

$$m = m_\mu \in (V^*)^* \cong V$$

called the *mean* of the Gaussian measure. Moreover

$$m_\mu = \int_V x \mu[dx] \in V.$$

(ii) Define $C = C_\mu : V^* \times V^* \to \mathbb{R}$

$$C(\xi, \eta) = \frac{1}{4}\bigl(v[\xi + \eta] - v[\xi - \eta]\bigr) = \mathbb{E}_\mu\bigl[(\xi - m[\xi])(\eta - m[\eta])\bigr].$$

Show that C is a bilinear form, it is symmetric and positive semidefinite. It is called the *covariance form* of the Gaussian measure μ.

(iii) Show that if μ_0, μ_1 are Gaussian measures on V_0 and respectively V_1, then the product $\mu_0 \otimes \mu_1$ is a Gaussian measure on $V_0 \oplus V_1$. Moreover,

$$m[\mu_0 \otimes \mu_1] = m_{\mu_0} \oplus m_{\mu_1}, \quad C_{\mu_0 \otimes \mu_1} = C_{\mu_0} \oplus C_{\mu_1}.$$

We set

$$\mathbf{\Gamma}_{\mathbf{1}_n} := \underbrace{\mathbf{\Gamma}_1 \otimes \cdots \otimes \mathbf{\Gamma}_1}_{n}.$$

$\mathbf{\Gamma}_{\mathbf{1}_n}$ is called the canonical Gaussian measure on $\mathbb{R}^n$. More explicitly

$$\mathbf{\Gamma}_{\mathbf{1}_n}[dx] = \frac{1}{(2\pi)^{n/2}} e^{-\frac{|x|^2}{2}} dx,$$

where $|x|$ denotes the Euclidean norm of the vector $x \in \mathbb{R}^n$.

(iv) Suppose that V_0, V_1 are real finite dimensional vector spaces, μ is a Gaussian measure on V_0 and $A : V_0 \to V_1$ is a linear map. Denote by μ_A the pushforward of μ via the map A, $\mu_A := A_\# \mu$. Prove that μ_A is a Gaussian measure on V_1 with mean $m_{\mu_A} = A m_\mu$ and covariance form

$$C_A : V_1^* \times V_1^* \to \mathbb{R}, \quad C_A(\xi_1, \eta_1) = C_\mu(A^* \xi_1, A^* \eta_1), \quad \forall \xi_1, \eta_1 \in V_1^*.$$

Above, $A^* : V_1 \to V_0^*$ is the dual (transpose) of the linear map A.

(v) Fix a basis $\{e_1, \ldots, e_n\}$ of V so we can identify V and V^* with $\mathbb{R}^n$ and C with a symmetric positive semidefinite matrix. Denote by A its unique positive semidefinite square root. Show that the pushforward $A_\# \mathbf{\Gamma}_{1_n}$ is a Gaussian measure on $\mathbb{R}^n$ with mean zero and covariance form $C = A^2$.

(vi) Define the Fourier transform of a measure $\mu \in \text{Prob}(V)$ to be the function

$$\widehat{\mu} : V^* \to \mathbb{C}, \quad \widehat{\mu}(\xi) = \mathbb{E}_\mu\big[e^{i\xi}\big] = \int_V e^{i\langle \xi, x \rangle} \mu[dx].$$

Show that if μ is a Gaussian measure on V with mean m covariance form C, then

$$\widehat{\mu}(\xi) = e^{im[\xi]} e^{-\frac{1}{2} C(\xi, \xi)}, \quad \forall \xi \in V^*.$$

(vii) Use the ideas in the proof of Theorem 2.39 to show that a Gaussian measure is uniquely determined by its mean and covariance form. We denote by $\mathbf{\Gamma}_{m,C}$ the Gaussian measure with mean m and covariance C.

(viii) Suppose that C is a symmetric positive definite $n \times n$ matrix. Prove that the Gaussian measure on $\mathbb{R}^n$ with mean 0 and covariance from C is

$$\mathbf{\Gamma}_{0,C}\big[dx\big] = \frac{1}{\big(\det(2\pi C)\big)^{n/2}} e^{-\frac{\langle C^{-1} x, x \rangle}{2}} dx$$

where $\langle -, - \rangle$ denotes the canonical inner product on $\mathbb{R}^n$. **Hint.** Analyze first the case when C is a diagonal matrix. □

Exercise 2.55. Let $(\Omega, \mathcal{S}, \mathbb{P})$ be a probability space and E a finite dimensional real vector space. Recall that a Borel measurable map $X : (\Omega, \mathcal{S}, \mathbb{P}) \to E$ is called a *Gaussian random vector* if its distribution $\mathbb{P}_X = X_\# \mathbb{P}$ is a Gaussian measure on E; see Exercise 2.54.

Suppose that $X_1, \ldots, X_n \in \mathcal{L}^0(\Omega, \mathcal{S}, \mathbb{P})$ are *jointly Gaussian* random variables, i.e., the random vector $\vec{X} = (X_1, \ldots, X_n) : \Omega \to \mathbb{R}^n$ is Gaussian.

(i) Prove that each of the variables $X_1, \ldots, X_n$ is Gaussian and the covariance form

$$C : \mathbb{R}^n \times \mathbb{R}^n \to \mathbb{R}$$

of the Gaussian measure $\mathbb{P}_{\vec{X}} \in \text{Prob}(\mathbb{R}^n)$ is given by the matrix $(c_{ij})_{1 \le i, j \le n}$

$$c_{ij} = \text{Cov}\big[X_i, X_j\big], \quad \forall 1 \le i, j \le n.$$

(ii) Prove that $X_1, \ldots, X_n$ are independent if and only if the matrix $(c_{ij})_{1 \leq i,j \leq n}$ is diagonal, i.e.,
$$\mathbb{E}[X_i X_j] = \mathbb{E}[X_i]\mathbb{E}[X_j], \quad \forall i \neq j.$$

Hint. Use the results in Exercise 2.54. □

Exercise 2.56 (Gaussian regression). Suppose that $X_0, X_1, \ldots, X_n$ are *jointly Gaussian* random variables with zero means. Let $\overline{X}_0$ denote the orthogonal projection of $X_0 \in L^2(\Omega, \mathcal{S}, \mathbb{P})$ onto the finite dimensional subspace
$$\mathrm{span}\{X_1, \ldots, X_n\} \subset L^2(\Omega, \mathcal{S}, \mathbb{P}).$$

(i) Prove that $\overline{X}_0 = \mathbb{E}[X_0 \,\|\, X_1, \ldots, X_n]$.
(ii) Suppose that the covariance matrix C of the Gaussian vector $(X_1, \ldots, X_n)$ is invertible. Denote by $L = [\ell_1, \ldots, \ell_n]$ the $1 \times n$ matrix
$$\ell_i = \mathbb{E}[X_0 X_i], \quad i = 1, \ldots, n.$$
Prove that
$$\overline{X}_0 = L \cdot C^{-1} \cdot \boldsymbol{X}, \quad \boldsymbol{X} := \begin{bmatrix} X_1 \\ \vdots \\ X_n \end{bmatrix}.$$

Hint. For (i) use Exercise 2.55(ii) and (1.4.10). □

Remark 2.93. The result in Exercise 2.56 is remarkable. Let us explain its typical use in statistics.

Suppose we want to understand the random quantity X_0 and all we truly understand are the random variables $X_1, \ldots, X_n$. A quantity of the form $f(X_1, \ldots, X_n)$ is called a predictor, and the simplest predictors are of the form $c_0 + c_1 X_1 + \cdots + c_n X_n$. These are called *linear predictors*. The conditional expectation $\mathbb{E}[X_0 \,\|\, X_1, \ldots, X_n]$ is the predictor closest to X_0. The *linear* predictor closest to X_0 is called the *linear regression*. The coefficients $c_0, c_1, \ldots, c_n$ corresponding to the linear regression are obtained via the least squares approximation.

The result in the above exercise shows that, when the random variables $X_0, X_1, \ldots, X_1$ are jointly Gaussian, the best predictor of X_0, given $X_1, \ldots, X_n$ is the linear predictor. This is another reason why the Gaussian variables are extremely convenient to work with in practice. □

Exercise 2.57 (Maxwell). Suppose that $(X_n)_{n \in \mathbb{N}}$ is a sequence of mean zero i.i.d. random variables. For each $n \in \mathbb{N}$ we denote by V_n the random vector $V_n := (X_1, \ldots, X_n)$. Prove that the following are equivalent.

(i) The random variables X_n are Gaussian.
(ii) For any $n \in \mathbb{N}$ and for any orthogonal map $T : \mathbb{R}^n \to \mathbb{R}^n$ the random vectors V_n and RV_n have identical distributions.

□

Exercise 2.58. Suppose that X is a standard normal random variable and Z is a Bernoulli random variable, independent of X, with success probability $p = \frac{1}{2}$.

(i) Prove that $Y = XZ$ is also a standard normal random variable.
(ii) Prove that $X + Y$ is not Gaussian.

□

Exercise 2.59. Suppose that $\mathbb{T}$ is a compact interval of the real axis, and $(X_t)_{t \in \mathbb{T}}$, $(Y_t)_{t \in \mathbb{T}}$, $(Z_t)_{t \in \mathbb{T}}$ are real valued stochastic processes such that (Y_t) and (Z_t) are modifications of (X_t) with a.s. continuous paths. Prove that the processes (Y_t) and (Z_t) are indistinguishable. □

Exercise 2.60. Fix a Brownian motion $(B_t)_{t \geq 0}$ defined on a probability space $(\Omega, \mathcal{S}, \mathbb{P})$. Denote by $\mathcal{E}$ the vector subspace of $L^2([0,1], \boldsymbol{\lambda})$ spanned by the functions $\boldsymbol{I}_{(s,t]}$, $0 \leq s < t \leq 1$.

(i) Prove that any function $f \in \mathcal{E}$ admits a convenient representation, i.e., a representation of the form

$$f = \sum_{k=1}^{n} c_k \boldsymbol{I}_{(s_k, t_k]}, \quad c_k \in \mathbb{R},$$

where the intervals $(s_j, t_j]$, $(s_k, t_k]$ are disjoint for $j \neq k$.

(ii) Let $f \in \mathcal{E}$ and consider two convenient representations of f

$$\sum_{k=1}^{n} c_k \boldsymbol{I}_{(s_k, t_k]} = f = \sum_{i=1}^{m} c'_k \boldsymbol{I}_{(s'_k, t'_k]}.$$

Show that

$$\sum_{k=1}^{n} c_k (B_{t_k} - B_{s_k}) = \sum_{i=1}^{m} c'_k (B_{t'_k} - B_{s'_k}) =: W(f).$$

(iii) Show that for any $f \in \mathcal{E}$ we have $W(f) \in L^2(\Omega, \mathcal{S}, \mathbb{P})$ and $\|W(f)\|_{L^2(\Omega)} = \|f\|_{L^2([0,1])}$.
(iv) Prove that the map $W : \mathcal{E} \to L^2(\Omega, \mathcal{S}, \mathbb{P})$ is linear and extends to a linear isometry $W : L^2([0,1]\boldsymbol{\lambda}) \to L^2(\Omega, \mathcal{S}, \mathbb{P})$ whose image consists of Gaussian random variables. In other words, this isometry is a Gaussian white noise. The map W is called the *Wiener integral*. It is customary to write

$$W(f) = \int_0^1 f(s) dB_s.$$

□

Exercise 2.61. The space $F := C\big([0,\infty)\big)$ of continuous functions $[0,\infty) \to \mathbb{R}$ is equipped with a natural metric d,

$$d(f,g) = \sum_{n\in\mathbb{N}} \frac{1}{2^n}\min\big(1, d_n(f,g)\big), \quad d_n(f,g) := \sup_{t\in[n-1,n]} |f(t) - g(t)|.$$

Denote by $\mathcal{B}_F$ the Borel algebra of F. For each $t \geq 0$ we define $E_t : F \to \mathbb{R}$, $E_t(f) = f(t)$. We set

$$\mathcal{S}_t = E_x^{-1}(\mathcal{B}_R), \quad \forall t \geq 0, \quad \mathcal{S} = \bigcup_{t\geq 0} \mathcal{S}_t.$$

(i) Prove that E_t is a continuous function on F, $\forall t \geq 0$.
(ii) Prove that $\mathcal{B}_F = \mathcal{S}$.
(iii) Suppose that $(\Omega, \mathcal{A})$ is a measurable space and $W : \Omega \to F$ is a map

$$\Omega \in \omega \mapsto W_\omega \in F.$$

Prove that W is $(\mathcal{A}, \mathcal{B}_F)$-measurable if and only if for any $t \geq 0$ the function

$$W^t : (\Omega, \mathcal{A}) \to \mathbb{R}, \quad \omega \mapsto W_\omega(t)$$

is measurable. **Hint.** Use (ii). $\square$

Chapter 3

Martingales

The usefulness of the martingale property was recognized by P. Lévy (condition (𝒞) in [105, Chap. VIII]), but it was J. L. Doob [47] who realized its full potential by discovering its most important properties: optional stopping/sampling, existence of asymptotic limits, maximal inequalities.

I have to admit that when I was first introduced to martingales they looked alien to me. Why would anyone be interested in such things? What are really these martingales?

I can easily answer the first question. Martingales are ubiquitous, they appear in the most unexpected of situations, though not always in an obvious way, and they are "well behaved". Since their appearance on the probabilistic scene these stochastic processes have found many applications.

As for the true meaning of this concept let me first remark that the name "martingale" itself is a bit unusual. It is French word that has an equestrian meaning (harness) but, according to [113], the term was used among the French gamblers when referring to a gambling system. I personally cannot communicate clearly, beyond a formal definition, what is the true meaning of this concept. I believe it is a fundamental concept of probability theory and I subscribe to R. Feynman's attitude: it is more useful to know how the electromagnetic waves behave than knowing what the look like. The same could be said about the concept of martingale and, why not, about the concept of probability. I hope that the large selection of examples discussed in this chapter will give the reader a sense of this concept.

This chapter is divided into two parts. The first and bigger part is devoted to discrete time martingales. The second and smaller part is devoted to continuous time martingales. I have included many and varied applications of martingales with the hope that they will allow the reader to see the many facets of this concept and convince him/her of its power and versatility. My presentation was inspired by many sources and I want to single out [75; 33; 47; 53; 102; 103; 135; 160] that influenced me the most.

3.1 Basic facts about martingales

We need to introduce some basic terminology.

3.1.1 Definition and examples

Suppose that $(\Omega, \mathcal{S}, \mathbb{P})$ is a probability space and $\mathbb{T} \subset \mathbb{R}$. Recall that a *random or stochastic process with parameter space* $\mathbb{T}$ is a family of random variables

$$X_t : (\Omega, \mathcal{S}, \mathbb{P}) \to \mathbb{R}, \ t \in \mathbb{T}.$$

A $\mathbb{T}$-*filtration* of the probability space $(\Omega, \mathcal{S}, \mathbb{P})$ is a family $\mathcal{F}_\bullet = (\mathcal{F}_t)_{t \in \mathbb{T}}$ of sub-σ-algebras of $\mathcal{S}$ such that

$$\mathcal{F}_s \subset \mathcal{F}_t, \ \forall s \leq t.$$

We set

$$\mathcal{F}_\infty := \bigvee_{t \in \mathbb{T}} \mathcal{F}_t.$$

A family of random variables $X_t : (\Omega, \mathcal{S}, \mathbb{P}) \to \mathbb{R}, \ t \in \mathbb{T}$, is said to be *adapted to the filtration* $\mathcal{F}_\bullet = (\mathcal{F}_t)_{t \in \mathbb{T}}$, if X_t is $\mathcal{F}_t$-measurable for any t.

Remark 3.1. If we think of a σ-algebra as encoding all the measurable information in a given random experiment, then we can think of a $\mathbb{T}$-filtration as an increasing flow of information. For example, if $\mathbb{T} = \mathbb{N}_0$, and $(X_n)_{n \geq 0}$ is a sequence of random variables, then the collection

$$\mathcal{F}_n = \sigma(X_0, X_1, \ldots, X_n), \ n \in \mathbb{N}_0,$$

is a filtration of σ-algebras. At epoch n, information about X_n becomes available to us, on top of the information about $X_0, X_1, \ldots, X_{n-1}$ that we have collected along the way. □

Definition 3.2. Suppose that $(\Omega, \mathcal{S}, \mathbb{P})$ equipped with a filtration $\mathcal{F}_\bullet = (\mathcal{F}_t)_{t \in \mathbb{T}}$. An $\mathcal{F}_\bullet$-*martingale* is a family of random variables $X_t : (\Omega, \mathcal{S}, \mathbb{P}) \to \mathbb{R}, \ t \in \mathbb{T}$, satisfying the following two conditions.

(i) The family is adapted to the filtration $\mathcal{F}_\bullet$ and X_t is <u>integrable</u> for any $t \in \mathbb{T}$.
(ii) For all $s, t \in \mathbb{T}, \ s < t$, we have $X_s = \mathbb{E}[X_t \| \mathcal{F}_s]$.

The family of *integrable* random variables $(X_t)_{t \in \mathbb{T}}$ is called a $\mathcal{F}_\bullet$-*submartingale* (resp. *supermartingale*) if it is adapted to the filtration and for any $s, t \in \mathbb{T}, \ s < t$, we have $X_s \leq \mathbb{E}[X_t \| \mathcal{F}_s]$ (resp. $X_s \geq \mathbb{E}[X_t \| \mathcal{F}_s]$).

When $\mathbb{T}$ is a discrete subset of $\mathbb{R}$ we say that the (sub- or super-)martingale is a *discrete time* (sub/super)martingale. □

Note that a sequence of random variables $(X_n)_{n\in\mathbb{N}_0}$ is a discrete time submartingale (resp. martingale) with respect to a filtration $(\mathcal{F}_n)_{n\in\mathbb{N}_0}$ of $\mathcal{F}$ if

$$\boxed{\mathbb{E}[\,X_{n+1}\|\mathcal{F}_n\,] \geq X_n, \quad (\text{resp. } \mathbb{E}[\,X_{n+1}\|\mathcal{F}_n\,] = X_n), \quad \forall n \in \mathbb{N}_0}.$$

Remark 3.3. Suppose that $(X_n)_{n\geq 0}$ is a sequence of integrable random variables and

$$\mathcal{F}_n = \sigma(X_1, \ldots, X_n).$$

Then $\mathbb{E}[\,X_{n+1}\,\|\,\mathcal{F}_n\,]$ is a measurable function of the variables $X_0, \ldots, X_n$,

$$\mathbb{E}[\,X_{n+1}\,\|\,\mathcal{F}_n\,] = f_{n+1}(X_0, X_1, \ldots, X_n), \quad f_{n+1} : \mathbb{R}^{n+1} \to \mathbb{R}.$$

The sequence $(X_n)_{n\geq 0}$ is a martingale if and only if $f_{n+1}(x_0, x_1, \ldots, x_n) = x_n$, $\forall n \geq 0$, $\forall x_0, \ldots, x_n \in \mathbb{R}$. If the joint distribution of $(X_0, \ldots, X_n)$ is described by a density $p_n(x_0, \ldots, x_n)$, then

$$f_{n+1}(x_0, \ldots, x_n) = \int_{\mathbb{R}} x_{n+1} \frac{p_{n+1}(x_0, \ldots, x_n, x_{n+1})}{p_n(x_0, \ldots, x_n)} dx_{n+1}.$$

□

Example 3.4 (Closed martingales). Suppose that $\mathcal{F}_\bullet = (\mathcal{F}_n)_{n\in\mathbb{N}_0}$ is a filtration of $\mathcal{S}$ and $X \in L^1(\Omega, \mathcal{S}, \mathbb{P})$. Then the sequence of random variables

$$X_n = \mathbb{E}[\,X\|\mathcal{F}_n\,] \in L^1(\Omega, \mathcal{F}_n, \mathbb{P}), \quad n \in \mathbb{N}_0,$$

is a martingale since

$$\mathbb{E}[\,X_{n+1}\,\|\,\mathcal{F}_n\,] = \mathbb{E}\Big[\mathbb{E}[\,X\,\|\,\mathcal{F}_{n+1}\,]\,\Big\|\,\mathcal{F}_n\Big] = \mathbb{E}[\,X\,\|\,\mathcal{F}_n\,] = X_m.$$

Such a martingale is called *closed* or *Doob martingale*. □

Example 3.5 (Unbiased random walk). Suppose that $(X_n)_{n\in\mathbb{N}}$ is a sequence of independent integrable random variables such that $\mathbb{E}[X_n] = 0$, $\forall n \in \mathbb{N}_0$.

One should think that X_n is the size of the n-th step so that the location after n steps is

$$S_n = X_1 + \cdots + X_n.$$

Set $\mathcal{F}_n := \sigma(X_1, \ldots, X_n)$, $S_0 := 0$. Then the sequence $(S_n)_{n\in\mathbb{N}_0}$ is a martingale adapted to the filtration $\mathcal{F}_n$. Indeed

$$\mathbb{E}[\,S_{n+1}\|\mathcal{F}_n\,] = \mathbb{E}[\,X_{n+1}\|X_1, \ldots, X_n\,] + \mathbb{E}[\,X_1 + \cdots + X_n\|X_1, \ldots, X_n\,]$$

$$= \mathbb{E}[\,X_{n+1}\,] + X_1 + \cdots + X_n = S_n.$$

□

Example 3.6 (Random products). Suppose that $(Y_n)_{n \in \mathbb{N}}$ are positive i.i.d. random variables such that $\mathbb{E}[Y_1] = 1$. Then the sequence of products

$$Z_n = Y_1 Y_2 \cdots Y_n, \quad n \in \mathbb{N},$$

is a martingale adapted to the filtration $\mathcal{F}_n = \sigma(Y_1, \ldots, Y_n)$. Indeed

$$\mathbb{E}[Z_{n+1} \| Y_1, \ldots, Y_n] = \mathbb{E}[Y_{n+1}] Y_1 \cdots Y_n = Z_n. \qquad \square$$

Example 3.7 (Biased random walk). Suppose that $(X_n)_{n \in \mathbb{N}}$ are i.i.d. random variables such that the moment generating function

$$\mu(\lambda) := \mathbb{E}[e^{\lambda X_n}]$$

is well defined for λ in some interval Λ. We set

$$S_n := X_1 + \cdots + X_n, \quad M_n = M_n(\lambda) := e^{\lambda S_n} \mu(\lambda)^{-n}, \quad \mathcal{F}_n := \sigma(X_1, \ldots, X_n).$$

If we define

$$Y_n := \frac{1}{\mu(\lambda)} e^{\lambda X_n},$$

then we deduce that

$$\mathbb{E}[Y_n] = 1, \quad M_n = Y_1 \cdots Y_n.$$

From the previous example we deduce that $\bigl(M_n(\lambda)\bigr)_{n \in \mathbb{N}}$ is a martingale.

As a concrete example, suppose that the random variables X_n are all binomial type

$$\mathbb{P}[X_n = 1] = p, \quad \mathbb{P}[X_n = -1] = q = 1 - p.$$

In this case $\mu(\lambda) = pe^\lambda + qe^{-\lambda}$. Note that if $e^\lambda = \frac{q}{p}$, then $\mu(\lambda) = 1$ and we deduce that

$$M_n = \left(\frac{q}{p}\right)^{S_n}$$

is a martingale. This is sometimes referred to as the *De Moivre's martingale*. $\qquad \square$

Example 3.8 (Galton-Watson/branching processes). Fix a probability measure μ on $\mathbb{N}_0$ such that

$$m := \sum_{k \in \mathbb{N}_0} k\mu[k] < \infty, \quad \mu[k] := \mu[\{k\}],$$

and $\mu[k_0] > 0$ for some $k_0 > 0$. Consider next a sequence $(X_{n,j})_{j,n \in \mathbb{N}_0}$ of i.i.d. $\mathbb{N}_0$-valued random variables with common probability distribution μ. Fix $\ell \in \mathbb{N}$, set $Z_0 := \ell$, and for any $n \in \mathbb{N}_0$ define

$$Z_{n+1} = \sum_{j=1}^{Z_n} X_{n,j}, \quad \mathcal{F}_n = \sigma\bigl(X_{k,j}; \, k \in \mathbb{N}_0, k < n\bigr).$$

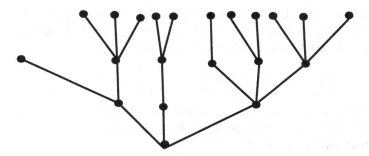

Fig. 3.1   *Three generations of a Galton-Watson (random) tree. Here $Z_1 = 3$, $Z_2 = 2+1+3 = 6$, $Z_3 = 3+2+1+2+3 = 11$.*

The random variable Z_n can be interpreted as the population of the n-th generation of a species that had ℓ individuals at $n = 0$ and such that the number of offsprings of a given individuals is a random variable with distribution μ. The j-th individual of the n-th generators has $X_{n,j}$ offsprings. We will refer to μ as the *reproduction law*.

The sequence $(Z_n)_{n \geq 0}$ is known as the *Galton-Watson process* or the *branching process* with reproduction law μ.

When $\ell = 1$ this process can be visualized as a random rooted tree. The root v_0 has $Z_1 = X_{0,1}$ successor vertices. $v_{1,1}, \ldots, v_{1,Z_1}$. The vertex $v_{1,i}$ has $X_{1,i}$ successors etc.; see Figure 3.1. For any $n \in \mathbb{N}_0$ we have

$$Z_{n+1} = \sum_{k=1}^{\infty} \left(\sum_{j=1}^{k} X_{n,j} \right) \boldsymbol{I}_{\{Z_n = k\}}$$

so

$$\mathbb{E}\left[Z_{n+1} \| \mathcal{F}_n \right] = \sum_{k=1}^{\infty} \mathbb{E}\left[\left(\sum_{j=1}^{k} X_{n,j} \right) \boldsymbol{I}_{\{Z_n = j\}} \,\middle\|\, \mathcal{F}_n \right]$$

$$= \sum_{k=0}^{\infty} \mathbb{E}\left[\left(\sum_{j=1}^{k} X_{n,j} \right) \,\middle\|\, \mathcal{F}_n \right] \boldsymbol{I}_{\{Z_n = k\}}$$

$(X_{n,j} \perp\!\!\!\perp \mathcal{F}_n, \forall n, j)$

$$= \sum_{k=1}^{\infty} \underbrace{\left(\sum_{j=1}^{k} \mathbb{E}\left[X_{n,j} \right] \right)}_{=km} \boldsymbol{I}_{\{Z_n = k\}} = m \sum_{k=0}^{\infty} k \boldsymbol{I}_{\{Z_n = k\}} = m Z_n.$$

This proves that the sequence $Y_n = m^{-n} Z_n$, $n \in \mathbb{N}_0$ defines a martingale. □

Example 3.9 (Polya's urn). An urn contains $r > 0$ red balls and $g > 0$ green balls. At each moment of time we draw a ball uniformly likely from the balls existing at that moment, we replace it by $c+1$ balls of the same color, $c \geq 0$. Denote by R_n and G_n the number of red and respectively green balls in the urn after the n-th draw. We denote by X_n the ratio of red balls after n draws, i.e.,

$$X_n := \frac{R_n}{R_n + G_n} = \frac{R_n}{r + g + cn}.$$

Note that when $c = 1$, the scheme can be alternatively described as randomly adding at each moment of time a red/green ball with probability equal to the fraction of red/green balls that exist at that moment in the urn.

We set

$$\mathcal{F}_n = \sigma(R_0, G_0, \cdots, R_n, G_n) = \sigma(X_0, X_1, \ldots, X_n).$$

We will show that $(X_\bullet)$ is an $\mathcal{F}_\bullet$-martingale. To see this observe that

$$X_n = \sum_{i,j>0} \frac{i}{i+j} I_{\{R_n=i, G_n=j\}}$$

so

$$\mathbb{E}\big[X_{n+1} \| \mathcal{F}_n\big] = \sum_{i,j>0} \frac{i}{i+j} \mathbb{E}\big[I_{\{R_{n+1}=j, G_{n+1}=j\}} \| \mathcal{F}_n\big].$$

Now observe that

$$\mathbb{E}\big[I_{\{R_{n+1}=i, G_{n+1}=j\}} \big\| \mathcal{F}_n\big]$$

$$= \sum_{k,\ell>0} \mathbb{P}\big[R_{n+1}=i, G_{n+1}=j \| R_n=k, G_n=\ell\big] I_{\{R_n=k, G_n=\ell\}}$$

$$= \frac{i-c}{i+j-c} I_{\{R_n=i-c, G_n=j-c\}} + \frac{j-c}{i+j-c} I_{\{R_n=i, G_n=j-c\}}.$$

We deduce

$$\mathbb{E}\big[X_{n+1} \| \mathcal{F}_n\big] = \sum_{i,j} \frac{i}{i+j} \cdot \frac{i-c}{i+j-c} I_{\{R_n=i-c, G_n=j\}}$$

$$+ \sum_{i,j} \frac{i}{i+j} \cdot \frac{j-c}{i+j-c} I_{\{R_n=i, G_n=j-c\}}$$

$$= \sum_{u,v} \frac{u+c}{u+v+c} \cdot \frac{u}{u+v} I_{\{R_n=u, G_n=v\}} + \sum_{u,v} \frac{u}{u+v+c} \cdot \frac{v}{u+v} I_{\{R_n=u, G_n=v\}}$$

$$= \sum_{u,v} \frac{u(u+v+c)}{(u+v)(u+v+c)} I_{\{R_n=u, G_n=v\}} = \sum_{u,v} \frac{u}{u+v} I_{\{R_n=u, G_n=v\}} = X_n. \quad \square$$

Example 3.10 (Random walks on graphs). Suppose that Γ is a connected simple graph with vertex set $\boldsymbol{V}_\Gamma$ and edges $\boldsymbol{E}_\Gamma$. Assume that there are no multiple edges between two vertices $u, v \in \boldsymbol{V}_\Gamma$. Assume that Γ is *locally finite* i.e., for any vertex $u \in \boldsymbol{V}_\Gamma$, its set of neighbors $\mathcal{N}(u)$ is finite. We set $\deg(u) := |\mathcal{N}(u)|$.

A function $F : \boldsymbol{V}_\Gamma \to \mathbb{R}$ is called *harmonic* if

$$F(u) = \frac{1}{\deg(u)} \sum_{v \in \mathcal{N}(u)} F(v).$$

Consider the simple random walk on Γ that starts at a given vertex v_0 and the probability of transitioning from a vertex u to a neighbor v is equal to $\frac{1}{\deg(u)}$. Denote by V_n the location after n steps of the walk. Suppose that $F : \boldsymbol{V}_\Gamma \to \mathbb{R}$ is a harmonic function. Then the sequence of random variables

$$X_n = F(V_n), \quad n \in \mathbb{N}_0,$$

is a martingale with respect to the filtration $\mathcal{F}_n = \sigma(V_0, V_1, \ldots, V_n)$. Moreover

$$\mathbb{E}[X_n] = F(v_0), \quad \forall n \in \mathbb{N}_0. \qquad \square$$

Example 3.11 (New (sub)martingales from old). Suppose that $(\Omega, \mathcal{S}, \mathbb{P})$ is equipped with a filtration $\mathcal{F}_\bullet = (\mathcal{F}_n)_{n \in \mathbb{N}_0}$ and $X_n \in \mathcal{L}^1(\Omega, \mathcal{F}, \mathbb{P})$ is a sequence of random variables adapted to the above filtration.

(i) If $(X_n)_{n \in \mathbb{N}_0}$ is a martingale and $\varphi : \mathbb{R} \to \mathbb{R}$ is a convex function such that $\varphi(X_n)$ is integrable $\forall n \in \mathbb{N}_0$, then the conditional Jensen inequality implies that the sequence $\varphi(X_n)$ is a submartingale. Indeed, Jensen's inequality implies

$$\mathbb{E}[\varphi(X_{n+1}) \| \mathcal{F}_n] \geq \varphi\Big(\mathbb{E}[X_{n+1} \| \mathcal{F}_n]\Big) = \varphi(X_n).$$

(ii) If $(X_n)_{n \in \mathbb{N}_0}$ is a submartingale and $\varphi : \mathbb{R} \to \mathbb{R}$ is a *nondecreasing* convex function such that $\varphi(X_n)$ is integrable $\forall n \in \mathbb{N}_0$, then the sequence $\varphi(X_n)$ is a submartingale. Indeed, follow the same argument as above where at the last step use the fact that φ is nondecreasing. In particular if $(X_n)_{n \geq 0}$ is a submartingale, then so is $(X_n^+)_{n \geq 0}$, $x^+ = \max(0, x)$.

(iii) If $(X_n)_{n \in \mathbb{N}_0}$ is a supermartingale and $\varphi : \mathbb{R} \to \mathbb{R}$ is a nondecreasing concave function such that $\varphi(X_n)$ is integrable $\forall n \in \mathbb{N}_0$, then the sequence $\varphi(X_n)$ is a supermartingale. Indeed

$$\mathbb{E}[\varphi(X_{n+1}) \| \mathcal{F}_n] \leq \varphi\Big(\mathbb{E}[X_{n+1} \| \mathcal{F}_n]\Big) \leq \varphi(X_n).$$

In particular, if $(X_n)_{n \in \mathbb{N}_0}$ is a supermartingale, then so is $(\min(X_n, c))_{n \geq 0}$, $\forall c \in \mathbb{R}$.

$\square$

3.1.2 Discrete stochastic integrals

Fix a probability space $(\Omega, \mathcal{S}, \mathbb{P})$ and an $\mathbb{N}_0$-filtration $\mathcal{F}_\bullet$ of $\mathcal{S}$. If $C_\bullet$ is an increasing $\mathcal{F}_\bullet$-adapted process, then obviously $C_\bullet$ is a submartingale. If we add to this process a martingale $M_\bullet$, then the resulting process $X_\bullet = M_\bullet + C_\bullet$ is a submartingale.

It turns out that all submartingales can be obtained in this fashion. In fact, the increasing process C_n can be chosen to be of a special type: the random variable C_{n+1} can be chosen to be $\mathcal{F}_n$-measurable, i.e., the value of $C_\bullet$ at time $n+1$ is predictable at time n, i.e., can be determined from the information available to us at time n encoded in the σ-algebra $\mathcal{F}_n$.

Definition 3.12. A sequence of random variable $\{H_n : \Omega \to \mathbb{R},\ n \in \mathbb{N}_0\}$ is called $\mathcal{F}_\bullet$-*previsible* or *predictable* if H_0 is $\mathcal{F}_0$-measurable, and H_n is $\mathcal{F}_{n-1}$-measurable $\forall n \in \mathbb{N}$. $\square$

The next result formalizes the discussion at the beginning of this subsection.

Proposition 3.13 (Doob decomposition of discrete submartingales). *Let $X_\bullet = (X_n)_{n \in \mathbb{N}_0}$ be an $(\mathcal{F}_n)_{n \in \mathbb{N}_0}$-adapted process such that $X_n \in L^1$, $\forall n \in \mathbb{N}_0$. Then the following statements are equivalent.*

(i) The process $X_\bullet$ is a submartingale.
(ii) There exists an $\mathcal{F}_\bullet$-martingale $M_\bullet$ and an $\mathcal{F}_\bullet$-predictable nondecreasing process $C_\bullet$ such that
$$M_0 = 0 = C_0,\quad X_n = X_0 + M_n + C_n,\quad \forall n \geq 0.$$

Moreover, when $X_\bullet$ is a submartingale, then the martingale $M_\bullet$ and the nondecreasing predictable process are uniquely determined by $X_\bullet$ up to indistinguishability. In this case $M_\bullet$ is called the martingale component *of the submartingale $X_\bullet$ and $C_\bullet$ is called the* compensator *of $X_\bullet$. We denote it by $C(X_\bullet)$. The decomposition $X_n = X_0 + M_n + C_n$ is called the* Doob decomposition *of the submartingale $X_\bullet$.*

Proof. *Existence.* We describe M_n and C_n in terms of their increments. More precisely
$$\begin{aligned}C_{n+1} - C_n &= \mathbb{E}\big[X_{n+1} - X_n \,\big\|\, \mathcal{F}_n\big] + \mathbb{E}\big[M_{n+1} - M_n \,\big\|\, \mathcal{F}_n\big]\\ &= \mathbb{E}\big[X_{n+1} \,\big\|\, \mathcal{F}_n\big] - X_n,\quad \forall n \in \mathbb{N}_0,\end{aligned} \tag{3.1.1a}$$
$$M_{n+1} - M_n = (X_{n+1} - X_n) - (C_{n+1} - C_n),\quad \forall n \in \mathbb{N}_0. \tag{3.1.1b}$$

Note that $C_{n+1} - C_n$ is $\mathcal{F}_n$ measurable so (C_n) is predictable. By construction $M_\bullet$ is an $\mathcal{F}_\bullet$-martingale. Clearly, if $X_\bullet$ is a submartingale then, tautologically, C_n is increasing.

Uniqueness. Suppose that $X_\bullet$ is a submartingale, $M'_\bullet$ is a martingale, and $C_\bullet$ is a nondecreasing predictable process such that
$$M_0 = C_0 = 0,\quad X_n = X_0 + M'_n + C'_n,\quad \forall n \in \mathbb{N}_0.$$

We deduce
$$\mathbb{E}\big[\,X_{n+1}\,\big\|\,\mathcal{F}_n\,\big] - X_n = \underbrace{\mathbb{E}\big[\,M'_{n+1}\,\big\|\,\mathcal{F}_n\,\big] - M'_n}_{=0} + \underbrace{\mathbb{E}\big[\,C'_{n+1}\,\big\|\,\mathcal{F}_n\,\big] - C'_n}_{=C'_{n+1}-C'_n}.$$

This shows that the increments of C'_n are given by (3.1.1a) so $C'_n = C_n$. In particular, $M'_n = M_n$, $\forall n \in \mathbb{N}$. □

Example 3.14. Suppose that $(X_n)_{n\geq 0}$ is a sequence of nonnegative integrable random variables and $X_0 = 0$. Then
$$S_n = X_1 + \cdots + X_n$$
is a submartingale with respect to the filtration $\mathcal{F}_n = \sigma(X_1, \ldots, X_n)$. Indeed
$$\mathbb{E}\big[\,S_n\,\big\|\,\mathcal{F}_{n-1}\,\big] = \mathbb{E}\big[\,X_n\,\big\|\,\mathcal{F}_{n-1}\,\big] + S_{n-1} \geq S_{n-1}.$$
Consider the Doob decomposition $S_n = M_n + C_n$. The compensator C_n satisfies
$$C_{n+1} - C_n = \mathbb{E}\big[\,S_{n+1}\,\big\|\,\mathcal{F}_n\,\big] - S_n = \mathbb{E}\big[\,X_{n+1}\,\big\|\,\mathcal{F}_{n-1}\,\big]$$
so
$$C_n = \sum_{k=1}^n \mathbb{E}\big[\,X_n\,\big\|\,\mathcal{F}_{n-1}\,\big]$$
and
$$M_n = S_n - \sum_{k=1}^n \mathbb{E}\big[\,X_n\,\big\|\,\mathcal{F}_{n-1}\,\big] = \sum_{k=1}^n \big(\,X_n - \mathbb{E}\big[\,X_n\,\big\|\,\mathcal{F}_{n-1}\,\big]\,\big).$$
If the variables X_n are independent, then
$$M_n = \sum_{k=1}^n \big(\,X_k - \mathbb{E}\big[\,X_k\,\big]\,\big).$$
□

Definition 3.15 (Quadratic variation). Suppose that $(X_n)_{n\geq 0}$ is a martingale adapted to the filtration $(\mathcal{F}_n)_{n\geq 0}$ such that $\mathbb{E}\big[\,X_n^2\,\big] < \infty$, $\forall n \geq 0$. The compensator of the submartingale $(X_n^2)_{n\geq 0}$ is called the *quadratic variation* and it is denoted by $\langle X_\bullet \rangle$. □

Example 3.16. Suppose that $(X_n)_{n\geq 1}$ are independent random variables with zero means and finite variances. We set $S_0 = 0$,
$$S_n = X_1 + \cdots + X_n.$$
Then
$$\mathbb{E}\big[\,S_n^2\,\big] = \sum_{k=1}^n \mathbb{E}\big[\,X_k^2\,\big] < \infty, \quad \forall n \geq 1.$$
Thus $(S_\bullet)$ is an L^2-martingale. From the computations in Example 3.14 we deduce
$$\langle S_\bullet \rangle_n = \sum_{k=1}^n \mathbb{E}\big[\,X_k^2\,\big] = \sum_{k=1}^n \mathbb{E}\big[\,(S_k - S_{k-1})^2\,\big].$$
This explains why we refer to $\langle S_\bullet \rangle$ as quadratic *quadratic variation*. □

Theorem 3.17 (Discrete Stochastic Integral). *Suppose that $(X_n)_{n\in\mathbb{N}_0}$ be an $\mathcal{F}_\bullet$-adapted process and $(H_n)_{n\in\mathbb{N}}$ is a <u>bounded predictable</u> process. Define the process $(H \cdot X)_\bullet$ by setting*

$$(H\cdot X)_0 := 0, \quad (H\cdot X)_n = H_1(X_1 - X_0) + \cdots + H_n(X_n - X_{n-1}), \quad \forall n \in \mathbb{N}. \quad (3.1.2)$$

Then the following hold.

(i) *If $(X_n)_{n\in\mathbb{N}_0}$ is a martingale, then the process $(H \cdot X)_n$, $n \in \mathbb{N}_0$ is also an $\mathcal{F}_\bullet$-adapted martingale.*
(ii) *If $(X_n)_{n\in\mathbb{N}_0}$ is a submartingale and $H_n \geq 0$, $\forall n \in \mathbb{N}$, then the process $(H \cdot X)_n$, $n \in \mathbb{N}_0$ is also an $\mathcal{F}_\bullet$-adapted submartingale.*

Proof. (i) Clearly $(H\cdot X)_n \in L^1(\Omega, \mathcal{F}_n)$. We have

$$\mathbb{E}\big[(H\cdot X)_{n+1}\|\mathcal{F}_n\big] = \mathbb{E}\big[H_{n+1}(X_{n+1} - X_n)\|\mathcal{F}_n\big] + (H\cdot X)_n$$

(H_{n+1} is $\mathcal{F}_n$-measurable)

$$= H_{n+1}\mathbb{E}\big[(X_{n+1} - X_n)\|\mathcal{F}_n\big] + (H\cdot X)_n = H_{n+1}\big(\mathbb{E}\big[X_{n+1}\|\mathcal{F}_n\big] - X_n\big) + (H\cdot X)_n$$

((X_n) is a martingale)

$$= (H \cdot X)_n.$$

The proof of (ii) is similar. □

Remark 3.18. (a) When $X_\bullet$ is a martingale the process $(H \cdot X)_\bullet$ is called the *discrete stochastic integral* of H with respect to X and it is alternatively denoted

$$\int^n H dX := (H \cdot X)_n.$$

One should think of X_n as a random signed measure assigning mass $X_n - X_{n-1}$ to the point n.

(b) The discrete stochastic integral has a stock-trading interpretation. Suppose that X_n represents the price of a stock at the end of the n-th trading day. A day trader buys H_n shares at the beginning of the n-th trading day, based on the information available then. This information is encoded by the sigma-algebra $\mathcal{F}_{n-1}$ and the price of a share at the beginning of the n-th trading day is X_{n-1}. He sells them at the end of the n-the trading day. The resulting profit at the end of day n is then $H_n(X_n - X_{n-1})$. We deduce that $(H \bullet X)_n$ is represents the profit of the day trader after n trading days.

(c) The special case Theorem 3.17 when the variables H_n are Bernoulli random variables was discovered by P. Halmos and is classically known as the *impossibility of systems* theorem. In this case H_n represents the decision of a gambler to play or not the next game based on the information gathered during the games he observed so far. □

The applicability of Theorem 3.17 depends on our ability of producing interesting predictable processes. We describe one very useful class of examples.

Example 3.19. Observe first that a discrete time process $(Y_n)_{n\in\mathbb{N}}$ on $(\Omega, \mathcal{S}, \mathbb{P})$ can be viewed as a map

$$Y : \mathbb{N} \times \Omega \to \mathbb{R}, \quad (n, \omega) \mapsto Y_n(\omega).$$

We equip $\mathbb{N} \times \Omega$ with the product σ-algebra. A measurable set $\mathscr{X} \subset \mathbb{N} \times \Omega$ defines a stochastic process

$$\boldsymbol{I}_{\mathscr{X}} : \mathbb{N} \times \Omega \to \{0,1\}, \quad \left(\boldsymbol{I}_{\mathscr{X}}\right)_n = \boldsymbol{I}_{\mathscr{X}_n}, \quad \mathscr{X}_n := \{\,\omega \in \Omega;\ (n,\omega) \in \mathscr{X}\,\}.$$

The set $\mathscr{X}$ is called $\mathcal{F}_\bullet$-*predictable* if the process $\boldsymbol{I}_{\mathscr{X}}$ is such. More precisely, this means that $\mathscr{X}_0 \in \mathcal{F}_0$ and, for any $n \in \mathbb{N}$, the set $\mathscr{X}_n$ is $\mathcal{F}_{n-1}$-measurable. □

3.1.3 Stopping and sampling: discrete time

We want to describe one technique that makes the martingales extremely useful in applications. Fix a probability space $(\Omega, \mathcal{S}, \mathbb{P})$.

Definition 3.20. A random variable $T : (\Omega, \mathcal{S}, \mathbb{P}) \to \mathbb{N}_0 \cup \{\infty\}$ is called a *stopping time* adapted to the filtration $\mathcal{F}_\bullet = (\mathcal{F}_n)_{n\geq 0}$, or an $\mathcal{F}_\bullet$-*stopping time* if,

$$\{T \leq n\} \in \mathcal{F}_n, \quad \forall n \in \mathbb{N}_0 \cup \{\infty\}.$$

If $(X_n)_{n\in\mathbb{N}}$ is an $\mathcal{F}_\bullet$-adapted process, and T is an $\mathcal{F}_\bullet$-stopping time, then the T-*sample of the process* is the <u>random variable</u>

$$X_T := \sum_{n\in\mathbb{N}_0} X_n \boldsymbol{I}_{\{T=n\}}. \tag{3.1.3}$$

Observe that $X_T = 0$ on the set $\{T = \infty\}$. □

Example 3.21. (a) For each $n \in \mathbb{N}_0$ the constant random variable equal to n is a stopping time.
(b) Suppose that $(X_n)_{n\in\mathbb{N}_0}$ is $\mathcal{F}_\bullet$-adapted and $C \subset \mathbb{R}$ is a Borel set. We define the *hitting time* of C to be the random variable

$$H_C : \Omega \to \mathbb{N}_0 \cup \{\infty\}, \quad H_C(\omega) := \min\{\,n \in \mathbb{N}_0;\ X_n(\omega) \in C\,\}.$$

This is a stopping time since

$$\{\,H_C \leq n\,\} = \bigcup_{k\leq n} \{X_k \in C\}$$

and the process (X_n) is $\mathcal{F}_\bullet$-adapted.
(c) If S, T are stopping times, then $S \wedge T = \min(S,T)$ and $S \vee T = \max(S,T)$ are also stopping times.
(d) If $(T_k)_{k\in\mathbb{N}}$ is a sequence of stopping times, then $\inf T_k$, $\sup T_k$, $\liminf T_k$ and $\limsup T_k$ are also stopping times. □

Definition 3.22. Let $X_\bullet = (X_n)_{n \in \mathbb{N}}$ be a process adapted to the filtration $(\mathcal{F}_n)_{n \geq 0}$. For any stopping time T we denote by $X_\bullet^T$ the *process stopped at T* defined by

$$X_n^T := X_{T \wedge n}, \text{ where } X_{T \wedge n} = X_{\min(T(\omega), n)}(\omega) = \begin{cases} X_n(\omega), & n \leq T(\omega), \\ X_{T(\omega)}, & T(\omega) < n. \end{cases} \quad (3.1.4)$$

□

Note that the process stopped at T is also adapted to the filtration $(\mathcal{F}_n)_{n \geq 0}$.

Proposition 3.23. *Suppose that S, T is are stopping times such that $S \leq T$. Define*

$$]]T, \infty[[:= \{ (n, \omega) \in \mathbb{N}_0 \times \Omega; \ T(\omega) < n \},$$

$$]]S, T]] := \{ (n, \omega) \in \mathbb{N}_0 \times \Omega; \ S(\omega) < n \leq T(\omega) \}.$$

Then $]]T, \infty[[$, $[[0, T]]$ and $]]S, T]]$ are predictable subsets of $\mathbb{N}_0 \times \Omega$.

Proof. We have $]]T, \infty[[_n = \{T < n\} = \{T \leq n-1\} \in \mathcal{F}_{n-1}$. Next observe that

$$\boldsymbol{I}_{[[0,T]]} = 1 - \boldsymbol{I}_{]]T,\infty[[}$$

so $\boldsymbol{I}_{[[0,T]]}$ is a predictable process as a linear combination of predictable processes. Finally observe that since $S \leq T$ we have

$$\boldsymbol{I}_{]]S,T]]} = \boldsymbol{I}_{[[0,T]]} - \boldsymbol{I}_{[[0,S]]},$$

so $\boldsymbol{I}_{]]S,T]]}$ is predictable as a linear combination of predictable processes. □

Suppose now that $(X_n)_{n \in \mathbb{N}}$ is a (sub)martingale and T is a stopping time. Then $S_0 = 0$ is also a stopping time, $S_0 \leq T$. As we have seen above, the process $\boldsymbol{I}_{]]S_0,T]]} = \boldsymbol{I}_{]]0,T]]} \cdot X$ is a submartingale.

For every $n \in \mathbb{N}$ we have

$$(\boldsymbol{I}_{]]0,T]]} \cdot X)_n = (\boldsymbol{I}_{]]0,T]]})_n (X_n - X_{n-1}) + \cdots + (\boldsymbol{I}_{]]0,T]]})_1 (X_1 - X_0)$$

$$= (\boldsymbol{I}_{\{T \geq n\}})(X_n - X_{n-1}) + \cdots + (\boldsymbol{I}_{\{T \geq 1\}})(X_1 - X_0) = X_{T \wedge n} - X_0 = X_n^T - X_0.$$

Thus

$$\boxed{X_\bullet^T = X_0 + (\boldsymbol{I}_{]]0,T]]} \cdot X)_\bullet.} \quad (3.1.5)$$

This proves the following result.

Theorem 3.24 (Optional Stopping Theorem). *Suppose that*

$$X_n : (\Omega, \mathcal{F}, \mathbb{P}) \to \mathbb{R}, \ n \geq 0,$$

is a (sub)martingale adapted to the filtration $\mathcal{F}_\bullet$ and T is an $\mathcal{F}_\bullet$-stopping time. Then $X_\bullet^T$, the process stopped at T, is also a (sub)martingale adapted to $\mathcal{F}_\bullet$. □

Suppose that $T : (\Omega, \mathcal{S}, \mathbb{P}) \to \mathbb{N}_0 \cup \{\infty\}$ is a stopping time adapted to the filtration $\mathcal{F}_\bullet$. We define

$$\begin{aligned}\mathcal{F}_T &:= \Big\{ E \in \mathcal{F} : \ E \cap \{T \leq n\} \in \mathcal{F}_n, \ \forall n \in \mathbb{N}_0 \cup \{\infty\} \Big\} \\ &= \Big\{ E \in \mathcal{F} : \ E \cap \{T = n\} \in \mathcal{F}_n, \ \forall n \in \mathbb{N}_0 \cup \{\infty\} \Big\}.\end{aligned} \quad (3.1.6)$$

Tautologically, the random variable T is $\mathcal{F}_T$-measurable.

Example 3.25. Suppose that T is the hitting time of a Borel set $C \subset \mathbb{R}$. Then the event E belongs to $\mathcal{F}_T$ if, at any moment of time n, we can decide using the information $\mathcal{F}_n$ available to us at time n whether, up to that moment, we have visited C and the event E has occurred. $\square$

A few remarks are in order.

- The collection $\mathcal{F}_T$ is a σ-subalgebra of $\mathcal{F}$. It is called the *past-until-T* σ-algebra.
- The random variable X_T is $\mathcal{F}_T$-measurable. Indeed,

$$\{X_T \leq c\} \cap \{T = n\} = \{X_n \leq c\} \cap \{T = n\} \in \mathcal{F}_n, \ \forall n.$$

- If S, T are stopping times such that $S \leq T$, then $\mathcal{F}_S \subset \mathcal{F}_T$.

Definition 3.26. Suppose that $(X_n)_{n \in \mathbb{N}_0}$ is an $\mathcal{F}_\bullet$-(sub)martingale and T is an $\mathcal{F}_\bullet$-stopping time. We say that the stopping time T satisfies the *Doob conditions*[1] if the following hold.

$$\mathbb{P}[T < \infty] = 1. \quad (3.1.7\text{a})$$

$$X_T \in L^1. \quad (3.1.7\text{b})$$

$$\lim_{n \to \infty} \mathbb{E}\big[\, \boldsymbol{I}_{\{T > n\}} |X_n| \,\big] = 0. \quad (3.1.7\text{c})$$

$\square$

Roughly speaking, the Doob conditions state that the random process $(X_n)_{n \geq 0}$ is not sampled "too late". In Proposition 3.66 we provide another characterization of the Doob conditions in terms of the asymptotic behavior of the stopped process $X_\bullet^T$.

Example 3.27. Suppose that T is a bounded $\mathcal{F}_\bullet$-stopping time. Then T satisfies the Doob conditions.

To see this, choose $N \in \mathbb{N}$ such that $T < N$ a.s. The stopped process $X_\bullet^T$ is a submartingale so $X_T = X_{T \wedge N} = X_N^T \in L^1$. As for the second condition (3.1.7c), note that since T is a.s. bounded, then $\boldsymbol{I}_{\{T > n\}} X_n^+$ is a.s. 0 for $n > N$. $\square$

---

[1] There is no consensus on terminology in the literature. We use the term *Doob conditions* since they were first spelled out by J. L. Doob in his influential monograph [47].

Theorem 3.28 (Optional Sampling Theorem). *Suppose that $X_n : (\Omega, \mathcal{F}, \mathbb{P}) \to \mathbb{R}$, $n \geq 0$, is a (sub)martingale adapted to the filtration $\mathcal{F}_\bullet$, and $S \leq T$ are stopping times adapted to the same filtration. If T satisfies Doob's conditions in Definition 3.26, then*

$$\mathbb{E}[X_T] \geq \mathbb{E}[X_S].$$

If $X_\bullet$ is a martingale, then

$$\mathbb{E}[X_T] = \mathbb{E}[X_S] = \mathbb{E}[X_0].$$

Proof. We follow the original approach in [47, VII.2]; see also [4, Thm. 6.7.4]. Suppose that $(X_n)_{n \geq 0}$ is a martingale. Set $A_m := \{S = m\}$. Then

$$\mathbb{E}[X_S] = \sum_{m \geq 0} \mathbb{E}[X_S \boldsymbol{I}_{A_m}]$$

so it suffices to show that

$$\forall m \geq 0: \quad \mathbb{E}[X_T \boldsymbol{I}_{A_m}] = [X_S \boldsymbol{I}_{A_m}].$$

We have

$$\mathbb{E}[X_S \boldsymbol{I}_{A_m}] = \mathbb{E}[X_m \boldsymbol{I}_{A_m} \boldsymbol{I}_{T=m}] + \mathbb{E}[X_S \boldsymbol{I}_{A_m} \boldsymbol{I}_{T>m}]$$

$$= \mathbb{E}[X_T \boldsymbol{I}_{T=m}] + \boxed{\mathbb{E}[X_m \boldsymbol{I}_{A_m} \boldsymbol{I}_{T>m}]}_*$$

$(A_m \cap \{T > m\} \in \mathcal{F}_m$, $X_\bullet$ martingale$)$

$$= \mathbb{E}[X_T \boldsymbol{I}_{A_m} \boldsymbol{I}_{\{T=m\}}] + \boxed{\mathbb{E}[X_{m+1} \boldsymbol{I}_{A_m} \boldsymbol{I}_{\{T>m\}}]}_*$$

$$= \mathbb{E}[X_T \boldsymbol{I}_{A_m} \boldsymbol{I}_{\{T=m\}}] + \mathbb{E}[X_{m+1} \boldsymbol{I}_{A_m} \boldsymbol{I}_{\{T=m+1\}}] + \mathbb{E}[X_{m+1} \boldsymbol{I}_{A_m} \boldsymbol{I}_{\{T>m+1\}}]$$

$$= \mathbb{E}[X_T \boldsymbol{I}_{A_m} \boldsymbol{I}_{\{m \leq T \leq m+1\}}] + \boxed{\mathbb{E}[X_{m+1} \boldsymbol{I}_{A_m} \boldsymbol{I}_{\{T>m+1\}}]}_\bullet$$

$(A_m \cap \{T > m+1\} \in \mathcal{F}_{m+1}$, $X_\bullet$ martingale$)$

$$= \mathbb{E}[X_T \boldsymbol{I}_{A_m} \boldsymbol{I}_{\{m \leq T \leq m+1\}}] + \boxed{\mathbb{E}[X_{m+2} \boldsymbol{I}_{A_m} \boldsymbol{I}_{\{T>m+1\}}]}_\bullet$$

$$= \mathbb{E}[X_T \boldsymbol{I}_{A_m} \boldsymbol{I}_{\{m \leq T \leq m+2\}}] + \mathbb{E}[X_{m+2} \boldsymbol{I}_{A_m} \boldsymbol{I}_{\{T>m+2\}}].$$

Iterating this procedure we deduce that, $\forall n > 0$, we have

$$\mathbb{E}[X_S \boldsymbol{I}_{A_m}] = \mathbb{E}[X_T \boldsymbol{I}_{A_m} \boldsymbol{I}_{\{m \leq T \leq m+n\}}] + \mathbb{E}[X_{m+n} \boldsymbol{I}_{\{T>m+n\}}].$$

The condition (3.1.7c) shows that

$$\lim_{n \to \infty} \mathbb{E}[X_{m+n} \boldsymbol{I}_{T>m+n}] = 0,$$

so

$$\mathbb{E}[X_S \boldsymbol{I}_{A_m}] = \mathbb{E}[X_T \boldsymbol{I}_{A_m} \boldsymbol{I}_{0 \leq T \leq \infty}] = \mathbb{E}[X_T \boldsymbol{I}_{A_m}].$$

The submartingale situation is dealt with similarly. $\square$

Remark 3.29. Suppose that T is an a.s. finite $\mathcal{F}_\bullet$-stopping time and $X_\bullet$ is an $\mathcal{F}_\bullet$-submartingale such that $X_T \in L^1$. Then T satisfies Doob's conditions if and only if

$$\lim_{n \to \infty} \mathbb{E}\left[X_n^+ \boldsymbol{I}_{T>n} \right] = 0. \tag{3.1.8}$$

Clearly (3.1.7c) implies (3.1.8). Let us show that (3.1.8) $\Rightarrow$ (3.1.7c). Assume first that $X_\bullet$ is a martingale. Fix $m, n \in \mathbb{N}_0$, $m < n$. Observing that $\{T > m\} \in \mathcal{F}_m$ we deduce

$$\mathbb{E}\left[X_m \boldsymbol{I}_{T>n} \right] = \mathbb{E}\left[X_{m+1} \boldsymbol{I}_{T>m} \right] = \mathbb{E}\left[X_{m+1} \boldsymbol{I}_{T=m+1} \right] + \mathbb{E}\left[X_{m+1} \boldsymbol{I}_{T>m+1} \right]$$

($\{T > m+1\} \in \mathcal{F}_{m+1}$)

$$= \mathbb{E}\left[X_{m+1} \boldsymbol{I}_{T=m+1} \right] + \mathbb{E}\left[X_{m+2} \boldsymbol{I}_{T>m+1} \right]$$

$$= \mathbb{E}\left[X_{m+1} \boldsymbol{I}_{T=m+1} \right] + \mathbb{E}\left[X_{m+2} \boldsymbol{I}_{T=m+2} \right] + \mathbb{E}\left[X_{m+2} \boldsymbol{I}_{T>m+2} \right]$$

$$= \cdots = \mathbb{E}\left[X_{m+1} \boldsymbol{I}_{T=m+1} \right] + \cdots + \mathbb{E}\left[X_N \boldsymbol{I}_{T=n} \right] + \mathbb{E}\left[X_n \boldsymbol{I}_{T>n} \right] = \mathbb{E}\left[X_T \boldsymbol{I}_{m<T\leq n} \right].$$

We deduce

$$\mathbb{E}\left[X_m \boldsymbol{I}_{T>m} \right] - \mathbb{E}\left[X_n \boldsymbol{I}_{T>n} \right] = \mathbb{E}\left[X_T \boldsymbol{I}_{m<T\leq n} \right], \quad \forall n > m.$$

Using the equality $X_\bullet = X_\bullet^+ - X_\bullet^-$ we deduce that, $\forall n > m$.

$$\mathbb{E}\left[X_n^- \boldsymbol{I}_{T>n} \right] - \mathbb{E}\left[X_m^- \boldsymbol{I}_{T>m} \right] = \mathbb{E}\left[X_T \boldsymbol{I}_{m<T\leq n} \right] - \left(\mathbb{E}\left[X_m^+ \boldsymbol{I}_{T>m} \right] - \mathbb{E}\left[X_n^+ \boldsymbol{I}_{T>n} \right] \right).$$

If we let $n \to \infty$ in the above equality and recall that $T < \infty$ a.s., $X_T \in L^1$ and $X_\bullet^+$ satisfies (3.1.8) we deduce

$$\lim_{n \to \infty} \mathbb{E}\left[X_n^- \boldsymbol{I}_{T>n} \right] - \mathbb{E}\left[X_m^- \boldsymbol{I}_{T>m} \right] = \mathbb{E}\left[X_T \boldsymbol{I}_{T>m} \right] - \mathbb{E}\left[X_m^+ \boldsymbol{I}_{T>m} \right].$$

Using the Optional Sampling Theorem 3.24 for the stopping times $S \equiv m$ and T we deduce

$$\mathbb{E}\left[X_T \boldsymbol{I}_{T>m} \right] - \mathbb{E}\left[X_m^+ \boldsymbol{I}_{T>m} \right] = \mathbb{E}\left[X_m \boldsymbol{I}_{T>m} \right] - \mathbb{E}\left[X_m^+ \boldsymbol{I}_{T>m} \right] = -\mathbb{E}\left[X_m^- \boldsymbol{I}_{T>m} \right].$$

Hence

$$\lim_{n \to \infty} \mathbb{E}\left[X_n^- \boldsymbol{I}_{T>n} \right] - \mathbb{E}\left[X_m^- \boldsymbol{I}_{T>m} \right] = -\mathbb{E}\left[X_m^- \boldsymbol{I}_{T>m} \right]$$

so that

$$\lim_{n \to \infty} \mathbb{E}\left[|X_n| \boldsymbol{I}_{T>n} \right] = \lim_{n \to \infty} \mathbb{E}\left[X_n^+ \boldsymbol{I}_{T>n} \right] + \lim_{n \to \infty} \mathbb{E}\left[X_n^- \boldsymbol{I}_{T>n} \right] = 0.$$

Suppose now that $(X_\bullet)$ is a submartingale. Consider its Doob decomposition $X_n = X_0 + M_n + C_n$. If $X_\bullet$ satisfies (3.1.8), then

$$0 \leq (X_0 + M_n)^+ \leq X_n^+$$

and we deduce that the *martingale* $Y_\bullet = X_0 + M_\bullet$ satisfies (3.1.8) and thus (3.1.8). Next, observe that

$$X_n^+ = (Y_n^+ + C_n^+) \boldsymbol{I}_{Y_n \geq 0} + (C_n - Y_n^-) \boldsymbol{I}_{0 < Y_n^- \leq C_n}.$$

This proves that $0 \leq C_n \leq X_n^+ + Y_n^-$, so

$$\lim_{n \to \infty} \mathbb{E}\left[C_n \boldsymbol{I}_{T>n} \right] = \lim_{n \to \infty} \mathbb{E}\left[(X_n^+ + Y_n^-) \boldsymbol{I}_{T>n} \right] = 0.$$

Hence

$$\lim_{n \to \infty} \mathbb{E}\left[|X_n| \boldsymbol{I}_{T>n} \right] = \lim_{n \to \infty} \mathbb{E}\left[|Y_n| + C_n \boldsymbol{I}_{T>n} \right] = 0. \qquad \square$$

3.1.4 Applications of the optional sampling theorem

It is time to give the reader a first taste of the versatility of the optional sampling theorem. After we present more properties of martingales we will be able to extend the range of applications of this theorem.

Example 3.30 (The Ballot Problem). Let us consider again the *ballot problem* first discussed in Example 1.60. Recall the setup.

Two candidates A and B run for an election. Candidate A received a votes while candidate B received b votes, where $b < a$. The votes were counted in random order, so any permutation of the $a + b$ votes cast is equally likely. We have shown in Example 1.60 that the probability that A was ahead throughout the count is

$$p = \frac{a-b}{a+b}.$$

We want to described an alternate proof using martingale methods. Our presentation is inspired from [117, Sec. 12.2].

Set $n := a + b$ and denote by D_k the denote the number votes by which A was ahead when the k-th voted was tabulated. Note that $S_n = a - b$. Let X_k denote the random variable indicating the k-th vote. Thus, $X_k = 1$, if the vote went for A, and $X_k = -1$ if the vote went for B so that

$$D_0 = 0, \quad D_k = X_1 + \cdots + X_k.$$

For $k = 0, 1, \ldots, n$ we denote by R_k the ratio

$$R_k := \frac{D_{n-k}}{n-k}.$$

In other words R_k is candidate's A the lead in percentages after the $(n-k)$-th counted vote. Let us first show that R_k is a martingale with respect to the filtration

$$\mathcal{F}_k = \sigma(R_0, \ldots, R_k) = \sigma(D_n, D_{n-1}, \ldots, D_{n-k}).$$

Thus, conditioning on $\mathcal{F}_k$ corresponds to conditioning on the results of the last $k+1$ votes. Observe that, given D_{n-k} the result D_{n-k-1} one vote earlier, is independent of the results at the later votes $D_{n-k+1}, \ldots, D_n$. In other words,

$$\mathbb{E}\big[\, D_{n-k-1} \,\|\, D_{n-k}, \ldots, D_n \,\big] = \mathbb{E}\big[\, D_{n-k-1} \,\|\, D_{n-k} \,\big].$$

One might be tempted to think of D_{n-k} as a random walk in reverse, but there is a silent trap: there is a condition at the n-th step in reverse namely $D_0 = 0$.

To compute the above conditional expectation denote by A_m (resp. B_m) the number of votes A (resp. B) has received after m votes. Thus

$$D_m = A_m - B_m, \quad m = A_m + B_m.$$

Note that A_m and B_m are determined by D_m via the equalities

$$A_m = \frac{D_m + m}{2}, \quad B_m = \frac{m - D_m}{2}.$$

Thus, if D_{n-k} is known, the $(n-k)$-th vote could have been either a vote for A, and the probability of such a vote is $\frac{A_{n-k}}{n-k}$, or it could have been a vote for B, and the probability of such a vote is $\frac{B_{n-k}}{n-k}$. Hence

$$\mathbb{E}\big[\,D_{n-k-1}\,\|\,D_{n-k}\,\big] = \big(\,D_{n-k}-1\,\big)\frac{A_{n-k}}{n-k} + \big(\,D_{n-k}+1\,\big)\frac{B_{n-k}}{n-k}$$

$$= D_{n-k} - \frac{D_{n-k}}{n-k} = \frac{n-k-1}{n-k}D_{n-k}.$$

Dividing by $(n-k-1)$ we deduce that $(R_k)_{0 \leq k \leq n-1}$ is indeed a martingale.

Now define the stopping times

$$S := \big\{\,0 \leq k \leq n-1;\ R_k = 0\,\big\},$$

where $\min \emptyset := \infty$ and $T := \min(S, n-1)$. The stopping time T is bounded and the Optional Sampling Theorem 3.28 implies

$$\mathbb{E}\big[\,R_T\,\big] = \mathbb{E}\big[\,R_0\,\big] = \frac{D_n}{n} = \frac{a-b}{a+b}.$$

Now observe that

$$\mathbb{E}\big[\,R_T\,\big] = \mathbb{E}\big[\,R_T \mathbf{I}_{S=\infty}\,\big] + \mathbb{E}\big[\,R_T \mathbf{I}_{S<\infty}\,\big].$$

Note $R_T = 0$ on $\{S < \infty\}$. Observe that if $S = \infty$, then $D_k > 0$, for all $1 \leq k \leq n$. Hence $T = (n-1)$ on $\{S = \infty\}$ so $R_T = D_1 = 1$ on $\{S = \infty\}$.

$$\frac{a-b}{a+b} = \mathbb{E}\big[\,R_T\,\big] = \mathbb{P}\big[\,S = \infty\,\big]$$

$=$ the probability that candidate A lead throughout the vote count.

$\square$

Example 3.31 (Expected time to observe a pattern). Suppose that we are given a finite set (alphabet) $\mathcal{A}$ and a probability distribution π on it so that

$$\pi(a) := \pi(\{a\}) > 0, \quad \forall a \in \mathcal{A}.$$

Define

$$f : \mathcal{A} \to (0, \infty), \quad f(a) = \frac{1}{\pi(a)}.$$

Fix a word (or pattern) of length $\ell > 0$ in this alphabet, $\boldsymbol{a} = (a_1, \ldots, a_\ell) \in \mathcal{A}^\ell$.

Suppose that $(A_n)_{n \geq 1}$ is a sequence of independent $\mathcal{A}$-valued random variables with common distribution π. We say that the *pattern $\boldsymbol{a}$ is observed at time n* if $n \geq \ell$ and

$$(A_{n-\ell+1}, A_{n-\ell+2}, \ldots, A_n) = (a_1, a_2, \ldots, a_n).$$

We let $T = T_{\boldsymbol{a}}$ denote the first time the pattern $\boldsymbol{a}$ is observed

$$T_{\boldsymbol{a}} := \min\big\{\,n \geq \ell;\ (A_{n-\ell+1}, A_{n-\ell+2}, \ldots, A_n) = (a_1, a_2, \ldots, a_\ell)\,\big\}.$$

To visualize this, think that we have an urn with balls labeled by the letters in $\mathcal{A}$ in proportions given by π. We sample with replacement the urn and we record in succession the labels we draw. We are interested in the moment we first observe the labels $a_1, \ldots, a_\ell$ in succession as we sample the urn. As a special case, think that we flip a fair coin and we stop the first we see T, H, T, H in succession. In this case $\mathcal{A} = \{H, T\}$, $\pi(H) = \pi(T) = \frac{1}{2}$, $\boldsymbol{a} = THTH$.

An amusing quote by Bertrand Russel comes to mind. "There is a special department of Hell for students of probability. In this department there are many typewriters and many monkeys. Every time that a monkey walks on a typewriter, it types by chance one of Shakespeare's sonnets."

We will compute $\mathbb{E}\big[T_{\boldsymbol{a}}\big]$ by using a clever martingale method due to Li [107]. The precise answer is contained in (3.1.11).

Let us first observe that $\mathbb{E}\big[T_{\boldsymbol{a}}\big] < \infty$. This follows from a very useful trick, [160, E10.5], generalizing the result in Example 1.167.

Lemma 3.32 ('Sooner-rather-than-later'). *Suppose that T is a stopping time adapted to the filtration $(\mathcal{F}_n)_{n \in \mathbb{N}_0}$ with the property that there exist $r_0 > 0$ and $N_0 \in \mathbb{N}$ such that*

$$\forall n \in \mathbb{N}_0, \quad \mathbb{P}\big[T \leq n + N_0 \| \mathcal{F}_n \big] > r_0. \qquad (3.1.9)$$

Then there exists $c \in (0, 1)$ such that $\mathbb{P}\big[T > n\big] < c^n$, $\forall n > N_0$. In particular,

$$\mathbb{E}\big[T\big] = \sum_{n \geq 0} \mathbb{P}\big[T > n\big] < \infty. \qquad \square$$

In Exercise 3.6 we ask the reader to provide a proof of this result. It is a nice application of various properties of the conditional expectation.

In the case at hand (3.1.9) is satisfied with $N_0 = \ell$ and $r = \Big(\min_{a \in \mathcal{A}} \pi(a)\Big)^\ell$.

Following [107] we consider the following betting game involving the House (casino) and a random number of players. At each moment of time $n = 1, 2, \ldots$ the House samples the alphabet $\mathcal{A}$ according to the probability distribution π. (The House runs a chance game with set of outcomes $\mathcal{A}$ and probability distribution π.) The outcome of this sampling is the sequence of i.i.d. random variables A_n.

The first player adopts the following $\boldsymbol{a}$-based strategy.

- At time 0 he bets his fortune $F_0^1 = 1$ that the outcome of the first game is $A_1 = a_1$. If $A_1 = a_1$ his fortune will change to $F_1^1 = f(a_1) = \frac{1}{a_1}$. Otherwise, he will lose his fortune F_0^1 to the house, so $F_1^1 = 0$ in this case.
- At time 1 he bets his fortune that $A_2 = a_2$. If he wins, i.e., $A_2 = a_2$, his fortune at time 2 will grow to $F_2^1 = f(a_2) F_1^1$. If he loses, he will have to turn all its fortune to the House.
- In general, if $k \leq \ell$ and his fortune at time $k-1$ is F_{k-1}^1 (the fortune could be 0 at that moment), the player bets all its fortune, $f(a_k)$ on a dollar, that $A_k = a_k$. If this happens, his fortune will grow to $F_k^1 = f(a_k) F_{k-1}^1$. Otherwise, he will surrender his fortune F_{k-1}^1 to the house, so $F_k^1 = 0$ in this case.

- At time ℓ the first gambler stops playing, so $F_n^1 = F_\ell^1$, $\forall n \geq \ell$.
- We denote by X_n^1 the profit of the first player at time n, $X_n^1 = F_n^1 - F_0^1 = F_n^1 - 1$.

Concisely, if we define
$$M_k^1 = \begin{cases} f(a_k)\mathbf{I}_{\{A_k=a_k\}}, & 1 \leq k \leq \ell, \\ 1, & k < 1 \text{ or } k > \ell, \end{cases}$$
then
$$F_n^1 = \prod_{k=1}^n M_k^1.$$

Since $\mathbb{E}[M_k^1] = 1$ we deduce that $F_\bullet^1$ and $X_\bullet^1 = F_\bullet^1 - 1$ are martingales.

In general, for $m = 1, 2, \ldots$, the m-th player also plays ℓ rounds using the same strategy as the first player, but with a delay of $m - 1$ units of times.

Thus, the second player skips game 1 and only starts betting before the 2nd game using the same betting strategy as if the game started when he began playing: at his j-th round he bets $f(a_j)$ on a dollar that the outcome is $A_{j+1} = a_j$. The third player skips the first two games etc.

In general, at his j-th round, the m-th player bets $f(a_j)$ on a dollar that the outcome is $A_{j+m-1} = a_j$. We denote by F_n^m the fortune of the m-th player at time n. More precisely, if we set
$$M_k^m := \begin{cases} f(a_{k-m+1})\mathbf{I}_{\{A_k=a_{k-m+1}\}}, & m \leq k \leq m+\ell-1, \\ 1, & k < m \text{ or } k \geq m+\ell, \end{cases}$$
then
$$F_n^m := \prod_{k=1}^n M_k^m, \quad X_n^m = F_n^m - 1 \quad n = 1, 2, \ldots.$$

Note that $F_n^m = 1$ for $n < m$ because the m-th player skips the games $n = 1, 2, \ldots, m - 1$. Define
$$S_n := \sum_{m \geq 1} X_n^m = \sum_{m=1}^n X_n^m = \sum_{m=1}^n F_n^m - n.$$

In other words, S_n is the sum of the profits of all the players after n games. The process $S_\bullet$ is obviously a martingale. Note that
$$S_T = \sum_{m \leq T} F_T^m - T, \quad T = T_{\boldsymbol{a}}.$$

Recall that T is the first moment of time such that
$$A_{T-\ell+1} = a_1, \ A_{T-\ell+2} = a_2, \ldots, A_T = a_\ell. \tag{3.1.10}$$

Thus the player $(T - \ell + 1)$ will be the first player to hit the jackpot, i.e., observes the pattern $\boldsymbol{a}$ during the first ℓ games he plays. This proves $F_T^m = 0$ for $m \leq T - \ell$. Indeed, the minimality of T implies
$$(A_m, \ldots, A_{m+\ell-1}) \neq (a_1, \ldots, a_\ell)$$
and thus $\mathbf{I}_{\{A_m=a_1\}} \cdots \mathbf{I}_{\{A_{m+\ell-1}=a_\ell\}} = 0.$

The fortune of the player $T - \ell + 1$ at time T is
$$F_T^{T-\ell+1} = f(a_1) \cdots f(a_\ell).$$
Using the equalities (3.1.10) we deduce that the fortune of the next player, $T - \ell + 2$, at time T is nonzero if and only if
$$(a_2, \ldots, a_\ell) = (a_1, \ldots, a_{\ell-1}).$$
In this case the fortune is $f(a_1) \cdots f(a_{\ell-1})$. Similarly,
$$F_T^{T-\ell+3} = \begin{cases} f(a_1) \cdots f(a_{\ell-2}), & (a_1, \ldots, a_{\ell-2}) = (a_3, \ldots, a_\ell), \\ 0, & (a_1, \ldots, a_{\ell-2}) \neq (a_3, \ldots, a_\ell). \end{cases}$$
More generally, denote by $\delta_{\alpha,\beta}$ the Kronecker symbol
$$\delta_{\alpha,\beta} := \begin{cases} 1, & \alpha = \beta, \\ 0, & \alpha \neq \beta. \end{cases}$$
We deduce
$$S_T + T = \underbrace{F_T^1 + \cdots + F_T^{T-\ell}}_{=0} + F_T^{T-\ell+1} + F_T^{T-\ell+2} + \cdots + F_T^T$$
$$= F_T^{T-\ell+1} + F_T^{T-\ell+2} + \cdots + F_T^T$$
$$= \underbrace{f(a_1) \cdots f(a_\ell)}_{F_T^T} + \underbrace{\prod_{j=1}^{\ell-1} f(a_j) \delta_{a_{j+1}, a_j}}_{F_T^{T-1}} + \underbrace{\prod_{j=1}^{\ell-2} f(a_j) \delta_{a_{j+2}, a_j}}_{F_T^{T-2}} + \cdots$$
$$= \underbrace{\sum_{k=0}^{\ell-1} \prod_{j=1}^{\ell-k} f(a_j) \delta_{a_{j+k}, a_j}}_{=:\tau(\boldsymbol{a})}.$$
Hence $S_T = \tau(\boldsymbol{a}) - T$. If we could show that T satisfies Doob's conditions (Definition 3.26), then we could invoke the Optional Sampling Theorem 3.28 and conclude that
$$0 = \mathbb{E}[S_0] = \mathbb{E}[S_T] = \tau(\boldsymbol{a}) - \mathbb{E}[T].$$
Let us show that indeed the stopping time satisfies Doob's conditions.

Since $\mathbb{E}[T] < \infty$ and $S_T = \tau(\boldsymbol{a}) - T$ we deduce $S_T \in L^1$. Arguing as above we deduce that if $n < T$, then
$$F_n^1 + \cdots + F_n^n \leq F^{n-\ell+1} + \cdots + F_n^n$$
$$\leq f(a_1) \cdots f(a_\ell) + \prod_{j=1}^{\ell-1} f(a_j) \delta_{a_{j+1}, a_j} + \prod_{j=1}^{\ell-2} f(a_j) \delta_{a_{j+2}, a_j} + \cdots = \tau(\boldsymbol{a}).$$

Hence
$$|S_n|\boldsymbol{I}_{\{T>n\}} \leq \big(\tau(\boldsymbol{a})+n\big)\boldsymbol{I}_{\{T>n\}} \leq \big(\tau(\boldsymbol{a})+T\big)\boldsymbol{I}_{\{T>n\}}.$$
Since $\mathbb{E}[T]<\infty$ we deduce
$$\lim_{n\to\infty}\mathbb{E}\big[\,|S_n|\boldsymbol{I}_{\{T>n\}}\,\big] = 0.$$
This shows that the stopping time $T_{\boldsymbol{a}}$ satisfies Doob's conditions so that
$$\mathbb{E}[T_{\boldsymbol{a}}] = \tau(\boldsymbol{a}) = \sum_{k=0}^{\ell-1}\prod_{j=1}^{\ell-k}\frac{\delta_{a_{j+k},a_j}}{\pi(a_j)}. \tag{3.1.11}$$
Let us describe this equality using a more convenient notation. Denote by $V(\mathcal{A})$ the vocabulary of the alphabet $\mathcal{A}$
$$V(\mathcal{A}) = \bigsqcup_{\ell\geq 0}\mathcal{A}^\ell,\ \ \mathcal{A}^0 := \{\emptyset\}.$$
We denote by $\ell(\boldsymbol{a})$ the length of a word $\boldsymbol{a}$.

We define a weight $w = w_\pi : V(\mathcal{A}) \to (0,\infty)$ by setting
$$w(a_1,\dots,a_\ell) = \prod_{k=1}^{\ell} f(a_k),\ \ w(\emptyset) = 1.$$
For $\boldsymbol{a} = (a_1,\dots,a_\ell) \in \mathcal{A}^\ell$, $\ell \geq 1$, and $j = 1,\dots,\ell$ we define the left/right tail maps
$$L_j, R_j : \mathcal{A}^\ell \to \mathcal{A}^j,\ \ L_j(\boldsymbol{a}) = (a_1,\dots,a_j),\ \ R_j(\boldsymbol{a}) = (a_{\ell-j+1},\dots,a_\ell).$$
Thus, R_j retains only the last j letters of a word while L_j retains the first j letters.

Given two words $\boldsymbol{a}, \boldsymbol{b} \in V(\mathcal{A})$ we set
$$\langle \boldsymbol{a}, \boldsymbol{b}\rangle := \begin{cases} 1, & \boldsymbol{a} = \boldsymbol{b}, \\ 0, & \boldsymbol{a} \neq \boldsymbol{b}. \end{cases}$$
Now define
$$\Phi : V(\mathcal{A}) \times V(\mathcal{A}) \to [0,\infty),\ \ \Phi(\boldsymbol{a},\boldsymbol{b}) = \sum_{j=1}^{\ell(\boldsymbol{a})\wedge\ell(\boldsymbol{b})} \langle R_j\boldsymbol{a}, L_j\boldsymbol{b}\rangle w(L_j\boldsymbol{b}). \tag{3.1.12}$$
We can rewrite (3.1.11) as
$$\mathbb{E}[T_{\boldsymbol{a}}] = \Phi(\boldsymbol{a},\boldsymbol{a}). \tag{3.1.13}$$
In the special case when $\mathcal{A} = \{1,2,\dots,6\}$, π is the uniform counting probability and
$$\boldsymbol{a} = \underbrace{6\cdots 6}_{k} \in \mathcal{A}^k,$$
then the waiting time $\tau(\boldsymbol{a})$ coincides with the waiting time T to observe the first occurrence of a k-run of 6-s discussed in Example 1.167. In this case we have
$$\mathbb{E}[T] = \sum_{j=1}^{k} 6^j = \frac{6^{k+1}-6}{5}.$$

We refer to Example A.22 for an R-code that simulates sampling an alphabet until a given pattern is observed.

Let us discuss in more detail the special case $\mathcal{A} = \{H, T\}$, $\pi(H) = \pi(T) = \frac{1}{2}$. Suppose that $\boldsymbol{a}$ is the pattern $\boldsymbol{a} = (TTHH)$, $\boldsymbol{b} = HHH$. Observe that $\langle L_j\boldsymbol{a}, R_j\boldsymbol{a}\rangle = 1$ for $j = 4$ and 0 otherwise. Hence $\mathbb{E}[T_{\boldsymbol{a}}] = 16$. A similar computation shows that $\mathbb{E}[T_{\boldsymbol{b}}] = 14$. Thus we have to wait a longer time for the pattern $\boldsymbol{a}$ to occur.

On the other hand, a formula of Conway (see Exercise 3.14) shows that

$$\frac{\mathbb{P}[T_{\boldsymbol{b}} < T_{\boldsymbol{a}}]}{\mathbb{P}[T_{\boldsymbol{a}} < T_{\boldsymbol{b}}]} = \frac{\Phi(\boldsymbol{a},\boldsymbol{a}) - \Phi(\boldsymbol{a},\boldsymbol{b})}{\Phi(\boldsymbol{b},\boldsymbol{b}) - \Phi(\boldsymbol{b},\boldsymbol{a})}.$$

We have $\langle L_i\boldsymbol{a}, R_j\boldsymbol{b}\rangle = 0$, $\forall j$ and $\langle R_j\boldsymbol{a}, L_j\boldsymbol{b}\rangle = 1$ for $j = 1, 2$ so that

$$\Phi(\boldsymbol{a},\boldsymbol{b}) = 0, \quad \Phi(\boldsymbol{b},\boldsymbol{a}) = 6, \quad \frac{\mathbb{P}[T_{\boldsymbol{b}} < T_{\boldsymbol{a}}]}{\mathbb{P}[T_{\boldsymbol{a}} < T_{\boldsymbol{b}}]} = \frac{5}{7}.$$

We have reached a somewhat surprising conclusion: although, on average, we have to wait a shorter amount of time to observe the pattern $\boldsymbol{b}$, it is less likely that we will observe $\boldsymbol{b}$ before $\boldsymbol{a}$. The odds that $\boldsymbol{b}$ will appear first versus that $\boldsymbol{a}$ will appear first are $5 : 7$.

There are other strange phenomena. We should mention M. Gardner's even stranger nontransitivity paradox [66, Chap. 5]. More precisely, given any pattern $\boldsymbol{a} \in \mathcal{A}^k$ there exists a pattern $\boldsymbol{b} \in \mathcal{A}^k$ such that $\boldsymbol{b}$ is more likely to occur before $\boldsymbol{a}$, i.e., $\mathbb{P}[T_{\boldsymbol{b}} < T_{\boldsymbol{a}}] > \frac{1}{2}$. As shown in by Guibas and Odlyzko [76], if $\boldsymbol{a} = (a_1, \ldots, a_k)$ we can choose $\boldsymbol{b}$ to be of the form $\boldsymbol{b} = (b, a_1, \ldots, a_{k-1})$. □

3.1.5 Concentration inequalities: martingale techniques

Hoeffding's inequality (2.3.12) has a martingale counterpart usually referred to as *Azuma's inequality*.

Theorem 3.33 (Azuma). *Suppose that $(X_n)_{n \geq 0}$ is a martingale adapted to a filtration $\mathcal{F}_\bullet = (\mathcal{F}_n)_{n \geq 0}$ of the probability space $(\Omega, \mathcal{S}, \mathbb{P})$. Assume that for any $n \in \mathbb{N}$ there exist constants $a_n < b_n$ such the differences $D_n = X_n - X_{n-1}$ satisfy*

$$a_n \leq D_n \leq b_n \text{ a.s.}$$

Then

$$\forall x > 0, \quad \mathbb{P}[|X_n - X_0| > x] \leq 2e^{-\frac{2x^2}{(s_1^2 + \cdots + s_n^2)}}, \quad s_k = b_k - a_k. \tag{3.1.14}$$

Proof. The strategy is a variation on the Chernoff's method. Set

$$D_n := X_n - X_{n-1}, \quad S_n^2 := s_1^2 + \cdots + s_n^2, \quad \forall n \in \mathbb{N}.$$

We will prove inductively that

$$X_n - X_0 \in \mathbb{G}(S_n^2/4), \text{ i.e., } \mathbb{E}[e^{\lambda(X_n - X_0)}] \leq e^{\frac{\lambda S_n^2}{8}}, \quad \forall n \in \mathbb{N}, \ \lambda \in \mathbb{R}. \tag{3.1.15}$$

Assuming this, the inequality (3.1.14) follows from (2.3.11b).

To prove (3.1.15) note that since (X_n) is a martingale we have
$$\mathbb{E}\big[\,e^{\lambda(X_n-X_0)}\,\|\,\mathcal{F}_{n-1}\,\big] = e^{\lambda(X_{n-1}-X_0)}\mathbb{E}\big[\,e^{\lambda D_n}\,\|\,\mathcal{F}_{n-1}\,\big].$$

We set
$$Z_n(\lambda) := \mathbb{E}\big[\,e^{\lambda D_n}\,\|\,\mathcal{F}_{n-1}\,\big], \quad \forall n \in \mathbb{N}, \quad \lambda \in \mathbb{R}.$$

We claim that
$$\forall n \in \mathbb{N}, \quad \forall \lambda \in \mathbb{R}, \quad Z_n(\lambda) \leq e^{\frac{\lambda s_n^2}{8}} \quad \text{a.s.} \tag{3.1.16}$$

Obviously this implies that
$$\mathbb{E}\big[\,e^{\lambda(X_n-X_0)}\,\big] \leq e^{\frac{\lambda s_n^2}{8}}\mathbb{E}\big[\,e^{\lambda(X_{n-1}-X_0)}\,\big],$$

from which we can conclude inductively that $X_n - X_0 \in \mathbb{G}(S_n^2/4)$.

To prove (3.1.16) observe that, by construction, $Z_n(\lambda)$ is $\mathcal{F}_{n-1}$-measurable. We have to show that for any $S \in \mathcal{F}_{n-1}$ such that $\mathbb{P}[\,S\,] \neq 0$
$$\mathbb{E}\big[\,Z_n(\lambda)\boldsymbol{I}_S\,\big] \leq \mathbb{P}[\,S\,]e^{\frac{\lambda s_n^2}{8}}.$$

Denote by D_n^S the random variable $D_n\big|_S$ defined on the probability space $\big(S, \mathcal{F}_{n-1}\big|_S, \mathbb{P}_S\big)$, where
$$\mathbb{P}_S[\,A\,] = \mathbb{P}[\,A\,\|\,S\,] = \frac{\mathbb{P}[\,A\,]}{\mathbb{P}[\,S\,]}, \quad \forall A \in \mathcal{F}_{n-1}\big|_S.$$

We denote by $\mathbb{E}_S$ the expectation on $\big(S, \mathcal{F}_{n-1} \cap S, \mathbb{P}_S\big)$. Since $\mathbb{E}\big[\,D_n\,\|\,\mathcal{F}_{n-1}\,\big] = 0$ we deduce
$$\mathbb{E}_S\big[\,D_n^S\,\big] = 0.$$

Clearly $a_n \leq D_n^S \leq b_n$. We deduce from Hoeffding's Lemma (Proposition 2.56) that
$$\mathbb{E}_S\big[\,e^{\lambda D_n^S}\,\big] \leq e^{\frac{\lambda s_n^2}{8}},$$

and therefore, $\forall S \in \mathcal{F}_{n-1}$ such that, $\mathbb{P}[\,S\,] \neq 0$ we have
$$\mathbb{E}\big[\,Z_n(\lambda)\boldsymbol{I}_S\,\big] = \mathbb{E}\big[\,e^{\lambda D_n}\boldsymbol{I}_S\,\big] = \mathbb{P}[\,S\,]\mathbb{E}_S\big[\,e^{\lambda D_n^S}\,\big] \leq \mathbb{P}[\,S\,]e^{\frac{\lambda s_n^2}{2}}.$$

This concludes the proof of Azuma's inequality. □

The strength of Azuma's inequality is best appreciated in concrete examples.

Example 3.34 (Longest common subsequence). We want to have another look at the problem of the longest common subsequence first discussed in Example 1.150. Let us briefly recall the set-up.

We are given a finite set (alphabet) $\mathcal{A}$, $|\mathcal{A}| = k$, and a family of independent $\mathcal{A}$-valued random variables
$$\big\{\,X_n, Y_n;\ m, n \in \mathbb{N}\,\big\}$$

all with the same distribution π. We denote by L_n the length of the longest common subsequence of two random words

$$(X_1,\ldots,X_n) \text{ and } (Y_1,\ldots,Y_n).$$

We set

$$R_n := \frac{1}{n}L_n, \quad R := \sup_n R_n.$$

In Example 1.150 we have shown

$$\frac{1}{n}L_n \to R \quad \text{a.s.},$$

and

$$\lim_{n\to\infty} \mathbb{E}[R_n] = r(\pi) := \mathbb{E}[R].$$

We will to show that R_n is highly concentrated around its mean r_n. We follow the presentation in [144, Sec. 1.3].

Set $\ell_n := \mathbb{E}[L_n]$, $Z_n = (X_n, Y_n)$. Consider the finite filtration

$$\mathcal{F}_0 := \sigma(\emptyset), \quad \mathcal{F}_j = \sigma(Z_1,\ldots,Z_j), \quad j=1,\ldots,n.$$

Form the Doob (closed) martingale $U_j := \mathbb{E}[L_n \| \mathcal{F}_j]$. Note that $U_0 = \ell_n$. The random variable L_n is a function of the Z_j's

$$L_n = L_n(Z_1,\ldots,Z_n),$$

and U_j is a function of $Z_1,\ldots,Z_j$, $U_j = F_j(Z_1,\ldots,Z_j)$. More precisely,

$$F_j(z_1,\ldots,z_j) = \mathbb{E}[L_n[Z_1=z_1,\ldots,Z_j=z_j]$$

$$= \mathbb{E}[L_n(z_1,\ldots,z_j,Z_{j+1},\ldots,Z_n)]$$

$$= \int_{(\mathcal{A}^2)^{n-j}} L_n(z_1,\ldots z_j, z_{j+1},\ldots,z_n) \pi^{\otimes 2(n-j)}[dz_{j+1}\cdots dz_n].$$

Note that for any $z_1,\ldots,z_{j-1},z_j,z_j',z_{j+1},\ldots,z_n \in \mathcal{A}^2$ we have

$$-1 \leq L_n(z_1,\ldots,z_{j-1},z_j',z_{j+1},\ldots,z_n) - L_n(z_1,\ldots,z_{j-1},z_j,z_{i+1},\ldots,z_n) \leq 1.$$

Integrating with respect to $z_j', z_{j+1},\ldots,z_n$ we deduce

$$-1 \leq F_{j-1}(z_1,\ldots,z_{j-1}) - F_j(z_1,\ldots,z_n) \leq 1.$$

Hence $|U_j - U_{j-1}| \leq 1$. From Azuma's inequality with $s_n = 2$ we deduce

$$\mathbb{P}[|L_n - \ell_n| \geq nx] \leq 2e^{-\frac{nx^2}{2}},$$

so that

$$\mathbb{P}[|R_n - r_n| \geq x] \leq 2e^{-\frac{nx^2}{2}}.$$

This proves that R_n is highly concentrated around its mean. Obviously

$$\forall \varepsilon > 0, \quad \sum_{n \geq 1} \mathbb{P}\big[\,|R_n - r_n| \geq \varepsilon\,\big] < \infty,$$

and Corollary 1.141 implies that $R_n - r_n \to 0$ a.s.

On the other hand, we know from Example 1.150 that

$$R_n \to R \text{ a.s. and } r_n \to r(\pi) = \mathbb{E}\big[\,R\,\big].$$

Hence $\frac{1}{n}L_n$ converges almost surely to a constant $r(\pi)$.

We write $r(k)$ instead of $r(\pi)$ when π is the uniform distribution on an alphabet of cardinality k. In this case one has additional information about the rate of convergence of r_n to $r(k)$. However, the exact value of $r(k)$ remains illusive, even for small k. □

Example 3.35 (Bin packing). The bin packing problem has a short formulation: pack n items of sizes $x_1, \ldots, x_n \in [0, 1]$ in as few bins of maximum capacity 1 each. We denote by $B_n(x_1, \ldots, x_n)$ the lowest numbers of bins we can use to pack the items of sizes $x_1, \ldots, x_n$.

As in the case of the longest common subsequence problem, the bin packing problem has a probabilistic counterpart. Consider independent random variables $X_n \sim \text{Unif}([0,1])$, $n \in \mathbb{N}$ defined on a probability space $(\Omega, \mathcal{S}, \mathbb{P})$. We will describe the behavior of $b_m := \mathbb{E}\big[\,B_n(X_1, \ldots, X_n)\,\big]$ as $n \to \infty$.

Note that

$$X_1 + \cdots + X_n \leq B_n(X_1, \ldots, X_n) \leq n.$$

By taking expectations we deduce

$$\frac{n}{2} \leq b_n \leq n, \quad \forall n \in \mathbb{N}, \tag{3.1.17}$$

showing that b_n has linear growth as $n \to \infty$. On the other hand,

$$\begin{aligned}&B_{n+m}(X_1, \ldots, X_n, X_{n+1}, \cdots X_{n+m}) \\ &\leq B_n(X_1, \ldots, X_n) + B_m(X_{n+1}, \cdots X_{n+m}),\end{aligned} \tag{3.1.18}$$

and thus

$$b_{n+m} \leq b_n + b_m, \quad \forall n, m \in \mathbb{N}.$$

Setting $r_n := \frac{b_n}{n}$, we deduce from Fekete's Lemma 1.151 that

$$\lim_{n \to \infty} r_n = r := \inf_n r_n.$$

The inequalities (3.1.17) show that $r \in \big[\,\frac{1}{2}, 1\,\big]$.

We set $R_n := \frac{B_n}{n}$. We deduce from (3.1.18) and Fekete's Lemma that

$$R_n \to R := \inf_n R_n \text{ a.s. and } r = \mathbb{E}\big[\,R\,\big].$$

We want to show that R_n is highly concentrated around its mean. We use the same approach as in Example 3.34.

We set
$$\mathcal{F}_j = \sigma(X_1,\ldots,X_j), \quad \mathcal{F}_0 = \{\emptyset, \Omega\}.$$
Fix $n \in \mathbb{N}$. For $j = 0, 1, \ldots, n$ we set
$$U_j = U_{n,j} := \mathbb{E}[\,B_n \,\|\, \mathcal{F}_j\,]$$
so the collection $(U_j)_{0 \leq j \leq n}$ is a martingale adapted to the filtration $(\mathcal{F}_j)_{0 \leq j \leq n}$.

There exist Borel measurable maps $F_j : [0,1]^j \to \mathbb{N}$ such that $U_j = F_j(X_1,\ldots,X_j)$. More precisely,
$$F_j(x_1,\ldots,x_j) = \int_{[0,1]^{n-j}} B_n(x_1,\ldots,x_j, x_{j+1},\ldots x_n)\, dx_{j+1}\cdots dx_n.$$
For any $x_1,\ldots,x_{j-1}, x_j, x_j', x_{j+1},\ldots,x_n \in [0,1]$ we have
$$-1 \leq B_n(x_1,\ldots,x_{j-1}, x_j', x_{j+1},\ldots,x_n) - B_n(x_1,\ldots,x_{j-1}, x_j, x_{j+1},\ldots,x_n) \leq 1.$$
Integrating with respect to $x_j', x_{j+1},\ldots,x_n$ we deduce $|U_j - U_{j-1}| \leq 1$. Invoking Azuma's inequality we deduce as in Example 3.34 that
$$\mathbb{P}[\,|R_n - r_n| > x\,] \leq 2e^{-\frac{nx^2}{2}}.$$
This shows that R_n is highly concentrated around its mean and that $R_n \to r$ a.s.

In this case it is known that $r = \frac{1}{2}$. More precisely, there is an algorithm called MATCH which takes as input the sizes $x_1, \ldots, x_n$ of the n items and packs them into $M_n = M_n(x_1, \ldots, x_n)$ boxes where
$$\frac{n}{2} \leq \mathbb{E}[\,B_n\,] \leq \mathbb{E}[\,M_n\,] \leq \frac{n}{2} + O(\sqrt{n}).$$
This is the best one can hope for since it is also known
$$\mathbb{E}[\,B_n\,] \geq \frac{n}{2} + (\sqrt{3} - 1)\sqrt{\frac{n}{24\pi}} + o(\sqrt{n}).$$
For details we refer to [34, Sec. 5.1]. $\square$

The tricks used in the above examples are generalized and refined in McDiarmid's inequality.

Definition 3.36 (Bounded difference property). Suppose that S is a set. A function $f : S^n \to \mathbb{R}$, $n \in \mathbb{N}$, is said to satisfy the *bounded difference property* if there exist $L_1, \ldots, L_n > 0$ such that,
$$\big|f(s_1,\ldots,s_{k-1}, s, s_{k+1},\ldots, s_n) - f(s_1,\ldots,s_{k-1}, s', s_{k+1},\ldots, s_n)\big| \leq L_k, \quad (3.1.19)$$
$\forall k = 1, \ldots, n$, $\forall s_1, \ldots, s_{k-1}, s, s', s_{k+1}, \ldots, s_n \in S$. $\square$

Let us observe that the above condition is satisfied if and only if f is Lipschitz with respect to the *Hamming distance* on S

$$d_H : S^n \times S^n \to [0, \infty), \quad d_H(\underline{s}, \underline{t}) := \sum_{k=1}^n \boldsymbol{I}_{\mathbb{R} \setminus \{0\}}(s_k - t_k). \tag{3.1.20}$$

Theorem 3.37 (McDiarmid's inequality).
Suppose that $X_1, \ldots, X_n : (\Omega, \mathcal{S}, \mathbb{P}) \to \mathbb{R}$ are independent random variables and $f : \mathbb{R}^n \to \mathbb{R}$ satisfies the bounded difference property with constants $L_1, \ldots, L_n$. If $Z = f(X_1, \ldots, X_n)$ is integrable, then

$$\mathbb{P}\big[\,|Z - \mathbb{E}[Z]| > t\,\big] \leq e^{-2t^2/L^2}, \quad L^2 = L_1^2 + \cdots + L_n^2. \tag{3.1.21}$$

Proof. Denote by $\mathbb{P}_k$ the distribution on X_k. Let $\mathcal{F}_0 = \{\emptyset, \Omega\}$, $\mathcal{F}_k := \sigma(X_1, \ldots, X_k)$ and set

$$Z_k := \mathbb{E}[Z \,\|\, \mathcal{F}_k], \quad k = 0, \ldots, n$$

so that $Z_n = Z$ and $Z_0 = \mathbb{E}[Z]$. Since $X_1, \ldots, X_n$ are independent we deduce that $\forall \omega \in \Omega$

$$Z_k(\omega) = g_k\big(X_1(\omega), \ldots, X_k(\omega)\big),$$

where

$$g_k(x_1, \ldots, x_k) = \int_{\mathbb{R}^{n-k}} f(x_1, \ldots, x_k, x_{k+1}, \ldots, x_n) \mathbb{P}_{k+1}[dx_{k+1}] \cdots \mathbb{P}_n[dx_n].$$

Note that

$$g_{k-1}(x_1, \ldots, x_{k-1}) = \mathbb{E}_k[g_k] := \int_{\mathbb{R}} g_k(x_1, \ldots, x_{k-1}, x_k) \mathbb{P}_k[dx_k].$$

Hence

$$D_k = g_k - \mathbb{E}_k g_k, \quad \mathbb{E}\big[e^{\lambda D_k} \,\|\, \mathcal{F}_{k-1}\big] = h_{k-1}(X_1, \ldots, X_{k-1}),$$

where $h_{k-1}(x_1, \ldots, x_{k-1}) := \mathbb{E}_k\big[e^{\lambda(g_k - \mathbb{E}_k[g_k])}\big]$. Fix $x_1, \ldots, x_{k-1}$ and set

$$a_k = a_k(x_1, \ldots, x_{k-1}) := \inf_{x_k} g(x_1, \ldots, x_{k-1}, x_k),$$

$$b_k = b_k(x_1, \ldots, x_{k-1}) := \sup_{y_k} g(x_1, \ldots, x_{k-1}, y_k).$$

We deduce that

$$0 \leq b_k - a_k \leq \sup_{x_k, y_k} \big|g(x_1, \ldots, x_{k-1}, y_k) - g(x_1, \ldots, x_{k-1}, x_k)\big| \leq L_k.$$

We deduce from Hoeffding's inequality (2.3.13) that

$$\mathbb{E}_k\big[e^{\lambda(g_k - \mathbb{E}_k g_k)}\big] \leq e^{\frac{\lambda L_k^2}{8}}, \quad \forall x_1, \ldots, x_{k-1}.$$

Hence

$$\mathbb{E}\big[e^{\lambda D_k} \,\|\, \mathcal{F}_{k-1}\big] \leq e^{\frac{\lambda L_k^2}{8}} \text{ a.s., i.e., } \mathbb{E}\big[e^{\lambda(Z_n - Z_0)}\big] \leq e^{\frac{\lambda(L_1^2 + \cdots + L_n^2)}{8}}.$$

$\square$

In Exercise 3.20 we outline an important application of McDiarmid's inequality.

3.2 Limit theorems: discrete time

We have seen in the previous section how the Optional Stopping Theorem combined with quite a bit of ingenuity can produce miraculous results. This section is devoted to another miraculous property of martingales, namely, their rather nice asymptotic behavior. The foundational results in this section are all due to J. L. Doob. To convince the reader of the amazing versatility of martingales we have included a large eclectic collection of concrete applications.

3.2.1 *Almost sure convergence*

Fix a probability space $(\Omega, \mathcal{S}, \mathbb{P})$ and a $\mathbb{N}_0$-filtration $\mathcal{F}_\bullet$ of $\mathcal{F}$. We will investigate the behavior of an $\mathcal{F}_\bullet$-submartingale $(X_n)_{n \in \mathbb{N}_0}$ as $n \to \infty$. The key to this investigation is *Doob's upcrossing inequality*.

Given real numbers $a < b$ and a sequence of real numbers $\alpha = (\alpha_n)_{n \geq 0}$ we define inductively the sequences

$$\bigl(S_k(\alpha) = S_k(\alpha; a, b)\bigr)_{k \geq 1} \text{ and } \bigl(T_k(\alpha) = T_k(\alpha; a, b)\bigr)_{k \geq 1}$$

in $\mathbb{N}_0 \cup \{\infty\}$ as follows. We set

$$S_1(\alpha) := \inf_{n \geq 0}\{\alpha_n \leq a\}, \quad T_1(\alpha) := \inf_{n \geq S_1}\{\alpha_n \geq b\}.$$

Thus, S_1 is the first moment the sequence α drops below a, and T_1 is the first moment after S_1 when the sequence α crosses the upper level b. We then define inductively

$$S_{k+1}(\alpha) := \inf_{n \geq T_k}\{\alpha_n \leq a\}, \quad T_{k+1}(\alpha) := \inf_{n \geq S_k}\{\alpha_n \geq b\},$$

where we set $\inf \emptyset = \infty$; see Figure 3.2.

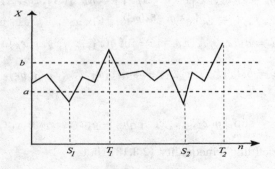

Fig. 3.2 *Up/downcrosssing of the interval $[a, b]$.*

The terms S_k are called *downcrossing times* while the terms T_k are called the *upcrossing times* of the sequence $(\alpha_n)_{n \geq 0}$. We define the *upcrossing numbers*

$$\begin{aligned} N_n\bigl([a,b], \alpha\bigr) &:= \#\{k \in \mathbb{N};\ T_k(\alpha) \leq n\}, \quad n \in \mathbb{N}, \\ N_\infty\bigl([a,b], \alpha\bigr) &:= \lim_{n \to \infty} N_n([a,b], \alpha). \end{aligned} \quad (3.2.1)$$

The importance of the upcrossing numbers in convergence problems is explained by the following elementary but rather clever result. In Exercise 3.8 we ask the reader to provide a proof.

Lemma 3.38. *Suppose that $\alpha = (\alpha_n)_{n \geq 0}$ is a sequence of real numbers. Then the following statements are equivalent.*

(i) The sequence α has a limit (possibly infinite).
(ii) For any rational numbers $a < b$ the total number of upcrossings $N_\infty([a,b], \alpha)$ is finite. □

Suppose now that $(X_n)_{n \in \mathbb{N}_0}$ is a process adapted to the filtration $\mathcal{F}_\bullet$. Then, for any $k \in \mathbb{N}$, the down/up-crossing times $S_k(X)$ and $T_k(X)$ are stopping times.

Theorem 3.39 (Doob's upcrossing inequality). *Assume that $X = (X_n)_{n \in \mathbb{N}_0}$ is a submartingale. Then for any real numbers $a < b$ we have*

$$(b-a)\mathbb{E}\big[\, N_n([a,b], X)\,\big] \leq \mathbb{E}\big[\,(X_n - a)^+\,\big] - \mathbb{E}\big[\,(X_0 - a)^+\,\big], \quad x^+ := \max(x, 0). \quad (3.2.2)$$

Proof. Since $(X - a)^+$ is a submartingale and

$$N_n\big([a,b], X\big) = N_n\big([0, b-a], (X-a)^+\big),$$

we see that it suffices to prove the result in the special case $X \geq 0$ and $a = 0 < b$. In other words, it suffices to prove that if $X \geq 0$, then

$$b\mathbb{E}\big[\, N_n\big([0,b], X\big)\,\big] \leq \mathbb{E}\big[\,(X_n - X_0)\,\big]. \quad (3.2.3)$$

The key fact underlying this inequality is the existence of a submartingale Y that lies above the random process $N_n\big(([0,b], X\big)$ and, in the mean, below the process X.

Consider the predictable process

$$H = \sum_{k=1}^{\infty} \boldsymbol{I}_{]]S_k(X), T_k(X)]]},$$

i.e.,

$$H_n = \sum_{k=1}^{\infty} \boldsymbol{I}_{\{S_k(X) < n \leq T_k(X)\}}.$$

Since the intervals

$$\big(S_1(X), T_1(X)\big], \; \big(S_2(X), T_2(X)\big], \ldots,$$

are pairwise disjoint (when finite) we have $H_n \leq 1$. Set $Y_n := (H \cdot X)_n$.

In stock market terms, think of the following investing strategy. Start buying a stock when its price cost hits zero, and sell it at the end the trading day. Continue buying (and selling) the stock as long as its price at the start of the trading day is below b. Once the price crosses b stop buying and wait until the price hits 0 again. The price of the stock at the beginning of the n-th trading day is X_{n-1} and changes to X_n at the end of the n-th trading day. Then Y_n is the profit following this strategy at the end of n days. Clearly the profit will be at least as big as $b\times$ the number of upcrossings of the interval $(0, b)$. This is the content of the following fundamental inequality.

$$\boxed{Y_n \geq bN_n\bigl([0,b], X\bigr)}. \qquad (3.2.4)$$

Here is a formal proof of this inequality. Let $M := N_n\bigl([0,b], X\bigr)$. Then

$$Y_n = \sum_{j=1}^n H_j \cdot (X_j - X_{j-1}) = \sum_{k=1}^M \sum_{j=S_k(X)+1}^{T_k(X)} (X_j - X_{j-1}) + \sum_{j=S_{M+1}+1}^n (X_j - X_{j-1})$$

$$= \sum_{k=1}^M (X_{T_k} - X_{S_k}) + I_{\{S_{M+1} < n\}}(X_n - X_{S_{M+1}})$$

(use the fact that $X_{S_{M+1}} = 0$ and $X_n \geq 0$)

$$\geq \sum_{k=1}^M \underbrace{(X_{T_k} - X_{S_k})}_{\geq b} \geq bM = bN_n([0,b], X).$$

Hence

$$b\mathbb{E}\bigl[N_n([0,b], X)\bigr] \leq \mathbb{E}[Y_n], \quad \forall n \in \mathbb{N}.$$

Note that the inequality (3.2.4) does not rely on the fact that X is a submartingale.

The process (H_n) is predictable and thus

$$\mathbb{E}\bigl[Y_k - Y_{k-1} \| \mathcal{F}_{k-1}\bigr] = \mathbb{E}\bigl[H_k(X_k - X_{k-1}) \| \mathcal{F}_{k-1}\bigr]$$

$$= H_k \mathbb{E}\bigl[(X_k - X_{k-1}) \| \mathcal{F}_{k-1}\bigr].$$

Since X is a *submartingale* we deduce

$$\mathbb{E}\bigl[(X_k - X_{k-1}) \| \mathcal{F}_{k-1}\bigr] \geq 0.$$

On the other hand $H_k \leq 1$ so that

$$H_k \mathbb{E}\bigl[(X_k - X_{k-1}) \| \mathcal{F}_{k-1}\bigr] \leq \mathbb{E}\bigl[(X_k - X_{k-1}) \| \mathcal{F}_{k-1}\bigr].$$

Hence

$$\mathbb{E}[Y_k - Y_{k-1}] \leq \mathbb{E}[X_k - X_{k-1}]. \qquad (3.2.5)$$

We deduce

$$b\mathbb{E}\bigl[N_n([0,b], X)\bigr] \leq \mathbb{E}[Y_n] = \sum_{k=1}^n \mathbb{E}[Y_k - Y_{k-1}] \leq \mathbb{E}[X_n - X_0].$$

$\square$

Remark 3.40. We should ponder why the inequality (3.2.5) is miraculous. We know that $H_k \in [0,1]$ so, whenever $X_k \leq X_{k-1}$, i.e., the price of stock goes down, we have $X_k - X_{k-1} \leq H_k(X_k - X_{k-1}) = Y_k - Y_{k-1}$. The inequality (3.2.5) shows that *this is not the expected behavior*. The fact that $X_\bullet$ is a *submartingale* biases the price in favor of increase. That is the reason why (3.2.5) holds. □

Theorem 3.41 (Submartingale Convergence Theorem). *Suppose that $(X_n)_{n \in \mathbb{N}_0}$ is a submartingale satisfying*

$$\sup_{n \in \mathbb{N}_0} \mathbb{E}\big[\, X_n^+ \,\big] < \infty. \tag{3.2.6}$$

Then X_n converges almost surely to an integrable *random variable X_∞.*

Remark 3.42. Observe that since X_n is a submartingale we have

$$\mathbb{E}\big[\,X_0\,\big] \leq \mathbb{E}\big[\,X_n\,\big] = \mathbb{E}\big[\,X_n^+\,\big] - \mathbb{E}\big[\,X_n^-\,\big], \quad x^- = \max(-x,0),$$

so that

$$\sup_{n \in \mathbb{N}_0} \mathbb{E}\big[\,X_n^-\,\big] < \infty$$

showing that (3.2.6) is equivalent to

$$\sup_{n \in \mathbb{N}_0} \mathbb{E}\big[\,|X_n|\,\big] < \infty. \tag{3.2.7}$$

□

Proof. Set

$$M := \sup_{n \in \mathbb{N}_0} \mathbb{E}\big[\,|X_n|\,\big].$$

Now let $a, b \in \mathbb{Q}$, $a < b$. Doob's upcrossing inequality shows that, for all $n \geq 1$, we have

$$(b-a)\mathbb{E}\big[\,N_n(a,b,X_\bullet)\,\big] \leq \mathbb{E}\big[\,(X_n - a)^+\,\big] \leq |a| + \mathbb{E}\big[\,|X_n|\,\big] \leq |a| + M.$$

Letting $n \to \infty$ we deduce $\mathbb{E}\big[\,N_\infty([a,b],X_\bullet)\,\big] < \infty$, and thus $N_\infty([a,b],X_\bullet) < \infty$ a.s. By removing a countable family of negligible sets (one for each pair of rational numbers a,b, $a < b$) we deduce that there exists a negligible set $\mathcal{N} \subset \Omega$ such that $\forall \omega \in \Omega \setminus \mathcal{N}$ we have

$$N_\infty\big([a,b], X_\bullet(\omega)\big) < \infty, \quad \forall a, b \in \mathbb{Q}, \quad a < b.$$

Lemma 3.38 implies that the sequence $X_\bullet$ converges a.s. to a random variable X_∞. The integrability of X_∞ follows from Fatou's lemma

$$\mathbb{E}\big[\,|X_\infty|\,\big] \leq \liminf_{n \to \infty} \mathbb{E}\big[\,|X_n|\,\big] < \infty.$$

□

Corollary 3.43. *Suppose that $(X_n)_{n \in \mathbb{N}_0}$ is a nonnegative supermartingale. Then X_n converges a.s. to an* integrable *random variable X_∞.*

Proof. Observe that $Y_n = -X_n$ is a submartingale and $Y_n^+ = 0$. The result now follows from the Submartingale Convergence Theorem. □

Corollary 3.44. *Suppose that $(X_n)_{n \in \mathbb{N}_0}$ is a submartingale adapted to the filtration $(\mathcal{F}_n)_{n \in \mathbb{N}_0}$ and T is an a.s. finite stopping time. If $\sup_n \mathbb{E}[|X_n|] < \infty$, then*

$$\lim_{n \to \infty} X_n^T = \lim_{n \to \infty} X_{n \wedge T} = X_T \text{ a.s.}$$

Proof. Note that $(X_\bullet^+)^T = (X_\bullet^T)^+$ so $(X_\bullet^T)^+$ is a submartingale. The Optional Sampling Theorem applied to the bounded stopping times $n \wedge T \leq n$ implies

$$\mathbb{E}[X_{n \wedge T}^+] \leq \mathbb{E}[X_n^+]$$

so that

$$\sup_n \mathbb{E}[X_{n \wedge T}^+] < \infty.$$

The conclusion now follows from the Submartingale Convergence Theorem. □

Example 3.45 (Galton-Watson/branching processes). [2] Consider again the branching process in Example 3.8 with reproduction law $\mu \in \text{Prob}(\mathbb{N}_0)$ and mean m,

$$0 < m := \sum_{n \geq 0} n \mu[n] < \infty.$$

As explained in Example 3.8, the sequence

$$W_n = \frac{1}{m^n} Z_n, \ n \in \mathbb{N}_0$$

is a nonnegative martingale so, according to Corollary 3.43, it converges a.s. to an integrable random variable W_∞.

If $m < 1$, the original sequence $Z_n = m^n W_m$ converges a.s. and in mean to 0. Moreover

$$\mathbb{E}[Z_n] = m^n \mathbb{E}[Z_0] = m^n \ell.$$

Thus, the expected population decays *exponentially* to zero. Something more dramatic holds.

Since $Z_n \geq 1$ if $Z_n > 0$ we deduce

$$\mathbb{P}[Z_n > 0] = \mathbb{P}[Z_n \geq 1] \leq \mathbb{E}[Z_n] = \ell m^n.$$

Hence

$$\sum_{n \geq 0} \mathbb{P}[Z_n > 0] < \infty.$$

The Borell-Cantelli Lemma implies that $\mathbb{P}[Z_n > 0 \text{ i.o.}] = 0$.

---

[2]**To the post pandemic reader**. I wrote most of this book during the great covid pandemic. I even taught this example to a group of masked students that were numbed by the news about the R-factor. The mean m is a close relative of this R-factor. This example explains the desirability of $R < 1$.

Thus, a population of bacteria that have on average less that one successor will die out, i.e., with probability 1 there exists $n \in \mathbb{N}$ such that $Z_n = 0$. If we set

$$E_n := \{ Z_k = 0, \ \forall k \geq n \} = \{ Z_n = 0 \},$$

then the event

$$E = \bigcup_{n \geq 0} E_n$$

is called the *extinction event*. Note that

$$E_0 \subset E_1 \subset \cdots \subset E_n \subset \cdots.$$

The probability of E is called *extinction probability*. We see that when $m < 1$, the extinction probability is 1. □

Remark 3.46. It is known that a random series with *independent* terms converges a.s. if and only if it converges in probability; see Exercise 2.2. However, there exist martingales that converge in probability, but not a.s.

Here is one such example, [53, Example 4.2.14]. Consider the following random walk $(X_n)_{n \geq 0}$ on $\mathbb{Z}$ where you should think of X_n as the location at time n. We set $X_0 = 0$. If X_{n-1} is known, then

$$\mathbb{P}[X_n = \pm 1 \,\|\, X_{n-1} = 0] = \frac{1}{2n}, \quad \mathbb{P}[X_n = 0 \,\|\, X_{n-1} = 0] = 1 - \frac{1}{n},$$

$$\mathbb{P}[X_n = 0 \,\|\, X_{n-1} = x \neq 0] = 1 - \frac{1}{n}, \quad \mathbb{P}[X_n = nx \,\|\, X_{n-1} = x \neq 0] = \frac{1}{n}.$$

The existence of such a process is guaranteed by Kolmogorov's theorem.

Denote by $\mathcal{F}_n$ the sigma-algebra generated by the random variables $X_0, X_1, \ldots, X_n$. From the construction we deduce that $\mathbb{E}[X_n \,\|\, X_{n-1}] = X_{n-1}$ so (X_n) is a martingale with respect to the filtration $\mathcal{F}_n$. Let $p_n := \mathbb{P}[X_n \neq 0]$. Note that

$$p_n = \mathbb{P}[X_n \neq 0 \,|\, X_{n-1} = 0]\mathbb{P}[X_{n-1} = 0] + \mathbb{P}[X_n \neq 0 \,|\, X_{n-1} \neq 0]\mathbb{P}[X_{n-1} \neq 0]$$

$$= \frac{1}{n}(1 - p_{n-1}) + \frac{1}{n}p_{n-1} = \frac{1}{n}.$$

Hence

$$\lim_{n \to \infty} \mathbb{P}[X_n \neq 0] = 0,$$

so that X_n converges in probability to 0. To show it does not converge a.s. it suffices to show that it does not converge a.s. to 0.

Denote by F_n the event $\{X_n \neq 0\}$. The random variables X_n have integer values so $F_n = \{|X_n| \geq 1\}$. Note that $F_n \in \mathcal{F}_n$ and

$$\mathbb{E}[\boldsymbol{I}_{F_n} \,\|\, \mathcal{F}_{n-1}] = \frac{1}{n}\boldsymbol{I}_{\{X_{n-1}=0\}} + \frac{1}{n}\boldsymbol{I}_{\{X_{n-1} \neq 0\}} = \frac{1}{n}.$$

Hence
$$\sum_{n\geq 1}\mathbb{E}\bigl[\,I_{F_n}\,\|\,\mathcal{F}_{n-1}\,\bigr] = \sum_{n\geq 1}\frac{1}{n} = \infty.$$
The conditional Borel-Cantelli result in Exercise 3.12 implies that
$$\mathbb{P}\bigl[\,|X_n|\geq 1 \text{ i.o.}\,\bigr] = \mathbb{P}\bigl[\,F_n \text{ i.o.}\,\bigr] = 1.$$
Thus (X_n) does not converge a.s. to 0.

Recently (2021) Iosif Pinelis gave another beautiful example of martingale converging in probability but not a.s. Here is briefly the construction.

Choose a sequence of independent geometric random variables
$$(T_n)_{n\geq 1}, \quad T_n \sim \text{Geom}(p_n).$$
We perform the following delayed and frequently stopped random walk on $\mathbb{Z}$. We start at $X_0 = 0$ and we wait for T_1 moments and we begin a standard random walk on $\mathbb{Z}$ until we first return to the origin. At that moment take a brake lasting T_2 moments and begin the standard walk until we return back to the origin etc. Denote by X_n the location after n moments. Then (X_n) is a martingale (with respect to an appropriate filtration). Moreover, if
$$\sum_{n\geq 1}\sqrt{p_n} < \infty,$$
then X_n converges in probability to 0 but not a.s. For details we refer to [128]. □

Example 3.47. The assumptions in the (sub)martingale convergence theorem are not strong enough to guarantee L^1-convergence. The following example shows what can happen.

Consider the standard random walk on $\mathbb{Z}$ that starts at 1. Each second the traveler takes a size 1 step forward or back with equal probability. More precisely, consider a sequence of i.i.d. Rademacher random variables $(X_n)_{n\in\mathbb{N}}$, $\mathbb{P}\bigl[\,X_1 = 1\,\bigr] = \mathbb{P}\bigl[\,X_n = -1\,\bigr] = \frac{1}{2}$. Then the sequence
$$S_0 = 1, \quad S_n = 1 + X_1 + \cdots + X_n, \quad n\in\mathbb{N}$$
is a martingale describing the evolution of the walk. Denote by N the first moment the walk reaches the origin, i.e.,
$$N := \inf\bigl\{\,n\in\mathbb{N};\ S_n = 0\,\bigr\}.$$
Observe that $N < \infty$ a.s.; see Exercise 3.13. Consider the random walk stopped at N
$$Y_n := S_n^N = S_{n\wedge N}.$$
From the Optional Stopping Theorem 3.24 we deduce that Y_n is a martingale which, by construction, is also nonnegative. Clearly $Y_n \to 0$ a.s. since $N < \infty$ a.s. This convergence is not L^1 since
$$\mathbb{E}\bigl[\,Y_n\,\bigr] = \mathbb{E}\bigl[\,Y_0\,\bigr] = 1, \quad \forall n\in\mathbb{N}.$$
□

3.2.2 Uniform integrability

We will describe in this subsection necessary and sufficient conditions guaranteeing that a sequence that converges in probability also converges in p-mean.

We begin with a basic fact.

Lemma 3.48. *Let $X \in L^1(\Omega, \mathcal{S}, \mathbb{P})$. Then*
$$\lim_{n \to \infty} \mathbb{E}\big[\, |X| \, \boldsymbol{I}_{\{|X| \geq n\}} \,\big] = 0.$$

Proof. The sequence $Z_n := |X| \boldsymbol{I}_{\{|X|>n\}}$ converges a.s. to 0 and $|Z_n| \leq |X|$, $\forall n$. The desired conclusion now follows from the Dominated Convergence theorem. $\square$

Definition 3.49 (Uniform integrability). A collection $\mathscr{X} \subset L^1(\Omega, \mathcal{S}, \mathbb{P})$ is called *uniformly integrable* (or UI for brevity) if
$$\lim_{r \to \infty} \mathbb{E}\big[\, |X| \, \boldsymbol{I}_{\{|X| \geq r\}} \,\big] = 0 \;\; \underline{\text{uniformly}} \text{ in } X \in \mathscr{X}. \tag{$\mathbf{UI_1}$}$$
$\square$

Remark 3.50. (a) Let $\mathscr{X} \subset L^1(\Omega, \mathcal{S}, \mathbb{P})$. Set
$$\chi(r) = \chi(r, \mathscr{X}) := \sup_{X \in \mathscr{X}} \mathbb{E}\big[\, |X| \boldsymbol{I}_{\{|X| \geq r\}} \,\big].$$
Then $\mathscr{X}$ is uniformly integrable iff $\lim_{r \to \infty} \chi(r) = 0$.

(b) A uniformly integrable family $\mathscr{X} \subset L^1(\Omega, \mathcal{S}, \mathbb{P})$ is bounded in the L^1-norm, i.e., $\chi(0) < \infty$. Indeed, $\forall X \in \mathscr{X}$ and r is sufficiently large so that $\chi(r) < 1$, we have
$$\mathbb{E}\big[\, |X| \,\big] = \mathbb{E}\big[\, |X| \boldsymbol{I}_{\{|X| < r\}} \,\big] + \mathbb{E}\big[\, |X| \boldsymbol{I}_{\{|X| \geq r\}} \,\big] \leq r + \chi(r) < \infty.$$
$\square$

Theorem 3.51. *Let $\mathscr{X} \subset L^1(\Omega, \mathcal{S}, \mathbb{P})$. Then the following statements are equivalent.*

(i) $\mathscr{X}$ is uniformly integrable.

(ii) The family $\mathscr{X}$ is L^1-bounded and, for any $\varepsilon > 0$, there exists $\delta = \delta(\varepsilon) > 0$ such that, $\forall X \in \mathscr{X}$ and any $S \in \mathcal{S}$, we have
$$\mathbb{P}[S] \leq \delta \Rightarrow \mathbb{E}\big[\, |X| \boldsymbol{I}_S \,\big] = \int_S |X(\omega)| \, \mathbb{P}[d\omega] < \varepsilon. \tag{$\mathbf{UI_2}$}$$

Proof. (i) $\Rightarrow$ (ii) Fix $\varepsilon > 0$. There exists $r_\varepsilon > 0$ such that $\chi(r_\varepsilon) < \varepsilon/2$. Now fix $\delta > 0$ such that $\delta r_\varepsilon < \frac{\varepsilon}{2}$. Then, for any $X \in \mathscr{X}$ and any $S \in \mathcal{S}$ such that $\mathbb{P}[S] < \delta$, we have
$$\mathbb{E}\big[\, |X| \boldsymbol{I}_S \,\big] = \mathbb{E}\big[\, |X| \boldsymbol{I}_{S \cap \{|X| < r_\varepsilon\}} \,\big] + \mathbb{E}\big[\, |X| \boldsymbol{I}_{S \cap \{|X| \geq r_\varepsilon\}} \,\big]$$
$$\leq r_\varepsilon \mathbb{P}[S] + \mathbb{E}\big[\, |X| \boldsymbol{I}_{\{|X| \geq r_\varepsilon\}} \,\big] \leq \delta r_\varepsilon + \chi(r_\varepsilon) < \varepsilon.$$

(ii) ⇒ (i) Set
$$B := \sup_{X \in \mathscr{X}} \mathbb{E}\big[|X|\big] < \infty.$$
Markov's inequality implies that for $r > 0$ we have
$$\mathbb{P}\big[|X| > r\big] \leq \frac{B}{r}, \quad \forall X \in \mathscr{X}.$$
Fix $\varepsilon > 0$ and $r_\varepsilon > 0$ such that $\frac{B}{r_\varepsilon} < \delta(\varepsilon)$. Then $\mathbb{P}\big[|X| > r_\varepsilon\big] < \delta(\varepsilon), \forall X \in \mathscr{X}$. Assumption (ii) implies
$$\chi(r_\varepsilon) = \sup_{X \in \mathscr{X}} \mathbb{E}\big[|X| \, \boldsymbol{I}_{\{|X| > r_\varepsilon\}}\big] < \varepsilon.$$
□

Remark 3.52. We should draw attention to the qualitatively different conditions $(\mathbf{UI}_1)$ and $(\mathbf{UI}_2)$. Condition $(\mathbf{UI}_1)$ involves only the probability distributions of the random variables $X \in \mathscr{X}$ with no mention of the probability space on which they are defined, whereas condition $(\mathbf{UI}_2)$ makes explicit reference to their domain of definition $(\Omega, \mathcal{S}, \mathbb{P})$. It is related to the *absolute continuity* condition in real analysis. □

Corollary 3.53. *Let $\mathscr{X} \in L^1(\Omega, \mathcal{S}, \mathbb{P})$ be a family of random variables such that there exists $Z \in L^1(\Omega, \mathcal{S}, \mathbb{P})$ with the property*
$$|X| \leq |Z| \text{ a.s.}, \quad \forall X \in \mathscr{X}.$$
Then $\mathscr{X}$ is UI.

Proof. The family $\mathscr{X}$ satisfies condition (ii) of Theorem 3.51. □

Theorem 3.54. *Let $\mathscr{X} \subset L^1(\Omega, \mathcal{S}, \mathbb{P})$. Then the following statements are equivalent.*

(i) $\mathscr{X}$ is UI.
(ii)
$$\lim_{r \to \infty} \sup_{X \in \mathscr{X}} \int_r^\infty \mathbb{P}\big[|X| > x\big] dx = 0.$$

(iii) There exists a convex increasing function $f : [0, \infty) \to [0, \infty)$ which is also superlinear,
$$\lim_{r \to \infty} \frac{f(r)}{r} = \infty$$
and satisfies
$$\sup_{X \in \mathscr{X}} \mathbb{E}\big[f(|X|)\big] < \infty. \tag{3.2.8}$$

Proof. (i) $\iff$ (ii) Proposition 1.126 shows that

$$\int_r^\infty \mathbb{P}\big[\,|X|>x\,\big]dx = \mathbb{E}\big[\,|X|\boldsymbol{I}_{\{|X|>r\}}\,\big], \quad \forall X\in\mathscr{X}.$$

(ii) $\Rightarrow$ (iii) Set

$$h(r):=\sup_{X\in\mathscr{X}}\int_r^\infty \mathbb{P}\big[\,|X|>x\,\big]dx.$$

Note that $h(0)\le r+h(r)<\infty$. Since $h(r)=o(1)$ as $r\to\infty$ we can find

$$0=r_0<r_1<r_2<\cdots$$

such that

$$h(r_n)\le \frac{h(0)}{2^n},\quad \forall n\in\mathbb{N}.$$

Now define

$$g(r):=\sum_{n\ge 0}\boldsymbol{I}_{[r_n,\infty)}(r),\quad f(x)=\int_0^x g(r)dr.$$

Note that $g(r)$ is nondecreasing and $\lim_{r\to\infty}g(r)=\infty$. This shows that f is increasing convex and superlinear. Using the Fubini-Tonelli theorem as in the proof of Proposition 1.126 we deduce

$$\mathbb{E}\big[\,f(|X|)\,\big]=\mathbb{E}\left[\int_0^{|X|}g(r)dr\right]=\mathbb{E}\left[\sum_{n\ge 0}\int_{r_n}^\infty \boldsymbol{I}_{|X|>r_n}(x)dx\right]$$

$$\le \sum_{n\ge 0}h(r_n)\le h(0)\sum_{n\ge 0}\frac{1}{2^n}.$$

(iii) $\Rightarrow$ (i) For every $n\in\mathbb{N}$ there exists $r_n>0$ such that

$$\forall x:\ x>r_n \Rightarrow x<\frac{f(x)}{n}.$$

We deduce that for any $X\in\mathscr{X}$ we have

$$\mathbb{E}\big[\,|X|\boldsymbol{I}_{|X|>r_n}\,\big]\le \frac{1}{n}\mathbb{E}\big[\,f(|X|)\boldsymbol{I}_{|X|>r_n}\,\big]\le \frac{1}{n}\mathbb{E}\big[\,f(|X|)\,\big].$$

The conclusion now follows from (3.2.8). $\square$

If in the above theorem we choose $f(r)=r^p$, $p>1$, we obtain the following result.

Corollary 3.55. *Let $\mathscr{X}\in L^1(\Omega,\mathcal{S},\mathbb{P})$ be a family of random variables such that there exist $p\in(1,\infty)$ with the property*

$$\sup_{X\in\mathscr{X}}\mathbb{E}\big[\,|X|^p\,\big]<\infty.$$

Then $\mathscr{X}$ is UI. $\square$

Corollary 3.56. *Let $X \in L^2(\Omega, \mathcal{F}, \mathbb{P})$ and suppose that $(\mathcal{F}_i)_{i \in I}$ is a family of sigma subalgebras. Set $X_i := \mathbb{E}[X \| \mathcal{F}_i]$, $i \in I$. Then the family $(X_i)_{i \in I}$ is UI.*

Proof. Lemma 3.48 shows that the family $\{X\}$ consisting of the integrable random variable X is uniformly integrable. Theorem 3.54 implies that there exists a superlinear, convex, increasing function $f : [0, \infty) \to [0, \infty)$ such that $\mathbb{E}[f(X)] < \infty$. From the conditional Jensen inequality in Theorem 1.166(ix) we deduce that
$$|X_i| = |\mathbb{E}[X \| \mathcal{F}_i]| \leq \mathbb{E}[|X| \| \mathcal{F}_i].$$
Since f is increasing and convex we deduce that
$$f(|X_i|) \leq f(\mathbb{E}[|X| \| \mathcal{F}_i]) \leq \mathbb{E}[f(|X|) \| \mathcal{F}_i].$$
Taking the expectations of both sides of this inequality we deduce
$$\mathbb{E}[f(|X_i|)] \leq \mathbb{E}[f(|X|)], \quad \forall i \in I.$$
Using Theorem 3.54 again we deduce that the family $(X_i)_{i \in I}$ is UI. $\square$

The next result clarifies the importance of the uniform integrability condition.

Theorem 3.57. *Consider a sequence (X_n) in $L^1(\Omega, \mathcal{S}, \mathbb{P})$ that converges in probability to X. Then the following statements are equivalent.*

(i) The sequence (X_n) is UI.
(ii) The limit X is integrable and sequence (X_n) converges to X in the L^1-norm.
(iii) The limit X is integrable and
$$\lim_{n \to \infty} \mathbb{E}[|X_n|] = \mathbb{E}[|X|].$$

Proof. We follow the approach in [53, Thm. 5.5.2].
(i) $\Rightarrow$ (ii). For every $M > 0$ we define
$$\Phi_M : \mathbb{R} \to \mathbb{R}, \quad \Phi_M(x) = \begin{cases} M, & x \geq M, \\ x, & |x| < M, \\ -M, & x \leq -M. \end{cases}$$
We have
$$\mathbb{E}[|X_n - X|] \leq \mathbb{E}[|X_n - \Phi_M(X_n)|] + \mathbb{E}[|\Phi_M(X_n) - \Phi_M(X)|]$$
$$+ \mathbb{E}[||\Phi_M(X) - X|]$$
$$\leq 2\mathbb{E}[|X_n|\boldsymbol{I}_{\{|X_n|>M\}}] + \mathbb{E}[|\Phi_M(X_n) - \Phi_M(X)|] + 2\mathbb{E}[|X|\boldsymbol{I}_{\{|X|>M\}}].$$
The sequence (X_n) is uniformly integrable and Theorem 3.51 implies that $\sup_n \mathbb{E}[|X_n|] < \infty$. Fatou's Lemma applied to an a.s. convergent subsequence of X_n implies that $X \in L^1$. We conclude that for any $\varepsilon > 0$ there exists $M = M(\varepsilon) > 0$ such that
$$2\mathbb{E}[|X_n|\boldsymbol{I}_{\{|X_n|>M\}}] + 2\mathbb{E}[|X|\boldsymbol{I}_{\{|X|>M\}}] < \frac{\varepsilon}{2}, \quad \forall n \in \mathbb{N}. \qquad (3.2.9)$$

From Corollary 1.145 we deduce that $\Phi_{M(\varepsilon)}(X_n)$ converges to $\Phi_{M(\varepsilon)}(X)$ in probability. Moreover,
$$\big|\Phi_{M(\varepsilon)}(X_n)\big| < M(\varepsilon), \quad \forall n \in \mathbb{N}.$$
The Bounded Convergence Theorem 1.153 implies that there exists $n = n(\varepsilon) > 0$ such that for any $n \geq n(\varepsilon)$ we have
$$\mathbb{E}\big[\,|\Phi_{M(\varepsilon)}(X_n) - \Phi_{M(\varepsilon)}(X)|\,\big] < \frac{\varepsilon}{2}.$$
From (3.2.9) we deduce that $\mathbb{E}\big[\,|X_n - X|\,\big] < \varepsilon$ for $n > n(\varepsilon)$.

Clearly (ii) $\Rightarrow$ (iii) since $X_n \to X$ in L^1 implies $\|X_n\|_{L^1} \to \|X\|_{L^1}$.

(iii) $\Rightarrow$ (i) For any $M > 0$ consider the *continuous* function
$$\Psi_M : [0, \infty) \to \mathbb{R}, \quad \Psi_M(x) = \begin{cases} x, & x \in [0, M-1], \\ 0, & x \geq M, \\ \text{linear}, & x \in (M-1, M). \end{cases}$$
The Dominated Convergence Theorem implies that $\Psi_M(|X|)$ converges to $|X|$ in L^1 as $M \to \infty$. Thus, there exists $M = M(\varepsilon)$ such that
$$\mathbb{E}\big[\,|X|\,\big] - \mathbb{E}\big[\,\Psi_M(|X|)\,\big] < \frac{\varepsilon}{2}, \quad \forall M \geq M(\varepsilon). \tag{3.2.10}$$
Using the Bounded Convergence Theorem as in the proof of the implication (i) $\Rightarrow$ (ii) we deduce that
$$\mathbb{E}\big[\,\Psi_M(|X_n|)\,\big] \to \mathbb{E}\big[\,\Psi_M(|X|)\,\big], \quad \forall M > 0. \tag{3.2.11}$$
Thus, for any $n \in \mathbb{N}$ we have
$$\mathbb{E}\big[\,|X_n|\boldsymbol{I}_{\{|X_n|>M(\varepsilon)\}}\,\big] \leq \mathbb{E}\big[\,|X_n|\,\big] - \mathbb{E}\big[\,\Psi_{M(\varepsilon)}(|X_n|)\,\big]$$
$$= \Big(\mathbb{E}\big[\,|X_n|\,\big] - \mathbb{E}\big[\,|X|\,\big]\Big)$$
$$+ \Big(\mathbb{E}\big[\,|X|\,\big] - \mathbb{E}\big[\,\Psi_{M(\varepsilon)}(|X|)\,\big]\Big) + \Big(\mathbb{E}\big[\,\Psi_{M(\varepsilon)}(|X|)\,\big] - \mathbb{E}\big[\,\Psi_{M(\varepsilon)}(|X_n|)\,\big]\Big)$$
$$\stackrel{(3.2.10)}{<} \Big(\mathbb{E}\big[\,|X_n|\,\big] - \mathbb{E}\big[\,|X|\,\big]\Big) + \Big(\mathbb{E}\big[\,\Psi_{M(\varepsilon)}(|X|)\,\big] - \mathbb{E}\big[\,\Psi_{M(\varepsilon)}(|X_n|)\,\big]\Big) + \frac{\varepsilon}{2}.$$
We can choose $n = n(\varepsilon, M(\varepsilon))$ so that for $n > n(\varepsilon)$ we have
$$\Big(\mathbb{E}\big[\,|X_n|\,\big] - \mathbb{E}\big[\,|X|\,\big]\Big) + \Big(\mathbb{E}\big[\,\Psi_{M(\varepsilon)}(|X|)\,\big] - \mathbb{E}\big[\,\Psi_{M(\varepsilon)}(|X_n|)\,\big]\Big) < \frac{\varepsilon}{2}.$$
Hence for any $M \geq M(\varepsilon)$
$$\sup_{n > n(\varepsilon)} \mathbb{E}\big[\,|X_n|\boldsymbol{I}_{\{|X_n|>M\}}\,\big] \leq \sup_{n > n(\varepsilon)} \mathbb{E}\big[\,|X_n|\boldsymbol{I}_{\{|X_n|>M(\varepsilon)\}}\,\big] < \varepsilon.$$
Now choose $M_1 > M(\varepsilon)$ such that
$$\mathbb{E}\big[\,|X_n|\boldsymbol{I}_{\{|X_n|>M_1\}}\,\big] < \varepsilon, \quad \forall n = 1, 2, \ldots, n(\varepsilon).$$
Hence for $M \geq M_1$ we have
$$\sup_{n \in \mathbb{N}} \mathbb{E}\big[\,|X_n|\boldsymbol{I}_{\{|X_n|>M\}}\,\big] < \varepsilon.$$
Thus (X_n) is uniformly integrable. $\square$

Remark 3.58. (a) The implication (iii) $\Rightarrow$ (ii) is sometimes referred to as *Scheffé's Lemma*.

(b) We used the Bounded Convergence Theorem to prove the implication (i) $\Rightarrow$ (ii). Obviously the Bounded Convergence Theorem is a special case of this implication. One can prove the equivalence (i) $\iff$ (ii) without relying on the Bounded Convergence Theorem; see [50, Thm. 10.3.6].

(c) The sequence in Example 2.27 converges in law, it is uniformly integrable yet it does not converge in probability. This shows that in the above theorem we cannot relax the convergence-in-probability condition to convergence in law. $\square$

3.2.3 Uniformly integrable martingales

We can now formulate and prove a refinement of Theorem 3.41.

Theorem 3.59. *If $(X_n)_{n \in \mathbb{N}_0}$ is a martingale adapted to the filtration $(\mathcal{F}_n)_{n \geq 0}$ of $(\Omega, \mathcal{F}, \mathbb{P})$. Set*

$$\mathcal{F}_\infty := \bigvee_{n \geq 0} \mathcal{F}_n = \sigma(\mathcal{F}_n, \, n \geq 0).$$

The following are equivalent.

(i) The collection $(X_n)_{n \in \mathbb{N}_0}$ is UI.
(ii) The sequence $(X_n)_{n \in \mathbb{N}_0}$ converges a.s. and L^1 to a random variable X_∞.
(iii) The sequence $(X_n)_{n \in \mathbb{N}_0}$ converges L^1 to a random variable X_∞.
(iv) There exists an integrable random variable X such that

$$X_n = \mathbb{E}[X \| \mathcal{F}_n], \quad \forall n \in \mathbb{N}_0.$$

If the above conditions are satisfied, then the limiting random variable X_∞ in (ii) and (iii) is related to the random variable X in (iv) via the equality $X_\infty = \mathbb{E}[X \| \mathcal{F}_\infty]$, i.e.,

$$\lim_{n \to \infty} \mathbb{E}[X \| \mathcal{F}_n] = \mathbb{E}[X \| \mathcal{F}_\infty]$$

a.s. and L^1.

Proof. Note that if a martingale (X_n) is UI, then it is bounded in L^1 and, according to Theorem 3.41, converges a.s. to an integrable random variable X_∞. In view of the previous discussion the statements (i)–(iii) are equivalent. The implication (iv) $\Rightarrow$ (i) follows from Corollary 3.56. The only thing left to prove is (iii) $\Rightarrow$ (iv).

More precisely, we will show that if $X_n \to X_\infty$ in L^1, then

$$X_n = \mathbb{E}[X_\infty \| \mathcal{F}_n], \quad \text{a.s.}, \quad \forall n \in \mathbb{N}_0.$$

In other words, we have to show that, for all $m \in \mathbb{N}_0$, and all $A \in \mathcal{F}_m$ we have

$$\mathbb{E}[X_m \mathbf{I}_A] = \mathbb{E}[X_\infty \mathbf{I}_A].$$

Since (X_n) is a martingale we deduce that, for $n > m$, we have
$$\mathbb{E}\big[X_m \boldsymbol{I}_A\big] = \mathbb{E}\Big[\mathbb{E}\big[X_n\,\|\,\mathcal{F}_m\big]\boldsymbol{I}_A\Big] = \mathbb{E}\Big[\mathbb{E}\big[X_n \boldsymbol{I}_A\,\|\,\mathcal{F}_m\big]\Big] = \mathbb{E}\big[X_n \boldsymbol{I}_A\big].$$
Now let $n \to \infty$.

Suppose now that for some integrable random variable X we have $X_n = \mathbb{E}\big[X\,\|\,\mathcal{F}_n\big]$. We want to show that
$$\lim_n X_n = X_\infty := \mathbb{E}\big[X\,\|\,\mathcal{F}_\infty\big]$$
i.e., for any $F \in \mathcal{F}_\infty$ we have
$$\mathbb{E}\big[X_\infty \boldsymbol{I}_F\big] = \mathbb{E}\big[X \boldsymbol{I}_F\big].$$
Denote by $\mathcal{Z} \subset \mathcal{F}_\infty$ the collection of $F \subset \mathcal{F}_\infty$ for which the above holds. Clearly $\mathcal{F}_n \subset \mathcal{Z}$. Clearly, $\mathcal{Z}$ is a λ-system and contains the π-system
$$\bigcup_{n \geq 0} \mathcal{F}_n.$$
Thus it contains $\mathcal{F}_\infty$, the σ-algebra generated by this system. $\square$

Theorem 3.59 implies that
$$\lim_{n \to \infty} \mathbb{E}\big[X\|\mathcal{F}_n\big] = \mathbb{E}\big[X\|\mathcal{F}_\infty\big] \text{ a.s. and } L^1,\ \forall X \in L^1(\Omega, \mathcal{F}, \mathbb{P}). \qquad (3.2.12)$$
In particular, we deduce

Corollary 3.60 (Lévy's 0-1 law). *For any set $A \in \mathcal{F}_\infty$, the random variables*
$$\mathbb{E}\big[\boldsymbol{I}_A\|\mathcal{F}_n\big],\ n \in \mathbb{N},$$
converge a.s. and L^1 to $\boldsymbol{I}_A$ as $n \to \infty$. $\square$

Corollary 3.61 (Kolmogorov's 0-1 law). *Suppose that $\mathcal{G}_1, \mathcal{G}_2, \ldots$ are independent σ-subalgebras of $\mathcal{F}$. We set*
$$\mathcal{T}_n := \sigma\big(\mathcal{G}_{n+1}, \mathcal{G}_{n+2}, \ldots\big)$$
and form the tail *σ-algebra*
$$\mathcal{T}_\infty = \bigcap_{n \geq 1} \mathcal{T}_n.$$
Then $\mathcal{T}_\infty$ is a 0-1 sigma-algebra,
$$H \in \mathcal{T}_\infty \Rightarrow \mathbb{P}\big[H\big] \in \{0,1\}.$$

Proof. Define $\mathcal{F}_n := \sigma\big(\mathcal{G}_1, \ldots, \mathcal{G}_n\big)$. Let $H \in \mathcal{F}_\infty$. By Levy's $0-1$ law we have
$$\mathbb{E}\big[\boldsymbol{I}_H\,\|\,\mathcal{F}_n\big] \to \boldsymbol{I}_H \text{ a.s.}$$
On the other hand, if $H \in \mathcal{T}$, then since $\mathcal{T} \perp\!\!\!\perp \mathcal{F}_n$ we deduce $\mathbb{E}\big[\boldsymbol{I}_H\,\|\,\mathcal{F}_n\big] = \mathbb{P}\big[H\big]$, so that $\mathbb{P}\big[H\big] = \boldsymbol{I}_H$ a.s. $\square$

Example 3.62. Consider again the Galton-Watson branching process in Example 3.8. Suppose that the reproduction law μ satisfies
$$m := \sum_{n \geq 0} n\mu[n] < \infty.$$
Assume $m \geq 1$. Consider the *extinction event* defined in Example 3.45
$$E := \bigcup_{n \geq 0} E_n, \quad E_n = \{ Z_k = 0, \ \forall k \geq n \}.$$
Consider next the event
$$U := \left\{ \sup_n Z_n = \infty \right\}.$$
We want to prove that if the probability that an individual has no successor is positive then, with probability 1, either the population extinguishes in finite time, or explodes. In particular it cannot stabilize to a finite nonzero limit. More precisely, we have the following dichotomy result.

If $p_0 = \mu[0] > 0$, then, with probability 1, the population either becomes extinct or explodes, i.e.,
$$E = U^c, \quad \mathbb{P}[E \cup U] = 1. \tag{3.2.13}$$

In particular
$$E = \left\{ \lim_n Z_n = 0 \right\}. \tag{3.2.14}$$
Note that
$$\forall \nu \in \mathbb{N}_0, \ \exists \delta(\nu) \in (0,1) : \ \forall n \in \mathbb{N}, \tag{3.2.15}$$
$$\mathbb{P}[E \,\|\, Z_1, \ldots, Z_n] \geq \delta(\nu) \text{ on } \{Z_n \leq \nu\}.$$
Indeed, if the population of the n-th generation has at most ν individuals then the probability that there will be no $(n+1)$-th generation is at least p_0^ν. More formally,
$$\mathbb{P}[E \,\|\, Z_1, \ldots, Z_n] \geq \mathbb{P}[E_{n+1} \,\|\, Z_1, \ldots, Z_n] = \mathbb{P}[E_{n+1} \,\|\, Z_n].$$
We have
$$\mathbb{P}[E_{n+1} \,\|\, Z_n] \boldsymbol{I}_{\{Z_n \leq \nu\}} = \sum_{k=0}^{\nu} \mathbb{P}[E_{n+1} \,\|\, Z_n = k] \boldsymbol{I}_{\{Z_n = k\}}$$
$$= \sum_{k=0}^{\nu} p_0^k \boldsymbol{I}_{\{Z_n = k\}} \geq p_0^\nu \boldsymbol{I}_{\{Z_n \leq \nu\}}.$$
This proves (3.2.15) with $\delta(\nu) = p_0^\nu$.

Since Z_n are integer valued we deduce that

Lemma 3.63. *Suppose that $(Z_n)_{n \geq 1}$ is a sequence of nonnegative random variables. Set*
$$E = \{ Z_n = 0 \text{ for some } n \}, \quad B := \left\{ \sup_n Z_n < \infty \right\}.$$
If (Z_n) satisfies (3.2.15), then
$$E \supset B. \tag{3.2.16}$$

Proof. Set $\mathcal{F}_n := \sigma(Z_1, \ldots, Z_n\}$, $B_\nu := \{ \sup_n Z_n \leq \nu \}$, so that
$$B_1 \subset B_2 \subset \cdots, \quad \bigcup_\nu B_\nu = B.$$
We have $\mathbb{E}[E \| \mathcal{F}_n] \geq \delta(\nu)$ on B_ν. Letting $n \to \infty$ we deduce from Lévy's 0-1 theorem (Corollary 3.60) that
$$\lim_{n \to \infty} \mathbb{E}[I_E \| \mathcal{F}_n] = I_E.$$
Hence $B_\nu \subset E$ for any ν. Hence $B \subset E$. □

In our special case, $B = U^c$. Note also that if the population dies at a time at a time n_0, then $Z_n = 0$, $\forall n \geq n_0$. Hence $E \subset B$ or, in view of (3.2.16), $E = B = U^c$. This proves the claimed dichotomy (3.2.13).

When $m = 1$, then $W_n = Z_n$ converges almost surely to an integrable random variable and we see that
$$\left\{ \lim_n Z_n < \infty \right\} \subset \{ \sup_n Z_n < \infty \} \subset E$$
and we deduce that
$$1 \geq \mathbb{P}[E] \geq \mathbb{P}\left[\lim_n Z_n < \infty\right] = 1.$$
Thus, when $m = 1$ and the probability having *no* successors is positive, i.e., $\mu[0] > 0$, then the extinction probability is also 1. One can show (see [6, Sec. I.9] or [89]) that if $m = 1$ and
$$\sigma^2 := \operatorname{Var}[X_{n,j}] = \sum_k k(k-1)\mu[k] < \infty,$$
then
$$\lim_{n \to \infty} n\mathbb{P}[Z_n > 0] = \frac{2}{\sigma^2}.$$
Thus, the probability of the population surviving more than n generations given that individuals have on average 1 successor is $O(1/n)$.

When $m > 1$ the extinction probability is still positive but < 1. Exercise 3.26 describes this probability and gives additional information about the distribution of W. For more details about branching processes we refer to [6; 80]. □

Suppose that $(X_n)_{n \in \mathbb{N}_0}$ is a process adapted to a filtration $\mathcal{F}_\bullet$ such that X_n converges a.s. to a random variable X_∞ as $n \to \infty$. If T is a stopping time adapted to the same filtration, finite or not, we set
$$\hat{X}_T := \sum_{n \in \mathbb{N}_0} X_n I_{\{T=n\}} + X_\infty I_{\{T=\infty\}} = X_T + X_\infty I_{\{T=\infty\}}. \qquad (3.2.17)$$
Note that
$$\mathbb{P}[T = \infty] = 0 \Rightarrow \hat{X}_T = X_T \text{ a.s.},$$
$$\boxed{\hat{X}_T = X_\infty^T := \lim_{n \to \infty} X_{T \wedge n} = \lim_{n \to \infty} X_n^T}. \qquad (3.2.18)$$

Theorem 3.64 (Optional sampling: UI martingales). *Suppose that* $X_\bullet = (X_n)_{n \in \mathbb{N}_0}$ *is a UI martingale and T is a stopping time, not necessarily a.s. finite. Then $\hat{X}_T \in L^1$ and*

$$\hat{X}_T = \mathbb{E}[X_\infty \| \mathcal{F}_T]. \tag{3.2.19}$$

Moreover, if S, T are stopping times such that $S \leq T$, then

$$\mathbb{E}[\hat{X}_T \| \mathcal{F}_S] = \hat{X}_S. \tag{3.2.20}$$

Proof. Let us first prove that $\hat{X}_T \in L^1$. We have

$$\mathbb{E}[|\hat{X}_T|] = \sum_{n \geq 0} \mathbb{E}[I_{\{T=n\}}|X_n|] + \mathbb{E}[I_{\{T=\infty\}}|X_\infty|]$$

$$= \sum_{n \geq 0} \mathbb{E}\Big[I_{\{T=n\}}\big|\mathbb{E}[X_\infty \| \mathcal{F}_n]\big|\Big] + \mathbb{E}[I_{\{T=\infty\}}|X_\infty|]$$

$$\Big(\,|\mathbb{E}[X\|\mathcal{F}]| \leq \mathbb{E}[|X|\,\|\mathcal{F}]\,\Big)$$

$$\leq \sum_{n \geq 0} \mathbb{E}\Big[I_{\{T=n\}}\mathbb{E}[|X_\infty|\,\|\mathcal{F}_n]\Big] + \mathbb{E}[I_{\{T=\infty\}}|X_\infty|]$$

(use the definition of conditional expectation)

$$= \sum_{n \geq 0} \mathbb{E}\Big[I_{\{T=n\}}|X_\infty|\Big] + \mathbb{E}[I_{\{T=\infty\}}|X_\infty|] = \mathbb{E}[|X_\infty|] < \infty.$$

Moreover, for $A \in \mathcal{F}_T$ we have

$$\mathbb{E}[I_A \hat{X}_T] = \sum_{n \in \mathbb{N}_0} \mathbb{E}[I_{A \cap \{T=n\}} X_n] + \mathbb{E}[I_{A \cap \{T=\infty\}} X_\infty]$$

$$\Big(\,I_{A \cap \{T=n\}} X_n = \mathbb{E}[I_{A \cap \{T=n\}} X_\infty \| \mathcal{F}_n]\,\Big)$$

$$= \sum_{n \in \mathbb{N}_0} \mathbb{E}[I_{A \cap \{T=n\}} X_\infty] + \mathbb{E}[I_{A \cap \{T=\infty\}} X_\infty] = \mathbb{E}[I_A X_\infty],$$

and thus $\hat{X}_T = \mathbb{E}[X_\infty \| \mathcal{F}_T]$. The since $\mathcal{F}_S \subset \mathcal{F}_T$, the equality (3.2.20) follows immediately from (3.2.19) and the properties of conditional expectation. □

Corollary 3.65 (Optional Stopping). *Suppose that $(X_n)_{n \in \mathbb{N}_0}$ is a UI martingale and T is any stopping time. Then the stopped martingale $X_n^T = X_{T \wedge n}$ is also a uniformly integrable martingale with respect to the filtration $\mathcal{F}_{T \wedge n}$.*

Proof. From Theorem 3.64 we deduce that $X_{T \wedge n} = \mathbb{E}[X_\infty \| \mathcal{F}_{T \wedge n}]$ and Corollary 3.56 implies it is UI. □

Doob's conditions in Definition 3.26 are closely related to uniform integrability.

Proposition 3.66. *Suppose that $(X_n)_{n \geq 0}$ is a martingale adapted to the filtration $(\mathcal{F}_n)_{n \geq 0}$ and T is an a.s. finite stopping time adapted to the same filtration. Then the following statements are equivalent.*

(i) The stopping time satisfies Doob's conditions (3.1.7b) and (3.1.7c).
(ii) The stopped martingale $X_n^T = X_{T \wedge n}$ is UI.

Proof. (i) $\Rightarrow$ (ii) Consider the submartingale $|X_n|$. Since T satisfies Doob's conditions we deduce from Theorem 3.28 that

$$\mathbb{E}[|X_T|] \geq \mathbb{E}[X_0] \geq \mathbb{E}[|X_{T \wedge n}|] \quad \forall n \geq 0.$$

Thus

$$\limsup_{n \to \infty} \mathbb{E}[|X_{T \wedge n}|]] \leq \mathbb{E}[|X_T|].$$

Since $\lim_{n \to \infty} X_{T \wedge n} = X_T$, a.s., we deduce from Fatou's Lemma that

$$\mathbb{E}[|X_T|] \leq \liminf_{n \to \infty} \mathbb{E}[|X_{T \wedge n}|]]$$

so that

$$\limsup_{n \to \infty} \mathbb{E}[|X_{T \wedge n}|]] = \mathbb{E}[|X_T|].$$

The desired conclusion now follows from Theorem 3.57.

(ii) $\Rightarrow$ (i) Observe first that $\lim_{n \to \infty} X_{T \wedge n} = X_T$ and since X_n^T is UI we deduce X^T is integrable. Now observe that

$$\mathbb{E}[|X_n|\boldsymbol{I}_{T>n}] = \mathbb{E}[|X_{T \wedge n}|\boldsymbol{I}_{T>n}].$$

Since $\mathbb{P}[T < \infty] = 1$ we deduce $\lim_{n \to \infty} \mathbb{P}[T > n] = 0$. Finally, using the fact that the stopped martingale X_n^T is UI we deduce

$$\lim_{n \to \infty} \mathbb{E}[|X_{T \wedge n}|\boldsymbol{I}_{T>n}] = 0.$$

$\square$

Corollary 3.67 (Optional Sampling Theorem). *Suppose that $(X_n)_{n \geq 0}$ is a martingale adapted to the filtration $(\mathcal{F}_n)$, $S \leq T$ are stopping times adapted to the same filtration and T satisfies the Doob conditions (3.1.7a, 3.1.7b, 3.1.7c). Then*

$$\mathbb{E}[X_T \| \mathcal{F}_S] = X_S.$$

Proof. Note that X^T is UI and, since $X^S = (X^T)^S$, we deduce from Theorem 3.64 that

$$\mathbb{E}[X_T \| \mathcal{F}_S] = \mathbb{E}[X_\infty^T \| \mathcal{F}_S] = X_S^T = X_S.$$

$\square$

3.2.4 Applications of the optional sampling theorem

Let us observe that the above discussion yields an alternate proof of the Optional Sampling Theorem 3.28. We restate it below.

Corollary 3.68 (Optional Sampling Theorem). *Suppose that $(X_n)_{n\geq 0}$ is a martingale adapted to the filtration $(\mathcal{F}_n)_{n\geq 0}$ and T is an a.s. stopping time such that the stopped martingale $X_n^T = X_{n \wedge T}$ is UI, i.e., T satisfies Doobs conditions. Then $\mathbb{E}[X_T] = \mathbb{E}[X_0]$.* □

The Optional Sampling Theorem is a versatile tool for computing expectations. Its applicability is greatly enhanced once we have simple criteria for recognizing when a stopped martingale is UI. We have the following result of J. L. Doob, [47, Thm. VII.2.2].

Proposition 3.69. *Suppose that $(M_n)_{n\geq 0}$ is a random process adapted to the filtration $(\mathcal{F}_n)_{n\geq 0}$ such that*

$$\mathbb{E}[M_n] < \infty, \ \forall n,$$

and T is a stopping time adapted to the same filtration. Suppose that

$$\mathbb{E}[T] < \infty, \tag{3.2.21a}$$

$$\exists C > 0 : \ \forall n \in \mathbb{N}, \ \mathbb{E}\big[|M_n - M_{n-1}| \,\|\, \mathcal{F}_{n-1}\big] \leq C. \tag{3.2.21b}$$

Then the stopped process $M_n^T = M_{T \wedge n}$ is UI.

Proof. We will show that there exists $Y \in L^1(\Omega, \mathcal{F}, \mathbb{P})$ such that

$$|M_{T \wedge n}| \leq Y, \ \forall n \in \mathbb{N}.$$

Note that

$$M_{T \wedge n} = \sum_{k=0}^{n-1} M_k \boldsymbol{I}_{\{T=k\}} + M_n \boldsymbol{I}_{\{T \geq n\}}$$

$$= \sum_{k=0}^{n-1} M_k \big(\boldsymbol{I}_{\{T \geq k\}} - \boldsymbol{I}_{\{T \geq k+1\}}\big) + M_n \boldsymbol{I}_{\{T \geq n\}} = M_0 + \sum_{k=1}^{n} (M_k - M_{k-1}) \boldsymbol{I}_{\{T \geq k\}}$$

so

$$\big[M_{T \wedge n}\big] \leq |M_0| + \sum_{k=1}^{n} |M_k - M_{k-1}| \boldsymbol{I}_{\{T \geq k\}}.$$

Set

$$Y := |M_0| + \sum_{k=1}^{\infty} |M_k - M_{k-1}| \boldsymbol{I}_{\{T \geq k\}}.$$

Clearly $|M_{T \wedge n}| \leq Y$, $\forall n \geq 0$. We will show that $\mathbb{E}[Y] < \infty$. We have

$$\mathbb{E}\big[\,|M_k - M_{k-1}|\boldsymbol{I}_{\{T \geq k\}}\,\big] = \mathbb{E}\Big[\,\mathbb{E}\big[\,|M_k - M_{k-1}|\boldsymbol{I}_{\{T \geq k\}}\,\|\,\mathcal{F}_{k-1}\big]\,\Big],$$

($\{T \geq k\} \in \mathcal{F}_{k-1}$)

$$\mathbb{E}\Big[\,\boldsymbol{I}_{\{T \geq k\}}\mathbb{E}\big[\,|M_k - M_{k-1}|\,\|\,\mathcal{F}_{k-1}\big]\,\Big] \overset{(3.2.21b)}{\leq} C\mathbb{E}\big[\,\boldsymbol{I}_{\{T \geq k\}}\,\big] = C\mathbb{P}\big[\,T \geq k\,\big].$$

Thus

$$\mathbb{E}[Y] \leq \mathbb{E}\big[\,|M_0|\,\big] + C\sum_{k=1}^{\infty}\mathbb{P}[T \geq k] = \mathbb{E}\big[\,|M_0|\,\big] + C\mathbb{E}[T] \overset{(3.2.21a)}{<} \infty.$$

$\square$

Theorem 3.70 (Wald's formula). *Suppose that $(Y_n)_{n \geq 0}$ is a sequence of i.i.d. integrable random variables with finite mean μ. Set*

$$S_n := \sum_{k=0}^{n} Y_k.$$

Let T be a stopping time adapted to the filtration $\mathcal{F}_n = \sigma(Y_0, \ldots, Y_n)$ and such that $\mathbb{E}[T] < \infty$. The following hold.

(i) $\mathbb{E}[S_T] = \mu\mathbb{E}[T]$.
(ii) Suppose additionally that

$$Y_n \in L^2, \quad \mu = 0, \quad \sigma^2 = \text{Var}[Y_n].$$

Then $\text{Var}[S_T] = \sigma^2\mathbb{E}[T]$.

Proof. (i) Set $\overline{Y}_n = Y_n - \mu$,

$$M_n := S_n - n\mu = \sum_{k=1}^{n}\overline{Y}_k.$$

Then

$$\mathbb{E}\big[\,M_n\,\|\,\mathcal{F}_{n-1}\,\big] = \mathbb{E}\big[\,\overline{Y}_n + M_{n-1}\,\|\,\mathcal{F}_{n-1}\,\big]$$

$$= \mathbb{E}\big[\,\overline{Y}_n\,\|\,\mathcal{F}_{n-1}\,\big] + \mathbb{E}\big[\,M_{n-1}\,\|\,\mathcal{F}_{n-1}\,\big] = \mathbb{E}\big[\,\overline{Y}_n\,\big] + M_{n-1} = M_{n-1}.$$

Observe that

$$\mathbb{E}\big[\,|M_n - M_{n-1}|\,\|\,\mathcal{F}_{n-1}\,\big] = \mathbb{E}\big[\,|\overline{Y}_n|\,\|\,\mathcal{F}_{n-1}\,\big] = \mathbb{E}\big[\,|\overline{Y}_n|\,\big] = \mathbb{E}\big[\,|\overline{Y}_0|\,\big]$$

so that (3.2.21b) is satisfied. We deduce from Proposition 3.69 that the stopped martingale M_n^T is UI and the Optional Sampling Theorem implies

$$0 = \mathbb{E}\big[\,M_0\,\big] = \mathbb{E}\big[\,M_T\,\big] = \mathbb{E}\big[\,S_T\,\big] - \mu\mathbb{E}[T].$$

(ii) From (i) we deduce $\mathbb{E}[S_T] = 0$ so $\mathrm{Var}[S_T] = \mathbb{E}[S_T^2]$. Set

$$Q_n := \sum_{k=1}^n Y_k^2.$$

We have

$$\mathbb{E}[S_n^2] = \sum_{k=1}^n \mathbb{E}[Y_k^2] + 2\sum_{1 \leq i < j \leq n} \mathbb{E}[Y_i Y_j] = \mathbb{E}[Q_n].$$

As in (i) we observe that $Z_n = Q_n - n\sigma^2$ is a martingale adapted to the filtration $\mathcal{F}_n$, the increments $Q_n - Q_{n-1}$ are independent of $\mathcal{F}_n$ and

$$\mathbb{E}\big[\,|Z_n - Z_{n-1}|\,\|\,\mathcal{F}_{n-1}\big] = \mathbb{E}\big[\,|Z_n - Z_{n-1}|\,\big] \leq \mathbb{E}[Y_n^2] + \sigma^2 = 2\sigma^2.$$

We deduce from Proposition 3.69 that the stopped martingale Z_n^T is UI and the Optional Sampling Theorem implies

$$0 = \mathbb{E}[Z_0] = \mathbb{E}[Z_T] = \mathbb{E}[Q_T] - \sigma^2 \mathbb{E}[T]$$

$$= \mathbb{E}[S_T^2] - \sigma^2 \mathbb{E}[T] = \mathrm{Var}[S_T] - \sigma^2 \mathbb{E}[T].$$

$\square$

Remark 3.71. In Exercise 1.16 we described a version (1.6.1) of Wald's formula that has a different nature than the one presented in Theorem 3.70. The random time T in (1.6.1) is independent of the variables X_n and the proof of (1.6.1) is a simple exercise in conditioning.

In Theorem 3.70 the random time T is quite dependent of these variables given that it is adapted to the filtration $\mathcal{F}_n = \sigma(X_1, \ldots, X_n)$ and the proof of the corresponding version of Wald's formula required the machinery of martingale theory.

We want to point out that without some assumptions on T we cannot expect the equality $\mathbb{E}[S_T] = \mu \mathbb{E}[T]$ to hold. Here is an example.

Suppose that the random variables X_n are exponential with parameter λ. For fix $t > 0$ we set

$$N(t) := \max\{n \geq 0;\; S_n \leq t\}.$$

The collection $(N(t))_{t>0}$ is the Poisson process introduced in Example 1.136. Thus, $N(t) \sim \mathrm{Poi}(\lambda t)$ so that

$$\mathbb{E}[N(t)] = \lambda t.$$

In this case

$$\mu = \mathbb{E}[\,\mathrm{Exp}(\lambda)\,] = \frac{1}{\lambda}.$$

For fixed t, the random variable $N(t)$ *is not adapted* to the filtration $\mathcal{F}_n = \sigma(X_1, \ldots, X_n)$. Indeed, knowing $S_1, \ldots, S_n$, we cannot conclude that

$S_{n+1} > t$, i.e., that n is the largest index k such that $S_k \leq t$. If Wald's formula were true in this case it would predict $\mathbb{E}[S_{N(t)}] = t$. However, we know from (1.3.50) that

$$\mathbb{E}[S_{N(t)}] = t - \frac{1}{\lambda} + \frac{e^{-\lambda t}}{\lambda}.$$

Let us observe that $T = N(t) + 1$ is adapted to the filtration $\mathcal{F}_n$. Indeed

$$T = n \Longleftrightarrow N(t) = n - 1$$

$$\Longleftrightarrow X_1 + \cdots + X_{n-1} \leq t \text{ and } X_1 + \cdots + X_n + X_{n+1} > t,$$

so that $\{T = n\} \in \sigma(X_1, \ldots, X_n)$. Wald's formula implies

$$\mathbb{E}[S_{N(t)+1}] = \mathbb{E}[N(t) + 1] \cdot \mathbb{E}[X_1] = \frac{\lambda t + 1}{\lambda} = t + \frac{1}{\lambda}.$$

This agrees with our earlier conclusion (1.3.51). □

Example 3.72 (Gambler's Ruin). Suppose that

$$X_n : (\Omega, \mathcal{F}, \mathbb{P}) \to \{-1, 1\}, \quad n \in \mathbb{N},$$

is a sequence of i.i.d. random variables with common probability distribution $\mathbb{P}[\,]X_n = 1\,] = p$, $\mathbb{P}[\,X_n = -1\,] = q = 1 - p$, $p \in (0, 1)$. Fix $k \in \mathbb{N}$ and set

$$S_0 =: k, \quad S_n := k + X_1 + \cdots + X_n, \quad \forall n \in \mathbb{N}.$$

Intuitively, k is the initial fortune of a gambler that plays a sequence of independent games where he wins \$1 with probability p and loses \$1 with probability q. Then S_n is the fortune of the gambler after n games. The game stops when the gambler is out of money, or his fortune reaches a prescribed threshold $N > k$.

The sequence $(S_n)_{n \in \mathbb{N}}$ is a random process adapted to the filtration

$$\mathcal{F}_n = \sigma(X_1, \ldots, X_n).$$

The random variable

$$T = T_k := \min\{n \in \mathbb{N}; \ S_n \in \{0, N\}\} \tag{3.2.22}$$

is a stopping time adapted to this filtration. It is the moment the gambler stops playing. The 'sooner-rather-than-later' Lemma 3.32 implies that $\mathbb{E}[T] < \infty$ since T satisfies (3.1.9)

$$\forall n \in \mathbb{N}_0, \ \mathbb{P}[T \leq n + N \| \mathcal{F}_n] > r, \ r = (\min(p, q))^N.$$

In particular, $\mathbb{P}[T < \infty] = 1$. We want to compute $p_k(N) := \mathbb{P}[S_T = N]$. We distinguish two cases.

A. $p = 1/2$ so that the game is fair. Then S_n is a martingale. Consider now the stopped process S^T. It is a UI martingale since its is uniformly bounded. We deduce from the Optional Sampling Theorem that

$$k = \mathbb{E}[S_0] = \mathbb{E}[S_T] = p_k(N)N \Rightarrow p_k(N) = \frac{k}{N}.$$

Thus, the ruin probability is $1 - p_k(N) = \frac{N-k}{N} = 1 - \frac{k}{N}$. In Example A.19 we describe R codes simulating this situation.

B. $p \neq 1/2$ so the game is biased. Consider the De Moivre's martingale M_n defined Example 3.7, i.e.,

$$M_n = \left(\frac{q}{p}\right)^{S_n}.$$

The stopped martingale M^T is UI since it is bounded. Hence

$$\mathbb{E}[M_T] = \mathbb{E}[M_0] = \left(\frac{q}{p}\right)^k.$$

If we set $p_k(N) := \mathbb{P}[S_T = N]$, then we deduce

$$\left(\frac{q}{p}\right)^k = \mathbb{P}[S_T = 0]\left(\frac{q}{p}\right)^0 + \mathbb{P}[S_T = N]\left(\frac{q}{p}\right)^N = (1 - p_k(N)) + p_k(N)\left(\frac{q}{p}\right)^N.$$

Hence

$$p_k(N) = \frac{(\frac{q}{p})^k - 1}{(\frac{q}{p})^N - 1}. \qquad \square$$

Example 3.73 (The coupon collector problem revisited). Let us recall the coupon collector problem we discussed in Example 1.112.

Suppose that each box of cereal contains one of m different coupons. Once you obtain one of every type of coupons, you can send in for a prize. Ann wants that prize and, for that reason, she buys one box of cereals everyday. Assuming that the coupon in each box is chosen independently and uniformly at random from the m possibilities and that Ann does not collaborate with others to collect coupons, how many boxes of cereal is she expected to buy before she obtain at least one of every type of coupon?

Let N denote the number of boxes bought until Ann has at least one of every coupon. We have shown in Example 1.112 that

$$\mathbb{E}[N] = mH_m, \quad H_m := \left(1 + \frac{1}{2} + \cdots + \frac{1}{m-1} + \frac{1}{m}\right).$$

Suppose now that Ann has a little brother, Bob, and, every time she collects a coupon she already has, she gives it to Bob. At the moment when she completed her collection, Bob is missing B coupons. What is the expectation of B?

To answer this question we follow the approach in [61, Sec. 12.5]. Assume that the coupons are labelled $1, \ldots, m$. We denote by C_k the label of the coupon Ann found in the k-th box she bought. Thus $(C_k)_{k \geq 1}$ are i.i.d., uniformly distributed in $\{1, \ldots, m\}$. We set

$$\mathcal{F}_n := \sigma(C_1, \ldots, C_n), \quad n \in \mathbb{N}.$$

Denote by X_n the number of coupons Ann is missing after she bought n cereal boxes and by Y_n the number of coupons that have appeared exactly one time in the first n boxes Ann bought.

From the equality

$$N = \min\{n \in \mathbb{N};\ X_n = 0\},$$

we deduce that N is a stopping time adapted to the filtration $\mathcal{F}_\bullet$. Moreover, $Y_N = B$, the number of coupons Bob is missing the moment Ann completed her collection.

Fix a function $f : \mathbb{N}_0^2 \to \mathbb{N}_0$ satisfying the difference equation

$$x\big(f(x-1, y+1) - f(x,y)\big) + yf(x, y-1) = 0, \ \forall\ x, y \geq 1 \qquad (3.2.23)$$

and form the process $Z_n := f(X_n, Y_n)$, $n \in \mathbb{N}$.

Lemma 3.74. *The process $Z_\bullet$ is a martingale adapted to the filtration $\mathcal{F}_n$.*

Proof. We set $\Delta Z_n := Z_{n+1} - Z_n$. Note that Z_n is $\mathcal{F}_n$-measurable so we have to show that

$$\mathbb{E}\big[\Delta Z_n \,\|\, \mathcal{F}_n\big] = 0. \qquad (3.2.24)$$

Let us observe that, when Ann buys a new cereal box there are only three, mutually exclusive possibilities,

$$\Delta X_n = -1, \ \Delta Y_n = -1, \ \Delta X_n = \Delta Y_n = 0.$$

The first possibility corresponds to Ann obtaining a new coupon, the second possibility corresponds to Bob obtaining a new coupon, and the third possibility occurs when the $(n+1)$-th coupon is owned by both Ann and Bob. Hence

$$\Delta Z_n = \boldsymbol{I}_{\{\Delta X_n = -1\}}\big(f(X_n - 1, Y_n + 1) - f(X_n, Y_n)\big)$$

$$+ \boldsymbol{I}_{\{\Delta Y_n = -1\}}\big(f(X_n, Y_n - 1) - (f(X_n, Y_n))\big).$$

Now observe that

$$\mathbb{E}\big[\boldsymbol{I}_{\{\Delta X_n = -1\}} \,\|\, \mathcal{F}_n\big] = \frac{X_n}{m} \ \text{and}\ \mathbb{E}\big[\boldsymbol{I}_{\{\Delta Y_n = -1\}} \,\|\, \mathcal{F}_n\big] = \frac{Y_n}{m}.$$

To understand the first equality observe that if Ann is missing X_n coupons at time n, then the probability of getting a new one in the new box is $\frac{X_n}{m}$. The second equality is proved in a similar fashion. Hence

$$\mathbb{E}\big[\Delta Z_n \,\|\, \mathcal{F}_n\big] = \frac{X_n}{m}\Big(f(X_n - 1, Y_n + 1) - f(X_n, Y_n)\Big)$$

$$+ \frac{Y_n}{m}\Big(f(X_n, Y_n - 1) - (f(X_n, Y_n))\Big) \stackrel{(3.2.23)}{=} 0.$$

$\square$

The martingale $(Z_n)_{n\geq 0}$ is bounded so it is uniformly integrable and we deduce from the Optional Sampling Theorem that

$$\mathbb{E}[\,f(0,Y_N)\,] = \mathbb{E}[\,Z_N\,] = \mathbb{E}[\,Z_1\,] = \mathbb{E}[\,f(X_1,Y_1)\,] = \mathbb{E}[\,f(m,0)\,].$$

This holds for any function f satisfying (3.2.23). Now observe that the function

$$f : \mathbb{N}_0 \to \mathbb{N}_0, \quad f(x,y) = H_x + \frac{y}{1+x},$$

where

$$H_0 = 0, \quad H_x = 1 + \frac{1}{2} + \cdots + \frac{1}{x}, \quad \forall x > 0,$$

satisfies (3.2.23), and we conclude

$$\mathbb{E}[\,Y_N\,] = H_m \sim \log m \text{ as } m \to \infty.$$

For example, if $m = 30$, then $H_m \approx 3.99$ so at the moment Ann has all the complete collection of 30 coupons, we expect that her little brother is missing only about 4 of them. Nearly there. □

3.2.5 Uniformly integrable submartingales

The proof of Theorem 3.59 yields the following submartingale counterpart.

Theorem 3.75. *If $(X_n)_{n\in\mathbb{N}_0}$ is a submartingale, then the following are equivalent.*

(i) *The collection $(X_n)_{n\in\mathbb{N}_0}$ is uniformly integrable.*
(ii) *The sequence $(X_n)_{n\in\mathbb{N}_0}$ converges a.s. and L^1 to a random variable X_∞.*
(iii) *The sequence $(X_n)_{n\in\mathbb{N}_0}$ converges L^1 to a random variable X_∞.* □

Corollary 3.76. *Suppose that $X_\bullet = (X_n)_{n\in\mathbb{N}_0}$ is a submartingale with Doob decomposition $X_n = X_0 + M_n + C_n$, where $(M_n)_{n\in\mathbb{N}_0}$ is the martingale component and $(C_n)_{n\in\mathbb{N}_0}$ is the predictable compensator. Then the following are equivalent.*

(i) *The submartingale $(X_n)_{n\in\mathbb{N}_0}$ is uniformly integrable.*
(ii) *The martingale $(M_n)_{n\in\mathbb{N}_0}$ and the compensator $(C_n)_{n\in\mathbb{N}_0}$ are uniformly integrable.*

Proof. Clearly (ii) ⇒ (i). To prove the converse note that

$$\mathbb{E}[\,|C_n|\,] = \mathbb{E}[\,C_n\,] = \mathbb{E}[\,X_n\,] - \mathbb{E}[\,X_0\,]$$

and since (X_n) is uniformly integrable we deduce

$$\sup_n \mathbb{E}[\,|C_n|\,] \leq \sup_n \mathbb{E}[\,|X_n|\,] - \mathbb{E}[\,X_0\,] < \infty.$$

The limit $C_\infty := \lim_{n\to\infty} C_n$ exists because (C_n) is nondecreasing. The Monotone Convergence theorem implies that C_∞ is integrable. Since $|C_n| = C_n \leq C_\infty, \forall n$, we deduce that the family (C_n) is UI. □

We can use Doob's decomposition to prove a submartingale version of Theorem 3.64.

Corollary 3.77. *If $X_\bullet = (X_n)_{n\in\mathbb{N}_0}$ is a uniformly integrable submartingale, then for any stopping time T the stopped submartingale $X_n^T = X_{T\wedge n}$ is a uniformly integrable submartingale.*

Proof. Consider the Doob decomposition of $X_\bullet$, $X_n = X_0 + M_n + C_n$. From Corollary 3.76 we deduce that $M_\bullet$ and $C_\bullet$ are UI. Moreover, the Doob decomposition of X^T is $X^T = M^T + C^T$. Corollary 3.65 shows that M^T is UI and C^T is UI since $0 \le C_n^T \le C_\infty \in L^1$. □

Theorem 3.78 (Optional Sampling: UI submartingales). *Suppose that*

$$X_\bullet = (X_n)_{n\in\mathbb{N}_0}$$

is a UI submartingale. Then for any stopping times S, T such that $S \le T$ we have

$$\hat{X}_S \le \mathbb{E}[\hat{X}_T \| \mathcal{F}_T]. \tag{3.2.25}$$

In particular, if we let $T = \infty$,

$$\hat{X}_S \le \mathbb{E}[X_\infty \| \mathcal{F}_S]. \tag{3.2.26}$$

Proof. If $X_\bullet = M_\bullet + C_\bullet$ is the Doob decomposition of X, then

$$X_\bullet^S = M_\bullet^S + C_\bullet^S, \quad X_\bullet^T = M_\bullet^T + C_\bullet^T.$$

In this case $X^S = (X^T)^S$ we deduce that $\hat{X}_S = \widehat{X^T}_S$. Then, since $\hat{C}_S$ is $\mathcal{F}_S$-measurable,

$$\hat{X}_S = \hat{M}_S + \hat{C}_S \stackrel{(3.2.19)}{=} \mathbb{E}[M_\infty^T \| \mathcal{F}_S] + \mathbb{E}[\hat{C}_S \| \mathcal{F}_S]$$

$$\le \mathbb{E}[M_\infty^T \| \mathcal{F}_S] + \mathbb{E}[\hat{C}_T \| \mathcal{F}_S] = \mathbb{E}[M_\infty^T \| \mathcal{F}_S] + \mathbb{E}[C_\infty^T \| \mathcal{F}_S] = \mathbb{E}[\hat{X}_T \| \mathcal{F}_S].$$

□

Corollary 3.79 (Optional Sampling). *Suppose that $Y_\bullet = (Y_n)_{n\in\mathbb{N}_0}$ is a uniformly integrable submartingale and S, T are a.s. finite stopping times such that $S \le T$. Then $Y_S \le \mathbb{E}[Y_T \| \mathcal{F}_S]$.*

Proof. Use (3.2.26) with the UI submartigale $X = Y^T$ and observe that $X_\infty = Y_\infty^T = \hat{Y}_T = Y_T$. □

Example 3.80 (Optimal Gambling Strategies). Consider a game of chance where the winning probability is $p < \frac{1}{2}$. For example, in the red-and-black roulette game one bets on black with winning probability $p = \frac{18}{38} \approx 0.473$. (The fair case $p = \frac{1}{2}$ is discussed in Exercise 3.19.)

Before each game the player bets a sum s, called *stake*, that cannot be larger than his fortune at that moment. If he wins, his fortune increases by the amount that he bet. Otherwise he loses his stake.

The player starts with a sum of money x and decides that he will play until the first moment his fortune goes above a set sum, the goal, say 1. His strategy is based on a function $\sigma(x)$. If his fortune after n games is X_n, then the amount he wagers for the next game depends on his current fortune X_n and is $\sigma(X_n)$. The player stops playing when, either he is broke, or he has reached (or surpassed) his goal. The function σ is known as the *strategy* of the gambler.

We denote by $\pi(x,\sigma)$ the probability that the gambler will reach his goal using the strategy σ, given that his initial fortune is x.

We want to show that the strategy that maximizes the winning probability $\pi(x,\sigma)$ is the "*go-bold*" strategy: if your fortune is less than half the goal, bet it all, and if your fortune is more than half the goal, bet as much as you need to reach your goal. Our presentation follows [64, §24.8]. To find out about gambling strategies for more complex games we refer to [49].

First let us introduce the appropriate formalism. The strategies will be chosen from a space $\boldsymbol{S}$, the collection of measurable functions $\sigma : [0,\infty) \to [0,\infty)$ such that

$$\sigma(x) \leq x, \ \forall x \in [0,1] \ \text{and} \ \sigma(x) = 0, \ \forall x > 1.$$

Note that the stopping rule is built in the definition of $\boldsymbol{S}$.

The sequence of games encoded by the sequence of i.i.d. random variables $(Y_n)_{n\in\mathbb{N}}$ such that

$$\mathbb{P}[Y_n = 1] = p, \ \mathbb{P}[Y_n = -1] = 1 - p, \ 0 < p < \frac{1}{2}.$$

For each $x \geq 0$ and each $\sigma \in \boldsymbol{S}$ define inductively a sequence of random variables $X_n = X_n^{x,\sigma}$,

$$X_0^x = x, \ X_{n+1} = X_n + \sigma(X_n)Y_{n+1}, \ n \geq 0. \tag{3.2.27}$$

We denote by $(\Omega, \mathcal{S}, \mathbb{P})$ the probability space where the random variables X_n and Y_n are defined. Thus $X_n^{x,\sigma}$ is the fortune of the player after n games starting with initial fortune x and using the strategy σ. Note that $\sigma(X_n)$ is the amount of money the player bets before the $(n+1)$-th game. It depends only on its fortune X_n at that time. If $Y_n = 1$ the player gains $\sigma(X_n)$ and if $Y_n = -1$, the player loses this amount. His strategy σ stays the same for the duration of the game.

Let us observe first that

$$X_\infty^{x,\sigma} := \lim_{n\to\infty} X_n^{x,\sigma}$$

exists a.s. and L^1. We will prove this by showing that $X_n^{x,\sigma}$ is a bounded supermartingale.

Since $\sigma(x) \leq x$ we deduce $x - \sigma(x) \geq 0$ and we deduce inductively that $X_n \geq 0$, a.s., $\forall n$. Next, we observe that if $x \leq 1 + x$ then $x + \sigma(x) \leq x + 1$. We deduce inductively that $X_n \leq x + 1$, a.s., $\forall n$.

We have $\mathbb{E}[Y_n] = 2p - 1 < 0$ and thus
$$\mathbb{E}[X_{n+1} \| \mathcal{F}_n] = X_n + \sigma(X_n)\mathbb{E}[Y_{n+1}] \leq X_n.$$
Thus (X_n) is a uniformly bounded supermartingale and thus UI. Set
$$h(x, \sigma) := \mathbb{E}[\min(X_\infty^{x,\sigma}, 1)] \text{ and } \pi(x, \sigma) := \mathbb{P}[X_\infty^{x,\sigma} \geq 1].$$
Observe that
$$x \geq h(x, \sigma) \geq \pi(x, \sigma), \quad \forall x \in [0, 1], \quad \sigma \in \boldsymbol{S}. \tag{3.2.28}$$
Since (X_n) is a supermartingale and the function $x \mapsto \min(x, 1)$ is concave and non-decreasing, the sequence $\min(X_n, 1)$ is also a supermartingale. Using the continuity of $x \mapsto \min(x, 1)$ we deduce from (i) that
$$\min(X_n^{x,\sigma}, 1) \to \min(X_\infty^{x,\sigma}, 1) \text{ a.s.}$$
Since $0 \leq \min(X_n, 1) \leq 1$ we deduce from the Dominated Convergence theorem that
$$x = \mathbb{E}[\min(X_0, 1)] \geq \mathbb{E}[\min(X_\infty^{x,\sigma}, 1)] \geq \mathbb{E}[\boldsymbol{I}_{X_\infty \geq 1}] \geq \mathbb{P}[X_\infty \geq 1] \geq \pi(x, \sigma).$$
Let us observe that if a strategy σ depends continuously on x, then
$$h(x, \sigma) = \pi(x, \sigma).$$
Set again $X_n = X_n^{x,\sigma}$. We will prove that $\mathbb{P}[0 < X_\infty < 1] = 0$. We argue by contradiction and assume
$$\mathbb{P}[0 < X_\infty < 1] > 0.$$
Thus, assume there exists $\omega \in \Omega$ such that $X_\infty(\omega) \in (0, 1)$ and
$$\lim_{n \to \infty} X_n(\omega) = X_\infty(\omega).$$
Thus
$$\lim_{n \to \infty} \sigma(X_n(\omega)) = \sigma(X_\infty(\omega)) > 0.$$
On the other hand,
$$\sigma(X_n) = |X_{n+1}(\omega) - X_n(\omega)| \to 0 \text{ as } n \to \infty.$$
Hence $\mathbb{P}[0 < X_\infty < 1] = 0$ so
$$\mathbb{E}[\min(X_\infty, 1)] = \mathbb{P}[X_\infty \geq 1].$$
We have the following *optimality criterion*.

Lemma 3.81. *Let $\sigma_0 \in \boldsymbol{S}$ and set $h_0(x) := h(x, \sigma_0)$, $\pi_0(x) = \pi(x, \sigma_0)$. If $h_0(x)$ is continuous and satisfies,*
$$h_0(x) \geq ph_0(x + s) + (1 - p)h_0(x - s), \tag{3.2.29}$$
then, for any $\sigma \in \boldsymbol{S}$, and any $x \in [0, 1]$ we have $\pi(x, \sigma_0) = h_0(x) \geq \pi(x, \sigma)$.

Proof. Fix $\sigma \in S$ and $x \in [0,1]$ and set $X_n = X_n^{x,\sigma}$. We set $h_0(x) = 1$, for $x \geq 1$. This is a natural condition: if the initial fortune is greater than the goal then the probability of achieving the goal is 1.

Observe that the random process $Y_n = h_0(X)$ is a supermartingale. Indeed,

$$\mathbb{E}\big[\, h_0(\overline{X}_{n+1})\,\|\,\mathcal{F}_n\,\big] = \mathbb{E}\big[\, h_0(\overline{X}_n + \sigma(\overline{X}_n)Y_{n+1})\,\|\,\mathcal{F}_n\,\big]$$

$$= \mathbb{E}\big[\, h_0(\overline{X}_n + \sigma(\overline{X}_n))\boldsymbol{I}_{\{Y_{n+1}=1\}} + h_0(\overline{X}_n - \sigma(\overline{X}_n))\boldsymbol{I}_{\{Y_{n+1}=-1\}}\,\|\,\mathcal{F}_n\,\big]$$

$$= p h_0(\overline{X}_n + \sigma(\overline{X}_n)) + (1-p) h_0(\overline{X}_n - \sigma(\overline{X}_n)) \stackrel{(3.2.29)}{\leq} h_0(\overline{X}_n).$$

Thus Y_n is a bounded supermartingale and thus

$$h_0(x) = \mathbb{E}\big[\, h_0(X_0^{x,\sigma})\,\big]\mathbb{E}\big[\, Y_0\,\big] \geq \mathbb{E}\big[\, Y_n\,\big].$$

Now observe that $\mathbb{E}\big[\, Y_0\,\big] = h_0(x)$.

On the other hand, since $h_0(x)$ is continuous and bounded we deduce that $h_0(X_n)$ converges a.s. and L^1 to $h_0(X_\infty)$. Thus

$$\mathbb{E}\big[\, Y_\infty\,\big] = \mathbb{E}\big[\, h_0(X_\infty^{x,\sigma})\,\big] \geq \mathbb{P}\big[\, X_\infty^{x,\sigma} \geq 1\,\big] \geq \pi(x,\sigma).$$

$\square$

Define $\sigma_0 \in S$

$$\sigma_0(x) := \begin{cases} \min(x, 1-x), & x \in [0,1], \\ 0, & x \geq 1, \end{cases}$$

and set $h_0(x) := h(x, \sigma_0)$, $\pi_0(x) = \pi(x, \sigma_0)$. We want to show that σ_0 satisfies all the conditions of Lemma 3.81.

Clearly σ_0 is a continuous strategy. By construction, for any $x \in [0,1]$ we have $0 \leq X_n^{x,\sigma_0} \leq 1$ a.s. so

$$\pi_0(x) = h_0(x) = \mathbb{E}\big[\, X_\infty^{x,\sigma_0}\,\big], \quad \forall x \in [0,1].$$

The functions

$$[0,1] \ni x \mapsto x + \sigma_0(x) \in [0,1], \quad [0,1] \ni x \mapsto x - \sigma_0(x) \in [0,1]$$

are non-decreasing. We deduce inductively that if $x \leq y$ then

$$\mathbb{E}\big[\, X_n^{x,\sigma_0}\,\big] \leq \mathbb{E}\big[\, X_n^{y,\sigma_0}\,\big]$$

and, by letting $n \to \infty$ we deduce that $h_0(x) \leq h_0(y)$ so that h_0 is non-decreasing.

By conditioning on Y_1 we deduce that

$$h_0(x) = \begin{cases} p h_0(2x), & x \leq 1/2, \\ p + (1-p) h_0(2x-1), & 1/2 \leq x \leq 1, \\ 1, & x > 1. \end{cases} \qquad (3.2.30)$$

Set
$$\mathcal{D} := \left\{ \frac{k}{2^n};\ n \in \mathbb{N}_0,\ 0 \leq k \leq 2^n \right\}.$$

We will prove by induction on n that (3.2.29) holds for x of the form $x = \frac{k}{2^n}$. Start with $n = 1$ so $x = \frac{1}{2}$. We have

$$h(1/2) - ph(1/2 + s) - (1-p)h(1/2 - s)$$

$$\stackrel{(3.2.30)}{=} p - p\big(p + (1-p)h(2s)\big) - (1-p)ph(1-2s)$$

$$= p(1-p)\Big(1 - \big(h(2s) + h(1-2s)\big)\Big) \geq 0,$$

where at the last step we used the fact that $h(x) \leq x$, $\forall x \in [0,1]$.

For the inductive step, assume that $n > 1$ and $x = \frac{k}{2^n}$, $k < 2^n$. Choose $s \in [0, x]$. We consider several cases.

Case 1. $x + s \leq \frac{1}{2}$. Using (3.2.30) and the induction hypothesis we deduce

$$ph(x+s) + (1-p)h(x-s) = p\big(ph(2x+2s) + (1-p)ph(2x-2s)\big)$$

$$\leq ph(2x) = h(x).$$

Case 2. $x - s \geq \frac{1}{2}$. Similar to **Case 1**.

Case 3. $x \leq \frac{1}{2}$ and $x + s \geq \frac{1}{2}$. Using (3.2.30) we have

$$A := h(x) - ph(x+s) - (1-p)h(x-s)$$

$$= ph(x2x) - p\big(p + (1-p)h(2x+2s-1)\big) - (1-p)ph(2x-2s)$$

$$= p\big(h(2x) - p - (1-p)h(2x+2s-1) - (1-p)h(2x-2s)\big).$$

Observe that since $\frac{1}{2} \leq x + s \leq 2x$. Using (3.2.30) we deduce

$$h(2x) = p + (1-p)h(4x - 1)$$

so that

$$A = p\big(p + (1-p)h(4x-1) - p - (1-p)h(2x+2s-1) - (1-p)h(2x-2s)\big)$$

$$= p(1-p)\big(h(4x-1) - h(2x+2s-1) - h(x-2s)\big)$$

$$= (1-p)\big(h(2x-1/2) - ph(x+2s-1) - p(x-2s)\big)$$

$(p \leq 1 - p)$

$$\geq (1-p)\big(h(2x-1/2) - ph(x+2s-1) - (1-p)(x-2s)\big).$$

The induction hypothesis implies $h(2x - 1/2) - ph(x+2s-1) - (1-p)(x-2s) \geq 0$.

Case 4. $x \geq \frac{1}{2}$ and $x - s \leq \frac{1}{2}$. This is similar to the previous case.

We can now prove that h_0 is continuous. Since h is nondecreasing we deduce that the right/left limits $h(x\pm)$ exist at each $x \in [0,1]$. Since (3.2.29) holds for every x in a dense set we deduce

$$h(x-) \geq ph\big((x+s)-\big) + (1-p)h\big((x-s)-\big)$$

$\forall 0 \leq s < x \leq 1$. Now let $s \searrow 0$ to conclude

$$h(x-) \geq ph(x+) + (1-p)h(x-) \Rightarrow ph(x-) \geq ph(x+)$$

so that $h(x-) = h(x+)$, i.e., h is continuous. Since $\mathcal{D}$ is dense in $[0,1]$ we deduce that h_0 satisfies (3.2.29) on $[0,1]$. We can now invoke Lemma 3.81 to deduce that

$$\pi(x, \sigma_0) = h_0(x) \geq \pi(x, \sigma), \quad \forall x \in [0,1], \ \sigma \in \boldsymbol{S},$$

i.e., σ_0 is an optimal gambling strategy.

Let us explain how to compute $h_0(x)$, $x \in D$. Every number $x \in \mathcal{D}$ has a binary expansion

$$x = 0.\epsilon_1 \epsilon_2 \cdots = \sum_{n \geq 1} \frac{\epsilon_n}{2^n}$$

where $\epsilon_n \in \{0,1\}$, and $\epsilon_n = 0$ for $n \gg 0$. Note that

$$x < \frac{1}{2} \iff \epsilon_1 = 0.$$

The first equation in (3.2.30) reads

$$h(0.0\epsilon_2 \cdots) = p \cdot h(0.\epsilon_2 \cdots).$$

In particular

$$h\big(0.\underbrace{0 \cdots 0}_{k} 1\epsilon_{k+2} \cdots\big) = p^k (0.1\epsilon_{k+2} \cdots).$$

The second equation in (3.2.30) reads

$$h(0.1\epsilon_2 \ldots) = p + (1-p)h(0.\epsilon_2 \cdots).$$

We define $f_0, f_1 : [0,1] \to [0,1]$ by

$$f_0(x) = px, \quad f_1(x) = p + (1-p)x.$$

The above discussion shows that

$$h(0.\epsilon_1 \cdots \epsilon_n) = f_{\epsilon_1}\big(h(0.\epsilon_2 \cdots \epsilon_n)\big).$$

Since $h(0) = h(1/2) = p$ we deduce by iteration that if,

$$x = 0.\epsilon_1 \epsilon_2 \cdots \epsilon_n 1,$$

then

$$h(x) = f_{\epsilon_1} \circ f_{\epsilon_2} \circ \cdots \circ f_{\epsilon_n}(p).$$

Thus h is uniquely determined on $\mathcal{D}$ and, since $\mathcal{D}$ is dense on $[0,1]$, the function h is uniquely determined on $[0,1]$. Let us emphasize that $h_0(x)$ depends on the winning probability p.

As an illustration let us compute $h_0(21/32)$. Note that $\frac{21}{32}$ has the binary expansion
$$\frac{21}{32} = 0.10101$$
so that
$$h(21/32) = f_1 \circ f_0 \circ f_1 \circ f_0(p) = f_1 \circ f_0 \circ f_1(p^2)$$
$$= f_1 \circ f_0(p + p^2 - p^3) = f_1(p^2 + p^3 - p^4) = p + (1-p)(p^2 + p^3 - p^4)$$
$$= p + p^2 + p^3 - p^4 - p^3 - p^4 + p^5 = p + p^2 - 2p^4 + p^5.$$

For example if the winning probability is $p = 0.4$, then $h_0(21/32) = 0.519 > 0.5$. Thus, although the winning probability $p < 0.5$, using this strategy with an initial fortune 21/32, the odds of increasing the fortune to 1 are better than 50 : 50.

If the initial fortune is $x = \frac{1}{4}$, then using its binary expansion $\frac{1}{4} = 0.01$ we deduce
$$h_0(1/4) = p h_0(1/2) = \frac{p}{2}.$$

In this case, if $p = 0.4$, the probability of reaching his goal is 0.2, substantially smaller. □

3.2.6 Maximal inequalities and L^p-convergence

The results in this subsection are wide ranging generalizations of Kolmogorov's one series theorem. They depend on Doob's maximal inequality which generalizes Kolmogorov's inequality (2.1.3).

Theorem 3.82 (Doob's maximal inequality). *Suppose that $(X_n)_{n \in \mathbb{N}_0}$ is a submartingale. Set*
$$\tilde{X}_n := \sup_{k \leq n} X_k.$$
Then, for any $a > 0$, we have
$$\boxed{a\mathbb{P}\big[\tilde{X}_n \geq a\big] \leq \mathbb{E}\big[X_n \boldsymbol{I}_{\{\tilde{X}_n \geq a\}}\big] \leq \mathbb{E}\big[X_n^+\big].} \qquad (3.2.31)$$

Proof. Let us introduce the stopping time
$$T := \inf\{\, n \geq 0;\ X_n \geq a\,\}.$$
Then
$$A := \big\{\, \tilde{X}_n \geq a\,\big\} = \Big\{\, \sup_{k \leq n} X_k \geq a\,\Big\} = \{\, T \leq n\,\}.$$
Applying the Optional Sampling Theorem 3.28 to the bounded stopping times $T \wedge n$ and n we deduce
$$\mathbb{E}\big[X_{T \wedge n}\big] \leq \mathbb{E}\big[X_n\big].$$

On the other hand,
$$X_{T\wedge n}(\omega) = X_{T(\omega)}\boldsymbol{I}_A(\omega) + X_n \boldsymbol{I}_{A^c}(\omega) \geq a\boldsymbol{I}_A(\omega) + X_n(\omega)\boldsymbol{I}_{A^c}(\omega),$$
so $X_{T\wedge n} \geq a\boldsymbol{I}_A + X_n \boldsymbol{I}_{A^c}$. We deduce
$$a\mathbb{P}[A] + \mathbb{E}[X_n \boldsymbol{I}_{A^c}] \leq \mathbb{E}[X_{T\wedge n}] \leq \mathbb{E}[X_n] = \mathbb{E}[X_n \boldsymbol{I}_A] + \mathbb{E}[X_n \boldsymbol{I}_{A^c}].$$
This implies the first inequality in (3.2.31). The second inequality is trivial. $\square$

Corollary 3.83. *Suppose that (Y_n) is a martingale. We set*
$$Y_n^* = \max_{k \leq n} |Y_n|.$$
Then for every $c > 0$ and any $p \in [1, \infty)$ we have
$$\mathbb{P}[Y_n^* > c] \leq \frac{1}{c^p} \mathbb{E}[|Y_n|^p].$$

Proof. Doob's maximal inequality applied to the submartingale $X_n = |Y_n|^p$ yields
$$\mathbb{P}[Y_n^* > c] = \mathbb{P}\left[\max_{0 \leq k \leq n} |Y_n|^p > c^p\right] \leq \frac{1}{c^p} \mathbb{E}[|Y_n|^p].$$
$\square$

Theorem 3.84 (Doob's L^p-inequality). *Let $p > 1$ and suppose that $(X_n)_{n \in \mathbb{N}_0}$ is a positive submartingale such that $X_n \in L^p$, $\forall n \geq 0$. Set*
$$\widetilde{X}_n := \sup_{k \leq n} X_k.$$
Then for any $n \geq 0$ we have
$$\mathbb{E}\big[(\widetilde{X}_n)^p\big]^{\frac{1}{p}} \leq q \mathbb{E}[X_n^p]^{\frac{1}{p}}, \qquad (3.2.32)$$
where
$$\frac{1}{p} + \frac{1}{q} = 1 \;\; or \;\; q = \frac{p}{p-1}.$$
In particular, if $(Y_n)_{n \in \mathbb{N}_0}$ is a martingale and if
$$Y_n^* := \max_{k \leq n} |Y_k|,$$
then for any $n \geq 0$ we have[3]
$$\|Y_n^*\|_{L^p} \leq q \|Y_n\|_{L^p}. \qquad (3.2.33)$$

---
[3] Note that $q = \frac{p}{p-1}$ is the exponent conjugate to p, $\frac{1}{p} + \frac{1}{q} = 1$.

Proof. Clearly (3.2.32) ⇒ (3.2.33). Note that $(X_n^p)_{n\geq 0}$ is also a submartingale and $\tilde{X}_n \in L^p$. From Doob's maximal inequality we deduce

$$a\mathbb{P}\big[\,\tilde{X}_n \geq a\,\big] \leq \mathbb{E}\big[\,X_n \mathbf{I}_{\{\tilde{X}_n \geq a\}}\,\big]$$

so

$$\frac{1}{p}\mathbb{E}\big[\,\tilde{X}_n^p\,\big] \stackrel{(1.3.43)}{=} \int_0^\infty a^{p-1}\mathbb{P}\big[\,\tilde{X}_n \geq a\,\big]da \stackrel{(3.2.31)}{\leq} \int_0^\infty a^{p-2}\mathbb{E}\big[\,X_n \mathbf{I}_{\{\tilde{X}_n \geq a\}}\,\big]da.$$

Switching the order of integration we deduce

$$\int_0^\infty a^{p-2}\mathbb{E}\big[\,X_n \mathbf{I}_{\{\tilde{X}_n \geq a\}}\,\big]da = \mathbb{E}\left[X_n \int_0^{\tilde{X}_n} a^{p-2}da\right] = \frac{1}{p-1}\mathbb{E}\big[\,X_n \tilde{X}_n^{p-1}\,\big]$$

(use Hölder's inequality with $\frac{1}{q} = 1 - \frac{1}{p}$)

$$\leq \frac{1}{p-1}\mathbb{E}\big[\,X_n^p\,\big]^{\frac{1}{p}}\mathbb{E}\big[\,\tilde{X}_n^p\,\big]^{\frac{p-1}{p}}.$$

Hence

$$\frac{1}{p}\mathbb{E}\big[\,\tilde{X}_n^p\,\big] \leq \frac{1}{p-1}\mathbb{E}\big[\,X_n^p\,\big]^{\frac{1}{p}}\mathbb{E}\big[\,\tilde{X}_n^p\,\big]^{\frac{p-1}{p}}.$$

This proves (3.2.32). □

Definition 3.85. Let $p \in [1, \infty)$. A martingale $(X_n)_{n \in \mathbb{N}_0}$ is called an L^p-*martingale* if

$$\mathbb{E}\big[\,|X_n|^p\,\big] < \infty, \ \ \forall n \in \mathbb{N}_0.$$

A *bounded L^p-martingale* is a martingale $(X_n)_{n \in \mathbb{N}_0}$ such that

$$\sup_{n \in \mathbb{N}_0} \mathbb{E}\big[\,|X_n|^p\,\big] < \infty.$$ □

Corollary 3.86 (L^p-**martingale convergence theorem**). *Suppose that $(X_n)_{n \in \mathbb{N}_0}$ is a* <u>*bounded*</u> *L^p-martingale for some $p > 1$. Set*

$$X_n^* := \max_{k \leq n}|X_k|, \ \ X_\infty^* = \sup_{k \geq 0}|X_k| = \lim_{n \to \infty} X_n^*.$$

Then $(X_n)_{n \in \mathbb{N}_0}$ is a UI martingale and X_n converges a.s. and L^p to a random variable

$$X_\infty \in L^p(\Omega, \mathcal{F}_\infty, \mathbb{P}).$$

Moreover

$$\mathbb{E}\big[\,(X_\infty^*)^p\,\big] \leq \left(\frac{p}{p-1}\right)^p \mathbb{E}\big[\,|X_\infty|^p\,\big].$$

Proof. From the Monotone Convergence Theorem we deduce

$$\mathbb{E}\big[\,(X_\infty^*)^p\,\big] = \lim_{n\to\infty} \mathbb{E}\big[\,(X_n^*)^p\,\big] \leq \left(\frac{p}{p-1}\right)^p \sup_{n\geq 0} \mathbb{E}\big[\,|X_n|^p\,\big] < \infty$$

so $X_\infty^* \in L^p$ and $|X_n| \leq X_\infty^*$, $\forall n \geq 0$. The desired conclusions now follow from the martingale convergence theorem and the Dominated Convergence Theorem. □

Example 3.87 (Kolmogorov's one series theorem). Suppose that $(X_n)_{n\geq 0}$ is a sequence of independent random variables such that

$$\mathbb{E}[X_n] = 0, \ \ \forall n \geq 0, \ \ \sum_{n\geq 0} \text{Var}[X_n] < \infty.$$

Then the random series $X_0 + X_1 + \cdots$ is a.s. and L^2-convergent. Indeed, the sequence of partial sums

$$S_n = X_0 + \cdots + X_n$$

is a bounded L^2-martingale and so it converges a.s. and L^2. □

Example 3.88 (Likelihood ratio). This example has origin in statistics. Suppose that we have a random quantity and we have reasons to believe that its probability distribution is either of the form $p(x)dx$ or $q(x)dx$ where $p, q : \mathbb{R} \to [0, \infty)$ are mutually absolutely continuous probability densities on $\mathbb{R}$

$$\int_\mathbb{R} p(x)dx = \int_\mathbb{R} q(x)dx = 1.$$

We want to describe a statistical test that helps deciding which is the real distribution. Our presentation follows [75, Sec. 12.8].

We take a large number of samples of the random quantity, or equivalently, suppose that we are given a sequence of i.i.d. random variables $(X_n)_{n\geq 1}$ with common probability density f, where f is one of the two densities p or q. Assume for simplicity that $p(x), q(x) > 0$, for almost any $x \in \mathbb{R}$.

The products

$$Y_n := \prod_{k=1}^n \frac{p(X_k)}{q(X_k)}$$

are called *likelihood ratios*. Note that if $f = q$, then $\mathbb{E}\big[\,Y_n\,\big] = 1$, $\forall n$.

To decide whether $f = q$ or $f = p$ we fix a (large) positive number a and a large $n \in \mathbb{N}$ and adopt the prediction strategy

$$\bar{f}_n := \begin{cases} p, & Y_n \geq a, \\ q, & Y_n < a. \end{cases}$$

We want to show that this strategy picks the correct density with high confidence, i.e., $\mathbb{P}\big[\,f = \bar{f}_n\,\big]$ is very close to 1 for large n and a.

If $f = q$, then Y_n is a product of i.i.d. nonnegative random variables with mean 1 and, as shown in Example 3.6, it is a martingale with respect to the filtration $\mathcal{F}_n = \sigma(X_1, \ldots, X_n)$.

The function log is *strictly* concave and we deduce from Jensen's inequality

$$\mathbb{E}\left[\log \frac{p(X_n)}{q(X_n)}\right] < \log \mathbb{E}\left[\frac{p(X_n)}{q(x_n)}\right] = 0.$$

The Strong Law of Large Numbers shows that

$$\frac{1}{n}\sum_{k=1}^{n} \log \frac{p(X_k)}{q(X_k)} \to \mathbb{E}[\log Y_1] < 0, \quad \text{a.s.}$$

Thus

$$\log Y_n = \sum_{k=1}^{n} \log \frac{p(X_k)}{q(X_k)} \to -\infty \quad \text{a.s.}$$

Thus, if $f = q$, then $Y_n \to 0$ a.s. In particular, $\mathbb{P}[f = \bar{f}_n] = \mathbb{P}[Y_n < a] \to 1$ as $n \to \infty$.

If $f = p$, then a similar argument shows that $\frac{1}{Y_n} \to 0$ a.s. We deduce that

$$Y_n \to \begin{cases} 0, & f = q, \\ \infty, & f = p. \end{cases}$$

Moreover, Doob's maximal inequality (3.2.31) shows that if $f = q$ so Y_n is a martingale, we have

$$\mathbb{P}\left[\max_{1 \le k \le n} Y_k \ge a\right] \le \frac{1}{a}.$$

Thus, if $f = q$ the probability Y_n overshoots the level $a \gg 1$ is small and this statistical test makes the right decision with high confidence. $\square$

Example 3.89. Consider again the branching process in Example 3.8. Suppose that the reproduction law μ satisfies

$$m = \sum_{k=0}^{\infty} k\mu(k) < \infty, \quad \sum_{k=0}^{\infty} k^2 \mu(k) < \infty.$$

We set

$$\sigma^2 := \text{Var}[\mu] = \sum_{k=0}^{\infty} k^2 \mu(k) - m^2.$$

Note that

$$Z_{n+1} = \sum_{k=1}^{\infty}\left(\sum_{j=1}^{k} X_{n,j}\right) \boldsymbol{I}_{\{Z_n = k\}} = \sum_{j=0}^{\infty} X_{n,j} \sum_{k \ge j} \boldsymbol{I}_{\{Z_n = k\}} = \sum_{j=0}^{\infty} X_{n,j} \boldsymbol{I}_{\{Z_n \ge j\}},$$

$$\mathbb{E}[Z_{n+1}^2 \| \mathcal{F}_n] = \mathbb{E}\left[\sum_{k,j=1}^{\infty} \boldsymbol{I}_{\{Z_n \ge j, Z_n \ge k\}} X_{n,j} X_{n,k} \| \mathcal{F}_n\right]$$

$(X_{n,j}, X_{n,k} \perp\!\!\!\perp \mathcal{F}_n)$

$$= \sum_{k,j=1}^{\infty} \boldsymbol{I}_{\{Z_n \geq j, Z_n \geq k\}} \mathbb{E}[X_{n,j} X_{n_k}] = \sum_{k,j=1}^{\infty} \boldsymbol{I}_{\{Z_n \geq j, Z_n \geq k\}} (m^2 + \delta_{jk}\sigma^2)$$

$$= m^2 \sum_{j,k=j=1}^{\infty} \boldsymbol{I}_{\{Z_n \geq j\}} \boldsymbol{I}_{\{Z_n \geq k\}} + \sigma^2 \sum_{k=1}^{\infty} \boldsymbol{I}_{\{Z_n \geq k\}}$$

$(\mathbb{E}[Z_n] = \sum_{k \geq 1} \mathbb{P}(Z_n \geq 1))$

$$= m^2 \left(\sum_{k=1}^{\infty} \boldsymbol{I}_{\{Z_n \geq k\}} \right)^2 + \sigma^2 \sum_{k=1}^{\infty} \boldsymbol{I}_{\{Z_n \geq k\}} = m^2 Z_n^2 + \sigma^2 Z_n.$$

Hence
$$\mathbb{E}[Z_{n+1}^2] = m^2 \mathbb{E}[Z_n^2] + m^2 \mathbb{E}[Z_n] = \sigma^2 \mathbb{E}[Z_n^2] + \sigma^2 m^n \mathbb{E}[Z_0] = m^2 \mathbb{E}[Z_n^2] + \sigma^2 m^n \ell.$$
We set
$$q_{n+1} := m^{-2n} \mathbb{E}[Z_n^2]$$
and we get from the above that
$$q_{n+1} = q_n + m^{-n-2} \sigma^2 \ell.$$

This shows that if $m > 1$, then the sequence (q_n) converges so the martingale $W_n := m^{-n} Z_n$ converges in L^2 and a.s. The limit W_∞ is nonzero if $\ell = \mathbb{E}[Z_0] > 0$ because
$$\mathbb{E}[W_\infty] = \mathbb{E}[W_0] = \mathbb{E}[Z_0] = \ell.$$
We refer to Exercise 3.26 for more details about W_∞. □

3.2.7 Backwards martingales

Suppose that the parameter set $\mathbb{T}$ is
$$\mathbb{T} = -\mathbb{N}_0 = \{0, -1, -2, \dots, \}.$$
In this case a $\mathbb{T}$-filtration $\mathcal{F}_n$, $n \in -\mathbb{N}_0$ is called a *backwards filtration*. We set
$$\mathcal{F}_{-\infty} := \bigcap_{n \leq 0} \mathcal{F}_n.$$

A *backwards martingale* (submartingale, supermartingale) is a martingale (resp. submartingale, supermartingale) adapted to a backwards filtration.

Theorem 3.90 (Convergence of backwards submartingales). *Suppose that $\mathcal{F}_\bullet = (\mathcal{F}_n)_{n \in -\mathbb{N}_0}$ is a backwards filtration of $(\Omega, \mathcal{F}, \mathbb{P})$ and $X_\bullet = (X_n)_{n \in -\mathbb{N}_0}$ is $\mathcal{F}_\bullet$-submartingale, i.e.,*
$$X_n \leq \mathbb{E}[X_m \| \mathcal{F}_n], \quad \forall n, m \in -\mathbb{N}_0, \; n \leq m,$$
and
$$C := \inf_{n \leq 0} \mathbb{E}[X_n] > -\infty.$$
Then the following hold.

(i) The family $(X_n)_{n \in -\mathbb{N}_0}$ is UI.
(ii) There exists $X_{-\infty} \in L^1(\Omega, \mathcal{F}_{-\infty}, \mathbb{P})$ such that $X_n \to X_{-\infty}$ a.s. and L^1 as $n \to -\infty$.

Moreover
$$X_{-\infty} \leq \mathbb{E}[X_n \| \mathcal{F}_{-\infty}], \tag{3.2.34}$$
with equality if $(X_n)_{n \in -\mathbb{N}_0}$ is a martingale.

Proof. Step 1. *Boundedness in L^1.* Observe that (X_n^+) is a submartingale and thus
$$\mathbb{E}[X_n^+] \leq \mathbb{E}[X_0^+], \quad \forall n \leq 0.$$
On the other hand, there exists $C \in \mathbb{R}$ such that
$$\mathbb{E}[X_n] = \mathbb{E}[X_n^+] - \mathbb{E}[X_n^-] \geq C, \quad \forall n \geq 0.$$
Hence
$$\mathbb{E}[X_n^-] \leq C + \mathbb{E}[X_n^+] \leq C + \mathbb{E}[X_0^+], \quad \forall n \leq 0,$$
and consequently,
$$Z := \sup_{n \leq 0} \mathbb{E}[|X_n|] < \infty. \tag{3.2.35}$$

Step 2. *Almost sure convergence.* For $K \in \mathbb{N}$ consider the (increasing) filtration
$$\mathcal{G}_n^K := \mathcal{F}_{(-K+n) \wedge 0}, \quad n \in \mathbb{N}_0,$$
and the $\mathcal{G}_n^K$-submartingale $Y_n^K = X_{(-K+n) \wedge 0}$. Thus
$$Y_0^K = X_{-K}, \; Y_1^K = X_{-K+1}, \ldots, Y_K^K = X_0, \; Y_{K+1}^K = X_0, \ldots .$$
Doob's upcrossing inequality applied to the submartingale Y_n^K shows that, for any rational numbers $a < b$ we have
$$(b-a)\mathbb{E}[N_K([a,b], Y^K)] \leq \mathbb{E}[(X_0 - a)^+] - \mathbb{E}[(X_{-K} - a)^+]$$
$$\leq \mathbb{E}[(X_0 - a)^+] \leq |a| + \mathbb{E}[|X_0|].$$
This proves that, for any rational numbers $a < b$, the nondecreasing sequence
$$K \mapsto N_K([a,b], Y^K)$$
is also bounded, and thus it has a finite limit $N_\infty([a,b], X)$ as $K \to \infty$. An obvious version of Lemma 3.38 shows that X_n has an a.s. limit a.s. $n \to -\infty$. The limit is a $\mathcal{F}_{-\infty}$-measurable random variable $X_{-\infty}$. Fatou's Lemma shows that
$$\mathbb{E}[|X_{-\infty}|] < \infty.$$

Step 3. *Uniform integrability.* This is obvious if $(X_n)_{n \leq}$ is a martingale since
$$X_n = \mathbb{E}[X_0 \| \mathcal{F}_n]$$
and the conclusion follows from Corollary 3.56.

In general, if $(X_n)_{n\leq 0}$ is a submartingale, we have
$$\mathbb{E}[X_n] \leq \mathbb{E}[X_m], \quad \forall n \leq m \leq 0.$$
Since the sequence $\mathbb{E}[X_n]$ is bounded below we deduce that it has a finite limit. Thus, for any $\varepsilon > 0$, there exists $K = K(\varepsilon) > 0$ such that
$$\mathbb{E}[X_{-n}] \geq \mathbb{E}[X_{-K}] - \frac{\varepsilon}{2}, \quad \forall n \geq K.$$
For $n > K$ and $a > 0$ we have
$$\mathbb{E}[|X_{-n}|\boldsymbol{I}_{\{|X_{-n}|>a\}}] = \mathbb{E}[(-X_{-n})\boldsymbol{I}_{\{X_{-n}<-a\}}] + \mathbb{E}[X_{-n}\boldsymbol{I}_{\{X_{-n}>a\}}]$$
$$= \boxed{-\mathbb{E}[X_{-n}]} + \mathbb{E}[X_{-n}\boldsymbol{I}_{\{X_{-n}\geq -a\}}] + \mathbb{E}[X_{-n}\boldsymbol{I}_{\{X_{-n}>a\}}]$$
$$\leq \boxed{-\mathbb{E}[X_{-K}] + \frac{\varepsilon}{2}} + \mathbb{E}[X_{-n}\boldsymbol{I}_{\{X_{-n}\geq -a\}}] + \mathbb{E}[X_{-n}\boldsymbol{I}_{\{X_{-n}>a\}}].$$
Now observe that, for any $H \in \mathcal{F}_n$, we have
$$X_{-n}\boldsymbol{I}_H \leq \mathbb{E}[X_{-K}\|\mathcal{F}_{-n}]\boldsymbol{I}_H = \mathbb{E}[X_{-K}\boldsymbol{I}_H\|\mathcal{F}_{-n}],$$
so
$$\mathbb{E}[X_{-n}\boldsymbol{I}_H] \leq \mathbb{E}[X_{-K}\boldsymbol{I}_H].$$
Hence, if $H = \{X_{-n} \geq -a\}$, or $H = \{X_{-n} > a\}$, then
$$\mathbb{E}[X_{-n}\boldsymbol{I}_{\{X_{-n}\geq -a\}}] + \mathbb{E}[X_{-n}\boldsymbol{I}_{\{X_{-n}>a\}}] \leq \mathbb{E}[X_{-K}\boldsymbol{I}_{\{X_{-n}\geq -a\}}] + \mathbb{E}[X_{-K}\boldsymbol{I}_{\{X_{-n}>a\}}],$$
and
$$\mathbb{E}[|X_{-n}|\boldsymbol{I}_{\{|X_{-n}|>a\}}] \leq -\mathbb{E}[X_{-K}] + \mathbb{E}[X_{-K}\boldsymbol{I}_{\{X_{-n}\geq -a\}}] + \mathbb{E}[X_{-K}\boldsymbol{I}_{\{X_{-n}>a\}}] + \frac{\varepsilon}{2}$$
$$= \mathbb{E}[|X_{-K}|\boldsymbol{I}_{\{|X_{-n}|\geq a\}}] + \frac{\varepsilon}{2}.$$
From Markov's inequality and (3.2.35) we deduce
$$\mathbb{P}[|X_{-m}| > a] \leq \frac{Z}{a}, \quad \forall m \in \mathbb{N}_0.$$
Since the family consisting of the single random variable X_{-K} is uniformly integrable, we deduce that there exists $\delta = \delta(\varepsilon) > 0$ such that, for any $A \in \mathcal{F}_K$ satisfying $\mathbb{P}[A] < \delta$ we have
$$\mathbb{E}[|X_{-K}|\boldsymbol{I}_A] < \frac{\varepsilon}{2}.$$
We deduce that for any $a > 0$ such that $\frac{Z}{a} < \delta(\varepsilon)$ we have
$$\mathbb{E}[|X_{-n}|\boldsymbol{I}_{\{|X_{-n}|>a\}}] \leq \mathbb{E}[|X_{-K}|\boldsymbol{I}_{\{|X_{-n}|\geq a\}}] < \frac{\varepsilon}{2}.$$
This proves that the family $(X_{-n})_{n\in\mathbb{N}_0}$ is UI.

Step 4. *Conclusion.* Finally, observe that for any $A \in \mathcal{F}_{-\infty}$ and any $n \leq m \leq 0$ we have $\mathbb{E}[X_n\boldsymbol{I}_A] \leq \mathbb{E}[X_m\boldsymbol{I}_A]$. If we let $n \to -\infty$ we deduce
$$\mathbb{E}[X_{-\infty}\boldsymbol{I}_A] \leq \mathbb{E}[X_m\boldsymbol{I}_A], \quad \forall m \leq 0, \ A \in \mathcal{F}_{-\infty}.$$

This is precisely the inequality (3.2.34). When (X_n) is a martingale all the above inequalities are equalities. $\square$

Corollary 3.91 (Backwards Martingale Convergence). *Suppose that* $(\mathcal{G}_n)_{n \in \mathbb{N}_0}$ *is a decreasing family of σ-subalgebras of $\mathcal{F}$ and*

$$Z \in L^1(\Omega, \mathcal{F}, \mathbb{P}).$$

Then the sequence $\mathbb{E}[Z \| \mathcal{G}_n]$ *converges a.s. and L^1 to* $\mathbb{E}[Z \| \mathcal{G}_\infty]$, *where*

$$\mathcal{G}_\infty = \bigcap_{n \geq 0} \mathcal{G}_n.$$

Proof. Apply the previous theorem to the backwards filtration $\mathcal{F}_n := \mathcal{G}_{-n}$, $n \leq 0$, and the martingale $Z_n := \mathbb{E}[Z \| \mathcal{F}_n]$, $n \leq 0$. □

3.2.8 Exchangeable sequences of random variables

An n-dimensional random vector $\boldsymbol{X} = (X_1, \ldots, X_n)$ is called *exchangeable* if, for any permutation π of $\{1, \ldots, n\}$ the random vectors $(X_1, \ldots, X_n)$ and $(X_{\pi(1)}, \ldots, X_{\pi(n)})$ have identical distributions.

A sequence of random variables $(X_k)_{k \in \mathbb{N}}$ is called *exchangeable* if for any $n \in \mathbb{N}$ the random vector $(X_1, \ldots, X_n)$ is exchangeable. One also refers to an exchangeable sequence as an *exchangeable process*.

Equivalently, if we denote by $\mathfrak{S}_n$ the subgroup of permutations φ of $\mathbb{N}$ such that $\varphi(r) = r$, $\forall r > n$, then the sequence $(X_n)_{n \geq 1}$ is exchangeable if for any $n \in \mathbb{N}$ and any $\varphi \in \mathfrak{S}_n$ the sequences $(X_n)_{n \in \mathbb{N}}$ and $(X_{\varphi(n)})_{n \in \mathbb{N}}$ are identically distributed.

Example 3.92. (a) A sequence of i.i.d. random variables $(X_n)_{n \geq 1}$ is exchangeable. (b) Suppose that $(\mu_\lambda)_{\lambda \in \Lambda}$ is a family of Borel probability measures on $\mathbb{R}$ parametrized by a probability space $(\Lambda, \mathcal{S}, \mathbb{P}_\Lambda)$ such that, for any Borel subset $B \subset \mathbb{R}$, the function

$$\Lambda \ni \lambda \mapsto \mu_\lambda[B]$$

is measurable. In other words, $\mu_\bullet$ is a *random probability measure*. In the language of kernels, the function $\mu_\bullet : \Lambda \to \mathcal{B}_\mathbb{R}$ is a Markov kernel $(\Lambda, \mathcal{S}) \to (\mathbb{R}, \mathcal{B}_\mathbb{R})$.

For each $\lambda \in \Lambda$ we have a product measure $\mu_\lambda^{\otimes n}$ on $\mathbb{R}^n$ equipped with its natural σ-algebra, $\mathcal{B}_n = \mathcal{B}_\mathbb{R}^{\otimes n}$. The *mixture* of the family (μ_λ^n) directed by $\mathbb{P}^\Lambda$ is the measure μ_Λ^n defined by the averaging formula

$$\mu_\Lambda^n[S] := \int_\Lambda \mu_\lambda^n[B] \, \mathbb{P}_\Lambda[d\lambda], \quad \forall B \in \mathcal{B}_n.$$

The collection $(\mu_\Lambda^n)_{n \in \mathbb{N}}$ forms a projective family. Kolmogorov's existence theorem shows that this family induces a unique probability measure μ_Λ^∞ on $\mathbb{R}^\mathbb{N}$. The random variables

$$X_n : \mathbb{R}^\infty \to \mathbb{R}, \quad X_n(x_1, x_2, \ldots) = x_n$$

form an exchangeable sequence. The measure μ_Λ^∞ is called a *mixture of* i.i.d. *directed by the random measure μ.*

For example, suppose that ν is a Borel probability measure on $\Lambda = [0,1]$. For any $p > 0$ define

$$\mu_p = \mathrm{Bin}(p) = (1-p)\delta_0 + p\delta_1 \in \mathrm{Prob}\big(\{0,1\}\big).$$

Then we obtain the mixtures $\mu_\nu^n \in \mathrm{Prob}\big(\{0,1\}^n\big)$ defined by,

$$\mu_\nu^n\big[\{\epsilon_1,\ldots,\epsilon_n\}\big] = \binom{n}{k}\int_{[0,1]} (1-p)^{n-k} p^k \nu\big[dp\big], \quad k = \epsilon_1 + \cdots + \epsilon_n.$$

The collection $\mu_\nu^n \in \mathrm{Prob}\big(\{0,1\}^n\big)$, $n \in \mathbb{N}$ is a projective family and thus it defines a measure μ_ν^∞ on $\{0,1\}^{\mathbb{N}}$.

The random vector $X = (X_1, X_2, \ldots)$ with distribution μ^∞ defines an exchangeable sequence of Bernoulli random variables. Observe that their common success probability is

$$\bar{p} := \mathbb{P}\big[X_n = 1\big] = \int_{[0,1]} p\nu\big[dp\big], \quad \forall n \in \mathbb{N}.$$

$\square$

Denote by $\mathcal{B}$ the Borel σ-algebra of $\mathbb{R}$. The groups $\mathfrak{S}_n$ act on $\mathbb{R}^{\mathbb{N}}$ by permuting the first n coordinates and we say that a function $\Phi : \mathbb{R}^{\mathbb{N}} \to \mathbb{R}$ is n-symmetric if it is $\mathfrak{S}_n$-invariant. We denote by $\mathcal{S}_n \subset \mathcal{B}^{\mathbb{N}}$ the sigma-subalgebra generated by the n-symmetric measurable functions $\Phi : \mathbb{R}^{\mathbb{N}} \to \mathbb{R}$. Equivalently,

$$S \subset \mathcal{S}_n \Longleftrightarrow \sigma(S) = S, \quad \forall \sigma \in \mathfrak{S}_n.$$

We set

$$\mathcal{S}_\infty := \bigcap_{n \geq 1} \mathcal{S}_n \subset \mathcal{B}^{\mathbb{N}}.$$

We will refer to $\mathcal{S}_\infty$ as the σ-algebra of *permutable or exchangeable events* associated to the exchangeable sequence $(X_n)_{n \in \mathbb{N}}$. Note that $\mathcal{S}_\infty \supset \mathfrak{T}_\infty$, where $\mathfrak{T}_\infty$ denotes the tail σ-algebra of the coordinate sequence $X_n : \mathbb{R}^{\mathbb{N}} \to \mathbb{R}$,

$$X_n(x_1, x_2, \ldots) = x_n, \quad n \in \mathbb{N}.$$

It turns out that exchangeable sequences have a rather nice structure.

Theorem 3.93 (de Finetti). *Suppose that $\underline{X} := (X_n)_{n \in \mathbb{N}}$ is an exchangeable sequence of integrable random variables defined on the same probability space $(\Omega, \mathcal{F}, \mathbb{P})$. Set*

$$\underline{\mathcal{S}}_n := \underline{X}^{-1}\mathcal{S}_n, \quad \forall n \in \mathbb{N} \cup \{\infty\}.$$

Then the following hold.

(i) *The random variables $(X_n)_{n \geq 1}$ are conditionally independent given $\underline{\mathcal{S}}_\infty$.*
(ii) *The random variables $(X_n)_{n \geq 1}$ are identically distributed given $\underline{\mathcal{S}}_\infty$, i.e., there exists a negligible subset $\mathcal{N} \in \mathcal{F}$ such that, on $\Omega \setminus \mathcal{N}$*

$$\mathbb{P}\big[X_i \leq x \,\|\, \underline{\mathcal{S}}_\infty\big] = \mathbb{P}\big[X_j \leq x \,\|\, \underline{\mathcal{S}}_\infty\big], \quad \forall i,j \in \mathbb{N}, \ \forall x \in \mathbb{R}.$$

(iii) The empirical means
$$\frac{X_1 + \cdots + X_n}{n}$$
converge a.s. and L^1 to $\mathbb{E}[\, X_1 \,\|\, \underline{\mathcal{S}}_\infty \,]$.

Proof. We follow the presentation in [91]. Without any loss of generality we can assume that $(\Omega, \mathcal{F}) = (\mathbb{R}^{\mathbb{N}}, \mathcal{B}^{\mathbb{N}})$ and $X_n(x_1, x_2, \dots) = x_n$. In this case $\underline{\mathcal{S}}_n = \mathcal{S}_n$. Observe that the exchangeability condition implies that the random variables X_n are identically distributed. Suppose that $f : \mathbb{R} \to \mathbb{R}$ is a measurable function such that $f(X_1) \in L^1$. We claim that

$$\forall k \in \mathbb{N}, \quad \frac{1}{n}\Big(f(X_1) + \cdots + f(X_n) \Big) = \mathbb{E}[\, f(X_k) \| \mathcal{S}_n \,]. \tag{3.2.36}$$

Note that $\mathcal{S}_n = \underline{X}^{-1}(\mathcal{B}_n)$, where $\mathcal{B}_n$ is the σ-subalgebra of $\mathcal{B}^{\mathbb{N}}$ consisting of $\mathfrak{S}_n$-invariant subsets. In particular, a function $g : \Omega \to \mathbb{R}$ is $\mathcal{S}_n$-measurable iff there exists an n-symmetric function Φ such that $g = \Phi(\underline{X})$.

Let $A \in \mathcal{S}_n$ and choose an n-symmetric function Φ such that $\boldsymbol{I}_A = \Phi(\underline{X})$. Then, for $1 \leq j \leq n$ we have

$$\mathbb{E}[\, f(X_j)\Phi(\underline{X}) \,] = \mathbb{E}[\, f(X_j)\Phi(X_j, X_2, \dots, X_{j-1}, X_1, X_{j+1}, \dots) \,]$$

$$= \mathbb{E}[\, f(X_1)\Phi(\underline{X}) \,],$$

so that

$$\mathbb{E}[\, f(X_1)\boldsymbol{I}_A \,] = \mathbb{E}\left[\frac{f(X_1) + \cdots + f(X_n)}{n} \boldsymbol{I}_A \right] = \mathbb{E}[\, f(X_j)\Phi(\underline{X}) \,].$$

The equality (3.2.36) follows by observing that $f(X_1) + \cdots + f(X_n)$ is $\mathcal{S}_n$-measurable.

The convergence theorem for backwards martingales (Corollary 3.91) shows that the empirical mean

$$\frac{f(X_1) + \cdots + f(X_n)}{n}$$

converges a.s. and L^1 to $\mathbb{E}[\, f(X_1) \,\|\, \mathcal{S}_\infty \,]$. By choosing $f(x) = x$ we obtain the statement (iii) of Theorem 3.93.

By choosing $f(x) = \boldsymbol{I}_{(-\infty, x]}$ we deduce

$$\lim_{n \to \infty} \frac{\#\{j \leq n;\ X_j \leq x\}}{n} = F(x) := \mathbb{P}[\, X_1 \leq x \,\|\, \mathcal{S}_\infty \,], \tag{3.2.37}$$

a.s. and L^1.

Let $k \in \mathbb{N}$. For $n \geq k$ we set $(n)_k := n(n-1)\cdots(n-k+1)$. Suppose that $f_1, \dots, f_k : \mathbb{R} \to \mathbb{R}$ are *bounded* and measurable. The above argument generalizes to prove that for $n \geq k$ we have

$$A_{k,n} := \frac{1}{(n)_k} \sum_{\substack{j_1, \dots, j_k \\ j_i \text{ distinct}}} f_1(X_{j_1}) \cdots f_k(X_{j_k}) = \mathbb{E}[\, f_1(X_1) \cdots f_k(X_k) \,\|\, \mathcal{S}_n \,].$$

Using the backwards martingale convergence theorem we deduce
$$\lim_{n\to\infty} A_{k,n} = \mathbb{E}\big[f_1(X_1)\cdots f_k(X_k) \,\|\, \mathcal{S}_\infty \big]. \qquad (3.2.38)$$
Consider now
$$B_{k,n} := \frac{1}{n^k} \sum_{j_1,\ldots,j_k=1}^{n} f_1(X_{j_1})\cdots f_k(X_{j_k}) = \prod_{i=1}^{k} \frac{f_i(X_1)+\cdots+f_i(X_n)}{n}.$$
We deduce from (3.2.36) that
$$\lim_{n\to\infty} B_{k,n} = \prod_{i=1}^{k} \mathbb{E}\big[f_i(X_i) \,\|\, \mathcal{S}_\infty \,\|.$$
Now observe that
$$A_{k,n} - B_{k,n} = O(1/n) \text{ as } n\to\infty,$$
since the contribution to $B_{k,n}$ corresponding to k-tuples with j_i non-distinct is $O(n^{k-1})$ and $n^k \sim (n)_k$ as $n\to\infty$. If we choose
$$f_i = \boldsymbol{I}_{(-\infty,x_i]}, \ 1\le i\le k,$$
we deduce from (3.2.37) that
$$\mathbb{P}\big[X_1\le x_1,\ldots,X_k\le x_k \,\|\, \mathcal{S}_\infty \big] = \prod_{i=1}^{k} \mathbb{P}\big[X_i\le x_i \,\|\, \mathcal{S}_\infty \big].$$
This proves (i) and (ii) of the theorem. □

Remark 3.94. Suppose that $(X_n)_{n\in\mathbb{N}}$ is an exchangeable sequence of random variables defined on the probability space $(\Omega,\mathcal{F},\mathbb{P})$. Denote by $\mathcal{S}_\infty$ the sigma-algebra of exchangeable events. Suppose that
$$Q:\Omega\times\mathcal{B}_\mathbb{R}\to[0,1], \ (\omega,B)\mapsto Q_\omega[B]$$
is a regular version of the conditional distribution $\mathbb{P}_{X_1}[dx\,\|\,\mathcal{S}_\infty]$, i.e.,
$$\forall B\in\mathcal{B}_\mathbb{R}, \ \mathbb{P}\big[X_1\in B\,\|\,\mathcal{S}_\infty\big] = Q_\square[B], \text{ a.s.}$$
De Finnetti's theorem implies that
$$\mathbb{P}\big[X_1\in B\,\|\,\mathcal{S}_\infty\big] = \lim_{n\to\infty} \frac{1}{n}\sum_{k=1}^{n} \boldsymbol{I}_B(X_k) = \mathbb{P}\big[X_m\in B\,\|\,\mathcal{S}_\infty\big], \ \forall m\in\mathbb{N}.$$
Thus the random variables (X_n) are equidistributed, conditional on $\mathcal{S}_\infty$.

Let us show that the distribution of the sequence $(X_n)_{n\in\mathbb{N}}$ is a mixture directed by the random measure $\omega\mapsto Q_\omega[-]$ as in Example 3.92(b).

Indeed, for any Borel subsets $B_1,\ldots,B_n\subset\mathbb{R}$ we have
$$\mathbb{P}\big[X_1\in B_1,\ldots,X_n\in B_n\big] = \mathbb{E}\Big[\mathbb{E}\big[\boldsymbol{I}_{B_1}(X_1)\cdots\boldsymbol{I}_{B_n}(X_n)\,\|\,\mathcal{S}_\infty\big]\Big]$$
(use the conditional independence given $\mathcal{S}_\infty$)
$$= \mathbb{E}\Big[\mathbb{E}\big[\boldsymbol{I}_{B_1}(X_1)\,\|\,\mathcal{S}_\infty\big]\cdots\mathbb{E}\big[\boldsymbol{I}_{B_n}(X_n)\,\|\,\mathcal{S}_\infty\big]\Big]$$
$$= \mathbb{E}\Big[Q[B_1]\cdots Q[B_n]\Big] = \int_\Omega Q_\omega^{\otimes n}[B_1\times\cdots B_n]\mathbb{P}[d\omega].$$
Thus the distribution of the sequence (X_n) is a mixture of i.i.d. driven by the random distribution Q. □

The σ-algebra $\mathcal{S}_\infty \subset \mathcal{B}^{\mathbb{N}}$ of permutable events of an exchangeable sequence $(X_n)_{n\in\mathbb{N}}$ contains its tail σ-algebra $\mathcal{T}_\infty$. It turns out that they are not so different. We have the following general result.

Proposition 3.95. *Suppose that $(X_n)_{n\in\mathbb{N}}$ is an exchangeable sequence of random variables. Then the $\mathbb{P}$-completion of $\mathcal{S}_\infty$ coincides with the completion of the tail $\mathcal{T}_\infty$.*

Proof. We follow the approach in [29, Sec. 7.3, Thm. 4]. For a different but related proof we refer to [1, Cor. (3.10)].

Denote by $\mathcal{S}_\infty^*$ and $\mathcal{T}_\infty^*$ the completions of $\mathcal{S}_\infty$ and respectively $\mathcal{T}_\infty$. We have

$$\lim_{n\to\infty} \frac{1}{n} \sum_{k=1}^n \mathbb{1}_{\{X_k \leq x\}} = \mathbb{P}[\, X_1 \leq x \,\|\, \mathcal{S}_\infty \,].$$

Clearly the limit in the left-hand side is $\mathcal{T}_\infty$-measurable since it is not affected by changing finitely many of the random variables. Hence

$$\mathbb{P}[\, X_1 \leq x \,\|\, \mathcal{S}_\infty \,] = \mathbb{P}[\, X_1 \leq x \,\|\, \mathcal{T}_\infty \,] = \mathbb{P}[\, X_1 \leq x \,\|\, \mathcal{T}_\infty^* \,]. \qquad (3.2.39)$$

Similarly, for any $x_1, \ldots, x_n \in \mathbb{R}$, the random variable

$$\prod_{k=1}^n \mathbb{P}[\, X_k \leq x_k \,\|\, \mathcal{S}_\infty \,]$$

is $\mathcal{T}_\infty$-measurable. Hence, for any $S \in \mathcal{S}_\infty$ we have

$$\mathbb{P}\left[S \cap \bigcap_{k=1}^n \{X_k \leq x_k\} \,\Big\|\, \mathcal{T}_\infty \right] = \mathbb{E}\left[\mathbf{I}_S \cdot \prod_{k=1}^n \mathbb{P}[\, X_k \leq x_k \,\|\, \mathcal{S}_\infty \,] \,\Big\|\, \mathcal{T}_\infty \right]$$

$$= \mathbb{E}\left[\prod_{k=1}^n \mathbb{P}[\, X_k \leq x_k \,\|\, \mathcal{S}_\infty \,] \,\Big\|\, \mathcal{T}_\infty \right] \mathbb{P}[\, S \,\|\, \mathcal{T}_\infty \,]$$

$$\stackrel{(3.2.39)}{=} \prod_{k=1}^n \mathbb{P}[\, X_k \leq x_k \,\|\, \mathcal{T}_\infty \,] \mathbb{P}[\, S \,\|\, \mathcal{T}_\infty \,].$$

Thus $\mathcal{S}_\infty$ and $X_1, \ldots, X_n$ are conditionally independent given $\mathcal{T}_\infty$ so $\mathcal{S}_\infty$ and $(X_n)_{n\in\mathbb{N}}$ are conditionally independent given $\mathcal{T}_\infty$. Since $\mathcal{S}_\infty \subset \sigma(X_n, n \in \mathbb{N})$ we deduce that for any $S \in \mathcal{S}_\infty$ is conditionally independent of itself given $\mathcal{T}_\infty$. We have

$$\mathbb{P}[\, S \,\|\, \mathcal{T}_\infty \,]^2 = \mathbb{P}[\, S \,\|\, \mathcal{T}_\infty \,]$$

so $\mathbb{P}[\, S \,\|\, \mathcal{T}_\infty \,] \in \{0, 1\}$. $\forall S \in \mathcal{S}_\infty$. Set

$$T = T_S = \{\, \omega : \mathbb{P}[\, S \,\|\, \mathcal{T}_\infty \,](\omega) = 1 \,\}.$$

Then $T \in \mathcal{T}_\infty$, $T \subset S$ and

$$\mathbb{P}[\, T \cap S \,] = \mathbb{E}\left[\, \mathbf{I}_T \mathbb{P}[\, S \,\|\, \mathcal{T}_\infty \,] \,\right] = \mathbb{P}[\, T \,],$$

$$\mathbb{P}[\, S \,] = \mathbb{E}\left[\, \mathbb{P}[\, S \,\|\, \mathcal{T}_\infty \,] \,\right] = \mathbb{E}[\, \mathbf{I}_T \,] = \mathbb{P}[\, T \,].$$

This concludes the proof. $\square$

Observe that a sequence of i.i.d. random variables $(X_n)_{n\geq 1}$ is exchangeable. The Kolmogorov 0-1 law and the above proposition imply the following result.

Theorem 3.96 (Hewitt-Savage 0-1 Law). *If $(X_n)_{n\geq 1}$ is a sequence of iid random variables and $A \in \mathcal{S}_\infty$, then $\mathbb{P}[A] \in \{0,1\}$.* □

Corollary 3.97 (The Strong Law of Large Numbers). *Suppose that $(X_n)_{n\in\mathbb{N}}$ is a sequence of i.i.d. integrable random variables. Then*

$$\bar{X}_n := \frac{1}{n}(X_1 + \cdots + X_n)$$

converges a.s. and L^1 to $\mathbb{E}[X_1]$.

Proof. From De Finetti's Theorem 3.93 we deduce that $\bar{X}_n$ converges a.s. and L^1 to $\mathbb{E}[X_1 \| \mathcal{S}_\infty]$. Proposition 3.95 that $\mathbb{E}[X_1 \| \mathcal{S}_\infty] = \mathbb{E}[X_1 \| \mathcal{T}_\infty]$ and Kolmogorov's 0-1 theorem shows that $\mathbb{E}[X_1 \| \mathcal{T}_\infty] = \mathbb{E}[X_1]$. □

Theorem 3.98. *Suppose that $\big(X_n : (\Omega, \mathcal{F}, \mathbb{P}) \to \{0,1\}\big)_{n\in\mathbb{N}}$ is an exchangeable sequence of Bernoulli random variables. Set*

$$S := \lim_{n\to\infty} \frac{1}{n}(X_1 + \cdots + X_n).$$

Then

$$S = \mathbb{P}[X_1 = 1 \| \mathcal{S}_\infty], \tag{3.2.40a}$$

$$\mathbb{P}[X_1 = \cdots = X_k = 1, X_{k+1} = \cdots = X_n = 0 \| S] = S^k(1-S)^{n-k}, \tag{3.2.40b}$$

$$\mathbb{P}[X_1 = \cdots = X_k = 1, X_{k+1} = \cdots = X_n = 0] = \mathbb{E}[S^k(1-S)^{n-k}]. \tag{3.2.40c}$$

In particular, the moment generating function of S is

$$\mathbb{E}[e^{tS}] = \sum_{n\geq 0} \mathbb{P}[X_1 = \cdots = X_n = 1] \frac{t^n}{n!}.$$

Proof. Using de Finetti's theorem we deduce that $S = \mathbb{E}[X_1 \| \mathcal{S}_\infty]$. Observe that $X_1 = \mathbf{I}_{\{X_1 = 1\}}$ so that

$$S = \mathbb{E}[X_1 \| \mathcal{S}_\infty] = \mathbb{E}[\mathbf{I}_{\{X_1 = 1\}} \| \mathcal{S}_\infty] = \mathbb{P}[X_1 = 1 \| \mathcal{S}_\infty].$$

Note that $0 \leq S \leq 1$ a.s. and

$$1 - S = \mathbb{E}[1 - \mathbf{I}_{\{X_1 = 1\}} \| \mathcal{S}_\infty] = \mathbb{P}[X_n = 0 \| \mathcal{S}_\infty].$$

Then, since $X_1, \ldots, X_n$ are conditionally i.i.d. given $\mathcal{S}_\infty$, we have

$$\mathbb{P}[X_1 = 1, \ldots, X_k = 1, X_{k+1} = 0, \ldots, X_n = 0 \| \mathcal{S}_\infty]$$

$$= \mathbb{P}[X_1 = 1 \| \mathcal{S}_\infty]^k \mathbb{P}[X_1 = 0 \| \mathcal{S}_\infty]^{n-k} = S^k(1-S)^{n-k}.$$

Since S is $\mathcal{S}_\infty$-measurable we have

$$\mathbb{P}\left[\, X_1 = 1, \ldots, X_k = 1, X_{k+1} = 0, \ldots, X_n = 0 \,\|\, S \,\right]$$
$$= \mathbb{E}\left[\, \mathbb{P}\left[\, X_1 = 1, \ldots, X_k = 1, X_{k+1} = 0, \ldots, X_n = 0 \,\|\, \mathcal{S}_\infty \,\right] \,\|\, S \,\right]$$
$$= \mathbb{E}\left[\, S^k(1 - S^k) \,\|\, S \,\right] = S^k(1 - S^k).$$

Clearly,

$$\mathbb{P}\left[\, X_1 = 1, \ldots, X_k = 1, X_{k+1} = 0, \ldots, X_n = 0 \,\right] = \mathbb{E}\left[\, S^k(1 - S)^{n-k} \,\right].$$

$\square$

Example 3.99 (Polya's urn revisited). We want to conclude this introduction to exchangeability with an application to Polya's urn problem introduced in Example 3.9. We recall this process.

We start with an urn containing $r > 0$ red balls and $g > 0$ green balls. At each moment of time we draw a ball uniformly likely from the balls existing at that moment, we replace it by $c + 1$ balls of the same color, $c \geq 0$. Denote by R_n and G_n the number of red and respectively green balls in the urn after the nth draw. As we have seen in Example 3.9 the ratio of red balls

$$Z_n = \frac{R_n}{R_n + G_n} = \frac{R_n}{r + g + cn}$$

is a bounded martingale and thus it has an a.s. and L^1 limit Z_∞. We will determine this limit using de Finetti's theorem. We discuss only the nontrivial case $c > 0$.

Introduce the $\{0,1\}$-valued random variables $(X_n)_{n\geq 1}$ where $X_n = 1$ if the n-drawn ball is red and $X_n = 0$ if it is green. Then

$$R_n = r + c S_n, \quad S_n := X_1 + \cdots + X_n,$$

and we deduce that

$$\lim_{n \to \infty} \frac{c S_n}{cn} = \lim_{n \to \infty} \frac{R_n}{R_n + G_n} = Z_\infty.$$

Let us observe that the sequence $(X_n)_{n \geq 1}$ is exchangeable. We prove by induction that $(X_1, \ldots, X_n)$ is exchangeable. For $n = 1$ the result is trivial.

Let $n > 1$ and $\epsilon_1, \ldots, \epsilon_n \in \{0,1\}$. We denote by r_k and g_k the number of red balls and respectively green balls after the k-th draw. We deduce

$$\mathbb{P}\left[\, X_1 = \epsilon_1, \ldots, X_n = \epsilon_n \,\right] = \begin{cases} \dfrac{\prod_{k=1}^n \left(\epsilon_k r_{k-1} + (1 - \epsilon_k) g_{k-1}\right)}{\prod_{k=1}^n (r + g + (k-1)c)}, & c > 0, \\[2ex] z_0^k (1 - z_0)^{n-k}, & c = 0, \end{cases}$$

where $z_0 = Z_0 = \frac{r}{r+g}$. When $c > 0$ the denominator above is independent of $\{\epsilon_1, \ldots, \epsilon_n\}$. We set $S_n := \epsilon_1 + \cdots + \epsilon_n$ and we rewrite the numerator in the form

$$\prod_{k=1}^n \left(\epsilon_k r_{k-1} + (1 - \epsilon_k) g_{k-1}\right) = \prod_{i=1}^{S_n} \left(r + c(i-1)\right) \prod_{j=1}^{n-S_n} \left(g + c(j-1)\right).$$

The last expression only depends on S_n which is obviously a symmetric function in the variables $\epsilon_1,\ldots,\epsilon_n$. If $c=0$, then this expression is equal to 0.

When $c>0$, we set

$$\rho := \frac{r}{c}, \quad \gamma := \frac{g}{c},$$

and we deduce

$$\mathbb{P}\big[X_1 = \cdots = X_k = 1,\ X_{k+1} = \cdots = X_n = 0\big] = \frac{\prod_{i=0}^{k-1}(\rho+i)\prod_{j=0}^{n-k-1}(\gamma+j)}{\prod_{k=0}^{n-1}(r+\gamma+k)}$$

$$= \frac{\Gamma(\rho+\gamma)}{\Gamma(\rho)\Gamma(\gamma)} \cdot \frac{\Gamma(\rho+k)\Gamma(\gamma+n-k)}{\Gamma(\rho+\gamma+n)} = \frac{B(\rho+k,\gamma+n-k)}{B(\rho,\gamma)},$$
(3.2.41)

where $B(x,y)$ denotes the Beta function

$$B(x,y) = \frac{\Gamma(x)\Gamma(y)}{\Gamma(x+y)} = \int_0^1 t^{x-1}(1-t)^{y-1}dt.$$

We now invoke Theorem 3.98. Note that

$$Z_\infty = \lim_{n\to\infty} \frac{1}{n}(X_1 + \cdots + X_n) = S = \mathbb{P}\big[X_1 = 1 \,\|\, \mathcal{S}_\infty\big]$$

is a $[0,1]$-valued random variable and (3.2.40c) with $k=n$ shows that, for any $n \geq 0$, we have

$$\int_0^1 z^n \mathbb{P}_{Z_\infty}[dz] = \mathbb{E}\big[Z_\infty^n\big] = \mathbb{P}\big[X_1 = \cdots = X_n = 1\big]$$

$$\stackrel{(3.2.41)}{=} \begin{cases} \frac{B(\rho+n,\gamma)}{B(\rho,\gamma)}, & c>0, \\ z_0^n, & c=0 \end{cases} = \begin{cases} \int_0^1 z^n \frac{z^{\rho-1}(1-z)^{\gamma-1}}{B(\rho,\gamma)}dz, & c>0, \\ \int_0^1 s^n \delta_{z_0}[dz] & c=0, \end{cases}$$

where δ_{z_0} is the Dirac measure concentrated at z_0. Hence

$$\int_0^1 z^n \mathbb{P}_{Z_\infty}[dz] = \begin{cases} \int_0^1 z^n \frac{z^{\rho-1}(1-z)^{\gamma-1}}{B(\rho,\gamma)}dz, & c>0, \\ \int_0^1 z^m \delta_{z_0}[dz], & c=0,\ \forall n \geq 0. \end{cases}$$

Since the probability measures on $[0,1]$ are uniquely determined by their momenta (see Corollary 1.108) we deduce

$$\mathbb{P}_{Z_\infty}[dz] = \begin{cases} \frac{z^{\rho-1}(1-z)^{\gamma-1}}{B(\rho,\gamma)}dz, & c>0 \\ \delta_{z_0}[dz], & c=0. \end{cases}$$

The distribution in the case $c>0$ is the *Beta distribution* with parameters ρ,γ discussed in Example 1.122. $\square$

3.3 Continuous time martingales

The study of martingales parametrized by $\mathbb{T} = [0, \infty)$ faces a few fundamental technical difficulties stemming from the fact that the space of parameters is not countable. To deal with these issues we need to introduce several new concepts.

3.3.1 *Generalities about filtered processes*

Suppose that $(\Omega, \mathcal{F}, \mathbb{P})$ is a probability space and $\mathcal{F}_\bullet = (\mathcal{F}_t)_{t \geq 0}$ is a filtration of sigma-subalgebras of $\mathcal{S}$. We denote by $\operatorname{Proc}(\mathcal{F}_\bullet)$ the collection of random processes (parametrized by $\mathbb{T}$) that are adapted to the filtration $\mathcal{F}_\bullet$. If no confusion is possible, we will use the simpler notation Proc when referring to adapted processes.

A function $f : [0, \infty) \to \mathbb{R}$ is called an *R-function*[4] if it is *right* continuous with left limits. It is called an *L-function*[5] if it is left continuous with right limits.

Definition 3.100. Let $X_\bullet = \{ X_t : (\Omega, \mathcal{F}, \mathbb{P}) \to \mathbb{R} \}_{t \in [0, \infty)}$ be a random process, not necessarily adapted to the filtration $\mathcal{F}_\bullet$.

(i) We say that the random process $X_\bullet$ is *measurable* if the map
$$X : [0, \infty) \times \Omega \to \mathbb{R}, \quad (t, \omega) \mapsto X_t(\omega)$$
is measurable with respect to the σ-algebra $\mathcal{B}_{[0,\infty)} \times \mathcal{F}$.

(ii) We say that the random process $X_\bullet$ is *progressively measurable* or *progressive* (with respect to the filtration $\mathcal{F}_\bullet$) if for any $t > 0$, the map
$$[0, t] \times \Omega \ni (s, \omega) \mapsto X_t(\omega) \in \mathbb{R}$$
is $\mathcal{B}_{[0,t]} \otimes \mathcal{F}_t$ measurable, where $\mathcal{B}_{[0,t]}$ denotes the σ-algebra of Borel subsets of $[0, t]$.

(iii) A subset $A \subset [0, \infty) \times \Omega$ is called *progressive* if the associated process
$$\boldsymbol{I}_A : [0, \infty) \times \Omega \to \mathbb{R}$$
is progressive.

(iv) We say that the adapted random process $X_\bullet$ is an *R-process* (resp. *L-process*) if there exists a negligible subset $N \subset \Omega$ such that, for any $\omega \in \Omega \setminus N$, the function $\mathbb{T} \ni t \mapsto X_t(\omega)$ is and R-function (resp. L-function).

$\square$

Remark 3.101. The progressive subsets of $[0, \infty) \times \Omega$ form a σ-subalgebra of $\mathcal{B}(\mathbb{R}) \otimes F$ that we denote by $\mathcal{F}_{\operatorname{prog}}$. Observe that a process is progressively measurable if and only if it is $\mathcal{F}_{\operatorname{prog}}$-measurable. For this reason we will denote by $\operatorname{Proc}(\mathcal{F}_{\operatorname{prog}})$ or $\operatorname{Proc}_{\operatorname{prog}}$ the collection of progressive processes.

An $\mathcal{F}_\bullet$-progressive process is also adapted to the filtration $\mathcal{F}_\bullet$ so
$$\operatorname{Proc}(\mathcal{F}_{\operatorname{prog}}) \subset \operatorname{Proc}(\mathcal{F}_\bullet).$$
$\square$

---

[4] A.k.a. *cadlag* function, *continue à droite limite à gauche*.
[5] A.k.a. *caglad* function, *continue à gauche limite à droite*.

Proposition 3.102. *Suppose that $X_\bullet \in \text{Proc}(\mathcal{F}_\bullet)$ is either an R-process or an L-process. Then $X_\bullet$ is a progressive process.*

Proof. Assume X is an R-process. The case of L-processes is similar. Fix $t \geq 0$. For each $n \in \mathbb{N}$, we subdivide the interval $[0,t]$ into n intervals of the same size. For $n \in \mathbb{N}$, define

$$X^n : [0,t] \times \Omega \to \mathbb{R}, \quad X^n_s(\omega) = \begin{cases} X_{kt/n}(\omega), & s \in [(k-1)t/n, kt/n), \; 1 \leq k \leq n, \\ X_t(\omega), & s = t. \end{cases}$$

Since $X_\bullet$ is an R-process we deduce that there exists a negligible subset $N \subset \Omega$ such that

$$\lim_{n \to \infty} X^n_s(\omega) = X_s(\omega), \quad \forall s \in [0,t], \; \omega \in \Omega \setminus N.$$

Clearly the function $X^n : [0,t] \times \Omega \to \mathbb{R}$ is $\mathcal{B}_{[0,t]} \otimes \mathcal{F}_t$-measurable. It follows that the a.s. limit $X : [0,t] \times \Omega \to \mathbb{R}$ is also $\mathcal{B}_{[0,t]} \otimes \mathcal{F}_t$-measurable, $\forall t \geq 0$. $\square$

We have the following nontrivial result, [30].

Theorem 3.103 (Chung-Doob). *Suppose that*

$$X_\bullet = \left\{ X_t : (\Omega, \mathcal{F}, \mathbb{P}) \to \mathbb{R} \right\}_{t \in [0,\infty)}$$

is a measurable process adapted to the filtration $\mathcal{F}_\bullet$. Then $X_\bullet$ admits a progressive modification. $\square$

Definition 3.104. Fix a filtration $\mathcal{F}_\bullet = (\mathcal{F}_t)_{t \geq 0}$ of the probability space $(\Omega, \mathcal{F}, \mathbb{P})$.

(i) An $\mathcal{F}_\bullet$-*stopping time* is a random variable $T : \Omega \to [0,\infty]$ such that

$$\{T \leq t\} \in \mathcal{F}_t, \quad \forall t \geq 0.$$

(ii) An $\mathcal{F}_\bullet$-*optional time* is a random variable $T : \Omega \to [0,\infty]$ such that

$$\{T < t\} \in \mathcal{F}_t, \quad \forall t > 0.$$

(iii) If $T : \Omega \to [0,\infty]$ is a stopping time, then the *past before* T is collection $\mathcal{F}_T \subset \mathcal{F}_\infty$ consisting of the sets $F \in \mathcal{F}$ satisfying the property $F \cap \{T \leq t\} \in \mathcal{F}_t$, $\forall t \geq 0$.

$\square$

Lemma 3.105. *For any stopping time T adapted to the filtration $\mathcal{F}_\bullet$ the collection $\mathcal{F}_T$ is a σ-algebra.* $\square$

The proof is left to the reader as an exercise.

Lemma 3.106. *Any stopping time T is an optional time.*

Proof. Indeed,
$$\{T < t\} = \bigcup_{n \geq 0} \{T \leq t - 1/n\},$$
and $\{T \leq t - 1/n\} \in \mathcal{F}_{t-1/n} \subset \mathcal{F}_t$. □

Definition 3.107. Fix a probability space $(\Omega, \mathcal{F}, \mathbb{P})$ and a filtration $\mathcal{F}_\bullet = (\mathcal{F}_t)_{t \geq 0}$ of $\mathcal{F}$. We set
$$\mathcal{F}_{t+} := \bigcap_{s > t} \mathcal{F}_s, \quad t \geq 0.$$

(i) We say that the filtration $\mathcal{F}_\bullet = (\mathcal{F}_t)_{t \geq 0}$ *right-continuous* if
$$\mathcal{F}_t = \mathcal{F}_{t+}, \quad \forall t \geq 0.$$

(ii) We say that the filtration $\mathcal{F}_t$ is $\mathbb{P}$-*complete* if the probability space $(\Omega, \mathcal{F}, \mathbb{P})$ is $\mathbb{P}$-complete[6] and the collection $\mathcal{N} \subset \mathcal{F}$ of $\mathbb{P}$-negligible events is contained in $\mathcal{F}_t$, $\forall t \geq 0$.

(iii) We say that the filtration $\mathcal{F}_t$ satisfies the *usual conditions* (or that it is *usual*) if it is both right-continuous and $\mathbb{P}$-complete.

□

Remark 3.108. If $(\mathcal{F}_t)_{t \geq 0}$ is a filtration of the *complete* probability space $(\Omega, \mathcal{F}, \mathbb{P})$, then the *usual augmentation* of $(\mathcal{F}_t)$ is the minimal filtration $(\hat{\mathcal{F}}_t)$ containing $(\mathcal{F}_t)$ and satisfying the usual conditions. More precisely if $\mathcal{N} \subset \mathcal{F}$ is the collection of probability zero events, then
$$\hat{\mathcal{F}}_t = \bigcap_{s \in (t, \infty)} \sigma(\mathcal{N}, \mathcal{F}_s). \quad \square$$

Proposition 3.109. *Consider a random variable* $T : \Omega \to [0, \infty]$. *Then the following statements are equivalent.*

(i) T *is an optional time for* $(\mathcal{F}_t)$.
(ii) T *is a stopping time for* $(\mathcal{F}_{t+})$.

In particular, if $\mathcal{F}_t$ is right-continuous, then T is a stopping time if and only if it is an optional time.[7] □

Example 3.110. Suppose that $(X_t)_{t \geq 0}$ is a process adapted to $\mathcal{F}_\bullet$ and $\Gamma \subset \mathbb{R}$. The $(\Gamma$-$)$ *début time* of (X_t) is the function
$$D_\Gamma : \Omega \to [0, \infty], \quad D_\Gamma(\omega) = \inf\{t \geq 0; \; X_t(\omega) \in \Omega\},$$

---

[6]Recall that this means that any set contained in a $\mathbb{P}$-null subset is measurable.

[7]This settles an inconsistency in the existence literature. Many authors refer to stopping times as optional times, while our optional times are sometimes referred to as weakly optional times. When the filtration is right continuous all these terms refer to the same concept, that of stopping time.

and the (Γ-)*hitting time* of (X_t) is the function

$$H_\Gamma : \Omega \to [0, \infty], \quad H_\Gamma(\omega) = \inf \{ t > 0; \; X_t(\omega) \in \Omega \}.$$

The following facts are not hard to prove; see [85, Lemma 9.6], [103, Prop. 3.9].

(i) If Γ is *open*, and the paths of X_t are right continuous, then the *début time* D_Γ is a stopping time of (X_t), while the *hitting time* H_Γ *is an optional time*.
(ii) If Γ is *closed*, and the paths of X_t are continuous, then the *début time* D_Γ is a stopping time of (X_t), while the *hitting time* H_Γ is an optional time.

We deduce from the above that if the filtration $\mathcal{F}_t$ is right-continuous and the paths of (X_t) are continuous, then both D_Γ and H_Γ are stopping times if Γ is either open or closed. $\square$

If the filtration $\mathcal{F}_\bullet$ satisfies the usual condition a much more general result is true. More precisely, we have the following highly nontrivial result of Dellacherie and Meyer [39, Thm. IV.50].

Theorem 3.111 (Début Theorem). *Suppose that the filtration $\mathcal{F}_\bullet$ satisfies the usual conditions and $(X_t)_{t\geq 0}$ is an $\mathcal{F}_\bullet$-progressive process. Then, for any Borel subset $\Gamma \subset \mathbb{R}$, the début time D_Γ is a stopping time.* $\square$

We list below a few elementary properties of stopping times.

Proposition 3.112. *Fix a filtered probability space $(\Omega, \mathcal{F}_\bullet, \mathbb{P})$.*

(i) *If T is a stopping time, then T is also $\mathcal{F}_T$-measurable.*
(ii) *If S is a stopping time and T is an $\mathcal{F}_S$-measurable random variable such that $T \geq S$, then T is also a stopping time and $\mathcal{F}_S \subset \mathcal{F}_T$.*
(iii) *Suppose that S, T are stopping times. Then $S \wedge T$ and $S \vee T$ are also stopping times and*

$$\mathcal{F}_{S \vee T} = \mathcal{F}_S \cap \mathcal{F}_T.$$

(iv) *An increasing limit of stopping times is a stopping time while a decreasing limit of stopping times is an optional time.*
(v) *Suppose that T is a stopping time. A function*

$$\{T < \infty\} \ni \omega \mapsto Y(\omega) \in \mathbb{R}$$

is $\mathcal{F}_T$-measurable if and only if, $\forall t \geq 0$, the restriction of Y to $\{T \leq t\}$ is $\mathcal{F}_t$-measurable.

Proof. We prove only (i). The rest are left to the reader as an exercise. To prove that the sublevel set $\{T \leq c\}$ is measurable we have to show that for any $t \geq 0$ the intersection

$$\{T \leq c\} \cap \{T \leq t\} = \{T \leq t \wedge c\}$$

is $\mathcal{F}_t$-measurable. This is a consequence of the fact that T is compatible with the filtration $\mathcal{F}_t$. □

Definition 3.113. Fix a filtered probability space $(\Omega, \mathcal{F}_\bullet, \mathbb{P})$. Given a random process $(X_t)_{t\geq 0}$ and an $\mathcal{F}_\bullet$-stopping time $T : \Omega \to [0, \infty]$ we denote by X_T the random variable

$$I_{\{T(\omega)<\infty\}} = X_T(\omega) = \begin{cases} X_{T(\omega)}(\omega), & T(\omega) < \infty \\ 0, & T(\omega) = \infty \end{cases}.$$

□

The proof of the following result is left to the reader as an exercise.

Proposition 3.114. *If $(X_t)_{t\geq 0}$ is a progressively measurable random process and T is a stopping time, then the random variable X_T is $\mathcal{F}_T$-measurable.* □

3.3.2 The Brownian motion as a filtered process

Let us illustrate the concepts introduced in the previous subsection on the stochastic process defined by the Brownian motion. The following result should be obvious from the definition of the Brownian motion.

Proposition 3.115. *Suppose that B is a Brownian motion. Then the following hold.*

Symmetry. The stochastic process $-B$ is also a Brownian motion.

Time rescaling. For any $c > 0$ the rescaled Brownian motion

$$B_t^c := \frac{1}{\sqrt{c}} B_{ct}$$

is another standard Brownian motion.

Time inversion. The stochastic process

$$X_t := \begin{cases} tB_{1/t}, & t > 0, \\ 0, & t = 0, \end{cases}$$

is another standard Brownian motion.

Proof. The statements (i) and (ii) are immediate. The last statement concerning time inversion requires a bit more work. We follow the approach in the proof of [33, Thm. VIII.1.6].

Observe that first X_t is a Gaussian process with mean zero and covariances

$$\mathbb{E}[X_s X_t] = \min(s,t), \quad \forall s, t \geq 0.$$

Thus it suffices to show that (X_t) is a.s. continuous, i.e.,

$$\lim_{t \searrow 0} X_t = 0 \quad \text{a.s.}$$

Equivalently, we will show that
$$\lim_{t\to\infty}\frac{1}{t}B_t = 0 \quad \text{a.s.}$$
Note that for $n \in \mathbb{N}$ and $t \in (n, n+1]$ we have
$$\frac{1}{t}|X_t| \leq \frac{1}{n}|X_n + (X_t - X_n)| \leq \frac{1}{n}|X_n| + \frac{1}{n}\sup_{s\in[0,1]}|X_{n+s} - X_n|$$
(the process (X_t) is a.s. continuous on $(0, \infty)$)
$$= \frac{1}{n}|X_n| + \frac{1}{n}\sup_{s\in[0,1]\cap\mathbb{Q}}|X_{n+s} - X_n|.$$
The Strong Law of Large Numbers shows that
$$\lim_{n\to\infty}\frac{1}{n}X_n = 0 \quad \text{a.s.}$$
For each $m = 1, 2, \ldots$, the process
$$D_k^n = X_{n+k/m} - X_n = \sum_{j=1}^m (X_{n+\frac{j}{m}} - X_{n+\frac{j-1}{m}}),$$
is a martingale since the above summands have mean zero and are independent. Applying Doob's maximal inequalities (3.2.31) to the discrete submartingales
$$Y^m = \{\,|D_n^m|^2,\ 0 \leq k \leq m\,\}, \quad m = 1, 2, \ldots,$$
we deduce that, for any $\varepsilon > 0$,
$$\mathbb{P}\left[\sup_{s\in[0,1]}|X_{n+s} - X_n| > n\varepsilon\right] = \mathbb{P}\left[\sup_{s\in[0,1]\cap\mathbb{Q}}|X_{n+s} - X_n|^2 > n^2\varepsilon^2\right]$$
$$\leq \frac{1}{n^2\varepsilon^2}\mathbb{E}\left[|X_{n+1} - X_n|^2\right] = \frac{1}{n^2\varepsilon^2}.$$
Since $\sum_{n\geq 1}\frac{1}{n^2} < \infty$ we deduce from the Borel-Cantelli Lemma that
$$\lim_{n\to\infty}\frac{1}{n}\sup_{s\in[0,1]}|X_{n+s} - X_n| = 0, \quad \text{a.s.}$$
$\square$

Theorem 3.116. *Suppose that $B : [0, \infty) \times \Omega \to \mathbb{R}$ is a Brownian motion and $(\Omega, \mathcal{F}, \mathbb{P})$ is a complete probability space. Let $\mathcal{N}$ denote the collection of $\mathbb{P}$-negligible events. We set*
$$\mathcal{F}_t = \sigma\big(\mathcal{N},\ B_s,\ 0 \leq s \leq t\big).$$
Then the filtration $(\mathcal{F}_t)_{t\geq 0}$ satisfies the usual conditions.

Proof. We follow the approach in the proof of [33, Thm. VII.3.20]. It suffices to prove that $(\mathcal{F}_t)$ is right-continuous, i.e.,

$$\mathcal{F}_{t_0} = \bigcap_{t>t_0} \mathcal{F}_t.$$

We set

$$\mathcal{G} = \mathcal{F}_{t_0}, \quad \mathcal{G}_n = \sigma\big(B_{t_0+2^{-n}} - B_{t_0+2^{-n-1}}\big), \quad n \in \mathbb{N}.$$

Clearly the σ-algebras $\mathcal{G}, \mathcal{G}_1, \ldots$, are independent. Set

$$\mathcal{T}_n := \sigma(\mathcal{G}, \mathcal{G}_{n+1}, \mathcal{G}_{n+2}, \ldots), \quad \mathcal{T}_\infty := \bigcap_{n \in \mathbb{N}} \mathcal{T}_n.$$

From Corollary 3.61 we deduce that $\mathcal{F}_{t_0} = \mathcal{T}_\infty$. On the other hand, $\mathcal{T}_\infty \supset \mathcal{F}_{t_0+}$ so $\mathcal{F}_{t_0+} = \mathcal{F}_{t_0}$. $\square$

Corollary 3.117 (Blumenthal's 0-1 law). *If $H \in \mathcal{F}_{0+}$ then, $\mathbb{P}[H] \in \{0,1\}$.* $\square$

Proposition 3.118. *Suppose that $(B_t)_{t \geq 0}$ is a standard Brownian motion and*

$$\mathcal{F}_t = \sigma\big(B_s, \ 0 \leq s \leq t\big).$$

Then the following hold.

(i) For any $\varepsilon > 0$ we have

$$\mathbb{P}\left[\sup_{s \in [0,\varepsilon]} B_s > 0\right] = \mathbb{P}\left[\inf_{s \in [0,\varepsilon]} B_s < 0\right] = 1.$$

(ii) For any $a \in \mathbb{R}$ we set

$$T_a := \inf_{t \geq 0} B_t = a.$$

Then

$$\mathbb{P}[T_a < \infty] = 1, \quad \forall a \in \mathbb{R}.$$

In particular, a.s.,

$$\limsup_{t \to \infty} B_t = \infty, \quad \liminf_{t \to \infty} B_t = -\infty.$$

Proof. (i) For any $c \neq 0$, the rescaled process

$$B^c(t) := \frac{1}{c} B_{c^2 t}, \quad t \geq 0$$

is also a standard Brownian motion. Note that since the paths of B_t are continuous we have

$$\sup_{t \in [0,1]} B_t = \sup_{t \in \mathbb{Q} \cap [0,1]} B_t.$$

Thus the set
$$\left\{\omega;\ \sup_{t\in[0,1]} B_t(\omega) > 0\right\}$$
is a Brownian event. The discussion in Remark 2.76 shows that
$$\mathbb{P}\left[\sup_{t\in[0,1]} B_t > 0\right] = \mathbb{P}\left[\sup_{t\in[0,1]} B_t^c > 0\right],\ \forall c \neq 0. \tag{3.3.1}$$
If we let $c = -1$ in the above equality we deduce,
$$\mathbb{P}\left[\sup_{t\in[0,1]} B_t > 0\right] = \mathbb{P}\left[\inf_{t\in[0,1]} B_t < 0\right]. \tag{3.3.2}$$
If we let $c = \sqrt{n}$, $n \in \mathbb{N}$ we deduce
$$\mathbb{P}\left[\sup_{t\in[0,1]} B_t > 0\right] = \mathbb{P}\left[\sup_{t\in[0,1/n]} B_t > 0\right],\ \forall n > 0. \tag{3.3.3}$$
We denote by E_n the Brownian event $\sup_{t\in[0,1/n]} B_t > 0$. Clearly
$$E_1 \supset E_2 \supset \cdots \supset E_n \supset \cdots$$
and $E_n \in \mathcal{F}_{1/n}$. We deduce from (3.3.3) that $\mathbb{P}[E_n] = \mathbb{P}[E_1]$, $\forall n$. If we set
$$E_\infty := \bigcap_n E_n,$$
then we deduce that $E_\infty \in \mathcal{F}_{0+}$ and $\mathbb{P}[E_\infty] = \mathbb{P}[E_1]$. Blumenthal's 0-1 theorem implies that
$$\mathbb{P}[E_n] = \mathbb{P}[E_\infty] \in \{0, 1\}.$$
Now observe that
$$\mathbb{P}[E_1] \subset \mathbb{P}[B_{1/2} > 0] = \frac{1}{2} > 0.$$
Hence
$$\mathbb{P}\left[\sup_{t\in[0,1/n]} B_t > 0\right] = \mathbb{P}\left[\inf_{t\in[0,1/n]} B_t < 0\right] = 1,\ \forall n \in \mathbb{N}. \tag{3.3.4}$$
This shows that a path of the Brownian motion oscillates wildly.

(ii) We have
$$1 = \mathbb{P}\left[\sup_{0\leq s\leq 1} B_s > 0\right] = \lim_{\delta \searrow 0} \mathbb{P}\left[\sup_{0\leq s\leq 1} B_s > \delta\right],$$
where the second is an increasing limit. The rescaling invariance of the Brownian motion implies
$$\mathbb{P}\left[\sup_{0\leq s\leq 1} B_s > \delta\right] = \mathbb{P}\left[\sup_{0\leq s\leq 1/\delta^2} B_s^\delta > 1\right].$$

We deduce

$$\mathbb{P}\left[\sup_{s\geq 0} B_s > 1\right] = \lim_{\delta \searrow} \mathbb{P}\left[\sum_{0\leq s \leq 1/\delta^2} B_s^\delta > 1\right] = 1.$$

Another rescaling argument shows that

$$\mathbb{P}\left[\sup_{s\geq 0} B_s > M\right] = 1, \quad \forall M > 0.$$

Replacing B by $-B$ we deduce

$$\mathbb{P}\left[\inf_{s\geq 0} B_s < -M\right] = 1, \quad \forall M > 0.$$

The conclusion (ii) is now obvious. □

Remark 3.119. The above result shows that, with probability 1 the Brownian motion has a zero on any arbitrarily small interval $[0, \varepsilon]$. As a matter of fact, the set of zeros of a Brownian motion is a large set: its Hausdorff dimension is a.s. $\frac{1}{2}$, [118, Thm. 4.24]. □

Let us observe that if $(B_t)_{t\geq 0}$ is a Brownian motion, then for any $t_0 \geq 0$, the process

$$\left(B_{t+t_0} - B_{t_0} \right)_{t\geq 0}$$

is also a Brownian motion, independent of $\sigma(B_s,\ 0 \leq s \leq t_0)$. We will refer to this elementary fact as the *simple Markov property*. We want to show that a stronger result holds where t_0 is allowed to be random.

Theorem 3.120 (The strong Markov property). *Suppose that $(B_t)_{t\geq 0}$ is a standard Brownian motion and T is a stopping time with respect to the filtration $\mathcal{F}_t = \sigma(B_s, 0 \leq s \leq t)$ such that $\mathbb{P}[T < \infty] > 0$. For every $t \geq 0$ we set*

$$B_t^{(T)} := \boldsymbol{I}_{\{T<\infty\}}\left(B_{T+t} - B_T \right).$$

Then, with respect to the probability measure $\mathbb{P}[-|T < \infty]$, the process $B_t^{(T)}$ is a standard Brownian motion, independent of $\mathcal{F}_T$.

Proof. We follow the approach in [103, Thm. 2.20].

Lemma 3.121. *Fix $A \in \mathcal{F}_T$. Let $F : \mathbb{R}^p \to \mathbb{R}$ be a <u>bounded</u> continuous function. Then, $\forall t_1, \ldots, t_p \geq 0$, we have*

$$\mathbb{E}\left[\boldsymbol{I}_A \boldsymbol{I}_{T<\infty} F\left(B_{t_1}^{(T)}, \ldots, B_{t_p}^{(T)} \right) \right] = \mathbb{P}\left[A \cap \{T < \infty\} \right] \mathbb{E}\left[F\left(B_{t_1}, \ldots, B_{t_p} \right) \right]. \quad (3.3.5)$$

□

Let us show first that conclusions of theorem follow from the above lemma. Set $S_\infty := \{T < \infty\}$. Assume first that $\mathbb{P}[S_\infty] = 1$. Then (3.3.5) reads

$$\mathbb{E}\big[\boldsymbol{I}_A F\big(B^{(T)}_{t_1}, \ldots, B^{(T)}_{t_p}\big)\big] = \mathbb{P}[A]\mathbb{E}\big[F(B_{t_1}, \ldots, B_{t_p})\big]. \qquad (3.3.6)$$

Indeed, if we set $A = \Omega$ in (3.3.6) we deduce that $B^{(T)}_t$ is a Brownian motion. In particular, for every choice of $t_1, \ldots, t_p \geq 0$, the vectors

$$\big(B^{(T)}_{t_1}, \ldots, B^{(T)}_{t_p}\big) \text{ and } \big(B_{t_1}, \ldots, B_{t_p}\big)$$

have the same distribution. Next, (3.3.6) implies that for every choice of $t_1, \ldots, t_p \geq 0$ the vector $(B^{(T)}_{t_1}, \ldots, B^{(T)}_{t_p})$ is independent of $\mathcal{F}_T$.

If $\mathbb{P}[S_\infty] < 1$, t and we denote by $\mathbb{E}_{S_\infty}$ the expectation with respect to the probability measure $\mathbb{P}[-|S_\infty]$, then (3.3.5) implies

$$\mathbb{E}_{S_\infty}\big[\boldsymbol{I}_A F\big(B^{(T)}_{t_1}, \ldots, B^{(T)}_{t_p}\big)\big] = \mathbb{P}[A|E_\infty]\mathbb{E}\big[F(B_{t_1}, \ldots, B_{t_p})\big].$$

Arguing as before we reach the conclusions of Theorem 3.120 assuming the validity of Lemma 3.121. $\square$

Proof of Lemma 3.121. For the clarity of exposition we discuss only the case $\mathbb{P}[S_\infty] = 1$. The case $\mathbb{P}[S_\infty] < 1$ requires no new ideas. The details can be safely left to the reader.

For every $t \geq 0$ and any $n \in \mathbb{N}$ we denote by $[t]_n$ the smallest rational number of the form $k/2^n$ and $\geq t$. Note that the quantities $[T]_n$ are stopping times: stopping the process at $[T]_n$ corresponds to stopping the process at the first time of the form $k/2^n$ after T. Then

$$\lim_{n \to \infty} [T]_n = T$$

and

$$F\big(B^{(T)}_{t_1}, \ldots, B^{(T)}_{t_p}\big) = \lim_{n \to \infty} F\big(B^{([T]_n)}_{t_1}, \ldots, B^{([T]_n)}_{t_p}\big).$$

From the Dominated Convergence theorem we deduce that

$$\mathbb{E}\big[\boldsymbol{I}_A F\big(B^{(T)}_{t_1}, \ldots, B^{(T)}_{t_p}\big)\big] = \lim_{n \to \infty} \mathbb{E}\big[\boldsymbol{I}_A F\big(B^{([T]_n)}_{t_1}, \ldots, B^{([T]_n)}_{t_p}\big)\big]$$

$$= \lim_{n \to \infty} \sum_{k=0}^{\infty} \mathbb{E}\big[\boldsymbol{I}_A \boldsymbol{I}_{(k-1)2^{-n} < T \leq k 2^{-n}} F\big(B^{([T]_n)}_{t_1}, \ldots, B^{([T]_n)}_{t_p}\big)\big].$$

Observe now that if $A \in \mathcal{F}_T$, then the event

$$A_{k,n} := A \cap \big\{(k-1)2^{-n} < T \leq k 2^{-n}\big\}$$

$$= A \cap \{T \leq k 2^{-n}\} \cap \{T > (k-1)2^{-n}\}$$

is $\mathcal{F}_{k 2^{-n}}$-measurable.

From the simple Markov property of the Brownian motion we deduce

$$\mathbb{E}\big[\boldsymbol{I}_{A_{k,n}} F\big(B^{([T]_n)}_{t_1}, \ldots, B^{([T]_n)}_{t_p}\big)\big]$$

$$= \mathbb{E}\big[\, \boldsymbol{I}_{A_{k,n}} F\big(B_{t_1+k2^{-n}} - B_{k2^{-n}}, \ldots, B_{t_p+k2^{-n}} - B_{k2^{-n}} \big) \,\big]$$

$$= \mathbb{P}\big[\, A_{k,n}\,\big] \mathbb{E}\big[\, F\big(B_{t_1}, \ldots, B_{t_p} \big) \,\big].$$

Observing that

$$\sum_{k=0}^{n} \mathbb{P}\big[\, A_{k,n}\,\big] = \mathbb{P}[A]$$

we deduce

$$\sum_{k=0}^{\infty} \mathbb{E}\big[\, \boldsymbol{I}_A \boldsymbol{I}_{(k-1)2^{-n} < T \leq k2^{-n}} F\big(B_{t_1}^{([T]_n)}, \ldots, B_{t_p}^{([T]_n)} \big) \,\big]$$

$$= \sum_{k=0}^{\infty} \mathbb{E}\big[\, \boldsymbol{I}_{A_{k,n}} F\big(B_{t_1}^{([T]_n)}, \ldots, B_{t_p}^{([T]_n)} \big) \,\big]$$

$$= \sum_{k=0}^{n} \mathbb{P}\big[\, A_{k,n}\,\big] \mathbb{E}\big[\, F\big(B_{t_1}, \ldots, B_{t_p} \big) \,\big] = \mathbb{P}\big[\, A \,\big] \mathbb{E}\big[\, F\big(B_{t_1}, \ldots, B_{t_p} \big) \,\big].$$

$\square$

Let us present some applications application of the strong Markov property. For $a \in \mathbb{R}$ we define the hitting time

$$T_a := \inf\{t > 0;\ B_t = a\}.$$

This is a stopping time for the standard Brownian motion B_t and Proposition 3.118(ii) shows that

$$\mathbb{P}\big[\, T_a < \infty \,\big] = 1.$$

Theorem 3.122 (Reflection Principle). *Fix $a \in \mathbb{R}$. If $(B_t)_{t \geq 0}$ is a standard Brownian motion, then the process*

$$\widetilde{B}_t = \begin{cases} B_t, & t < T_a \\ 2a - B_t, & t \geq T_a \end{cases} \quad (3.3.7)$$

is also a standard Brownian motion.

Proof. We follow the approach in [135, I.13]. Consider the processes

$$Y_t = B_t \boldsymbol{I}_{[[0, T_a]]}, \quad Z_s = B_{s+T_a} - a, \quad s \geq 0.$$

By the strong Markov property, Z is a standard Brownian motion, independent of Y. The process $-Z$ is also a Brownian motion independent of Y. Thus, the processes (Y, Z) and $(Y, -Z)$ have the same distribution. The map

$$(Y, Z) \mapsto \varphi(Y, Z) := Y_t \boldsymbol{I}_{[[0, T_a]]} + \big(a + Z_{t - T_a}\big) \boldsymbol{I}_{]]T_a, \infty[[}$$

produces the continuous process which will therefore have the same law as $\varphi(Y, -Z)$. Now observe that $\varphi(Y, Z) = B$ and $\varphi(Y, -Z) = \widetilde{B}$.

$\square$

Remark 3.123. The above result is called the reflection principle for a simple reason. In the region $t \geq T_a$ the graph of the function $t \to \tilde{B}_t$, viewed as a curve in the Cartesian plane with coordinates (t, x), is the reflection of the graph of B_t in the horizontal line $x = a$. This reflection principle is intimately related to André's reflection trick. □

Corollary 3.124. *Define*

$$S_t := \sup_{u \leq t} B_u.$$

Then, for any $a, y, t \geq 0$ we have

$$\mathbb{P}[S_t \geq a, B_t \leq a - y] = \mathbb{P}[B_t \geq a + y]. \qquad (3.3.8)$$

In particular, S_t has the same distribution as $|B_t|$.

Proof. Note that $S_t \geq a$ if and only if $T_a \leq t$. We have

$$\mathbb{P}[S_t \geq a, B_t \leq a - y] = \mathbb{P}[T_a \leq t, B_t \leq a - y] \stackrel{(3.3.7)}{=} \mathbb{P}[\tilde{B}_t \geq a + y]$$

(use the Reflection Principle)

$$= \mathbb{P}[B_t \geq a + y].$$

Now observe that

$$\mathbb{P}[S_t \geq a] = \underbrace{\mathbb{P}[S_t \geq a, B_t \geq a]}_{=\mathbb{P}[B_t \geq a]} + \mathbb{P}[S_t \geq a, B_t \leq a]$$

$$\stackrel{(3.3.8)}{=} 2\mathbb{P}[B_t \geq a] = \mathbb{P}[B_t \geq a] + \mathbb{P}[B_t \leq -a] = \mathbb{P}[|B_t| \geq a].$$

□

Corollary 3.125. *For every $a > 0$ the stopping time T_a has the same distribution as $\frac{a^2}{B_1^2}$ and has density*

$$f_a(t) = \frac{a}{\sqrt{2\pi t^3}} \exp\left(-\frac{a^2}{2t}\right) \boldsymbol{I}_{\{t > 0\}}.$$

Proof. Note that

$$\mathbb{P}[T_a \leq t] = \mathbb{P}[S_t \geq a] = \mathbb{P}[|B_t| \geq a] = \mathbb{P}[B_t^2 \geq a^2]$$

$$= \mathbb{P}[tB_1^2 \geq a^2] = \mathbb{P}\left[\frac{a^2}{B_1^2} \leq t\right].$$

The statement about f_a now follows from the fact that B_1 is a standard normal random variable. □

3.3.3 Definition and examples of continuous time martingales

Fix a filtered probability space $\left(\Omega, \mathcal{F}, (\mathcal{F}_t)_{t\geq 0}, \mathbb{P}\right)$.

Definition 3.126. A random process $(X_t)_{t\geq 0}$ adapted to the filtration $(\mathcal{F}_t)_{t\geq 0}$ such that $X_t \in L^1$, $\forall t$, is called a

- *martingale* if,
$$\mathbb{E}\left[X_t \| \mathcal{F}_s\right] = X_s, \quad \forall 0 \leq s < t,$$

- *submartingale* if,
$$\mathbb{E}\left[X_t \| \mathcal{F}_s\right] \geq X_s, \quad \forall 0 \leq s < t,$$

- *supermartingale* if,
$$\mathbb{E}\left[X_t \| \mathcal{F}_s\right] \leq X_s, \quad \forall 0 \leq s < t.$$

□

Example 3.127 (Uniformly integrable martingales). To any integrable random variable X we can associate the martingale $X_t := \mathbb{E}\left[X \| \mathcal{F}_t\right]$. □

Example 3.128 (Processes with independent increments). Suppose that the random process $(Z_t)_{t\geq}$ has *independent increments*, i.e., for any $n \in \mathbb{N}$ and any
$$0 \leq s_1 < t_1 \leq s_2 < t_2 \leq \cdots \leq s_n < t_n,$$
the increments
$$Z_{t_1} - Z_{s_1}, \; Z_{t_2} - Z_{s_2}, \; \ldots, \; Z_{t_n} - Z_{s_n}$$
are independent. The process (Z_t) is adapted to the natural filtration
$$(\mathcal{F}_t)_{t\geq 0}, \quad \mathcal{F}_t = \sigma\left(Z_s, \; s \leq t\right).$$
We deduce that, $\forall 0 \leq s < t$, the increment $Z_t - Z_s$ is independent of $\mathcal{F}_s$ so
$$\mathbb{E}\left[X_t \| \mathcal{F}_s\right] - X_s = \mathbb{E}\left[(X_t - X_s) \| \mathcal{F}_s\right] = \mathbb{E}\left[X_t - X_s\right].$$
Hence
$$\mathbb{E}\left[X_t - \mathbb{E}\left[X_t \| \mathcal{F}_s\right]\right] = X_s - \mathbb{E}\left[X_s\right], \quad \forall 0 \leq s < t. \tag{3.3.9}$$
Then

(i) if $Z_t \in L^1$, $\forall t \geq 0$, then $\widetilde{Z}_t := Z_t - \mathbb{E}\left[Z_t\right]$ is a martingale;
(ii) if $Z_t \in L^2$, $\forall t \geq 0$, then $Y_t := \widetilde{Z}_t^2 - \mathbb{E}\left[\widetilde{Z}_t^2\right]$ is a martingale;
(iii) if, for some $\theta \in \mathbb{R}$, we have $\mathbb{E}\left[e^{\theta Z_t}\right] < \infty$, $\forall t \geq 0$, then
$$X_t := \frac{e^{\theta Z_t}}{\mathbb{E}\left[e^{\theta Z_t}\right]}$$
is a martingale.

The case (i) follows from (3.3.9). The case (iii) is the continuous time analogue of Example 3.7 and the proof is similar. To prove (ii) note that

$$\mathbb{E}\big[\,\widetilde{Z}_t^2\|\mathcal{F}_s\,\big] = \mathbb{E}\big[\,(\widetilde{Z}_s + \widetilde{Z}_t - \widetilde{Z}_s)^2\,\|\mathcal{F}_s\,\big]$$

$$= \widetilde{Z}_s^2 + 2\widetilde{Z}_s\underbrace{\mathbb{E}\big[\,(\widetilde{Z}_t - \widetilde{Z}_s)\,\|\mathcal{F}_s\,\big]}_{=0} + \mathbb{E}\big[\,(\widetilde{Z}_t - \widetilde{Z}_s)^2\|\mathcal{F}_s\,\big] = \widetilde{Z}_s^2 + \mathbb{E}\big[\,(\widetilde{Z}_t - \widetilde{Z}_s)^2\,\big]$$

$$= \widetilde{Z}_s^2 + \mathbb{E}\big[\,\widetilde{Z}_t^2\,\big] - 2\mathbb{E}\big[\,\widetilde{Z}_s\widetilde{Z}_t\,\big] + \mathbb{E}\big[\,\widetilde{Z}_s^2\,\big]$$

$$= \widetilde{Z}_s^2 + \mathbb{E}\big[\,\widetilde{Z}_t^2\,\big] - 2\mathbb{E}\big[\,\mathbb{E}\big[\,\widetilde{Z}_s\widetilde{Z}_t\|\mathcal{F}_s\,\big]\,\big] + \mathbb{E}\big[\,\widetilde{Z}_s^2\,\big] = \widetilde{Z}_s^2 + \mathbb{E}\big[\,\widetilde{Z}_t^2\,\big] - \mathbb{E}\big[\,\widetilde{Z}_s^2\,\big].$$

Hence

$$\mathbb{E}\Big[\,\widetilde{Z}_t^2 - \mathbb{E}\big[\,\widetilde{Z}_t^2\,\big]\,\|\mathcal{F}_s\,\Big] = \widetilde{Z}_s^2 - \mathbb{E}\big[\,\widetilde{Z}_s^2\,\big].$$

Classical examples of processes with independent increments are the Brownian motion, the Poisson process, or more generally the Lévy processes, [33, Chap. VII].

If B_t is a 1-dimensional Brownian motion started at 0, adapted to $\mathcal{F}_t$, then B_t is a normal random variable with mean 0 and variance t, for each $t > 0$. The moment generating function of B_t is

$$M_{B_t}(\theta) = \mathbb{E}\big[\,e^{\theta B_t}\,\big] = e^{\frac{\theta^2 t}{2}}.$$

We deduce from the above that

$$B_t,\ \ B_t^2 - t,\ \ e^{\theta B_t - \frac{\theta^2}{2}t}$$

are martingales, $\forall \theta \in \mathbb{R}$. The martingale

$$\left(e^{\theta B_t - \frac{\theta^2}{2}t}\right)_{t \geq 0},$$

is called the *exponential martingale* of the Brownian motion.

Note that if we set $\lambda := \theta\sqrt{t}$, and $X = \frac{B_t}{\sqrt{t}}$, then

$$e^{\theta B_t - \frac{\theta^2}{2}t} = e^{\lambda X - \lambda^2/2} \stackrel{(1.6.5)}{=} \sum_{n \geq 0} H_n(X)\frac{\lambda^n}{n!},$$

where $H_n(x)$ is the n-th Hermite polynomial (1.6.4). We can rewrite the above equality as

$$e^{\theta B_t - \frac{\theta^2}{2}t} = \sum_{n \geq 0} M_n(t)\frac{\theta^n}{n!},\ \ M_n(t) = t^{n/2}H_n\big(B_t/\sqrt{t}\,\big).$$

Each of the coefficients $M_n(t)$ is a continuous time martingale. Note that

$$M_1(t) = B_t.\ \ M_2(t) = B_t^2 - t.$$

$\square$

Example 3.129 (New submartingales from old). If $(X_t)_{t \geq 0}$ is a martingale and $f : \mathbb{R} \to \mathbb{R}$ is a convex function such that $f(X_t) \in L^1$, $\forall t \geq 0$, then $\big(f(X_t)\big)_{t \geq 0}$ is a submartingale. If $(X_t)_{t \geq 0}$ is only a submartingale and additionally, f is nondecreasing, then $\big(f(X_t)\big)_{t \geq 0}$ is a submartingale. $\square$

3.3.4 Limit theorems

Fix a filtered probability space $\big(\Omega, (\mathcal{F}_t)_{t\geq 0}, \mathcal{F}, \mathbb{P}\big)$.

Definition 3.130. An R-(sub/super)martingale is a (sub/super)martingale $(X_t)_{t\geq 0}$ adapted to the filtration $(\mathcal{F}_t)_{t\geq 0}$ such that the paths of X_t are a.s. R-functions. □

Remark 3.131. Suppose that $(X_t)_{t\geq 0}$ is an R-submartingale. Fix a negligible set $\mathcal{N} \subset \Omega$ such that $t \mapsto X_t(\omega)$ is an R-function for any $\omega \in \Omega \setminus \mathcal{N}$. Fix a dense countable subset D of $[0, \infty)$.

Note that for every open interval $I \subset [0, \infty)$ we have

$$\sup_{t \in D \cap I} X_t(\omega) = \sup_{t \in I} X_t(\omega), \quad \inf_{t \in D \cap I} X_t(\omega) = \inf_{t \in I} X_t(\omega), \; \forall \omega \in \Omega \setminus \mathcal{N}. \qquad (3.3.10)$$

This shows that $(X_t)_{t \geq}$ is a *separable process* in the sense of Doob, [47, II.2]. This means that there exist

- a countable dense subset $D \subset [0, \infty)$, and
- a negligible subset $\mathcal{N} \subset \Omega$,

such that, for any closed interval $I \subset \mathbb{R}$, and any open subset $\mathcal{O}$ of $[0, \infty)$, the sets

$$\{\, \omega; \; X_s(\omega) \in I, \; \forall s \in D \cap \mathcal{O} \,\} \text{ and } \{\, \omega; \; X_t(\omega) \in I, \; \forall t \in \mathcal{O} \,\}$$

differ by a subset of $\mathcal{N}$. A dense countable subset D with the above property is called a *separability set*. □

Before we proceed investigating the properties of R-submartingales we want to understand how restrictive is the assumption that the paths are a.s. R-functions. The proof of Theorem 3.90 shows that if $(X_t)_{t\geq 0}$ is an R-submartingale, then, for any bounded set $S \subset [0, \infty)$ the family $(X_s)_{s \in S}$ is UI. This implies that the function $t \mapsto \mathbb{E}[X_t]$ is an R-function. We have a more precise result, [103, Sec. 3.3], [135, II.65-67].

Theorem 3.132 (Doob's regularization theorem). *If the filtration $(\mathcal{F}_t)_{t\geq 0}$ satisfies the usual conditions, then a submartingale $(X_t)_{t\geq 0}$ adapted to this filtration admits an R-submartingale modification if and only if the function $t \mapsto \mathbb{E}[X_t]$ is right continuous.* □

Theorem 3.133 (Doob's maximal inequality). *Suppose that $(X_t)_{t\geq 0}$ is an R-submartingale. Then, for any $a, t > 0$ we have*

$$a\mathbb{P}\left[\sup_{s \in [0,t]} |X_s| > a\right] \leq \mathbb{E}\big[\,|X_t^+|\,\big] \leq \mathbb{E}\big[\,|X_t|\,\big] + \mathbb{E}\big[\,|X_0|\,\big]. \qquad (3.3.11)$$

Proof. For any $m \in \mathbb{N}$ we set
$$D_m := \left\{0, \frac{t}{m}, \ldots, \frac{(m-1)t}{m}, t\right\}, \quad D := \bigcup_{m \in \mathbb{N}} D_m.$$
The discrete Doob maximal inequality (3.2.31) implies that
$$a\mathbb{P}\left[\sup_{s \in D_m} |X_s| > a\right] \leq \mathbb{E}[|X_t^+|] \quad \text{and} \quad a\mathbb{P}\left[\sup_{s \in D} |X_s| > a\right] \leq \mathbb{E}[|X_t^+|].$$
As observed in Remark 3.131 $(X_\bullet)$ is a separable process so (3.3.10)
$$\mathbb{P}\left[\sup_{s \in D} |X_s| > a\right] = \mathbb{P}\left[\sup_{s \in [0,t]} |X_s| > a\right].$$
$\square$

Theorem 3.134 (Doob's L^p-inequality). *Suppose that $(X_t)_{t \geq 0}$ is an R-martingale. Then, for any $t > 0$ and $p > 1$ we have*

$$\boxed{\mathbb{E}\left[\sup_{s \in [0,t]} |X_s|^p\right]^{\frac{1}{p}} \leq q \|X_t\|_{L^p}, \quad \frac{1}{q} = 1 - \frac{1}{p}.} \qquad (3.3.12)$$

Proof. Argue as in the proof of Theorem 3.133 by relying on the separability of $(X_\bullet)$ and the discrete L^p-inequality (3.2.33). $\square$

Theorem 3.135. *Suppose that $(X_t)_{t \geq 0}$ is an R-submartingale and*
$$\sup_{t > 0} \mathbb{E}[|X_t|] < \infty. \qquad (3.3.13)$$
Then there exists an integrable random variable X_∞ such that
$$\lim_{t \to \infty} X_t = X_\infty \quad a.s.$$

Proof. For any $m \in \mathbb{N}$ we set
$$D_m := \frac{1}{2^m}\mathbb{N}, \quad m \in \mathbb{N}, \quad D = \bigcup_{m \in \mathbb{N}} D_m.$$
For any function $f : [0, \infty) \to \mathbb{R}$, any rational numbers $a < b$ and any $S \subset [0\infty)$ we denote by $N(f, S, [a, b])$ the supremum of the set of integers k such that there exist
$$s_1 < t_1 < \cdots s_k < t_k$$
in S such that $f(s_i) \leq a$, $f(t_i) \geq b$, $\forall i = 1, \ldots, k$.

For $m \in \mathbb{N}$ we set $N_m(f, [a, b]) := N(f, D_m, [a, b])$. Equivalently, $N_m(f, [a, b])$ is the number of upcrossings of the strip $[a, b]$ by the function $f\big|_{D_m}$. Note that
$$N_m(X, [a, b]) \leq N_{m+1}(X, [a, b]), \quad \forall m,$$
and
$$N(f, D, [a, b]) = \lim_{m \to \infty} N_m(X, [a, b]).$$

Doob's upcrossing inequality (3.2.2) implies

$$(b-a)\mathbb{E}\big[\,N_m\big(\,X,[a,b]\,\big)\,\big] \leq \sup_{t>0}\mathbb{E}\big[\,(X_t-a)^+\,\big] - \mathbb{E}\big[\,(X_0-a)^+\,\big], \quad \forall m \in \mathbb{N}.$$

Letting $m \to \infty$ we deduce from the Monotone Convergence Theorem

$$(b-a)\mathbb{E}\big[\,N\big(\,X,D,[a,b]\,\big)\,\big] \leq \sup_{t>0}\mathbb{E}\big[\,(X_t-a)^+\,\big] - \mathbb{E}\big[\,(X_0-a)^+\,\big] < \infty.$$

Thus $N\big(\,X,D,[a,b]\,\big) < \infty$ a.s. so the limit

$$X_\infty := \lim_{\substack{t \to \infty \\ t \in D}} X_t$$

exists a.s. We leave the reader convince her/himself that since the process $X_\bullet$ is separable (see Remark 3.131) the limit

$$X_\infty = \lim_{t \to \infty} X_t$$

exists a.s. The boundedness assumption (3.3.13) coupled with Fatou's lemma implies that X_∞ is integrable. □

The above theorem implies immediately the following continuous time counterpart of Theorem 3.59.

Theorem 3.136 (UI martingales). *Suppose that $(X_t)_{t \geq 0}$ is an UI R-martingale. Then*

$$X_\infty = \lim_{t \to \infty} X_t$$

exists a.s. and L^1 and

$$X_t = \mathbb{E}\big[\,X_\infty\|\,\mathcal{F}_t\,\big], \quad \forall t > 0.$$ □

3.3.5 Sampling and stopping

Suppose that $(X_t)_{t \geq 0}$ is an R-submartingale such that

$$X_\infty = \lim_{t \to \infty} X_t$$

exists a.s. Let $T : \Omega \to [0, \infty]$ be a stopping time adapted to the filtration $(\mathcal{F}_t)$. The *optional sampling* of $X_\bullet$ at T is the random variable

$$X_T(\omega) = \boldsymbol{I}_{T<\infty}X_{T(\omega)}(\omega) + \boldsymbol{I}_{T=\infty}X_\infty(\omega).$$

Theorem 3.137 (Optional sampling). *Suppose that $(X_t)_{t \geq 0}$ is an UI R-martingale and S, T are stopping times such that $S \leq T$. Then the following hold.*

(i) The random variables X_S, X_T are integrable.
(ii) $X_S = \mathbb{E}\big[\,X_T\|\,\mathcal{F}_S\,\big] = \mathbb{E}\big[\,X_\infty\|\,\mathcal{F}_S\,\big].$
(iii) $\mathbb{E}[X_S] = \mathbb{E}\big[\,X_\infty\,\big] = \mathbb{E}\big[\,X_0\,\big].$

Proof. We set

$$S_n = \sum_{k=0}^{\infty} \frac{k+1}{2^n} I_{\{k2^{-n} < S \leq (k+1)2^{-n}\}} + \infty I_{S=\infty},$$

$$T_n = \sum_{k=0}^{\infty} \frac{k+1}{2^n} I_{\{k2^{-n} < T \leq (k+1)2^{-n}\}} + \infty I_{T=\infty}.$$

Observe that $S_n \geq S$, $T_n \geq T$ and $S_n \leq T_n$, $\forall n$.

Let us show that S_n is $\mathcal{F}_S$ measurable and T_n is T-measurable. In other words, we have to show that

$$\{S_n \leq c\} \cap \{S \leq s\} \in \mathcal{F}_s, \quad \forall c, s \geq 0.$$

Note that

$$\{S \leq s\} \cap \{S_n \leq c\} = \{S \leq s\} \cap \left(\bigcup_{(k+1)2^{-n} \leq c} \left\{ k2^{-n} < S \leq (k+1)2^{-n} \right\} \right) \in \mathcal{F}_c$$

$$= \bigcup_{(k+1)2^{-n} \leq c} \left\{ k2^{-n} < S \leq \min\bigl(s, (k+1)2^{-n}\bigr) \right\} \in \mathcal{F}_s.$$

Proposition 3.112(ii) now implies that S_n is a stopping time. A similar argument shows that T_n is a stopping time. Note that

$$S_n \searrow S \text{ and } T_n \searrow T \text{ as } n \to \infty.$$

For $n \in \mathbb{N}_0$ set $D_n = 2^{-n}\mathbb{N}_0$. For each $n \in \mathbb{N}_0$ the stochastic process

$$X^n := (X_t)_{t \in D_n},$$

is a UI *discrete* martingales with respect to the filtration $\mathcal{F}_\bullet^n := (\mathcal{F}_t)_{t \in D_n}$. The above arguments show that S_n and T_n are stopping times with respect to these filtrations. We deduce from the discrete Optional Sampling Theorems 3.64 that

$$X_{S_n} = X_{S_n}^n = \mathbb{E}\bigl[\,X_{T_n}^n \,\|\, \mathcal{F}_{S_n}^n\,\bigr] = \mathbb{E}\bigl[\,X_{T_n} \,\|\, \mathcal{F}_{S_n}\,\bigr],$$

and

$$X_{S_n} = \mathbb{E}\bigl[\,X_\infty \,\|\, \mathcal{F}_{S_n}\,\bigr], \quad X_{T_n} = \mathbb{E}\bigl[\,X_\infty \,\|\, \mathcal{F}_{T_n}\,\bigr].$$

Now observe that since (X_t) is a.s. right continuous we have

$$X_S = \lim_{n \to \infty} X_{S_n} \text{ and } X_T = \lim_{n \to \infty} X_{T_n} \text{ a.s.}$$

The families (X_{S_n}) and (X_{T_n}) are UI so the above convergences also hold in L^1. Since $\mathcal{F}_S \subset \mathcal{F}_{S_n} \subset \mathcal{F}_{T_n}$ and the conditional expectation map

$$\mathbb{E}\bigl[\,-\,\|\mathcal{F}_S\,\bigr] : L^1(\Omega,,\mathcal{F},\mathbb{P}) \to L^1(\Omega, \mathcal{F}_S, \mathbb{P})$$

is a contraction we deduce

$$X_S = \mathbb{E}\bigl[\,X_S\|\mathcal{F}_S\,\bigr] = \lim_{n \to \infty} \mathbb{E}\bigl[\,X_{S_n}\|\mathcal{F}_S\,\bigr] = \lim_{n \to \infty} \mathbb{E}\bigl[\,X_{T_n}\|\mathcal{F}_S\,\bigr] = \mathbb{E}\bigl[\,X_T\|\mathcal{F}_S\,\bigr],$$

where the above converges are in L^1. □

Corollary 3.138. *Suppose that $(X_t)_{t\geq 0}$ is an R-martingale and S, T are bounded stoping times such that $S \leq T$ a.s. Then the following hold.*

(i) The random variables X_S, X_T are integrable.
(ii) $X_S = \mathbb{E}\big[\, X_T \|\, \mathcal{F}_S \,\big] = \mathbb{E}\big[\, X_\infty \|\, \mathcal{F}_S \,\big].$

Proof. Fix $t_0 > 0$ such $S, T \leq t_0$ a.s. Then the stopped process $X_{t \wedge t_0}$ is an UI R-martingale. The conclusions now follow from Theorem 3.137 applied to this stopped martingale. □

Corollary 3.139 (Optional stopping). *Let $(X_t)_{t\geq 0}$ be an R-martingale compatible with the filtration $(\mathcal{F}_t)_{t\geq 0}$. Then the following hold.*

(i) The stopped process
$$X_t^T := X_{T \wedge t}$$
is an R-martingale compatible with the same filtration $(\mathcal{F}_t)_{t\geq 0}$.
(ii) If additionally $(X_t)_{t\geq 0}$ is UI, then so is the stopped process and we have

$$X_{T \wedge t} = \mathbb{E}\big[\, X_T \| \mathcal{F}_t \,\big], \tag{3.3.14}$$

$$X_T = \lim_{t \to \infty} X_t \quad a.s.\ and\ L^1. \tag{3.3.15}$$

Proof. We begin by proving (ii). For $s < t$, the stopping times $s \wedge T$ and $t \wedge T$ are bounded and $s \wedge T \leq t \wedge T$. The random variables $X_{t \wedge T}$ are $\mathcal{F}_{t \wedge T}$-measurable and thus $\mathcal{F}_t$-measurable since $\mathcal{F}_{t \wedge T} \subset \mathcal{F}_t$. To prove (3.3.14) it suffices to check that for any $A \in \mathcal{F}_t$ we have

$$\mathbb{E}\big[\, X_T \boldsymbol{I}_A \,\big] = \mathbb{E}\big[\, X_{t \wedge T} \boldsymbol{I}_A \,\big].$$

Decompose $\boldsymbol{I}_A = \boldsymbol{I}_{A \cap \{T \leq t\}} + \boldsymbol{I}_{A \cap \{T > t\}}$. We have

$$X_T \boldsymbol{I}_{A \cap \{T \leq t\}} = X_{t \wedge T} \boldsymbol{I}_{A \cap \{T \leq t\}}$$

so that

$$\mathbb{E}\big[\, X_T \boldsymbol{I}_{A \cap \{T \leq t\}} \,\big] = \mathbb{E}\big[\, X_{t \wedge T} \boldsymbol{I}_{A \cap \{T \leq t\}} \,\big]. \tag{3.3.16}$$

On the other hand, we deduce from Theorem 3.137 that

$$X_{t \wedge T} = \mathbb{E}\big[\, X_T \| \mathcal{F}_{t \wedge T} \,\big].$$

Now observe that

$$A \cap \{T > t\} \in \mathcal{F}_t \ \text{and}\ A \cap \{T > t\} \in \mathcal{F}_T,$$

so $A \cap \{T > t\} \in \mathcal{F}_t \cap \mathcal{F}_T = \mathcal{F}_{t \wedge T}$. Hence

$$X_{t \wedge T} \boldsymbol{I}_{A \cap \{T > t\}} = \mathbb{E}\big[\, X_T \boldsymbol{I}_{A \cap \{T > t\}} \| \mathcal{F}_{t \wedge T} \,\big],$$

$$\mathbb{E}\big[\, X_{t \wedge T} \boldsymbol{I}_{A \cap \{T > t\}} \,\big] = \mathbb{E}\big[\, X_T \boldsymbol{I}_{A \cap \{T > t\}} \,\big]. \tag{3.3.17}$$

The desired conclusion follows by adding (3.3.16) and (3.3.17). The assertion (3.3.15) follows from the fact that the stopped martingale X^T is UI. Part (i) now follows from (ii) applied to the sequence of UI martingales

$$(X_t^n)_{t\geq 0} := (X_{n\wedge t})_{t\geq 0}, \; n \in \mathbb{N}.$$

Indeed, the martingales X^n are compatible with $\mathcal{F}_t$ and for $s < t$ we have

$$\mathbb{E}\big[X_{T\wedge t}^n \big\| \mathcal{F}_s\big] = \mathbb{E}\big[\mathbb{E}\big[X_T^n \|\mathcal{F}_t\big]\big\| \mathcal{F}_s\big] = \mathbb{E}\big[X_T^n \| \mathcal{F}_s\big] = X_{T\wedge s}^n.$$

Now let $n \to \infty$ and observe that for $n > t$ we have $X_{T\wedge t}^n = X_{T\wedge n}$. □

Example 3.140. Suppose that $(B_t)_{t\geq 0}$ is a Brownian motion started at 0 and $(\mathcal{F}_t)_{t\geq 0}$ is its canonical filtration. For any $a \in \mathbb{R}$ we set

$$T_a := \inf\{t \geq 0 : \; B_t = a\}.$$

According to Proposition 3.118(ii), $\mathbb{P}[T_a < \infty] = 1$.

(a) We want to show that if $a < 0 < b$, then

$$\mathbb{P}[T_a < T_b] = \frac{b}{b-a}, \;\; \mathbb{P}[T_a > T_b] = \frac{-a}{b-a}. \quad (3.3.18)$$

Consider the stopping time $T = T_a \wedge T_b$ and the stopped martingale $M_t = B_{T\wedge t}$. This martingale is UI since $|M_t| \leq |a| \vee |b|$. We deduce

$$0 = \mathbb{E}[M_0] = \mathbb{E}[M_\infty] = \mathbb{E}[B_T] = a\mathbb{P}[T_a < T_b] + b\mathbb{P}[T_b < T_a].$$

The equalities (3.3.18) follow by observing that the probabilities $\mathbb{P}[T_a < T_b]$ and $\mathbb{P}[T_a > T_b]$ satisfy a second linear constraint

$$\mathbb{P}[T_a < T_b] + \mathbb{P}[T_a > T_b] = 1.$$

(b) For $a > 0$ we set

$$U_a := \inf\{t \geq 0 : \; |B_t| = a\} = T_a \wedge T_{-a}.$$

We want to show that

$$\mathbb{E}[U_a] = a^2. \quad (3.3.19)$$

To see this consider the martingale of Example 3.3.9(ii), $M_t = B_t^2 - t$. The stopped process $M_{t\wedge U_a}$ is still a martingale so

$$\mathbb{E}[M_{t\wedge U_a}] = \mathbb{E}[M_0] = 0 \; \text{ and } \; \mathbb{E}[B_{t\wedge U_a}^2] = \mathbb{E}[t \wedge U_a].$$

The Monotone Convergence Theorem implies that

$$\lim_{t\to\infty} \mathbb{E}[t \wedge U_a] = \mathbb{E}[U_a].$$

The martingale $B_{t\wedge U_a}$ is bounded, $|B_{t\wedge U_a}| \leq a$, $\forall t \geq 0$ and we deduce from the Dominated Convergence Theorem that

$$\mathbb{E}[U_a] = \lim_{t\to\infty} \mathbb{E}[t \wedge U_a] = \lim_{t\to\infty} \mathbb{E}[B_{t\wedge U_a}^2] = \mathbb{E}[B_{U_a}^2] = a^2.$$

(c) Fix $a > 0$. We want to compute the moment generating function of T_a. To this aim, we consider for any $\lambda \in \mathbb{R}$ the martingale of Example 3.3.9(iii)

$$X_t^\lambda := \exp\left(\lambda B_t - \frac{\lambda^2 t}{2}\right). \tag{3.3.20}$$

For $\lambda > 0$ the stopped martingale $Y_y^\lambda = X_{t \wedge T_a}^\lambda$ is bounded thus UI and we deduce

$$1 = \mathbb{E}[Y_0^\lambda] = \mathbb{E}[Y_\infty^\lambda] = e^{\lambda a}\mathbb{E}\left[e^{-\frac{\lambda^2 T_a}{2}}\right].$$

Replacing λ with $\sqrt{2\lambda}$ we deduce

$$\mathbb{E}[e^{-\lambda T_a}] = e^{-a\sqrt{2\lambda}}.$$

This can be alternatively verified using the distribution of T_a computed in Corollary 3.125.

(d) We want to compute the Laplace transform of U_a (or moment generating function). Consider the stopped martingale $Z_t^\lambda := X_{t \wedge U_a}^\lambda$, where X_t^λ is defined as in (3.3.20). We deduce as above that

$$1 = \mathbb{E}[e^{\lambda B_{U_a}} e^{-\lambda^2 U_a/2}].$$

The computations in (a) show that

$$\mathbb{P}[B_{U_a} = a] = \mathbb{P}[B_{U_a} = -a] = \frac{1}{2}.$$

Note that

$$\mathbb{P}[U_a \leq u] = \mathbb{P}[B_{U_a} = a, U_a \leq u] + \mathbb{P}[B_{U_a} = -a, U_a \leq u].$$

Using the symmetry $B_t \mapsto -B_t$ we deduce

$$\mathbb{P}[B_{U_a} = a, U_a \leq u] = \mathbb{P}[B_{U_a} = -a, U_a \leq u] = \frac{1}{2}\mathbb{P}[U_a \leq u]$$

$$= \mathbb{P}[B_{U_a} = a]\mathbb{P}[U_a \leq u] = \mathbb{P}[B_{U_a} = -a]\mathbb{P}[U_a \leq u],$$

proving that B_{U_a} and U_a are independent. Hence

$$1 = \mathbb{E}[e^{\lambda B_{U_a}}]\mathbb{E}[e^{-\lambda^2 U_a/2}] = \cosh(\lambda a)\mathbb{E}[e^{-\lambda^2 U_a/2}]. \qquad \square$$

3.4 Exercises

Exercise 3.1. Suppose that $(X_n)_{n\geq 0}$ is a sequence of integrable random variables and $(q_n)_{n\geq 1}$ is a sequence of nonzero real numbers such that, for any $n \in \mathbb{N}$
$$\mathbb{E}[X_n \| \mathcal{F}_{n-1}] = q_n X_{n-1}, \quad \mathcal{F}_{n-1} := \sigma(X_0, \ldots, X_{n-1}).$$
Define $Q_0 = 1$, $Q_n = q_1 \cdots q_n$, $\forall n \in \mathbb{N}$ and set $Y_n := \frac{1}{Q_n} X_n$. Prove that $(Y_n)_{n\geq 0}$ is a martingale with adapted to the filtration $(\mathcal{F}_n)_{n\geq 0}$. □

Exercise 3.2. Suppose that $(X_n)_{n\geq 0}$ is a martingale with respect to a filtration $(\mathcal{F}_n)_{n\geq 0}$ such that $X_0 = 0$ and $\mathbb{E}[|X_n|^2] < \infty$, $\forall n$. Using the sequence of differences $D_n = X_n - X_{n-1}$, $n \geq 1$ we construct two new processes, the *optional quadratic variation*
$$Q_n = \sum_{k=1}^n D_k^2$$
and the *predictable quadratic variation*
$$V_n = \sum_{k=1}^n \mathbb{E}[D_k^2 \| \mathcal{F}_{k-1}].$$
Prove that the processes
$$A_n = X_n^2 - Q_n \quad B_n = X_n^2 - V_n$$
are martingales with respect to the $(\mathcal{F}_n)_{n\geq 0}$. □

Exercise 3.3. Let $x_1, \ldots x_r \in \mathbb{R}$. Fix a family $\{I_n, J_n; \ n \in \mathbb{N}\}$ of independent random variables such that I_n, J_n are uniformly distributed on $\{1, \ldots, n-1\}$, $\forall n \geq 2$. Define inductively
$$X_n := \begin{cases} x_r, & n \leq r \\ X_{I_n} + X_{J_n}, & n > r, \end{cases}$$
and set
$$Y_n := \frac{1}{n(n+1)} \sum_{k=1}^n X_k.$$
Prove that the sequence (Y_n) is a martingale with respect to the filtration $\sigma(X_1, \ldots, X_n)$. □

Exercise 3.4. Prove all the claims in Example 3.21. □

Exercise 3.5 (Optional switching). Suppose that $\mathcal{F}_\bullet := (\mathcal{F}_n)_{n\geq 0}$ is a filtration of the probability space $(\Omega, \mathcal{S}, \mathbb{P})$ and $(X_n)_{n\geq 0}$, $(Y_n n \geq 0$ are two $\mathcal{F}_\bullet$-martingales. Let $T : \Omega \to \mathbb{N}_0$ be a stopping time adapted to $\mathcal{F}_\bullet$. For $n \in \mathbb{N}_0$ define
$$Z_n : \Omega \to \mathbb{R}, \quad Z_n(\omega) = \begin{cases} X_n(\omega), & n \leq T(\omega), \\ Y_n(\omega), & n > T(\omega). \end{cases}$$
Prove that $(Z_n)_{n\geq 0}$ is a martingale adapted to $\mathcal{F}_\bullet$. □

Exercise 3.6. Prove Lemma 3.32. □

Exercise 3.7 (Dubins' inequality). Let $X_\bullet = (X_n)_{n\geq 0}$ be a nonnegative supermartingale adapted to the filtration $\mathcal{F}_\bullet$ of a probability space $(\Omega, \mathcal{S}, B)$. For $0 \leq a < b$ denote by $N_n([a,b], X)$ the number of upcrossings of $[a,b]$ by $X_\bullet$ up to time n; see (3.2.1). Prove that for any $k = 1, 2, \ldots, n$

$$\mathbb{P}\big[\, N_n(a,b,X) \geq k \,\big] \leq \Big(\frac{a}{b}\Big)^k \mathbb{E}\big[\, \min(1, X_0/a) \,\big].$$

Exercise 3.8. Prove Lemma 3.38. □

Exercise 3.9. Suppose that $X_n \in L^1(\Omega, \mathcal{S}, \mathbb{P})$, $n \in \mathbb{N}$, is a uniformly integrable sequence of random variables that converges in law to the random variable X, $X_n \Rightarrow X$. Then $X \in L^1(\Omega, \mathcal{S}, \mathbb{P})$ and

$$\lim_{n\to\infty} \mathbb{E}\big[\, X_n^\pm \,\big] = \mathbb{E}\big[\, X^\pm \,\big], \quad \lim_{n\to\infty} \mathbb{E}\big[\, |X_n| \,\big] = \mathbb{E}\big[\, |X| \,\big],$$

$$\lim_{n\to\infty} \mathbb{E}\big[\, X_n \,\big] = \mathbb{E}\big[\, X \,\big]. \qquad \square$$

Exercise 3.10 (Pratt's Lemma). Let (X_n), (Y_n), (Z_n) be three sequences of integrable random variables with the following properties.

(i) $X_n \leq Y_n \leq Z_n$, $\forall n$.
(ii) $X_n \xrightarrow{P} X$, $Y_n \xrightarrow{P} Y$, $Z_n \xrightarrow{P} Z$.
(iii) $\mathbb{E}[X_n] \to \mathbb{E}[X]$, $\mathbb{E}[Z_n] \to \mathbb{E}[Z]$.

Prove that $\mathbb{E}\big[\, Y_n \,\big] \to \mathbb{E}\big[\, Y \,\big]$. □

Exercise 3.11. Suppose that $(X_n)_{n\geq 0}$ is a martingale defined on a probability space $(\Omega, \mathcal{S}, \mathbb{P})$ such that $\exists M > 0$,

$$\forall n \in \mathbb{N} \quad |X_n - X_{n-1}| \leq M, \text{ a.s.}$$

Define

$$A := \Big\{\omega \in \Omega;\ \lim_{n\to\infty} X_n(\omega) \text{ exists and is finite}\Big\},$$

$$B := \Big\{\omega \in \Omega;\ \liminf_{n\to\infty} X_n(\omega) = -\infty,\ \limsup_{n\to\infty} X_n(\omega) = \infty \Big\}.$$

Prove that $\mathbb{P}\big[\, A \cup B \,\big] = 1$. In other words, when a martingale (with bounded increments) does not have a limit, it oscillates wildly.

Hint. For $C > 0$ look at $T_C^\pm = \min\{n;\ \pm X_n > C\}$. □

Exercise 3.12 (P. Lévy). Suppose that $(\Omega, \mathcal{S}, \mathbb{P})$ is a probability space and $(\mathcal{F}_n)_{n\geq 1}$ is a filtration of sigma-subalgebras. Let (F_n) be a sequence of events such that $F_n \in \mathcal{F}_n$, $\forall n$. We set

$$X_n = \sum_{k=1}^n \big(\mathbf{I}_{F_k} - \mathbb{E}\big[\, \mathbf{I}_{F_k} \,\|\, \mathcal{F}_{k-1} \,\big] \big).$$

(i) Prove that X_n is a martingale and $|X_n - X_{n-1}| \leq 4$, $\forall n$. **Hint.** Have a look at Example 3.14.

(ii) Prove that
$$\{F_n \text{ i.o.}\} = \left\{\sum_{n\geq 1} \mathbb{E}\left[\,\boldsymbol{I}_{F_n} \,\|\, \mathcal{F}_{n-1}\right] = \infty\right\}.$$

Hint. Use Exercise 3.11.

(iii) Deduce from (ii) the second Borel-Cantelli Lemma, Theorem 1.139(ii).

□

Exercise 3.13. Suppose that $(X_n)_{n\in\mathbb{N}}$ is a sequence of independent Rademacher random variables
$$\mathbb{P}[X_n = 1] = \mathbb{P}[X_n = -1] = \frac{1}{2}, \quad \forall n.$$

Set
$$S_n := X_1 + \cdots + X_n, \quad p_n := \mathbb{P}[\exists k = 1,\ldots,n, \ S_k < 0].$$

(i) Compute p_n. **Hint.** Use the André's reflection trick in Example 1.60.

(ii) Show that $p_n \to 1$ as $n \to \infty$.

□

Exercise 3.14. Consider the situation in Example 3.31. We have a finite set $\mathcal{A}$ called alphabet, a probability distribution π on $\mathcal{A}$ such that $\pi[a] \neq 0$, $\forall a \in \mathcal{A}$. Fix two words
$$\boldsymbol{a} = (a_1,\ldots,a_k) \in \mathcal{A}^k, \quad \boldsymbol{b} = (b_1,\ldots,b_\ell) \in \mathcal{A}^\ell$$
and assume that $\boldsymbol{b}$ is not a subword of $\boldsymbol{a}$, i.e.,
$$(a_{i+1},\ldots,a_{i+\ell}) \neq (b_1,\cdots,b_\ell), \quad \forall i = 0,\ldots,k-\ell.$$

Let $(A_n)_{n\geq 1}$ be i.i.d. $\mathcal{A}$ valued random variable with common distribution π. As in Example 3.31 we denote by $T_{\boldsymbol{b}}$ the time to observe the pattern $\boldsymbol{b}$.

(i) Prove that
$$\mathbb{E}[T_{\boldsymbol{b}} \,\|\, A_1 = a_1,\ldots,A_k = a_k] - k = \Phi(\boldsymbol{b},\boldsymbol{b}) - \Phi(\boldsymbol{a},\boldsymbol{b})$$
where Φ is defined by (3.1.12).

(ii) Set $p_{\boldsymbol{a}} := \mathbb{P}[T_{\boldsymbol{a}} < T_{\boldsymbol{b}}]$, $p_{\boldsymbol{b}} := \mathbb{P}[T_{\boldsymbol{b}} < T_{\boldsymbol{a}}]$, $T = \min(T_{\boldsymbol{a}}, T_{\boldsymbol{b}})$. Prove that
$$p_{\boldsymbol{a}}\Phi(\boldsymbol{a},\boldsymbol{a}) + p_{\boldsymbol{b}}\Phi(\boldsymbol{b},\boldsymbol{a}) = \mathbb{E}[T] = p_{\boldsymbol{a}}\Phi(\boldsymbol{a},\boldsymbol{b}) + p_{\boldsymbol{b}}\Phi(\boldsymbol{b},\boldsymbol{b}).$$

(iii) Show that
$$\frac{p_{\boldsymbol{b}}}{p_{\boldsymbol{a}}} = \frac{\Phi(\boldsymbol{a},\boldsymbol{a}) - \Phi(\boldsymbol{a},\boldsymbol{b})}{\Phi(\boldsymbol{b},\boldsymbol{b}) - \Phi(\boldsymbol{a},\boldsymbol{a})}.$$

Hint. Consider the same martingale (X_n) as in Example 3.31. Observe that $X_k = \Phi(a,b) - k$ given that $A_j = a_j$, $j = 1, \ldots, k$. (ii) Note that $\mathbb{E}[T_b] = \mathbb{E}[T_b] + \mathbb{E}[T_b - T]$ and (i) gives a formula for $\mathbb{E}[T_b - T \| T = T_a]$. □

Exercise 3.15 (Kakutani). Let (X_n) be a sequence of independent positive random variables such that $\mathbb{E}[X_n] = 1$. Consider the product martingale

$$Y_n = \prod_{k=1}^{n} X_k.$$

Doob's convergence theorem shows that Y_n converges a.s. to a random variable Y_∞ satisfying $\mathbb{E}[Y_\infty] \leq 1$. Set $a_n := \mathbb{E}[X_n^{1/2}]$. Prove that the following are equivalent.

(i) $\mathbb{E}[Y_\infty] = 1$.
(ii) $Y_n \to Y_\infty$ in L^1.
(iii) The martingale $(Y_n)_{n \in \mathbb{N}}$ is UI.
(iv) $\prod_n a_n > 0$.
(v) $\sum_n (1 - a_n) > 0$.

□

Exercise 3.16. Consider the unbiased random walk in Example 3.5

$$S_0 = a \in \mathbb{Z}, \quad S_n = X_1 + \cdots + X_n, \quad n \geq 1,$$

where $(X_n)_{n \geq 1}$ are i.i.d. random variables such that $\mathbb{E}[X_n] = 0$, $\forall n$. Set $\mathcal{F}_n = \sigma(X_1, \ldots, X_n)$, $n \in \mathbb{N}$.

(i) Assume $\sigma^2 := \mathbb{E}[X_1^2] < \infty$. Show that the sequence $(S_n^2 - n\sigma^2)_{n \geq 0}$ is a martingale with respect to the filtration $\mathcal{F}_n$.
(ii) Assume that $M(t) = \mathbb{E}[e^{tX_1}]$ exists for all $|t| < t_0$, $t_0 > 0$. For $|t| < t_0$ and $n \in \mathbb{N}$ we set

$$Z_n(t) := \frac{1}{M(t)^n} e^{tS_n}$$

is a martingale with respect to the filtration $\mathcal{F}_n$.
(iii) Set $D = \frac{d}{dx}$. We define $M(D) : \mathbb{R}[x] \to \mathbb{R}[x]$ by the equality

$$M(D)[P](x) = \sum_{k \geq 0} \frac{M^{(k)}(0)}{k!} D^k P(x).$$

Prove that $M(D)$ is bijective and for any polynomial P the sequence

$$Y_n = M(D)^{-n} P(S_n), \quad n \geq 1,$$

is a martingale. Find Y_n when $P(x) = x$ and $P(x) = x^2$.

Hint. Set $P_n := \mathcal{T}^{-n}[P]$ and express $\mathbb{E}[P_{n+1}(S_n + X_{n+1}) \| X_1, \ldots, x_n]$ using the operator $M(D)$.

□

Exercise 3.17. Suppose that $(X_n)_{n\geq 0}$ is a martingale with respect to the filtration $\mathcal{F}_\bullet = (\mathcal{F}_n)_{\geq 0}$ such that $\mathbb{E}[X_n^2] < \infty$, $\forall n$. The sequence $(X_n^2)_{n\geq 0}$ is a submartingale and thus, according to Proposition 3.13 it admits a Doob decomposition $X_n^2 = X_0 + M_n + C_n$, where $(M_n)_{n\geq 0}$ is a martingale and the compensator (C_n) is a predictable, nondecreasing process. Set
$$A_n = X_0 + C_n, \quad A_\infty = \lim_{n\to\infty} A_n = \sup_{n\in\mathbb{N}} A_n.$$

(i) Prove that $\mathbb{E}[\sup_{n\geq 0} X_n] \leq 4\mathbb{E}[A_\infty]$. **Hint.** Use Doob's L^2-maximal inequality.

(ii) Prove that $\lim_{n\to\infty} X_n$ exists and is finite a.s. on the set $\{A_\infty < \infty\}$. **Hint.** For $a > 0$ we set $N_a = \min\{n; \ A_{n+1} > a^2\}$. Show that it is adapted to the filtration $\mathcal{F}_\bullet$. Apply (i) to the stopped martingale $X_{n\wedge N_a}$.

(iii) Suppose that $f : [0, \infty) \to [1, \infty)$ is an increasing function such that
$$\int_0^\infty \frac{f(t)}{t^2} dt < \infty.$$
Prove that $\frac{X_n}{f(A_n)} \to 0$ a.s. on the set $\{A_\infty = \infty\}$. **Hint.** Set $H_n = \frac{1}{f(A_n)}$, $\forall n \in \mathbb{N}$. Let $Y_\bullet$ denote the martingale defined by the discrete stochastic integral $(H \cdot X)_\bullet$; see (3.1.2). Use the Doob decomposition of Y_n to prove that Y_n converges L^2 a.s. Conclude using Kronecker's lemma, Lemma 2.10.

$\square$

Exercise 3.18 (Dubins-Freedman). Suppose that $(\Omega, \mathcal{S}, \mathbb{P})$ is a probability space and $(\mathcal{F}_n)_{n\geq 1}$ is a filtration of sigma-subalgebras. Let (F_n) be a sequence of events such that $F_n \in \mathcal{F}_n$, $\forall n$. We set
$$X_n = \sum_{k=1}^n (I_{F_n} - f_n), \quad f_n := \mathbb{E}[I_{F_n} \| \mathcal{F}_{n-1}].$$

(i) Prove that $(X_n)_{\geq 0}$ is a martingale and $\mathbb{E}[X_n^2] < \infty$, $\forall n \geq 0$.

(ii) Define $S = \{\sum_n f_n = \infty\}$. Prove that
$$\frac{\sum_{k=0}^n I_{F_k}}{\sum_{k=0}^n f_k} \to 1, \quad \text{a.s. on } S. \tag{3.4.1}$$

(iii) Deduce from (3.4.1) the conclusion of Exercise 3.12(ii). Thus (3.4.1) is a generalization of the second Borel-Cantelli lemma, Theorem 1.139(ii).

$\square$

Exercise 3.19 (Conservation of fairness). A fair coin is flipped repeatedly and independently. A gambler starts with an initial fortune $f_0 > 0$. Before the n-th flip, his fortune is F_{n-1}. Based only on the information available to him at that moment, the gambler bets a sum $B_n \in (0, b)$, $0 \leq B_n \leq F_{n_1}$. If the n-th flip shows Heads he earns B_n dollars and if its shows Tails, he loses B_n dollars. The gambler stops gambling when he is broke or at the first moment when he reaches his goal, i.e., $F_n \geq g$ where $g > 0$ set in advance of his gambling.

(i) Prove that the probability p_g that he reaches his goal is $\leq \frac{f_0}{g}$.
(ii) Prove that if $B_n \leq \min(F_{n-1}, g - F_{n-1})$, $\forall n \geq 1$, then $p_g = \frac{f_0}{g}$.
(iii) Find p_g if $B_n = \frac{1}{2}F_{n-1}$.

□

Remark 3.141. Note that if $f_0, g \in \mathbb{N}$ and the gambling strategy is $B_n = 1$ whenever his fortune is $< g$ the above problem reduces to the classical Gambler's ruin problem discussed in Example 3.72. The name *"conservation of fairness"* seems appropriate: whatever gambling strategy satisfying (ii) and based only on the information available at each moment, the probability of reaching the goal is the same, $\frac{f_0}{g}$.

□

Exercise 3.20. Suppose that we are given a sequence of i.i.d. random vectors

$$X_n : (\Omega, \mathcal{S}, \mathbb{P}) \to \mathbb{X} := \mathbb{R}^N$$

and a collection $\mathcal{F}$ of uniformly bounded measurable functions $f : \mathbb{X} \to \mathbb{R}$, i.e., there exists $C > 0$ such that $\|f\|_{L^\infty} \leq C$, $\forall f \in \mathcal{F}$. For $n \in \mathbb{N}$ we set

$$D_n(\mathcal{F}) := \sup_{f \in \mathcal{F}} \frac{1}{n} \left| \sum_{k=1}^n \overline{f}(X_k) \right|, \quad \overline{f}(x) := f(x) - \mathbb{E}[X_j],$$

$$\mathcal{R}_n(\mathcal{F}) = \sup_{f \in \mathcal{F}} \mathbb{E}\left[\frac{1}{n} \left| \sum_{k=1}^n R_k f(X_k) \right| \right],$$

where $(R_n)_{n \geq 1}$ is a sequence of independent Rademacher random variables that are also independent of $(X_n)_{n \geq 1}$. Assume that D_n is measurable.

(i) Prove that the function $G_n : \mathbb{X}^n \to \mathbb{R}$

$$G_n(x_1, \ldots, x_n) = \sup_{f \in \mathcal{F}} \frac{1}{n} \left| \sum_{k=1}^n \overline{f}(x_k) \right|$$

satisfies the bounded difference property (Definition 3.36) with $L_k = \frac{2C}{n}$, $\forall k = 1, \ldots, n$.

(ii) Prove that

$$\mathbb{P}[D_n - \mathbb{E}[D_n] \leq t] \geq 1 - e^{-\frac{nt^2}{2C^2}}, \quad \forall t > 0.$$

(iii) Prove that for any $\delta > 0$

$$\mathbb{P}[D_n \leq 2\mathcal{R}_n + \delta] \geq 1 - e^{-\frac{n\delta^2}{2C^2}}.$$

Hint. Use (2.4.13) in Remark 2.66.

□

Exercise 3.21. Consider the standard random walk $(S_n)_{n\geq 0}$ on $\mathbb{Z}$ started at 0, i.e.,
$$S_0 = a \in \mathbb{Z}, \quad S_n = X_1 + \cdots + X_n,$$
where $(X_n)_{n\geq 1}$ are i.i.d. with $\mathbb{P}[X_n = \pm 1] = \frac{1}{2}$. Fix $a, g \in \mathbb{N}_0$, $a < g$ and set
$$T_a := \min\{n \in \mathbb{N}; \; S_n = 0 \text{ or } S_n = g\}.$$

(i) Show $\mathbb{P}[S_{T_a} = 0] = \frac{g-a}{g}$ and $\mathbb{P}[S_{T_a} = g] = \frac{a}{g}$.
(ii) Show that $\mathbb{E}[T_a] = a(g-a)$.
(iii) Compute the pgf of T_a
$$PG_{T_a}(s) := \sum_{n=0}^{\infty} \mathbb{P}[T_a = n] s^n.$$

□

Exercise 3.22. Suppose that $(X_n)_{n\geq 0}$ adapted to the filtration $(\mathcal{F}_n)_{n\geq 0}$ and T is a stopping time adapted to the same filtration such that $\mathbb{P}[T < \infty] = 1$ and $X_T \in L^1$. Prove that
$$E[X_T \,\|\, \mathcal{F}_n] = X_n \text{ on } \{T \geq n\}.$$

Hint. Have a look at the Proof of Theorem 3.28.

Exercise 3.23. Suppose that $(X_n)_{n\geq 1}$ is a sequence of i.i.d., *nonnegative, integer valued* random variables with finite mean. Set
$$S_n := X_1 + \cdots + X_n.$$
For $k = 1, \ldots, n$, set $\mathcal{F}_{-k} = \sigma(S_k, S_{k+1}, \ldots, S_n)$, $Y_{-k} = S_k/k$.

(i) Prove that for $j \leq k$ we have
$$\mathbb{E}[X_j \,\|\, \mathcal{F}_{-k}] = X_k.$$
(ii) Prove that $(Y_{-k})_{1\leq k\leq n}$ is a martingale with respect to the filtration $(\mathcal{F}_{-k})_{1\leq k\leq n}$. (Compare with Example 3.30.)
(iii) Show that
$$\mathbb{P}[S_k < k, \; \forall 1 \leq k \leq n \,\|\, S_n] = (1 - S_n/n)^+.$$

Hint. (iii) Set $T = \inf\{-n \leq k \leq -1; \; Y_k \geq 1\}$, where we define $\inf \emptyset = -1$. Use Exercise 3.22.

□

Exercise 3.24. Suppose that $f : [0,1] \to \mathbb{R}$ is a Lebesgue integrable function. For any $n \in \mathbb{N}_0$ we define the step function $f_n : [0,1] \to \mathbb{R}$ by setting $f_n(0) = 0$ and
$$f_n(x) = \frac{1}{2^n} \int_{(k-1)/2^n}^{k/2^n} f(x)dx, \text{ if } 0 \leq \frac{(k-1)}{2^n} < x \leq \frac{k}{2^n} \leq 1.$$

Prove that f_n converges a.s. and L^1 to f as $n \to \infty$.

□

Exercise 3.25. Suppose that $(X_n)_{n\geq 0}$ is a supermartingale such that, there exist $f_0, g > 0$ with the property
$$X_0 = f_0 \text{ a.s.}, \quad 0 \leq X_n \leq g \text{ a.s.}, \quad \forall n \in \mathbb{N}.$$
Prove that for any stopping time T such that $\mathbb{P}[T < \infty] = 1$ we have $\mathbb{P}[X_T = g] \leq \frac{f_0}{g}$. □

Exercise 3.26. Consider the branching process $(Z_n)_{n\geq 0}$ with initial condition $Z_0 = 1$ and reproduction law $\mu \in \text{Prob}(\mathbb{N}_0)$ such that
$$m := \mathbb{E}[\mu] = \sum_{n\geq 0} n\mu_n < \infty, \quad \mu_n := \mu[n].$$
Assume $\mu_0 > 0$. Denote by $f(s)$ the probability generating function (pgf) of μ
$$f(s) = \sum_{n\geq 0} \mu_n s^n.$$
We set
$$f_n(s) := \underbrace{f \circ \cdots \circ f}_{n}(s), \quad n \in \mathbb{N}.$$

(i) Show that if $m > 1$ the equation $f(s) = s$ has a unique solution $r = r(\mu)$ in the interval $(0,1)$. Compute $r(\mu)$ when
$$\mu_n = qp^n, \quad n \in \mathbb{N}_0,$$
where $p \in (1/2, 1)$, $q = 1 - p$.

(ii) Prove that $\mathbb{P}[Z_n = 0] = f_n(0)$.

(iii) Denote by E the extinction event
$$E = \bigcup_{n\geq 0} \{Z_n = 0\}.$$
Prove that
$$\mathbb{P}[E] = \begin{cases} 1, & m \leq 1, \\ r(\mu), & m > 1. \end{cases}$$

(iv) Assume $m > 1$. Prove that the sequence $(r^{Z_n})_{n\geq 0}$ is a martingale.

(v) Set
$$W_n := \frac{1}{m^n} Z_n.$$
Assume
$$m > 1, \quad \mathbb{E}[Z_1^2] = \sum_{n\geq} n^2 \mu_n < \infty,$$
and set
$$W := \lim_{n\to\infty} W_n.$$

Denote by $\mathbb{P}_W$ the probability distribution of W and by $\varphi(\lambda)$ its Laplace transform

$$\varphi(\lambda) = \mathbb{E}\big[e^{-\lambda W}\big] = \int_\mathbb{R} e^{-\lambda w}\mathbb{P}_W[dw], \quad \lambda \in \mathbb{C}, \; \mathbf{Re}\,\lambda \geq 0.$$

Prove that

$$\varphi'(0) = 1, \;\; \varphi(\lambda) = f\big(\varphi(\lambda/m)\big) = \sum_{n=0}^\infty \mu_n \varphi(\lambda/m))^n, \;\; \forall\, \mathbf{Re}\,\lambda \geq 0. \quad (3.4.2)$$

(vi) Prove that there exists at most one probability measure $\nu \in \mathrm{Prob}\big([0,\infty)\big)$ such that

$$\int_0^\infty t^2 \nu[dt] < \infty$$

and its Laplace transform

$$\varphi_\nu(\lambda) := \int_0^\infty e^{-\lambda t}\nu[dt], \quad \lambda \in \mathbb{C}, \; \mathbf{Re}\,\lambda \geq 0,$$

satisfies (3.4.2).

Hint. Consider two such measures ν_k, $k=0,1$, denote by $\Phi_k(t)$ their characteristic functions. Set $\Phi(t) = \Phi_1(t) - \Phi_0(t)$, $\gamma(t) = \Phi(t)/t$, $t \neq 0$. Prove that $|\gamma(mt)| \leq |\gamma(t)|$ and conclude that $\Phi \equiv 0$.

$\square$

Exercise 3.27. Let $\mathfrak{S}_n$ denote the group of permutations of $\mathbb{I}_n := \{1,\ldots,n\}$. We equip it with the uniform probability measures. A *run* of a permutation π is a pair $(s,r) \in \mathbb{I}_n$, $s < t$ such that

$$\pi_{s-1} > \pi_s < \pi_{s+1} < \cdots < \pi_r > \pi_{r+1},$$

where $\pi_0 := n+1$ and $\pi_{n+1} := 0$. We denote by $R_n(\pi)$ the number of runs of $\pi \in \mathfrak{S}_n$. Set

$$X_n := nR_n - \frac{1}{2}n(n+1).$$

(i) For $\pi \in \mathfrak{S}_{n+1}$ we set $k_\pi := \pi^{-1}(n+1)$ and denote by φ_π the unique increasing bijection

$$\varphi_\pi : \mathbb{I}_n \to \mathbb{I}_{n+1} \setminus \{k_\pi\}.$$

Set $\bar{\pi} := \pi \circ \varphi_\pi$. Show that the random maps

$$\mathfrak{S}_{n+1} \ni \pi \mapsto k_\pi \in \mathbb{I}_{n+1}, \;\; \mathfrak{S}_{n+1} \ni \pi \mapsto \bar{\pi} \in \mathfrak{S}_n$$

are independent and uniformly distributed on their ranges.

(ii) Prove that (X_n) is a martingale.
(iii) Compute $\mathbb{E}\big[R_n\big]$ and $\mathbb{E}\big[R_n^2\big]$.

(iv) Show that
$$\lim_{n \to \infty} \mathbb{E}\left[\left(\frac{R_n}{n} - \frac{1}{2}\right)^2\right] = 0.$$

□

Exercise 3.28. Suppose that $(X_n)_{n \geq 0}$ is an L^2-martingale adapted to the filtration $(eF_n)_{n \geq 0}$ and $\langle X_\bullet \rangle$ is its quadratic variation; see Definition 3.15. Fix a bounded predictable process $(H_n)_{n \geq 0}$ and form the discrete stochastic integral $(H \bullet X)$ (see Theorem 3.17.

(i) Show that
$$\mathbb{E}[X_n^2] - \mathbb{E}[X_0^2] = \mathbb{E}[\langle X \rangle_n].$$

(ii) Prove that the martingale $(H \bullet X)$ is an L^2 martingale.

(iii) Prove that
$$\langle H \bullet X \rangle_m = (H^2 \bullet \langle X \rangle)_n := \sum_{k=1}^{n} H_k^2 (\langle X \rangle_k - \langle X \rangle_{k-1}), \ \forall n \geq 1.$$

(iv) Prove that
$$\mathbb{E}[(H \bullet X)_n^2] = \mathbb{E}\left[\sum_{k=1}^{n} H_k^2 (X_k - X_k-1)^2\right], \ l \ \forall n \geq 1.$$

Exercise 3.29. Suppose that $(X_n)_{n \in \mathbb{N}}$ is an exchangeable sequence of random variables and T is a stopping time adapted to the filtration $\mathcal{F}_n = \sigma(X_1, \ldots, X_n)$. Prove that if $T < N$ a.s., then X_{T+1} has the same distribution as X_1. □

Exercise 3.30. Suppose that $(X_n)_{n \in \mathbb{N}}$ is a sequence of random variables such that for any $n \in \mathbb{N}$ the distribution of the random vector $(X_1, \ldots, X_n)$ is orthogonally invariant, i.e., for any $T \in O(n)$, $T_\# \mathbb{P}_{X_1, \ldots, X_n} = \mathbb{P}_{X_1, \ldots, X_n}$. Prove that $(X_n)_\mathbb{N}$ are conditionally i.i.d. $N(0, \sigma^2)$ given a random variable $\sigma^2 \geq 0$. □

Exercise 3.31. Prove Lemma 3.105. □

Exercise 3.32. Finish the proof of Proposition 3.112. □

Exercise 3.33. Prove Proposition 3.114. □

Exercise 3.34. Let $N(t)$ be a Poisson process with intensity λ as described in Example 1.136. Denote by $(\mathcal{F}_t)$ the natural filtration, $\mathcal{F}_t = \sigma(N(s), \ s \leq t)$.

(i) Prove that $N(t)$ is an R-process.
(ii) Prove that $\mathcal{F}_{t+} = \mathcal{F}_t, \forall t \geq 0$.
(iii) Prove that $\mathbb{E}[N(t) \| \mathcal{F}_s] = \mathbb{E}[N(t) \| N(s)], \forall 0 \leq s < t$. □

Exercise 3.35. Suppose that $W : L^2([0,\infty)) \to L^2(\Omega, \mathcal{S}, \mathbb{P})$ is a Gaussian white noise; see Example 2.78. Fix $f \in L^2([0,\infty))$ and consider the Wiener integral (see Example 2.78 and Exercise 2.60)

$$X_t = \int_0^t f(s) dB(s) := W(\mathbf{I}_{[0,t]} f), \quad t \geq 0.$$

(i) Prove that (X_t) is an L^2 martingale adapted to the filtration $\mathcal{F}_t := \sigma(X_s, \ s \leq t)$.
(ii) Use Kolmogorov's Continuity Theorem 2.81 to show that $(X_t)_{t \geq 0}$ admits a continuous modification.

$\square$

Exercise 3.36. Let $B(t)$, $t \geq 0$ be a one-dimensional Brownian motion started at 0. For each $n \in \mathbb{N}$ and each $t \geq 0$ we set

$$X_t^n := \sum_{k=1}^n B\big((k-1)t/n\big)\Big(B\big(kt/n\big) - B\big((k-1)t/n\big)\Big).$$

(i) Prove that for any $n \in \mathbb{N}$ the stochastic process (X_t^n) is an L^2-martingale.
(ii) Prove that for each $t \geq 0$ X_t^n converges to $B(t)^2 - t$ in L^2 as $n \to \infty$.

$\square$

Exercise 3.37. Suppose that $(W_t)_{t \geq 0}$ is a pre-Brownian motion defined on a probability space $(\Omega, \mathcal{S}, \mathbb{P})$; see Definition 2.71. Let $t_0, \delta \geq 0$. Set

$$R(t_0, \delta) = \sup_{t \in \mathbb{Q} \cap [t_0, t_0+\delta]} |B(t) - B(t_0)|.$$

(i) Prove that

$$\mathbb{P}\big[R(t_0, \delta) > \varepsilon\big] \leq \frac{3\delta^2}{\varepsilon^4}, \quad \forall \varepsilon, \delta > 0.$$

Hint. Use Doob's maximal inequalities.

(ii) Prove that W_t is a a.s. *uniformly* continuous on $\mathbb{Q}_{\geq 0}$ and conclude that (W_t) admits a modification continuous on $[0, \infty)$.

$\square$

Exercise 3.38. Let $(B_t)_{t \geq 0}$ be a standard Brownian motion and $-a < 0 < b$. Set $T = \min(T_{-a}, T_b)$ where for $c \in \mathbb{R}$, we set $T_c = \inf\{t \geq 0; \ B_t = c\}$. Prove that

$$\mathbb{E}[T] = \mathbb{E}[B_T^2] = ab.$$

$\square$

Exercise 3.39 (P. Lévy). Let $(B_t)_{t \geq 0}$ be a standard Brownian motion and $c > 0$. For $a \in \mathbb{R}$ we denote by r_a the reflection $r_a : \mathbb{R} \to \mathbb{R}$, $r_a(x) = 2a - x$.

(i) Prove that for any Borel subsets $U_- \subset (-\infty, -c]$, $U_+ \subset [c, \infty)$ we have

$$\mathbb{P}[T_c < T_{-c}, B_1 \in U_-] + \mathbb{P}[T_c > T_{-c}, B_1 \in r_c(U_-)] = \mathbb{P}[B_1 \in r_c(U_-)]$$

$$\mathbb{P}[T_c > T_{-c}, B_1 \in U_+] + \mathbb{P}[T_c < T_{-c}, B_1 \in r_{-c}(U_+)] = \mathbb{P}[B_1 \in r_{-c}(U_+)].$$

(ii) Denote by J the interval $[-c, c]$. Prove that

$$\mathbb{P}[T_c \leq T_{-c} \wedge 1, B_t \in J] = \mathbb{P}[B_1 \in r_c(J)] - \mathbb{P}[T_c > T_{-c}, B_1 \in r_c(J)],$$

$$\mathbb{P}[T_{-c} \leq T_c \wedge 1, B_t \in J] = \mathbb{P}[B_1 \in r_{-c}(J)] - \mathbb{P}[T_c < T_{-c}, B_1 \in r_{-c}(J)].$$

(iii) Prove that

$$\mathbb{P}\left[\sup_{t \in [0,1]} |B_t| < c\right] = \mathbb{P}[B_1 \in J]$$

$$-\Big(\mathbb{P}[T_c \leq T_{-c} \wedge 1, B_t \in J] + \mathbb{P}[T_{-c} \leq T_c \wedge 1, B_t \in J]\Big).$$

(iv) Prove that

$$\mathbb{P}[|B_t| < c] = \mathbb{P}[|B_1| \leq c] - \mathbb{P}[c \leq |B_1| \leq 3c] + \mathbb{P}[3c \leq |B_1| \leq 5c] - \cdots.$$

$\square$

Remark 3.142. Exercise 3.39 is a special case of a more general result called the *support theorem*. For any continuous function $f : [0,1] \to \mathbb{R}$ such that $f(0) = 0$ and any $\varepsilon > 0$ we have

$$\mathbb{P}\left[\sup_{t \in [0,1]} |B_t - f(t)| \leq \varepsilon\right] > 0. \tag{3.4.3}$$

For a proof we refer to [63, Ch. 1, Thm. (38)].

Let us describe and amusing application of this fact. Suppose that $(B_t^i)_{t \geq 0}$, $i = 1, 2$, are two independent Brownian motions and $f^i : [0,1] \to \mathbb{R}$, $i = 1, 2$ are two continuous functions such that $f^i(0) = 0$. The equality (3.4.3) implies immediately that for any $\varepsilon > 0$ we have

$$\mathbb{P}\left[\max_i \sup_{t \in [0,1]} |B_t^i - f^i(t)| \leq \varepsilon\right]$$

$$= \mathbb{P}\left[\sup_{t \in [0,1]} |B_t^1 - f^1(t)| \leq \varepsilon\right] \mathbb{P}\left[\sup_{t \in [0,1]} |B_t^2 - f^2(t)| \leq \varepsilon\right] > 0. \tag{3.4.4}$$

The pair of functions (f^1, f^2) defines a path

$$F : [0,1] \to \mathbb{R}^2, \quad F(t) = (f^1(t), f^2(t)).$$

Think of $F(t)$ as tracing the motion of the tip of an infinitesimally fine pen as you sign a planar piece of paper, starting at the origin.

Any other path $G = (g^1, g^2) : [0,1] \to \mathbb{R}^2$ satisfying

$$\left| g^i(t) - f^i(t) \right| < \varepsilon, \ \forall t \in [0,1], \ i = 1, 2,$$

will follow closely the original motion of the fine pen, producing a curve essentially indistinguishable with the naked eye from the original signature. In fact, if $\varepsilon > 0$, one cannot distinguish the two curves, even using a magnifying glass.

The random path (B_t^1, B_t^2) is the so called *planar Brownian motion* started at the origin. The equality (3.4.3) shows that the probability p_0 that this random path follows closely the motion of the tip of the fine pen is positive. For this reason the inequality (3.4.3) is sometimes referred to as *Lévy's forgery theorem*. □

Chapter 4

Markov chains

The Markov chains form a special but sufficiently general class of examples of stochastic processes. Their investigation requires a diverse arsenal of techniques, probabilistic and not only, and they reveal important patterns arising in many other instances.

The foundations of this theory were laid by the Russian mathematician A. A. Markov at the beginning of the twentieth century. By most accounts, Markov was a rather unconventional individual. He discovered what we now know as Markov chains in his attempts to contradict Pavel Nekrasov, a mathematician/theologian of that time who maintained on a theological basis that the Law of Large Numbers was specific to independent events/random variables and cannot be seen in other contexts. Markov succeeded in proving Nekrasov wrong and in the process laid the foundations of the theory of Markov chains. For more on this history of this concept we refer to the very readable article [82].

So what did Markov discovered? Think of a Markov chain as a random walk on a finite set $\mathscr{X}$. From a given location x the walker can go to a location x' with probability $q_{x,x'}$. Suppose that at some location $x_0 \in \mathscr{X}$ we placed a pile of sand consisting of giddy grains of sand: every second one of them starts this random walk and performs a billion steps (think of a fixed but very large number of steps). After all the grains of sand performed this ritual, the initial pile of sand is redistributed at various points of $\mathscr{X}$. Denote by m_x^1 the mass of the pile of send relocated at x. Next, collect the piles from their locations and move them back to the initial location x_0.

Run the above experiment again we get a new distribution of piles of sand at the points of $\mathscr{X}$. Denote the mass at x by m_x^2. Markov observed that

$$\frac{m_x^1}{m_x^2} \approx 1, \ \forall x.$$

Run the experiment a third time to obtain a third distribution of mass $(m_x^3)_{x \in \mathscr{X}}$ and the conclusion is the same

$$\frac{m_x^1}{m_x^3} \approx 1, \ \forall x.$$

To put it differently, if m is the mass of the pile of sand at x_0, then, for any $x \in X$,

$$\frac{m_x^1}{m} \approx \frac{m_x^2}{m} \approx \frac{m_x^3}{m} \approx \cdots.$$

This phenomenon is one manifestation of the Law of Large Numbers for Markov chains.

During this more than a century since its creation, the theory of Markov chains has witnessed dramatic growth and generalizations, and has found applications in unexpected problems. For example, Googly's Pageant algorithm is a special application of the Law of Large Numbers for Markov chains.

The present chapter is an introduction to the theory of Markov chains. We present the classical results and spend some time on some more recent developments. As always, we try to illustrate the power of the theory on many concrete example. Needless to say, we barely scratch the surface of this subject.

4.1   Markov chains

In the sequel $\mathscr{X}$ will denote a finite or countable set equipped with the discrete topology. We will refer to it as the *state space*. The Borel sigma-algebra of $\mathscr{X}$ coincides with the sigma-algebra $2^{\mathscr{X}}$ of all subsets of $\mathscr{X}$.

4.1.1   *Definition and basic concepts*

Definition 4.1. A *Markov chain with state space* $\mathscr{X}$ is a sequence of random variables

$$X_n : (\Omega, \mathcal{S}, \mathbb{P}) \to (\mathscr{X}, 2^{\mathscr{X}}), \ \ n \in \mathbb{N}_0,$$

satisfying the *Markov property*

$$\mathbb{P}\big[X_{n+1} = x_{n+1} \big| X_n = x_n\big] = \mathbb{P}\big[X_{n+1} = x_{n+1} \big| X_n = x_n, \ldots, X_0 = x_0\big], \quad (4.1.1)$$

$\forall n \in \mathbb{N}, \, x_0, x_1, \ldots, x_n, x_{n+1} \in \mathscr{X}$.

The filtration associated to the Markov chain is the sequence of sigma-subalgebras

$$\mathcal{F}_n := \sigma(X_0, \ldots, X_n), \ \ n \in \mathbb{N}_0.$$

The probability distribution of X_0 is called the *initial distribution* of the system.

The Markov chain is called *homogeneous* if, for any $x, x' \in \mathscr{X}$, and any $n \in \mathbb{N}$ we have

$$\mathbb{P}\big[X_{n+1} = x' \big| X_n = x\big] = \mathbb{P}\big[X_1 = x' \big| X_0 = x\big].$$

In this case the function

$$Q : \mathscr{X} \times \mathscr{X} \to [0,1], \ \ Q(x_0, x_1) = Q_{x_0, x_1} = \mathbb{P}\big[X_1 = x_1 \big| X_0 = x_0\big]$$

is called the *transition matrix*[1] of the homogeneous Markov chain. We denote by Markov($\mathscr{X}, \mu, Q$) the collection of HMC-s with state space $\mathscr{X}$, initial distribution μ and transition matrix Q. □

Remark 4.2. (a) Let us observe that the Markov property can be written in the more compact form

$$\mathbb{P}[X_{n+1} = x \,\|\, X_n] = \mathbb{P}[X_{n+1} = x \,\|\, \mathcal{F}_n], \;\; \forall n \in \mathbb{N}, \; x \in \mathscr{X}. \tag{4.1.2}$$

In view of Proposition 1.172, the last property is equivalent to the conditional independence

$$X_{n+1} \perp\!\!\!\perp_{X_n} \mathcal{F}_{n-1}, \;\; \forall n \in \mathbb{N}. \tag{4.1.3}$$

Exercise 1.51 shows that this is also equivalent to the condition

$$X_{n+1} \perp\!\!\!\perp_{X_n} \mathcal{F}_n, \;\; \forall n \in \mathbb{N}. \tag{4.1.4}$$

One can show that this further equivalent to that

$$\sigma(X_{n+1}, X_{n+2}, \dots) \perp\!\!\!\perp_{X_n} \mathcal{F}_n. \tag{4.1.5}$$

This is colloquially expressed as saying that *the future is conditionally independent of the past given the present.*

(b) It is convenient to think of a Markov chain with state space $\mathscr{X}$ as describing the random walk of a grasshopper hopscotching on the elements of $\mathscr{X}$. The decision where to jump next is not influenced by the past, but only by the current location and the current time. For a homogeneous Markov chain the decision where to jump next depends only on the current location and not on the "time" n when the grasshopper reaches that state. Thus Q_{x_0, x_1} is the probability that the grasshopper, currently located at x_0, will jump to x_1.

We can represent an HMC with state space $\mathscr{X}$ and transition matrix Q as a directed graph (loops allowed) with vertex set $\mathscr{X}$ constructed as follows: there is a directed edge from x_0 to x_1 if and only if $Q_{x_0, x_1} > 0$. □

If $(X_n)_{n \geq 0}$ is a homogeneous Markov chain (or HMC for brevity), then its transition matrix Q is *stochastic*, i.e.,

$$Q_{x_0, x_1} \geq 0, \;\; \sum_{x \in \mathscr{X}} Q_{x_0, x} = 1, \;\; \forall x_0, x_1 \in \mathscr{X}. \tag{4.1.6}$$

In other words, the entries of the matrix Q are nonnegative and the sum of the entries in each row is equal to 1.

If μ_n is the distribution of X_n, then, for any $x \in \mathscr{X}$ we have

$$\mathbb{P}[X_{n+1} = x] = \sum_{x' \in \mathscr{X}} \mathbb{P}[X_n = x']Q_{x', x} = \sum_{x' \in \mathscr{X}} \mu_n[x']Q_{x', x}.$$

---

[1] I made the decision to break with the tradition and use the letter Q to denote the transition matrix after teaching this topic and realizing that there were too many P's on the blackboard and this sometimes confused the audience.

Think of μ_n and μ_{n+1} as matrices consisting of a single *row*. We can rewrite the above equality as an equality of matrices $\mu_{n+1} = \mu_n Q$. In particular,

$$\mu_n = \mu_0 Q^n, \tag{4.1.7}$$

where Q^n denotes the n-th power of the matrix Q, $Q^n = \left(Q^n_{x,y}\right)_{x,y \in \mathscr{X}}$. From (4.1.7) we deduce that

$$\mathbb{P}[X_n = x_n \mid X_0 = x_0] = Q^n_{x_0, x_n}. \tag{4.1.8}$$

For this reason the matrix Q^n is also known as the n-th step transition matrix.

Let us show that given any matrix $Q : \mathscr{X} \times \mathscr{X} \to [0,1]$ satisfying (1.2.19) and any probability measure μ on $\mathscr{X}$, there exists a homogeneous Markov chain, with state space $\mathscr{X}$, initial distribution μ and transition matrix Q, i.e., $\mathrm{Markov}(\mathscr{X}, \mu, Q) \neq \emptyset$.

Observe that we can view Q as a kernel or random probability measure

$$\hat{Q} : \mathscr{X} \times 2^{\mathscr{X}} \to [0,1], \quad (x, A) \mapsto \hat{Q}_x[A] = \sum_{a \in A} Q_{x,a}.$$

Note that $\hat{Q}_x[-]$ is a probability measure on $\mathscr{X}$. It is described by row x of the matrix Q.

Consider the set $\mathscr{X}^{\mathbb{N}_0}$ equipped with the natural product sigma algebra $\mathcal{E}$; see Definition 1.192. In this case it coincides with the sigma algebra generated by π-system consisting of the *cylinders*

$$C_{s_0, s_1, \ldots, s_k} := \left\{ \underline{x} = (x_n)_{n \in \mathbb{N}_0} \in \mathscr{X}^{\mathbb{N}_0}; \; x_i = s_i, \; \forall i = 0, \ldots, k \right\}.$$

Let us observe that there exists a probability measure $\mathbb{P}_\mu : \mathcal{E} \to [0,1]$ uniquely determined by the conditions

$$\mathbb{P}_\mu\left[C_{s_0, s_1, \ldots, s_k}\right] = \mu[s_0] \prod_{i=1}^k Q_{s_{i-1}, s_i}. \tag{4.1.9}$$

To prove that such a measure does indeed exist for any μ and Q we will rely on Kolmogorov's existence theorem, Theorem 1.195.

The equalities (4.1.9) define probability measures $\mathbb{P}_k = \mathbb{P}_k^{\mu, Q}$ on the product spaces $\mathscr{X}^{\{0, 1, \ldots, k\}}$ by setting

$$\mathbb{P}_k\left[(s_0, \ldots, s_k)\right] := \mu[s_0] \prod_{i=1}^k Q_{s_{i-1}, s_i}. \tag{4.1.10}$$

Note that for $f : \mathscr{X}^{\{0, 1, \ldots, k\}} \to \mathbb{R}$ we have

$$\begin{aligned}&\int_{\mathscr{X}^{\{0,1,\ldots,k\}}} f(x_0, \ldots, x_k) \mathbb{P}_k[dx_0 \cdots dx_k] \\ &= \sum_{x_0 \in \mathscr{X}} \sum_{x_1 \in \mathscr{X}} \cdots \sum_{x_k \in \mathscr{X}} \mu[x_0] Q_{x_0, x_1} \cdots Q_{x_{k-1}, x_k} f(x_0, \ldots, x_k).\end{aligned} \tag{4.1.11}$$

The family of measures $(\mathbb{P}_k)_{k\geq 0}$ is projective since the transition matrix Q is stochastic. Indeed,

$$\mathbb{P}_{k+1}\big[\,(s_0,\ldots,s_k)\times \mathscr{X}\,\big] = \sum_{x\in\mathscr{X}} \mathbb{P}_{k+1}\big[\,(s_0,\ldots,s_k,x)\,\big]$$

$$= \left(\mu[s_0]\prod_{i=1}^{k} Q_{s_{i-1},s_i}\right)\underbrace{\sum_x Q_{s_k,x}}_{=1} \qquad (4.1.12)$$

$$= \mu[s_0]\prod_{i=1}^{k} Q_{s_{i-1},s_i} = \mathbb{P}_k\big[\,(s_0,\ldots,s_k)\times \mathscr{X}\,\big].$$

Kolmogorov's existence theorem, then implies the existence of $\mathbb{P}_\mu \in \mathrm{Prob}\,\big(\,\mathscr{X}^{\mathbb{N}_0}\,\big)$ satisfying (4.1.9). Note that

$$\mathbb{P}_\mu = \sum_{x\in\mathscr{X}} \mu\big[\,x\,\big]\mathbb{P}_x, \quad \mathbb{P}_x := \mathbb{P}_{\delta_x}. \qquad (4.1.13)$$

For $n\in\mathbb{N}_0$ we denote by $\mathcal{E}_n$ the sub-sigma-algebra of $\mathcal{E}$ generated by X_0, $X_1,\ldots,X_n$. Note that $\mathbb{P}_n$ can be identified with the restriction of $\mathbb{P}_\mu$ to $\mathcal{E}_n$.

For $\mu \in \mathrm{Prob}(\mathscr{X})$ we denote by $\mathbb{E}_\mu$ the expectation (integral) with respect to $\mathbb{P}_\mu$

$$\mathbb{E}_\mu : L^1(\mathscr{X}^{\mathbb{N}_0},\mathcal{E},\mathbb{P}_\mu)\to\mathbb{R}, \quad \mathbb{E}_\mu\big[\,F\,\big] = \int_{\mathscr{X}^{\mathbb{N}_0}} F(\underline{x})\mathbb{P}_\mu\big[\,d\underline{x}\,\big]. \qquad (4.1.14)$$

For $x\in\mathscr{X}$ we set

$$\mathbb{E}_x := \mathbb{E}_{\delta_x}. \qquad (4.1.15)$$

We have a shift operator

$$\Theta : \mathscr{X}^{\mathbb{N}_0}\to\mathscr{X}^{\mathbb{N}_0}, \quad \Theta(x_0,x_1,x_2,\ldots) = (x_1,x_2,\ldots).$$

Note that $X_n = X_0\circ\Theta^n$, $\Theta^n = \underbrace{\Theta\circ\cdots\circ\Theta}_{n}$.

Theorem 4.3. *Consider the random variables*

$$X_n : \mathscr{X}^{\mathbb{N}_0}\to\mathscr{X}, \quad X_n(\underline{x}) = x_n, \quad n\in\mathbb{N}_0.$$

*Then the stochastic process $(X_n)_{n\in\mathbb{N}_0}$ is an HMC, defined on $(\mathscr{X}^{\mathbb{N}_0},\mathcal{E},\mathbb{P}_\mu)$ with transition state space $\mathscr{X}$ matrix Q and initial distribution μ. The probability space $(\mathscr{X}^{\mathbb{N}_0},\mathcal{E},\mathbb{P}_\mu)$ is called the **path space** of this HMC.*

Moreover, if $F\in L^1(\mathscr{X}^{\mathbb{N}_0},\mathcal{E},\mathbb{P}_\mu)$, then

$$\mathbb{E}_\mu\big[\,F\circ\Theta^n\,\|\,\mathcal{E}_n\,\big] = \mathbb{E}_\mu\big[\,F\,\|\,X_n\,\big]. \qquad (4.1.16)$$

Proof. For each x we have a probability measure Q_x on $\mathscr{X}$ given by
$$Q_x[\{x'\}] = Q_{x,x'}, \ \forall x' \in \mathscr{X}.$$
We will show that for any $A \subset \mathscr{X}$ we have the equality of random variables
$$\mathbb{P}[X_{n+1} \in A \,\|\, \mathcal{E}_n] = Q_{X_n}[A] = \sum_{a \in A} Q_{X_n,a}. \qquad (4.1.17)$$
Let $B \in \mathcal{E}_n$. It is a cylinder of the form
$$B = \{X_0 \in B_0, \ldots, X_n \in B_n\}, \ B_0, B_1, \ldots, B_n \subset \mathscr{X}\}.$$
Then
$$\boldsymbol{E}[\boldsymbol{I}_A(X_{n+1})\boldsymbol{I}_B] = \mathbb{P}_\mu[\{X_{n+1} \in A\} \cap B] = \mathbb{P}_\mu[X_0 \in B_0, \ldots, X_n \in B_n\, X_{n+1} \in A]$$
$$\stackrel{(4.1.11)}{=} \int_B Q_{X_n}[A] d\mathbb{P}_\mu.$$
This proves (4.1.17).

The random measure Q_{X_n} is a regular version of the conditional probability $\mathbb{P}[X_{n+1} \in -\,\|\, X_n]$, i.e.,
$$Q_{X_n}[S] = \mathbb{P}[X_{n+1} \in S \,\|\, X_n], \ \forall S \subset \mathscr{X}.$$
Using Proposition 1.178 we deduce that for every bounded function $f : \mathscr{X} \to \mathbb{R}$ we have
$$\mathbb{E}[f(X_{n+1}) \,\|\, \mathcal{E}_n] = \sum_{x \in \mathscr{X}} Q_{X_n, x} f(x). \qquad (4.1.18)$$
Let $\mathcal{M} \subset L^1(\mathscr{X}^{\mathbb{N}_0}, \mathcal{E}, \mathbb{P}_\mu)$ denote the collection of functions F satisfying (4.1.16). Clearly $\mathcal{M}$ is a vector space and if F_n is a sequence in $\mathcal{M}$ such that $F_n \nearrow F$, $F \in L^1$, then $F \in \mathcal{M}$. To show that $\mathcal{M} = L^1$ we use Monotone Class Theorem so it suffices to show that there exists a π-system $\mathcal{C} \subset \mathcal{E}$ that generates $\mathcal{E}$ such that $\boldsymbol{I}_C \in \mathcal{M}$.

Denote by $\mathcal{C}$ the set of cylinders
$$C_{A_0, A_1, \ldots, A_N} := \{\underline{x} \in \mathscr{X}^{\mathbb{N}_0};\ x_i \in A_i,\ i = 1, \ldots, N\}.$$
Note that
$$\boldsymbol{I}_{C_{A_0, \ldots, A_N}} = \prod_{k=0}^{N} \boldsymbol{I}_{\{X_k \in A_k\}},$$
and
$$\boldsymbol{I}_{C_{A_0, \ldots, A_N}} \circ \Theta^n = \prod_{k=0}^{N} \boldsymbol{I}_{\{X_{n+k} \in A_k\}}.$$
By definition $\mathcal{C}$ generates $\mathcal{E}$. Since $\mathcal{M}$ is a vector space it suffices to check that $\boldsymbol{I}_C \in \mathcal{M}$ for $C \in \mathcal{C}$ of the form
$$C = C_{A_0, \ldots, A_N},\ A_k = \{x_k\},\ x_k \in \mathscr{X},\ k = 0, 1, \ldots, N.$$

To verify (4.1.16) for sets of this form and arbitrary n we argue by induction on N. For $N=1$ this follows from (4.1.17). For the inductive step note that

$$\mathbb{E}\big[\, I_{\{X_n=x_0, X_{n+1}=x_1,\ldots,X_{n+N}=x_N\}} \,\|\, \mathcal{E}_n \,\big] = \mathbb{E}\left[\prod_{k=0}^{N} I_{\{X_{n_k}=x_k\}} \,\Big\|\, \mathcal{E}_n\right]$$

$$= \mathbb{E}\left[\, I_{\{X_n=x_0\}} \mathbb{E}\left[\prod_{k=1}^{N} I_{\{X_{n+k}=x_k\}} \,\Big\|\, \mathcal{E}_{n+1}\right] \,\Big\|\, \mathcal{E}_n\right]$$

$$= \mathbb{E}\left[\, I_{\{X_n=x_0\}} \underbrace{\mathbb{E}\left[\prod_{k=1}^{N} I_{\{X_{n+k}=x_k\}} \,\Big\|\, X_{n+1}\right]}_{=:f(X_{n+1})} \,\Big\|\, \mathcal{E}_n\right]$$

(use the inductive assumption)

$$= \mathbb{E}\big[\, I_{\{X_n=x_0\}} f(X_{n+1}) \,\|\, X_n \,\big] = \mathbb{E}\left[\, I_{\{X_n=x_0\}} \underbrace{\mathbb{E}\left[\prod_{k=1}^{N} I_{\{X_{n+k}=x_k\}} \,\Big\|\, \mathcal{E}_{n+1}\right]}_{=f(X_{n+1})} \,\Big\|\, X_n\right]$$

($\sigma(X_n) \subset \mathcal{E}_{n+1}$)

$$= \mathbb{E}\left[\prod_{k=0}^{N} I_{\{X_{n+k}=x_k\}} \,\Big\|\, X_n\right].$$

$\square$

Remark 4.4. We have deduced (4.1.16) relying on the Markov property. The above proof shows that the Markov property (4.1.17) is a special case of (4.1.16). For this reason we can take (4.1.16) as definition of Markov's property. $\square$

Given a homogeneous Markov chain $X_n : (\Omega, \mathcal{S}, \mathbb{P}) \to \mathcal{X}$, $n \geq 0$, with state space $\mathcal{X}$, initial distribution μ and transition matrix Q, we obtain a measurable map

$$\vec{X} : (\Omega, \mathcal{S}) \to (\mathcal{X}^{\mathbb{N}_0}, \mathcal{E}), \quad \omega \mapsto \vec{X}(\omega) = \big(X_n(\omega)\big)_{n \geq 0}.$$

The *distribution of the Markov chain* is the pushforward measure

$$\mathbb{P}_{\vec{X}} := \vec{X}_\# \mathbb{P} \in \mathrm{Prob}\big(\mathcal{X}^{\mathbb{N}_0}, \mathcal{E}\big).$$

It is uniquely determined by the equalities

$$\mathbb{P}_{\vec{X}}\big[\, C_{s_0,s_1,\ldots,s_k} \,\big] := \mathbb{P}\big[\, X_0=s_0, \ldots, X_k=s_k \,\big] = \mu_0[s_0] \prod_{i=1}^{k} Q_{s_{i-1},s_i}. \quad (4.1.19)$$

We deduce
$$\mathbb{P}_{\vec{X}} = \mathbb{P}_\mu.$$
For every, $F \in L^1(\mathscr{X}^{\mathbb{N}_0}, \mathcal{E}, \mathbb{P}_\mu)$, we have
$$\mathbb{E}_{\mathbb{P}}[F(X_0, X_1, \dots)] = \mathbb{E}_\mu[F] = \int_{\mathscr{X}^{\mathbb{N}_0}} F(\underline{x}) \mathbb{P}_\mu[d\underline{x}].$$
This is a special case of the change in variables formula (1.2.21).

This shows that the distribution of the Markov chain is uniquely determined by the initial distribution $\mu \in \text{Prob}(\mathscr{X})$ and the transition matrix Q.

Remark 4.5. One can define any HMC on probability spaces other than $\mathscr{X}^{\mathbb{N}_0}$. Here is a such a construction corresponding to state space $\mathscr{X}$ transition matrix Q and initial probability distribution μ. We set $\mu_x := \mu[x]$.

First, a little bit of terminology. We say that an interval is convenient if it either empty or the form $[a, b)$, $a < b$. If $[a, b), [c, d)$ are nonempty convenient intervals, then we say that $[a, b)$ precedes $[c, d)$ and we write $[a, b) \prec [c, d)$ if $b \leq c$. The empty set is allowed to precede or succeed any nonempty convenient interval. Assume that $\mathscr{X}$ is a subset of $\mathbb{N}$. As such it is equipped with a total order.

The probability space is the unit interval $[0, 1)$ equipped with the Lebesgue measure. The random variables X_n, depend on the choice of initial distribution, and are defined inductively as follows.

- Partition $[0, 1)$ into convenient intervals $I_x = I_x^0$, $x \in \mathscr{X}$ of Lebesgue measures $\mu_x = \boldsymbol{\lambda}[I_x^0]$, such that
$$x < x' \Rightarrow I_x \prec I_{x'}.$$
- Partition each interval $I_{x_0}^0$ into convenient intervals I_{x_0,x_1}^1 of sizes $\mu_{x_0} Q_{x_0,x_1}$, $x_0, x_1 \in \mathscr{X}$, such that
$$x < x' \Rightarrow I_{x_0,x}^1 \prec I_{x_0,x'}^1.$$
- Inductively, partition each interval $I_{x_0,x_1,\dots,x_n}^n$ into convenient intervals $I_{x_0,x_1,\dots,x_n,x_{n+1}}^{n+1}$ of sizes
$$\boldsymbol{\lambda}[I_{x_0,x_1,\dots,x_n}^n] Q_{x_n,x_{n+1}} = \mu_{x_0} \prod_{j=0}^{n} Q_{x_j,x_{j+1}},$$
such that
$$x < x' \Rightarrow I_{x_0,\dots,x_n,x}^{n+1} \prec I_{x_0,x_1,\dots,x_n,x'}^{n+1}.$$

Now define $X_n : [0, 1) \to \mathscr{X}$ by setting
$$X_n(t) = x_n \text{ if } t \in \bigcup_{x_0,x_1,\dots,x_{n-1} \in \mathscr{X}} I_{x_0,x_1,\dots,x_n}^n.$$
Note that these random variables are defined on the same probability space
$$([0, 1], \mathcal{B}_{[0,1]}, \boldsymbol{\lambda}),$$
but *they depend on the choice of the initial distribution*.

This is different from the construction based on path spaces. In that case we are given measurable maps defined on the same measurable space and we obtain different HMC's by choosing *different* probability measures. $\square$

4.1.2 Examples

The homogeneous Markov chains appear in many and diverse situations. According to the discussion in the previous subsection, to describe an HMC it suffices to describe the state space $\mathscr{X}$ and the transition matrix Q. We will remain vague about the initial distribution μ.

Example 4.6 (Gambler's ruin). Consider the gambler's ruin problem discussed in Example 3.72. The state space is $\mathscr{X} = \{0, 1, \ldots, N\}$. Then X_n is the fortune of a gambler at time n. The gambler flips a fair coin with two faces labeled ± 1. If its fortune is strictly in between 0 and N, then its fortune changes by the amount shown on the face of the coin. The game stops when its fortune reaches either 0 or N. Concretely

$$Q_{N,k} = 0, \quad \forall k < N, \quad Q_{N,N} = 1,$$

$$Q_{0,j} = 0, \quad \forall j > 0, \quad Q_{0,0} = 1,$$

$$Q_{k,k+1} = Q_{k,k-1} = \frac{1}{2}, \quad Q_{k,j} = 0, \text{ if } |k-j| > 1, \ 0 < k, j < N.$$

The directed graph describing this HMC is depicted in Figure 4.1 where, for clarity, we have omitted the loops at 0 and N. □

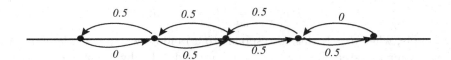

Fig. 4.1 *The gambler's ruin chain.*

Example 4.7 (The Ehrenfest Urn). Consider the following situation. There are B balls in two urns. Equivalently, think of an urn with two chambers. Pick one of these B balls uniformly at random and move it in the other box/chamber. Denote by X_n the number of balls in the left box at time n. Then $(X_n)_{n \geq 0}$ is an HMC with transition probabilities

$$Q_{i,i+1} = \frac{B-i}{B}, \ i = 0, 1, \ldots, B-1, \quad Q_{i,i-1} = \frac{i}{B}, \ i = 1, \ldots, B,$$

$$Q_{i,j} = 0, \ |i-j| > 1.$$

This HMC is known as the *Ehrenfest urn*. Note that during this process it is more likely that a ball moves from the more crowded box to the less crowded one, similarly to what happens in diffusion processes. □

Example 4.8 (Random placement of balls). Consider a sequence of independent trials each consisting in randomly placing a ball in one of r given urns. We say that the system is in state k if exactly k urns are occupied.

We obtain an HMC with state space $\{0, 1, \ldots, r\}$ with transition probabilities

$$Q_{j,j} = \frac{j}{r}, \quad Q_{j,j+1} = \frac{r-j}{r}, \quad 0 \leq j < r,$$

and of course $Q_{j,k} = 0$ for any other pairs (j, k). If $X_0 = 0$, so initially all boxes are empty, then $X_n = r - N_{r,n}$, where $N_{r,n}$ is the number of empty boxes investigated in Exercise 2.19. □

Example 4.9 (Random walk on $\mathbb{Z}^d$). Suppose that $(X_n)_{n \geq 1}$ are i.i.d. $\mathbb{Z}^d$-valued random variables. Denote by π their common distribution. Set

$$S_0 = 0, \quad S_n = X_1 + \cdots + X_n.$$

Then the random process $(S_n)_{n \in \mathbb{N}_0}$ is an HMC with transition matrix

$$Q_{\boldsymbol{m},\boldsymbol{n}} = \mathbb{P}[X_1 = \boldsymbol{n} - \boldsymbol{m}] = \pi[\boldsymbol{n} - \boldsymbol{m}], \quad \boldsymbol{m}, \boldsymbol{n} \in \mathbb{Z}^d.$$

One can imagine this process as a person starting at the origin of $\mathbb{Z}^d$ and walking with random step sizes, with X_n the size of the n-th step.

A standard random walk is obtained as follows. Denote by $\boldsymbol{e}_1, \ldots, \boldsymbol{e}_d$ the canonical basis $\mathbb{Z}^d$ and choose π to be uniformly distributed on the set $\{\pm\boldsymbol{e}_1, \ldots, \pm\boldsymbol{e}_d\}$, i.e.,

$$\pi[\pm\boldsymbol{e}_k] = \frac{1}{2d}, \quad k = 1, \ldots, d.$$

For example, when $d = 1$, this corresponds to a random walk on $\mathbb{Z}$ where, at each moment, going one step ahead or one step back is decided by flipping a fair coin. □

Example 4.10 (Simple random walk on a graph). Consider an undirected graph $G = (V, E)$, where V is the set of vertices, and E denotes the set of edges. We do not allow for multiple edges connecting two vertices. For every vertex v we denotes by $\deg(v)$ is degree, i.e., the number of edges of E at v. We assume that the graph is locally finite, i.e., $\deg(v) < \infty$, $\forall v \in V$.

Suppose now that a grasshopper hopscotches on the set vertices V according to the following rule: if situated at a vertex v_0, the grasshopper will jump to one of the neighbors of v_0 in V chosen uniformly randomly. Denote by X_n the location of the grasshopper at time n. Then $(X_n)_{n \geq 0}$ is an HMC with state space V with transition matrix

$$Q_{v_0, v_1} = \begin{cases} \frac{1}{\deg(v_0)}, & \text{if } v_0 \text{ and } v_1 \text{ are neighbors}, \\ 0, & \text{otherwise}. \end{cases}$$

□

Example 4.11 (The branching process). Consider again the branching process with reproduction law μ described in Example 3.8. Recall that it deals with the evolution of a population of individuals of a species with $\mu[j]$ denoting the probability that a given individual will have $j \in \mathbb{N}_0$ offsprings.

Denote by Z_n the size of the n-th generation population. We assume that $Z_0 = 1$. Then $(Z_n)_{n \geq 0}$ is an HMC with state space $\mathbb{N}_0$.

To see this, choose a sequence of i.i.d. random variables $(\xi_k)_{k \in \mathbb{N}}$ with common distribution μ. Then

$$\mathbb{P}[Z_{n+1} = j \mid Z_n = i] = \mathbb{P}[\xi_1 + \cdots + \xi_i = j].$$

The distribution of the random variable $\xi_1 + \cdots + \xi_i$ is the convolution of μ^{*i}, the convolution of i copies of μ. More precisely,

$$\mu^{*i}[j] = \sum_{k_1 + \cdots + k_i = j} \mu[k_1] \cdots \mu[k_i].$$

The transition matrix is then $Q_{i,j} = \mu^{*i}[j]$. $\square$

Example 4.12 (Queing). Customers arrive for service and take their place in a waiting line. During each period of time one customer is served, if at least one customer is present. During a service period new customers may arrive. We assume that the number of customers that arrive during the n-th service period is a random variable ξ_n, and that the random variables $\xi_1, \xi_2, \ldots$ are i.i.d. with common distribution $\mu \in \mathrm{Prob}(\mathbb{N}_0)$. We set $\mu_i := \mu[i]$, $i \in \mathbb{N}_0$. For notation convenience we set $\mu_n = 0$ for $n < 0$.

We denote by X_n the number of customers in line at the end of the n-th period. Note that

$$X_{n+1} = (X_n - 1)^+ + \xi_n.$$

The sequence $(X_n)_{n \geq 0}$ is an HMC with state space $\mathbb{N}_0$ and transition matrix

$$Q_{i,j} = \begin{cases} \mu_j, & i = 0, \\ \mu_{j-i+1}, & i > 0. \end{cases}$$

$\square$

Example 4.13 (Noisy dynamical systems). Suppose that $T: \mathscr{X} \to \mathscr{X}$ is a selfmap of an at most countable set $\mathscr{X}$. This defines a dynamical system $(T^n)_{n \in \mathbb{N}}$ which can be viewed as a trivial Homogeneous Markov Chain with transition matrix

$$Q_{x,x'} = \begin{cases} 1, & x' = T(x), \\ 0, & x' \neq T(x). \end{cases}$$

We can obtain more general Markov chains if work with "noisy" selfmaps.

More precisely suppose that $(S, \mathcal{S})$ is a measurable space and

$$T: \mathscr{X} \times S \to \mathscr{X}, \quad (x, s) \mapsto T_s(x)$$

is a measurable map. In other words, (T_s) is a measurable family of maps $\mathscr{X} \to \mathscr{X}$.

Fix an S-valued "noise", i.e., a sequence

$$Z_n : (\Omega, \mathcal{F}, \mathbb{P}) \to (S, \mathcal{S}), \quad n \in \mathbb{N}$$

of i.i.d. S-valued random variables and an $\mathscr{X}$-valued random variable X_0 independent of the Z's we obtain a *noisy dynamical system*

$$X_{n+1} = T_{Z_{n+1}}(X_n), \quad \forall n \in \mathbb{N}.$$

Hence

$$X_n = T_{Z_n} \circ \cdots \circ T_{Z_1}(X_0).$$

The sequence $(X_n)_{n \geq 0}$ is an HMC. Indeed,

$$\mathbb{P}\big[X_{n+1} = x_{n+1} \big| X_n = x_n, \ldots, X_0 = x_0 \big]$$

$$= \mathbb{P}\big[T_{Z_{n+1}}(x_n) = x_{n+1} \big| X_n = x_n, \ldots, X_0 = x_0 \big]$$

$$= \mathbb{P}\big[T_{Z_{n+1}}(x_n) = x_{n+1} \big]$$

since the event $\{X_n = x_n, \ldots, X_0 = x_0\}$ belongs to the sigma algebra generated by $X_0, Z_1, \ldots, Z_n$ and thus is independent of Z_{n+1}. On the other hand, obviously

$$\mathbb{P}\big[T_{Z_{n+1}}(x_n) = x_{n+1} \big] = \mathbb{P}\big[X_{n+1} = x_{n+1} \big| X_n = x_n \big].$$

This Markov chain is *homogenous* since the random variables Z_n are identically distributed.

The standard random walk on $\mathbb{Z}$ is a Markov system generated in an obvious way by a random dynamical system defined by the map

$$T : \mathbb{Z} \times \mathbb{Z} \to \mathbb{Z}, \quad (m, z) \mapsto T_z(m) = m + z$$

and the noise described by a sequence of i.i.d. Rademacher random variables $(Z_n)_{n \in \mathbb{N}}$. Then $X_{n+1} = X_n + Z_{n+1}$. One can show that any Markov chain can be produced in this fashion, as iterates of random maps. $\square$

4.2 The dynamics of homogeneous Markov chains

In this section we will consistently adopt the dynamical point of view on Markov chains described in Remark 4.2(b) and extract some useful consequences.

4.2.1 Classification of states

Suppose that $(X_n)_{n \geq 0}$ is an HMC with state space $\mathscr{X}$ and transition matrix Q.

Definition 4.14. (a) A state $x_1 \in \mathscr{X}$ is said to be *accessible* from a state $x_0 \in \mathscr{X}$, and we denote this by $x_0 \to x_1$, if $Q^n_{x_0, x_1} > 0$ for some $n \in \mathbb{N}_0$.

(b) The states x_0 and x_1 *communicate* if $x_0 \to x_1$ and $x_1 \to x_1$. We indicate this using the notation $x_0 \leftrightarrow x_1$. $\square$

Recall that to an HMC with state space $\mathscr{X}$ we can associate a directed graph with vertex set $\mathscr{X}$; see Remark 4.2(b). A *walk* from x to x' in this graph is a sequence of vertices
$$x = x_0, \, x_1, \ldots, x_n = x'$$
such that, for any $i = 1, \ldots, n$, there exists a directed edge from x_{i-1} to x_i. If $x \neq x'$, then x' is accessible from x if there is a walk from x to x'.

Proposition 4.15. *The communication relation "$\leftrightarrow$" is an equivalence relation.*

Proof. Reflexivity. $x \leftrightarrow x$ since $Q^0_{x,x} = 1$.

Symmetry. The relation is symmetric by definition.

Transitivity. Suppose that $x_0 \leftrightarrow x_1$ and $x_1 \leftrightarrow x_2$. Then, there exist $m, n \in \mathbb{N}_0$ such that
$$Q^m_{x_0, x_1} > 0 \text{ and } Q^n_{x_1, x_2} > 0.$$
Observe that
$$Q^{m+n}_{x_0, x_2} = Q^m_{x_0, x_1} Q^n_{x_1, x_2} + \underbrace{\sum_{x \in \mathscr{X} \setminus \{x_1\}} Q^m_{x_0, x} Q^n_{x, x_2}}_{\geq 0} > 0.$$
Hence $x_0 \to x_2$. The opposite relation $x_2 \to x_0$ is proved in identical fashion. □

Definition 4.16. The equivalence classes of the relation $\leftrightarrow$ are called the *communication classes* of the given HMC. □

Example 4.17. (a) Consider the HMC associated to the gamblers's ruin problem described in Example 4.6. The state space is $\{0, 1, \ldots, N\}$ and there are three communication classes
$$C_0 = \{0\}, \ C = \{1, \ldots, N-1\}, \ C_N = \{N\}.$$
Note that no state in C is accessible from C_0 or C_N.

(b) The HMC associated to the Ehrenfest urn model in Example 4.7 has state space $\{0, 1, \ldots, N\}$ and any two states communicate so that there is only a single communication class.

(c) The HMC corresponding to the random placement of balls problem in Example 4.8 has state space $\{0, 1, \ldots, r\}$ and communication classes
$$C_0 = \{0\}, \ \{1\}, \ldots, C_r = \{r\}.$$
Note that for $j > i$, the class C_j is accessible from the class C_i. □

Definition 4.18. Let $(X_n)_{n \in \mathbb{N}_0}$ be an HMC with state space $\mathscr{X}$ and transition matrix Q.

(i) A subset $C \subset \mathscr{X}$ is *closed* with respect to this HMC if no state outside C is accessible from a state in C.
(ii) A subset of $\mathscr{X}$ is called *irreducible* if its is closed and contains no proper closed subset.
(iii) A state $x \in \mathscr{X}$ is called *absorbing* if the set $\{x\}$ is irreducible.
(iv) The HMC is called *irreducible* if its state space is irreducible.

□

Example 4.19. For the HMC corresponding to the random placement of balls problem in Example 4.8 with state space $\{0, \ldots, r\}$, all the subsets $\{k, k+1, \ldots, r\}$ are closed and the state r is absorbing. This is not an irreducible Markov chain. □

Note that a subset $C \subset \mathscr{X}$ is closed if and only if for any $x \in C$ we have

$$\sum_{y \in C} Q_{x,y} = 1.$$

Equivalently, this means that $\mathbb{P}[X_n \in C \mid X_0 \in C] = 1$.

Using the intuition of the randomly hopping grasshopper, this says that, once the grasshopper steps in a closed set it will be trapped there. In particular, this argument proves the following result.

Proposition 4.20. *A closed subset of the state space of an HMC is a union of communication classes.*

□

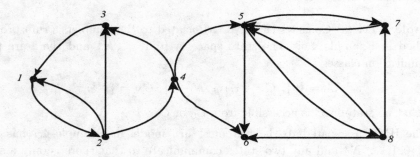

Fig. 4.2 An *HMC* with a single irreducible subset.

Example 4.21. Consider an HMC with associated digraph depicted in Figure 4.2. It consists of three communication classes

$$C_1 := \{1, 2, 3, 4\}, \quad C_2 := \{5, 7, 8\}, \quad C_3 := \{6\}.$$

The communication class C_3 is closed while C_1 and C_2 are not. The only irreducible set is C_3. In particular the state 6 is absorbing.

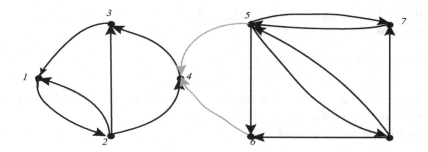

Fig. 4.3 *Another HMC with a single irreducible subset.*

Suppose now that we change the directions of the edges $4 \to 5$ and $4 \to 6$ as depicted in Figure 4.3. This HMC has the same communication classes C_1, C_2, C_3, but this time only C_1 is closed. □

Lemma 4.22. *Let $C \subset \mathscr{X}$ be a <u>closed</u> subset. Then the following are equivalent.*

(i) C is irreducible.
(ii) C is a communication class.

In particular, an HMC is irreducible if and only if it consists of a single communication class.

Proof. The implication (ii) ⇒ (i) follows from Proposition 4.20: a closed set is a union of communication classes.

(i) ⇒ (ii) Suppose that C is an irreducible subset of $\mathscr{X}$. In particular, C is a union of communication classes

$$C = \bigcup_{j=1}^{N} C_j, \ N \in \mathbb{N} \cup \{\infty\}.$$

Suppose that $N \geq 2$. Set $j_0 := 1$. Since C_{j_0} is not closed, there exists $j_1 \neq j_0$, $x_{j_0} \in C_{j_0}$ and $x_{j_1} \in C_{j_1}$ such that $x_{j_0} \to x_{j_1}$. In fact, any class in C_{j_1} is accessible from any class in C_{j_0}. We write this $C_{j_0} \to C_{j_1}$.

Next, we can find $j_2 \notin \{j_0, j_1\}$ such that $C_{j_1} \to C_{j_2}$. Clearly no class in $C_{j_0} \cup C_{j_1}$ is accessible from C_{j_2}. Iterating, we obtain a (possible finite) subsequence in $\mathbb{N} \cap [1, N]$

$$j_0, j_1, \ldots, j_k, \ldots,$$

where the j_k's are pairwise distinct, such that

$$C_{j_0} \to C_{j_1} \to C_{j_2} \to \cdots$$

and no state in $C_{j_0} \cup \cdots \cup C_{j_k}$ is accessible from $C_{j_{k+1}}$. Note that

$$C' = \bigcup_{k \geq 1} C_{j_k} \subset C \setminus C_{j_0}$$

is a proper closed subset of C, contradicting the fact that C is irreducible. □

Definition 4.23. Suppose that $(X_n)_{n \in \mathbb{N}_0}$ is an HMC with state space $\mathscr{X}$ and transition matrix Q.

(i) The set of *periods* of a state $x \in \mathscr{X}$ is
$$\mathcal{P}_x := \{ n \in \mathbb{N}; \ Q^n_{x,x} > 0 \}.$$
(ii) The *period* of a state x is $d = d(x) := \gcd \mathcal{P}_x$, where "gcd" stands for greatest common divisor. When $\mathcal{P}_x = \emptyset$ we set $d(x) := \infty$.
(iii) A state x is called *aperiodic* if $d(x) = 1$.

$\square$

Lemma 4.24. *Let $(X_n)_{n \geq 0}$ be an HMC with state space $\mathscr{X}$ and transition matrix Q. Suppose that $x \in \mathscr{X}$ and $d(x) < \infty$. Then the following hold.*

(i) The set $\mathcal{P}_x$ is a semigroup of the additive semigroup $(\mathbb{N}, +)$.
(ii) There exists $N \in \mathbb{N}$ such that $nd(x) \in \mathcal{P}_x$, $\forall n \geq N$.
(iii) If $x \leftrightarrow y$, then $d(x) = d(y)$.

Proof. (i) Follows from the fact that $Q^{m+n}_{x,x} \geq Q^m_{x,x} Q^n_{x,x}$.

(ii) We claim that there exist $k \geq 2$ and $m_1, \ldots, m_k \in \mathcal{P}_x$ such that
$$d(x) = \gcd(m_1, m_2, \ldots, m_k).$$
Pick $m_1, m_2 \in \mathcal{P}_x$ and set $d_1 := \gcd(m_1, m_2)$. Then $d \leq d_1$. If $d < d_1$ define
$$d_2 := \min \{ \gcd(m_1, m_2, m); \ m \in \mathcal{P}_x \}.$$
Then $d \leq d_2 \leq d_1$ and $d_2 = d_1$ iff $d = d_2 = d_1$. If $d_2 < d_1$ choose $m_3 \in \mathcal{P}_x$ such that
$$d_2 = \gcd(m_1, m_2, m_3) \geq d.$$
If $d < d_2$ define
$$d_3 := \min \{ \gcd(m_1, m_2, m_3, m); \ m \in \mathcal{P}_x \}.$$
If $d_3 = d_2$ we stop because $d_3 = d$. If $d_3 < d_2$, we iterate the above procedure. Clearly this procedure will stop after finitely many iterations.

An old result of I. Schur [159, Thm. 3.15.2] implies that the set
$$\{ m_1 n_1 + \cdots + m_k n_k; \ n_1, \ldots, n_k \in \mathbb{N} \}$$
contains all the sufficiently large multiples of d.

(iii) For $x, y \in \mathscr{X}$ we set
$$\mathcal{P}_{x,y} := \{ n \in \mathbb{N}; \ Q^n_{x,y} > 0 \}.$$
Thus $\mathcal{P}_x = \mathcal{P}_{x,x}$. Suppose $x \leftrightarrow y$. Note that
$$\mathcal{P}_{x,y} + \mathcal{P}_{y,x} \subset \mathcal{P}_x.$$
Hence $d(x) | (\mathcal{P}_{x,y} + \mathcal{P}_{y,x})$. From the inclusion
$$\mathcal{P}_{x,y} + \mathcal{P}_y + \mathcal{P}_{y,x} \subset \mathcal{P}_x$$
we deduce that $d(x) | \mathcal{P}_y$ so $d(x) | d(y)$. Reversing the roles of x, y in the above argument we deduce $d(y) | d(x)$ so $d(x) = d(y)$.

$\square$

According to the above result, all the states of an irreducible HMC have the same period so we can speak of the period of that HMC.

Definition 4.25. An irreducible HMC is called *aperiodic* if each of its states has period 1. □

Example 4.26. (i) Each state in the standard random walk on $\mathbb{Z}$ locally finite graph has period 2.

More generally, given a vertex v in a locally finite, connected graph, its set of periods with respect to the standard random walk coincides with the set of lengths of paths in the graph that start and end at v. Since there is such a path of length 2 we deduce that the vertex is aperiodic if and only if there exists a path of odd length starting and ending at x.

(ii) The Ehrenfest urn in Example 4.7 is irreducible with period 2. □

Proposition 4.27. *Let $(X_n)_{n\geq 0}$ be an irreducible HMC with state space $\mathscr{X}$, transition matrix Q, and period $d < \infty$. Fix $x_0 \in \mathscr{X}$. Consider the HMC $(Y_n)_{n\geq 0}$ with state space $\mathscr{X}$, initial state $Y_0 = x_0$ and transition matrix $T = Q^d$. Denote by $\mathcal{C}_T$ the set of communication classes of T. For each $x \in \mathscr{X}$ we denote by $[x]_T$ the T-communication class of x. Then the following hold.*

(i) There exists a bijection $r = r_{x_0} : \mathcal{C}_T \to \mathbb{Z}/d\mathbb{Z}$ such that $r([x]_T) = k \bmod d$ iff there exists $n \in \mathbb{N}_0$ such that $Q_{x_0,x}^{nd+k} > 0$.

(ii) If $Q_{x,y} > 0$, then $r(y) \equiv r(x) + 1 \bmod d$.

(iii) Each T-communication class is T-closed.

Proof. As in the proof of Lemma 4.24 we set
$$\mathcal{P}_{x,y} := \{n \in \mathbb{N};\ Q_{x,y}^n > 0\}.$$
Let us observe that
$$\forall x,y \in \mathscr{X},\ \forall n,m \in \mathcal{P}_{x,y}:\ n \equiv m \bmod d. \tag{4.2.1}$$
The claim is obviously true if $n = m$. Suppose that $n > m$. Then $n - m \in \mathcal{P}_{y,y}$ so $d = d(y)$ divides $n - m$.

Thus, for any $x,y \in \mathscr{X}$ there exists $r = r(x,y) \in \{0,1,\ldots,d-1\}$ such that
$$\mathcal{P}_{x,y} \subset r(x,y) + d\mathbb{N}_0.$$
For any $x \in \mathscr{X}$ we set $r(x) := r(x_0, x)$. We want to prove that
$$[x]_T = [y]_T \Longleftrightarrow r(x) = r(y). \tag{4.2.2}$$
Indeed, suppose that $[x]_T = [y]_T$. Then there exists n such that $T_{x,y}^n > 0$, i.e., $Q_{x,y}^{nd} > 0$. Fix $m \in \mathbb{N}_0$ such that $Q_{x_0,x}^m > 0$. Then $Q_{x_0,y}^{m+nd} > 0$. Clearly
$$r(y) \equiv m + nd \equiv m \equiv r(x) \bmod d.$$

Conversely, suppose that $r(x) = r(y)$. Fix $n_x, n_y \in \mathbb{N}_0$ such that $Q_{x_0,x}^{n_x}, Q_{x_0,y}^{n_y} > 0$. Choose N large enough such that $Nd > n_x$ and $Nd \in \mathcal{P}_{x_0}$. Then $Nd - n_x \in \mathcal{P}_{x,x_0}$ and $n_y \in \mathcal{P}_{x_0,x}$ so $Nd - n_x + n_y \in \mathcal{P}_{x,y}$. Hence

$$0 = r(y) - r(x) \equiv n_y - n_x \bmod d.$$

We deduce that there exists $m \in \mathbb{N}_0$ such that $md = Nd - n_x + n_y$. Hence $T_{x,y}^m > 0$. In other words y is T-accessible from x. A symmetric argument shows that x is T-accessible from y so that $[x]_T = [y]_T$. This proves (4.2.2) and (i).

The statements (ii) and (iii) follow immediately from the equality

$$r(x,z) \equiv r(x,y) + r(y,z) \bmod d,$$

so $r(y) = r(x) + r(x,y)$. $\square$

Remark 4.28. Suppose that $(X_n)_{n \geq 0}$ is an HMC as in the above proposition and

$$C_0, C_1, \ldots, C_{d-1} \subset \mathcal{X}$$

are the communication classes of Q^d. If $X_0 \in C_i$, then $X_n \in C_{i+n \bmod d}$, for any n. Thus a grasshopper hopscotching following the prescriptions of this Markov chain will jump from a region C_i to somewhere in the next region C_{i+1} and so on, returning after d jumps to the region where he started.

Observe also that the map $r : \mathcal{C}_T \to \mathbb{Z}/d\mathbb{Z}$ depends on the choice of x_0. On the other hand, the action of $\mathbb{Z}/d\mathbb{Z}$ on $\mathcal{C}_T$ is independent of x_0 and it is free and transitive. Thus $\mathcal{C}_T$ is naturally a $\mathbb{Z}/d\mathbb{Z}$-torsor. $\square$

4.2.2 The strong Markov property

Suppose that $X_n : (\Omega, \mathcal{S}, \mathbb{P}) \to \mathcal{X}$, $n \geq 0$ is an HMC with state space $\mathcal{X}$ and transition matrix Q. As usual, we denote by $(\mathcal{F}_n)_{n \geq 0}$ the filtration of $\mathcal{S}$ determined by this random process, i.e., $\mathcal{F}_n = \sigma(X_0, \ldots, X_n)$, $n \in \mathbb{N}_0$.

Suppose now that T is a stopping time adapted to this filtration. In (3.1.6) we defined the sigma-algebra $\mathcal{F}_T$ associated to T by the requirements

$$S \in \mathcal{F}_T \Longleftrightarrow S \cap \{T \leq n\} \in \mathcal{F}_n, \quad \forall n \in \mathbb{N}_0.$$

Example 4.29 (Return times). Let $(X_n)_{n \in \mathbb{N}_0}$ be an HMC with state space $\mathcal{X}$. For $A \subset \mathcal{X}$ we define

$$T_A := \min\{n \geq 1; \; X_n \in A\}.$$

We will refer to T_A as the *return time* to A. This is a stopping time with respect to the canonical filtration $\mathcal{F}_n$. For $x \in \mathcal{X}$ we set $T_x := T_{\{x\}}$.

Note that the event S belongs to $\mathcal{F}_{T_A}$ if at any moment n we can decide using the information collected up to that point in $\mathcal{F}_n$ whether S occurred and we have returned to A up to that moment. $\square$

Example 4.30 (Hitting times). Let $(X_n)_{n \in \mathbb{N}_0}$ be an HMC with state space $\mathscr{X}$. For $A \subset \mathscr{X}$ we define
$$H_A := \min\{n \geq 0;\ X_n \in A\}.$$
We will refer to H_A as the *hitting time* of A. This is a stopping time with respect to the canonical filtration $\mathcal{F}_n$. For $x \in \mathscr{X}$ we set $H_x := H_{\{x\}}$. □

Theorem 4.31. *Let $X_n : (\Omega, \mathcal{S}, \mathbb{P}) \to \mathscr{X}$, $n \in \mathbb{N}_0$, be an HMC with initial distribution μ and transition matrix Q. Suppose that T is a stopping time adapted to the canonical filtration $(\mathcal{F}_n)_{n \in \mathbb{N}_0}$. Conditional on $X_T = x \in \mathscr{X}$ and $T < \infty$ the stochastic process*
$$Y_n := X_{T+n},\ n \in \mathbb{N}_0,$$
is in Markov$(\mathscr{X}, \delta_x, Q)$ *and independent of $\mathcal{F}_T$. More explicitly, if Λ is the event*
$$\Lambda = \{T < \infty,\ X_T = x\},$$
and $\mathbb{P}_\Lambda : \mathcal{S} \to [0,1]$ is the probability measure $\mathbb{P}_\Lambda[S] = \mathbb{P}[S|\Lambda]$, then the stochastic process
$$Y_n : (\Omega, \mathcal{S}, \mathbb{P}_\Lambda) \to \mathscr{X}$$
is Markov$(\mathscr{X}, \delta_x, Q)$ *and independent of $\mathcal{F}_T$.*

Proof. For $n \in \mathbb{N}$ denote by T_n the stopping time $T_n := T + n$. Denote by S the event
$$\{T < \infty\} \cap \{X_T = x_0 = x,\ X_{T+1} = x_1, \ldots, X_{T+n} = x_n\}$$
$$= \Lambda \cap \{X_{T+1} = x_1, \ldots, X_{T+n} = x_n\}.$$
Note that $S \in \mathcal{F}_{T_n}$. We have to show that
$$\mathbb{P}_\Lambda[Y_{T+n+1} = x_{n+1} | S] = Q_{x_n, x_{n+1}},$$
i.e.,
$$\frac{\mathbb{P}_\Lambda[\{X_{T+n+1} = x_{n+1}\} \cap S \cap \Lambda]}{\mathbb{P}_\Lambda[S]} = Q_{x_n, x_{n+1}}$$
or
$$\frac{\mathbb{P}[\{X_{T+n+1} = x_{n+1}\} \cap S \cap \Lambda]}{\mathbb{P}[S \cap \Lambda]} = \frac{\mathbb{P}[\{X_{T+n+1} = x_{n+1}\} \cap S]}{\mathbb{P}[S]} = Q_{x_n, x_{n+1}}.$$
We have
$$\mathbb{P}[\{X_{T+n+1} = x_{n+1}\} \cap S] = \sum_{k=0}^{\infty} \mathbb{P}[\{X_{k+n+1} = x_{n+1}\} \cap S \cap \{T = k\}]$$
(use the Markov property (4.1.2) and the definition of S)
$$= \sum_{k=0}^{\infty} \mathbb{P}[\{T = k\} \cap S] Q_{x_n, x_{n+1}}$$

$$= Q_{x_n,x_{n+1}} \sum_{k=0}^{\infty} \mathbb{P}[\{T=k\} \cap S] = Q_{x_n,x_{n+1}} \mathbb{P}[S].$$

To prove the independence of $\mathcal{F}_T$ given $\Lambda = \{X_T = x,\ T < \infty\}$ it suffices to show that each of the events

$$\Gamma_0 = \{X_T = x\}, \ldots, \Gamma_n = \{X_T = x_0 = x, X_{T+1} = x_1, \ldots, X_{T+n} = x_n\}, \ldots$$

are independent of $\mathcal{F}_T$ given $\Lambda = \{X_T = x, T < \infty\}$. Let $S \in \mathcal{F}_T$. We have

$$\mathbb{P}[S \cap \Gamma_n \cap \Lambda] = \sum_{k=0}^{\infty} \mathbb{P}[S \cap \Gamma_n \cap \{T = k\}]$$

(use the Markov property repeatedly)

$$= \sum_{k=0}^{\infty} \mathbb{P}[\{T=k\} \cap S \cap \Gamma_0] Q_{x_0,x_1} \cdots Q_{x_{n-1},x_n}$$

$$= \mathbb{P}[S \cap \Gamma_0 \cap \{T < \infty\}] Q_{x_0,x_1} \cdots Q_{x_{n-1},x_n} = \mathbb{P}[S \cap \Lambda] Q_{x_0,x_1} \cdots Q_{x_{n-1},x_n},$$

i.e.,

$$\mathbb{P}[S \cap \Gamma_n \cap \Lambda] = \mathbb{P}[S \cap \Lambda] Q_{x_0,x_1} \cdots Q_{x_{n-1},x_n}.$$

Then

$$\mathbb{P}[S \cap \Gamma_n | \Lambda] = \frac{\mathbb{P}[S \cap \Lambda]}{\mathbb{P}[\Lambda]} \cdot Q_{x_0,x_1} \cdots Q_{x_{n-1},x_n},\quad x_0 = x.$$

Since the stochastic process $Y_n : (\Omega, \mathcal{S}, \mathbb{P}_\Lambda) \to \mathcal{X}$ is Markov$(\mathcal{X}, \delta_x, Q)$ we deduce

$$Q_{x_0,x_1} \cdots Q_{x_{n-1},x_n} = \mathbb{P}[\Gamma_n | \Lambda].$$

Hence $\mathbb{P}[S \cap \Gamma_n | \Lambda] = \mathbb{P}[S | \Lambda] \cdot \mathbb{P}[\Gamma_n | \Lambda]$. □

In the following subsections we will have plenty of opportunities to see the strong Markov principle at work.

4.2.3 Transience and recurrence

Suppose that $(X_n)_{n \in \mathbb{N}_0}$ is an HMC with state space $\mathcal{X}$ and transition matrix Q. For any $x \in \mathcal{X}$ we denote by T_x the return time to x, i.e.,

$$T_x := \min\{n \geq 1;\ X_n = x\}.$$

Moreover,

$$\mathbb{P}_x := \mathbb{P}[- | X_0 = x], \quad \mathbb{E}_x := \mathbb{E}[- | X_0 = x].$$

Definition 4.32. A state $x \in \mathcal{X}$ is called *recurrent* or *persistent* if $\mathbb{P}_x[T_x < \infty] = 1$. Otherwise it is called *transient*. □

Example 4.33. If $\mathscr{X}$ is *finite* and irreducible, then any state of $\mathscr{X}$ is recurrent. Indeed the 'sooner-rather-than-later' Lemma 3.32 implies that $\mathbb{E}_x[T_x] < \infty, \forall x \in \mathscr{X}$.

□

We set $T_x^1 := T_x$ and we define inductively

$$T_x^{k+1} := \min\{n > T_x^k;\ X_n = x\},$$

$$N_x := \#\{k \in \mathbb{N};\ T_x^k < \infty\} = \#\{n \in \mathbb{N};\ X_n = x\} \in \mathbb{N} \cup \{\infty\}.$$

Thus T_x^k is the time of the k-th return to x. We will refer to N_x the *number of returns to x*. We set

$$p = p_x := \mathbb{P}_x[T_x < \infty].$$

Lemma 4.34. *For any $n \in \mathbb{N}_0$ we have $\mathbb{P}_x[N_x \geq n] = p^n$. In particular, if x is recurrent, i.e., $p = 1$, then $N_x = \infty$ a.s. and, if X is transient, then*

$$\mathbb{E}_x[N_x] = \frac{p}{1-p}.$$

Proof. Set $p_n := \mathbb{P}_x[N_x \geq n]$. We will prove inductively that $p_n = p^n$.

Clearly $\mathbb{P}[N_x \geq 1] = \mathbb{P}[T_x^1 < \infty] = p$. Suppose that $p_n = p^n$. The post-T_x^k process $Y_n = X_{T_x^n + n}$ starts at x and the strong Markov property implies that it is a HMC with the same transition matrix. In particular, the probability that it returns to x is p. On the other hand, Y_n returns to x if and only if $N_x \geq k+1$. Since the post-T_x^k process is independent of $\mathcal{F}_{T_x^k}$ we deduce

$$\mathbb{P}[N_x \geq n+1] = p\mathbb{P}[N_x \geq n] = p^{n+1}.$$

□

Corollary 4.35. *Assume that $X_0 = x$ a.s. Then the following hold*

$$x \text{ is recurrent} \iff N_x = \infty \text{ a.s.} \iff \mathbb{E}_x[N_x] = \infty.$$

$$x \text{ is transient} \iff \mathbb{E}_x[N_x] < \infty.$$

□

Clearly the recurrence/transience of a state depends only on the transition matrix. The next result characterizes these features in terms of the transition matrix.

Theorem 4.36. *Let $x \in \mathscr{X}$. The following statements are equivalent.*

(i) The state x is recurrent.
(ii)

$$\sum_{n \in \mathbb{N}} Q_{x,x}^n = \infty.$$

Proof. Observe that
$$N_x = \sum_{n \in \mathbb{N}} I_{\{X_n = x\}}$$
and
$$\mathbb{E}_x[N_x] = \sum_{n \in \mathbb{N}} \mathbb{E}_x[I_{\{X_n = x\}}] = \sum_{n \in \mathbb{N}} Q^n_{x,x}.$$
The result now follows from Corollary 4.35. □

Corollary 4.37. *Let $x, y \in \mathscr{X}$. If $x \to y$ and x is recurrent, then*

(i) $x \leftrightarrow y$,
(ii) $\mathbb{P}_y[T_x < \infty] = 1$,
(iii) *the state y is recurrent.*

Proof. The state x is recurrent so $N_x = \infty$ a.s. Since $x \to y$ we deduce that
$$\mathbb{P}_x[T_y < \infty] = \mathbb{P}[T_y < \infty \,|\, X_0 = x] > 0.$$
The post-T_y chain $Y_n = X_{n+T_y}$, $n \geq 0$, will almost surely reach x since $N_x = \infty$ a.s. Using the strong Markov property at T_y we deduce that Y_n has the same transition matrix Q. Hence $y \to x$, i.e. $x \leftrightarrow y$. In particular, the original chain, started at y will almost surely reach x, i.e., $\mathbb{P}[T_x < \infty \,|\, X_0 = y] = 1$.

Since $x \leftrightarrow y$ there exist $j, k \in \mathbb{N}$ such that
$$c = \min\{Q^j_{x,y}, Q^k_{y,x}\} > 0.$$
We deduce
$$Q^{n+j+k}_{y,y} \geq Q^k_{y,x} Q^n_{x,x} Q^j_{x,y} \geq c^2 Q^n_{x,x}, \quad \forall n \in \mathbb{N}.$$
Hence
$$\sum_{m \geq 1} Q^m_{y,y} \geq \sum_{m > j+k} Q^m_{y,y} \geq c^2 \sum_{n \geq 1} Q^n_{x,y} = \infty.$$
□

The above result shows that if C is a communication class then, either all classes in C are recurrent, or all classes in C are transient. In the first case C is called a *recurrence class* and in the second case C is called a *transience class*. An irreducible HMC consists of a single communication class. Accordingly an irreducible HMC can be either transient, or recurrent.

Proposition 4.38. *Suppose that $(X_n)_{n \geq 0}$ is an irreducible transient HMC with state space $\mathscr{X}$, transition matrix Q and initial distribution μ. Then,*
$$\mathbb{E}_\mu[N_x] < \infty, \quad \forall x \in \mathscr{X}.$$

Proof. We first prove that given $x_0 \in \mathscr{X}$ there exists $C = C_{x_0} > 0$ such that
$$\mathbb{E}_y[\,N_{x_0}\,] \leq C, \quad \forall y \in \mathscr{X}.$$
Indeed, using the strong Markov property as in the proof of Lemma 4.34 we deduce that for any $y \in \mathscr{X}$ we have
$$\mathbb{E}_y[\,N_{x_0}\,] = \sum_{n \geq 1} \mathbb{P}_y[\,N_{x_0} \geq n\,] = \sum_{n \geq 1} \mathbb{P}_{x_0}[\,N_{x_0} \geq n-1\,]\mathbb{P}_y[\,T_{x_0} < \infty\,]$$
$$= \underbrace{\mathbb{P}_{x_0}[\,N_{x_0} \geq 0\,]}_{=1}\mathbb{P}_y[\,T_{x_0} < \infty\,] + \mathbb{P}_y[\,T_{x_0} < \infty\,]\underbrace{\sum_{m \geq 1}\mathbb{P}_{x_0}[\,N_{x_0} \geq m\,]}_{\mathbb{E}_{x_0}[\,N_{x_0}\,]}$$
$$= \mathbb{P}_y[\,T_{x_0} < \infty\,]\bigl(1 + \mathbb{E}_{x_0}[\,N_{x_0}\,]\bigr) \leq \underbrace{1 + \mathbb{E}_{x_0}[\,N_{x_0}\,]}_{C_{x_0}}.$$

Now observe that
$$\mathbb{E}_\mu[\,N_{x_0}\,] = \sum_{y \in \mathscr{X}} \mu[\,y\,]\mathbb{E}_y[\,N_{x_0}\,] \leq C_{x_0}.$$
$\square$

Theorem 4.39. *Suppose that C is a recurrence class and $X_0 = x \in C$ a.s. Then,*
$$\mathbb{P}[\,N_y = \infty\,|\,X_0 = x\,] = 1, \quad \forall y \in C.$$
In particular, with probability one, the chain visits every state of C infinitely often, i.e.,
$$\mathbb{P}[\,\forall y \in C,\ N_y = \infty,\,|\,X_0 \in C\,] = 1.$$

Proof. Let $x, y \in C$. We have
$$\mathbb{P}[\,N_x = \infty\,|\,X_0 = x\,] = 1, \quad \mathbb{P}[\,T_y < \infty\,|\,X_0 = x\,] = 1.$$
The strong Markov property shows that the post-T_y chain has the same distribution as the chain started at y. Since y is recurrent we have $\mathbb{P}[\,N_y = \infty\,|\,X_0 = x\,] = 1$ and we deduce that
$$\mathbb{P}[\,N_y < \infty\,|\,X_0 = x\,] = 0, \quad \forall x, y \in C,$$
$$\mathbb{P}[\,N_y < \infty\,|\,X_0 \in C\,] = 0, \quad \forall y \in C.$$
In particular,
$$\mathbb{P}[\,\exists y \in C,\ N_y < \infty\,|\,X_0 \in C\,] \leq \sum_{y \in C}\mathbb{P}[\,N_y < \infty\,|\,X_0 \in C\,] = 0.$$
Hence
$$\mathbb{P}[\,\forall y \in C,\ N_y = \infty,\,|\,X_0 \in C\,] = 1 - \mathbb{P}[\,\exists y \in C,\ N_y < \infty\,|\,X_0 \in C\,] = 1.$$
$\square$

Proposition 4.40. *A recurrence communication class C is closed.*

Proof. Suppose that C is not closed. Then there exist $c \in C$ and $x \in \mathscr{X} \setminus C$ such that $c \to x$. Since C is a communication class x does not communicate with any $y \in C$. Fix $n_0 \in \mathbb{N}$ such that $p := \mathbb{P}[X_{n_0} = x | X_0 = c] > 0$. In particular, since x does not communicate with C we deduce that $\mathbb{P}[X_n \in \mathscr{X} \setminus X, \; \forall n \geq n_0 | X_0 = c] \geq p$. In particular $\mathbb{P}[N_c < n_0 | X_0 = c] \geq p$. This contradicts the fact that $\mathbb{P}[N_c = \infty | X_0 = c] = 1$. $\square$

The set of communication classes $\overline{\mathscr{X}} := \mathscr{X}/\leftrightarrow$ is itself the state space of an HMC with transition matrix

$$\overline{Q}_{C,C'} = \mathbb{P}[X_1 \in C' | X_0 \in C].$$

Each state of $\overline{\mathscr{X}}$ is in itself a communication class of the new Markov chain. The state space $\overline{\mathscr{X}}$ is partitioned into two types: transient states and recurrent states. The recurrent states are closed, i.e., they are absorbing as states in $\overline{\mathscr{X}}$. Given a recurrent state $R \in \overline{\mathscr{X}}$, no other communication class is accessible from R.

Example 4.41. Consider for example the gambler's ruin problem with total fortune $N \in \mathbb{N}$; see Example 4.6. This can be viewed as a Markov chain with state space $\{0, 1, \ldots, N\}$ and transition probabilities

$$q_{i,i\pm 1} = \frac{1}{2}, \; \forall 0 < i < N, \; q_{0,0} = q_{N,N} = 1.$$

The communication classes of this Markov chain are

$$\{0\}, \; \{N\}, \; \{1, 2, \ldots, N-1\}.$$

The first two are recurrent while the third is transient. $\square$

If $\mathscr{X}$ is finite, the argument in Example 4.33 shows that a communication class is closed iff it is recurrent.

Example 4.42 (G. Polya). (a) Consider the standard random walk on $\mathbb{Z}$. We denote by Q the transition matrix. This is an irreducible Markov chain and each state has period 2. To decide whether it is transient or recurrent it suffices to verify if the origin is such. Note that $Q_{0,0}^{2n-1} = 0, \; \forall n \in \mathbb{N}$. To compute $Q_{0,0}^{2n}$ we observe that a path of length $2n$ starts and ends at the origin if and only if it consists of exactly n steps to the right and n steps to the left. Since each such step occurs with probability $\frac{1}{2}$ we deduce

$$Q_{0,0}^{2n} = \frac{1}{2^{2n}} \binom{2n}{n} = \frac{(2n)!}{2^{2n}(n!)^2}.$$

Using Stirling's formula (A.1.6) we deduce that, as $n \to \infty$, we have

$$\frac{(2n)!}{2^{2n}(n!)^2} \sim \frac{\sqrt{4\pi n}}{2\pi n} \sim \frac{1}{\sqrt{\pi n}},$$

so
$$\sum_{n\in\mathbb{N}} Q_{0,0}^n = \infty.$$

Thus, the 1-dimensional standard random walk is recurrent.

(b) Consider the standard walk on $\mathbb{Z}^2$. It is irreducible. We want to decide if the origin is recurrent or transient. To compute $Q_{0,0}^{2n}$ we observe that a path of length $2n$ starts and ends at the origin if and only if the number of steps up is equal to the number of steps down and the number of steps to the right is equal to the number of steps to the left. We deduce that

$$Q_{0,0}^{2n} = \sum_{k=0}^{n} \frac{(2n)!}{4^{2n}(k!)^2((n-k)!)^2} = \frac{(2n)!}{4^{2n}(n!)^2} \sum_{k=0}^{n} \binom{n}{k}^2.$$

Using Newton's binomial formula in the equality

$$(x+y)^{2n} = (x+y)^n (x+y)^n$$

and identifying the coefficients of $x^n y^n$ on either side of the above equality we deduce

$$\binom{2n}{n} = \sum_{k=0}^{n} \binom{n}{k}\binom{n}{n-k} = \sum_{k=0}^{n} \binom{n}{k}^2,$$

so that

$$Q_{0,0}^{2n} = \left(\frac{1}{2^{2n}} \binom{2n}{n}\right)^2 \sim \frac{1}{\pi n} \text{ as } n \to \infty.$$

Hence, again

$$\sum_{n\in\mathbb{N}} Q_{0,0}^n = \infty$$

so the standard 2-dimensional random walk is also recurrent.

(c) Consider the standard random walk on $\mathbb{Z}^3$. Arguing as in the 2-dimensional case we deduce

$$Q_{0,0}^{2n} = \frac{1}{6^{2n}} \sum_{j+k+\ell=n} \frac{(2n)!}{(j!)^2(k!)^2(\ell!)^2} = \frac{1}{2^{2n}}\binom{2n}{n} \sum_{j+k+\ell=n} \left(\frac{n!}{j!k!\ell!3^n}\right)^2.$$

Now observe that

$$\sum_{j+k+\ell=n} \underbrace{\frac{n!}{j!k!\ell!3^n}}_{=:p_{jk\ell}} = \left(\frac{1}{3}+\frac{1}{3}+\frac{1}{3}\right)^n = 1.$$

Hence

$$\sum_{j,k,\ell} p_{jk\ell}^2 \leq \max p_{j,k,\ell} \sum_{j,k,\ell} p_{j,k,\ell} = \max p_{j,k,\ell},$$

so
$$Q_{0,0}^{2n} \leq \frac{1}{2^{2n}} \binom{2n}{n} \max p_{j,k,\ell}.$$

Let us observe that the maximum value of $p_{j,k,\ell}$ is achieved when j,k,ℓ are as close to $n/3$ as possible. Indeed, if $j \leq k \leq \ell$, $j < \ell$, then
$$(j+1)!(\ell-1)! = \frac{j+1}{\ell} j!\ell! \leq j!\ell!$$
so
$$p_{j+1,k,\ell-1} \geq p_{j,k,\ell}.$$

Assume now that $n = 3m$. We deduce
$$Q_{0,0}^{2n} \leq \frac{1}{2^{2n}} \binom{2n}{n} \frac{(3m)!}{(m!)^3 3^{3m}}.$$

Using again Stirling's formula we deduce that, as $m \to \infty$ we have
$$\frac{(3m)!}{(m!)^3 3^{3m}} \sim \frac{\sqrt{6\pi m}}{(2\pi m)^{3/2}} = \frac{\sqrt{3}}{2\pi m}.$$

On the other hand
$$\frac{1}{2^{2n}} \binom{2n}{n} \sim \frac{1}{\sqrt{\pi n}} = \frac{1}{\sqrt{3\pi m}}.$$

We deduce that
$$Q_{0,0}^{6m} = O(m^{-3/2}) \text{ as } m \to \infty.$$

Arguing in a similar fashion we deduce
$$Q_{0,0}^{6m+2},\ Q_{0,0}^{6m+4} = O(m^{-3/2}) \text{ as } m \to \infty.$$

We conclude that
$$\sum_{n \in \mathbb{N}} Q_{0,0}^n = \sum_{n \in \mathbb{N}} Q_{0,0}^{2n} < \infty,$$

so the standard 3-dimensional random walk is transient! □

4.2.4 Invariant measures

Suppose that $(X_n)_{n \in \mathbb{N}_0}$ is an HMC with state space $\mathscr{X}$ and transition matrix Q. We will identify a σ-finite measure λ on $\mathscr{X}$ with function
$$\lambda : \mathscr{X} \to [0,\infty),\ x \mapsto \lambda_x = \lambda[\{x\}].$$

Definition 4.43. An *invariant* or *stationary measure* for the HMC $(X_n)_{n \in \mathbb{N}_0}$ is a σ-finite measure λ on $\mathscr{X}$ such that $\lambda = \lambda Q$, i.e.,
$$\lambda_x = \sum_{y \in \mathscr{X}} \lambda_y Q_{y,x},\ \forall x \in \mathscr{X}. \tag{4.2.3}$$

An *invariant* or *stationary distribution* is an invariant *probability* measure. □

Example 4.44 (Time reversal). Suppose that π is an invariant *probability* distribution for $(X_n)_{n\geq 0}$ such that $\pi_x > 0$, $\forall x \in \mathscr{X}$. Suppose additionally that

$$\mathbb{P}[X_0 = x] = \pi_x, \quad \forall x \in \mathscr{X}.$$

Then

$$\mathbb{P}[X_1 = x] = \sum_{y \in \mathscr{X}} \mathbb{P}[X_1 = x \mid X_0 = y]\pi_y = \sum_y \pi_y Q_{y,x} = \pi_x.$$

Iterating we deduce that the random variables $(X_n)_{n\in\mathbb{N}_0}$ are identically distributed. For $x, y \in \mathscr{X}$ we set

$$R_{y,x} := \mathbb{P}[X_0 = x \mid X_1 = y] = \frac{\mathbb{P}[X_1 = y \mid X_0 = x]\mathbb{P}[X_0 = x]}{\mathbb{P}[X_1 = y]} = \frac{\pi_x}{\pi_y} Q_{x,y}.$$

Note that for every $x, y \in \mathscr{X}$ we have

$$\sum_x R_{y,x} = \frac{1}{\pi_y} \sum_x \pi_x Q_{x,y} = 1,$$

so $(R_{y,x})_{x,y\in\mathscr{X}}$ is a stochastic matrix describing the so called *time reversed chain*.

Suppose now that π is a probability distribution on $\mathscr{X}$ such that $\pi_x > 0$, $\forall x \in \mathscr{X}$ and satisfying

$$Q_{y,x} = R_{y,x} = \frac{\pi_x}{\pi_y} Q_{x,y}, \quad \forall x, y \in \mathscr{X}. \tag{4.2.4}$$

From the equality

$$1 = \sum_x Q_{y,x} = \sum_x \frac{\pi_x}{\pi_y} Q_{x,y}$$

we deduce that π is a stationary distribution and the time reversed chain coincides with the initial chain. This is the reason why the chains satisfying (4.2.4) are called *reversible*. □

Definition 4.45. An irreducible HMC with state space $\mathscr{X}$ and transition matrix is called *reversible* if there exists a function $\lambda: \mathscr{X} \to (0, \infty)$ satisfying the *detailed balance equations*

$$\lambda_y Q_{y,x} = \lambda_x Q_{x,y}, \quad \forall x, y \in \mathscr{X}. \tag{4.2.5}$$

□

Observe that if Q satisfies the detailed balance equation, then an argument as in Example 4.44 shows λ defines a Q-invariant measure

Example 4.46. (a) If $Q_{x,y} = Q_{y,x}$ for any $x, y \in \mathscr{X}$, then the corresponding chain is reversible and any uniform measure on X is invariant. This happens for example if $(X_n)_{n\geq 0}$ describes the standard random walk on $\mathbb{Z}^d$.

(b) In the case the standard random walk on a locally finite connected graph we have
$$Q_{x,y} = \frac{1}{\deg x}, \quad \deg x \cdot Q_{x,y} = 1 = \deg y \cdot Q_{y,x}.$$
Hence Q is in detailed balance with invariant measure $x \mapsto \deg x$. If, additionally, $\mathscr{X}$ is finite, then the probability measure
$$\pi_x = \frac{\deg x}{\sum_y \deg y}$$
is invariant. □

Example 4.47 (The Ehrenfest urn). Consider the Ehrenfest urn model detailed in Example 4.7. We recall that the state space is $\mathscr{X} = \{0, 1, \ldots, B\}$, $B \in \mathbb{N}$ and the only nontrivial transition probabilities are
$$Q_{k,k+1} = \frac{B-k}{B}, \quad Q_{k,k-1} = \frac{k}{B}.$$
Note that
$$\frac{Q_{k,k+1}}{Q_{k+1,k}} = \frac{B-k}{k+1} = \frac{\binom{B}{k+1}}{\binom{B}{k}}.$$
Then the measure $k \to \lambda_k = \binom{B}{k}$ is invariant and
$$\pi_k = \frac{1}{2^B}\binom{B}{k}, \quad k = 0, 1, \ldots, B,$$
is an invariant probability distribution. □

Theorem 4.48. *Suppose that the HMC $(X_n)_{n \in \mathbb{N}_0}$ is irreducible and recurrent. Fix $x_0 \in \mathscr{X}$, and denote by T_0 the time of first return to x_0, i.e.,*
$$T_0 := T_{x_0} = \min\{n \geq 1; \ X_n = x_0\}.$$
For any $x \in X$, define
$$N_x = \sum_{n \in \mathbb{N}} I_{\{X_n = x\}} I_{\{n \leq T_0\}} = \sum_{n=1}^{T_0} I_{\{X_n = x\}}, \quad (4.2.6a)$$
$$\lambda_x = \lambda_{x, x_0} = \begin{cases} \mathbb{E}_{x_0}[N_x], & x \neq x_0, \\ 1, & x = x_0. \end{cases} \quad (4.2.6b)$$
In other words, λ_x is the expected number of visits to x before returning to x_0 when starting from x_0. Then, the following hold.

(i) $\lambda_x \in (0, \infty)$, $\forall x \in \mathscr{X}$ *and the associated measure λ on $\mathscr{X}$, given by*
$$\lambda[\{x\}] = \lambda_x, \quad \forall x \in \mathscr{X}$$
is invariant.

(ii) $\lambda[\mathscr{X}] = \mathbb{E}_{x_0}[T_{x_0}]$.

(iii) The measure λ is the unique invariant measure such that $\lambda_{x_0} = 1$.

Proof. (i) We follow the approach in [20, Thm. 3.2.1]. Clearly $\lambda_{x_0} = 1$. For $x \in \mathscr{X} \setminus \{x_0\}$ and $n \in \mathbb{N}$ we set

$$p_x(n) := \mathbb{P}[X_n = x, \ n \le T_0].$$

Thus, $p_x(n)$ is the probability of visiting state x at time n before returning to x_0. The equality (4.2.6a) implies that

$$\lambda_x = \sum_{n \in \mathbb{N}} p_x(n), \ \forall x \ne x_0. \tag{4.2.7}$$

Let us prove that λ satisfies (4.2.3). Observe first that $p_x(1) = Q_{x_0,x}$. From the Markov property we deduce

$$p_x(n) = \sum_{y \ne x_0} p_y(n-1) Q_{y,x}. \tag{4.2.8}$$

We deduce that

$$\lambda_x \stackrel{(4.2.7)}{=} \sum_{n \in \mathbb{N}} p_x(n) = p_x(1) + \sum_{y \ne x_0} \underbrace{\left(\sum_{n \in \mathbb{N}} p_y(n)\right)}_{=\lambda_y} Q_{y,x}$$

$(\lambda_{x_0} = 1, \ p_x(1) = Q_{x_0,x})$

$$= \lambda_0 Q_{x_0,x} + \sum_{y \ne x_0} \lambda_y Q_{y,x} = \sum_y \lambda_y Q_{y,x}.$$

This proves (4.2.3). Let us now show that the λ_x defined by (4.2.6b) are positive.

Suppose that $\lambda_x = 0$ for some $x \in \mathscr{X}$. Obviously $x \ne x_0$. Moreover, from the equality $\lambda = \lambda Q^n$, $\forall n \in \mathbb{N}$ we deduce

$$0 = \lambda_x = Q^n_{x_0,x} + \sum_{y \ne x_0} \lambda_y Q^n_{y,x}.$$

Thus $Q^n_{x_0,x} = 0$, $\forall n \in \mathbb{N}$, which contradicts the fact that x_0 and x communicate.

Finally, let us prove that $\lambda_x < \infty$, $\forall x$. Observe that

$$1 = \lambda_{x_0} = \sum_{x \in \mathscr{X}} \lambda_x Q^n_{x,x_0}. \tag{4.2.9}$$

Suppose that $\lambda_x = \infty$ for some $x \ne x_0$. Since the chain is irreducible, the state x_0 communicates with x so there exists $n = n(x)$ such that $Q^n_{x_0,x} \ne 0$. The equality (4.2.9) implies $\lambda_x \le \frac{1}{Q^n_{x,x_0}}$.

(ii) We have

$$\sum_{x \in \mathscr{X}} \sum_{n \ge 1} I_{\{X_n = x\}} I_{\{n \le T_0\}} = \sum_{n \ge 1} \sum_{x \in \mathscr{X}} I_{\{X_n = x\}} I_{\{n \le T_0\}} = \sum_{x \in \mathscr{X}} I_{\{n \le T_0\}} = T_0.$$

Hence
$$\lambda[\mathscr{X}] = \sum_{x \in \mathscr{X}} \lambda_x = \sum_{x \in \mathscr{X}} \sum_{n \geq 1} \mathbb{E}_{x_0}\big[I_{\{X_n = x\}} I_{\{n \leq T_0\}}\big] = \mathbb{E}_{x_0}[T_0].$$

(iii) We follow the approach in [2; 123]. Consider the matrix $K : \mathscr{X} \times \mathscr{X} \to [0,1]$
$$K_{x,y} = \begin{cases} Q_{x,y}, & y \neq x_0, \\ 0, & y = x_0. \end{cases}$$
Consider the sequence $(\mu_n)_{n \geq 0}$ of measures on $\mathscr{X}$ defined by
$$\mu_0 = \delta_{x_0}, \quad \mu_n[x] = p_x(n) = \mathbb{P}[X_n = x, \ n < T_0], \quad x \in \mathscr{X}.$$
Note that $\mu_n[x_0] = 0$, $\forall n$. The equality (4.2.8) implies that
$$\mu_n = \mu_{n-1} K, \quad \forall n \geq 1$$
so $\mu_n = \delta_{x_0} K^n$. Observe that
$$\lambda = \sum_{n \geq 0} \mu_n = \sum_{n \geq 0} \delta_{x_0} K^n.$$
Fix an invariant measure ν such that $\nu_{x_0} = 1$. The invariance condition reads $\nu = \delta_{x_0} + \nu K$. We deduce
$$\nu = \delta_{x_0} + \big(\delta_{x_0} + \nu K\big) K = \delta_{x_0} + \delta_{x_0} K + \nu K^2.$$
Arguing inductively we deduce
$$\nu = \sum_{m=0}^{n} \delta_{x_0} K^m + \nu K^{n+1} \geq \sum_{m=0}^{n} \delta_{x_0} K^m, \quad \forall n \in \mathbb{N}.$$
Letting $n \to \infty$ we deduce $\nu \geq \mu$. The difference $\sigma = \nu - \mu$ is also an invariant measure such that $\sigma[x_0] = 0$. Set $\sigma_x := \sigma[x]$.

Fix $x \in \mathscr{X} \setminus \{x_0\}$. Since the Markov chain is irreducible there exists $n \in \mathbb{N}$ such that $Q^n_{x,x_0} > 0$. From the equality $\sigma = \sigma Q^n$ we deduce
$$0 = \sigma_{x_0} = \sigma_x Q^n_{x,x_0} + \sum_{y \neq x} \sigma_x Q^n_{y,x_0} \geq \sigma_{x_1} Q^n_{x,x_0}.$$
Hence $\sigma_x = 0$, $\forall x \in \mathscr{X}$ so that $\nu = \lambda$. $\square$

Remark 4.49. The example of the standard random walk on $\mathbb{Z}^3$ shows that even transient chains can admit invariant measures. $\square$

Suppose that $(X_n)_{n \geq 0}$ is irreducible and recurrent. For each $x \in \mathscr{X}$ we denote by π^x the unique invariant measure on $\mathscr{X}$ such that $\pi^x[x] = 1$. We know that for $x, y \in \mathscr{X}$ the measure π^y is a positive multiple of π^x,
$$\pi^y = c_{y,x} \pi^x.$$
From the equality $\pi^x[x] = 1$ we deduce $c_{y,x} = \pi^y[x]$ so that
$$\pi^y = \pi^y[x] \pi^x. \tag{4.2.10}$$
From Theorem 4.48(ii) we deduce that
$$\pi^x[\mathscr{X}] = \mathbb{E}_x[T_x]. \tag{4.2.11}$$
In particular this shows that the following statements are equivalent

(i) $\exists x \in \mathcal{X}$ such that $\mathbb{E}_x[T_x] < \infty$.
(ii) $\forall x \in \mathcal{X}$, $\mathbb{E}_x[T_x] < \infty$.

Definition 4.50. Let $(X_n)_{n\geq 0}$ be an irreducible recurrent HMC.

(i) The chain is called *positively recurrent* if $\mathbb{E}_x[T_x] < \infty$ for some $x \in \mathcal{X}$.
(ii) The chain is called *null recurrent* if $\mathbb{E}_x[T_x] = \infty$ for all $x \in \mathcal{X}$.

$\square$

Corollary 4.51. *An irreducible recurrent HMC is positively recurrent if for some (for all) $x \in \mathcal{X}$ we have*

$$\pi^x[\mathcal{X}] < \infty. \tag{4.2.12}$$

In particular, a positively recurrent irreducible HMC admits a unique invariant probability measure π_∞ described by

$$\boxed{\pi_\infty = \frac{1}{\mathbb{E}_x[T_x]}\pi^x, \ \forall x \in \mathcal{X}}.$$

In other words

$$\boxed{\pi_\infty[x] = \frac{1}{\mathbb{E}_x[T_x]}, \ \forall x \in \mathcal{X}}. \tag{4.2.13}$$

$\square$

Proof. The equality (4.2.13) follows from the equality $\pi^x[x] = 1$. $\square$

Corollary 4.52. *Any irreducible HMC with finite state space $\mathcal{X}$ admits a unique stationary probability measure.* $\square$

Proof. As shown is Example 4.33 this chain is recurrent since the state space is finite. The finiteness of $\mathcal{X}$ implies (4.2.12). $\square$

Example 4.53 (Random knight moves). Consider a regular 8×8 chess table and a knight that starts in the lower left-hand corner and then moves randomly along, making each permissible move with equal probability.

This is an example of random walk on a graph with vertex set $\mathcal{X}$ consisting of the centers of the 64 squares of the board, where two vertices are connected by as many edges as possibilities for the knight to go from a square to the other in one move.

It is easily seen that this is a connected graph so the corresponding random walk is irreducible and has a unique invariant probability distribution given by

$$\pi[x] = \frac{\deg x}{Z}, \quad Z = \sum_{y \in \mathcal{X}} \deg y.$$

Now observe that Z is twice the number of edges of this graph.

To count them observe that each 2×3 sub-rectangle of the chess table determines four edges in this graph, two for each diagonal. The same is true for the 3×2 rectangle. Moreover, any knight move is corresponds to a diagonal of unique such rectangle. If $N_{2\times 3}$ and respectively $N_{3\times 2}$ denote the number of 2×3 rectangles respectively 3×2 rectangles, then

$$N_{2\times 3} = N_{3\times 2} =: N,$$

so that $Z = 16N$. Now observe that since a 3×2 rectangle is uniquely determined by the location of its lower left corner we have $N = 6 \times 7 = 42$ so $Z = 16 \cdot 42 = 672$.

If x corresponds to the left-hand square, then $\deg x = 4$ so

$$\mathbb{E}_x[T_x] = \frac{1}{\pi[x]} = \frac{16 \cdot 42}{4} = 4 \cdot 42 = 168.$$

Thus, given that the knight starts at x, the expected time to return to x is 168. □

Theorem 4.54. *Suppose that (X_n) is an irreducible HMC with state space $\mathscr{X}$ and transition matrix Q. Then the following are equivalent.*

(i) The chain is positively recurrent.
(ii) There exists an invariant probability measure.

Proof. We have already shown that (i) $\Rightarrow$ (ii). To prove the implication (ii) $\Rightarrow$ (i) fix an invariant probability measure π. Thus $\pi = Q^n \pi$, $\forall n \in \mathbb{N}$, so that

$$\pi[y] = \sum_{x \in \mathscr{X}} \pi[x] Q^n_{x,y}, \quad \forall y \in \mathscr{X}, \; n \in \mathbb{N}. \tag{4.2.14}$$

Fix $y_0 \in \mathscr{X}$ such that $\pi[y_0] \neq 0$. We prove first that if the chain is recurrent then it has to be positively recurrent. Denote by λ_{y_0} the unique invariant measure such that $\lambda_{y_0}[y_0] = 1$. The measure λ_{y_0} is a constant multiple of π so it is finite. Hence

$$\mathbb{E}_{y_0}[T_{y_0}] = \lambda_{y_0}[\mathscr{X}] < \infty,$$

showing that the chain in fact positively recurrent.

We now argue by contradiction that the chain is indeed recurrent. Assume that our Markov chain is transient. Proposition 4.38 implies that for any $x \in \mathscr{X}$ we have

$$\infty > \mathbb{E}_\pi[N_x] = \sum_n \mathbb{E}_\pi[\boldsymbol{I}_{N_x=n}] = \sum_n \sum_{x' \in \mathscr{X}} Q^n_{x,x'} \pi[x'] \geq \pi[y_0] \sum_n Q^n_{x,y_0}.$$

Hence

$$\sum_n Q^n_{x,y_0} < \infty \quad \text{and} \quad \lim_{n \to \infty} Q^n_{x,y_0} = 0, \; \forall x \in \mathscr{X}. \tag{4.2.15}$$

Set $q_n(x) = Q^n_{x,y_0}$, $\forall x \in \mathscr{X}$. The equality (4.2.14) implies that

$$\int_{\mathscr{X}} q_n(x) \pi[dx] = \pi[y_0] > 0, \; \forall n \in \mathbb{N}.$$

On the other hand, the equality (4.2.15) coupled with the Dominated Convergence theorem implies
$$\lim_{n\to\infty}\int_{\mathscr{X}} q_n(x)\pi[dx] = 0.$$
This contradiction completes the proof. □

Example 4.55. We have shown in Example 4.42 that the standard random walks on $\mathbb{Z}$ and $\mathbb{Z}^2$ are recurrent. Let us show that they are null recurrent.

Note that for $k = 1, 2$, the measure on $\mathbb{Z}^k$ defined by $\lambda[x] = 1$, $\forall x \in \mathbb{Z}^k$ is invariant. By Theorem 4.48, λ is the unique invariant measure such that $\lambda[0] = 1$. Since $\lambda[\mathbb{Z}^k] = \infty$ we deduce that there is no invariant finite measure. □

Proposition 4.56. *Suppose that $(X_n)_{n\geq 0}$ is an irreducible, positively recurrent HMC with state space $\mathscr{X}$ and transition matrix Q. Then*
$$\mathbb{E}_y[T_x] < \infty, \quad \forall x, y \in \mathscr{X}.$$

Proof. Fix $x \in \mathscr{X}$. Let $\mathcal{Y} \subset \mathscr{X}$ denote the set of $y \in \mathscr{X}$ such that $\mathbb{E}_y[T_x] < \infty$. Note that $x \in \mathcal{Y}$.

For $y \in \mathcal{Y}$ we have
$$\mathbb{E}_y[T_x] = \mathbb{E}_y[T_x | X_1 = y]Q_{y,y} + \sum_{z\neq y}\mathbb{E}_y[T_x | X_1 = z]Q_{y,z}.$$
Hence
$$y \neq z, \ Q_{y,z} > 0 \Rightarrow \mathbb{E}_y[T_x | X_1 = z] < \infty.$$
Now observe that for $z \neq y$
$$\mathbb{E}_y[T_x | X_1 = z] = \begin{cases} 1, & z = x, \\ 1 + \mathbb{E}_z[T_x], & z \neq x. \end{cases}$$
We deduce that $y \in \mathscr{X}$ and $Q_{y,z} > 0$, then $z \in \mathcal{Y}$. We conclude iteratively that
$$y \in \mathcal{Y}, \quad \forall y, \ x \to y.$$
Since the chain is irreducible we deduce $\mathcal{Y} = \mathscr{X}$. □

Let $(X_n)_{n\geq 0}$ be an HMC with state space and transition matrix Q. Recall that for any set $A \subset \mathscr{X}$ we denoted by T_A the time of first return to A
$$T_A := \inf\{n \geq 1; \ X_n \in A\}.$$
Note that $T_A \leq T_a$, $\forall a \in A$, so
$$\mathbb{E}_x[T_A] \leq \mathbb{E}_x[T_a], \quad \forall x \in \mathscr{X}, \ \forall a \in A.$$
We deduce that if the chain is irreducible and positively recurrent then
$$\mathbb{E}_x[T_A] < \infty, \quad \forall x \in \mathscr{X}, \ \forall A \subset \mathscr{X}.$$
We have a sort of converse.

Proposition 4.57. *Suppose that $(X_n)_{n\geq 0}$ is an irreducible HMC with state space $\mathscr{X}$ and transition matrix Q. If there exists a finite subset $A \subset \mathscr{X}$ such that*
$$\mathbb{E}_a[T_A] < \infty, \quad \forall a \in A,$$
then $(X_n)_{n\geq 0}$ is positively recurrent.

Proof. We follow the approach in [20, Chap. 5, Sec. 1.1]. Set
$$M_A := \max_{a \in A} \mathbb{E}_a[T_A].$$
Consider the epochs of return to A,
$$T^1 := T_A, \quad T^{k+1} := \min\{n > T^k; \ X_n \in A\}.$$
Fix $a_0 \in A$ and suppose that $(X_n)_{n \geq 0}$ starts at a_0, $X_0 = a_0$ a.s. We set
$$Y_0 := X_0, \quad Y_k := X_{T^k}, \quad k \in \mathbb{N}.$$
The strong Markov property shows that $(Y_k)_{k \geq 0}$ is an HMC with state space A. Since (X_n) is irreducible we deduce that $(Y_k)_{k \geq 0}$ is such. Since A is finite, the chain (Y_k) is positively recurrent. Denote by $\widehat{T}_0$ the time of first return to a_0 of the chain $(Y_k)_{k \geq 0}$. Set
$$S_0 = T^1, \quad S_k = T^{k+1} - T^k.$$
If T_{a_0} denotes the time of first return to a_0 of the original chain, then
$$T_{a_0} = \sum_{k=0}^{\infty} S_k \boldsymbol{I}_{\{k < \widehat{T}_0\}}, \quad \mathbb{E}_{a_0}[T_{a_0}] = \sum_{k=0}^{\infty} \mathbb{E}_{a_0}[S_k \boldsymbol{I}_{\{k < \widehat{T}_0\}}].$$
On the other hand,
$$\mathbb{E}_{a_0}[S_k \boldsymbol{I}_{\{k < \widehat{T}_0\}}] = \sum_{a \in A} \mathbb{E}_{a_0}[S_k \boldsymbol{I}_{\{k < \widehat{T}_0\}} \boldsymbol{I}_{\{X_{T^k} = a\}}].$$
Observe that the event $\{k < \widehat{T}_0\}$ belongs to $\mathcal{F}_{T^k}$. We deduce
$$\mathbb{E}_{a_0}[S_k \boldsymbol{I}_{\{k < \widehat{T}_0\}} \boldsymbol{I}_{\{X_{T^k} = a\}}] = \mathbb{E}_{a_0}[S_k | k < \widehat{T}_0, X_{T^k} = a] \mathbb{P}_{x_0}[k < \widehat{T}_0, X_{T^k} = a]$$
(use the strong Markov property for T^k)
$$= \mathbb{E}_{a_0}[S_k | X_{T^k} = a] \mathbb{P}_{x_0}[k < \widehat{T}_0, X_{T^k} = a] = \mathbb{E}_a[T_A] \mathbb{P}_{x_0}[k < \widehat{T}_0, X_{T^k} = a]$$
$$\leq M_A \mathbb{P}_{a_0}[k < \widehat{T}_0, X_{T^k} = a].$$
Hence
$$\mathbb{E}_{a_0}[T_{a_0}] \leq M_A \sum_{k=0}^{\infty} \left(\sum_{a \in A} \mathbb{P}_{x_0}[\widehat{T}_0 > k, X_{T^k} = a] \right)$$
$$= M_A \sum_{k=0}^{\infty} \mathbb{P}_{a_0}[\widehat{T}_0 > k] = M_A \mathbb{E}_{a_0}[\widehat{T}_0] < \infty,$$
since the chain (Y_k) is positively recurrent. $\square$

4.3 Asymptotic behavior

Suppose that $(X_n)_{n \in \mathbb{N}_0}$ is an *irreducible, positively recurrent* HMC with state space $\mathscr{X}$ and transition matrix Q. It thus has a unique stationary probability measure $\pi_\infty \in \text{Prob}(\mathscr{X})$. In this section we will provide a dynamical description of π_∞ and prove a Law of Large Numbers that involves this measure.

4.3.1 The ergodic theorem

Fix an arbitrary state $x_0 \in \mathscr{X}$, let $T_0 := T_{x_0}$ denote the time of first return to x_0 and denote by π^0 the unique invariant measure on $\mathscr{X}$ such that $\pi^0[x_0] = 1$. The results in the previous section show that

$$\pi^0[x] = \mathbb{E}_{x_0}\left[\sum_{n\geq 1} \boldsymbol{I}_{\{X_n=x\}}\boldsymbol{I}_{\{n\leq T_0\}}\right] = \mathbb{E}_{x_0}\left[\sum_{n=1}^{T_0} \boldsymbol{I}_{\{X_n=x\}}\right], \quad \forall y \in \mathscr{X}. \quad (4.3.1)$$

For $n \in \mathbb{N}$ we set

$$\nu(n) = \nu_{x_0}(n) := \sum_{k=1}^{n} \boldsymbol{I}_{\{X_k=x_0\}}.$$

In other words, the random variable $\nu_{x_0}(n)$ is the number of returns to x_0 during the interval $[1, n]$.

Proposition 4.58. *Suppose that $f \in L^1(\mathscr{X}, \pi^0)$. Then*

$$\lim_{n\to\infty} \frac{1}{\nu_{x_0}(n)} \sum_{k=1}^{n} f(X_k) = \int_{\mathscr{X}} f(x)\pi^0[dx] = \sum_{x\in\mathscr{X}} f(x)\pi^0[x], \quad \mathbb{P}_{x_0} - \text{a.s.} \quad (4.3.2)$$

Proof. We follow the proof in [20, Prop. 3.4.1]. Using the decomposition $f = f^+ - f^-$ we see that it suffices to consider only the case when f is nonnegative. Let $T_0 = \tau_1 \leq \tau_2 \leq \cdots$ denote the successive times of return to x_0. We set

$$U_p := \sum_{k=\tau_{p-1}+1}^{\tau_p} f(X_k).$$

The strong Markov property shows that the random variables $U_1, U_2, \ldots$ are i.i.d. We have

$$\mathbb{E}_{x_0}[U_1] = \sum_{k=1}^{T_0} \mathbb{E}_{x_0}[f(X_k)] = \mathbb{E}_{x_0}\left[\sum_{k=1}^{T_0} \sum_{x\in\mathscr{X}} f(x)\boldsymbol{I}_{\{X_k=x\}}\right]$$

$$= \sum_{x\in\mathscr{X}} f(x)\mathbb{E}_{x_0}\left[\sum_{k=1}^{T_0} \boldsymbol{I}_{\{X_k=x\}}\right] \stackrel{(4.3.1)}{=} \sum_{x\in\mathscr{X}} f(x)\pi^0[x].$$

The Strong Law of Large Numbers implies

$$\lim_{p\to\infty} \frac{1}{p} \sum_{k=1}^{p} U_k = \sum_{x\in\mathscr{X}} f(x)\pi^0[x], \quad \mathbb{P}_{x_0} - \text{a.s.}$$

In other words

$$\frac{1}{p} \sum_{k=1}^{\tau_p} f(X_k) \to \sum_{x\in\mathscr{X}} f(x)\pi^0[x], \quad \mathbb{P}_{x_0} - \text{a.s.}$$

Observing that $\tau_{\nu(n)} \leq n < \tau_{\nu(n)+1}$, we deduce that for nonnegative of f we have

$$\frac{1}{\nu(n)} \sum_{k=1}^{\tau_{\nu(n)}} f(X_k) \leq \frac{1}{\nu(n)} \sum_{k=1}^{n} f(X_k) \leq \frac{1}{\nu(n)} \sum_{k=1}^{\tau_{\nu(n)+1}} f(X_k)$$

$$= \frac{\nu(n)+1}{\nu(n)} \frac{1}{\nu(n)+1} \sum_{k=1}^{\tau_{\nu(n)+1}} f(X_k).$$

Since the chain is recurrent we have $\nu(n) \to \infty$ so

$$\lim_{n \to \infty} \frac{\nu(n)+1}{\nu(n)} = 1.$$

Hence the extremes in the last inequality converge to $\sum_{x \in \mathscr{X}} f(x) \pi^0 [x]$, $\mathbb{P}_{x_0}$ – a.s. □

Corollary 4.59. *We have*

$$\frac{\nu(n)}{n} \to \frac{1}{\mathbb{E}_{x_0}[T_{x_0}]} = \pi_\infty [x_0], \ \mathbb{P}_{x_0} - \text{a.s.} \tag{4.3.3}$$

Proof. Let $f = 1$ in Proposition 4.58. This f is integrable so

$$\frac{n}{\nu(n)} \to \pi^0 [\mathscr{X}] \stackrel{(4.2.11)}{=} \mathbb{E}_{x_0}[T_{x_0}].$$

□

Corollary 4.60 (Ergodic Theorem). *Suppose that $(X_n)_{n \geq 0}$ is a positively recurrent irreducible HMC with state space $\mathscr{X}$, transition matrix Q and stationary distribution π_∞. Let $f \in L^1(\mathscr{X}, \pi_\infty)$. Then, for any $\mu \in \text{Prob}(\mathscr{X})$ we have*

$$\lim_{n \to \infty} \frac{1}{n} \sum_{k=1}^{n} f(X_k) = \int_{\mathscr{X}} f(x) \pi_\infty [dx], \ \mathbb{P}_\mu - \text{a.s.} \tag{4.3.4}$$

Proof. Assume first that (X_n) are defined on the path space $(\mathscr{X}^{\mathbb{N}_0}, \mathcal{E}, \mathbb{P}_\mu)$.

Suppose $\mu = \delta_{x_0}$. If we divide both sides of (4.3.2) by $\mathbb{E}_{x_0}[T_{x_0}]$ we deduce

$$\lim_{n \to \infty} \frac{1}{\nu(n) \mathbb{E}_{x_0}[T_{x_0}]} \sum_{k=1}^{n} f(X_k) = \int_{\mathscr{X}} f(x) \pi_\infty [dx].$$

Now observe that

$$\frac{1}{n} \sum_{k=1}^{n} f(X_k) = \frac{n}{\nu(n) \mathbb{E}_{x_0}[T_{x_0}]} \frac{1}{n} \sum_{k=1}^{n} f(X_k),$$

and (4.3.3) implies

$$\frac{n}{\nu(n) \mathbb{E}_{x_0}[T_{x_0}]} \to 1.$$

More generally, for any $\mu \in \text{Prob}(\mathscr{X})$,

$$\mathbb{P}_\mu = \sum_{x \in \mathscr{X}} \mu[x] \mathbb{P}_x \in \text{Prob}(\mathscr{X}^{\mathbb{N}_0}, \mathcal{E})$$

we denote by $\mathcal{C}_x$ the set

$$\mathcal{C} := \left\{ \underline{x} = (x_0, x_1, \ldots) \in \mathscr{X}^{\mathbb{N}_0} : \lim_{n \to \infty} \frac{1}{n} \big(f(x_1) + \cdots + f(x_n) \big) = \int_{\mathscr{X}} f(x) \pi_\infty[dx] \right\}.$$

From the above we deduce that $\mathbb{P}_x[\mathcal{C}] = 1$, $\forall x \in \mathscr{X}$. Then

$$\mathbb{P}_\mu[\mathcal{C}] \stackrel{(4.1.13)}{=} \sum_{x \in \mathscr{X}} \mu[x] \mathbb{P}_x[\mathcal{C}] = 1.$$

Suppose that random maps (X_n) are defined on a probability space $(\Omega, \mathcal{S}, \mathbb{P})$, not necessarily the path space. Using the map $\vec{X} : \Omega \to \mathscr{X}^{\mathbb{N}_0}$ we reduce this case to the situation we have discussed above. □

The Ergodic Theorem is a Law of Large Numbers for a sequence of, not necessarily independent, random variables.

4.3.2 Aperiodic chains

When $(X_n)_{n \geq 0}$ is an irreducible, *aperiodic*, positively recurrent HMC the Ergodic Theorem can be considerably strengthened. We need to introduce some terminology.

The *variation distance* $d_v(\mu, \nu)$ between two probability measures μ, $\nu \in \text{Prob}(\mathscr{X})$ is defined by

$$d_v(\mu, \nu) := \frac{1}{2} \sum_{x \in \mathscr{X}} | \mu[x] - \nu[x] |.$$

If X, Y are $\mathscr{X}$-valued random variables, then the variation distance between them is defined to be the variation distance between their distributions $\mathbb{P}_X, \mathbb{P}_Y$,

$$d_v(X, Y) := d_v(\mathbb{P}_X, \mathbb{P}_Y).$$

Lemma 4.61. *Let* $\mu, \nu \in \text{Prob}(\mathscr{X})$. *Then*

$$d_v(\mu, \nu) = \sup_{A \subset \mathscr{X}} \big| \mu[A] - \nu[A] \big| = \sup_{A \subset \mathscr{X}} \big(\mu[A] - \nu[A] \big).$$

Proof. The second equality follows from the elementary observation

$$\big| \mu[A] - \nu[A] \big| = \max\big\{ (\mu[A] - \nu[A]), (\mu[A^c] - \nu[A^c]) \big\}.$$

Now define

$$B := \big\{ x \in \mathscr{X};\ \mu[x] \geq \nu[x] \big\}.$$

Observe that for any $A \in \mathscr{X}$ we have
$$\mu[A] - \nu[A] \le \mu[A \cap B] - \nu[A \cap B] \le \mu[B] - \nu[B].$$
The first inequality follows from the fact that, for $x \in A \cap B^c$, we have $\mu[x] < \nu[x]$. A similar reasoning shows that
$$\nu[A] - \mu[A] \le \nu[B^c] - \mu[B^c].$$
Observe that
$$\mu[B] - \nu[B] + \mu[B^c] - \nu[B^c] = \mu[\mathscr{X}] - \nu[\mathscr{X}] = 0.$$
Hence
$$\sup_{A \subset \mathscr{X}} (\mu[A] - \nu[A]) = \mu[B] - \nu[B] = \frac{1}{2} \sum_{x \in \mathscr{X}} |\mu[x] - \nu[x]|.$$
$\square$

Remark 4.62. Suppose that μ, ν are two probability measures on a measurable spaces $(\Omega, \mathcal{S})$. Then their difference $\lambda = \mu - \nu$ is a signed measure on $(\Omega, \mathcal{S})$. As such, λ has a positive part, λ_+ a negative part, and a variation $|\lambda| = \lambda_+ + \lambda_-$. We set
$$\|\lambda\| := |\lambda|[\Omega].$$
For example, if $\Omega = \mathbb{R}$ equipped with the Borel sigma-algebra and μ, ν are given by densities p and respectively q, then
$$\|\mu - \nu\| = \int_\mathbb{R} |p(x) - q(x)| dx.$$
If μ, ν are probability measures on a finite or countable set $\mathscr{X}$, then
$$d_v(\mu, \nu) = \frac{1}{2} \|\mu - \nu\|.$$
$\square$

One technique for estimating the variation distance between two probability measures on $\mathscr{X}$ is based on the idea of coupling.

Definition 4.63. Let $\mu, \nu \in \text{Prob}(\mathscr{X})$. A *coupling* of μ with ν is a probability measure λ on $\mathscr{X} \times \mathscr{X}$ whose marginals equal μ and respectively ν, i.e.,
$$\lambda[\{x_0\} \times \mathscr{X}] = \mu[x_0], \quad \lambda[\mathscr{X} \times \{x_1\}] = \nu[x_1], \quad \forall x_0, x_1 \in \mathscr{X}.$$
We will use the notation $\mu \overset{\lambda}{\leftrightsquigarrow} \nu$ to indicate that λ is a coupling of μ with ν. We will denote by $\text{Couple}(\mu, \nu)$ the set of couplings of μ with ν.

A coupling of a pair of $\mathscr{X}$-valued random variables is defined to be an $\mathscr{X} \times \mathscr{X}$ random variable Z whose distribution is a coupling of the distributions of X and Y.
$\square$

The next result explains the relevance of couplings in estimating the variation distance between two measures.

Proposition 4.64. *Let $\mu, \nu \in \text{Prob}(\mathscr{X})$. Set*
$$\mathscr{X}^{(2)} := \{(x_0, x_1) \in \mathscr{X}^2;\ x_0 \neq x_1\}.$$
Then,
$$d_v(\mu, \nu) \leq \lambda[\mathscr{X}^{(2)}], \quad \forall \lambda \in \text{Couple}(\mu, \nu). \tag{4.3.5}$$

Proof. For any $A \subset \mathscr{X}$ we have
$$\mathscr{X}^2 \supset A \times A^c = A \times \mathscr{X} \setminus A \times A$$
so
$$\lambda[\mathscr{X}^{(2)}] \geq \lambda[A \times \mathscr{X}] - \lambda[A \times A]$$
$$\geq \lambda[A \times \mathscr{X}] - \lambda[\mathscr{X} \times A] = \mu[A] - \nu[A].$$
Hence
$$\lambda[\mathscr{X}^{(2)}] \geq \sup_{A \subset \mathscr{X}} (\mu[A] - \nu[A]) = d_v(\mu, \nu).$$
$\square$

Remark 4.65. The inequality (4.3.5) is optimal in the sense that there exists a coupling $\lambda \in \text{Couple}(\mu, \nu)$ such that $d_v(\mu, \nu) = \lambda[\mathscr{X}^{(2)}]$. For details we refer to [20, Sec. 4.1.2] or [104, Sec. 4.2]. $\square$

Definition 4.66. Two $\mathscr{X}$-valued stochastic processes $(X_n)_{n \in \mathbb{N}_0}$ and $(Y_n)_{n \in \mathbb{N}_0}$ are said to *couple* if the random variable
$$T = \min\{n \in \mathbb{N};\ X_m = Y_m,\ \forall m \geq n\}$$
is a.s. *finite*. The random variable T is called the *coupling time*. $\square$

Lemma 4.67. *Suppose that the $\mathscr{X}$-valued processes $(X_n)_{n \in \mathbb{N}_0}$ and $(Y_n)_{n \in \mathbb{N}_0}$ couple with coupling time T. Then*
$$d_v(X_n, Y_n) \leq \mathbb{P}[T > n].$$
In particular,
$$\lim_{n \to \infty} d_v(X_n, Y_n) = 0.$$

Proof. For all $A \subset \mathscr{X}$ we have
$$\mathbb{P}[X_n \in A] - \mathbb{P}[Y_n \in A] = \mathbb{P}[X_n \in A,\ T \leq n] + \mathbb{P}[X_n \in A,\ T > n]$$
$$- \mathbb{P}[Y_n \in A, T \leq n] - \mathbb{P}[Y_n \in A, T > n]$$
$(X_n = Y_n, \forall n \leq T)$
$$= \mathbb{P}[X_n \in A,\ T > n] - \mathbb{P}[Y_n \in A, T > n] \leq \mathbb{P}[X_n \in A,\ T > n] \leq \mathbb{P}[T > n].$$
$\square$

Theorem 4.68. *Suppose that Q is a probability transition matrix on the state space $\mathscr{X}$ such that the associated HMC's are irreducible,* aperiodic *and positively recurrent. Denote by π the unique invariant probability measure on $\mathscr{X}$. Then, for any $\mu \in \text{Prob}(\mathscr{X})$*

$$\lim_{n\to\infty} d_v(\mu Q^n, \pi) = 0. \qquad (4.3.6)$$

In particular if $\mu = \delta_{x_0}$ we deduce that

$$\pi[x] = \lim_{n\to\infty} Q^n_{x_0, x}. \qquad (4.3.7)$$

Proof. Consider two independent HMCs $(X_n)_{n \in \mathbb{N}}$ and $(Y_n)_{n \geq 0}$ with state space $\mathscr{X}$, transition matrix Q such that initial distribution of (X_n) is μ and the initial distribution of Y_n is π. Since π is stationary, the probability distribution of Y_n is π, $\forall n$. According to Lemma 4.67 it suffices to show that the stochastic processes (X_n), (Y_n) couple.

Consider the stochastic process (X_n, Y_n). This is an HMC with state space $\mathscr{X} \times \mathscr{X}$ and transition matrix $\widehat{Q}$

$$\widehat{Q}_{(x_0,y_0),(x_1,y_1)} = Q_{x_0,x_1} \cdot Q_{y_0,y_1}.$$

Since Q is irreducible and *aperiodic* we deduce that for any $x_0, x_1, y_0, y_1 \in \mathscr{X}$, $\exists N_0 > 0$ such that $Q^n_{x_1,x_1} \cdot Q^n_{y_1,y_1} > 0$, $\forall n \geq N_0$. We deduce that there exists $N \geq N_0$ such that

$$Q^n_{x_0,x_1} \cdot Q^n_{y_0,y_1} > 0, \ \forall n \geq N.$$

This shows that $\widehat{Q}$ is irreducible.

The product measure $\pi \otimes \pi$ is an invariant probability measure of the chain (X_n, Y_n). Theorem 4.54 implies that $\widehat{Q}$ is positively recurrent. Proposition 4.57 thus, for any $x \in \mathscr{X}$, the chain (X_n, Y_n) will almost surely return to (x, x) in finite time. $\square$

Remark 4.69. Suppose that Q is the transition matrix of an irreducible, positively recurrent Markov chain with state space $\mathscr{X}$ and invariant probability measure π. We form a new stochastic matrix

$$\widetilde{Q} = \frac{1}{2}(1 + Q).$$

The chain with this transition matrix is called *the lazy version* of the original chain. It is irreducible and π is the invariant probability measure of the lazy version as well. However, the lazy chain is also *aperiodic*, even if the original chain is not. This follows from the equality

$$\widetilde{Q}^n_{x,y} = \sum_{k=0}^{n} \frac{1}{2^n} \binom{n}{k} Q^k_{x,y}.$$

This shows that if $Q^k_{x,y} \neq 0$, then $\widetilde{Q}^n_{x,y} > 0$, $\forall n \geq k$. Using the terminology of generalized convergence in [79], we can say that the *Euler means* of the sequence $\left(Q^n_{x,y} \right)_{n \geq 0}$ converge to the invariant measure. $\square$

4.3.3 Martingale techniques

Suppose that $(X_n)_{n\geq 0}$ is an HMC with state space $\mathscr{X}$, transition matrix Q and initial distribution π_0. We assume that all the random variables X_n are defined on the same probability space $(\Omega, \mathcal{S}, \mathbb{P})$. Set
$$\pi_n := \pi_0 \cdot Q^n, \quad \forall n \in \mathbb{N}.$$
We denote by $\mathcal{F}_n$ the filtration
$$\mathcal{F}_n = \sigma(X_0, X_1, \ldots, X_n) \subset \mathcal{S}, \quad n \geq 0.$$
We want to investigate the (sub/super)martingales with respect to this filtration and show some of their applications to the dynamics of the underlying HMC.

Note that any function $\mathscr{X} \to \mathbb{R}$ is measurable with respect to the sigma-algebra $2^{\mathscr{X}}$. For this reason we will denote by $\mathcal{L}^0(\mathscr{X})$ the space of functions $\mathscr{X} \to \mathbb{R}$. We think of a function $f \in \mathcal{L}^0(\mathscr{X})$ as a *column* vector $\bigl(f(x)\bigr)_{x\in\mathscr{X}}$ and we denote by Qh the function described by the multiplication of the matrix Q with the column vector f. More precisely,
$$(Qf)(x) = \sum_{y\in\mathscr{X}} Q_{x,y} f(y), \quad \forall x \in \mathscr{X}.$$
There is a small problem with this definition namely, if $\mathscr{X}$ is infinite, then the above series may by divergent. Since the rows of Q define probability distributions on $\mathscr{X}$ we see that the above sums are finite if f is bounded. We obtain in this fashion a linear map
$$Q : \mathcal{L}^\infty(\mathscr{X}) \to \mathcal{L}^\infty(\mathscr{X}), \quad f \to Qf.$$
We say that the transition matrix is *locally finite* if each of its rows has only finitely many nonzero entries. Equivalently, at each state $x \in \mathscr{X}$ the system can transition only to finitely many states. In this case Q defines a linear map
$$Q : \mathcal{L}^0(\mathscr{X}) \to \mathcal{L}^0(\mathscr{X}).$$
Note
$$Q\boldsymbol{I}_{\mathscr{X}} = \boldsymbol{I}_{\mathscr{X}}, \quad f \geq 0 \Rightarrow Qf \geq 0.$$
If we think of π_n as a row vector, then for any $g \in \mathcal{L}^1(\mathscr{X}, \pi_n)$ we have
$$\int_{\mathscr{X}} g\, d\pi_n = \pi_n \cdot g,$$
where the "$\cdot$" denotes the multiplication of a one-row matrix π_n with a one-column matrix g. We deduce
$$\pi_n \cdot g = (\pi_0 \cdot Q^n) \cdot g = \pi_0 \cdot (Q^n g).$$
Thus
$$g \in L^1(\mathscr{X}, \pi_n), \quad \forall n \geq 0 \Longleftrightarrow Q^n g \in L^1(\mathscr{X}, \pi_0), \quad \forall n \geq 0.$$

Definition 4.70. A function $h \in \mathcal{L}^1(\mathscr{X}, \pi_0)$ is called a *Lyapunov function* of the HMC $(X_n)_{n\geq 0}$ if the stochastic process $\bigl(h(X_n)\bigr)_{n\geq 0}$ is a supermartingale adapted to the filtration $\mathcal{F}_n$, i.e.,
$$f(X_n) \in L^1(\Omega, \mathcal{S}, \mathbb{P}), \quad \mathbb{E}\bigl[\, f(X_{n+1}) \,\|\, \mathcal{F}_n \,\bigr] \leq h(X_n), \quad \forall n \geq 0. \qquad \square$$

Since the distribution of X_n is π_n, we deduce that
$$f(X_n) \in L^1(\Omega, \mathcal{S}, \mathbb{P}) \Longleftrightarrow f \in L^1(\mathcal{X}, \pi_n).$$
The Markov condition implies $\mathbb{E}\bigl[\,f(X_{n+1}) \,\|\, \mathcal{F}_n\,\bigr] = \mathbb{E}\bigl[\,f(X_{n+1}) \,\|\, X_n\,\bigr]$. Let us observe that
$$\mathbb{E}\bigl[\,f(X_{n+1}) \,\|\, \mathcal{F}_n\,\bigr] = \mathbb{E}\bigl[\,f(X_{n+1}) \,\|\, X_n\,\bigr] = Qf(X_n), \quad \forall n \geq 0. \tag{4.3.8}$$
Indeed,
$$\mathbb{E}\bigl[\,f(X_{n+1})\bigm| X_n = x\,\bigr] = \sum_{y \in \mathcal{X}} f(y)\mathbb{P}\bigl[\,X_{n+1} = y \bigm| X_n = x\,\bigr]$$
$$= \sum_{y \in \mathcal{X}} Q_{x,y} f(y) = (Qf)(x).$$

Proposition 4.71. *Let $f \in \mathcal{L}^\infty(\mathcal{X})$. Then the sequence $\bigl(f(X_n)\bigr)_{n \geq 0}$ is a martingale (resp. supermartingale) iff $Qf = f$ (resp. $Qh \leq h$).* □

Definition 4.72. The operator $\Delta := \mathbb{1} - Q : \mathcal{L}^\infty(\mathcal{X}) \to \mathcal{L}^\infty(\mathcal{X})$ is called the *Laplacian*[2] of the HMC. □

Observe that for any $f \in \mathcal{L}^\infty(\mathcal{X})$ we have
$$(\Delta f)(x) = \sum_{y \in \mathcal{X}} Q_{x,y}\bigl(f(x) - f(y)\bigr).$$
Thus $f(X_n)$ martingale iff $\Delta f = 0$, i.e., h is *harmonic* with respect to this Laplacian. This sequence is a supermartingale iff $\Delta f \geq 0$, i.e., f is superharmonic with respect to the Laplacian Δ.

A function $f : \mathcal{X} \to \mathbb{R}$ is said to be *harmonic on a subset* $U \subset \mathcal{X}$ if
$$\Delta f(u) = 0, \ \forall u \in U \Longleftrightarrow f(u) = \sum_{x \in \mathcal{X}} Q_{u,x} f(x), \ \forall u \in U.$$

Example 4.73. Fix a nonempty subset $Y \subset \mathcal{X}$ and $x_0 \in \mathcal{X} \setminus Y$. We denote by H_Y the hitting time of Y, and by H_{x_0} the hitting time x_0. For $x \in \mathcal{X}$ we set
$$f(x) = \mathbb{P}_x\bigl[\,H_{x_0} < H_Y\,\bigr],$$
i.e., $f(x)$ is the probability that the system started at x hits x_0 before it hits E. Note that
$$0 = f(y) \leq f(x) \leq 1 = f(x_0), \ \forall x \in \mathcal{X}, \ \forall y \in Y.$$
Note that for any $x \notin \{x_0\} \cup Y$ we have
$$f(x) = \mathbb{P}_x\bigl[\,H_{x_0} < H_Y\,\bigr] = \sum_{x' \in \mathcal{X}} \underbrace{\mathbb{P}_x\bigl[\,H_{x_0} < H_Y \bigm| X_1 = x'\,\bigr]}_{=f(x')} Q_{x,x'} = Qf(x).$$
Thus f is harmonic on $\mathcal{X} \setminus \bigl(\{x_0\} \cup Y\bigr)$. □

---
[2] Here we are using the geometers' convention. As defined, the Laplacian is nonnegative definite.

Proposition 4.74 (Lévy's martingale). *Suppose that $f \in \mathbb{B}(\mathscr{X})$. For each $n \in \mathbb{N}_0$ we set*

$$M_n^f := f(X_n) - f(X_0) + \sum_{k=0}^{n-1} \Delta f(X_k)$$

$$= f(X_n) - f(X_0) + \sum_{k=0}^{n-1} \big(X_k - \mathbb{E}\big[\, f(X_{k+1}) \,\|\, X_k \,\big] \big).$$

Then the sequence $\big(M_n^f \big)_{n \geq 0}$ is a martingale.

Proof. Note that

$$M_{n+1}^f - M_n^f = f(X_{n+1}) - f(X_n) + \Delta f(X_n)$$

and

$$\mathbb{E}\big[\, M_{n+1}^f - M_n^f \,\|\, \mathcal{F}_n \,\big] = \mathbb{E}\big[\, f(X_{n+1}) \,\|\, \mathcal{F}_n \,\big] - f(X_n) - \Delta f(X_n)$$

$$\stackrel{(4.3.8)}{=} Qf(X_n) - f(X_n) + \Delta f(X_n) = 0.$$

$\square$

Let $\boldsymbol{I} : \mathscr{X} \to \mathbb{R}$ denote the indicator of $\mathscr{X}$, $\boldsymbol{I}(x) = 1$, $\forall x \in \mathscr{X}$. Since Q is a stochastic matrix we have $Q\boldsymbol{I} = \boldsymbol{I}$, so that the constant functions are harmonic.

Theorem 4.75. *Suppose that the HMC $(X_n)_{n \geq 0}$ is irreducible and recurrent. Then any bounded Lyapunov function is constant.*

Proof. We argue by contradiction. Suppose that h is non-constant bounded Lyapunov function on $\mathscr{X}$. There exist $x_0, x_1 \in \mathscr{X}$ such that $h(x_0) \neq h(x_1)$.

Suppose that $\pi_0 = \delta_{x_0}$. The sequence $h(X_n)$ is a bounded supermartingale. The Submartingale Convergence Theorem implies that the sequence $h(X_n)$ converges a.s.

On the other hand, since (X_n) is recurrent we deduce

$$\mathbb{P}\big[\, X_n = x_0 \text{ i.o.} \,\big] = \mathbb{P}\big[\, X_n = x_1 \text{ i.o.} \,\big] = 1.$$

Thus, the sequence $h(X_n)$ has a.s. two different limit points $h(x_0)$ and $h(x_1)$ and thus $h(X_n)$ is a.s. divergent! $\square$

Corollary 4.76. *If the irreducible HMC $(X_n)_{n \geq 0}$ admits a nonconstant, bounded Lyapunov function then it must be transient.* $\square$

Example 4.77. Suppose that $(X_n)_{n \geq 0}$ describes the simple random walk on a locally finite connected graph $G = (V, E)$ with vertex set V and edge set E. A function $f : V \to \mathbb{R}$ is then superharmonic with respect to this Markov chain if its value at each vertex is at least the average of the values at neighbors

$$f(v) \geq \frac{1}{\deg v} \sum_{u \sim v} f(u), \ \forall v \in V,$$

where $u \sim v$ indicates that the vertices u and v are neighbors, i.e., connected by an edge.

Suppose that G is a rooted binary tree. This means that G is a tree, it has a unique vertex v_0 of degree 1, and every other vertex has degree 3. One can think that any vertex other than the root has a unique direct ancestor and two direct successors. The root has a unique successor

One can think of v_0 as the generation zero vertex. It has a unique successor. This is the generation 1 vertex. Its two successors form the second generation of vertices. Their 4 successors determine the third generation etc. Equivalently, a vertex belongs to the n-th generation, $n > 1$, if its predecessor is in the $(n-1)$-th generation. We obtain in this fashion a generation function

$$g : V \to \mathbb{N}_0, \ g(v) := \text{the generation of the vertex } v.$$

Define

$$f : V \to [0,1], \ f(v) = 2^{-g(v)}.$$

Any vertex $v \neq v_0$ has two vertices of generation $g(v)+1$ and one vertex of generation $g(v) - 1$. Hence

$$\sum_{u \sim v} f(u) = 2^{-g(v)+1} + 2 \cdot 2^{-f(v)-1} = 3 \cdot 2^{-g(v)} = 3f(v),$$

so that

$$f(v) = \frac{1}{3} \sum_{u \sim v} f(u), \forall v \in V \setminus \{v_0\}.$$

Obviously $f(v_0) > f(v), \ \forall v \in V \setminus \{v_0\}$. This proves that f is superharmonic, nonconstant and bounded so the random walk on G is transient. $\square$

Definition 4.78. A function $f \in \mathcal{L}^0(\mathscr{X})$ is called *coercive* if, for any $C > 0$, the set $\{f \leq C\}$ is a *finite* subset of $\mathscr{X}$. $\square$

Proposition 4.79. *Let $(X_n)_{n \geq 0}$ be an irreducible HMC with state space $\mathscr{X}$ and transition matrix Q. Suppose that there exists a nonnegative coercive function $f : \mathscr{X} \to [0,\infty)$ and a finite set $A \subset \mathscr{X}$ such that*

$$\sum_{y \in \mathscr{X}} Q_{x,y} f(y) \leq f(x), \ \forall x \in \mathscr{X} \setminus A. \quad (4.3.9)$$

Then $(X_n)_{n \geq 0}$ is recurrent.

Proof. We follow [56, Sec. 2.2]. The condition (4.3.9) is equivalent to
$$\mathbb{E}_x\big[f(X_{n+1}) - f(X_n)\big| X_n = x\big] \leq 0, \quad \forall x \in \mathscr{X} \setminus A.$$
Denote by T_A the time of first return to A. For $x \in \mathscr{X} \setminus A$ we denote by (Y_n^x) the process started at x and stopped upon entry in A, $Y_n := X_{n \wedge T_A}$.

The sequence $F_n^x = f(Y_n^x)$ is a bounded below supermartingale adapted to $\sigma(X_0, \ldots, X_n)$. From the submartingale convergence theorem we deduce that F_n^x converges a.s. to F_∞^x. Moreover, Fatou's Lemma implies
$$\mathbb{E}_x\big[F_\infty^x\big] \leq \mathbb{E}_x\big[F_0\big] = f(x).$$
In particular, $\mathbb{P}_x\big[F_\infty^x = \infty\big] = 0$, $\forall x \in \mathscr{X} \setminus A$. Hence
$$\mathbb{P}_x\big[\lim f(X_{n \wedge T_A}) = \infty\big] = \mathbb{P}_x\big[F_\infty^x = \infty\big] = 0.$$
We can now argue by contradiction. Suppose that the chain (X_n) is transient. Then, with probability 1, the chain X_n will exit any finite set never to return; see Exercise 4.11. Hence, for $x \in \mathscr{X} \setminus A$, with probability 1, the chain X_n exits the finite set $\{f < N\}$, never to return so that
$$\mathbb{P}_x\Big[\lim_{n \to \infty} f(X_n) = \infty\Big] = 1.$$
We deduce
$$\mathbb{P}_x\big[T_A < \infty\big] = 1, \quad \forall x \in \mathscr{X} \setminus A.$$
Indeed, if it does not return to A in finite time, then
$$\mathbb{P}\big[f(X_n) = f(X_{n \wedge T_A}), \ \forall n\big] > 0$$
so that
$$\mathbb{P}_x\big[\lim f(X_{n \wedge T_A}) = \infty\big] > 0.$$
Since (X_n) is transient, with probability 1, it will exit A in finite time, never to return. This is impossible because we have just shown that if outside A it will return to A in finite time. $\square$

Remark 4.80. We want to mention that the condition (4.3.9) is also necessary for recurrence. For a proof we refer to [56, Sec. 2.2]. $\square$

Theorem 4.81 (Foster). *Let $(X_n)_{n \geq 0}$ be an irreducible HMC with state space $\mathscr{X}$ and transition matrix Q. Suppose that there exists a function $f : \mathscr{X} \to [0, \infty)$, a finite set $A \subset \mathscr{X}$ and $\varepsilon > 0$ such that*
$$\sum_{y \in \mathscr{X}} Q_{x,y} f(y) \leq f(x) - \varepsilon, \quad \forall x \in \mathscr{X} \setminus A. \tag{4.3.10a}$$
$$\sum_{y \in \mathscr{X}} Q_{x,y} f(y) < \infty, \quad \forall x \in A. \tag{4.3.10b}$$
Then $(X_n)_{n \geq 0}$ is positively recurrent.

Proof. We follow [56, Sec. 2.2]. Denote by T_A the time of first return to A and set $Y_n := X_{n \wedge T_A}$. Suppose $X_0 = x \in \mathcal{X} \setminus A$. Then (4.3.10a) reads
$$\mathbb{E}_x\big[f(Y_{n+1}) - f(Y_n) \,\|\, Y_n\big] = \mathbb{E}_x\big[f(Y_{n+1}) \,\|\, Y_n\big] - Y_n = \leq -\varepsilon \mathbf{I}_{\{T_A > n\}}.$$
Thus $f(Y_n)$ is a nonnegative supermartingale and
$$\mathbb{E}_x\big[f(Y_{n+1})\big] - \mathbb{E}_x\big[f(Y_n)\big] \leq -\varepsilon \mathbb{P}_x\big[T_A > n\big].$$
Hence
$$\mathbb{E}_x\big[f(Y_{n+1})\big] - f(x) = \mathbb{E}_x\big[f(Y_{n+1})\big] - \mathbb{E}_x\big[f(Y_0)\big] \leq \varepsilon \sum_{k=0}^{n} \mathbb{P}_x\big[T_A > k\big]$$
so that
$$\sum_{k=0}^{n} \mathbb{P}_x\big[T_A > k\big] \leq \frac{1}{\varepsilon} f(x).$$
Letting $n \to \infty$ we deduce
$$\mathbb{E}_x\big[T_A\big] \leq \frac{1}{\varepsilon} f(x), \quad \forall x \in \mathcal{X} \setminus A. \tag{4.3.11}$$
Now let $a \in A$. Then
$$\mathbb{E}_a\big[T_A\big] = \sum_{b \in A} Q_{a,b} + \sum_{x \in \mathcal{X} \setminus A} Q_{a,x}\big(1 + \mathbb{E}_x\big[T_A\big]\big)$$
$$= 1 + \sum_{x \in \mathcal{X} \setminus A} Q_{a,x} \mathbb{E}_x\big[T_A\big] \stackrel{(4.3.11)}{\leq} 1 + \frac{1}{\varepsilon} \sum_{x \in \mathcal{X} \setminus A} Q_{a,x} f(x) \stackrel{(4.3.10b)}{<} \infty.$$
Thus $\mathbb{E}_a\big[T_A\big] < \infty$, $\forall a \in A$ and Proposition 4.57 implies that $(X_n)_{n \geq 0}$ is positively recurrent. $\square$

Remark 4.82. (a) Note that condition (4.3.10a) reads
$$\Delta f(x) \geq \varepsilon, \quad \forall x \in \mathcal{X} \setminus \mathcal{A}.$$
Moreover, condition (4.3.10b) is automatically satisfied Q is locally finite, i.e., on each row there are only finitely many nonzero entries.

(b) If $(X_n)_{n \geq 0}$ positively recurrent, $x_0 \in \mathcal{X}$, then the function $f : \mathcal{X} \to [0, \infty)$
$$f(x) = \begin{cases} \mathbb{E}_x\big[T_{x_0}\big], & x \neq x_0, \\ 0, & x = x_0 \end{cases}$$
satisfies the conditions of Theorem 4.81 with $A = \{x_0\}$, $\varepsilon = 1$. $\square$

Example 4.83. Consider the biased random walk on $\mathbb{N}_0 = \{0, 1, \dots\}$ with transition probabilities
$$Q_{0,1} = 1, \quad Q_{n,n+1} = p_n, \quad Q_{n,n-1} = q_n := 1 - p_n, \quad \forall n \in \mathbb{N}.$$
Above $p_n, q_n > 0$, $\forall n \in \mathbb{N}$, so that the corresponding Markov chain is irreducible. Consider the coercive function
$$f : \mathbb{N}_0 \to [0, \infty), \quad f(n) = n.$$
Then, $\forall n \geq 1$ we have
$$\Delta f(n) = n - \big(p_n(n+1) + q_n(n-1)\big) = q_n - p_n.$$
Thus, if $q_n \geq p_n$, this random walk is recurrent. Moreover if
$$\inf_{n \in \mathbb{N}} (q_n - p_n) > 0$$
then this random walk is positively recurrent. $\square$

4.4 Electric networks

4.4.1 Reversible Markov chains as electric networks

Suppose that $(X_n)_{n\geq 1}$ is an irreducible, reversible, *locally finite* HMC with state space $\mathscr{X}$ and transition matrix Q. We recall that local finiteness means that

$$\forall x \in \mathscr{X}, \quad \#\{\, y \in \mathscr{X};\ Q_{x,y} \neq 0\,\} < \infty.$$

The reversibility means that there exists a function $c : \mathscr{X} \to (0, \infty)$ such that

$$c(y) Q_{y,x} = c(x) Q_{x,y} \quad \forall x, y \in \mathscr{X}. \tag{4.4.1}$$

Note that any positive multiple of c also satisfies (4.4.1). We set

$$c(x, y) := c(x) Q_{x,y}, \quad \forall x, y \in \mathscr{X}.$$

The detail balance condition (4.4.1) shows that $c(x, y) = c(y, x)$ and $c(x, y) \neq 0$ iff $Q_{x,y} > 0$. Note that

$$Q_{x,y} = \frac{c(x,y)}{c(x)}, \quad c(x) = \sum_{y \sim x} c(x, y). \tag{4.4.2}$$

It is convenient to visualize this Markov chain as a random walk on an undirected graph with vertex set $\mathscr{X}$ and *weighted* edges. Two vertices x, y are connected by an edge if and only if $Q_{x,y} > 0$. Since the Markov chain is irreducible, this graph is connected.

We use the notation $x \sim y$ to indicate that the vertices/nodes x, y are connected by an edge. We say that two vertices x, y are neighbors if $x \sim y$. For $x \in \mathscr{X}$ we denote by $N(x)$ the set of neighbors of x. If $y \in N(x)$, then we weigh the connecting edge with the weight $c(x, y) = c(y, x)$.

We will assume $c(x, x) = 0$, $\forall x \in \mathscr{X}$, i.e., the associated graph has no loops.

The Markov chain dynamics has the following equivalent description: if the system is at the state/vertex x it will transition to a neighbor y with a probability proportional to the weight $c(x, y)$. The weights $\{\, c(x)\,\}_{x \in \mathscr{X}}$ define a Q-invariant measure μ_c on $\mathscr{X}$

$$\mu_c[S] = \sum_{s \in S} c(s). \tag{4.4.3}$$

Formally, an electric network is a triplet $(\mathscr{X}, E, c)$, where $(\mathscr{X}, E)$ is a locally finite, connected, unoriented graph and $c : \mathscr{X} \times \mathscr{X} \to [0, \infty)$. The set of vertices $\mathscr{X}$ is assumed to be at most countable. We regard the set of edges E as a symmetric subset of $\mathscr{X} \times \mathscr{X}$, i.e., $(x, y) \in E \iff (y, x) \in E$. We assume there are no loops, i.e., $\forall (x, y) \in E$, $x \neq y$. We will frequently use the notation $x \sim y$ to indicate that $(x, y) \in E$.

The function $c : \mathscr{X} \times \mathscr{X} \to [0, \infty)$ satisfies

- $c(x, y) > 0 \iff (x, y) \in E$.
- $c(x, y) = c(y, x)$, $\forall (x, y) \in E$.

We have seen that a reversible Markov chain determines an electric network.

Conversely, an electric network $(\mathscr{X}, E, c)$ determines a reversible Markov chain with state space $\mathscr{X}$ and transition matrix $Q : \mathscr{X} \times \mathscr{X} \to [0,1]$

$$Q_{x,y} = \frac{c(x,y)}{c(x)}, \quad c(x) = \sum_{y \in N(x)} c(x,y).$$

An *electric network* corresponds to a real electric network in which an edge between two nodes x, y corresponds to a resistor between these two nodes with *resistance*

$$r(x,y) = \frac{1}{c(x,y)}.$$

The quantity $c(x,y)$ is called *conductance*.

4.4.2 Sources, currents and chains

The connection between electric networks and the dynamics associated Markov chain is through the classical physical laws of Kirchhoff and Ohm. As shown in the pioneering work of Nash-Williams [119], this point of view can shed remarkable insight into the behavior of the Markov chains. In the remainder of this section we will highlight some of this fruitful interplay between probability and physics. For more about this we refer to [48; 74; 109; 142] which served our sources of inspiration.

First, a matter of notation. For every pair of elements s, s' of a set S we denote by $\delta_{s,s'}$ the Kronecker symbol

$$\delta_{s,s'} = \begin{cases} 1, & s = s', \\ 0, & s \neq s'. \end{cases}$$

As observed by R. Bott and by H. Weyl, see e.g. [16], the physical laws of electric networks have simple geometric interpretations, best expressed in the language of Hodge theory.

The main objects in Hodge theory are the *chain/cochain complexes*. To define them we need to make some choices.

Consider a locally finite graph $(\mathscr{X}, E)$. An *orientation* of the graph is a subset $E_+ \subset E$ such that for any edge $(x,y) \in E$ either $(x,y) \in E_+$ or $(y,x) \in E_+$, but not both.

One can obtain such an E_+ by assigning orientations (arrows) along the edges. Define E_+ as the collection of *positively oriented* edges. More precisely $(x,y) \in E_+$, if and only if the arrow of the oriented edge goes from x to y.

The vector space of *0-chains*, denoted by C_0, consists of formal sums of the type

$$\boldsymbol{j} := \sum_{x \in \mathscr{X}} j(x)[x], \quad j(x) \in \mathbb{R}, \quad \forall x \in \mathscr{X}.$$

Equivalently, $C_0 = \mathbb{R}^{\mathscr{X}}$. In physics a 0-chain is known as a *source* (of current) and $j(x) = 0$ for all but finitely many x.

The vector space C_1 of 1-chains consists of skew-symmetric functions
$$\boldsymbol{i}: E \to \mathbb{R}, \quad (x,y) \mapsto i(x,y).$$
For any $(x,y) \in E$, define $[x] \oslash [y] : E \to \mathbb{R}$ by setting
$$[x] \oslash [y](\boldsymbol{e}) = \begin{cases} 1, & \boldsymbol{e} = (x,y) \\ -1, & \boldsymbol{e} = (y,x), \\ 0 & \text{otherwise}. \end{cases}$$
If we fix an orientation E_+, we will identify an oriented edge $\boldsymbol{e} = (x,y) \in E_+$ with the current $[x] \oslash [y]$ and we write $\boldsymbol{i_e} := [x] \oslash [y]$.

Once we fix an orientation E_+ we can describe each current as a formal sum of the type
$$\boldsymbol{i} = \sum_{(x,y) \in E_+} i(x,y)[x] \oslash [y] = \frac{1}{2} \sum_{(x,y) \in E} i(x,y)[x] \oslash [y]$$
where E_+ is an orientation of the edges. In physics, 1-chains are known as *currents*.

One should think of $[x] \oslash [y]$ as representing the edge (x,y) oriented from x to y. A current can then visualized as an assignment of arrows and weights on edges with the understanding that we get the same current if we reverse any of the arrows and change its weight to the opposite.

A 0-chain $\boldsymbol{j}$ is called *compactly supported* if $j(x) = 0$ for all but finitely many x. Similarly, a 1-chain $\boldsymbol{i}$ is *compactly supported* if $i(x,y) = 0$ for all but finitely many edges $(x,y) \in E$. For $k = 0, 1$ we denote by C_k^{cpt} the space of compactly supported k-chains. There are *boundary operators*
$$\partial : C_1 \to C_0, \quad \partial : C_0^{\text{cpt}} \to \mathbb{R}$$
defined as follows.

- If $\boldsymbol{i} \in C_1$, then
$$\partial \boldsymbol{i} := \sum_{x \in \mathscr{X}} w(x)[x], \quad w(x) = \sum_{y \in N(x)} i(x,y) = - \sum_{y \in N(x)} i(y,x), \quad \forall x \in \mathscr{X}.$$
In particular, for $x_0, x_1 \in \mathscr{X}$,
$$\partial [x_0] \oslash [x_1] = [x_1] - [x_0].$$

- If $\boldsymbol{j} \in C_0^{\text{cpt}}$, then
$$\partial \boldsymbol{j} = \sum_{x \in \mathscr{X}} j(x) \in \mathbb{R}.$$

Let us observe that for any *compactly supported* current $\boldsymbol{i}$ we have $\partial^2 \boldsymbol{i} = 0$. Indeed
$$\partial(\partial \boldsymbol{i}) = \sum_{x \in \mathscr{X}} \sum_{y \in N(x)} i(x,y) = \sum_{(x,y) \in E} i(x,y) = 0$$
since $i(x,y) + i(y,x) = 0$ whenever $x \sim y$.

Remark 4.84. If $\mathscr{X}$ is infinite, then there could exist 1-chains $\boldsymbol{i}$ such that $\partial \boldsymbol{i} \in C_0^{\text{cpt}}$ yet $\partial^2 \boldsymbol{i} \neq 0$. $\square$

The (finite) paths in the graph are special examples of compactly supported 1-chains. By a path of length n we understand a sequence of neighbors

$$x_0, x_1, \ldots, x_n, \quad x_{k-1} \sim x_k, \quad x_{k-1} \neq x_k, \quad \forall k = 1, \ldots, n.$$

The associated 1-chain $i = i_{x_0, x_1, \ldots, x_n}$ is

$$i_{x_0, x_1, \ldots, x_n} = \sum_{k=1}^n [x_{k-1}] \oslash [x_k].$$

Note that

$$\partial i_{x_0, x_1, \ldots, x_n} = [x_n] - [x_0].$$

The path is closed if $x_0 = x_n$ or, equivalently, $\partial i_{x_0, x_1, \ldots, x_n} = 0$.

4.4.3 Kirkhoff's laws and Hodge theory

The actual sources and currents in real electric network are governed by Kirchhoff's laws. We refer to [8, Chap. 12] for a more detailed description of the physical aspects. Fix an electrical network $(\mathscr{X}, E, c)$.

Kirchhoff's first law states that the source of a (physical) current $i \in C_1$ is the 0-chain $j = -\partial i$. More explicitly, this means that

$$j(x) + \sum_{y \in N(x)} i(x, y) = 0, \quad \forall x \in \mathscr{X}. \tag{4.4.4}$$

This is a purely topological condition in the sense that it is independent of the choice of conductance function.

The physics/geometry enters the scene through the conductance function. More precisely, in physics each current i in an electric network has *finite energy*[3] defined by

$$\mathcal{E}_r[i] := \frac{1}{2} \sum_{(x,y) \in E} r(x,y) i(x,y)^2. \tag{4.4.5}$$

If we fix an orientation E_+ of the edges we obtain the equivalent description

$$\mathcal{E}_r[i] = \sum_{(x,y) \in E_+} r(x,y) i(x,y)^2 = \sum_{(x,y) \in E_+} \frac{i(x,y)^2}{c(x,y)}. \tag{4.4.6}$$

We denote by C_1^∞ the space of finite energy 1-chains. The space C_1^∞ is endowed with a (resistor) inner product

$$\langle i_1, i_2 \rangle_r := \sum_{(x,y) \in E_+} r(x,y) i_1(x,y) i_2(x,y). \tag{4.4.7}$$

Thus, a physical current is an element of C_1^∞.

---

[3]The physical units of the expression in (4.4.6) are indeed the units for energy, Joules.

To formulate Kirkhoff's second law we need to introduce the concept of cochain. The *cochains* are objects dual to chains. The space of 0-cochains (resp. 1-cochains) is the dual vector space of C_0 (resp. C_1)

$$C^0 = C_0^* = \mathrm{Hom}(C_0, \mathbb{R}), \;\; C^1 := C_1^* = \mathrm{Hom}(C_1, \mathbb{R}).$$

One we can think of a 0-cochain as a function $u : \mathscr{X} \to \mathbb{R}$. Physicists call such functions *potentials*. For each $x \in \mathscr{X}$ denote by $\delta^x \in C^0$ the elementary 0-cochain defined by

$$\delta^x([y]) = \delta_{x,y}, \;\; \forall y \in \mathscr{X}.$$

A 0-cochain is then a formal sum

$$u = \sum_{x \in \mathscr{X}} u(x) \delta^x.$$

For each $(x,y) \in E$ denote by $dx \otimes dy : E \to \mathbb{R}$ the elementary 1-cochain defined by

$$dx \otimes dy(x', y') = \begin{cases} 1, & (x', y') = (x, y), \\ -1, & (x', y') = (y, x), \\ 0, & \text{otherwise.} \end{cases}$$

A 1-cochain should be viewed as a formal sum

$$v = \sum_{(x,y) \in E_+} v(x,y) dx \otimes dy = \frac{1}{2} \sum_{(x,y) \in E} v(x,y) dx \otimes dy, \;\; v(x,y) = -v(y,x).$$

More concretely, we identify a 1-cochain with a skew-symmetric function on $v : E \to \mathbb{R}$. In physics such a function is called *voltage* and it is measured in Volts.

We define the "integral" of a 1-cochain v along a path

$$\gamma = x_0, x_1, \ldots, x_n,$$

to be the real number

$$\int_\gamma v := \sum_{k=1}^n v(x_{k-1}, x_k).$$

There exists a *coboundary operator* $d : C^0 \to C^1$ that associates to each function $u : \mathscr{X} \to \mathbb{R}$ its "differential"

$$du = \sum_{(x,y) \in E_+} \big(u(y) - u(x) \big) dx \otimes dy. \tag{4.4.8}$$

A 1-cochain v is called *exact* if it is the differential of a 0-cochain.

The following fact is left to the reader as an exercise.

Lemma 4.85. *A 1-cochain v is exact if and only if its integral along any closed path is 0. Equivalently, this means that the integral along a path depends only on the endpoints of the path.* □

The *energy* of a 1-cochain
$$v = \sum_{(x,y)\in E_+} v(x,y) dx \otimes dy$$
is
$$\mathcal{E}_c[v] := \sum_{(x,y)\in E_+} c(x,y)v(x,y)^2 = \frac{1}{2}\sum_{(x,y)\in E} c(x,y)v(x,y)^2.$$

We denote by C^1_∞ the space of finite energy 1-cochains. It is a Hilbert space with (conductance) inner product
$$\langle v_1, v_2\rangle_c := \sum_{(x,y)\in E_+} c(x,y)v_1(x,y)v_2(x,y). \tag{4.4.9}$$

Hence
$$\mathcal{E}_c[v] = \langle v,v\rangle_c.$$

We have a "resistor" duality map $\mathcal{R}: C_1^\infty \to C^1$, $C_1 \ni i \to \mathcal{R}i = i^*$,
$$\mathcal{R}\left(\sum_{(x,y)\in E_+} i(x,y)[x]\otimes [y]\right) = \sum_{(x,y)\in E_+} r(x,y)i(x,y) dx\otimes dy.$$

Note that since $r(x,y) = \frac{1}{c(x,y)}$ we have
$$\mathcal{E}_c[\mathcal{R}i] = \sum_{(x,y)\in E_+} c(x,y)r(x,y)^2 i(x,y)^2 = \sum_{(x,y)\in E_+} r(x,y)i(x,y)^2 = \mathcal{E}_r[i],$$

so that $\mathcal{R}$ induces a (bijective) isometry of Hilbert space $C_1^\infty \to C_\infty^1$. In fact C_∞^1 can be identified with the topological dual of C_1^∞ with the induced inner product and norm. For this reason we will refer to $\mathcal{R}i$ as the dual of i and, when no confusion is possible, we will write $i^* := \mathcal{R}(i)$.

Ohm's law states that for any current i in an electric network there is a difference of potential/voltage $u(x,y)$ between any two neighbors $x \sim y$ related to $i(x,y)$ via the equality
$$u(x,y) = r(x,y)i(x,y). \tag{4.4.10}$$

In other words, the collection of voltages associated to the current i is the dual 1-cochain $\mathcal{R}i$.

Kirchhoff's second law states that a finite energy currents i generated by a source $-j = \partial i$ has a special property: *the dual 1-chain of voltages is exact*. In other words, there exists a function $u: \mathscr{X} \to \mathbb{R}$ such that $du = -i^* = -\mathcal{R}i$, i.e.,
$$c(x,y)\bigl(u(y) - u(x)\bigr) = \frac{1}{r(x,y)}\bigl(u(y) - u(x)\bigr) = i(x,y), \quad \forall (x,y)\in E_+.$$

Note that
$$\mathcal{E}_c[du] := \langle du, du\rangle_c = \langle i, i\rangle_r = \mathcal{E}_r[i] < \infty. \tag{4.4.11}$$

Definition 4.86. A *Kirkhoff current* is a finite energy current i such that its dual $i^* = \mathcal{R}i$ is exact. A function $u \in C_0$ such that $i^* = -du$ is called a *potential* of the Kirchhoff current. □

Suppose $\boldsymbol{i}$ is a Kirckhoff current and u is a potential of $\boldsymbol{i}$. If the graph is connected, then any other potential of $\boldsymbol{i}$ differs from u by an additive constant. The source $\boldsymbol{j} = -\partial \boldsymbol{i}$ of $\boldsymbol{i}$ can be described explicitly in terms of a potential u of $\boldsymbol{i}$. The equality $\partial \boldsymbol{i} = -\boldsymbol{j}$ reads

$$\sum_{y \in N(x)} c(x,y)\bigl(u(y) - u(x)\bigr) = -j(x), \quad \forall x \in \mathscr{X}.$$

Since $c(x,y) = c(x)Q_{x,y}$ we deduce

$$\sum_{y \in N(x)} Q_{x,y}\bigl(u(y) - u(x)\bigr) = -\frac{1}{c(x)} j(x), \quad \forall x \in \mathscr{X}.$$

Equivalently, this means that

$$\Delta u(x) = \frac{1}{c(x)} j(x), \quad \forall x \in \mathscr{X}, \tag{4.4.12}$$

where $\Delta = \mathbb{1} - Q$ is the Laplacian of the HMC with transition matrix Q; see Definition 4.72.

Denote by C_∞^0 space of finite energy 0-chains, i.e., 0-chains u satisfying

$$\sum_{x \in \mathscr{X}} c(x) u(x)^2.$$

This defines an inner product on C_∞^0

$$\langle u_1, u_2 \rangle_c = \sum_{x \in \mathscr{X}} c(x) u_1(x) u_2(x). \tag{4.4.13}$$

As such, C_∞^0 can be identified with the Hilbert space $L^2(\mathscr{X}, \mu_c)$ where μ_c is the Q-invariant measure on $\mathscr{X}$ determined by the detailed balance equations; see (4.4.3).

We denote by C_{cpt}^0 the spaces of functions $u : \mathscr{X} \to \mathbb{R}$ vanishing outside a finite set. Let us observe that if α_1, α_2 are k-cochains and at least one of them is compactly supported, then we can define $\langle \alpha_1, \alpha_2 \rangle_c$ using the same expressions (4.4.9), (4.4.13) as above.

Proposition 4.87 (Discrete integration by parts). *For any $u \in C^0$ and any $v \in C_{\mathrm{cpt}}^0$ we have*

$$\langle \Delta u, v \rangle_c = \langle du, dv \rangle_c = \langle u, \Delta v \rangle_c. \tag{4.4.14}$$

Proof. We have

$$\langle \Delta u, v \rangle_c = \sum_{x \in \mathscr{X}} c(x) \Delta u(x) v(x)$$

$$= \sum_{x \in \mathscr{X}} \left(\sum_{y \in Y} c(x) Q_{x,y} \bigl(u(x) - u(y)\bigr) \right) v(x) = \sum_{x,y \in \mathscr{X}} c(x,y)\bigl(u(x) - u(y)\bigr) v(x)$$

$$= \sum_{(x,y) \in E_+} c(x,y)\bigl(u(x) - u(y)\bigr) v(x) + \sum_{(y,x) \in E_+} c(x,y)\bigl(u(x) - u(y)\bigr) v(x)$$

(change variables $x \leftrightarrow y$ in the second sum)

$$= \sum_{(x,y)\in E_+} c(x,y)\big(u(x)-u(y)\big)v(x) + \sum_{(x,y)\in E_+} c(x,y)\big(u(y)-u(x)\big)v(y)$$

$$= \sum_{(x,y)\in E_+} c(x,y)\big(u(x)-u(y)\big)\big(v(x)-v(y)\big) = \langle du, dv\rangle_c.$$

The same argument, with the roles of u and v reversed show that

$$\langle du, dv\rangle_c = \langle u, \Delta v\rangle_c.$$

The above expressions are well defined since both dv and Δv are compactly supported because the graph is locally finite. □

Here is a simple consequence. We say that a set $S \subset \mathscr{X}$ is *cofinite* if $\mathscr{X} \setminus S$ is finite.

Corollary 4.88. *Suppose that S is a nonempty cofinite set and $u \in \mathscr{X}$ is a solution of the boundary value problem*

$$\Delta u(x) = 0, \ \ \forall x \in \mathscr{X} \setminus S, \ \ u(s) = 0, \ \ \forall s \in S.$$

Then $u(x) = 0$, $\forall x \in \mathscr{X}$.

Proof. We have $0 = \langle \Delta u, u\rangle_c = \langle du, du\rangle_c$. Hence $du = 0$ and since $\mathscr{X}$ is connected, we deduce that u is constant. Since $S \neq \emptyset$, we deduce that u is identically zero. □

4.4.4 A probabilistic perspective on Kirchoff laws

Denote by $(X_n)_{n\geq 0}$ the random walk on the weighted graph defined by the electric network $(\mathscr{X}, E, c)$.

Let $S \subset \mathscr{X}$ be a nonempty subset. Recall that we denote by H_S, respectively T_S, the hitting and respectively return time to S. Fix a bounded function $\varphi : S \to \mathbb{R}$ and define

$$u = u_\varphi : \mathscr{X} \to \mathbb{R}, \ \ u(x) = \mathbb{E}_x\big[\varphi(X_{H_S})\big].$$

Conditioning on neighbors we deduce u is harmonic on $\mathscr{X} \setminus S$ and $u = \varphi$ on S. Corollary 4.88 shows that if S is cofinite, then u is the unique function on $\mathscr{X}$ that is harmonic on $\mathscr{X} \setminus S$ and equal to φ on S.

Let us investigate a special case of this construction. Consider a cofinite set S_- and $x_+ \in \mathscr{X} \setminus S_-$. Set $S := \{x_+\} \cup S_-$. If $\varphi = I_{\{x_+\}} : S \to \mathbb{R}$, then the computation in Example 4.73 shows that u_φ is

$$u(x) = u_{x_+, S_-}(x) := \mathbb{P}_x\big[H_S = H_{x_+}\big] = \mathbb{P}_x\big[H_{x_+} < H_{S_-}\big]. \tag{4.4.15}$$

Thus $u(x)$ is the probability that the random walk started at x reaches x_+ before S_-. Clearly this function has finite energy since it has compact support. To this function we associate a current i defined by $\mathcal{R}i = -du$. More precisely

$$i = \sum_{(x,y)} c(x,y)\big(u(y)-u(x)\big)[x] \otimes [y],$$

and its source is
$$j = j_{x_+, S_-} : \mathcal{X} \to \mathbb{R}, \quad j(x) \stackrel{(4.4.12)}{=} c(x)\Delta u(x).$$

The current i has compact support contained in the finite set of edges with one end in the finite set $\mathcal{X} \setminus S_-$. Hence $\partial^2 i = 0$ so

$$0 = \sum_{x \in \mathcal{X}} j(x) = \sum_{x \in \mathcal{X}} c(x)\Delta u(x). \qquad (4.4.16)$$

The energy of u is

$$\langle du, du \rangle_c = \langle u, \Delta u \rangle_c = \sum_{x \in \mathcal{X}} c(x) u(x) \Delta u(x)$$
$$= c(x_+) u(x_+) \Delta u(x_+) = u(x_+) j(x_+). \qquad (4.4.17)$$

Now observe that $u(x_+) = 1$ so that

$$\Delta u(x_+) = 1 - \sum_{x \in N(x_+)} Q_{x_+, x} u(x)$$

$$= 1 - \sum_{x \in N(x_+)} Q_{x_+, x} \mathbb{P}_x[H_{S_-} > H_{x_+}] = \mathbb{P}_{x_+}[T_{x_+} > H_{S_-}].$$

Hence

$$\mathcal{E}_c[du] \stackrel{(4.4.17)}{=} c(x_+)\Delta u(x_+) = j_{x_+, S_+}(x_+) \qquad (4.4.18)$$
$$= c(x_+)\mathbb{P}_{x_+}[T_{x_+} > H_{S_-}] =: \kappa(x_+, S_-).$$

The quantity $\kappa(x_+, S_-)$ is called the *effective conductance* from x_+ to S_-. Its inverse is called *effective resistance* between x_+ and S_- and it is denoted by $\mathcal{R}_{\text{eff}}(x_+, S_-)$. Thus

$$\mathcal{R}_{\text{eff}}(x_+, S_-) = \frac{1}{c(x_+)\mathbb{P}_{x_+}[T_{x_+} > H_{S_-}]}.$$

We set

$$\boxed{\bar{u} = \bar{u}_{x_+, S_-} := \frac{1}{\kappa(x_+, S_-)} u_{x_+, S_-} = \frac{1}{\kappa(x_+, S_-)} \mathbb{P}_x[H_{x_+} < H_{S_-}]}.$$

This is the potential of the compactly supported Kirchhoff current $\bar{i}_{x_+, S_-}$ such that

$$\boxed{\mathcal{R} \bar{i}_{x_+, S_-} = d \bar{u}_{x_+, S_-}}, \qquad (4.4.19)$$

with source

$$\bar{j} = \bar{j}_{x_+, S_-} = \frac{1}{\kappa(x_+, S_-)} j_{x_+, S_+} = \frac{c(x)}{c(x_+)\mathbb{P}_{x_+}[T_{x_+} > H_{S_-}]} \Delta u(x), \qquad (4.4.20)$$

where u is defined in (4.4.15), $u(x) = \mathbb{P}_x[H_{x_+} < H_{S_-}]$. Note that

$$\bar{j}_{x_+, S_-}(x_+) = 1.$$

Its energy is
$$E_{x_+,S_-} := \frac{1}{\kappa(x_+,S_-)^2}\mathcal{E}_c\big[\,du_{x_+,S_-}\,\big] = \frac{1}{\kappa(x_+,S_-)} = \bar{u}_{x_+,S_-}(x_+). \qquad (4.4.21)$$

Since
$$u_{x_+,S_-}(x) = \mathbb{P}_x\big[\,H_{S_-} < T_{x_+}\,\big] \leq 1 = u_{x_+,S_-}(x_+),\ \forall x \in \mathscr{X},$$

we deduce
$$0 \leq \bar{u}_{x_+,S_-}(x) \leq \bar{u}_{x_+,S_-}(x_+) = E_{x_-,S_-},\ \forall x \in \mathscr{X}. \qquad (4.4.22)$$

Let us observe that if $\mathscr{X}$ is finite and $S_- = \{x_-\}$, then the equality (4.4.16) shows that
$$0 = \sum_{x \in \mathscr{X}} \bar{j}_{x_+,x_-}(x) = \bar{j}_{x_+}(x_+) + \bar{j}_{x_-}(x_-)$$

and thus
$$\bar{j}_{x_+,x_-}(x) = \begin{cases} \pm 1, & x = x_\pm, \\ 0, & x \neq x_\pm. \end{cases}$$

Definition 4.89. A *flow from x_+ to S_- on the electric network* is a *finite energy current* i such that $\partial i = -\bar{j}_{x_+,S_-}$. The source $\bar{j}_{x_+,S_-}$ defined in (4.4.20) is called the *unit dipole* with source x_+ and sink S_-. □

A flow from x_+ to S_- satisfies the second Kirchhoff law if and only if it has finite energy and i^* is the differential of a function $u : \mathscr{X} \to \mathbb{R}$. We will refer to such flows as *Kirchhoff flows*.

Lemma 4.90. *Suppose that i is a compactly supported current such that $\partial i = 0$. Then for any $u : \mathscr{X} \to \mathbb{R}$ we have $\langle i^*, du \rangle_c = 0$.*

Proof. We have
$$\sum_{y \in N(x)} i(x,y) = 0,\ \forall x \in \mathscr{X}.$$

We recall that $i(x,y) = -i(y,x), \forall (x,y) \in E$. We have
$$\langle i^*, du \rangle_c = \sum_{(x,y) \in E_+} r(x,y)c(x,y)i(x,y)\big(u(y) - u(x)\big)$$
$$= \sum_{(x,y) \in E_+} i(x,y)\big(u(y) - u(x)\big) = 2\sum_{(x,y) \in E} i(x,y)\big(u(y) - u(x)\big)$$
$$= -2\sum_{x \in \mathscr{X}} u(x)\left(\sum_{y \in N(x)} i(x,y)\right) + 2\sum_{y \in \mathscr{X}} u(y)\left(\sum_{x \in N(y)} i(x,y)\right) = 0.$$

All the above sums involve only finitely many terms since i is compactly supported. □

Theorem 4.91. *Suppose S_- is cofinite and $x_+ \in \mathscr{X} \setminus S_-$. Then the following hold.*

(i) *The current $i_0 := \bar{i}_{x_+,S_-}$ defined by (4.4.19) is the unique compactly supported Kirchhoff current with source the dipole $j_0 = \bar{j}_{x_+,S_-}$. In particular it is a Kirchhoff flow from x_+ to S_-.*

(ii) *The voltage function $\bar{u} = \bar{u}_{x_+,S_-}$ that determines i_0 is the unique solution of the boundary value problem*

$$\Delta v(x) = 0, \quad \forall x \in \mathscr{X} \setminus (S_- \cup \{x_+\})$$
$$v(x) = \begin{cases} \frac{1}{c(x_+)}, & x = x_+, \\ 0, & x \in S_-, \end{cases} \quad (4.4.23)$$

(iii) *The energy of i_0 is*

$$\mathcal{E}[i_0] = \boldsymbol{E}_{x_+,S_-} = \bar{u}_{x_+,S_-}(x_+) = \frac{1}{c(x_+)\mathbb{P}_{x_+}[T_{x_+} > H_{S_-}]} = \mathscr{R}_{\mathrm{eff}}(x_+, S_-).$$

(iv) *If i_1 is another compactly supported flow from x_+ to S_-, then*

$$\mathcal{E}[i_1] \geq \mathcal{E}[\bar{i}_{x_+,S_-}].$$

Proof. (i) Set $u_0 := \bar{u}_{x_+,S_-}$. Recall that u_0 has compact support. Suppose that i_1 is another compactly supported Kirchhoff flow from x_+ to S_-. Then there exists a function $u_1 : \mathscr{X} \to \mathbb{R}$ such that $i_1^* = du_1$. We deduce from (4.4.12) that the functions u_k, $k = 0, 1$ are solutions of the same equation

$$\Delta u_k(x) = \frac{1}{c(x)} j_0(x), \quad \forall x \in \mathscr{X}.$$

If we write $u = u_1 - u_0$, then $\Delta u = 0$ on $\mathscr{X}$. The function u may not have compact support, but du does. We have

$$\langle du, du \rangle_c = \frac{1}{2} \sum_{(x,y) \in E} c(x,y)(u(x) - u(y))(u(x) - u(y))$$

$$= \frac{1}{2} \sum_{(x,y) \in E} c(x,y)(u(x) - u(y))u(x) - \frac{1}{2} \sum_{(x,y) \in E} c(x,y)(u(x) - u(y))u(y)$$

$$= \frac{1}{2} \sum_{x \in \mathscr{X}} u(x) \underbrace{\sum_{y \in N(x)} c(x,y)(u(x) - u(y))}_{=c(x)\Delta u(x) = 0} + \frac{1}{2} \sum_{y \in \mathscr{X}} u(y) \underbrace{\sum_{x \in N(y)} c(y,x)(u(y) - u(x))}_{=c(y)\Delta u(y) = 0}$$

$$= 0.$$

Hence $du = 0$ so that $i_0 = i_1$.

(ii) If v_1, v_2 are two compactly supported solutions of (4.4.23), then the argument above shows that $\langle dv, dv \rangle_c = 0$ and, since v is compactly supported, we deduce that $v = 0$.

The equality (iii) follows from (4.4.21).

(iv) Set $i = i_1 - i_0$. Then
$$\mathcal{E}[i_1] = \mathcal{E}[i + i_1] = \langle i_0^*, i_0^2 \rangle_c + 2\langle i_0^*, i^* \rangle_c + \underbrace{\langle i^*, i^* \rangle_c}_{\geq 0}$$

$(i_0^* = du)$

$$\geq \mathcal{E}[i_0] + 2\langle du, i^* \rangle_c.$$

Lemma 4.90 shows that $\langle du, i^* \rangle_c = 0$ since i has compact support and $\partial i = 0$. $\square$

Remark 4.92. (a) Part (iv) of the theorem is known as the *Thompson* or *Dirichlet Principle*. It classically states that the Kirchoff flow is *the* least energy compactly supported flow sourced by the dipole $\bar{j}_{x_+, S_-}$. Observe that the energy of the Kirchhoff flow carries information about the dynamics of the Markov chain associated to the electric network.

(b) The Kirchhoff flow from x_+ to S_- is the unique compactly supported current i such that

- $\partial i(x_+) = -1$.
- There exists a function $u : \mathscr{X} \to \mathbb{R}$, identically zero on S_-, such that $i^* = du$.

$\square$

4.4.5 Degenerations

To proceed further we perform a reduction to a finite network. We set
$$S_+ := \mathscr{X} \setminus S_-, \quad \partial S_+ := \{ s_- \in S_-; \ N(s_-) \cap S_+ \neq \emptyset \}.$$

For simplicity we assume that x_+ does not have any neighbor in S_-. We obtain a new finite electric network $\mathscr{X}/S_-$ described as follows.

- Its vertex set is $S_+ \cup \{x_-\}$. Think that we have identified all the vertices in S_- with a single point x_-.
- The conductances $c_*(x, y)$ of $\mathscr{X}/S_-$ are defined according to the rule

$$c_*(x, y) = \begin{cases} c(x, y), & x, y \in S_+, \\ \sum_{s_+ \in \partial S_+} c(x, s_+), & y = x_- \\ \sum_{s_+ \in \partial S_+} c(s_+, y), & x = x_-. \end{cases}$$

Note that $c_*(x) = c(x)$, $\forall x \in S_+$. We denote by Δ_* the Laplacian determined by these conductances. The function $\bar{u} = \bar{u}_{x_0, S_-}$ is identically zero on S_- and thus descends to a function u_* on $\mathscr{X}/S_-$ such that $u_*(x_-) = 0$. The set S_+ is also a subset of $\mathscr{X}/S_-$.

$$\Delta_* u = \Delta \bar{u} \text{ on } S_+.$$

Moreover
$$c_*(x_\pm)\Delta_* u_*(x_+) = \pm 1.$$
Thus u_* is the potential of the Kirchoff flow on $\mathscr{X}/S_-$ from x_+ to x_-. We denote by $\boldsymbol{E}_{x_+,x_-}$ its energy.

Note that the induced Kirchoff flow on $\mathscr{X}/S_-$ has the same energy as the original Kirchhoff flow on $\mathscr{X}$, i.e.,
$$\boldsymbol{E}_{x_+,x_-} = \boldsymbol{E}_{x_+,S_-}. \qquad (4.4.24)$$
On the finite graph $\mathscr{X}/S_-$ the flows from x_+ to x_- can be thought of as paths from x_+ and x_-. They all have finite energy. The Kirchhoff flow is the path with minimal energy from x_+ to x_-.

In view of this reduction to finite graphs we concentrate on finite electric networks. Suppose $(\mathscr{X}, E, c)$ is such a network and $x_+, x_- \in \mathscr{X}$, $x_+ \neq x_-$. For finite graphs the finite energy condition is automatically satisfied and a flow from x_+ to x_- is simply a 1-chain $\boldsymbol{i}$ such that
$$\partial \boldsymbol{i} = [x_-] - [x_+].$$
The source $[x_+] - [x_-]$ is called a *dipole* with source x_+ and sink x_-.

☞ *The flow condition involves only the topology of graph and is independent of the physics/geometry of the network encoded by the conductance function. However, the Kirchhoff flow depends on the physics/geometry of the network.*

Denote by $\boldsymbol{i} = \boldsymbol{i}_{x_+,x_-}$ the Kirchhoff flow with source x_+ and sink x_-. Its *potential grounded at* x_- is the function $u : \mathscr{X} \to \mathbb{R}$ uniquely determined by the equations
$$\Delta u = 0 \in \mathscr{X} \setminus \{x_+, x_-\}, \quad u(x_-) = 0, \quad c(x_+)\Delta u(x_+) = 1. \qquad (4.4.25)$$
Then $\boldsymbol{i}_{x_+,x_-} = \mathcal{R}^{-1}du_{x_+,x_-}$. The energy of this flow is
$$\boldsymbol{E}_{x_+,x_-} = u_{x_+,x_-}(x_+) = \frac{1}{c(x_+)\mathbb{P}_{x_+}[T_{x_+} > T_{x_-}]}. \qquad (4.4.26)$$
This quantity is an invariant of the quadruplet $(\mathscr{X}, c, x_+, x_-)$.

Clearly if we vary the conductance function the energy changes, and a flow that is minimal for a choice of conductance may fail to be so for another choice. In particular, a flow that has minimal energy with respect to a conductance function may not have this property if we change the conductance or, equivalently, the resistance function $r(x,y) = \frac{1}{c(x,y)} \in (0, \infty]$. We will indicate the dependence of $\boldsymbol{E}_{x_+,x_-}$ on r using the notation $\boldsymbol{E}_{x_+,x'}(r)$.

Suppose we change the conductance function to a new function c' that is bigger or, equivalently, such that $r'(x,y) \leq r(x,y)$. Then for any current $\boldsymbol{i}$ we have
$$\mathcal{E}_r[\boldsymbol{i}] = \frac{1}{2}\sum_{(x,y)\in E} r(x,y)\boldsymbol{i}(x,y)^2 \geq \frac{1}{2}\sum_{(x,y)\in E} r'(x,y)\boldsymbol{i}(x,y)^2 = \mathcal{E}_{r'}[\boldsymbol{i}].$$
This implies the following result known as the *Raleigh Principle*.

Theorem 4.93 (Raleigh). *The energy of the Kirchoff flow with given source and sink increases with the increase of the resistance function or, equivalently, if the conductance function decreases.*

Proof. Suppose that we decrease the resistance of an edge from $r(x,y)$ to $r'(x,y)$. Denote by $\boldsymbol{i}(r)$ the Kirchoff flow with source x_+, sink x_- and choice of resistance r. Define $\boldsymbol{i}(r')$ in a similar fashion.

We have
$$\boldsymbol{E}_{x_+,x_-}(r) = \mathcal{E}_r\big[\boldsymbol{i}(r)\big] \geq \mathcal{E}_{r'}\big[\boldsymbol{i}(r)\big] \geq \mathcal{E}_{r'}\big[\boldsymbol{i}(r')\big] = \boldsymbol{E}_{x_+,x_-}(r').$$
□

We can use this principle to produce estimates for $\boldsymbol{E}_{x_+,x_-}(r)$ in terms $\boldsymbol{E}_{x_+,x_-}(r')$ if r' is chosen wisely making $\boldsymbol{E}_{x_+,x_-}(r')$ easier to compute. One way to simplify the computation of $\boldsymbol{E}_{x_+,x_-}$ is to modify the topology of the graph. We can achieve this by pushing r to extreme values. Let describe two such degenerations.

Suppose $y_0, y_1 \in \mathscr{X} \setminus \{x_+, x_-\}$ are two nodes connected by an edge. Upon rescaling c we can assume that $c(y_0, y_1) = 1 = r(y_0, y_1)$. We have a family of deformed resistances
$$r_t : E \to (0,\infty), \ t > 0, \ r_t(x,x') = \begin{cases} t, & (x,x') = (y_0,y_1) \text{ or } (y_1,y_0), \\ r(x,x'), & \text{otherwise}. \end{cases}$$

We denote by $\boldsymbol{i}^t$ the Kirchhoff flow with source x_+, sink x_- and resistances r_t, by $\boldsymbol{E}^t$ its energy $\boldsymbol{E}^t = \boldsymbol{E}_{x_+,x_-}(r_t)$ and by u^t its potential grounded at x_- defined by (4.4.25).

The Raleigh Principle shows that $\boldsymbol{E}^t$ is an increasing function of t. We want to describe what happens with $\boldsymbol{E}^t$ and u^t as $t \to 0, \infty$.

Cutting. The behavior as $t \to \infty$ is described by the electric network $(\mathscr{X}^\infty, c^\infty, E^\infty)$ obtained by *cutting* the edge (y_0, y_1). More precisely
$$\mathscr{X}^\infty = \mathscr{X}, \ E^\infty = E \setminus \{(y_0,y_1),(y_1,y_0)\},$$

$$c^\infty(x,x') = \lim_{t\to\infty} c^t(x,x') = \begin{cases} 0, & (x,x') = (y_0,y_1) \text{ or } (y_1,y_0), \\ c(x,x'), & \text{otherwise}. \end{cases}$$

Shorting. The behavior as $t \to 0$ is described by the network $(\mathscr{X}^0, E^0, c^0)$ obtained by *shorting* the edge (y_0, y_1). Intuitively, the shorted network is obtained by collapsing the vertices y_0, y_1 to single point $*$; see Figure 4.4. More precisely

- $\mathscr{X}^0 = \big(\mathscr{X} \setminus \{y_0, y_1\}\big) \cup \{*\}$.
- If $x, x' \in \mathscr{X} \setminus \{y_0, y_1\}$, then $c^0(x,x') = c(x,x')$ so that $(x,x') \in E \Longleftrightarrow (x,x') \in E^0$.
- $x \in \mathscr{X} \setminus \{y_0, y_1\}$, then $c(x,*) = c(x,y_0) + c(x,y_1)$ so that $(x,*) \in E^0$ if and only if $(x,y_0) \in E$ or $(x,y_1) \in E$.

Note that we have a natural projection $p: \mathscr{X} \to \mathscr{X}^0$

$$p(x) = \begin{cases} x, & x \neq y_0, y_1, \\ *, & x = y_0, y_1. \end{cases}$$

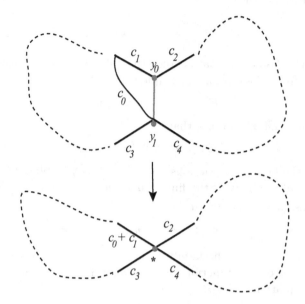

Fig. 4.4 Shorting an electric network along the edge (y_0, y_1).

For $\epsilon \in \{0, \infty\}$ by u^0 (resp. u^∞) the potential grounded at x_- of the Kirchhoff flow in $(\mathscr{X}^\epsilon, E^\epsilon, c^\epsilon)$ with source x_+ and sink x_-. Denote by U^ϵ the energy of u^ϵ

$$E^\epsilon = \frac{1}{2} \sum_{(x,y) \in E^\epsilon} c^\epsilon(x,y) \big(u^\epsilon(x) - u^\epsilon(u)\big)^2.$$

Theorem 4.94 (Maxwell-Raleigh). *Suppose that $y_0, y_1 \in \mathscr{X} \setminus \{x_+, x_-\}$ have the property that the removal of the edge connecting them does not disconnect the graph $(\mathscr{X}, E)$. Then*

$$\lim_{t \to \infty} u^t(x) = u^\infty(x), \quad \forall x \in \mathscr{X},$$

$$\lim_{t \to 0} u^t(x) = u^0\big(p(x)\big), \quad \forall x \in \mathscr{X},$$

and

$$\lim_{t \to \epsilon} \boldsymbol{E}^t = \boldsymbol{E}^\epsilon, \quad \epsilon = 0, \infty.$$

In particular, $\boldsymbol{E}^0 \leq \boldsymbol{E}^t \leq \boldsymbol{E}^\infty$, $\forall t > 0$. Thus the energy of the Kirchhoff flow from x_+ to x_- is increased by cutting and decreased by shorting.

Proof. We will carry the proof in several steps. We set $r = r^1$.

1. Compactness. Fix a path γ in $\mathscr{X}$ from x_+ to x_- that avoids the edge (y_0, y_1),
$$\gamma = x_-, x_0, x_1, \ldots, x_n = x_-.$$
The r^t-energy of this path is
$$\mathcal{E}_{r^t}[\gamma] = \sum_{k=1}^{n} r(x_{k-1}, x_k) = \mathcal{E}_r[\gamma].$$
It is independent of t since the path avoids the only edge whose resistance depends on t. We deduce from Thompson's principle that
$$\boldsymbol{E}^t \leq \mathcal{E}_r[\gamma], \quad \forall t > 0.$$
The local estimate (4.4.22) implies that
$$0 \leq u^t(x) \leq \boldsymbol{E}^t \leq \mathcal{E}_r[\gamma], \quad \forall t > 0.$$
This shows that the family of functions $u^t : \mathscr{X} \to [0, \infty)$ is relatively compact with respect to the usual topology of the finite dimensional vector space $\mathbb{R}^{\mathscr{X}}$.

2. $t \to \infty$. In this case observe that
$$\lim_{t \to \infty} c^t(x, y) = c^\infty(x, y), \quad \forall x, y \in \mathscr{X}.$$
We will show that as $t \to \infty$ the family u^t has only one limit point. Suppose that for a sequence $t_n \to \infty$ the functions u^{t_n} converge to a function v. The function u^{t_n} satisfies the equation
$$\sum_{y \in \mathscr{X}} c^{t_n}(x, y) \big(u^{t_n}(x) - u^{t_n}(y) \big) = \begin{cases} 0, & x \neq x_\pm, \\ \pm 1, & x = x_\pm, \end{cases} \quad u^{t_n}(x_-) = 0.$$
Letting $n \to \infty$ we deduce that v satisfies
$$\sum_{y \in \mathscr{X}} c^\infty(x, y) \big(v(x) - v(y) \big) = \begin{cases} 0, & x \neq x_\pm, \\ \pm 1, & x = x_\pm, \end{cases} \quad v(x_-) = 0.$$
According to Theorem 4.91(ii) the above equation has a unique solution, the potential u^∞ of the Kirchhoff flow from x_+ to x_- grounded at x_- in $(\mathscr{X}^\infty, c^\infty)$ proving that
$$\lim_{t \to \infty} u^t = u^\infty.$$
The equality
$$\lim_{t \to \infty} \boldsymbol{E}^t = \boldsymbol{E}^\infty$$
is obvious.

3. $t \to 0$. The above argument fails in this case because $c^t(y_0, y_1) = \frac{1}{t}$. Pick a sequence $t_n \nearrow 0$ such that u^{t_n} has a limit u^0 as $t_n \to 0$. To simplify the presentation we will write u^t instead of u^{t_n}. We will show that
$$u^0(y_0) = u^0(y_1) \qquad (4.4.27)$$

and the induced function $\bar{u}^0$ on $\mathscr{X}^0$,

$$\bar{u}^0(x) = \begin{cases} u^0(x), & x \neq *, \\ u^0(y_0) = u^0(y_1), & x = *, \end{cases}$$

satisfies

$$\bar{u}^0(x_-) = 0, \tag{4.4.28a}$$

$$\sum_{y \in N^0(x)} c^0(x,y)\big(\bar{u}^0(x) - \bar{u}^0(y)\big) = \begin{cases} 0, & x \in \mathscr{X}^0 \setminus \{x_+, x_-\} \\ 1, & x = x_+. \end{cases} \tag{4.4.28b}$$

We set

$$N_*(y_0) := N(y_0) \setminus \{y_1\}, \quad N_*(y_1) := N(y_1) \setminus \{y_0\},$$

$$c_*(y_i) := \sum_{y \in N_*(y_i)} c(y_0, y), \quad i = 0, 1.$$

Denote by $N^0(*)$ the set of neighbors of $*$ in the graph $(\mathscr{X}^0, E^0)$. Note that

$$N^0(*) = N_*(y_0) \cup N_*(y_1), \quad c^0(*) = c_*(y_0) + c_*(y_1).$$

Since $\Delta_{c^t} u^t(y_0) = 0$, $i = 0, 1$ we deduce

$$\frac{1}{t}\big(u^t(y_0) - u^t(y_1)\big) + \sum_{y \in N_*(y_0)} c(y_0, y)\big(u^t(y_0) - u^t(y)\big)$$

so that

$$\big(1 + t c_*(y_0)\big) u^t(y_0) - u^t(y_1) = t \sum_{y \in N_*(y_0)} c(y_0, y) u^t(y).$$

A similar computation shows that

$$-u^t(y_0) + \big(1 + t c_*(y_1)\big) u^t(y_1) = t \sum_{y \in N_*(y_1)} c(y_1, y) u^t(y).$$

Thus $\big(u^t(y_0), u^t(y_1)\big)$ is the solution of the 2×2 non-homogeneous linear system

$$\underbrace{\begin{bmatrix} a_0(t) & -1 \\ -1 & a_1(t) \end{bmatrix}}_{=:A(t)} \cdot \begin{bmatrix} u^t(y_0) \\ u^t(y_1) \end{bmatrix} = t \cdot \underbrace{\begin{bmatrix} c_0(t) \\ c_1(t) \end{bmatrix}}_{\alpha(t)},$$

where

$$a_i(t) = 1 + t c_*(y_i), \quad c_i(t) = \sum_{y \in N_*(y_i)} c(y_i, y) u^t(y), \quad i = 0, 1.$$

Note that

$$\det A(t) = a_0(t) a_1(t) - 1 = t\big(c_*(y_0) + c_*(y_1)\big) + O(t^2) = t c^0(*) + O(t^2).$$

Set
$$A_0(t) = \begin{bmatrix} c_0(t) & -1 \\ c_1(t) & a_1(t) \end{bmatrix}, \quad A_1(t) = \begin{bmatrix} a_0(t) & c_0(t) \\ -1 & c_1(t) \end{bmatrix}.$$

Using Cramer's rule we deduce
$$u^t(y_0) = \frac{t \det A_0(t)}{\det A(t)} = \frac{a_1(t)c_0(t) + c_1(t)}{c^0(*) + O(t)}$$

$$u^t(y_1) = \frac{a_0(t)c_1(t) + c_0(t)}{c^0(*) + O(t)}.$$

Now observe that
$$\lim_{t \to 0} a_i(t) = 1$$
and, since $N^0(*) = N_*(y_0) \cup N_*(y_1)$
$$\lim_{t \to 0} \big(c_0(t) + c_1(t) \big) = \sum_{y \in N^0(*)} c^0(*, y) u^0(y).$$

Hence
$$u^0(y_0) = u^0(y_1) = \bar{u}^0(*) := \frac{\sum_{y \in N^0(*)} c^0(*, y) u^0(y)}{c(*)}.$$

This proves (4.4.27). The equality (4.4.28a) is obvious. Observe that
$$\bar{u}^0(*) \sum_{y \in N^0(*)} c(*, y) = \sum_{y \in N^0(*)} c^0(*, y) \bar{u}^0(y),$$
i.e.,
$$\sum_{y \in N^0(*)} c^0(*, y) \big(\bar{u}^0(*) - \bar{u}^0(y) \big) = 0.$$

This proves (4.4.28b) for $x = *$.

If $x \in \mathscr{X} \setminus \{*, x_-\}$, then
$$\sum_{y \in N(x)} c^t(x, y) \big(u^t(x) - u^t(y) \big) = \begin{cases} 1, & x = x_+, \\ 0, & x \neq x_+. \end{cases}$$

The equality (4.4.28b) for $x \neq *, x_-$ follows by letting $t \to 0$ above and observing that
$$\lim_{t \to 0} u^0(y_i) = \bar{u}^0(*), \ i = 0, 1, \ \lim_{t \to 0} \big(c^t(x, y_0) + c^t(x, y_1) \big) = c^0(x, *)$$
and
$$N^0(x, *) = \big(N(x) \setminus \{y_0, y_1\} \big) \cup \{*\}.$$

This proves the equality (4.4.28b). This determines $\bar{u}^0$ uniquely and shows that
$$\lim_{t \to 0} u^t = \bar{u}^0 \big(p(x) \big).$$

It remains to verify only the claim
$$\lim_{t\to 0} \boldsymbol{E}^t = \boldsymbol{E}^0.$$

Note that
$$\boldsymbol{E}^t = \frac{1}{2} \sum_{(x,y)\in E} c^t(x,y)\big(u^t(x) - u^t(y)\big)^2.$$

There are two problematic terms in the above sum corresponding to $(x,y) = (y_0, y_1)$ or (y_1, y_0) and their contribution to the energy is
$$\frac{1}{t}\big(u^t(y_0) - u^t(y_1)\big)^2.$$

Now observe that
$$u^t(y_0) - u^t(y_1) = \frac{c_0(t)\big(a_1(t) - 1\big) - c_1(t)\big(a_0(t) - 1\big)}{c^0(*) + O(t)} = t \frac{c_*(y_1)c_0(t) - c_*(y_0)c_1(t)}{c^0(*) + O(t)}.$$

Hence
$$\frac{1}{t}\big(u^t(y_0) - u^t(y_1)\big)^2 = O(t) \text{ as } t \to 0,$$
so
$$\lim_{t\to 0} \boldsymbol{E}^t = \frac{1}{2} \lim_{\to 0} \sum_{(x,y)\in E\setminus\{(y_0,y_1),(y_1,y_0)\}} c^t(x,y)\big(u^t(x) - u^t(y)\big)^2$$
$$= \frac{1}{2} \sum_{(x,y)\in E^0} c^0(x,y)\big(u^0(x) - u^0(y)\big)^2 = \boldsymbol{E}^0.$$

□

Remark 4.95. (i) Let us explain what happens if the edge (y_0, y_1) disconnects the graph but x_+, x_- lie in the same connected component of the resulting graph. Denote by $(\mathscr{X}_0, E_0)$ the connected component containing x_+, x_- and by $(\mathscr{X}_*, E_*)$ in the other component. The compactness part of the argument still works since the energy of u^t is bounded by the energy of a path in $(\mathscr{X}_0, E_0)$ connecting x_+ to x_+.

Denote by u_0^t the restriction of u to $\mathscr{X}_0$ and by u_*^t its restriction of u to $\mathscr{X}^*$. Then
$$\mathcal{E}[du^t] = \underbrace{\frac{1}{2} \sum_{(x,y)\in E_0} c(x,y)\big(u^t(x) - u^t(y)\big)^2 + t\big(u^t(y_0) - u^t(y_1)\big)^2}_{=:\mathcal{E}_0^t}$$
$$+ \underbrace{\frac{1}{2} \sum_{(x,y)\in E_*} c(x,y)\big(u^t(x) - u^t(y)\big)^2}_{=:\mathcal{E}_*^t}.$$

Note that
$$\lim_{t \to 0} t\big(u^t(y_0) - u^t(y_1)\big)^2.$$

Arguing exactly as in Step 2 of the proof of Theorem 4.94 one can show that u_0^t converges to u_{x_+,x_-}^0 the potential grounded at 0 of the Kirchhoff flow in $\mathscr{X}^0$ from x_+ to x_-. If u_* is any limit point of u_*^t then u_* satisfies $\Delta_* u_* = 0$ so
$$\langle du_*, du_* \rangle_* = \langle \Delta_* u_*, u_* \rangle_* = 0$$
so 0 is the only limit point of $\mathcal{E}_*^t$ as $t \to 0$.
$$\mathcal{E}_c\big[du\big]_c \leq \lim_{t \to 0} \mathcal{E}_{c^t}\big[du^t\big] = \mathcal{E}_{c^0}\big[du^0\big]$$
so the energy of the Kirchhoff flow from x_+, x_- in $\mathscr{X}$ is not greater than the energy of the similar flow in $\mathscr{X}_0$.

(ii) To understand why shorting is tricky recall that $\mathscr{X}$ is finite so the Markov chain defined by the conductance c_t has an invariant probability measure is given by
$$\pi_t(x) = \pi(x) = \frac{c_t(x)}{Z_t}, \quad Z_t = \sum_{x \in \mathscr{X}} c_t(x).$$

If we let $t = c_t(y_0, y_1) \to \infty$ and leave the other conductances unchanged, then
$$\pi_t(x) \to 0, \ \forall x \neq y_0, y_1, \ \pi_t(x) \to \frac{1}{2}, \ x = y_0, y_1.$$

(iii) In view of the conservation of energy equality (4.4.24), the cutting and shorting procedures can be used in infinite graphs to estimate the energy $\boldsymbol{E}_{x_+,S_-}$ by reducing, them to cutting/shorting procedure on the collapsed graph X/S_-. Cutting has to be performed with care so that while cutting edges we do not disconnect x_+ from S_-. □

4.4.6 *Applications*

We want to illustrate the usefulness of the above results on some concrete example.

When the graph $(\mathscr{X}, E)$ is finite and all the edges have the same conductances, the Kirchhoff flow from x_+ to x_- can be described explicitly in terms certain counts of spanning trees, [74, Thm. 1.16]. In particular, its energy $\mathcal{K}(x_+, x_-)$ is a topological invariant of the quadruplet $(\mathscr{X}, E, x_+, x_-)$ described explicitly in terms of spanning trees.

If we now assign conductances c to the edges, the energy $\boldsymbol{E}_{x_+,x_-}(c)$ of the Kirchhoff flow from x_-, x_+ satisfies
$$\frac{1}{\sup c(x,y)} \mathcal{K}(x_+, x_-) \leq \boldsymbol{E}_{x_+,x_-}(c) \leq \frac{1}{\inf c(x,y)} \mathcal{K}(x_+, x_-).$$

The computation of $\mathcal{K}(x_+, x_-)$ is impractical for complicated graphs, but the above rather rough estimate expresses in a simple fashion the fact that $\boldsymbol{E}_{x_+,x_-}(c)$ depends on both the topology and the geometry of the electrical network.

Example 4.96. Suppose that $(\mathscr{X}, E, c)$ is a finite electric network such that the underlying graph is a tree. Then for any pair of points x_+, x_- there exists a unique 1-chain $\boldsymbol{i}$ such that
$$\partial \boldsymbol{i} = [x_-] - [x_+].$$
It is described by a minimal path
$$x_+ = x_0, x_1, \ldots, x_n = x_-.$$
This is the Kirchhoff flow from x_+ to x_- and its energy is
$$\boldsymbol{E}_{x_+, x_-} = \sum_{i=1}^{n} r(x_{i-1}, x_i) = \sum_{i=1}^{n} \frac{1}{c(x_{i-1}, x_i)}.$$
As a special case of this consider the Ehrenfest urn model. Recall that the state space is the set $\mathscr{X} := \{0, 1, \ldots, B\}$, $B \in \mathbb{N}$ and transition matrix Q given by
$$Q_{k,k-1} = \frac{k}{B}, \ \forall k \geq 1, \quad Q_{j,j+1} = \frac{B-j}{B}, \ \forall j < B.$$
As explained in Example 4.47, this can be described as an electric network whose underlining graph is a path
$$0 \to 1 \to \cdots \to B,$$
and conductances
$$c(j, j+1) = \binom{B}{j} \frac{B-j}{B} = \binom{B-1}{j}.$$
In particular,
$$c(j) = \binom{B-1}{j} + \binom{B-1}{j-1} = \binom{B}{j}.$$
If B is even, $B = 2N$, then
$$\boldsymbol{E}_{0,N} = \boldsymbol{E}_{N,0} = \sum_{j=0}^{N=-1} \frac{1}{\binom{2N-1}{j}}.$$
Thus
$$\mathbb{P}_N[T_N > T_0] = \frac{1}{c(N)\boldsymbol{E}_{N,0}}, \quad \mathbb{P}_0[T_0 > T_N] = \frac{1}{c(0)\boldsymbol{E}_{N,0}}.$$
Hence
$$\frac{\mathbb{P}_0[T_0 > T_N]}{\mathbb{P}_N[T_N > T_0]} = \frac{c(N)}{c(0)} = \binom{2N}{N} \sim \frac{4^N}{\sqrt{\pi N}}.$$
In particular, this shows that $\mathbb{P}_N[T_N > T_0]$ is extremely small for large N. Thus if initially in the two chambers there equal numbers of balls, the probability that during the random transfers of balls between them, one of the chambers will continuously have less than half the balls until it empties, is extremely small. In fact, the expected time of emptying the left chamber while starting with equal numbers of balls in both is (see [87, Sec. VII.3, p. 175] with $s = 2N$)
$$\mathbb{E}_N[T_0] \sim 4^N(1 + A/N) \text{ as } N \to \infty, \ 1 \leq A \leq 2. \tag{4.4.29}$$
This example is historically important because it was used to explain an apparent contradiction between Boltzmann's kinetic theory of gases and classical thermodynamics. We refer to [13; 84] for more details. □

Remark 4.97. There is a discrepancy between the estimate (4.4.29) proved in [87] and the estimate for $\mathbb{E}_N[T_0]$ proved in [9, Sec. III.5] which states that

$$\mathbb{E}_N[T_0] = \frac{4^N}{N}\bigl(1 + O(/N)\bigr) \text{ as } N \to \infty. \tag{4.4.30}$$

The estimate (4.4.30) also contradicts the estimates [88, Eq. (4.27)] and [125, Eq. (7)]. □

Example 4.98 (Random walks on infinite graphs). Let us investigate the standard random walk on an infinite, locally finite graph $(\mathscr{X}, E, c)$. Thus we think of an electric network in which all edges have the same conductance 1. For $x, y \in \mathscr{X}$ define $\mathrm{dist}(x, y)$ the minimal length of a path joining x and y. Fix $x_+ \in \mathscr{X}$ and set

$$B_n := \{x \in \mathscr{X};\ \mathrm{dist}(x_+, x) \leq n\},$$

$$\Sigma_n = \{x \in \mathscr{X};\ \mathrm{dist}(x_+, x) = n\} = B_n \setminus B_{n-1},\ S_n^- = \mathscr{X} \setminus B_n.$$

Note that the balls B_n are finite. For $n \in \mathbb{N}$ we denote by $C(n)$ the total number of edges connecting a point in Σ_{n-1} to a point in Σ_n.

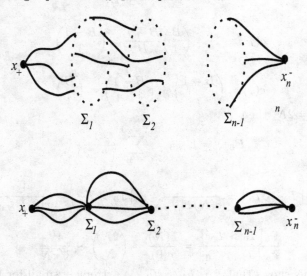

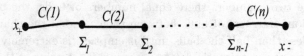

Fig. 4.5 *Shorting an infinite electric network inside spheres.*

Form the collapsed electric network $(\mathscr{X}^n, E^n, c^n)$, $\mathscr{X}^n := X/S_n^-$. The set S_n^- corresponds to a unique vertex x_n^- in $\mathscr{X}^n$; see the top of Figure 4.5. Denote by E_{x_+, x_n^-} the energy of the Kirchhoff flow in $\mathscr{X}^n$ from x_+ to x_n-.

As we have seen
$$\frac{1}{c(x_+)\mathbb{P}_{x_+}[T_{x_+} > H_{S_n^-}]} = \boldsymbol{E}_{x_+, x_n^+}.$$

Observe that the collapsed network $\mathscr{X}/S_n^-$ is obtained from the collapsed network $\mathscr{X}/S_{n+1}^-$ by first shorting the edges in $\Sigma_n \subset \mathscr{X}/S_{n+1}^-$ and then shorting the edge (x_n^-, x_{n+1}^-). Hence
$$\boldsymbol{E}_{x_+, x_n^-} \leq \boldsymbol{E}_{x_+, x_{n+1}^-}.$$

We set
$$\boldsymbol{E}_{x_+, \infty} := \lim_{n \to \infty} \boldsymbol{E}_{x_+, x_n} = \lim_{n \to \infty} \frac{1}{c(x_+)\mathbb{P}_{x_+}[T_{x_+} > S_n^-]}.$$

Thus
$$\lim_{n \to \infty} \mathbb{P}_{x_+}[T_{x_+} > S_n^-] = \frac{1}{c(x_+)\boldsymbol{E}_{x_+, \infty}}.$$

We deduce that the associated Markov chain is recurrent if and only if $\boldsymbol{E}_{x_+, \infty} = \infty$ and transient otherwise.

To estimate $\boldsymbol{E}_{x_+, x_n^-}$ from below we short edges in $\mathscr{X}/S_n^-$. First we short the edges between points in Σ_k, $k = 1, \ldots, n-1$. We obtain the electric network $\mathscr{X}_*^n$ at the bottom of Figure 4.5. As explained in Example 4.96, energy of the Kirchhoff flow in $\mathscr{X}_*^n$ from x_+ to x_n^- is
$$E^n = \sum_{k=1}^{n} \frac{1}{C(k)} \leq \boldsymbol{E}_{x_+, x_n^-}.$$

Hence
$$\boldsymbol{E}_{x_+, \infty} \geq \sum_{k=1}^{\infty} \frac{1}{C(k)}.$$

We deduce that if
$$\sum_{k=1}^{\infty} \frac{1}{C(k)} = \infty,$$
then the corresponding Markov chain is recurrent.

To estimate $\boldsymbol{E}_{x_+, \infty}$ from above we use the cutting trick. We gradually remove edges such that the component containing x_+ has infinitely many vertices. Restricting to the component containing x_+ we obtain a electric network with bigger $\boldsymbol{E}_{x_+, \infty}$ according to Theorem 4.94 and Remark 4.95(iii).

Thus if the graph $(\mathscr{X}, E)$ contains a connected subgraph $(\mathscr{X}_0, E_0)$ such that the random walk on $\mathscr{X}_0$ is transient, then the random walk on $(\mathscr{X}, E)$ is also transient. □

Example 4.99 (Random walk on $\mathbb{Z}^2$). Suppose that $(\mathscr{X}, E, c)$ corresponds to the standard random walk on $\mathbb{Z}^2$. Observe that the sphere Σ_{n-1}, $n-1 > 0$, is the square
$$\Sigma_{n-1} = \{(x, y) \in \mathbb{Z}^2; \ |x| + |y| = n-1\}.$$

Each of the four vertices if this square is connected to Σ_n through 3 edges. The interior of each of the four edges contains $(n-2)$ lattice points and each of them is connected to Σ_n through 2-edges. Thus

$$C(n) = 12 + 8(n-2) = 8n - 4, \quad \forall n \in \mathbb{N}.$$

Since

$$\sum_{n \geq 1} \frac{1}{8n-4} = \infty.$$

We deduce again that the random walk on $\mathbb{Z}^2$ is recurrent. □

Example 4.100 (Random walks on symmetric trees). Consider the unbiased random walk on an infinite locally finite tree $(\mathscr{X}, E)$. Fix $x_+ \in \mathscr{X}$ and think of x_+ as the root of the tree. As such every vertex has a unique predecessor and a number $s(x)$ of successors so the degree is

$$d(x) = \begin{cases} s(x) + 1, & x \neq x_+, \\ s(x_+), & x = x_+. \end{cases}$$

Define B_n, Σ_n, S_n^- as in the previous example. We assume that the tree is *radially symmetric* about the root i.e., for any $n \in \mathbb{N}$ the vertices on the sphere Σ_n have the same number s_n of successors. Set

$$\sigma_k := |\Sigma_k|.$$

Note that for any $k \geq 0$ we have

$$\sigma_{k+1} = s_0 s_1 \cdots s_k.$$

One can think of σ_k as the "volume" of the sphere Σ_k.

We want to investigate the unbiased random walk on this tree. Equivalently, this means assigning conductance 1 to every edge. We want to solve the equation

$$\Delta u(x) = \begin{cases} 0, & x \in B_n \setminus \{x_+\}, \\ \frac{1}{d(x_+)}, & x = x_+, \end{cases}$$

subject to the boundary condition

$$u(x) = 0, \quad \forall x \in S_n^- := \mathscr{X} \setminus B_n.$$

We know that this equation has a unique solution. We can invoke the symmetry of the graph and show that this solution must be constant along the spheres Σ_n but we do not really need to do this. If we can find a solution with this property then it has to be *it*. So make use of this Ansatz and seek a solution that is constant on the spheres.

Denote by u_k the value of u on Σ_k. We set $u_0 := u(x_+)$

$$\Delta_k = u_k - u_{k+1}, \quad \forall k \geq 0.$$

Note that $\Delta_n = u_n$. For $k \in \{1, n\}$ we have
$$u_k = \frac{s_k u_{k+1} + u_{k-1}}{s_k + 1}$$
so that
$$(s_k + 1)u_k = s_k u_{k+1} + u_{k-1} \iff \Delta_{k-1} = s_k \Delta_k.$$
Iterating we deduce
$$\Delta_{k-1} = s_{k+1} \cdots s_n \Delta_n = \frac{s_0 s_1 \cdots s_n}{s_0 \cdots s_k} \Delta_n = \frac{\sigma_n}{\sigma_k} \Delta_n = \frac{\sigma_n}{\sigma_k} u_n.$$
Hence
$$u_0 = u_0 - u_{n+1} = \sum_{k=0}^{n} \Delta_k = \sigma_n \left(\sum_{k=0}^{n} \frac{1}{\sigma_k} \right) u_n.$$
The equation
$$\Delta u(x_+) = \frac{1}{s_0}$$
is equivalent to $\Delta_0 = \frac{1}{s_0}$ so that
$$\frac{1}{s_0} = \frac{\sigma_n}{s_0} u_n, \quad u_n = \frac{1}{\sigma_n}, \quad \boldsymbol{E}_{x_+, S_n^-} = u_0 = \sum_{k=0}^{n} \frac{1}{\sigma_k}.$$
Hence,
$$\boldsymbol{E}_{x_+, \infty} = \sum_{k=0}^{\infty} \frac{1}{\sigma_k}.$$
This shows that if the number of vertices on Σ_n growth fast the random walk is transient and if it growth slow, the walk is recurrent. Intuitively, the more vertices far away, more opportunities to get lost. As an example fix $d \in \mathbb{N}$, $d \geq 2$. We denote by $\mathcal{T}_d$ the rooted radially symmetric tree with successor sequence (s_n) given by
$$s_n = \begin{cases} d, & n = 2^k - 1, \ k \geq 0, \\ 1, & \text{otherwise.} \end{cases}$$
Thus
$$\sigma_n = d^{k+1} \quad 2^k \leq n < 2^{k+1}$$
and
$$\sum_{n=0}^{\infty} \frac{1}{\sigma_n} = \frac{1}{d} + \sum_{n=2}^{3} \frac{1}{d^2} + \sum_{n=4}^{7} \frac{1}{d^3} + \cdots = \frac{1}{d} \sum_{k=0}^{\infty} \left(\frac{2}{d} \right)^k = \begin{cases} \frac{1}{d-2}, & d \geq 3, \\ \infty, & d = 2. \end{cases}$$
Thus, the random walk on $\mathcal{T}_d$ is transient if $d \geq 3$ and recurrent if $d = 2$.

We can obtain a more striking example of recurrent random walk by choosing the successor sequence to be
$$s_n = \begin{cases} k, & n = k!, \ k \geq 2 \\ 1, & \text{otherwise.} \end{cases}$$
For more information about random walks on trees we refer to the very comprehensive monograph [110]. □

4.5 Finite Markov chains

For HMC-s with finite state space the theory simplifies somewhat and new techniques are available.

4.5.1 *The Perron-Frobenius theory*

Consider a homogenous Markov chain with finite state space

$$\mathscr{X} = \mathbb{I}_m := \{1, 2, \ldots, m\}.$$

In this case the transition matrix Q is an $m \times m$ *stochastic matrix*, i.e., a matrix with nonnegative entries such that the sum of the entries on each row is 1. If we set

$$e := \begin{bmatrix} 1 \\ \vdots \\ 1 \end{bmatrix} \in \mathbb{R}^m$$

then we see that an $m \times m$ matrix Q with nonnegative entries is stochastic iff

$$Qe = e.$$

We view measures on $\mathscr{X}$ as *row* vectors $\mu = [\mu_1, \ldots, \mu_m]$.

For convenience we will denote by $\mathcal{R}_m$ the space of *row* vectors and by $\mathcal{C}_m$ the space of *column* vectors. We will denote the row vectors using Greek letters and we will think of them as *signed* measures on $\mathscr{X}$. The matrix Q acts on row vectors by right multiplication $\mu \to \mu \cdot Q$, and on column vectors by left multiplication, $v \mapsto Q \cdot v$.

A signed measure $\mu \in \mathcal{R}_m$ is a probability measure if

$$\mu_k \geq 0, \quad \forall k \in \mathbb{I}_m, \quad \mu \cdot e = 1.$$

Let $\text{Prob}_m \subset \mathcal{R}_m$ denote the space of probability measures on $\mathbb{I}_m$. We equip $\mathcal{R}_m$ with the variation norm

$$\|\alpha\|_v := \sum_{k=1}^m |\alpha_k|.$$

Observe that if $\mu, \nu \in \text{Prob}_m$, then

$$d_v(\mu, \nu) = \frac{1}{2}\|\mu - \nu\|_v.$$

Note that a *column* vector

$$z = \begin{bmatrix} z_1 \\ \vdots \\ z_m \end{bmatrix} \in \mathbb{C}^m$$

is a (left) eigenvector of Q^T corresponding to an eigenvalue $\lambda \in \mathbb{C}$ if and only if the row vector $z^\top$ is a (right) eigenvector of Q since

$$z^\top \cdot Q = \lambda z^\top.$$

The matrix Q and its transpose $Q^\top$ have the same eigenvalues.[4] The vector e is a (left) eigenvector of Q corresponding to the eigenvalue 1. We deduce that there exists a row vector $\alpha \in \mathcal{R}_m$ such that

$$\alpha \cdot Q = \alpha.$$

If α had nonnegative entries, then it would be an invariant measure for the HMC defined by Q. The classical Perron-Frobenius theory explains when this is the case and much more.

Observe that the HMC defined by Q is irreducible if and only if

$$\forall i, j \in \mathscr{X}, \; \exists \; n > 0 \; \text{such that} \; Q_{i,j}^n > 0.$$

Additionally, it is aperiodic if and only if Q is *primitive*, i.e., there exists $n_0 \in \mathbb{N}$ such that

$$\forall \; n > n_0, \; \forall i, j \in \mathscr{X} \; \text{such that} \; Q_{i,j}^n > 0, \; \forall 1 \le i, j \le m.$$

For a proof of the following result we refer to [65, Chap. XIII] or [138, Chap. 8].

Theorem 4.101 (Perron-Frobenius). *Suppose that Q is a stochastic $m \times m$ matrix. Then the following hold.*

(i) All the eigenvalues of $Q^\top$ are contained in the unit disk.
(ii) If Q is irreducible, then there exists $p \in \mathbb{N}$ such that

$$\lambda \in \mathrm{Spec}(Q) \; and \; |\lambda| = 1 \Longleftrightarrow \lambda^p = 1.$$

Moreover, every eigenvalue on the unit circle is simple.
(iii) The matrix Q is primitive if and only if $p = 1$. In this case

$$\rho := \max\{\,|\lambda|; \; \lambda \in \mathrm{Spec}(Q), \lambda \ne 1\,\} < 1.$$

□

Suppose that Q is primitive and denote by π the unique invariant probability distribution of Q, i.e., the unique row vector

$$\pi = (\pi_1, \ldots, \pi_m)$$

such that

$$\pi_k > 0, \; \forall k, \; \pi_1 + \cdots + \pi_m = 1.$$

Denote by $\Delta(\lambda)$ the characteristic polynomial of Q, $\Delta(\lambda) = \det(\lambda \mathbb{1} - Q)$. Set $B(\lambda) = \frac{1}{\lambda - 1} \Delta(\lambda)$.

Since 1 is a simple eigenvalue of Q the polynomials $\lambda - 1$ and $B(\lambda)$ have no common divisor and thus we have a decomposition of the space $\mathcal{R}_m$ (see [99, Thm. XI. 4.1]) as a direct sum of (right) Q-*invariant* subspaces

$$\mathcal{R}_m = \ker_r(\mathbb{1} - Q) \oplus \ker_r B(Q),$$

---

[4]$\det(\lambda \mathbb{1} - Q) = \det(\lambda \mathbb{1} - Q)^\top$.

where
$$\ker_r(\mathbb{1} - Q) = \{\alpha \in \mathcal{R}_m;\ \alpha \cdot (\mathbb{1} - Q) = 0\} = \mathrm{span}(\pi),$$

$$\ker_r B(Q) := \{\alpha \in \mathcal{R}_m;\ \alpha \cdot B(Q) = 0\}.$$

Thus any $\alpha \in \mathcal{R}_m$ admits a unique decomposition
$$\alpha = \alpha^0 + \alpha^\perp,\ \alpha^0 \in \ker_r(\mathbb{1} - Q),\ \alpha^\perp \in \ker_r B(Q).$$

More explicitly, choose polynomials $u(\lambda), v(\lambda)$ such that
$$u(\lambda)(\lambda - 1) + v(\lambda)B(\lambda) = 1.$$

Then
$$\alpha^\perp = \alpha \cdot u(Q)(Q - 1) \in \ker B(Q)\ \alpha^0 = \alpha \cdot v(Q)B(Q).$$

Note that
$$\alpha^\perp \cdot e = \alpha \cdot \alpha(Q)(Q - 1) \cdot e = 0.$$

If $\mu \in \mathcal{R}_m$ is a probability measure, then it has a canonical decomposition
$$\mu = c\pi + \mu^\perp,\ \mu^\perp \in \ker_r B(Q).$$

Since $\mu \cdot e = 1$ and $\mu^\perp \cdot e = 0$ we deduce $c = 1$ so $\mu = \pi + \mu^\perp$ and thus
$$\mu \cdot Q^n = \pi + \mu^\perp \cdot Q^n,$$

i.e.,
$$\mu \cdot Q^n - \pi = \mu^\perp \cdot Q^n.$$

Since $\ker_r B(Q)$ is Q-invariant we deduce from Theorem 4.101 that there exist $C > 0$, $r \in (0, 1)$ such that
$$\|\alpha \cdot Q\|_v \leq r\|\alpha\|_v \leq Cr\|\alpha\|_v,\ \forall \alpha \in \ker_r B(Q).$$

Hence
$$\|\mu \cdot Q^n - \pi\|_1 = \|\mu^\perp \cdot Q^n\|_v \leq Cr^n \|\mu\|_v = Cr^n,\ \forall \mu \in \mathrm{Prob}_m.$$

In particular, if we choose μ to be the Dirac measure concentrated at $k \in \mathbb{I}_m$, then $\delta^k \cdot Q^n$ is the k-th row of the matrix Q^n and we deduce
$$\sum_{\ell=1}^{m} |Q_{k,\ell}^n - \pi_\ell| \leq Cr^n,\ \forall k \in \mathbb{N}.$$

Theorem 4.101 allows us sharpen the above estimate. If
$$\Delta(\lambda) = \det(\lambda - Q) = \lambda^m + \sum_{j=0}^{m-1} a_j \lambda^j$$

denotes the characteristic polynomial of Q, then Cayley-Hamilton theorem implies that the sequence of matrices $(Q^n)_{n \in \mathbb{N}_0}$ satisfies the linear recurrence relation

$$Q^{n+m} + \sum_{j=0}^{m-1} a_j Q^{n+j} = 0, \quad \forall n \in \mathbb{N}_0.$$

Let $1, \lambda_2, \ldots, \lambda_s$ be the eigenvalues of Q,

$$1 > \rho = |\lambda_2| \geq \cdots \geq |\lambda_s|.$$

The eigenvalue λ_2 is usually referred to as the *second largest eigenvalue (or SLE)* of the transition matrix.

Denote by m_i is the size of the largest Jordan cell corresponding to the eigenvalue i. We assume that m_2 is the largest Jordan cell size the eigenvalues of norm ρ. The above recurrence relation shows that, for any $1 \leq i,j \leq m$, the sequence $(Q^n_{i,j})_{n \geq 0}$ admits a description of the form

$$Q^n_{i,j} = c_{ij} + \sum_{k=2}^{r} C^k_{i,j}(n) \lambda_k^n$$

where $C^k_{i,j}(z)$ is a complex polynomial of degree $\leq m_k - 1$. We deduce that

$$|Q^n_{ij} - c_{ij}| = O(n^{m_2 - 1} \rho^n).$$

We conclude that $c_{i,j} = \pi_j$ and thus

$$|Q^n_{ij} - \pi_j| = O(n^{m_2 - 1} \rho^n). \tag{4.5.1}$$

If the Markov chain is reversible, i.e.,

$$\pi_i Q_{ij} = \pi_j Q_{ji}, \quad \forall i,j \in \mathbb{I}_m,$$

then the operator $Q : \mathcal{C}_m \to \mathcal{C}_m$ is symmetric with respect to the $L^2(\pi)$-inner product $\langle -, - \rangle_\pi$ on $\mathcal{C}_m = \mathbb{R}^{\mathcal{X}}$

$$\langle x, y \rangle_\pi = \sum_{i=1}^{n} x_i y_i \pi_i, \quad \forall x, y \in \mathcal{C}_m.$$

Indeed

$$\langle Qx, y \rangle_\pi = \sum_i \sum_j Q_{ij} x_j y_i \pi_i = \sum_j \sum_i \pi_j Q_{ji} x_j y_i$$

$$= \sum_j \left(\sum_i Q_{ji} y_i \right) \pi_j x_j = \langle x, Qy \rangle_\pi.$$

In this case all the eigenvalues are real and the operator Q is diagonalizable and (4.5.1) improves to

$$|Q^n_{ij} - \pi_j| = O(\rho^n). \tag{4.5.2}$$

In general finding or estimating the SLE can be a daunting task. If some symmetry is present this is sometimes manageable.

Example 4.102 (Random walks on groups). Suppose that G is a finite group and $H \subset G$ is a set of generators. The set H determines a random walk on G. From g one can transition to $h \cdot g$, $h \in H$, with probability $\frac{1}{|H|}$.

A frequently encountered case is when H is symmetric, i.e.,

$$x \in H \Longleftrightarrow x^{-1} \in H.$$

The directed graph corresponding to this random walk is symmetric, i.e., there is a directed edge from g to g' if and only if there is a directed edge from g' to g. The resulting undirected graph is called the *Cayley graph* determined by the symmetric set of generators. The random walk on the groups is then the standard walk on the Cayley graph. The group structure behind the Cayley graph adds a lot of symmetry that we can use to our advantage. For a detailed presentation of this technique and many interesting applications we refer to the beautiful monograph [41].

We want illustrate this principle on a simpler situation. Suppose that G is the discrete torus

$$G := \left(\mathbb{Z}/n\mathbb{Z}\right)^d.$$

We will denote by $x = (x_1, \ldots, x_d)$ the elements of G, $x_k \in \mathbb{Z}/n\mathbb{Z}$. As generators e choose

$$\pm e_k \bmod n\mathbb{Z}, \quad k = 1, \ldots, d,$$

where

$$e_1 = (1, 0, \ldots, 0), \ldots, e_d = (0, \ldots, 0, 1).$$

For $d = 2$ this random walk can be visualised as a random walk on the vertices of the square grid $S_n = [0, n]^2 \cap \mathbb{Z}^2$ where the opposite edges are identified. Thus from $(0, y)$ we can transition $(0, y \pm 1 \bmod n)$ or $(\pm 1 \bmod n, y)$ with equal probabilities. Note that when n is odd, the random walk is irreducible and aperiodic.

When $n = 2$ this becomes a random walk on the set of vertices of the hypercube $[0, 1]^d$ or, equivalently, on the set of subsets of $\{1, \ldots, d\}$.

The invariant probability measure π is, up to a multiplicative constant, the uniform counting measure. We write

$$L^2(G) = L^2(G, \pi), \quad \|f\| = \|f\|_{L^2} = \frac{1}{|G|^{1/2}} \left(\sum_{x \in \mathbb{T}_n^d} |f(x)|^2 \right)^{1/2}.$$

Here we work with complex valued functions so the inner product is

$$\langle f, g \rangle = \frac{1}{|G|} \sum_{x \in G} f(x) \overline{g(x)}.$$

If Q denotes the transition matrix of this Markov chain, then for any $f \in L^2(G)$ we have

$$Qf(x) = \sum_{x' \in G} Q_{x, x'} f(x') = \frac{1}{d} \sum_{k=1}^{d} \frac{f(x + e_k) + f(x - e_k)}{2}, \tag{4.5.3a}$$

$$\Delta f(x) = f(x) - Qf(x) = -\frac{1}{d}\sum_{k=1}^{d} \frac{f(x+e_k) - 2f(x) + f(x-e_k)}{2}. \quad (4.5.3b)$$

One can verify that the induced operator $Q: L^2(G) \to L^2(G)$ is symmetric since Q is reversible but we will not rely on this fact in this example.

To compute the eigenvalues of $Q: L^2(G) \to L^2(G)$ we use Fourier analysis. This requires a little bit of representation theory and we will refer to [149] for the proofs of all the claims below.

A *character* of G is a group morphism

$$\chi: G \to S^1 := \{\, z \in \mathbb{C};\ |z| = 1\,\}.$$

The set $\widehat{G}$ of characters is a group itself with respect to the pointwise multiplication of characters. It is called the *dual group*.

Denote by $\mathcal{R}_n$ the group of n-th roots of unity

$$\mathcal{R}_n := \{\, z \in \mathbb{C}^*;\ z^n = 1\,\}.$$

Observe that for any character χ, the complex numbers $\chi(e_k)$ are n-th roots of 1. In fact, the map

$$\rho: \widehat{G} \to \mathcal{R}_n^d,\ \widehat{G} \ni \chi \mapsto (\rho_1, \ldots, \rho_d) = (\chi(e_1), \ldots, \chi(e_d)) \in \mathcal{R}_n^d$$

is a group isomorphism. The collection of functions

$$\chi: G \to \mathbb{C},\ \chi \in \widehat{G}$$

is an *orthonormal basis* of $L^2(G)$ and thus, for any $f \in L^2$ we have an orthogonal decomposition

$$f = \sum_{\chi \in \widehat{G}} \langle f, \chi \rangle \chi. \quad (4.5.4)$$

The function

$$\widehat{G} \ni \chi \mapsto \widehat{f}(\chi) := \langle f, \chi \rangle \in \mathbb{C}$$

is called the *Fourier transform* of f. More explicitly,

$$\widehat{f}(\chi) = \frac{1}{|G|} \sum_{x \in G} f(x)\overline{\chi(x)}.$$

The equality (4.5.4) can be rewritten

$$f(x) = \sum_{\chi \in \widehat{G}} \widehat{f}(\chi)\chi(x),\ \forall x \in G, \quad (4.5.5)$$

and, as such, it is known as the *Fourier inversion formula*.

If we identify $\chi \in \widehat{G}$ with $\rho(\chi) = (\rho_1, \ldots, \rho_d) \in \mathcal{R}_n^d$, then we can view the Fourier transform $\widehat{f}$ as a function on $\mathcal{R}_n^d$

$$\widehat{f}(\rho_1, \ldots, \rho_d) = \frac{1}{|G|} \sum_{x \in G} f(x)\rho_1^{-x_1} \cdots \rho_d^{-x_d},$$

and Fourier inversion formula reads
$$f(x) = \sum_{\substack{\rho_k^n = 1, \\ k=1,\ldots,d}} \widehat{f}(\rho_1,\ldots,\rho_d) \rho_1^{x_1} \cdots \rho_d^{x_d}.$$

Using (4.5.3a) and (4.5.5) we deduce
$$Qf(x) = \sum_\chi \widehat{f}(\chi) \cdot \underbrace{\frac{1}{2d}\left(\sum_{k=1}^d \bigl(\chi(e_k) + \chi(-e_k)\bigr)\right)}_{=:m(\chi)} \cdot \chi(x).$$

Thus
$$Qf = Q\left(\sum_\chi \widehat{f}(\chi)\chi\right) = \sum_\chi m(\chi)\widehat{f}(\chi)\chi.$$

In other words, the orthonormal basis $\{\chi;\ \chi \in \widehat{G},\}$ diagonalizes Q and
$$\operatorname{Spec} Q = \{\, m(\chi),\ \chi \in \widehat{G}\,\}.$$

If we write
$$\chi(e_k) = \rho_k = \cos\theta_k + i\sin\theta_k \in \mathcal{R}_n,$$
then $\chi(e_k) + \chi(-e_k) = \rho_k + \bar\rho_k = 2\cos\theta_k$ and
$$m(\chi) = \frac{1}{d}\sum_{k=1}^d \cos\theta_k, \quad \theta_k \in \left\{0, \frac{2\pi}{n}, \ldots, \frac{2\pi(n-1)}{n}\right\}.$$

Thus $\operatorname{Spec} Q \subset [-1,1]$ and $1 \in \operatorname{Spec} Q$. The SLE is
$$\lambda_2 = \lambda_2(d,n) = \frac{d - 1 + \cos 2\pi/n}{d} = 1 - \frac{2\sin^2 \pi/n}{d}.$$

Note that
$$\lambda_2(d,n) \sim 1 - \frac{2\pi^2}{dn^2} \quad \text{as } n \to \infty. \tag{4.5.6}$$

If $n = 2$ all the characters/eigenfunctions are real valued. More precisely, for every
$$\vec\epsilon = (\epsilon_1,\ldots,\epsilon_d) \in \{-1,1\}^d$$
we have an eigenfunction $\chi_{\vec\epsilon}$ given by
$$\chi_{\vec\epsilon}(x) = \prod_{k=1}^d \epsilon_k^{x_k}, \quad \forall x = (x_1,\ldots,x_d) \in \{0,1\}^d. \tag{4.5.7}$$

The corresponding eigenvalue is
$$\lambda_{\vec\epsilon} = \frac{1}{d}\bigl(\epsilon_1 + \cdots + \epsilon_d\bigr), \quad \epsilon_k = \pm 1.$$

Hence
$$\operatorname{Spec}(Q) = \left\{\, -1 + \frac{2k}{d},\ k = 0, 1, \ldots, d\,\right\}.$$

In this case the SLE is

$$\lambda_2(d,2) = 1 - \frac{2}{d}. \tag{4.5.8}$$

A probability measure μ on G can be identified with a continuous linear functional on $L^2(G,\pi)$ and, as such, can be identified with a function $\mu^* \in L^2(G,\pi)$

$$\mu^* = \sum_\chi \mu[\chi]\chi, \quad \mu[\chi] = \sum_{x \in \mathbb{T}_n^d} \mu[x]\chi(x) = \int_{\mathbb{T}_n^d} \chi d\mu.$$

Then

$$\mu \cdot Q^N = \sum_\chi m(\chi)^n \chi.$$

□

Example 4.103 (The Ehrenfest urn revisited). The random walk $(X_n)_{n \geq 0}$ on

$$\mathcal{V}_d := \{0,1\}^d,$$

the set of vertices of the hypercube $[0,1]^d$ is intimately related to Ehrenfest urn; see Example 4.7.

To see this, consider the *states*

$$s_k := \{\, x = (x_1, \ldots, x_d) \in \mathcal{V}_d;\ |x| := x_1 + \cdots + x_d = k\,\}, \quad k = 0, 1, \ldots, d.$$

If we think of the vertices $x \in \mathcal{V}_d$ as vectors of bits 0/1, then the random walk has a simple description: if located at $x \in \mathcal{V}_d$, pick a random component of x and flip it to the opposite bit. Note that

$$\mathbb{P}[\,X_{n+1} \in s_{k+1}\,\big|\, X_n \in s_k\,] = \frac{d-k}{d}, \quad \mathbb{P}[\,X_{n+1} \in s_{k-1}\,\big|\, X_n \in s_k\,] = \frac{k}{d}.$$

We recognize here the transition rules for the Ehrenfest urn model with d particles/balls. Thus, if on our walk along the vertices of the hypercube, we only keep track of the state we are in, we obtain the Markov chain defined by Ehrenfest's urn model.

For concrete computations it is convenient to have an alternate description of this phenomenon. Denote by $\mathfrak{S}_d$ the group of permutations of $\{1, \ldots, d\}$. There is an obvious *left* action of $\mathfrak{S}_d$ on $\mathcal{V}_d$,

$$\varphi \cdot (x_1, \ldots, x_d) = (x_{\varphi(1)}, \ldots, x_{\varphi(d)}), \quad \forall \varphi \in \mathfrak{S}_d,\ (x_1, \ldots, x_d) \in \{0,1\}^d.$$

On the other hand, $\mathcal{V}_d$ is equipped with a metric, the so called *Hamming distance*,

$$\delta(x,y) = \sum_{i=1}^d |x_i - y_i|, \quad x, y \in \mathcal{V}_d.$$

Two vertices $x, y \in \mathcal{V}_d$ are neighbors (connected by an edge of the cube) iff $\delta(x,y) = 1$. Since the above action of $\mathfrak{S}_d$ preserves the Hamming distance we deduce that $\mathfrak{S}_d$ is a group of graph isomorphisms, i.e.,

$$\forall x, y \in \mathcal{V}_d,\ \varphi \in \mathfrak{S}_d:\quad x \sim y \Longleftrightarrow \varphi \cdot x \sim \varphi \cdot y.$$

Observe also that the states s_k, $k = 0, 1, \ldots, d$, are the orbits of the above action of $\mathfrak{S}_d$. Thus, the state space of the Ehrenfest urn model can be identified with $\bar{\mathcal{V}}_d := \mathfrak{S}_d \backslash \mathcal{V}_d$, the space of orbits of the above left action. Denote by π the invariant probability measure of the random walk on $\mathcal{V}_d$ and $\bar\pi$ the invariant measure of the Ehrenfest urn model

$$\bar\pi[k] = \frac{1}{2^d}\binom{d}{k}.$$

If $\mathrm{Proj} : \mathcal{V}_d \to \mathfrak{S}_d \backslash \mathcal{V}_d$ is the natural projection, then

$$\mathrm{Proj}_\# \pi = \bar\pi.$$

The left action of $\mathfrak{S}_d$ on $\mathcal{V}_d$ induces a *right* action on the space $L^2(\mathcal{V}_d, \pi)$

$$(f \cdot \varphi)(x) = f(\varphi \cdot x), \quad \forall f : \mathcal{V}_d \to \mathbb{R}, \; x \in \mathcal{V}_d, \; \varphi \in \mathfrak{S}_d.$$

We denote by $L^2(\mathcal{V}_d, \pi)^{\mathfrak{S}_d}$ the subspace consisting of invariant functions, i.e., functions constant along the orbits of $\mathfrak{S}_d$. The pullback

$$\mathrm{Proj}^* : L^2(\mathfrak{S}_d \backslash \mathcal{V}_d, \bar\pi) \to L^2(\mathcal{V}_d, \pi), \; f \mapsto f \circ \mathrm{Proj}$$

is an isometry onto $L^2(\mathcal{V}_d, \pi)^{\mathfrak{S}_d}$.

Let us observe that the induced linear operator

$$Q : L^2(\mathcal{V}_d, \pi) \to L^2(\mathcal{V}_d, \pi)$$

is $\mathfrak{S}_d$-equivariant, i.e., for any $f \in L^2(\mathcal{V}_d, \pi)$, $\varphi \in \mathfrak{S}_d$,

$$Q(f \cdot \sigma) = (Qf) \cdot \sigma. \tag{4.5.9}$$

In particular, this shows that

$$Q\big(L^2(\mathcal{V}_d, \pi)^{\mathfrak{S}_d}\big) \subset L^2(\mathcal{V}_d, \pi)^{\mathfrak{S}_d}.$$

If $\bar{Q}$ denotes the transition matrix of the Ehrenfest model, then

$$\begin{array}{ccc}
L^2(\mathcal{V}_d, \pi)^{\mathfrak{S}_d} & \xrightarrow{Q} & L^2(\mathcal{V}_d, \pi)^{\mathfrak{S}_d} \\
\uparrow{\scriptstyle\mathrm{Proj}^*} & & \uparrow{\scriptstyle\mathrm{Proj}^*} \\
L^2(\bar{\mathcal{V}}_d, \bar\pi) & \xrightarrow{\bar{Q}} & L^2(\bar{\mathcal{V}}_d, \bar\pi)
\end{array}$$

If $\lambda \in \mathrm{Spec}\, Q$ and $\chi \in \ker(\lambda - Q)$ is an eigenfunction of Q, then (4.5.9) implies that $\chi \cdot \varphi \in \ker(\lambda - Q)$, $\forall \varphi \in \mathfrak{S}_d$.

For every $\vec\epsilon \in \{-1, 1\}^d$ we set

$$w(\vec\epsilon) = \#\{j; \; \epsilon_j = -1\}.$$

Note that

$$\sum_j \epsilon_j = d - 2w(\vec\epsilon), \quad \lambda(\vec\epsilon) = 1 - \frac{2w(\vec\epsilon)}{d}.$$

If $\lambda_j = 1 - \frac{2j}{d}$, then
$$\ker(\lambda_j - Q) = \operatorname{span}\{\chi_{\vec{\epsilon}};\ w(\vec{\epsilon}) = j\}.$$
The orthogonal projection Π onto $L^2(\mathcal{V}_d, \pi)^{\mathfrak{S}_d}$ is the symmetrization operator
$$L^2(\mathcal{V}_d) \ni f \mapsto \Pi f = \frac{1}{d!} \sum_{\varphi \in \mathfrak{S}_d} f \cdot \varphi \in L^2(\mathcal{V}_d, \pi)^{\mathfrak{S}_d}.$$
The above description shows that
$$\Pi \ker(\lambda - Q) \subset \ker(\lambda - Q), \quad \forall \lambda \in \operatorname{Spec} Q,$$
so that
$$\operatorname{Spec} \overline{Q} \subset \operatorname{Spec} Q.$$
Since
$$\chi_{\varphi \cdot \vec{\epsilon}}(x) = \chi_{\vec{\epsilon}}(\varphi^{-1} \cdot x), \quad \forall \varphi \in \mathfrak{S}_d,\ x \in \{0,1\}^d,$$
we deduce
$$\Pi \chi_{\vec{\epsilon}} = \Pi \chi_{\varphi \cdot \vec{\epsilon}}, \quad \forall \varphi \in \mathfrak{S}_d.$$
Thus $\Pi \chi_{\vec{\epsilon}}$ depends only on $w(\vec{\epsilon})$. We set
$$\Psi_j := \Psi \Pi \chi_{\vec{\epsilon}},\ w(\vec{\epsilon}) = j.$$
Note that
$$\Psi_j = \frac{1}{\binom{d}{j}} \sum_{w(\vec{\epsilon}) = j} \chi_{\vec{\epsilon}}. \tag{4.5.10}$$
Since the eigenfunctions $\chi_{\vec{\epsilon}}$ with fixed weight $w(\vec{\epsilon}) = j$ span the eigenspace of Q corresponding to the eigenvalue λ_j we deduce that
$$\ker(\lambda_j - \overline{Q}) = \operatorname{span}(\Psi_j)$$
so $\dim \ker(\lambda - \overline{Q}) \leq 1$, $\forall \lambda \in \operatorname{Spec} \overline{Q} \subset \operatorname{Spec} Q$. Hence
$$\#\operatorname{Spec}\overline{Q} = \dim \overline{\mathcal{V}}_d = d+1 = \#\operatorname{Spec} Q$$
and thus
$$\operatorname{Spec} Q = \operatorname{Spec} \overline{Q} \text{ and } \dim \ker(\lambda - \overline{Q}) = 1,\ \forall \lambda \in \operatorname{Spec} \overline{Q}.$$
Define $K : \mathcal{V}_d \times \mathbb{C} \to \mathbb{C}$,
$$K(x, z) := \prod_{i=1}^d \left(1 + (-1)^{x_i} z\right) = (1-z)^{|x|}(1+z)^{d-|x|}, \tag{4.5.11}$$
$$|x| = \sum_i x_i = \#\{i;\ x_i = 1\}.$$

Observe that
$$K(x,z) = \sum_{j=0}^{d}\left(\sum_{w(\vec{\varepsilon})=j}\chi_{\vec{\varepsilon}}(x)\right)z^j \stackrel{(4.5.10)}{=} \sum_{j=0}^{d}\binom{d}{j}\Psi_j(x)z^j.$$

Thus
$$(1-z)^{|x|}(1+z)^{d-|x|} = \sum_{j}\binom{d}{j}\Psi_j(x)z^j. \quad (4.5.12)$$

Integrating the equality $K(x,z)^2 = (1-z)^{2|x|}(1+z)^{2(d-|x|)}$ over $\mathcal{V}_d$ with the uniform probability measure π we deduce
$$\int_{\mathcal{V}_d} K(x,z)^2 \pi[dx] = \frac{1}{2^n}\sum_{k=1}^{d}\binom{d}{k}(1-z)^{2k}(1+x)^{2(d-k)}$$

$$= \frac{1}{2^d}\big((1-z)^2 + (1+z)^2\big)^d = (1+z^2)^d.$$

This shows that
$$\|\Psi_j\|^2_{L^2(\pi)} = \frac{1}{\binom{d}{j}}.$$

Identify $L^2(\overline{\mathcal{V}}_d, \overline{\pi})$ with the space $\mathbb{R}^{d+1}$
$$L^2(\overline{\mathcal{V}}_d, \overline{\pi}) \ni f \mapsto \begin{bmatrix} f(0) \\ \vdots \\ f(d) \end{bmatrix} \in \mathbb{R}^{1+d}$$

with the inner product
$$\langle u, v\rangle_\pi := \frac{1}{2^d}(Bu, v),$$

where $(-,-)$ denotes the canonical inner product on $\mathbb{R}^{d+1}$,
$$(u,v) = \sum_{i=0}^{d} u_i v_i,$$

and B is the diagonal matrix
$$B = \text{Diag}\left(\binom{d}{0},\ldots,\binom{d}{d}\right).$$

We denote by c_{kj} the coefficient of z^j in $(1-z)^k(1+z)^{d-k}$. If we think of the invariant eigenfunction Ψ_j as a function on $\overline{\mathcal{V}}_d$, $\Psi_j(k) := \Psi_j(x)$, $|x| = k$, then we have
$$\binom{d}{j}\Psi_j(k) \stackrel{(4.5.12)}{=} c_{kj}, \quad \binom{d}{j}\Psi_j = \underbrace{\begin{bmatrix} c_{0j} \\ \vdots \\ c_{dj} \end{bmatrix}}_{=:C_j}.$$

Denote by C the $(d+1) \times (d+1)$ matrix with columns C_j and by Λ the diagonal matrix

$$\Lambda = \mathrm{Diag}(\lambda_0, \lambda_1, \ldots, \lambda_d).$$

If we regard the columns C_j as functions in $L^2(\overline{\mathcal{V}}_d, \overline{\pi})$, then each is a multiple of an eigenfunction Ψ_j of $\overline{Q}$ so that

$$\overline{Q} C_j = \lambda_j C_j, \quad \forall j = 0, 1, \ldots, d.$$

Hence $\overline{Q} C = C \Lambda$ so that C diagonalizes $\overline{Q}$,

$$C^{-1} Q C = \Lambda, \quad \text{i.e., } Q = C \Lambda C^{-1}.$$

Remarkably, the inverse of C can be described explicitly.

From the equalities

$$\binom{d}{j} \Psi_j = C_j, \quad \|\Psi_j\|^2_{L^2(\pi)} = \frac{1}{\binom{d}{j}}$$

we deduce

$$\langle C_i, C_j \rangle_\pi = \binom{d}{i} \delta_{ij}, \quad \forall i = 0, 1, \ldots, d.$$

In other words,

$$\frac{1}{2^d}(BCx, Cy) = (Bx, y), \quad \forall x, y \in \mathbb{R}^{d+1}.$$

Hence

$$C^\top B C = 2^d B. \tag{4.5.13}$$

The matrix C has another miraculous symmetry. To prove it we need to get back to the definition of the entries c_{kj},

$$(1-z)^k (1+z)^{d-k} = \sum_j c_{kj} z^j.$$

Consider the function

$$F(u, z) = \sum_k \binom{d}{k} u^k (1-z)^k (1+z)^{d-k} = ((1+u) + (1-u)z)^n.$$

On one hand, we have

$$F(u, z) = \sum_k \binom{d}{k} u^k \boxed{(1-z)^k (1+z)^{d-k}}$$

$$= \sum_k \binom{d}{k} u^k \boxed{\sum_j c_{kj} z^j u^k} = \sum_{k,j} \binom{d}{k} c_{kj} z^j u^k.$$

On the other hand, the binomial formula yields

$$F(u, z) = ((1+u) + (1-u)z)^n = \sum_j \binom{d}{j} z^j \boxed{(1-u)^j u^{d-j}}$$

$$= \sum_j \binom{d}{j} z^j \boxed{\sum_k c_{jk} u^k} = \sum_{k,j} \binom{d}{j} c_{jk} z^j u^k.$$

Hence
$$\binom{d}{k} c_{kj} = \binom{d}{j} c_{jk}, \quad \forall j, k.$$

This can be written in more compact form as
$$(BC)_{kj} = (BC)_{jk} \iff BC = (BC)^\top = C^\top B.$$

Using this in (4.5.13) we deduce $BC^2 = 2^d B$ so that
$$C^{-1} = \frac{1}{2^d} C.$$

Hence
$$Q^n = C\Lambda^n C^{-1} = \frac{1}{2^d} C\Lambda^n C, \quad \forall n \geq 0. \tag{4.5.14}$$

The above formula was first obtained by M. Kac [84]. Since then, many different proofs were offered [87; 88; 147]. For more about the rich history and the ubiquity of the Ehrenfest urn we refer to [13; 147]. As a curiosity, we want to mention that the spectrum of $\overline{Q}$ was known to J. J. Sylvester in the 19th century.

One can use (4.5.14) to obtain important information about the dynamics of the Ehrenfest urn such that the return or first passage times T_i, $i = 0, 1, \ldots, d$. We refer to [13; 84; 87; 88] for more details.

The above "miraculous" properties of the matrix C are manifestations of the remarkable symmetries of the Krawtchouk polynomials. We refer to [43; 44] for more about these polynomials and their applications in probability. □

4.5.2 Variational methods

Consider a reversible, irreducible Markov chain with finite state space $\mathscr{X}$ and transition matrix Q. Set $N := |\mathscr{X}|$. Denote by π the invariant probability distribution. We have seen that Q is symmetric as a linear operator
$$L^2(\mathscr{X}, \pi) \to L^2(\mathscr{X}, \pi).$$

We denote by $\langle -, - \rangle_\pi$ the inner product in $L^2(\mathscr{X}, \pi)$ and by $\|-\|_\pi$ the associated norm. We identify $L^2(\mathscr{X}, \pi)$ with $\mathbb{R}^N$ equipped with the inner product
$$\langle u, v \rangle_\pi = \sum_{i=1}^N u_i v_i \pi_i.$$

The eigenvalues have variational characterizations. We order the eigenvalues of Q decreasingly
$$1 = \lambda_1 > \lambda_2 \geq \lambda_2 \geq \cdots \geq \lambda_N \geq -1.$$

Above, each eigenvalue of Q appears as often as its multiplicity. The eigenspace corresponding to the eigenvalue 1 is spanned by the constant function $e = 1$ or, equivalently the column vector $e \in \mathbb{R}^N$ with all the coordinates equal to 1.

As we have seen, the second largest eigenvalue (or SLE) λ_2 controls the rate of convergence of the Markov chain. It has the variational description

$$\lambda_2 = \sup_{\substack{u \in \mathbb{R}^N \setminus \{0\}, \\ \langle u, e \rangle_\pi = 0}} \frac{\langle Qu, u \rangle}{\|u\|_\pi^2}.$$

We will use this variational characterization to provide upper estimates for λ_2.

It is more convenient to work with the Laplacian $\Delta := \mathbb{1} - Q$. Note that $\ker \Delta = \operatorname{span}\{e\}$. Its eigenvalues are $\mu_k = 1 - \lambda_k$,

$$0 = \mu_1 < \mu_2 \leq \mu_3 \leq \cdots \leq \lambda_N \leq 2.$$

Note that lower estimates for μ_2 are equivalent with upper estimates for λ_2.

The first positive eigenvalue μ_2 has a variational characterization in terms of the *Dirichlet form*

$$\mathcal{E}(-, -) : L^2(\mathscr{X}, \pi) \times L^2(\mathscr{X}, \pi) \to \mathbb{R}, \quad \mathcal{E}(u, v) = \langle \Delta u, v \rangle_\pi.$$

Lemma 4.104.

$$\mathcal{E}(u, v) = \frac{1}{2} \sum_{x,y \in \mathscr{X}} \pi_x Q_{x,y} \big(u(x) - u(y) \big) \big(v(x) - v(y) \big).$$

Proof.

$$\sum_{x,y \in \mathscr{X}} \pi_x Q_{x,y} \big(u(x) - u(y) \big) \big(v(x) - v(y) \big)$$

$$= \underbrace{\sum_{x,y \in \mathscr{X}} Q_{x,y} \big(u(x) - u(y) \big) v(x) \pi_x}_{=:A} - \underbrace{\sum_{x,y \in \mathscr{X}} \pi_x Q_{x,y} \big(u(x) - u(y) \big) v(y)}_{=:B}.$$

Note that

$$A = \sum_{x \in \mathscr{X}} \sum_{y \in Y} Q_{x,y} \big(u(x) - u(y) \big) v(x) \pi_x$$

$$= \sum_{x \in \mathscr{X}} \big(u(x) - (Qu)(x) \big) v(x) \pi_x = \langle \Delta u, v \rangle_\pi.$$

Using the detailed balance equations $\pi_x Q_{x,y} = \pi_y Q_{y,x}$ we deduce

$$B = \sum_{y \in X} \left(\sum_{x \in \mathscr{X}} Q_{y,x} \big(u(x) - u(y) \big) \right) v(y) \pi_y$$

$$= \sum_{y \in \mathscr{X}} \big((Qu)(y) - u(y) \big) v(y) \pi_y = -\langle \Delta u, v \rangle_\pi.$$

□

Let us observe that the reversible Markov chain is defined by an electric network with conductances $c(x,y)$, where

$$c(x,y) := \pi_x Q_{x,y}.$$

Then $\forall u, v \in L^2(\mathscr{X}, \pi)$

$$\mathcal{E}(u,v) = \frac{1}{2} \sum_{x,y \in \mathscr{X}} c(x,y)\big(u(x) - u(y)\big)\big(v(x) - v(y)\big) = \langle du, dv \rangle_c,$$

where $\langle -, - \rangle_c$ is the inner product (4.4.9) on 1-cochains and d is the coboundary operator (4.4.8).

The classical Ritz-Raleigh description of eigenvalues of a symmetric operator shows that

$$\mu_2 := \inf\big\{\, \mathcal{E}(u,u);\ \|u\|_\pi = 1,\ \langle u, e \rangle_\pi = 0 \,\big\}.$$

Note that for any λ in $\mathbb{R}$ we have

$$\mathcal{E}(u + \lambda, u + \lambda) = \mathcal{E}(u, u).$$

If we think of $u \in L^2(\mathscr{X}, \pi)$ as a random variable defined on the probability space $(\mathscr{X}, \pi)$, then the above characterization of μ_2 can be rewritten as

$$\mu_2 := \inf_{\substack{\mathbb{E}_\pi[u]=0 \\ u \neq 0}} \frac{\mathcal{E}(u,u)}{\mathrm{Var}[u]}.$$

Lower bounds of μ_2 are classically known as *Poincaré inequalities*. Thus, a lower bund $\mu_2 > m > 0$ is equivalent to a statement of the form

$$\frac{1}{2} \sum_{x,y} \pi_x Q_{x,y}\big(u(x) - u(y)\big)^2 \geq m \sum_{x \in \mathscr{X}} \pi_x u(x)^2 \text{ if } \sum_x u(x)\pi_x = 0$$

$$\iff \frac{1}{2} \sum_{x,y} c(x,y)\big(u(x) - u(y)\big)^2 \geq m \sum_{x \in \mathscr{X}} c(x) u(x)^2 \text{ if } \sum_x u(x) c(x) = 0.$$

To state our first Poincaré type inequality we need a few geometric preliminaries.

To our reversible Markov chain we associate a graph G with vertex set $\mathscr{X}$. Two vertices x, y are connected by an edge iff $Q(x, y) \neq 0$. We write $x \sim y$ if x and y are connected by an edge in G. This graph could have loops. It is connected since the Markov chain is irreducible. We set

$$\widehat{E} := \big\{\, (x,y) \in \mathscr{X} \times \mathscr{X};\ x \sim y \,\big\}. \tag{4.5.15}$$

We think of the elements of $\widehat{E}$ as edges of G equipped with an orientation. For any $u : \mathscr{X} \to \mathbb{R}$ and $e = (x', x'') \in \widehat{E}$ we set

$$\delta_e u := u(x'') - u(x').$$

We can speak of the conductance $c(e)$ of any oriented edge $e = (x, y)$,

$$c(e) := c(x,y) = \pi_x Q_{x,y}.$$

Note that
$$\mathcal{E}(u,u) = \frac{1}{2}\sum_{e\in\widehat{E}} c(e)(\delta_e u)^2. \tag{4.5.16}$$

A path in G between two vertices x, y is a succession of vertices
$$\gamma: \quad x = x_0 \sim x_1 \sim \cdots x_{\ell-1} \sim x_\ell = y,$$
where we do not allow repeated edges. The number ℓ is called the *length* of γ and it is denoted by $\ell(\gamma)$. The path γ determines a collection of oriented edges
$$e_i = (x_{i-1}, x_i), \ i = 1, \ldots, \ell.$$
We will use the notation $e \in \gamma$ to indicate that e is one of the *oriented* edges determined by γ.

We denote by Γ the collection of paths in G. It comes with an obvious equivalence relation: two paths are equivalent if they have the same initial and final points. Fix a collection $\mathcal{C}$ of representatives of this equivalence relation. Thus, $\mathcal{C}$ contains exactly one path for $\gamma_{x,y}$ every pair (x, y) of vertices and this path connects x to y. Following [46] we set
$$K(\mathcal{C}) := \sup_{e\in E} K(\mathcal{C}, e), \quad K(\mathcal{C}, e) := \frac{1}{c(e)} \sum_{\mathcal{C} \ni \gamma_{x,y} \ni e} \ell(\gamma_{x,y}) \pi_x \pi_y. \tag{4.5.17}$$

If an oriented edge e is not contained in any path $\gamma \in \mathcal{C}$ we set $K(e) = 0$.

Theorem 4.105 (Diaconis-Stroock). *For any $u \in L^2(\mathcal{X}, \pi)$ we have*
$$\mathrm{Var}[\,u\,] \leq K(\mathcal{C})\mathcal{E}(u,u). \tag{4.5.18}$$
Thus $\mu_2 \geq \frac{1}{K(\mathcal{C})}$ so that
$$\lambda_2(Q) \leq 1 - \frac{1}{K(\mathcal{C})}.$$

Proof. We follow the approach in the proof of [46, Proposition 1]. Set $K = K(\mathcal{C})$. Let $u \in L^2(\mathcal{X}, \pi)$. For any $x, y \in \mathcal{X}$ we have the telescoping equality
$$u(y) - u(x) = \sum_{e\in\gamma_{x,y}} \delta_e u.$$
Using the Cauchy Schwartz inequality we deduce
$$\bigl(u(y) - u(x)\bigr)^2 = \Biggl(\sum_{e\in\gamma_{x,y}} \delta_e u\Biggr)^2 \leq \ell(\gamma_{x,y}) \sum_{e\in\gamma_{x,y}} (\delta_e u)^2.$$
Now observe that
$$\mathrm{Var}[\,u\,] = \frac{1}{2}\sum_{x,y} \bigl(u(y)-u(x)\bigr)^2 \pi_x \pi_y \leq \frac{1}{2}\sum_{x,y} \ell(\gamma_{x,y}) \sum_{e\in\gamma_{x,y}} (\delta_e u)^2$$
$$= \frac{1}{2}\sum_{e\in\widehat{E}} (\delta_e u)^2 \sum_{\gamma_{x,y} \ni e} \gamma_{x,y} = \frac{1}{2}\sum_{e\in E} c(e)(\delta_e u)^2 \underbrace{\frac{1}{c(e)} \sum_{\gamma_{x,y}\ni e} \gamma_{x,y}}_{\leq K}$$
$$\leq \frac{K}{2}\sum_{e\in E} c(e)(\delta_e u)^2 \stackrel{(4.5.18)}{=} K\mathcal{E}(u,u).$$
$\square$

Example 4.106. Suppose that our Markov chain corresponds to the random walk on the Cayley graph of the cyclic group $\mathbb{Z}/n\mathbb{Z}$, n odd; see Example 4.102. Equivalently, it is the random walk on the set

$$\mathscr{X} = \{x_i\}_{i \in \mathbb{Z}/n\mathbb{Z}}$$

of vertices of a regular n-gon, where at each vertex we are equally likely to move to one of its two neighbors. In this case we have

$$\pi_x = \frac{1}{n}, \quad Q_{x_i,x_{i+1}} = Q_{x_i,x_{i-1}} = \frac{1}{2}, \quad \forall i \in \mathbb{Z}/n\mathbb{Z},$$

$$c(x_i, x_j) = \frac{1}{2n} \times \begin{cases} 1, & i = j \pm 1, \\ 0, & \text{otherwise.} \end{cases}$$

As collection $\mathcal{C}$, we choose geodesics (shortest paths) connecting the pair of vertices. Since n is odd, for every $x, y \in \mathscr{X}$ there exists a unique such geodesic $\gamma_{x,y}$ and it has length $< \frac{n}{2}$. Due to the symmetry of the graph the quantity

$$K(e) := \frac{1}{c(e)} \sum_{\gamma_{x,y} \ni e} \ell(\gamma_{x,y}) \pi_x \pi_y = \frac{2}{n} \sum_{\gamma_{x,y} \ni e} \ell(\gamma_{x,y})$$

is independent of e so

$$K(\mathcal{C}) = K(e), \quad \forall e \in E.$$

Averaging over the n edges of the graph we deduce

$$K(\mathcal{C}) = \frac{1}{n} \sum_e K(e) = \frac{2}{n^2} \sum_e \sum_{\gamma_{x,y} \ni e} \ell(\gamma_{x,y})$$

$$= \frac{2}{n^2} \sum_{x,y} \sum_{e \in \gamma_{x,y}} \ell(\gamma_{x,y}) = \frac{2}{n^2} \sum_{x,y} \ell(\gamma_{x,y})^2$$

($n = 2m+1$)

$$= \frac{2}{n} \sum_{i=1}^n \ell(\gamma_{x_1,x_i})^2 = \frac{4}{n} \sum_{i=1}^m i^2 = \frac{n^2}{6} + O(n), \quad \text{as } n \to \infty.$$

Hence

$$\lambda_2 \leq 1 - \frac{6}{n^2} + O(n^{-3}), \quad \text{as } n \to \infty.$$

Thus, for large n this lower estimate is of the same order, as the precise estimate (4.5.6) with $d = 1$. □

We want to describe another geometric estimate for μ_2 of the type first described in Riemannian geometry by J. Cheeger [27].

The volume of a set $S \subset \mathscr{X}$ is computed using the stationary measure π,

$$V(S, Q) := \pi[S] = \sum_{s \in S} \pi_s.$$

The "boundary" of the set S is the collection of oriented edges
$$\partial S := \{ (s,s') \in \widehat{E};\ s \in S,\ s' \in S^c \}.$$
The "area" of the boundary of $S \subset \mathscr{X}$ is
$$A(\partial S, Q) := \sum_{e \in \partial S} c(e) = \sum_{(s,s') \in S \times S^c} \pi_s Q_{s,s'},\ S^c = \mathscr{X} \setminus S.$$
Note that $A(\partial S, Q) = A(\partial S^c, Q)$. The ratio
$$h(S, Q) = \frac{A(\partial S, Q)}{V(S, Q)}$$
is the conditional probability that the Markov chain will transition from a state in S to a state in S^c given that initial distribution is the equilibrium distribution.

Remark 4.107. If Q is associated to an electric network with arbitrary conductances $\widetilde{c}(x,y)$, then there exists $Z > 0$ such that
$$\widetilde{c}(x) = \sum_y \widetilde{c}(x,y) = Z\pi_x,\ \forall x \in \mathscr{X}.$$
Note that if we define
$$\widetilde{V}(S,Q) := \sum_{s \in S} \widetilde{c}(s),\ \widetilde{A}(\partial S, Q) := \sum_{e \in \partial S} \widetilde{c}(e),$$
then
$$\frac{\widetilde{A}(\partial S, Q)}{\widetilde{V}(S, Q)} = \frac{A(\partial S, Q)}{V(S, Q)}. \qquad \square$$

Now define the *Cheeger isoperimetric constant* or the *conductance* of $(\mathscr{X}, Q)$ to be
$$\begin{aligned} h(Q) &:= \inf \left\{ h(S,Q);\ 0 < \mu[S] < \frac{1}{2} \right\} \\ &= \inf \left\{ \max(h(S,Q), h(S^c,Q));\ \emptyset \neq S \subsetneq \mathscr{X} \right\}. \end{aligned} \qquad (4.5.19)$$

To get a feeling of the meaning of $h(Q)$ suppose that Q corresponds to the unbiased random walk on a connected graph G with vertex set $\mathscr{X}$. For any $S \subset \mathscr{X}$, the area $A(\partial S)$ is, up to a multiplicative constant, the number of edges connecting a vertex in S with a vertex outside S. The volume $V(S)$ is, up to a multiplicative constant the sum of degrees of vertices in S, or equivalently, $V(S) - A(\partial S)$ is twice the number of edges with both endpoints in S. Thus, a "large" $h(Q)$ signifies that, for any subset of $\mathscr{X}$, a large fraction of the edges with at least one endpoint in S have the other endpoint outside S.

As an example of graph with small h think of a "bottleneck", i.e., a graph obtained by connecting with a single edge two disjoint copies of a complete graph.

Various versions of Cheeger's isoperimetric constant of a (connected) graph play a key role in the definition of *expander families* of graphs, [96; 108]. It was in that context that the connection with random walks on graphs was discovered. For general reversible Markov chains we have the following result due to Jerrum and Sinclair [83].

Theorem 4.108. *Let Q denote the transition matrix of a reversible Markov chain with finite state space $\mathscr{X}$. Then*

$$\mu_2 \geq \frac{h(Q)^2}{2}.$$

In particular,

$$\lambda_2 \leq 1 - \frac{h(Q)^2}{2}.$$

Proof. We follow the presentation in [46]. Let $u \in L^2(\mathscr{X}, \pi)$. Set $u_+ := \max(u, 0)$. We set

$$S_u := \{u > 0\} \subset \mathscr{X}, \quad h(u) = \inf_{S \subset S_u} \frac{A(\partial S, Q)}{V(S)}.$$

Lemma 4.109. *If $u \in L^2(\mathscr{X}, \pi)$ and $u_+ \neq 0$, then*

$$\mathcal{E}(u_+, u_+) \geq \frac{h(u)^2}{2} \|u_+\|_\pi^2. \qquad (4.5.20)$$

Proof. We can assume without any loss of generality that $u = u_+$. Then

$$2 \sum_{u(x) < u(y)} (u(y)^2 - u(x)^2) c(x, y) = \sum_{x,y} |u(x)^2 - u(y)^2| c(x, y)$$

$$\leq \left(\underbrace{\sum_{x,y} (u(x) - u(y))^2 c(x, y)}_{=2\mathcal{E}(u,u)} \right)^{\frac{1}{2}} \left(\sum_{x,y} \underbrace{(u(x) + u(y))^2}_{\leq 2(u(x)^2 + u(y)^2)} c(x, y) \right)^{\frac{1}{2}}$$

$$\leq 2\mathcal{E}(u,u)^{1/2} \left(\underbrace{\sum_{x,y} (u(x)^2 + u(y)^2) c(x, y)}_{=2\|u\|_\pi^2} \right)^{\frac{1}{2}} = 2^{3/2} \mathcal{E}(u,u)^{1/2} \|u\|_\pi.$$

We deduce

$$2^{3/2} \mathcal{E}(u,u)^{1/2} \|u\|_\pi \geq 2 \sum_{u(x) < u(y)} (u(x)^2 - u(y)^2) c(x, y)$$

$$= 4 \sum_{x,y} \left(\int_{u(x)}^{u(y)} t \, dt \right) c(x, y) = 4 \int_0^\infty t \left(\sum_{u(x) \leq t < u(y)} c(x, y) \right) dt.$$

If we write $S_t := \{u > t\} \subset S_u$ and observe that
$$\sum_{u(x) \leq t < u(y)} c(x,y) = A(\partial S_t, Q) \geq h(u)\pi[S_t].$$
We deduce
$$\int_0^\infty t \left(\sum_{u(x) \leq t < u(y)} c(x,y) \right) dt \geq h(u) \int_0^\infty t\pi[u > t] dt \stackrel{(1.3.43)}{=} \frac{h(u)}{2} \|u\|_\pi^2.$$
$\square$

Observe now that for any $x, y \in \mathscr{X}$ we have
$$\big(u_+(x) - u_+(y)\big)\big(u(x) - u(y)\big) \geq \big(u_+(x) - u_+(y)\big)^2.$$
To see this, note first that above we have equality if both $u(x)$ and $u(y)$ are nonnegative or both nonpositive. We have strict inequality if one is positive and the other negative, say $u(x) > 0 > u(y)$. Indeed,
$$\big(u_+(x) - u_+(y)\big)\big(u(x) - u(y)\big) = u(x)\big(u(x) - u(y)\big) > u(x)^2 = \big(u_+(x) - u_+(y)\big)^2.$$
In particular, we deduce that
$$\mathcal{E}(u_+, u) \geq \mathcal{E}(u_+, u_+),$$
and thus,
$$\mu > 0, \ \Delta u \leq \mu u \text{ on } \{u > 0\} \Rightarrow \mu\|u_+\|_\pi^2 \geq \mathcal{E}(u_+, u_+). \tag{4.5.21}$$
Indeed,
$$\mu\|u_+\|_\pi^2 \geq \lambda \langle u_+, \Delta u \rangle_\pi = \mathcal{E}(u_+, u) \geq \mathcal{E}(u_+, u_+).$$
Combining (4.5.20) and (4.5.21) we deduce that $\mu \geq \frac{h(u)^2}{2}$ if $\Delta u \leq \mu u$ on $\{u > 0\} \neq \emptyset$.

Suppose now that u is a nontrivial eigenfunction corresponding to the eigenvalue μ_2 of Δ. Since
$$\sum_x u(x)\pi_x = 0$$
we deduce that $\{u > 0\} \neq \emptyset$ and we conclude that
$$\mu_2 \geq \frac{h(u)^2}{2} \geq \frac{h(Q)^2}{2}$$
as claimed. $\square$

The quantity $h(Q)$ is rather difficult to compute but lower estimates are easier to obtain. Consider a collection $\mathcal{C}$ of paths in G as in the definition (4.5.17). We set
$$\kappa(\mathcal{C}) = \sup_{e \in E} \kappa(\mathcal{C}, e), \quad \kappa(\mathcal{C}, e) = \frac{1}{c(e)} \sum_{\mathcal{C} \ni \gamma_{x,y} \ni e} \pi_x \pi_y.$$

If an oriented edge e is not contained in any path $\gamma \in \mathcal{C}$ we set $\kappa(e) = 0$. We have the following result, [46; 141].

Proposition 4.110. *We have*

$$h(Q) \geq \frac{1}{2\kappa(\mathcal{C})},$$

so that

$$\lambda_2 \leq 1 - \frac{1}{8\kappa(\mathcal{C})^2}, \quad \forall \mathcal{C}.$$

Proof. Let $S \subset \mathscr{X}$ be a set of vertices with $V(S) = \pi[S] \leq \frac{1}{2}$. We set

$$W(S) = \sum_{\substack{\gamma_{x,y} \in \mathcal{C} \\ x \in S,\, y \in S^c}} \pi_x \pi_y.$$

Clearly

$$W(S) = \pi[S]\pi[S^c] \geq \frac{1}{2}\pi[S] = \frac{1}{2}V(S).$$

On the other hand

$$W(S) \leq \sum_{e \in \partial S} \sum_{\gamma_{x,y} \ni e} \pi_x \pi_y = \sum_{e \in \partial S} c(e)\kappa(e) \leq \kappa \sum_{e \in \partial S} c(e) = \kappa A(\partial S),$$

and we deduce

$$\kappa A(\partial S) \geq \frac{1}{2}V(S).$$

$\square$

4.5.3 *Markov Chain Monte Carlo*

Since this is only an invitation to this subject we do not attempt to formulate the most general situation or technique. Suppose that we want to sample a very large but finite set $\mathscr{X}$ according to a probability measure on it. The information we have about the set and the given distribution is not complete but "obtainable".

The probability measure π is known only up to a multiplicative constant. More precisely, we know only a weight $w : \mathscr{X} \to (0, \infty)$ that is proportional to π, i.e.,

$$\pi[x] = \frac{w(x)}{Z}, \quad Z = \sum_{x \in \mathscr{X}} w(x).$$

For all intents and purposes, the normalizing constant Z is not effectively available to us. Still, we would like to produce an $\mathscr{X}$-valued random variable with distribution π.

The theory of Markov chains will allow us to produce, for any given $\varepsilon > 0$ an $\mathscr{X}$-valued random variable with distribution ν within $\varepsilon > 0$ (in total variation distance) from the desired but unknowable distribution π.

The *Metropolis algorithm* will allow us to achieve this. The input of the algorithm is a pair (G, w) where G is a with vertex set $\mathscr{X}$ w is a weight on its set of vertices, i.e., a function $w : \mathscr{X} \to (0, \infty)$ such that

$$\sum_{x \in \mathscr{X}} w(x) < infty.$$

The graph G is called the *candidate graph*. Often the candidate graph is suggested by the problem at hand.

A good example to have in mind is the set $\mathscr{X}$ of Internet nodes and we want to sample the set of nodes uniformly. In this case the weight w is a constant function. To simplify the presentation we assume that the graph is connected and the standard random walk on it is primitive.

The output of the algorithm is the transition matrix Q of a reversible, irreducible and aperiodic Markov chain with state space $\mathscr{X}$ and whose equilibrium probability π is proportional to w. We will refer to this Markov chain as the *Metropolis chain* with candidate graph G and equilibrium distribution π. If we run this Markov chain starting from an initial vertex $x_0 \in \mathscr{X}$, then for n sufficiently large, the state X_n reached after n steps will have a distribution close to π.

The transitions of this Markov chain are described by an *acceptance-rejection strategy* based on the standard random walk on the graph G. More precisely, the transitions from a vertex x to one of its neighbors follows these rules.

(i) Pick one of the neighbors y of x equally likely among its $d(x)$ neighbors. (This is what we would do if we were to perform a standard random walk on G.) This the *acceptance* part.

(ii) The transition to y is decided by a comparison between the weight $w(y)$ at y and the weight $w(x)$ at x. More precisely, we accept the move to y with probability $\min\left(1, \frac{w(y)/d(y)}{w(x)/d(x)}\right)$. Otherwise we reject the move and stay put at x. This is the *rejection* part.

In other words, the transition matrix Q of this Markov chain is given by

$$Q_{x,y} = \begin{cases} 0, & y \notin N(x), \\ \frac{1}{d(x)} \min\left(1, \frac{w(y)/d(y)}{w(x)/d(x)}\right), & y \in N(x), \\ 1 - \frac{1}{d(x)} \sum_{x' \in N(x)} \min\left(1, \frac{w(x')/d(x')}{w(x)/d(x)}\right), & y = x. \end{cases}$$

Above, $N(x)$ denotes the set of neighbors of x in the candidate graph. Let us show that

$$w(x) Q_{x,y} = w(y) Q_{y,x}, \quad \forall x, y \in \mathscr{X},$$

so that Q is reversible and its equilibrium distribution is proportional to w.

Indeed, for $x \neq y$, we have

$$w(x)Q_{x,y} = \frac{w(x)}{d(x)} \min\left(1, \frac{w(y)/d(y)}{w(x)/d(x)}\right)$$

$$= \begin{cases} w(y)/d(y), & w(y)/d(y) < w(x)/d(x), \\ w(x)d(x), & w(y)/d(y) \geq w(x)/d(x) \end{cases}$$

$$= \frac{w(y)}{d(y)} \min\left(1, \frac{w(x)/d(x)}{w(y)/d(y)}\right) = w(y)Q_{y,x}.$$

If the random walk on the candidate graph G is primitive, then so is the Metropolis chain. If not, we replace the Metropolis chain with its lazy version; see Remark 4.69.

We refer to [77; 141] for applications of this algorithm to combinatorics. In general, it is difficult to estimate the SLE or the rate of converges of the Metropolis chain, but in practice it works well. We refer to [77; 141] for applications of this algorithm to combinatorics.

Example 4.111. A few years ago (2016–17) I asked *Mike McCaffrey*, at that time a student writing his senior thesis under my supervision, to read Diaconis' excellent survey [42] and then to try to implement numerically the decryption strategy described in that paper, based on the Metropolis algorithm. I want to report some of McCaffrey's nice findings. For more details I refer to his senior thesis [116].

Let me first outline the encryption problem and the decryption strategy proposed in [42]. The encryption method is a simple substitution cipher. Scramble the 26 letters of the English alphabet $\mathcal{E}$. The encryption is captured by a permutation φ of the set $\mathcal{E}$, or equivalently, an element of φ the symmetric group $\mathfrak{S}_{26}$.

The decryption problem asks to determine the decoding permutation φ^{-1} given a text encoded by the (unknown) permutation φ. Thus, we need to find one element in a set of 26! elements. To appreciate how large 26! is, it helps to have in mind that a pile of 26! grains of sand will cover the continental United States with a layer of sand 0.6 miles (approx. 1 kilometer) thick. We are supposed to find a single grain of sand in this huge pile. Needle in a haystack sounds optimistic!

The strategy outlined in [42] goes as follows. There are 26^2 pairs of letters in the English alphabet $\mathcal{E}$, and they appear as adjacent letters in English texts with a certain frequency. E.g., one would encounter quite frequently the pair "*th*", less so pairs such as "*tt*" or "*tw*". We denote by $f(s_1, s_2)$ the frequency of the pair of letters (s_1, s_2). More precisely $f(s_1, s_2)$ is the conditional probability that in an English text the letter s_1 is followed by s_2. To any text of length n, viewed as string of n letters, $\underline{x} = x_1 \ldots, x_n$ we associate the weight

$$w(\underline{x}) := \prod_{i=2}^{n} f(x_{i-1}, x_i).$$

We can use a given encrypted text $\underline{x}$ to define a weight on $\mathfrak{S}_{26}$

$$w(\varphi) := w\big(\varphi(x_1)\ldots\varphi(x_n)\big).$$

If $\underline{x}$ is obtained from a genuine English text $a_1,\ldots,a_n$ via a permutation φ_0, $x_i = \varphi_0^{-1}(a_i)$, then φ_0 is the decoder $x_i \mapsto \varphi_0(x_i) = a_i$.

$$w(\varphi_0) := w\big(a_1\ldots a_n\big).$$

The hope is that permutations with higher weight are closer to the decoding permutation since they mimic closely the frequencies of adjacent pairs of letters in written English. In other words

$$\varphi = \mathrm{argmax}_{\sigma \in \mathfrak{S}_{26}}\, w(\sigma).$$

The weight function w defines a probability measure on $\mathfrak{S}_{26}$ highly concentrated around the decoding permutation. If we sample this probability measure there is a high probability that we will land near the decoding permutation.

To sample this probability measure we rely on the Metropolis algorithm. The symmetric group is generated by its $\binom{26}{2}$ transpositions and as candidate graph we take the associated Cayley graph defined by this set of generators. As initial state we take the identity permutation.

The question is how well does this work in practice. First, one needs to find the relative frequencies of adjacent pairs in English texts. One can do this by analyzing a large text. In [42] Diaconis suggested using "War and Peace". Mike McCaffrey used "Moby Dick" for this purpose. The table in Figure 4.6 (borrowed from [116]) depicts these relative frequencies.[5]

He then proceeded to test[6] this method using first a shorter text

THE PROBABILITY THAT WE MAY FAIL IN THE STRUGGLE OUGHT NOT TO DETER US FROM THE SUPPORT OF A CAUSE WE BELIEVE TO BE JUST

The scrambled version looked like

OVB CTEAJADKDOM OVJO SB RJM HJDK DN OVB WOTYXXKB EYXVO NEO OE ZBOBT YW HTER OVB WYCCETO EH J QJYWB SB ABKDBLB OE AB UYWO

We expect the weight of the decoded text w_{true} to be a lot higher than the weight of the encoded text. In the above example, the weight of the original text is $\underline{2.6 \times 10^{115}}$ higher than that of the cyphered text!!!

After 3,000 steps in the random walk governed by the above Metropolis algorithm, the output was close to the original text:

THE PROLALINITY THAT WE MAY FAIN ID THE STRUGGNE OUGHT DOT TO KETER US FROM THE SUPPORT OF A JAUSE WE LENIEVE TO LE BUST

---

[5]He actually used an alphabet consisting of 27 symbol, the 26 letters of the alphabet and a 27-th representing any symbol that is not a letter.

[6]An R-code implementing this algorithm was and is publicly available.

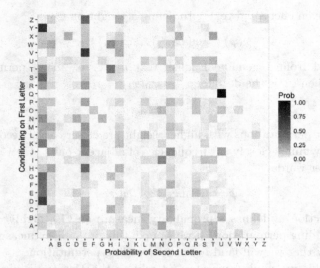

Fig. 4.6 Moby Dick Transition Matrix.

Mike then tested this algorithm on a bigger text. He chose the easily recognizable Gettysburg address by Abraham Lincoln.

The most vivid confirmation of the power of this method came when he presented his results to a mixed group of students in the College of Science of the University of Notre Dame. He began his presentation by projecting the ciphered Gettysburg address, but the audience was left in the dark about the nature of original text. While Mike was describing the problem and the decoding strategy, his laptop was running the algorithm in the background and every few seconds the text on the screen would scramble revealing a new text resembling more and more an English text. Ten minutes or so into his presentation the audience was able to recognize without difficulty the Gettysburg address. It took about 120 steps in the Metropolis random walk to reach an easily recognizable albeit misspelled text! □

4.6 Exercises

Exercise 4.1. Consider the construction in Remark 4.5 of an HMC with initial distribution μ and transition matrix Q as a sequence of random variables defined on $[0,1)$ equipped with the Lebesgue measure $\boldsymbol{\lambda}$. For every $t \in [0,1)$ there exists

$$\underline{x} = \underline{x}(t) \in \mathscr{X}^{\mathbb{N}_0}$$

uniquely determined by

$$t \in \bigcap_{n \geq 0} I^n_{x_0,\ldots,x_n}.$$

(i) Prove that the resulting map $\Psi : [0,1) \to \mathscr{X}^{\mathbb{N}_0}$ given by $t \mapsto \underline{x}(t)$ is measurable and $\Psi_{\#}\boldsymbol{\lambda} = \mathbb{P}_\mu$. **Hint.** Use the π-λ theorem.

(ii) Prove that the map Ψ is injective and its image is shift-invariant and has $\mathbb{P}_\mu$-negligible complement.

(iii) Describe the map $t \mapsto \underline{x}(t)$ when $\mathscr{X} = \{0,1\}$, $\mu[\,0\,] = \mu[\,1\,] = \frac{1}{2}$ and

$$Q = \begin{bmatrix} \frac{1}{2} & \frac{1}{2} \\ \frac{1}{2} & \frac{1}{2} \end{bmatrix}.$$

Describe explicitly the random variables

$$X_n : [0,1) \to \mathbb{R}, \quad X_n(t) = x_n(t), \quad \text{where } \underline{x}(t) = (x_0(t), x_1(t), \ldots) \in \{0,1\}^{\mathbb{N}_0}.$$

$\square$

Exercise 4.2. Two people A, B play the following game. Two dice are tossed. If the sum of the numbers showing is less than 7, A collects a dollar from B. If the total is greater than 7, then B collects a dollar from A. If a 7 appears, then the person with the fewest dollars collects a dollar from the other. If the persons have the same amount, then no dollars are exchanged. The game continues until one person runs out of dollars. Let A's number of dollars represent the states. We know that each person starts with 3 dollars.

(i) Show that the evolution of A is governed by a Markov chain. Describe its transition matrix.
(ii) If A reaches 0 or 6, then he stays there with probability 1. What is the probability that B loses in 3 tosses of the dice?
(iii) What is the probability that A loses in 5 or fewer tosses?

$\square$

Exercise 4.3. Prove that (4.1.4) is equivalent to (4.1.5). $\square$

Exercise 4.4. Let $\mathscr{X}$ be a finite or countable subset. Construct a Markov chain with state space $\mathscr{X}$ such that any subset of $\mathscr{X}$ is a closed set of this Markov chain.

$\square$

Exercise 4.5. Suppose that $(Y_n)_{n\geq 0}$ is a sequence of i.i.d., $\mathbb{N}_0$-valued variables with common probability generating function

$$G(s) = \sum_{k\geq 0} p_k s^n, \quad p_k := \mathbb{P}[Y_n = k], \forall k, n \in \mathbb{N}_0.$$

Let X_n the amount of water in a reservoir at noon on day n. During the 24 hour period beginning at this hour a quantity Y_n flows into reservoir, and just before noon a quantity of one unit of water is removed, if this amount can be found. The maximum capacity of the reservoir is K, excessive inflows are spilled and lost. Show that $(X_n)_{n\geq 0}$ is an HMC, and describe the transition matrix and its stationary distribution in terms of G. □

Exercise 4.6. Denote by X_n the capital of gambler at the end of the n-th game. He relies on the following gambling strategy. If his fortune is $\geq \$4$ he gambles $\$2$ expecting to win $\$4$, $\$3$, $\$2$ with respective probabilities 0.25, 0.30, 0.45. If his capital is 1, 2 or 3 dollars he bets $\$1$ expecting him to earn $\$2$ and $\$0$ with probabilities 0.45 and respectively 0.45 and 0.55. When his fortune is 0 he stops gambling.

(i) Show that $(X_n)_{n\geq 0}$ is a homogeneous Markov chain, compute its transition probabilities and classify its states.
(ii) Set

$$T := \inf\{n \in \mathbb{N};\ X_n = 0\}.$$

Show that $\mathbb{P}[T < \infty] = 1$.
(iii) Compute $\mathbb{E}[T]$.

□

Exercise 4.7. Suppose that $(X_n)_{n\geq 1}$ is a sequence of nonnegative i.i.d., continuously distributed random variables. Consider the sequence of *records* $(R_n)_{n\in\mathbb{N}}$ defined inductively by the rule

$$R_1 = 1, \quad R_n = \inf\{n > 1;\ X_n > \max(X_1, \ldots, X_{n-1})\}.$$

Show that the sequence (R_n) is an Markov chain with state space $\mathbb{N}$ and then compute its transition probabilities. Is this a homogeneous chain? □

Exercise 4.8. At an office served by a single clerk arrives a Poisson stream of clients. More precisely, the n-th client arrives client arrives at time $T_n = S_1 + \cdots + S_n$ where $(S_n)_{n\in\mathbb{N}}$ is a sequence of i.i.d. random variables, $S_n \sim \text{Exp}(\lambda)$. The time to process the n-th client is Z_n, where $(Z_n)_{n\geq 1}$ is a sequence of i.i.d. nonnegative random variables with common distribution $\mathbb{P}_Z$. We assume that the random variables Z_n are independent of the arrival times T_m. For $n \geq 0$ we denote by X_n the number of customers waiting in line immediately after the n-th arrived customer was served.

(i) Show that $(X_n)_{n\geq 0}$ is a homogeneous Markov chain with transition probabilities

$$\mathbb{P}\big[X_{n+1} = k \,\|\, X_n = j\big] = \begin{cases} q_{k-j}, & k \geq j, \\ 0, & k < j, \end{cases}$$

where

$$q_j = \int_0^\infty e^{-\lambda z} \frac{(\lambda z)^j}{j!} \mathbb{P}_Z\big[dz\big].$$

Hint. Use Exercise 1.47.

(ii) Set $\mu = \mathbb{E}\big[Z\big]$, $r := \lambda\mu$. Prove that the above chain is positively recurrent if and only if $r < 1$.

(iii) Assume that $r < 1$ and $c^2 := \mathbb{E}\big[Z^2\big] < \infty$. Prove that

$$\lim_{n\to\infty} \mathbb{E}\big[X_n\big] = r + \frac{\lambda^2 c^2}{2(1-r)}.$$

□

Exercise 4.9. Suppose that $(X_n)_{n\in\mathbb{N}_0}$ is an irreducible HMC with state space $\mathscr{X}$ and transition matrix Q. Prove that the following statements are equivalent.

(i) The chain is recurrent.
(ii) There exist $x, y \in \mathscr{X}$ such that

$$\sum_{n\in\mathbb{N}} Q^n_{x,y} = \infty.$$

(iii) For any $x, y \in \mathscr{X}$ we have

$$\sum_{n\in\mathbb{N}} Q^n_{x,y} = \infty.$$

□

Exercise 4.10. Suppose that $(X_n)_{n\geq 0}$ is an irreducible Markov chain with state space $\mathscr{X}$ transition probability matrix Q and $x_0 \in \mathscr{X}$.

(i) For $n \in \mathbb{N}$ set

$$\tau_x(n) = \mathbb{P}_x\big[T_{x_0} > n\big], \quad \tau_x = \lim_{n\to\infty} \tau_x(n).$$

Prove that

$$\tau_x = \sum_{y \neq x_0} Q_{x,y} \tau_y, \quad \forall x \in \mathscr{X} \setminus \{x_0\}. \tag{4.6.1}$$

(ii) Show that if x_0 is transient, then there exists $x \in \mathscr{X} \setminus \{x_0\}$ such that $\tau_x \neq 0$.
(iii) Suppose there exists a function $\alpha : \mathscr{X} \setminus \{x_0\} \to [-1, 1]$, not identically zero, satisfying (4.6.1). Prove that x_0 is transient.

Exercise 4.11. Suppose that $(X_n)_{n\geq 0}$ is a transient irreducible Markov chain with state space $\mathscr{X}$. Prove that, with probability 1 the chain will exit any finite subset $F \subset \mathscr{X}$, never to return, i.e.,

$$\mathbb{P}\Big[\lim_{n\to\infty} \boldsymbol{I}_F(X_n) = 0\Big] = 1.$$

□

Exercise 4.12. Bobby's business fluctuates in successive years between three sates between three states: $0 =$ bankruptcy, $1 =$ verge of bankruptcy, $2 =$ solvency. The transition matrix giving the probability of evolving from state to state is

$$Q = \begin{bmatrix} 1 & 0 & 0 \\ 0.5 & 0.25 & 0.25 \\ 0.5 & 0.25 & 0.25 \end{bmatrix}.$$

(i) What is the expected number of years until Bobby's business goes bankrupt, assuming it starts in solvency.
(ii) Bobby's rich father, deciding that it is bad for the family name if his son goes bankrupt. Thus, when state 0 is entered, his father infuses Bobby's business with cash returning him to solvency with probability 1. Thus the transition matrix for this Markov chain is

$$P = \begin{bmatrix} 0 & 0 & 1 \\ 0.5 & 0.25 & 0.25 \\ 0.5 & 0.25 & 0.25 \end{bmatrix}.$$

Show that this Markov chain irreducible aperiodic and find the expected number of years between cash infusions from his father. □

Exercise 4.13. Let Q be a stochastic $n \times n$ matrix and denote by C the $n \times n$ matrix such that $C_{i,j} = \frac{1}{n}, \forall i,j$.

(i) Prove that for any $r \in (0,1)$ the Markov chain defined by the stochastic matrix $Q(r) = (1-r)Q + rC$ is irreducible and aperiodic. Denote by π_r the unique stationary probability measure.
(ii) Prove that π_r converges as $r \to 0$ to a stationary probability measure π_0 of the HMC defined by Q.
(iii) Describe π_0 in the special case when the HMC determined by Q consists of exactly two communication classes C_1 and C_2 and there exist $x_i \in C_i$, $i=1,2$ such that $Q_{x_1,x_2} > 0$.

□

Exercise 4.14. The random walk of a chess piece on a chess table is govern by the rule: the feasible moves are equally likely. Suppose that a rook and a bishop start at the same corner of a 4×4 chess table and perform these random walks. Denote by T the time they meet again at the same corner. Find $\mathbb{E}[T]$. □

Exercise 4.15. Consider the HMC with state space $\mathscr{X} = \{0, 1, 2, \dots\}$ and transition matrix Q defined by
$$Q_{n,n+k} = \frac{1}{2^{n+1}} \binom{n}{k}, \quad \forall 0 \leq k \leq n, \ n \geq 1,$$
$$Q_{0,1} = 1, \ Q_{n,0} = \frac{1}{2}, \ \forall n \geq 1.$$
Prove that the chain is irreducible, positively recurrent and aperiodic and find $\mathbb{E}_0[T_0]$. □

Exercise 4.16. Let K_{n+1} denote the complete graph with $n+1$ vertices $v_0, v_1, \dots, v_n$. Denote by $(X_n)_{n \geq 0}$ the random walk on K_{n+1} transition rules
$$Q_{v_i, v_j} = \frac{1}{n}, \ \forall i > 0, \ j \geq 0, \ Q_{v_0, v_i} = 0, \ Q_{v_0, v_0} = 1.$$
Thus the vertex v_0 is absorbent. For $i > 0$ we denote that the time to reach the vertex v_0 starting at v_i,
$$H_i := \min\{j \geq 0 : X_0 = i, \ X_j = 0\}.$$
Prove that $\mathbb{E}[H_i] = n, \forall i > 0$. □

Exercise 4.17. A particle performs a random walk on the *nonnegative* integers with transition probabilities
$$p_{0,0} = q, \ p_{i,i+1} = p, \ p_{j,j-1} = q, \ i \geq 0, \ j > 0,$$
where $p \in (0, 1)$ and $q = 1 - p$.

Prove that the random walk is transient if $p > q$, null recurrent if $p = q$, and positively recurrent if $p < q$. In the last case determine the unique invariant probability distribution. □

Exercise 4.18. We generate a sequence B_n of bits, i.e., 0's and 1's, as follows. The first two bits are chosen randomly and independently with equal probabilities. (Flip a fair 0/1 coin twice and record the results.) If $B_1, \dots, B_n$ are generated, then we generate B_{n+1} according to the rules
$$\mathbb{P}[B_{n+1} = 0 \| B_n = B_{n-1} = 0] = \frac{1}{2} = \mathbb{P}[B_{n+1} = 0 \| B_n = 0, B_{n-1} = 1]$$
$$\mathbb{P}[B_{n+1} = 0 \| B_n = 1, B_{n-1} = 0] = \frac{1}{4} = \mathbb{P}[B_{n+1} = 0 \| B_n = B_{n-1} = 1].$$
What is the proportion of 0's in the long run? □

Exercise 4.19. Consider the Markov chain with state space $\mathscr{X} = \mathbb{N}_0$ and transition probabilities
$$Q_{n,n-1} = 1, \ \forall n \in \mathbb{N},$$
$$Q_{0,n} = p_n, \ \forall n \in \mathbb{N}_0, \ \sum_{n \geq 0} p_n = 1.$$
Find a necessary and sufficient condition on the distribution $(p_n)_{n \geq 0}$ guaranteeing that the above HMC is positively recurrent. □

Exercise 4.20. Suppose that a gambles plays a fair game with winning probability $p = \frac{1}{2}$. He starts with an initial fortune $X_0 = 1$ dollar. His goal is to reach a fortune of g dollars, $g \in \mathbb{N}$. He stops if he reaches this fortune or he is broke and he is employing a bold strategy: at every game he stakes the largest of money that will get him closest to but not above g. He cannot bet a sum greater that his fortune at that moment. Denote by X_n his fortune after the n-th game.

(i) Prove that $(X_n)_{n \geq 0}$ is an HMC. Describe its state space and it transition matrix.
(ii) Prove that the player reaches his goal with probability $\frac{1}{g}$ and goes broke with probability $\frac{g-1}{g}$.

$\square$

Exercise 4.21. Consider the standard random walk in $\mathbb{Z}^2$ started at the origin. For each $m \in \in \mathbb{Z}$ we denote by T_m the first moment the random walk reaches the line $x + y = m$ and we denote by (U_m, V_m) the point where this walk intersects the above line. Find the probability distributions of T_m, U_m and V_m. $\square$

Exercise 4.22. A top-to-bottom shuffle of a deck of k cards consist of choosing a position j uniformly random in $\{1, \ldots, k\}$ and moving the top card to occupy position j in the deck, counted from top to bottom. This defines a random walk $\left(\Phi_n\right)_{n \geq 0}$ on the symmetric group $\mathfrak{S}_k$. Assume that Φ_0 is the identity permutation. Denote by T the first moment when the card that used to be at the bottom of the deck reaches the top, i.e.,

$$T = \min\{n \geq 1;\ \Phi_n(1) = k\}.$$

(i) Compute $\mathbb{E}[T]$.
(ii) Denote by Ψ_T the random permutation $\mathfrak{S}_{k-1}$ given by $\Psi_T(j) = \Phi_T(j+1)$, $j = 1, \ldots, k-1$. In other words, Ψ_T is the permutation of the cards below the top card at time T. Prove that Ψ_T is uniformly distributed on $\mathfrak{S}_{k-1}$.
(iii) Prove that Φ_{T+1} is uniformly distributed on $\mathfrak{S}_k$.

$\square$

Exercise 4.23. Suppose that $\mathscr{X}$ is an at most countable set equipped with the discrete topology $\mu \in \mathrm{Prob}(\mathscr{X})$ and $Q : \mathscr{X} \times \mathscr{X} \to [0, 1]$ is a stochastic matrix. Let $X_n : (\Omega, \mathcal{S}, \mathbb{P}) \to \mathscr{X}$ be a sequence of measurable maps.

(i) Prove that $(X_n)_{n \geq 0} \in \mathrm{Markov}(\mu, Q)$ if and only if

$$\mathbb{E}\big[f(X_{n+1}) \,\|\, X_n, \ldots, X_0\big] = Q^* f(X_n)$$

for any bounded function $f : \mathscr{X} \to \mathbb{R}$.

(ii) (Lévy) Prove that $(X_n)_{n\geq 0} \in \text{Markov}(\mu, Q)$ if and only if, for any $f \in L^\infty(\mathscr{X}, \mu)$ the sequence
$$Y_0 = X_0, \quad Y_n = f(X_n) - \sum_{k=0}^{n-1}\bigl(Qf(X_k) - f(X_k)\bigr)$$
is a martingale with respect to the filtration $\mathcal{F}_n = \sigma(X_0, X_1, \ldots, X_n)$.

□

Exercise 4.24. Consider an irreducible HMC with finite state space $\mathscr{X}$ and transition matrix Q. We denote by π the invariant probability distribution. For every $x \in \mathscr{X}$ we denote by H_x the hitting time of x,
$$H_x := \min\{n \geq 0;\ X_n = x\}.$$

(i) Show that
$$\tau(x) := \sum_{y \in \mathscr{X}} \mathbb{E}_x[H_y]\pi[y]$$
is independent of x.

(ii) Prove that
$$\tau(x) = \sum_{x} \mathbb{E}_\pi[H_y]\pi[y].$$

□

Exercise 4.25. Suppose that $(X_n)_{n \in \mathbb{N}_0}$ is an HMC with state space $\mathscr{X}$ and transition matrix Q. Suppose that $B \subset \mathscr{X}$ and H_B is the *hitting* time of B
$$H_B := \min\{n \geq 0,\ X_n \in B\}.$$
We define
$$h_B : \mathscr{X} \to [0,1], \quad h_B(x) = \mathbb{P}_x[H_B < \infty] = \mathbb{P}[H_B < \infty \| X_0 = x],$$
$$k_B : \mathscr{X} \to [0,\infty], \quad k_B(x) = \mathbb{E}_x[H_B].$$

(i) Show that h_B satisfies the linear system
$$h_B(x) = 1, \quad \forall x \in B,$$
$$h_B(x) = \sum_{y \in \mathscr{X}} Q_{x,y} h_B(y), \quad x \in \mathscr{X} \setminus B. \qquad (4.6.2)$$

(ii) Show that if $h : \mathscr{X} \to [0, \infty)$ is a solution of (4.6.2), then
$$h_B(x) \leq h(x), \quad \forall x \in \mathscr{X}.$$

(iii) Show that k_B satisfies the linear system
$$k_B(x) = 0, \quad \forall x \in B,$$
$$k_B(x) = 1 + \sum_{y \in \mathscr{X}} Q_{x,y} k_B(y), \quad x \in \mathscr{X} \setminus B. \qquad (4.6.3)$$

(iv) Show that if $k : \mathscr{X} \to [0, \infty]$ satisfies (4.6.3), then $k_B(x) \leq k(x)$, $\forall x \in \mathscr{X}$.

□

Exercise 4.26. Suppose that $(X_n)_{n \in \mathbb{N}_0}$ is an HMC with state space $\mathscr{X}$ and transition matrix Q. For $x \in \mathscr{X}$ we denote by T_x the return time to x

$$T_x := \min\{ n \geq 1; \ X_n = x \}.$$

We set

$$f_{x,y}(n) := \mathbb{P}_x[T_y = n],$$

$$F_{x,y}(s) := \sum_{n \geq 0} f_{x,y}(n) s^n, \quad P_{x,y}(s) := \sum_{n \geq 0} Q_{x,y}^n s^n,$$

$$f_{x,y} := F_{x,y}(1) = \sum_{n \geq 0} f_{x,y}(n) = \mathbb{P}_x[T_y < \infty].$$

(i) Prove that

$$P_{x,y}(s) = \delta_{x,y} + F_{x,y}(s) P_{y,y}(s), \ \forall x, y \in \mathscr{X},$$

where

$$\delta_{x,y} = \begin{cases} 1, & x = y, \\ 0, & x \neq y. \end{cases}$$

(ii) Deduce from (i) that

$$\sum_{n \geq 0} Q_{x,x}^n < \infty \Longleftrightarrow \mathbb{P}_x[T_x < \infty] < 1.$$

(iii) Set $T_x^{(1)} := T_x$ and define inductively $T_x^{(k)} := \min\{, n > T_x^{(k-1)}; \ X_n = x\}$, $k > 1$. Prove that

$$\mathbb{P}[T_x^{(k-1)} < \infty] = f_{x,y} f_{yy}^{(k-1)}.$$

□

Exercise 4.27. Suppose that $\{X_n : (\Omega, \mathcal{S}, \mathbb{P}) \to \mathscr{X}\}_{n \geq 0}$ is an irreducible, recurrent HMC with state space $\mathscr{X}$ and transition matrix Q defined on a probability space $(\Omega, \mathcal{S}, \mathbb{P})$. Fix $x \in \mathscr{X}$, assume $\mathbb{P}[X_0 = x] = 1$ and denote by T_k the time of k-th return to x. More precisely

$$T_0 = 0, \ T_1 := \min\{n > 0; \ X_n = x\}, \ T_{k+1} = \min\{n > T_k; \ X_n = x\}.$$

We set

$$Y_k = (X_{t_k}, X_{T_k+1}, \ldots, X_{T_{k+1}-1}).$$

(i) Realize the quantities Y_k as random maps $Y_k \to \mathcal{Y}$ where $\mathcal{Y}$ is a countable set equipped with the sigma-algebra $2^{\mathcal{Y}}$.

(ii) Show that the resulting random maps are i.i.d.

□

Exercise 4.28. Consider a positively recurrent HMC $(X_n)_{n\geq 0}$ with state space $\mathscr{X}$ and transition matrix Q. Denote by π the stationary distribution. For $x \in \mathscr{X}$ we denote by T_x the first return time to x and for $y \in \mathscr{X}$ we set

$$N_{x,y} := \{n \in \mathbb{N};\ n \leq T_x,\ X_n = y\}, \quad \mathcal{G}(x,y) := \mathbb{E}_x[N_{x,y}].$$

In other words, $\mathcal{G}(x,y)$ is the expected number of visits to y before returning to x.

(i) Prove that $\mathcal{G}(x,y) = \pi[y]\mathbb{E}_x[T_y]$.
(ii) Prove that

$$\mathbb{P}_x[T_y < T_x] = \frac{1}{\pi[y](\mathbb{E}_x[T_y] + \mathbb{E}_y[T_x])}.$$

□

Exercise 4.29. Consider a positively recurrent HMC $(X_n)_{n\geq 0}$ with state space $\mathscr{X}$, transition matrix Q and stationary distribution π. Suppose that T is a stopping time adapted to $(X_n)_{n\geq 0}$ and let $x \in \mathscr{X}$ be such that $\mathbb{E}_x[T] < \infty$. We denote $\mathcal{G}_T(x,y)$ denote the expected number of visits to y before T, when started at x, i.e.,

$$\mathcal{G}_T(x,y) = \mathbb{E}_x[N_{x,y}^T], \quad N_{x,y}^T = \#\{n \geq 0;\ X_0 = x,\ X_n = y,\ n \leq T\}.$$

Prove that $\mathcal{G}_T(x,y) = \pi[y]\mathbb{E}_x[T]$. □

Exercise 4.30 (LeCam). Suppose that $(X_r)_{1\leq r\leq n}$ is a family of independent Bernoulli random variables with success probabilities p_r. Set

$$\lambda := p_1 + \cdots + p_n, \quad S = X_1 + \cdots + X_n.$$

(i) Show that measure μ_r on $\mathbb{N}_0 \times \mathbb{N}_0$ given by

$$\mu_r[(m,n)] = \begin{cases} 1 - p_r, & m = n = 0, \\ e^{-p_r} - 1 + p_r, & m = 1,\ n = 0, \\ \frac{p_r^n}{n!}e^{-p_r}, & m = 1,\ n \geq 1, \\ 0, & \text{elsewhere} \end{cases}$$

is a coupling of $\text{Bin}(p_r)$ with $\text{Poi}(p_r)$ and conclude that

$$d_v(\text{Bin}(p_r), \text{Poi}(p_r)) \leq p_r^2.$$

(ii) Denote by $\mathbb{P}_S$ the probability distribution of S. Prove that

$$d_v(\mathbb{P}_S, \text{Poi}(\lambda)) \leq \sum_{r=1}^n p_r^2.$$

□

Exercise 4.31. Let $(X_n)_{n\geq 0}$ be an irreducible Markov chain with finite state space $\mathscr{X}$, transition matrix Q and invariant probability measure $\mu \in \text{Prob}(\mathscr{X})$. Assume that the initial distribution is also μ, i.e., $\mathbb{P}_{X_0} = \mu$. For $n \in \mathbb{N}_0$ we set (see Exercise 2.50 for notation)

$$H_n = \frac{1}{n+1} \text{Ent}_2[X_0, X_1, \ldots X_n], \quad L_n = \text{Ent}_2[X_n | X_{n-1}, \ldots, X_0].$$

(i) Prove that the sequence $(L_n)_{n\geq 0}$ is non-increasing and nonnegative. Denote by L its limit.

(ii) Prove that

$$H_n = \frac{1}{n+1} \sum_{k=0}^{n} L_k.$$

(iii) Prove that the sequence (H_n) is convergent and its limit is L.

(iv) Prove that

$$L = -\sum_{x\in\mathscr{X}} \mu[x] \text{Ent}_2[Q_{x,-}] = -\sum_{x,y\in\mathscr{X}} \mu[x] Q_{x,y} \log_2 Q_{x,y}.$$

The number L is called *entropy rate* of the irreducible Markov chain. We denote it by $\text{Ent}_2[\mathscr{X}, Q]$.

□

Exercise 4.32. Let Q denote the $n \times n$ transition matrix describing the random walk on a complete graph with n vertices. Find the spectrum of Q. □

Exercise 4.33 (Doeblin). Suppose that $(X_n)_{\geq 0}$ is an HMC with state space $\mathscr{X}$, initial distribution μ and transition matrix Q satisfying the *Doeblin condition*

$$\exists \varepsilon > 0, \exists x_0 \in \mathscr{X}: \ Q_{x,x_0} > \varepsilon, \ \forall x \in \mathscr{X}.$$

Denote $\mathcal{M}$ the space of finite signed measures ρ on $\mathscr{X}$. For $\rho \in \mathcal{M}$ we set

$$\|\rho\|_1 := \sum_{x\in\mathscr{X}} |\rho_x| < \infty, \quad \rho_x := \rho[\{x\}].$$

(i) Prove that for any $\rho \in \mathcal{M}$ we have $\rho Q \in \mathcal{M}$ and

$$\sum_{x\in\mathscr{X}} \rho_x = \sup_{y\in\mathscr{X}} (\rho Q)_y.$$

If $\rho \in \mathcal{M}$ and

$$\sum_{x\in\mathscr{X}} \rho_x = 0,$$

then

$$\|\rho Q\|_1 \leq (1-\varepsilon)\rho \|\rho\|_1.$$

(ii) Set $\mu_n := \mu \cdot Q^n$. Prove that
$$\|\mu_n - \mu_m\|_1 \leq 2(1-\varepsilon)^m, \quad \forall n \geq m \geq 1.$$

(iii) Prove that the HMC is irreducible, positively recurrent and the unique invariant probability measure π satisfies
$$\|\mu_n - \pi\|_1 \leq 2(1-\varepsilon)^n, \quad \forall n \in \mathbb{N}.$$

$\square$

Exercise 4.34. Suppose that $(X_n)_{\geq 0}$ is an HMC with state space $\mathscr{X}$, initial distribution μ and transition matrix Q. For each $n \in \mathbb{N}$ we set
$$A_n := \frac{1}{n+1} \sum_{k=0}^{k} Q^k.$$
Suppose that there exist $N \in \mathbb{N}$, $x_0 \in \mathscr{X}$ and $\varepsilon > 0$ such that
$$(A_N)_{x,x_0} > \varepsilon, \quad \forall x \in \mathscr{X}.$$
Prove that the HMC is irreducible, positively recurrent and the unique invariant probability measure π satisfies
$$\|\mu A_n - \pi\|_1 \leq \frac{N}{(n+1)\varepsilon}, \quad \forall n \in \mathbb{N}.$$

$\square$

Exercise 4.35. Prove Lemma 4.85. $\square$

Exercise 4.36. Suppose that $(\mathscr{X}, E, c)$ is a finite connected electric network and x_+, x_- are distinct vertices. The *commute time* between x_+, x_- is the quantity
$$K_{x_+,x_-} = \mathbb{E}_{x_+}[T_{x_-}] + \mathbb{E}_{x_-}[T_{x_+}].$$
Set
$$C(\mathscr{X}) := \sum_{e \in E} c(e).$$

(i) Prove that
$$K_{x_+,x_-} = 2C\boldsymbol{E}_{x_+,x_-}$$
where $\boldsymbol{E}_{x_+,x_-}$ denotes the energy of the Kirchhoff flow with source x_+ and sink x_-; see (4.4.26). **Hint.** Use Exercise 4.28.

(ii) Consider the Ehrenfest urn with B balls defined in Example 4.7. Use (i) to compute $\boldsymbol{E}_0[T_B]$. In other words, given that initially all the B balls were in the right chamber, find the expected time until all of them move in the left chamber.

$\square$

Chapter 5

Elements of Ergodic Theory

Ergodic theory is a rather eclectic subject with applications in many areas of mathematics, including probability. The ergodicity feature first appeared in the works of L. Boltzmann on statistical mechanics, [23]. The modern formulation of this hypothesis, due to Y. Sinai, came much later, in 1963, and it took a few more decades to be adjudicated mathematically.

Our rather modest goal in this chapter is to describe enough of the fundamentals of this theory so we can shed new light on some of the fundamental limit theorems we have proved in the previous chapters. For more details we refer to [3; 11; 36; 93; 131; 154] that served as our main sources of inspiration.

5.1 The ergodic theorem

5.1.1 *Measure preserving maps and invariant sets*

Suppose that $(\Omega, \mathcal{S}, \mathbb{P})$ is a probability space. A measurable map $T : (\Omega, \mathcal{S}) \to (\Omega, \mathcal{S})$ is said the be *measure preserving* if $T_\# \mathbb{P} = \mathbb{P}$, i.e.,

$$\mathbb{P}\big[\, T^{-1}(S)\, \big] = \mathbb{P}\big[\, S\, \big], \ \ \forall S \in \mathcal{S}. \tag{5.1.1}$$

The measure preserving map T is called an *automorphism* of the probability space if it is bijective, and its inverse is also measure preserving.

Proposition 1.29 shows that (5.1.1) is satisfied if and only if there exists a π-system $\mathcal{C}$ that generates $\mathcal{S}$ such that

$$\mathbb{P}\big[\, T^{-1}(C)\, \big] = \mathbb{P}\big[\, C\, \big], \ \ \forall C \in \mathcal{C}. \tag{5.1.2}$$

Example 5.1. (a) Let $\mathbb{P}_{S^1}$ denote the Euclidean probability measure on S^1, the unit circle in $\mathbb{R}^2$, i.e. (see Example 1.87)

$$\mathbb{P}_{S^1}\big[\, d\theta\, \big] = \frac{1}{2\pi} d\theta.$$

We denote by R_φ the counterclockwise rotation by an angle φ about the center of this circle. Then R_φ is measure preserving. If we think of S^1 as the set of complex numbers of norm 1,

$$S^1 := \big\{\, z \in \mathbb{C};\ |z| = 1\, \big\},$$

then $R_\varphi(z) = e^{i\varphi} z$.

(b) Consider the n-dimensional torus $\mathbb{T}^n := \mathbb{R}^n/\mathbb{Z}^n$. Set $I = [0,1]$ and observe that the natural projection $\pi : I^n \to \mathbb{T}^n$ is Borel measurable. We denote by $\mathbb{P}_{\mathbb{T}^n}$ the push-forward by π of the Lebesgue measure on I^n. Let observe that the resulting probability space is isomorphic to the product of n copies of $(S^1, \mathcal{B}_{S^1}, \mathbb{P}_{S^1})$. Suppose that $A \in \mathrm{SL}_n(\mathbb{Z})$, i.e., A is an $n \times n$ matrix with integer coefficients and determinant 1. Then $A(\mathbb{Z}^n) = \mathbb{Z}^n$ and thus we have a well defined induced map

$$T_A : \mathbb{R}^n/\mathbb{Z}^n \to \mathbb{R}^n/\mathbb{Z}^n.$$

This map is clearly bijective and Borel measurable. It is also measure preserving since $\det C = 1$.

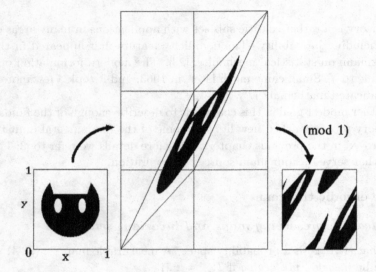

Fig. 5.1 Arnold's cat map.

In [3] Arnold and Avez memorably depicted the action of the map T_A for

$$A = \begin{bmatrix} 1 & 1 \\ 1 & 2 \end{bmatrix} \in \mathrm{SL}_2(\mathbb{Z})$$

as in Figure 5.1. This map is popularly known as *Arnold's cat map*.

(c) In the previous examples, the maps where automorphisms of the corresponding probability spaces. Here is an example of a measure preserving map that is not bijective. More precisely define

$$Q : S^1 \to S^1, \quad Q(z) = z^2.$$

Then the Lebesgue measure $\frac{1}{2\pi}d\theta$ is Q-invariant. If we identify S^1 with $\mathbb{R}$ mod $\mathbb{Z}$, then we can describe Q as the map $Q : [0,1) \to [0,1)$ given by

$$Q(x) = 2x \bmod 1.$$

If $x \in [0,1)$ has binary expansion

$$x = \sum_{n=1}^{\infty} \frac{\epsilon_n}{2^n}, \quad \epsilon_n \in \{0,1\},$$

then

$$Q(x) = \sum_{n=1}^{\infty} \frac{\epsilon_{n+1}}{2^n}.$$

(d) Consider the *tent map* $T : [0,1] \to [0,1]$, $T(x) = \min(2x, 2-2x)$. Equivalently, this is the unique continuous map such that $T(0) = T(1) = 0$, $T(1/2) = 1$ and it is linear on each of the intervals $[0, 1/2]$ and $[1/2, 1]$. Its graph looks like a tent with vertices $(0,0)$, $(1/2, 1)$ and $(1, 0)$.

This map preserves the Lebesgue measure. Indeed, if $I \subset [0,1]$ is a compact interval then $T^{-1}(I)$ consists of two intervals $I_{\pm}$, symmetrically located with respect to the midpoint $1/2$ of $[0,1]$, and each having half the size of I.

(e) Suppose that X is a compact metric space and $T : X \to X$ is a continuous map. Denote by $\mathrm{Prob}(X)$ the set of Borel probability measures on X. Then map T induces a push-forward map $T_\# : \mathrm{Prob}(X) \to \mathrm{Prob}(X)$. The T-invariant measures are precisely the fixed points of $T_\#$. One can show (see Exercise 5.2) that the set $\mathrm{Prob}_T(X)$ of T-invariant measures is nonempty, convex and closed with respect to the weak convergence. □

Example 5.2 (Stationary sequences). Let $(\mathbb{X}, \mathcal{F})$ be a measurable space and suppose that $X_n : \Omega \to \mathbb{X}$, $n \in \mathbb{N}$, is a sequence of random maps defined on the same probability space $(\Omega, \mathcal{S}, \mathbb{P})$. The sequence is said to be *stationary* if for any $m, k \in \mathbb{N}$ the random vectors

$$(X_1, \ldots, X_m) : \Omega \to \mathbb{X}^m \text{ and } (X_{k+1}, \ldots, X_{k+m}) : \Omega \to \mathbb{X}^m$$

have the same distribution.

(i) For example, a sequence of i.i.d. random variables is stationary. More generally, an exchangeable sequence of random variables is stationary.
(ii) Suppose that $(X_n)_{n \geq 0}$ is an HMC with state space $\mathscr{X}$ and transition matrix Q and initial distribution μ. The sequence $(X_n)_{n \geq 0}$ is stationary if and only if μ is an invariant distribution, i.e., $\mu = \mu \cdot Q$.

To any stationary sequence $X_n : \Omega \to \mathbb{X}$, we can canonically associate a measure preserving map as follows. Consider the *path space* $\mathbb{U} = \mathbb{U}_{\mathbb{X}} := \mathbb{X}^{\mathbb{N}}$. It consists of sequences of points in $\mathbb{X}$, $\underline{u} = (u_1, u_2, \ldots)$. We have natural coordinate maps $U_n : \mathbb{U} \to \mathbb{X}$, $U_n(\underline{u}) = u_n$.

For $n \in \mathbb{N}$ denote by $\mathcal{U}_n$ the sigma-algebra generated by $U_1, \ldots, U_n$ and we set

$$\mathcal{U} = \bigvee_{n=1}^{\infty} \mathcal{U}_n.$$

Note that we have a $\mathcal{U}$-measurable map *shift map*
$$\Theta : \mathbb{U} \to \mathbb{U}, \quad \Theta(u_1, u_2, \dots) = (u_2, u_3, \dots).$$
A sequence of random variables $X_n : (\Omega, \mathcal{S}, \mathbb{P}) \to (\mathbb{X}, \mathcal{F})$, $n \in \mathbb{N}$, defines a measurable map
$$\vec{X} : (\Omega, \mathcal{S}) \to (\mathbb{U}, \mathcal{U}), \quad \omega \mapsto (X_1(\omega), X_2(\omega), \dots).$$
The distribution of this sequence is the push-forward probability measure $\mathbb{P}_{\vec{X}} := \vec{X}_{\#}\mathbb{P}$. Note that
$$X_n = U_n \circ \vec{X} \text{ and } U_{k+1} = U_1 \circ \Theta^k, \quad \forall n, k \in \mathbb{N}.$$
Since the measure $\mathbb{P}_{\vec{X}}$ is uniquely determined by its restrictions to the sigma-subalgebras $\mathcal{U}_n$ we deduce that the sequence $(X_n)_{n \in \mathbb{N}}$ is stationary iff the shift Θ preserves the distribution $\mathbb{P}_{\vec{X}}$ on the path space.

When $\mathbb{X}$ is finite or countable, and the sequence $(X_n)_{n \in \mathbb{N}}$ is i.i.d., the resulting shift is known as *Bernoulli shift*.

Conversely, if T is a measurable map $T : (\Omega, \mathcal{S}, \mathbb{P}) \to (\Omega, \mathcal{S}, \mathbb{P})$, then it is measure preserving if and only if, for any measurable function $f : \Omega \to \mathbb{R}$ the sequence
$$f, f \circ T, f \circ T^2, \dots$$
is stationary. □

Definition 5.3. Suppose that $(\Omega, \mathcal{S})$ is a measurable space and $T : (\Omega, \mathcal{S}) \to (\Omega, \mathcal{S})$ is a measurable map.

(i) A measurable function $f : \Omega \to \mathbb{R}$ is called *T-invariant* if $f \circ T = f$.
(ii) A measurable set $S \in \mathcal{S}$ is called *T-invariant* if its indicator $\boldsymbol{I}_S$ is an invariant function.
(iii) We denote by $\mathcal{I} = \mathcal{I}_T$ the collection of invariant sets. □

Remark 5.4. (a) Note the definition of $\mathcal{I}_T$ involves no measure on $\mathcal{S}$.

(b) Note that if $S \in \mathcal{S}$, then $\boldsymbol{I}_S \circ T = \boldsymbol{I}_{T^{-1}(S)}$ so the set S is T-invariant iff $S = T^{-1}(S)$. Observe that
$$S \subset T^{-1}(S) \Longleftrightarrow T(S) \subset S, \tag{5.1.3a}$$
$$T^{-1}(S) \subset S \Longleftrightarrow \forall \omega \in \Omega, \ T(\omega) \in S \Rightarrow \omega \in S. \tag{5.1.3b}$$
We can give a dynamic description of invariance. For $\omega \in \Omega$ we denote by $\mathcal{O}_T(\omega)$ the orbit of ω with respect to the action of T
$$\mathcal{O}_T(\omega) := \{\omega, T(\omega), T^2(\omega), \dots\}.$$
A set S is invariant if and only if
$$\omega \in S \Rightarrow \mathcal{O}_T(\omega) \subset S \text{ and } \omega \in \Omega \setminus S \Rightarrow \mathcal{O}_T(\omega) \subset \Omega \setminus S.$$
In the universal case $(\mathbb{U}_{\mathbb{X}}, \mathcal{U})$, a subset $S \in \mathcal{U}_{\mathbb{X}}$ is Θ-invariant if
$$\underline{s} = (s_1, s_2, \dots) \in S \Rightarrow (s_2, s_3, \dots) \in S,$$
$$(s_2, s_3, \dots) \in S \Rightarrow \forall s_1 \in \mathbb{X} : \ (s_1, s_2, s_3, \dots) \in S.$$
Note that if T is an automorphism, then a set S is invariant iff $T(S) = S$. □

Proposition 5.5. *Suppose that $(\Omega, \mathcal{S})$ is a measurable space and $T : (\Omega, \mathcal{S}) \to (\Omega, \mathcal{S})$ is a measurable map. Then the following hold.*

(i) The collection $\mathcal{I} = \mathcal{I}_T$ of T-invariant measurable sets is a sigma-subalgebra of $\mathcal{S}$.

(ii) An $\mathcal{S}$-measurable function $f : \Omega \to \mathbb{R}$ is T-invariant if and only if it is $\mathcal{I}_T$-measurable.

Proof. (i) Thus follows from the fact that $S \in \mathcal{I}_T$ if and only if $S = T^{-1}(S)$.

(ii) Suppose that f is T-invariant. Then for any $x \in \mathbb{R}$ the set $S = \{f \leq x\}$ is T-invariant since $\boldsymbol{I}_S \circ T = \boldsymbol{I}_{\{f \circ T \leq x\}} = \boldsymbol{I}_S$.

Conversely, if f is $\mathcal{I}_T$-measurable, then $f^{-1}(\{y\}) \in \mathcal{I}_T$, $\forall y \in \mathbb{R}$ and
$$(f \circ T)^{-1}(\{y\}) = T^{-1}\big(f^{-1}(\{y\})\big) = f^{-1}(\{y\}).$$
If $f(x) = y$, then $x \in f^{-1}(\{y\}) = (f \circ T)^{-1}(\{y\})$ so that $f \circ T(x) = y = f(x)$. □

Remark 5.6. Consider the path space $\mathbb{U}_{\mathbb{X}} = \mathbb{X}^{\mathbb{N}}$. We have the *tail* sigma-subalgebra
$$\mathcal{T}_\infty = \bigcap_{m \geq 1} \mathcal{T}_m, \quad \mathcal{T}_m = \sigma(U_m, U_{m+1}, \ldots).$$
Note that
$$S \in \mathcal{T}_{m+1} \iff \Theta^m S \in \mathcal{T}_1 = \mathcal{U}, \quad \forall m \geq 0.$$
The shift map Θ is surjective and if S is Θ-invariant, then (5.1.3a) and (5.1.3b) imply that $\Theta S = S$. In particular, $\Theta^m S = S \in \mathcal{T}_1$, so $S \in \mathcal{T}_m$, $\forall m$. Hence, in the universal case $\mathcal{I} = \mathcal{I}_\Theta \subset \mathcal{T}_\infty$.

Observe that the sigma-algebras $\mathcal{I}_\Theta$ and $\mathcal{T}$ do not depend on any choice of probability measure on $\mathbb{U}_{\mathbb{X}}$. □

Definition 5.7. Suppose that $T : (\Omega, \mathcal{S}, \mathbb{P}) \to (\Omega, \mathcal{S}, \mathbb{P})$ is a measure preserving map. A *measurable function* $f : (\Omega, \mathcal{S}) \to \mathbb{R}$ is said to be *quasi-invariant* if
$$f = f \circ T \quad \mathbb{P} - \text{a.s.}$$
A subset $S \in \mathcal{S}$ is said to be *quasi-invariant* if $\boldsymbol{I}_S$ is quasi-invariant, i.e.,
$$\mathbb{P}\big[S \Delta T^{-1} S\big] = 0,$$
where $A \Delta B := (A \setminus B) \cup (B \setminus A)$ is the symmetric difference of two sets. We denote by $\mathcal{J} = \mathcal{J}_T$ the collection of T-quasi-invariant sets. □

Proposition 5.8. *Suppose that $T : (\Omega, \mathcal{S}, \mathbb{P}) \to (\Omega, \mathcal{S}, \mathbb{P})$ is a measure preserving map. Then the following hold.*

(i) The collection $\mathcal{J}_T$ of quasi-invariant sets is a sigma-algebra.

(ii) The $\mathbb{P}$-completions of $\mathcal{I}$ and $\mathcal{J}$ coincide, i.e., for any $S \in \mathcal{J}$ there exists $S' \in \mathcal{I}$ such that $\mathbb{P}\big[S \Delta S'\big] = 0$.

(iii) A measurable function is T-quasi-invariant if and only if it is $\mathcal{J}_T$-measurable.

Proof. (i) The fact that $\mathcal{J}_T$ is a sigma-algebra follows immediately from the definition of a quasi-invariant.

(ii) Denote by $\bar{\mathcal{J}}$ the $\mathbb{P}$-completion of $\mathcal{J}$. Let $\bar{S} \in \bar{\mathcal{J}}$. There exists $S \in \mathcal{J}$ such that $\mathbb{P}[\bar{S}\Delta S] = 0$. Since T is measure preserving we deduce

$$0 = \mathbb{P}[T^{-1}(\bar{S}\Delta S)] = \mathbb{P}[T^{-1}(\bar{S})\Delta S]$$

and thus

$$\mathbb{P}[T^{-1}(\bar{S})\Delta \bar{S}] = \mathbb{E}[|I_{T^{-1}(\bar{S})} - I_{\bar{S}}|]$$

$$\leq \mathbb{E}[|I_{T^{-1}(\bar{S})} - I_S|] + \mathbb{E}[|I_S - I_{\bar{S}}|] = 0.$$

Conversely, if $S \in \mathcal{J}$ define

$$\bar{S} := \bigcap_{n \in \mathbb{N}} S_n, \quad S_n := \bigcup_{k \geq n} T^{-k}(S).$$

Note that $\bar{S} = T^{-1}(\bar{S})$ so that $\bar{S}$ is invariant. Since $S_1 \supset S_2 \supset \cdots$, we have

$$I_{\bar{S}} = \lim_{n \to \in \infty} I_{S_n}.$$

On the other hand

$$I_{S_n} = \sup_{k \geq n} I_{T^{-k}(S)} = \sup_{k \geq n} I_S \circ T^n = I_S \text{ a.s.}$$

since S is quasi-invariant and thus $I_S \circ T^n = I_S$ a.s. Hence $I_{\bar{S}} = I_S$ a.s., so that $S \in \bar{\mathcal{J}}$.

(iii) Clearly, if f is quasi-invariant, then so are the sublevel sets $\{f \leq x\}$, $\forall x \in \mathbb{R}$ and thus f is $\mathcal{J}_T$-measurable.

Conversely if f is $\mathcal{J}_T$-measurable, then so are $f^\pm$ and it suffices to show that if $f \geq 0$ is $\mathcal{J}$-measurable, then f is quasi-invariant. Clearly any $\mathcal{J}$-measurable elementary function is quasi-invariant. Since f is an increasing limit of $\mathcal{J}$-measurable elementary functions, it is therefore an increasing limit of quasi-invariant elementary functions and thus it is quasi-invariant. □

Definition 5.9. Let $(\Omega, \mathcal{S}, \mathbb{P})$ be probability space. A measure preserving map $T : \Omega \to \Omega$ is said to be *ergodic* if any T-invariant set has measure 0 or 1, i.e., the sigma-algebra of invariant sets is a zero-one algebra. □

From Proposition 5.8 we deduce the following equivalent characterization of ergodicity.

Proposition 5.10. *The map T is ergodic if and only if any $\mathbb{P}$-quasi-invariant set is a zero-one event, i.e., has measure 0 or 1.* □

Remark 5.11. Suppose that T is an ergodic automorphism of the probability space $(\Omega, \mathcal{S}, \mathbb{P})$. If $S \in \mathcal{S}$, then the set

$$\widehat{S} = \bigcup_{n \geq 0} T^n(S)$$

is quasi-invariant. Indeed $T(\widehat{S}) \subset \widehat{S}$ and $\mathbb{P}[T(\widehat{S})] = \mathbb{P}[\widehat{S}]$ invariant. If $\mathbb{P}[S] > 0$, then $\mathbb{P}[\widehat{S}] > 0$ and the ergodicity of T implies that

$$\mathbb{P}[\Omega \setminus \widehat{S}] = 0.$$

The set $\widehat{S}$ is a union or orbits $\mathcal{O}_T(\omega) = \{T^n(\omega)\}_{n \geq 0}$ of the dynamical system on Ω determined by the iterates of T. The ergodicity shows that the orbits originating in a set S of positive measure reach almost any point in Ω; the unreachable ones form a negligible set. This shows that the dynamics of an ergodic automorphism is quite chaotic: orbits want to fill the space. □

Definition 5.12. Suppose that $X_n : (\Omega, \mathcal{S}, \mathbb{P}) \to (\mathbb{X}, \mathcal{F})$, $n \in \mathbb{N}$ is a sequence of measurable maps. We say that $(X_n)_{n \in \mathbb{K}}$ is a *Kolmogorov sequence* if its tail algebra

$$\mathcal{T}_\infty := \bigcap_{m \in \mathbb{N}} \mathcal{T}_m, \quad \mathcal{T}_m := \sigma(X_n, \; n \geq m)$$

is a zero-one algebra. □

As shown in Remark 5.6 the sigma-algebra $\mathcal{I}$ of Θ-invariant sets is contained in the tail algebra. Hence, if $(X_n)_{n \in \mathbb{N}}$ is a stationary Kolmogorov sequence, then the shift map Θ on the associated path space $\mathbb{X}^\mathbb{N}$ is ergodic. In particular, if $(X_n)_{n \in \mathbb{N}}$ is a sequence of i.i.d. random variables, then Kolmogorov's 0-1 theorem shows that the shift map on the path space is ergodic.

Example 5.13. Consider the map $Q : [0, 1) \to [0, 1)$ discussed in Example 5.1(iii). The interval $[0, 1)$ embeds in $\{0, 1\}^\mathbb{N}$

$$[0, 1) \ni x = \sum_{n=1}^\infty \frac{\epsilon_n}{2^n} \mapsto (\epsilon_1, \epsilon_2, \dots) \in \{0, 1\}^\mathbb{N}.$$

The image of the map is a shift-invariant subset of $\{0, 1\}^\mathbb{N}$. Its complement is negligible with respect to the product measure on $\{0, 1\}^\mathbb{N}$ and the restriction of the product measure on the image of this embedding coincides with the Lebesgue measure; see Exercise 1.3(vii). The space $\{0, 1\}^\mathbb{N}$ equipped with the product measure is the path space corresponding to an i.i.d. sequence of Bernoulli random variables with success probability $\frac{1}{2}$. Hence the shift map is ergodic, proving that the map Q is also ergodic. □

We have the following characterization of Kolmogorov sequences due to Blackwell and Freedman [14].

Theorem 5.14. *Suppose that $X_n : (\Omega, \mathcal{S}, \mathbb{P}) \to (\mathbb{X}, \mathcal{F})$, $n \in \mathbb{N}$ is a sequence of measurable maps. The following are equivalent.*

(i) The sequence is a Kolmogorov sequence.

(ii) For any $A \in \mathcal{S}$

$$\lim_{m \to \infty} \sup_{B \in \mathcal{T}_m} |\mathbb{P}[A \cap B] - \mathbb{P}[A]\mathbb{P}[B]| = 0, \qquad (5.1.4)$$

where we recall that $\mathcal{T}_m = \sigma(X_n, \ n \geq m)$.

Proof. (i) $\Rightarrow$ (ii) Then for any $B \in \mathcal{T}_\infty$, $A \in \mathcal{F}$ we have

$$|\mathbb{P}[A \cap B] - \mathbb{P}[A]\mathbb{P}[B]| = \left|\mathbb{E}[I_A I_B] - \mathbb{E}[I_A]\mathbb{E}[I_B]\right|$$

$$= \left|\int_B (\mathbb{E}[I_A \,\|\, \mathcal{T}_m] - \mathbb{P}[A])\right| \leq \int_\Omega |\mathbb{E}[I_A \,\|\, \mathcal{T}_m] - \mathbb{P}[A]|.$$

Hence

$$\sup_{B \in \mathcal{T}_m} |\mathbb{P}[A \cap B] - \mathbb{P}[A]\mathbb{P}[B]| \leq \int_\Omega |\mathbb{E}[I_A \,\|\, \mathcal{T}_m] - \mathbb{P}[A]|. \qquad (5.1.5)$$

The Backwards Martingale Convergence Theorem implies that

$$\mathbb{E}[I_A \,\|\, \mathcal{T}_m] \to \mathbb{E}[A \,\|\, \mathcal{T}_\infty] \text{ a.s. and } L^1.$$

Since $\mathcal{T}_\infty$ is a zero-one algebra, we deduce that

$$\lim_{m \to \infty} \mathbb{E}[A \,\|\, \mathcal{T}_\infty] = \mathbb{E}[I_A] = \mathbb{P}[A].$$

Using this in (5.1.5) we obtain (5.1.4).

(ii) $\Rightarrow$ (i) Let $A \in \mathcal{T}_\infty$. Then, $\forall m$, $A \in \mathcal{T}_m$ and thus

$$0 \leq \mathbb{P}[A] - \mathbb{P}[A]^2 = |\mathbb{P}[A \cap A] - \mathbb{P}[A]\mathbb{P}[A]|$$

$$\leq \sup_{B \in \mathcal{T}_m} |\mathbb{P}[A \cap B] - \mathbb{P}[A]\mathbb{P}[B]| \to 0.$$

Hence $\mathbb{P}[A] = \mathbb{P}[A]^2$ so that $\mathbb{P}[A] \in \{0, 1\}$ so that $\mathcal{T}_\infty$ is a zero-one algebra. $\square$

5.1.2 Ergodic theorems

Let $(\Omega, \mathcal{S}, \mathbb{P})$ be probability space and suppose that $T : \Omega \to \Omega$ is measure preserving map. For any measurable function $f : (\Omega, \mathcal{S}, \mathbb{P}) \to \mathbb{R}$ we denote by $\widehat{T}f$ its pullback by T

$$\widehat{T}f := f \circ T.$$

This is a measurable function and, since T is measure preserving, we deduce from the change-in-variables formula (Theorem 1.75) that

$$\int_\Omega \widehat{T}f d\mathbb{P} = \int_\Omega f dT_\# \mathbb{P} = \int_\Omega f d\mathbb{P}, \ \ \forall f \in \mathcal{L}^1(\Omega, \mathcal{S}, \mathbb{P}).$$

Note also that

$$(\widehat{T}f)^p = \widehat{T}f^p \ \forall f \in \mathcal{L}^0_+(\Omega, \mathcal{S}), \ p \geq 1.$$

We will denote by $\|-\|_p$ the norm of $L^p(\Omega, \mathcal{S}, \mathbb{P})$. We deduce that $\widehat{T}$ defines isometries

$$\widehat{T}: L^p(\Omega, \mathcal{S}, \mathbb{P}) \to L^p(\Omega, \mathcal{S}, \mathbb{P}), \quad \forall p \geq 1.$$

The operator $\widehat{T}$ is referred to as the *Koopman operator*.

We denote by $\mathcal{J} = \mathcal{J}_T$ the σ-subalgebra of quasi-invariant measurable subsets. Thus $S \in \mathcal{J}$ if and only if $\widehat{T} \boldsymbol{I}_S = \boldsymbol{I}_S$, $\mathbb{P}$-a.s.

More generally, if $f \in L^1(\Omega, \mathcal{S}, \mathbb{P})$ and $S \in \mathcal{J}$, then

$$\widehat{T} \boldsymbol{I}_S \in L^1(\Omega, \mathcal{S}, \mathbb{P}) \text{ and } \widehat{T}(f \boldsymbol{I}_S) = (\widehat{T}f) \cdot (\widehat{T} \boldsymbol{I}_S) = (\widehat{T}f) \cdot \boldsymbol{I}_S,$$

so that

$$\int_\Omega f \boldsymbol{I}_S d\mathbb{P} = \int_\Omega \widehat{T}(f \boldsymbol{I}_S) d\mathbb{P} = \int_\Omega (\widehat{T}f) \cdot \boldsymbol{I}_S d\mathbb{P}, \quad \forall S \in \mathcal{J}, \ f \in L^1(\Omega, \mathcal{S}, \mathbb{P}). \tag{5.1.6}$$

For each $p \geq 1$ we set

$$\mathcal{Q}_{T,p} := \{ f \in L^p(\Omega, \mathcal{S}, \mathbb{P}); \ \widehat{T}f = f \}. \tag{5.1.7}$$

In other words, $\mathcal{Q}_{T,p}$ consists of quasi-invariant L^p-functions, i.e.,

$$\mathcal{Q}_{T,p} = L^p(\Omega, \mathcal{J}, \mathbb{P}) = L^p(\Omega, \mathcal{J}, \mathbb{P}).$$

We set

$$\mathcal{Q}_T := \mathcal{Q}_{T,2} = L^2(\Omega, \mathcal{J}, \mathbb{P}) \tag{5.1.8}$$

and we denote by P_T the orthogonal projection onto $\mathcal{Q}_T$. In the proof of Theorem 1.162 we have shown that

$$P_T f = \mathbb{E}[f \| \mathcal{J}].$$

The space $\mathcal{Q}_T$ contains the constant functions so $\dim \mathcal{Q}_T \geq 1$.

Proposition 5.15. *Suppose that* $T : (\Omega, \mathcal{J}, \mathbb{P}) \to (\Omega, \mathcal{J}, \mathbb{P})$ *is a measure preserving map. Then the following statements are equivalent.*

(i) The map T is ergodic.
(ii) For any $p \geq 1$ $\dim \mathcal{Q}_{T,p} = 1$.
(iii) There exists $p \geq 1$ such that $\dim \mathcal{Q}_T^p = 1$.

Proof. (i) $\Rightarrow$ (ii) Assume that T is ergodic so $\mathcal{J}$ is a zero-one sigma-subalgebra. Hence, any $\mathcal{J}$-measurable elementary function is constant. Hence any L^p function must be a.s.-constant as a limit of elementary functions.

Clearly (ii) $\Rightarrow$ (iii). To prove the implication (iii) $\Rightarrow$ (i) note any $\mathcal{J}$-measurable function belongs to any L^p and thus must be a.s. constant. $\square$

To summarize

$$\boxed{T \text{ is ergodic} \iff \dim \mathcal{Q}_T = 1.} \tag{5.1.9}$$

For each n we denote by A_n the n-th *temporal average/mean operator*

$$f \mapsto A_n f = \frac{1}{n}\bigl(1 + \widehat{T} + \widehat{T}^2 + \cdots \widehat{T}^{n-1}\bigr)f.$$

Note that A_n defines linear operators

$$A_n : L^p(\Omega, \mathcal{S}, \mathbb{P}) \to L^p(\Omega, \mathcal{S}, \mathbb{P}), \ \ p \geq 1$$

satisfying

$$\|A_n f\|_p \leq \|f\|_p, \ \ \forall f \in L^p. \tag{5.1.10}$$

Remark 5.16. Let me briefly explain the intuition of the temporal averages $A_n(f)$. Think of Ω as the space of states of a physical system that evolves in discrete time. Thus, if the system was initially in the state ω, it will be in the state $T^n(\omega)$ after n units of time.

A function $f : \Omega \to \mathbb{R}$ can be viewed as a macroscopic quantity that associates to each state ω a measurable numerical quantity $f(\omega)$. Note that for each $n \in \mathbb{N}$ and each $\omega \in \Omega$ we have

$$(A_{n+1}f)(\omega) = \frac{f(\omega) + f(T\omega) + \cdots + f(T^n\omega)}{n+1}$$

is the average value of the macroscopic quantity f as the system evolves for n units of time. $\square$

We have the following *mean ergodic theorem* due to John von Neumann.

Theorem 5.17 (L^2-**Mean ergodic theorem**)**.** *Suppose that $(\Omega, \mathcal{S}, \mathbb{P})$ is a probability space and $T : \Omega \to \Omega$ is a measure preserving map. Then, $\forall f \in L^2(\Omega, \mathcal{S}, \mathbb{P})$, the temporal averages $A_n f$ converge in L^2 to the orthogonal projection of f onto the space $\mathcal{Q}_T$ of quasi-invariant functions, i.e.,*

$$\frac{1}{n}\bigl(1 + \widehat{T} + \widehat{T}^2 + \cdots + \widehat{T}^{n-1}\bigr)f \to P_T f = \mathbb{E}[f \,\|\, \mathcal{J}].$$

In particular, if T is ergodic we have

$$\frac{1}{n}\bigl(1 + \widehat{T} + \widehat{T}^2 + \cdots + \widehat{T}^{n-1}\bigr)f \to \mathbb{E}[f]\mathbf{I}_\Omega \ \text{ in } L^2.$$

Proof. Denote by $\mathcal{X}_2$ the collection of functions $f \in L^2(\Omega, \mathcal{S}, \mathbb{P})$ such that $A_n f$ converges in L^2 to some function $A_\infty f$. Clearly $\mathcal{X}_2$ is a vector space. We will gradually show that $\mathcal{X}_2 = L^2(\Omega, \mathcal{S}, \mathbb{P})$ and $A_\infty = P_T$.

1. $\mathcal{Q}_T \subset \mathcal{X}_2$ and $A_\infty f = f, \forall f \in \mathcal{Q}_T$.

Indeed $A_n f = f, \forall f \in \mathcal{Q}_T$.

2. $\forall f \in \mathcal{X}_2$, we have $\widehat{T} f \in \mathcal{X}_2$ and $A_\infty f \in \mathcal{Q}_T$.

Let $f \in \mathscr{X}_2$. Note first that $\widehat{T}$ commutes with A_n. Since $\widehat{T}$ is continuous we deduce
$$\lim_{n\to\infty} A_n \widehat{T} f = \lim_{n\to\infty} \widehat{T} A_n f = \widehat{T} A_\infty f,$$
i.e., $\widehat{T} f \in \mathscr{X}_2$ and $A_\infty \widehat{T} f = \widehat{T} A_\infty f$.

On the other hand,
$$n A_n \widehat{T} f = (n+1) A_{n+1} f - f \implies A_n \widehat{T} f = \frac{n+1}{n} A_{n+1} f - \frac{1}{n} f,$$
so that
$$\widehat{T} A_\infty f = A_\infty \widehat{T} f = \lim_{n\to\infty} \frac{n+1}{n} A_{n+1} f = A_\infty f.$$
Hence $A_\infty f \in \mathfrak{Q}_T$.

3. $\widehat{T} f - f \in \mathscr{X}_2$, $\forall f \in L^2(\Omega, \mathcal{S}, \mathbb{P})$.

Indeed
$$A_n(\widehat{T} f - f) = \frac{1}{n}(\widehat{T}^n f - f)$$
and, since $\widehat{T}$ is unitary, we deduce that
$$\left\| A_n(\widehat{T} f - f) \right\|_2 \leq \frac{1}{n}(\|\widehat{T} f\|_2 + \|f\|_2) = \frac{2}{n}\|f\|_2 \to 0.$$

4. $\forall f \in L^2(\Omega, \mathcal{S}, \mathbb{P})$, $\forall k \in \mathbb{N}$ we have $\widehat{T}^k f - f \in \mathfrak{Q}_T^\perp$.

We first prove the claim for $k = 1$. Indeed, for any $g \in \mathfrak{Q}_T$ we have $\widehat{T} g = g$ and
$$(\widehat{T} f - f, g) = (\widehat{T} f, g) - (f, g) = (\widehat{T} f, \widehat{T} g) - (f, g) = 0,$$
where at the last step we used the fact that $\widehat{T}$ is unitary. In general
$$\widehat{T}^k f - f = \sum_{j=1}^{k} (\widehat{T}^j f - \widehat{T}^{j-1} f) = \sum_{j=1}^{k} (\widehat{T} f_j - f_j), \quad f_j := (\widehat{T}^{j-1} f),$$
and $\widehat{T} f_j - f_j \in \mathfrak{Q}_T^\perp$, $\forall j$.

5. $A_\infty f = P_T f$, $\forall f \in \mathscr{X}_2$.

We have
$$f - A_n f = \frac{1}{n} \sum_{k=1}^{n-1} (f - \widehat{T}^k f) \in \mathfrak{Q}_T^\perp.$$

Letting $n \to \infty$ and using the fact that $\mathfrak{Q}_T^\perp$ is a closed subspace of L^2 we deduce from **4** $f - A_\infty f \in \mathfrak{Q}_T^\perp$ so that $A_\infty f = P_T f$.

6. $\mathscr{X}_2$ is closed.

Let $(f_k)_{k \in \mathbb{N}}$ be a sequence in $\mathscr{X}_2$ that converges in L^2 to f. To show that $f \in \mathscr{X}_2$ we will show that the sequence $A_n f$ is Cauchy. Fix $\varepsilon > 0$. We have
$$\|A_n f - A_m f\|_2 \leq \|A_n f - A_n f_k\|_2 + \|A_n f_k - A_m f_k\|_2 + \|A_m f_k - A_m f\|_2$$

(use (5.1.10), i.e., $\|A_n\| \leq 1$ as operator $L^2 \to L^2$)
$$\leq \|f - f_k\|_2 + \|A_n f_k - A_m f_k\|_2 + \|f - f_k\|_2.$$

Hence
$$\|A_n f - A_m f\|_2 \leq 2\|f - f_k\|_2 + \|A_n f_k - A_m f_k\|_2, \quad \forall k, m, n.$$

Fix k such that
$$\|f - f_k\|_2 < \frac{\varepsilon}{3}.$$

The sequence $(A_n f_k)_{n \in \mathbb{N}}$ is convergent since $f_k \in \mathscr{X}_2$. It is thus Cauchy, so there exists $N = N(\varepsilon, k)$ such that $\forall m, n > N$
$$\|A_n f_k - A_m f_k\|_2 < \frac{\varepsilon}{3}.$$

Hence
$$\|A_n f - A_m f\|_2 \leq \varepsilon, \quad \forall m, n > N(\varepsilon, k).$$

7. $\mathscr{X}_2 = L^2(\Omega, \mathcal{S}, \mathbb{P})$.

We know that $\mathcal{Q}_T \subset \mathscr{X}_2$ and
$$\text{Range}\,(\widehat{T} - 1) \subset \mathcal{Q}_T^\perp \cap \mathscr{X}_2.$$

At this point we invoke a classical result of functional analysis: if $S : H \to H$ is a bounded linear operator on a Hilbert space, then the closure of the range of S is $(\ker S^*)^\perp$; see e.g. [22, Cor. 2.18].

The operator $\widehat{T}$ is unitary, so that $\widehat{T}^* = \widehat{T}^{-1}$. Hence if we let $S = \widehat{T} - 1$, then
$$S^* = \widehat{T}^* - 1 = \widehat{T}^{-1} - 1,$$
since $\widehat{T}$ is unitary. We deduce
$$\text{closure}\,\Big(\text{Range}\,(\widehat{T} - 1)\Big) = \big(\ker(\widehat{T}^{-1} - 1)\big)^\perp = \mathcal{Q}_T^\perp.$$

Since $\mathscr{X}_2$ is closed we deduce $\mathcal{Q}_T^\perp \subset \mathscr{X}_2$. This completes the proof of Theorem 5.17.
$\square$

Corollary 5.18. *Suppose that $(\Omega, \mathcal{S}, \mathbb{P})$ is a probability space and $T : \Omega \to \Omega$ is a measure preserving map. Then, $\forall f \in L^1(\Omega, \mathcal{S}, \mathbb{P})$ the temporal averages $A_n f$ converge in L^1 to $\mathbb{E}[\,f \,\|\, \mathcal{I}\,]$, i.e.,*
$$\frac{1}{n}\big(1 + \widehat{T} + \widehat{T}^2 + \cdots \widehat{T}^{n-1}\big)f \to \mathbb{E}[\,f \,\|\, \mathcal{I}\,] \quad \text{in } L^1 \text{ as } n \to \infty.$$

Proof. Denote by $\mathscr{X}_1$ the collection of functions $f \in L^1(\Omega, \mathcal{S}, \mathbb{P})$ such that $A_n f$ converges in L^1 to some function $A_\infty f$. Since $\|-\|_1 \leq \|-\|_2$ we deduce that $\mathscr{X}_2 \subset \mathscr{X}_1$. The argument in Step **6.** in the proof of Theorem 5.17 extends without change to the L^1 since, according to (5.1.10), the operators A_n are contractions

$A_n : L^1 \to L^1$. This proves that $\mathscr{X}_1$ is a closed subspace of L^1 that contains L^2 and thus $\mathscr{X}_1 = L^1$.

From (5.1.6) we deduce that for any $f \in L^1(\Omega, \mathcal{S}, \mathbb{P})$, and any $S \in \mathcal{J}$ we have

$$\int_\Omega f \boldsymbol{I}_S d\mathbb{P} = \int_\Omega (A_n f) \boldsymbol{I}_S d\mathbb{P}.$$

Letting $n \to \infty$ we deduce

$$\int_\Omega f \boldsymbol{I}_S d\mathbb{P} = \int_\Omega (A_\infty f) \boldsymbol{I}_S d\mathbb{P}, \quad \forall S \in \mathcal{J}.$$

Hence (see Definition 1.157) $A_\infty f = \mathbb{E}\big[\, f \,\|\, \mathcal{J}\,\big]$. □

We can now formulate and prove *Birkhoff's ergodic theorem*.

Theorem 5.19 (Birkhoff's ergodic theorem).
Let $(\Omega, \mathcal{S}, \mathbb{P})$ be probability space. Suppose that $T : \Omega \to \Omega$ a measure preserving map. If $f \in \mathcal{L}^1(\Omega, \mathcal{S}, \mathbb{P})$, then the temporal averages

$$A_n(f) = \frac{1}{n+1}\Big(f + f \circ T + \cdots f \circ T^n\Big) = \frac{1}{n+1}\big(f + \widehat{T}f + \cdots + \widehat{T}^n f\,\big]$$

converge a.s. to $\mathbb{E}\big[\, f \,\|\, \mathcal{J}\,\big]$.

Proof. Denote by $\mathscr{X}_0$ the set of functions $f \in L^1(\Omega, \mathcal{S}, \mathbb{P})$ such that $A_n f$ converges a.s. to a function $A_\infty f \in L^1$. Corollary 5.18 shows that in this case $A_\infty f = \mathbb{E}\big[\, f \,\|\, \mathcal{J}\,\big]$. Clearly $\mathscr{X}_0$ is a vector subspace of $L^1(\Omega, \mathcal{S}, \mathbb{P})$.

We will show that $\mathscr{X}_0 = L^1(\Omega, \mathcal{S}, \mathbb{P})$ in two steps.

(i) The set $\mathscr{X}_0$ is a closed subspace of $L^1(\Omega, \mathcal{S}, \mathbb{P})$.
(ii) $\widehat{T}f - f \in \mathscr{X}_0$, $\forall f \in L^1(\Omega, \mathcal{S}, \mathbb{P})$.

The claim (i) is the difficult one. Temporarily assuming its validity we will show how it implies (ii) and the conclusion of the theorem.

Proof of (ii) assuming (i). Observe that for any $f \in L^1$ we have

$$A_n\big(\widehat{T}f - f\big) = \frac{1}{n}\big(\widehat{T}^n f - f\big).$$

In particular, if $f \in L^\infty$ we deduce

$$\|A_n(\widehat{T}f - f)\|_\infty \leq \frac{2}{n}\|f\|_\infty$$

so $A_n\big(\widehat{T}f - f\big) \to 0$ a.s. so $(\widehat{T}f - f) \in \mathscr{X}_0$ if $f \in L^\infty$.

Suppose now that $f \in L^1$ then $f = f_+ - f_-$ and $(\widehat{T}f)_\pm = \widehat{T}f_\pm$. Thus it suffices to show that $\widehat{T}f - f \in \mathscr{X}_0$ if $f \in L^1$ and $f \geq 0$ a.s.

In this case we can find a sequence of elementary functions f_n such that $f_n \nearrow f$. Hence

$$\widehat{T}f_n - f_n \to \widehat{T}f - f \text{ in } L^1.$$

Since the functions f_n are bounded, so are the functions $\widehat{T}f_n - f_n$ we deduce that $\widehat{T}f_n - f_n \in \mathscr{X}_0$. We know from (i) that $\mathscr{X}_0$ is L^1-closed. This proves (ii).

From (ii) we deduce that $\widehat{T}f - f \in \mathscr{X}_0$, $\forall f \in L^2 \subset L^1$. Since $\mathscr{X}_0$ is closed in L^1 we deduce from the proof of Theorem 5.17 that

$$\mathrm{closure}_{L^2}\big(\mathrm{range}(\widehat{T}-1)\big) \subset \mathrm{closure}_{L^1}\big(\mathrm{range}(\widehat{T}-1)\big) \subset \mathscr{X}_0.$$

On the other hand, $Q_{T,2} \subset \mathscr{X}_0$, so that

$$L^2 = \mathscr{X}_2 = Q_{T,2} + Q_{T,2}^\perp = Q_{T,2} + \mathrm{closure}_{L^2}\big(\mathrm{range}(\widehat{T}-1)\big) \in \mathscr{X}_0.$$

Since L^2 is dense in L^1, and $\mathscr{X}_0$ is closed in L^1 we conclude that $\mathscr{X}_0 = L^1$.

Proof of (i) The proof of this result is based on a technical inequality similar in spirit to Doob's maximal inequality (3.2.31). For $f \in L^1(\Omega, \mathcal{S}, \mathbb{P})$ we define $\mathbb{M}[f] \in \mathcal{L}^0(\Omega, \mathcal{S})$,

$$\mathbb{M}[f](\omega) := \sup_{n \geq 1} A_n f(\omega) = \sup_{n \geq 1} \frac{1}{n}\big(f(\omega) + f(T\omega) + \cdots + f(T^{n-1}\omega)\big).$$

Lemma 5.20 (Maximal Ergodic Lemma). $\forall \lambda > 0$, $\forall f \in L^1(\Omega, \mathcal{S}, \mathbb{P})$

$$\forall \lambda > 0, \ f \in L^1(\Omega, \mathcal{S}, \mathbb{P}): \ \lambda \mathbb{P}\big[\{\mathbb{M}[|f|] > \lambda\}\big] \leq \|f\|_1. \tag{5.1.11}$$

$\square$

Let us first explain why the Maximal Ergodic Lemma implies the claim (i).

Suppose that the sequence (f_k) in $\mathscr{X}_0$ converges in L^1 to a function f. We want to show that the sequence $A_n(f)$ is a.s. Cauchy, i.e., for every $\varepsilon > 0$, the set

$$\bigcup_N \underbrace{\bigcap_{m,n > N} \big\{|A_n(f) - A_m(f)| < \varepsilon\big\}}_{=: X_N(f,\varepsilon)}$$

has measure 1. Since $X_N(f, \varepsilon) \subset X_{N'}(f, \varepsilon)$ for $N < N'$ this is equivalent to

$$\lim_{N \to \infty} \mathbb{P}\big[X_N(f, \varepsilon)\big] = 1. \tag{5.1.12}$$

Fix $\varepsilon > 0$. Note that

$$\big|A_n(f) - A_m(f)\big| \leq \big|A_n(f) - A_n(f_k)\big| + \big|A_n(f_k) - A_m(f_k)\big| + \big|A_m(f_k) - A_m(f)\big|$$

$$\leq \big|A_n(|f - f_k|)\big| + \big|A_n(f_k) - A_m(f_k)\big| + \big|A_m(|f - f_k|)\big|$$

$$\leq 2\mathbb{M}\big[|f_k - f|\big] + \big|A_n(f_k) - A_m(f_k)\big|.$$

We deduce

$$X_N(f, \varepsilon) \supset \big\{2\mathbb{M}[|f_k - f|] < \varepsilon/2\big\} \cap X_N(f_k, \varepsilon/2), \ \forall N, k.$$

Letting $N \to \infty$ we deduce

$$\lim_{N \to \infty} \mathbb{P}\big[\, X_N(f, \varepsilon)\,\big] \geq \lim_{N \to \infty} \mathbb{P}\big[\,\big\{\, 2\mathbb{M}[\,|f_k - f|\,] < \varepsilon/2\,\big\} \,\cap\, X_N(f_k, \varepsilon/2)\,\big].$$

From the inclusion-exclusion principle we deduce that

$$\mathbb{P}\big[\,\big\{\, 2\mathbb{M}[\,|f_k - f|\,] < \varepsilon/2\,\big\} \cap X_N(f_k, \varepsilon/2)\,\big] = \mathbb{P}\big[\,\big\{\, 2\mathbb{M}[\,|f_k - f|\,] < \varepsilon/2\,\big\}\,\big]$$

$$+ \mathbb{P}\big[\, X_N(f_k, \varepsilon/2)\,\big] - \mathbb{P}\big[\,\big\{\, 2\mathbb{M}[\,|f_k - f|\,] < \varepsilon/2\,\big\} \cup X_N(f_k, \varepsilon/2)\,\big].$$

Since $f_k \in \mathscr{X}_0$, the sequence $\big(A_n(f_k)\big)_{n \geq 1}$ is a.s. Cauchy so, for any k, so

$$\lim_{N \to \infty} \mathbb{P}\big[\,\big\{\, 2\mathbb{M}[\,|f_k - f|\,] < \varepsilon/2\,\big\} \cup X_N(f_k, \varepsilon/2)\,\big] = \lim_{N \to \infty} \mathbb{P}\big[\, X_N(f_k, \varepsilon/2)\,\big] = 1.$$

Hence, $\forall k$,

$$\lim_{N \to \infty} \mathbb{P}\big[\,\big\{\, 2\mathbb{M}[\,|f_k - f|\,] < \varepsilon/2\,\big\} \cap X_N(f_k, \varepsilon/2)\,\big] = \mathbb{P}\big[\,\big\{\, 2\mathbb{M}[\,|f_k - f|\,] < \varepsilon/2\,\big\}\,\big].$$

We deduce that

$$\lim_{N \to \infty} \mathbb{P}\big[\, X_N(f, \varepsilon)\,\big] \geq \mathbb{P}\big[\, 2\mathbb{M}[\,|f_k - f|\,] > \varepsilon/2\,\big] \overset{(5.1.11)}{\geq} 1 - \frac{4}{\varepsilon}\|f - f_k\|_1, \;\; \forall k.$$

Letting $k \to \infty$ we obtain (5.1.12).

Proof of the Maximal Lemma Let us observe that the inequality (5.1.11) follows from

$$\int_{\{\mathbb{M}[g] > 0\}} g\, d\mathbb{P} \geq 0 \;\; \forall g \in L^1(\Omega, \mathcal{S}, \mathbb{P}). \tag{5.1.13}$$

Indeed, if in (5.1.13) we let $g = f - \lambda$, $\lambda > 0$, then

$$\|f\|_1 \geq \int_{\{\mathbb{M}[f] > \lambda\}} f\, d\mathbb{P} \overset{(5.1.13)}{\geq} \lambda \mathbb{P}\big[\,\{\mathbb{M}[f] > \lambda\}\,\big].$$

We will present two proofs of (5.1.13). The first proof, due to F. Riesz, is a bit longer but a bit more intuitive. The second proof, due to A. Garsia [67] is a lot shorter but less intuitive.

Set

$$X := \{\, \mathbb{M}[\,g\,] > 0\,\} \subset \Omega.$$

Define

$$S_n(g) := \sum_{j=0}^{n-1} g \circ T^j, \;\; \mathbb{M}_n[\,g\,] := \max_{1 \leq k \leq n} S_k(g), \;\; X_k := \{\, \mathbb{M}_k[\,g\,] > 0\,\}. \tag{5.1.14}$$

First proof of (5.1.13). Note that $X_k \subset X_{k+1}$ and

$$\int_X g\, d\mathbb{P} = \lim_{n \to \infty} \int_{X_n} g\, d\mathbb{P} = \lim_{n \to \infty} \frac{1}{n} \sum_{k=1}^n \int_{X_k} g\, d\mathbb{P}.$$

At the last step we used the fact that the Cèsaro means of a convergent sequence have the same limit as the sequence; see Exercise 2.3 with $p_{n,k} = \frac{1}{n}$. Thus, it suffices to show that

$$\sum_{k=1}^{n} \int_{X_k} g d\mathbb{P} \geq 0, \quad \forall n \geq 0. \tag{5.1.15}$$

Fix n. We have

$$\sum_{k=1}^{n} \int_{X_k} g d\mathbb{P} = \sum_{j=0}^{n-1} \int_{X_{n-j}} g d\mathbb{P} = \sum_{j=0}^{n-1} \int_{T^{-j}(X_{n-j})} g \circ T^j d\mathbb{P},$$

where at the last step we used the change-in-variables formula (1.2.21) and the fact that T is measure preserving. We set $Y_j := T^{-j}(X_{n-j})$. Hence

$$\sum_{k=1}^{n} \int_{X_k} g d\mathbb{P} = \int_{\Omega} \left(\sum_{j=0}^{n-1} g(T^j \omega) \mathbf{I}_{Y_j}(\omega) \right) \mathbb{P}[d\omega].$$

We will prove the stronger fact

$$h(\omega) := \sum_{j=0}^{n-1} g(T^j \omega) \mathbf{I}_{Y_j}(\omega) \geq 0, \quad \forall \omega \geq 0. \tag{5.1.16}$$

Let $\omega \in \Omega$. Set $x_j = x_j(\omega) := g(T^j \omega)$. Note that $\omega \in Y_j$ if and only if $T^j(\omega) \in X_{n-j}$, i.e., at least one of the numbers

$$x_j, x_j + x_{j+1}, \ldots, x_j + x_{j+1} + \cdots + x_{n-1}$$

is positive. The inequality (5.1.16) is a special case of the following cute combinatorial lemma of F. Riesz [133].

Lemma 5.21. *Suppose are given a finite sequence of real numbers*

$$\underline{x} := x_0, \ldots, x_{n-1}.$$

We say that x_j is a leading term of $\underline{x}$ if there exists $\ell \geq j$ such that $x_j + \cdots + x_\ell > 0$. Then the sum of the leading terms is ≥ 0.

Proof. The lemma is easily proved by induction on n. For $n = 1$ this is obviously true. Assume that it is true for any $m < n$ and any sequence of m real numbers. Denote by L the set of indices $j = 0, 1, \ldots, n-1$ such that x_j is a leading terms. If $L = \emptyset$ the conclusion is trivially true.

Suppose $L \neq \emptyset$, set $j_0 := \min L$ and denote by ℓ_0 the smallest $\ell \geq j_0$ such that

$$x_{j_0} + \cdots + x_\ell > 0.$$

If $\ell_0 = j_0$, then $x_{j_0} > 0$. Suppose that $\ell_0 > j_0$. The minimality of ℓ_0 implies that for any j, such that $j_0 \leq j < \ell_0$ we have $x_{j_0} + \cdots + x_j < 0$ so that, for $j_0 < k \leq \ell_0$ we have

$$x_k + \cdots x_{\ell_0} \geq 0.$$

This proves that each of the terms $x_{j_0}, x_{j_0+1}, \ldots, x_{\ell_0}$ is a leading term. Their sum is obviously nonnegative.

Consider now the (shorter) sequence

$$\underline{y}: \quad y_0 = x_{\ell_0+1}, \ldots, y_{m-1} := x_n, \quad m := n - 1 - \ell_0 < n - 1.$$

The induction assumption implies that the sum of the leading terms of $\underline{y}$ is ≥ 0. The minimality of j_0 implies that the leading terms of $\underline{x}$ are $x_{j_0}, \ldots, x_{\ell_0}$ together with the leading terms of $\underline{y}$. This proves Lemma 5.21 and completes the proof of Theorem 5.19. $\square$

Second proof of (5.1.13). We continue using the notations (5.1.14). Set $G_n := \mathbb{M}_n[g]$.

Since $X_n \nearrow X$, it suffices to show that

$$\int_{X_n} g \, d\mathbb{P} \geq 0, \quad \forall n.$$

The operator $f \mapsto \widehat{T}f$ is monotone, i.e., $f_0 \leq f_1 \Rightarrow \widehat{T}f_0 \leq \widehat{T}f_1$, and we deduce that for $1 \leq k \leq m$ we have

$$S_{k-1}(g) \leq \max_{1 \leq j \leq m-1} S_j(g) = G_{m-1} \leq G_{m-1}^+$$

and

$$S_k(g) = g + \widehat{T} S_{k-1}(g) \leq g + \widehat{T} G_{m-1} \leq g + \widehat{T} G_{m-1}^+$$

so that

$$G_{m-1} \leq G_m \leq g + \widehat{T} G_{m-1}^+, \quad \forall m \in \mathbb{N},$$

or equivalently

$$g \geq G_n - \widehat{T} G_n^+, \quad \forall n.$$

We deduce

$$\int_{X_n} g \geq \int_{X_n} G_n - \int_{X_n} \widehat{T} G_n^+$$

($\widehat{T} G_n^+ \geq 0$ on Ω, $G_n = G_n^+$ on X_n, $G_n^+ = 0$ on $\Omega \setminus X_n$)

$$\geq \int_{X_n} G_n^+ - \boxed{\int_\Omega \widehat{T} G_n^+} = \int_\Omega G_n^+ - \boxed{\int_\Omega G_n^+} = 0$$

where, at the last step, the equality of the boxed terms is due to the fact that T is measure preserving. $\square$

Remark 5.22. In Remark 5.11 we suggested that the ergodicity condition points to a chaotic behavior of the dynamics of the iterates of T. The ergodic theorem makes this much more precise.

Suppose that T is a measure preserving self-map of the probability space $(\Omega, \mathcal{S}, \mathbb{P})$. If T is ergodic, then for any subset $S \in \mathcal{S}$ there exists a negligible subset $\mathcal{N} \in \mathcal{S}$ such that
$$\forall \omega \in \Omega \setminus \mathcal{N}, \quad \lim_{n \to \infty} \frac{1}{n} \#\{ k; \ T^k \omega \in S, \ 0 \leq k < n \} = \mathbb{P}[S]. \tag{5.1.17}$$
Indeed, the left-hand-side of (5.1.17) is the temporal average $A_n[\boldsymbol{I}_S](\omega)$ so (5.1.17) follows from the Ergodic Theorem 5.19. Observe that (5.1.17) states that for most ω, the orbit $\mathcal{O}_T(\omega)$ spends equal amounts of time in sets of equal measures. In other words, most orbits are equidistributed. The equidistribution phenomenon characterizes ergodicity.

Let us observe that, conversely, if a measure preserving map T satisfies the above equidistribution property, then it has to be ergodic. Indeed, suppose that S is a T-invariant set. Let $\mathcal{N}$ be a negligible set as in (5.1.17). Then for any $\omega \in (\Omega \setminus S) \setminus \mathcal{N}$ the orbit $\mathcal{O}_T(\omega)$ does not intersect S since S is invariant. In this case the left-hand-side of (5.1.17) is equal to zero so $\mathbb{P}[S] = 0$. Thus if S is invariant and its complement $\Omega \setminus S$ is not negligible, then S must be so. Hence T is ergodic.

If we partition Ω into a finite number of measurable sets $S_1, \ldots, S_N$, $p_k = \mathbb{P}[S_k] > 0$, $\forall k$, then there exists a negligible set $N \subset \Omega$ so that for any $\omega \in \Omega \setminus N$ the orbit $\mathcal{O}_T(\omega)$ will be located at each moment of time in one of the chambers S_k of this partition. Moreover, it spends a fraction p_k of the time in the chamber S_k. From this point of view, we can regard the dynamics as hopscotching randomly from one chamber to another, and each chamber is frequented as often as its size. We want to warn that this hopscotching need not have a Markovian nature. □

5.2 Applications

Ergodicity is the unifying principle behind some of the limit theorems we have discussed in the previous chapters and it is the source of many interesting non-probabilistic results.

5.2.1 *Limit theorems*

The Strong Law of Large Numbers is a consequence of the Ergodic Theorem.

Example 5.23 (I.i.d. random variables). Suppose that $(X_n)_{n \in \mathbb{N}}$ is a sequence of i.i.d. integrable random variables defined on the same probability space $(\Omega, \mathcal{S}, \mathbb{P})$. Kolmogorov's 0-1 theorem shows that this is a Kolmogorov family, thus ergodic. Consider the coordinate maps on the path space
$$U_n : \mathbb{R}^{\mathbb{N}} \to \mathbb{R}, \quad U_m(u_1, u_2, \ldots) = u_n.$$
The Ergodic Theorem implies
$$\frac{1}{n}(U_1 + \cdots + U_n) \to \mathbb{E}[U_1] \quad \mu - \text{a.s.}$$
Observing that $X_n = U_n \circ \vec{X}$ we deduce the Strong Law of Large Numbers. □

Example 5.24 (Markov chains). Consider a HMC $(X_n)_{n\geq 0}$ with state space $\mathscr{X}$, transition matrix Q and initial distribution μ. The path space of this Markov chain (see Theorem 4.3) is the probability space

$$\mathbb{U} = \mathbb{U}_\mu = (\mathscr{X}^{\mathbb{N}_0}, \mathcal{E}, \mathbb{P}_\mu),$$

where

$$U_n(u_0, u_1, u_2, \ldots) = u_n.$$

Recall that for any $x \in \mathscr{X}$ we set $\mathbb{P}_x = \mathbb{P}_{\delta_x}$ where δ_x is the Dirac measure on $\mathscr{X}$ concentrated at x then

$$\mathbb{P}_\mu = \sum_{x \in \mathscr{X}} \mu_x \mathbb{P}_x, \quad \mu_x = \mu[\{x\}]. \tag{5.2.1}$$

Denote by $\mathcal{I} \subset \mathcal{E}$ the sigma-algebra of sets that are Θ-invariant sets, where Θ denotes the shift on $\mathbb{U}$. Fix $x \in \mathscr{X}$ and let $A \in \mathcal{I}_x$.

The Markov property (4.1.16) implies that

$$\mathbb{E}_x[\mathbf{I}_A \circ \Theta^n \,\|\, \mathcal{E}_n] = \mathbb{E}_{X_n}[\mathbf{I}_A], \quad \mathcal{E}_n = \sigma(X_0, \ldots, X_n).$$

Since A is invariant we deduce $\mathbf{I}_A = \mathbf{I}_A \circ \Theta^n$ a.s. so

$$\mathbb{E}_x[\mathbf{I}_A \,\|\, \mathcal{E}_n] = \mathbb{E}_{X_n}[\mathbf{I}_A].$$

Lévy's 0-1 theorem implies that

$$\mathbb{E}_x[\mathbf{I}_A \,\|\, \mathcal{E}_n] \to \mathbf{I}_A \quad \text{a.s.}$$

Hence

$$\mathbb{P}_{X_n}[A] = \underbrace{\mathbb{E}_{X_n}[\mathbf{I}_A]}_{=:f_n} \to \mathbf{I}_A \quad \text{a.s.}$$

On the other hand, $\mathbb{P}_x[X_n = x \text{ i.o.}] = 1$, since the chain is recurrent. Thus

$$\mathbb{P}\Big[f_n = \mathbb{P}_x[A] \text{ i.o.}\Big] = 1.$$

Hence $\mathbf{I}_A = \mathbb{P}_x[A]$ a.s. so $\mathbb{P}_x[A] \in \{0,1\}$. Using (5.2.1) we deduce that $\mathbb{P}_\mu[A] \in \{0,1\}$ for any initial distribution μ.

If the chain is positively recurrent and π_∞ is the invariant distribution, then Θ is measure preserving and we deduce that $\mathcal{I}$ is a zero-one algebra so Θ is ergodic. We see that the Ergodic Theorem for Markov chains (Corollary 4.60) is a special case of Birkhoff's Ergodic Theorem because any $f \in L^1(\mathscr{X}, \pi)$i induces a function $\bar{f} = f \circ U_0 \in L^1(\mathscr{X}^{\mathbb{N}_0}\mathcal{E}, \mathbb{P}_{\pi_\infty})$.

The fact that the shift map is ergodic allows us to state results stronger than Corollary 4.60. For any finite set $B \subset \mathscr{X} \times \mathscr{X}$ we obtain a function $F_B \in L^1(\mathscr{X}^{\mathbb{N}_0}\mathcal{E}, \mathbb{P}_{\pi_\infty})$

$$F_B(u_0, u_1, u_2, \ldots) = \mathbf{I}_B(u_0, u_1)$$

and a corresponding Law of Large Numbers

$$\sum_{k=0}^{n-1} I_B(X_k, X_{k+1}) = \mathbb{E}_{\pi_\infty}[F_B] \to \sum_{(x_0,x_1)\in B} \pi_\infty[x_0] Q_{x_0,x_1}. \qquad (5.2.2)$$

One should think of B as a collection of directed edges, a "bridge". In the left-hand-side we have the fraction of time a path of the Markov chain "crosses the bridge B".

Here is an amusing simple illustration of this result. Consider the graph G obtained from by connecting two disjoint connected graphs G_0, G_1 with a single edge from a vertex u_0 in G_0 to a vertex u_1 in G_1. For a vertex v_i of G_i we denote by $\deg_i(v_i)$ its degree in G_i. We denote by E_i the number of edges of G_i.

Let B be the set consisting of the single oriented edge (u_0, u_1). In this case

$$Q_{u_0,u_1} = \frac{1}{\deg_0(u_0)+1}, \quad \pi_\infty[u_0] = \frac{\deg_0(u_0)+1}{2E_0+2E_1+2}.$$

Formula (5.2.2) shows that the standard random walk on G crosses the bridge from u_0 to u_1 roughly a fraction $\frac{1}{2E_0+2E_1+2}$ of the time. $\square$

Example 5.25 (Weyl's equidistribution theorem). Fix $\varphi \in (0, 2\pi)$ and denote by R_φ the planar counterclockwise of angle φ about the origine. This induces a transformation of the unit circle

$$S^1 := \{z \in \mathbb{C}; \ |z| = 1\}.$$

This preserves the canonical probability measure μ on S^1

$$\mu[d\theta] = \frac{1}{2\pi} d\theta.$$

As in the previous section this induces a unitary operator

$$\widehat{R}_\varphi : L^2(S^1, \mu) \to L^2(S^1, \mu), \quad \widehat{R}_\varphi f(\theta) = f(\theta + \varphi).$$

Above the functions in $L^2(S^1, \mu)$ are *complex valued*. For $n \in \mathbb{Z}$ we set

$$e_n(\theta) = e^{in\theta} \in L^2(S^1, \mu).$$

Note that $\widehat{R}_\varphi e_n = e^{in\varphi} e_n$. Since the collection $(e_n)_{n \in \mathbb{Z}}$ is a complete orthonormal sistem we deduce that the eigenspace corresponding to the eigenvalue 1 of $\widehat{R}_\varphi$

$$\ker(1 - \widehat{R}_\varphi) = \mathrm{span}\Big\{e_n;\ \frac{n\varphi}{2\pi} \in \mathbb{Z}\Big\}.$$

We deduce that $\ker(1 - \widehat{R}_\varphi)$ is 1-dimensional iff $\frac{\varphi}{2\pi}$ is irrational. In this case $\widehat{R}_\varphi$ is ergodic and we deduce from (5.1.17) that if $A \subset S^1$ then for almost any $\theta \in S^1$ we have the asymptotic equidistribution equality

$$\frac{1}{n}\sum_{k=0}^{n-1} I_k(\theta + k\varphi) = \frac{\theta_1 - \theta_0}{2\pi}, \quad \text{a.s.} \qquad (5.2.3)$$

With a little bit more work one can show that (5.2.3) holds *for any* θ. This is *Weyl's equidistribution theorem*, [157]. The reader interested in more details on the equidistribution problem can consult [97]. $\square$

5.2.2 Mixing

Suppose that T is a measure preserving transformation of a probability space $(\Omega, \mathcal{S}, \mathbb{P})$. Note that if T is ergodic, then the L^2 ergodic theorem implies that

$$\frac{1}{n}\sum_{k=0}^{n-1} f\circ T^k \xrightarrow{L^2} \mathbb{E}[\,f\,]\boldsymbol{I}_\Omega, \quad \forall f \in L^2(\Omega, \mathcal{S}, \mathbb{P}).$$

If we take the inner product with $g \in L^2$ of both sides in the above equality we deduce

$$\frac{1}{n}\sum_{k=0}^{n-1}\int_\Omega (f\circ T^k) g\, d\mathbb{P} \to \mathbb{E}[\,f\,]\mathbb{E}[\,g\,], \quad \forall f, g \in L^2(\Omega, \mathcal{S}, \mathbb{P}). \qquad (5.2.4)$$

In particular, if we let $f = \boldsymbol{I}_A$, $g = \boldsymbol{I}_B$, $A, B \in \mathcal{S}$, we deduce

$$\lim_{n\to\infty} \frac{1}{n}\sum_{k=0}^{n-1} \mathbb{P}[\,T^{-k}(A)\cap B\,] = \mathbb{P}[\,A\,]\mathbb{P}[\,B\,]. \qquad (5.2.5)$$

Let us observe that the above condition is equivalent with ergodicity. Indeed if we let A quasi-invariant and $B = X \setminus A$, then $\mathbb{P}[\,T^{-k}(A)\cap B\,] = 0$, $\forall k$ and we deduce

$$\mathbb{P}[\,A\,](1 - \mathbb{P}[\,A\,]) = 0$$

so any quasi-invariant set has measure 0 or 1.

Since convergent sequences are also Cèsaro convergent we deduce that condition (5.2.5) follows from the stronger requirement

$$\lim_{n\to\infty} \mathbb{P}[\,T^{-n}(A)\cap B\,] = \mathbb{P}[\,A\,]\mathbb{P}[\,B\,], \quad \forall A, B \in \mathcal{S}. \qquad (5.2.6)$$

A measure preserving map T satisfying this condition is said to be *mixing*.

When T is an automorphism one can give a more visual interpretation of the mixing condition. In this case mixing is also equivalent to the condition

$$\lim_{n\to\infty} \mathbb{P}[\,A\cap T^n(B)\,] = \mathbb{P}[\,A\,]\mathbb{P}[\,B\,], \quad \forall A, B \in \mathcal{S}. \qquad (5.2.7)$$

Assume that the region B is occupied molecules of black ink in a glass of crystalline water. These molecules occupy a fraction $\mathbb{P}[\,B\,]$ of the entire space. Flow the black region B using T. Thus, $T^n(B)$ represents the location of the black region after n units of time.

The mixing condition shows that after a while, the fraction $\mathbb{P}[\,A\cap T^n(B)\,]/\mathbb{P}[\,A\,]$ of a region A occupied by these moving molecules of black ink is equal to $\mathbb{P}[\,B\,]$. Thus in the long run, all the regions will have the same fraction of black ink. To use a very apt analogy in Arnold and Avez [3], this is what happens when we mix well a cocktail.

The mixing condition (5.2.6) can be rewritten as

$$\lim_{n\to\infty}(\widehat{T}^n \boldsymbol{I}_A, \boldsymbol{I}_B) = \lim_{n\to\infty} \mathbb{E}[\,(\boldsymbol{I}_A\circ T^n)\cdot \boldsymbol{I}_B\,] = \mathbb{E}[\,\boldsymbol{I}_A\,]\cdot\mathbb{E}[\,\boldsymbol{I}_B\,], \quad \forall A, B \in \mathcal{S}. \qquad (5.2.8)$$

This implies that for any elementary functions $f, g \in \text{Elem}(\Omega, \mathcal{S})$ we have

$$\lim_{n \to \infty} \int_\Omega (f \circ T^n) g \, d\mathbb{P} = \left(\int_\Omega f \, d\mathbb{P} \right) \left(\int_\Omega g \, d\mathbb{P} \right).$$

Since $\text{Elem}(\Omega, \mathcal{S})$ is dense in $L^2(\Omega, \mathcal{S}, \mathbb{P})$ we deduce that if T is mixing, then

$$\forall f, g \in L^2(\Omega, \mathcal{S}, \mathbb{P}): \lim_{n \to \infty} \int_\Omega (f \circ T^n) g \, d\mathbb{P} = \left(\int_\Omega f \, d\mathbb{P} \right) \left(\int_\Omega g \, d\mathbb{P} \right). \quad (5.2.9)$$

Clearly, if a measure preserving map satisfies (5.2.9), then it is mixing. The above argument has the following immediate generalization.

Proposition 5.26. *Suppose that $T : (\Omega, \mathcal{S}, \mathbb{P}) \to (\Omega, \mathcal{S}, \mathbb{P})$ is a measure preserving map and $\mathcal{C} \subset L^2(\Omega, \mathcal{S}, \mathbb{P})$ is a collection of functions such that $\text{span}(\mathcal{C})$ is dense in $L^2(\Omega, \mathcal{S}, \mathbb{P})$. Then the following are equivalent.*

(i) The map T is mixing.
(ii) For every $f, g \in \mathcal{C}$,

$$\lim_{n \to \infty} (\widehat{T}^n f, g)_{L^2(\Omega)} = \left(\int_\Omega f \, d\mathbb{P} \right) \left(\int_\Omega g \, d\mathbb{P} \right). \quad (5.2.10)$$

$\square$

Let us give a few examples of mixing maps.

Proposition 5.27. *Suppose that $X_n : (\Omega, \mathcal{S}, \mathbb{P}) \to (\mathbb{X}, \mathcal{F})$, $n \in \mathbb{N}$, is a Kolmogorov stationary sequence of measurable maps. Then the shift map on the path space is mixing.*

Proof. We will show that the shift map Θ satisfies (5.2.6). Denote by U_n the coordinate maps on the path space $U_n : \mathbb{X}^\mathbb{N} \to \mathbb{X}$, and set

$$\mathcal{T}_n := \sigma(U_n, U_{n+1}, \dots).$$

For $B \in \mathcal{F}$ and $m \in \mathbb{N}$ we set

$$\varepsilon_m(B) := \sup_{S \in \mathcal{T}_m} |\mathbb{P}[S \cap B] - \mathbb{P}[S]\mathbb{P}[B]|.$$

Since $(X_n)_{n \in BN}$ is a Kolmogorov sequence we deduce from Theorem 5.14 $\varepsilon_m(B) \to 0$ as $m \to \infty$.

Observe that if $A \in \mathcal{F}$, then $T^{-m}(A) \in \mathcal{T}_n$ so that

$$| \mathbb{P}[T^{-m}(S) \cap B] - \mathbb{P}[T^{-m}(A)]\mathbb{P}[B]| \leq \varepsilon_m(B) \to 0.$$

This implies (5.2.6) since T is measure preserving so $\mathbb{P}[T^{-m}(A)] = \mathbb{P}[A]$. $\square$

Proposition 5.28. *Suppose that $(X_n)_{n \geq 0}$ is an irreducible, positively recurrent HMC with state space $\mathcal{X}$, transition matrix Q and stationary distribution π. Then the following are equivalent.*

(i) The HMC is aperiodic.
(ii) The shift map on the path space $(\mathscr{X}^{\mathbb{N}_0}, \mathcal{E}, \mathbb{P}_\pi)$ is mixing.

Proof. We follow the approach in [21, Sec. 16.1.2].

(i) $\Rightarrow$ (ii) Suppose that our HMC is aperiodic. Consider the path space $\mathbb{U} = \mathscr{X}^{\mathbb{N}_0}$. Denote by $\mathcal{C}$ the collection of cylindrical subsets of $\mathbb{U}$ of the form

$$C_{x_{i_1},\ldots,x_{i_k}} := \{\underline{u} \in \mathbb{U};\ u_{i_j} = x_{i_j} \in \mathscr{X},\ \forall 1 \leq j \leq k\},\ 0 \leq i_1 < \cdots < i_k,\ k \in \mathbb{N}.$$

In view of Proposition 5.26 it suffices to show that (5.2.6) is satisfied for any $A, B \in \mathcal{C}$. Suppose that

$$A = C_{x_{i_1},\ldots,x_{i_k}},\quad B = C_{x_{j_1},\ldots,x_{j_m}}.$$

For $n > j_m$ we have

$$\Theta^{-n}(A) \cap B = C_{x_{j_1},\ldots,x_{j_m},x_{i_1+n},\ldots,x_{i_k+n}},\quad x_{ij} = x_{i_j+n},$$

and

$$\mathbb{P}_\pi\big[\Theta^{-n}(A) \cap B\big] = \pi\big[x_{j_1}\big]Q^{j_2-j_1}_{x_{j_1},x_{j_2}} \cdots Q^{j_m-j_{m-1}}_{x_{j_{m-1}},x_{j_m}} \boxed{Q^{n+i_1-j_m}_{x_{j_m},x_{i_1}}} \qquad (5.2.11)$$
$$\times Q^{i_2-i_1}_{x_{i_1},x_{i_2}} \cdots Q^{i_k-i_{k-1}}_{x_{i_{k-1}},x_{i_k}}.$$

Since the HMC is aperiodic we deduce from (4.3.7) we deduce that

$$\lim_{n \to \infty} Q^{n+i_1-j_m}_{x_{j_m},x_{i_1}} = \pi\big[x_{i_1}\big].$$

Using this in (5.2.11) we deduce that

$$\lim_{n \to \infty} \mathbb{P}_\pi\big[\Theta^{-n}(A) \cap B\big] = \underbrace{\pi\big[x_{j_1}\big]Q^{j_2-j_1}_{x_{j_1},x_{j_2}} \cdots Q^{j_m-j_{m-1}}_{x_{j_{m-1}},x_{j_m}}}_{\mathbb{P}_\pi[B]}$$
$$\times \underbrace{\pi\big[x_{i_1}\big]Q^{i_2-i_1}_{x_{i_1},x_{i_2}} \cdots Q^{i_k-i_{k-1}}_{x_{i_{k-1}},x_{i_k}}}_{\mathbb{P}_\pi[A]}.$$

(ii) $\Rightarrow$ (i) Suppose that the shift map is mixing. To prove that it is aperiodic we argue by contraction and assume the period d is bigger than 1. As in Proposition 4.27 consider the communication classes of Q^d,

$$C_1, C_2, \ldots, C_d \subset \mathscr{X}.$$

Hence

$$\mathbb{P}\big[X_{n+1} \in C_{i+1 \bmod d}\ \|\ X_n \in C_{i \bmod d}\big] = 1,\ \forall n \geq 0,\ i = 1,\ldots,d.$$

Consider the sets

$$A_i = \{\underline{u} \in \mathscr{X}^{\mathbb{N}_0};\ u_0 \in C_i\},\ i = 1, 2, \ldots.$$

Then $\Theta^{-n}(A_i) = A_{i+n \bmod d}$, $A_i \cap A_j = \emptyset$ if $i \not\equiv j \bmod d$. We deduce that for any $n \in \mathbb{N}$ we have

$$\mathbb{P}_\pi\big[\Theta^{-nd}(A_0) \cap A_1\big] = 0,\ \mathbb{P}_\pi\big[\Theta^{-nd-1}(A_0) \cap A_1\big] = \mathbb{P}_\pi\big[A_1\big] = \pi\big[A_1\big] \neq 0.$$

This contradicts the fact that Θ is mixing. $\square$

Remark 5.29. Suppose that $(X_n)n \geq 0$ is an HMC as in the above proposition. We know that if the sequence $(X_n)_{n\geq 0}$ is Kolmogorov, then it is mixing. A theorem of Blackwell and Freedman [14] shows that the converse is also true so $(X_n)_{n\geq 0}$ is mixing if and on only if it is Kolmogorov. In fact, the following properties are equivalent.

(i) The HMC $(X_n)_{n\geq 0}$ is aperiodic.
(ii) For any probability measures $\mu, \nu \in \text{Prob}(\mathscr{X})$
$$\lim_{n\to\infty} d_v(\mu Q^n - \nu Q^n) = 0.$$
(iii) The HMC $(X_n)_{n\geq 0}$ is mixing.
(iv) The HMC $(X_n)_{n\geq 0}$ is Kolmogorov.

For a proof we refer to [85, Thm. 26.10]. □

Example 5.30 (The tent map). Consider the tent map $T : [0,1] \to [0,1]$ introduced in Example 5.1(d). Recall that T is the continuous map $[0,1] \to [0,1]$ such that $T(0) = 0 = T(1)$, $T(1/2) = 1$ and T is linear on each of the intervals $[0,1/2]$ and $[1/2, 1]$. We want to show that T is mixing.

Consider the Haar basis of $L^2([0,1])$. Recall its definition. It consists of the *Haar functions*

$$H_0 = 1, \quad H_1 = H_{0,0} = I_{[0,1/2]} - I_{[1/2,1]}$$

$$H_{n,k}(x) = 2^{n/2} H_{0,0}(2^n x - k)$$

$$= 2^{n/2} I_{\left[\frac{k}{2^n}, \frac{k}{2^n} + \frac{1}{2^{n+1}}\right)} - 2^{n/2} I_{\left[\frac{k}{2^n} + \frac{1}{2^{n+1}}, \frac{k+1}{2^n}\right)}, \quad 0 \leq k < 2^n.$$

Define

$$\boldsymbol{H}_{-1} = \text{span}\, \boldsymbol{I}_{[0,1]}, \quad \boldsymbol{H}_n := \text{span}\{H_{n,k};\ 0 \leq k < 2^n\}, \quad n \geq 0,$$

$$\mathcal{H} = \{H_0\} \cup \{H_{n,k};\ n \geq 0,\ 0 \leq k < 2^n\}.$$

The subspaces $\boldsymbol{H}_n$ are mutually orthogonal and the collection $\mathcal{H}$ spans a dense subspace of $L^2([0,1])$; see [24, Sec. 9.2]. Moreover

$$\widehat{T}\boldsymbol{H}_n \subset \boldsymbol{H}_{n+1}, \quad \forall n \geq 0.$$

Thus if $m,n \geq 0$, $0 \leq j \leq 2^m$, $0 \leq k \leq 2^n$ we have

$$\left(\widehat{T}^\ell H_{m,j}, H_{n,k}\right)_{L^2} = 0 = \left(\int_0^1 H_{m,j}(x)dx\right)\left(\int_0^1 H_{n,k}(x)dx\right), \quad \forall \ell > n - m.$$

Clearly $\widehat{T}H_0 = H_0$. Proposition 5.26 applied to the collection $\mathcal{H}$ implies that T is mixing, hence ergodic. □

Example 5.31 (Arnold's cat). We consider a slightly more general situation. Let $d > 1$ and denote by $\mathbb{T}^d$ the d-dimensional torus

$$\mathbb{T}^m = \underbrace{S^1 \times \cdots \times S^1}_{m}$$

equipped with the invariant probability measure

$$\mu[\,d\theta\,] = \frac{1}{(2\pi)^d} d\theta^1 \cdots d\theta_d,$$

where θ_i are the standard angular coordinates on the torus. Suppose that $A \in \mathrm{SL}_d(\mathbb{Z})$, i.e., A is a $d \times d$ matrix with integral entries and determinant 1. Since $A\mathbb{Z}^d = \mathbb{Z}^d$ we deduce that A defines a measure preserving map of $\mathbb{T}^d$

$$\theta = \begin{bmatrix} \theta_1 \\ \vdots \\ \theta_m \end{bmatrix} \mapsto T_A\theta := A \cdot \theta \bmod (2\pi\mathbb{Z})^d.$$

Denote by $\langle -, - \rangle$ the canonical inner product in $\mathbb{R}^d$ and by A^* the transpose of A. Clearly

$$A^* \in \mathrm{SL}_d(\mathbb{Z}) \text{ and } A^*(\mathbb{Z}^d) = \mathbb{Z}^d.$$

For each $\vec{m} \in \mathbb{Z}^d$ we denote by $\mathcal{O}_{\vec{m}}$ the orbit of the action of A^* on $\mathbb{Z}^d$, i.e., the set

$$\mathcal{O}_{\vec{m}} = \{\, (A^*)^n \vec{m};\ n \geq 0 \,\}.$$

For any $\vec{m} \in \mathbb{Z}^d$ we set define the *character*[1] $\chi_{\vec{m}} \in L^2(\mathbb{T}^d, \mu)$

$$\chi_{\vec{m}}(\theta) = e^{i\langle \vec{m}, \theta \rangle} = e^{i(m_1\theta_1 + \cdots + m_d\theta_d)}, \ i := \sqrt{-1}.$$

The set of characters

$$\mathcal{C}_d := \{ \chi_{\vec{m}};\ \vec{m} \in \mathbb{Z}^d \} \subset L^2(\mathbb{T}^d, \mu) \tag{5.2.12}$$

is an orthonormal family that spans a vector subspace dense in $L^2(\mathbb{T}^d, \mu)$.

The unitary operator $\widehat{T}_A : L^2(\mathbb{T}^d, \mu) \to L^2(\mathbb{T}^d, \mu)$ has the explicit description

$$\widehat{T}_A f(\theta) = f(A\theta).$$

In particular,

$$\widehat{T}_A \chi_{\vec{m}}(\theta) = e^{i\langle \vec{m}, A\theta \rangle} = e^{i\langle A^*\vec{m}, \theta \rangle} = \chi_{A^*\vec{m}}(\theta).$$

We have the following result.

Theorem 5.32. *Let $A \in \mathrm{SL}_d(\mathbb{Z})$, $d > 1$. The following are equivalent.*

(i) The map $A : \mathbb{T}^d \to \mathbb{T}^d$ is ergodic.
(ii) For any $\vec{m} \in \mathbb{Z}^d \setminus \{0\}$ the orbit $\mathbf{O}_{\vec{m}}$ is infinite.
(iii) The map $A : \mathbb{T}^d \to \mathbb{T}^d$ is mixing.

---
[1] Any continuous group morphism $\chi : \mathbb{T}^d \to S^1$ has the form $\chi_{\vec{m}}$ for some $\vec{m} \in \mathbb{Z}^d$.

Proof. We follow the approach in [36, Sec. 4.3]. We only need to prove (i) ⇒ (ii) ⇒ (iii).

(i) ⇒ (ii) We argue by contradiction. Suppose there exists $\vec{m} \in \mathbb{Z}^d \setminus \{0\}$ such that $O_{\vec{m}}$ is finite. Denote by n the smallest $n \in \mathbb{N}$ such that $(A^*)^n \vec{m} = \vec{m}$. Then the function
$$f = \chi_{\vec{m}} + \cdots + \chi_{(A^*)^{n-1}\vec{m}}$$
is $\widehat{T}_A$-invariant and nonconstant since the functions $1, \chi_{\vec{m}} \cdots \chi_{(A^*)^{n-1}\vec{m}}$ are linearly independent. Hence A is not ergodic.

(ii) ⇒ (ii) We apply Proposition 5.26 to the set of characters $\mathcal{C}_d$ in (5.2.12). Note that if
$$\int_{\mathbb{T}^d} \chi_{\vec{m}} d\mu = \begin{cases} 1, & \vec{m} = 0, \\ 0, & \vec{m} \neq 0. \end{cases}$$
Clearly if $f = g = 1$, then (5.2.10) holds trivially. Suppose $f \neq 1$. Then
$$\left(\int_\Omega f \, d\mathbb{P} \right) \left(\int_\Omega g \, d\mathbb{P} \right) = 0.$$
Assumption (ii) implies that $\widehat{T}_A^n f$ is a character different from g for all n sufficiently large and thus
$$\left(\widehat{T}_A^n f, g \right)_{L^1} = 0, \ \forall n \gg 0.$$
We deduce that A is mixing. □

The condition (ii) above holds if and only if none of the eigenvalues of A are roots of 1. Observe that if one eigenvalue of $A \in \mathrm{SL}_2(\mathbb{Z})$ is a root of 1 then all eigenvalues are roots of 1. We deduce that the only matrices in $\mathrm{SL}_2(\mathbb{Z})$ are
$$\pm \begin{bmatrix} 1 & 0 \\ 0 & 1 \end{bmatrix} \text{ and } \pm \begin{bmatrix} 0 & -1 \\ 1 & 0 \end{bmatrix}.$$
In particular, this shows that Arnold's cat map is mixing. □

Remark 5.33. There is another condition that intermediates between mixing and ergodicity. More precisely, a measure preserving self-map of a probability space $(\Omega, \mathcal{S}, \mathbb{P})$ is called *weakly mixing* if,
$$\lim_{n \to \infty} \frac{1}{n} \sum_{k=0}^{n-1} \left| \mathbb{P}[T^{-k}(A) \cap B] - \mathbb{P}[A]\mathbb{P}[B] \right| = 0 \quad (5.2.13)$$
for any $A, B \in \mathcal{S}$. Clearly (5.2.13) implies (5.2.5) so weakly mixing are ergodic.

Since convergent sequences are Cèsaro convergent we deduce that (5.2.6) implies (5.2.13) so mixing maps are weakly mixing.

It turn out that most weakly mixing automorphisms of a probability space $(\Omega, \mathcal{S}, \mathbb{P})$ are not mixing. More precisely the mixing operators form a meagre (first Baire category) subset in the set of weakly mixing automorphisms [78, p. 77].

5.3 Exercises

Exercise 5.1. Suppose that $(\Omega, \mathcal{S})$ is a measurable space and $T : (\Omega, \mathcal{S}) \to (\Omega, \mathcal{S})$ a measurable map. Denote by $\mathrm{Prob}_T(\Omega, \mathcal{S})$ the set of T-invariant probability measures $\mathbb{P} : \mathcal{S} \to [0, 1]$.

(i) Prove that $\mathrm{Prob}_T(\Omega, \mathcal{S})$ is a convex subset of the space of finite measures on $\mathcal{S}$.
(ii) Prove that T is ergodic with respect to a probability measure $\mathbb{P}$ if and only if $\mathbb{P}$ is an extremal point of $\mathrm{Prob}_T(\Omega, \mathcal{S})$ i.e., $\mathbb{P}$ cannot be written as a convex combination $\mathbb{P} = (1-t)\mathbb{P}_0 + t\mathbb{P}_1$, $t \in (0, 1)$, $\mathbb{P} \neq \mathbb{P}_0, \mathbb{P}_1$.

□

Exercise 5.2. Suppose that (X, d) is a compact metric spaces and $T : X \to X$ is a continuous map. Let $\mathbb{P}$ be a Borel probability measure on X. For $n \in \mathbb{N}$ we set

$$\mathbb{P}_n := \frac{1}{n} \sum_{k=01}^{n} T_{\#}^k \mathbb{P}.$$

(i) Prove that the sequence $(\mathbb{P}_n)_{n \in \mathbb{N}}$ contains a subsequence $(\mathbb{P}_{n_k})$ that converges weakly to a Borel probability measure $\mathbb{P}_*$ on X, i.e.,

$$\lim_{k \to \infty} \int_X f(x) \mathbb{P}_{n_k}[\,dx\,] = \int_{\mathbb{P}_*} f(x) \mathbb{P}_*[\,dx\,], \quad \forall f \in C(X).$$

Hint. Use Banach-Alaoglu compactness theorem.

(ii) Prove that $\mathbb{P}_*$ is T-invariant.
(iii) Prove that the set $\mathrm{Prob}_T(X)$ of T-invariant Borel probability measures on X is convex and closed with respect to the weak convergence.

□

Exercise 5.3. Let $(\Omega, \mathcal{S}, \mathbb{P})$ be a probability space and $T : \Omega \to \Omega$ a measure preserving map. We say that T is *quasi-mixing* if there exist $c_1, c_2 > 0$ such that $\forall A, B \in \mathcal{S}$

$$c_1 \mathbb{P}[\,A\,]\mathbb{P}[\,B\,] \leq \mathbb{P}[\,T^{-1}(A) \cap B\,] \leq c_2 \mathbb{P}[\,A\,]\mathbb{P}[\,B\,]. \tag{5.3.1}$$

(i) Suppose that $\mathcal{A} \subset \mathcal{S}$ is a collection of measurable subsets that generates $\mathcal{S}$, $\sigma(\mathcal{A}) = \mathcal{S}$. Show that T is quasi-mixing if (5.3.1) holds for all $A, B \in \mathcal{A}$.
(ii) Prove that if T is quasi-mixing, then it is ergodic.

□

Exercise 5.4. Let $(\Omega, \mathcal{S}, \mathbb{P})$ be a probability space and $T : \Omega \to \Omega$ a measure preserving map. Suppose that $(\mathcal{F}_n)_{n \geq 1}$ is a filtration of sigma-subalgebras with the following properties.

(i)
$$\bigvee_{n\geq 1} \mathcal{F}_n = \mathcal{S}.$$

(ii) $T^{-1}(\mathcal{F}_n) \subset \mathcal{F}_n$, $\forall n \in \mathbb{N}$.

(iii) For any $k \in \mathbb{N}$ the intersection
$$\bigcap_{n\geq} (T^k)^{-1}(\mathcal{F}_n)$$
is a 0-1-sigma subalgebra.

Prove that T is mixing. □

Exercise 5.5. Let $(\Omega, \mathcal{S}, \mathbb{P})$ be a probability space, $T : \Omega \to \Omega$ a measure preserving map and $g \in L^1(\Omega, \mathcal{S}, \mathbb{P})$. Prove that the following are equivalent.

(i) The function g is T-invariant, i.e., $g \circ T = g$ a.s.
(ii) For any $f \in L^\infty(\Omega, \mathcal{F}, \mathbb{P})$, $\mathbb{E}[gf] = \mathbb{E}[g(f \circ T)]$. □

Exercise 5.6 (Poincaré). Suppose that $(\Omega, \mathcal{S}, \mathbb{P})$ is a probability space and $T : \Omega \to \Omega$ is a measure preserving measurable map. Prove that for any $S \in \mathcal{S}$ such that $\mathbb{P}[S] > 0$ we have
$$\mathbb{P}[\{\omega \in \Omega; \; T^n\omega \in S \text{ i.o.}\}] = 1.$$
□

Exercise 5.7 (Kac). Suppose that $(\Omega, \mathcal{S}, \mathbb{P})$ is a probability space and $T : \Omega \to \Omega$ is a measure preserving measurable map. For $S \in \mathcal{S}$ such that $\mathbb{P}[S] > 0$ we define the first return map
$$T_S : \Omega \to \mathbb{N} \cup \{\infty\}, \quad T_S(\omega) = \min\{n \in \mathbb{N} : \; T^n\omega \in S\}.$$
Set
$$\Omega_S := \{\omega \in \Omega \setminus S; \; T^n\omega \notin S, \; \forall n \geq 1\}.$$

(i) Prove that
$$\int_S T_S(\omega) \; \mathbb{P}[d\omega] = 1 - \mathbb{P}[\Omega_S].$$

(ii) Prove that if T is ergodic then $\mathbb{P}[\Omega_S] = 0$. □

Exercise 5.8. Consider an irreducible HMC $(X_n)_{n\geq 0}$ with finite state space $\mathscr{X}$, transition matrix Q and whose initial distribution is the stationary distribution μ. The path space of this Markov chain is (see Theorem 4.3)
$$\mathbb{U}_\mu = (\mathscr{X}^{\mathbb{N}_0}, \mathcal{E}, \mathbb{P}_\mu).$$
For $n \in \mathbb{N}_0$ we denote by U_n the n-th coordinate map $U_n(u_0, u_1, \dots) = u_n$. Let
$$f : \mathbb{U}_\mu \to \mathbb{R}, \quad f((u_0, u_1, \dots)) = -\log_2 Q_{u_0, u_1}.$$

(i) Prove that
$$\int_{\mathbb{U}_\mu} f(\underline{u}) \mathbb{P}_\mu [d\underline{u}] = \mathrm{Ent}_2[\mathscr{X}, Q],$$
where $\mathrm{Ent}_2[\mathscr{X}, Q]$ denotes the entropy rate of the Markov chain described in Exercise 4.31.

(ii) Prove that
$$-\lim_{n \to \infty} \frac{1}{n} \log_2 (Q_{U_0, U_1} \cdots Q_{U_{n-1}, U_n}) = \mathrm{Ent}_2[\mathscr{X}, Q] \quad \text{a.s.}$$
□

Exercise 5.9 (Kac). Consider the map $Q : [0,1) \to [0,1)$, $x \mapsto Qx := 2x \bmod 1$. Show that Q it is mixing with respect to the Lebesgue measure. **Hint.** See Example 5.13.
□

Exercise 5.10. Consider the tent map $T : [0,1] \to [0,1]$, $T(x) = \min(2x, 2-2x)$ and the *logistic map* $L : [0,1] \to [0,1]$, $L(x) = 4x(1-x)$.

(i) Prove that the map $\Phi : [0,1] \to [0,1]$, $\Phi(x) = 1 - \cos(x/\pi)$ is a homeomorphism and $L \circ \Phi = \Phi \circ T$.
(ii) Describe the measure $\mu := \Phi_\# \lambda$, where λ is the Lebesgue measure on $[0,1]$.
(iii) Prove that the logistic map preserves μ and it is mixing with respect to this measure.
□

Exercise 5.11. Fix $m \in \mathbb{N}$, $m \geq 2$. For any $\vec{\epsilon} \in \{0,1\}^m$ define
$$F_{\vec{\epsilon}} : [0,1] \to [0,1], \quad F_{\vec{\epsilon}}(x) = \begin{cases} (-1)^{\epsilon_k} m \left(x - \frac{k-1}{m} \right) + \epsilon_k, & \frac{k-1}{m} \leq x < \frac{k}{m}, \\ 0, & x = 1. \end{cases}$$
Prove that $F_{\vec{\epsilon}}$ is mixing for any $\vec{\epsilon} \in \{0,1\}^m$.
□

Exercise 5.12. Consider the Haar functions $H_{n,k}$ used in Example 5.30. We define the *Rademacher functions*,
$$R_n : [0,1] \to \mathbb{R}, \quad R_n = 2^{-n/2} \sum_{0 \leq k < 2^n} H_{n,k}, \quad n \geq 0.$$

(i) Prove that
$$\sum_{n=0}^\infty \frac{1}{2^{n+1}} R_n(x) = 1 - 2x, \quad \forall x \in [0,1].$$

(ii) Prove that the functions $(R_n)_{n \geq 0}$, viewed as random variables defined on the probability space $([0,1], \lambda)$, are i.i.d.
□

Exercise 5.13. Suppose that X is a finite set and π is a probability measure on X given by the weights $\pi_x := \pi[\{x\}] > 0$, $\forall x \in X$. Consider the Cartesian product $\mathbb{U}_\pi := X^{\mathbb{Z}}$ equipped with the product sigma-algebra and product measure $\boldsymbol{\pi}_\infty := \pi^{\otimes \mathbb{Z}}$. The elements of $\mathbb{X}$ are functions $\boldsymbol{u} : \mathbb{Z} \to X$. Consider the shift $\Theta : \mathbb{U}_\pi \to \mathbb{U}_\pi$, $\Theta \boldsymbol{u}(n) = \boldsymbol{u}(n+1)$.

(i) Prove that Θ is mixing with respect to the measure $\boldsymbol{\pi}_\infty$.
(ii) Denote for $S \subset X$ by $\mathcal{N}_\pi^S$ the subset of $\mathbb{U}_\pi$ consisting of functions $\boldsymbol{u} : \mathbb{Z} \to X$ such that $\lim_{n \to \infty} \boldsymbol{u}(n)$ exists. Prove that $\boldsymbol{\pi}_\infty[\mathcal{N}_\pi^S] = 0$ and that the complement $\mathbb{U}_\pi^S$ is Θ-invariant. □

Exercise 5.14. Consider the *baker's transform* $B : [0,1]^2 \to [0,1]^2$,

$$B(x,y) = \begin{cases} (q(2x), q(y/2)), & x \leq 1/2, \\ (q(2x), q((y+1)/2)), & x > 1/2, \end{cases}$$

where $q(t)$ denotes the fractional part of the real number t, $q(t) = t - \lfloor t \rfloor$. Prove that B is mixing with respect to the Lebesgue measure. Consider the map $\Phi : \{0,1\}^{\mathbb{Z}} \to [0,1]^2$ given by $\Phi(x(\boldsymbol{u}), y(\boldsymbol{u}))$,

$$x(\boldsymbol{u}) = \sum_{n=0}^{\infty} \frac{\boldsymbol{u}(-n)}{2^{n+1}}, \quad y(\boldsymbol{u}) = \sum_{n=1}^{\infty} \frac{\boldsymbol{u}(n)}{2^n}.$$

Denote by π the uniform measure on $\{0,1\}$ and by $\boldsymbol{\pi}_\infty$ the induced product measure on $\{0,1\}^{\mathbb{Z}}$.

(i) Prove that $\Phi_\# \boldsymbol{\pi}_\infty = \boldsymbol{\lambda}$, where $\boldsymbol{\lambda}$ is the Lebesgue measure on the square $[0,1]^2$.
(ii) Show that $B \circ \Phi = \Phi \circ \Theta$, where Θ is the shift defined in Exercise 5.13.
(iii) Prove that the baker's transform is mixing with respect to the Lebesgue measure. □

Exercise 5.15 (Gauss). Consider the map $G : [0,1] \to [0,1]$ given by

$$G(x) = \begin{cases} 0, & x = 0, \\ \frac{1}{x} - \lfloor \frac{1}{x} \rfloor, & x \in (0,1]. \end{cases}$$

For $k \in \mathbb{N}$ we set $I_k := (1/(k+1), 1/k)$. Any $x \in (0,1]$ has a continuous fraction decomposition

$$x = [0 : a_1 : a_2 : \cdots] := 0 + \cfrac{1}{a_1 + \cfrac{1}{a_2 + \cfrac{1}{\ddots}}}, \quad a_n = a_n(x) \in \mathbb{N}_0, \; \forall n \in \mathbb{N}.$$

(The number x is rational if and only if $a_n = 0$ for all n sufficiently large.) Set $[0,1]_* := [0,1] \setminus \mathbb{Q}$.

(i) Let $x = [0 : a_1 : a_2 : \cdots] \in [0,1]_*$. Prove that $G(x) = [0 : a_2 : a_3 : \cdots]$ and, for any $n \in \mathbb{N}$, we have

$$x = [0 : a_1 : \cdots : a_{n-1} : a_n + G^n(x)] = \cfrac{1}{a_1 + \cfrac{1}{a_2 + \cfrac{1}{\ddots + a_{n-1} + \cfrac{1}{a_n + G^n(x)}}}}.$$

(ii) Let $x \in [0,1]_*$. Prove that $a_n(x) = k$ iff $G^n(x) \in I_k$, $\forall k, n \in \mathbb{N}$.

(iii) For each $a \in \mathbb{R}$ we set

$$T_a = \begin{bmatrix} a & 1 \\ 1 & 0 \end{bmatrix}.$$

Prove that

$$[0 : a_1 : \cdots : a_n] = \frac{p_n}{q_n}, \quad \begin{bmatrix} p_n \\ q_n \end{bmatrix} = T_0 \cdot T_{a_1} \cdots T_{a_n} \begin{bmatrix} 1 \\ 0 \end{bmatrix}. \quad (5.3.2)$$

(iv) Let $x := [0 : a_1 : a_2 : \cdots] \in [0,1]_*$. Prove that for any $n \in \mathbb{N}$ we have

$$x = \frac{p_n(x) + p_{n-1}(x) G^n(x)}{q_n(x) + q_{n-1}(x) G^n(x)}$$

where $p_n(x), q_n(x)$ are defined in terms of the $a_n(x)$'s by (5.3.2).

(v) Prove that $q_n(x) \geq 2^{\frac{n-2}{2}}$, $\forall x \in [0,1]_*$, $n \in \mathbb{N}$.

(vi) Prove that the restriction of G to I_k is a diffeomorphism onto $(0,1)$.

(vii) Fix $c > 0$ and set $\rho : [0,1] \to [0, \infty)$

$$\rho(x) = \frac{c}{x+1}.$$

Prove that for any $x \in [0,1]_*$ we have

$$\rho(x) = \sum_{G(y) = x} \frac{\rho(y)}{|G'(y)|}.$$

(viii) Prove that the probability measure μ on defined by

$$\mu[\,dx\,] = \frac{1}{\log 2 (x+1)} \boldsymbol{\lambda}[\,dy\,]$$

is G-invariant.

(ix) Prove that for any $n \in \mathbb{N}$ the map $A_n : [0,1]_* \to \mathbb{N}$, $x \mapsto a_n(x)$ is measurable and the sigma-algebra generated by these random variables coincides with the Borel sigma algebra. **Hint.** Show that the set $I_{a_1, \ldots, a_m} := \{A_k = a_k, \ 1 \leq k \leq m\}$ is an interval with endpoints expressible in terms of the fractions $\frac{p_k}{q_k}$ defined as in (5.3.2).

(x) Show that G is quasi-mixing (see Exercise 5.3) hence ergodic. □

Exercise 5.16. Suppose that T is an automorphism of a probability space $(\Omega, \mathcal{S}, \mathbb{P})$. Define

$$T^{\times 2} : \Omega \to \Omega \to \Omega \times \Omega, \quad T^{\times 2}(\omega_1, \omega_2) = (T\omega_1, T\omega_2).$$

Prove that $T^{\times 2}$ is ergodic (with respect to $\mathbb{P}^{\otimes 2}$) if and only if T is weakly mixing, i.e., satisfies (5.2.13). □

Appendix A

A few useful facts

A.1 The Gamma function

Definition A.1 (Gamma and Beta functions). The *Gamma function* is the function

$$\Gamma : (0, \infty) \to \mathbb{R}, \quad \Gamma(x) = \int_0^\infty t^{x-1} e^{-t} dt. \tag{A.1.1}$$

The *Beta function* is the function of two positive variables

$$B(x, y) := \frac{\Gamma(x)\Gamma(y)}{\Gamma(x+y)}, \quad x, y > 0. \tag{A.1.2}$$

□

We gather here a few basic facts about the Gamma and Beta functions used in the text. For proofs we refer to [100, Chap. 1] or [158, Chap. 12].

Proposition A.2. *The following hold.*

(i) $\Gamma(1) = 1$.
(ii) $\Gamma(x+1) = x\Gamma(x)$, $\forall x > 0$.
(iii) For any $n = 1, 2, \ldots$ *we have*

$$\Gamma(n) = (n-1)!. \tag{A.1.3}$$

(iv) $\Gamma(1/2) = \sqrt{\pi}$.
(v) For any $x, y > 0$ *we have* **Euler's formula**

$$B(x, y) = \int_0^1 s^{x-1}(1-s)^{y-1} ds = \int_0^\infty \frac{u^{x-1}}{(1+u)^{x+y}} du. \tag{A.1.4}$$

□

The equality (iv) above reads

$$\sqrt{\pi} = \Gamma(1/2) = \int_0^\infty e^{-t} t^{-1/2} dt$$

$(t = x^2, t^{-1/2} = x^{-1} \ dt = 2x dx)$

$$= 2 \int_0^\infty e^{-x^2} dx = \int_{-\infty}^0 e^{-x^2} dx + \int_0^\infty e^{-x^2} dx = \int_{-\infty}^\infty e^{-x^2} dx.$$

If we make the change in variables $x = \frac{s}{\sqrt{2}}$ so that $x^2 = \frac{s^2}{2}$ and $dx = \frac{1}{\sqrt{2}} ds$, then we deduce

$$\sqrt{\pi} = \frac{1}{\sqrt{2}} \int_{-\infty}^\infty e^{-\frac{x^2}{2}} dx.$$

From this we obtain the fundamental equality

$$\frac{1}{\sqrt{2\pi}} \int_{-\infty}^\infty e^{-\frac{x^2}{2}} dx = 1. \tag{A.1.5}$$

The function $\Gamma(x)$ grows very fast as $x \to \infty$. Its asymptotics is governed by the *Stirling's formula*

$$x\Gamma(x) \sim \sqrt{2\pi x} \left(\frac{x}{e} \right)^x \quad \text{as } x \to \infty. \tag{A.1.6}$$

Note that for $n \in \mathbb{N}$ the above estimate reads

$$n! \sim \sqrt{2\pi n} \left(\frac{n}{e} \right)^n \quad \text{as } n \to \infty. \tag{A.1.7}$$

There are very sharp estimates for the ratio

$$q_n = \frac{n!}{\sqrt{2\pi n} \left(\frac{n}{e} \right)^n}.$$

More precisely we have (see [58, II.9])

$$\frac{1}{12n+1} < \log q_n < \frac{1}{12n}. \tag{A.1.8}$$

We denote by ω_n the volume of the n-dimensional Euclidean unit ball

$$B^n := \{ \boldsymbol{x} \in \mathbb{R}^n; \ \|\boldsymbol{x}\| \leq 1 \}, \quad \|\boldsymbol{x}\| = \sqrt{x_1^2 + \cdots + x_n^2},$$

and by σ_{n-1} the "area" of the unit sphere in $\mathbb{R}^n$

$$S^{n-1} = \{ \boldsymbol{x} \in \mathbb{R}^n; \ \|\boldsymbol{x}\| = 1 \}.$$

Then

$$\sigma_{n-1} = \frac{2\Gamma(1/2)^n}{\Gamma(n/2)}, \quad \omega_n = \frac{1}{n} \sigma_{n-1} = \frac{\Gamma(1/2)^n}{\Gamma((n+1)/2)}. \tag{A.1.9}$$

A.2 Basic invariants of frequently used probability distributions

$$X \sim \text{Bin}(n,p) \iff \mathbb{P}[X=k] = \binom{n}{k} p^k q^{n-k}, \ \ k=0,1,\ldots,n, \ \ q=1-p.$$

$$\text{Ber}(p) \sim \text{Bin}(1,p).$$

$$X \sim \text{NegBin}(k,p) \iff \mathbb{P}[X=n] = \binom{n-1}{k-1} p^k q^{n-k}, \ \ n=k, k+1,\ldots$$

$$\text{Geom}(p) \sim \text{NegBin}(1,p).$$

$$X \sim \text{HGeom}(w,b,n), \ \ \mathbb{P}[X=k] = \frac{\binom{w}{k}\binom{b}{n-k}}{\binom{w+b}{n}}, k=0,1,\ldots,w.$$

$$X \sim \text{Poi}(\lambda), \ \ \lambda > 0 \iff \mathbb{P}[X=n] = e^{-\lambda}\frac{\lambda^n}{n!}, \ \ n=0,1,\ldots$$

$$X \sim \text{Unif}(a,b) \iff \mathbb{P}_X = \frac{1}{b-a} \boldsymbol{I}_{[a,b]} dx.$$

$$X \sim \text{Exp}(\lambda), \ \ \lambda > 0 \iff \mathbb{P}_X = \lambda e^{-\lambda x} \boldsymbol{I}_{[0,\infty)} dx$$

$$X \sim N(\mu,\sigma^2), \ \ \mu \in \mathbb{R}, \ \ \sigma > 0 \iff \mathbb{P}_X = \frac{1}{\sigma\sqrt{2\pi}} e^{-\frac{(x-\mu)^2}{2\sigma^2}} dx, \ \ x \in \mathbb{R}.$$

$$X \sim \text{Gamma}(\nu,\lambda) \iff p_X(x) = \frac{\lambda^\nu}{\Gamma(\nu)} x^{\nu-1} e^{-\lambda x} \boldsymbol{I}_{[0,\infty)} dx$$

$$X \sim \text{Beta}(a,b) \iff p_X = \frac{1}{B(a,b)} x^{a-1} (1-x)^{b-1} \boldsymbol{I}_{(0,1)} dx.$$

$$X \sim \text{Stud}_p \iff p_X = \frac{1}{\sqrt{p\pi}} \frac{\Gamma(\frac{p+1}{2})}{\Gamma(\frac{p}{2})} \frac{1}{(1+x^2/p)^{(p+1)/2}} dx, x \in \mathbb{R}.$$

Name	Mean	Variance	pgf	mgf
Ber(p)	p	pq	$(q+ps)$	pe^t
Bin(n,p)	np	npq	$(q+ps)^n$	$p^n e^{nt}$
Geom(p)	$\frac{1}{p}$	$\frac{q}{p^2}$	$\frac{ps}{1-qs}$	$\frac{pe^t}{1-qe^t}$
NegBin(k,p)	$\frac{k}{p}$	$\frac{kq}{p^2}$	$\left(\frac{ps}{1-qs}\right)^k$	$\left(\frac{pe^t}{1-qe^t}\right)^k$
Poi(λ)	λ	λ	$e^{\lambda(s-1)}$	$e^{\lambda(e^t-1)}$
HGeom(w,b,n)	$\frac{w}{w+b}\cdot n$	$*$	$*$	$*$
Unif(a,b)	$\frac{a+b}{2}$	$\frac{(b-a)^2}{12}$	NA	$\frac{e^{tb}-e^{ta}}{tb-ta}$
Exp(λ)	λ^{-1}	λ^{-2}	NA	$\frac{\lambda}{\lambda-t}$
$N(\mu,\sigma^2)$	μ	σ^2	NA	$\exp\left(\frac{\sigma^2}{2}t^2+\mu t\right)$
Gamma(ν,λ)	$\frac{\nu}{\lambda}$	$\frac{\nu}{\lambda^2}$	NA	$\left(\frac{\lambda}{\lambda-t}\right)^\nu$
Beta(a,b)	$\frac{a}{a+b}$	$\frac{ab}{(a+b)^2(a+b+1)}$	NA	$*$
Stud$_p$	$0,\ p>1$	$\frac{p}{p-2},\ p>2$	NA	NA

A.3 A glimpse at R

This section is merely an invitation to programming in R. It is not meant as a serious guide to learning R. It mainly lists a few basic tricks that will get the curios reader started and cover many with simple probability simulations I used in my classes.

First, here is how you install R on your computers.
For Mac users
    https://cran.r-project.org/bin/macosx/
For Windows users
    https://cran.r-project.org/bin/windows/base/
Next, install R Studio (the Desktop version). This is a very convenient interface for using R.
    https://www.rstudio.com/products/RStudio/
(*Install first R and then R Studio.*) You can also access RStudio and R in the cloud
    https://www.rollapp.com/app/rstudio
The site
    http://www.people.carleton.edu/~rdobrow/Probability/
has a repository of many simple R programs (or R scripts) that you can use as models.

The reader familiar with the basics of programming will have no problems learning the basics of R. This section is addressed to such a reader. We list some of the commands and objects most frequently used in probability and we have included several examples to help the reader get started. R-Studio comes with links to various freely available web sources for R-programming. A commercial source that I find very useful is *"The Book of R"*, [38]. Often I ask GOOGLE how to do this or that in R and I receive many satisfactory solutions.

Example A.3 (Operations with vectors). The workhorse of R is the object called *vector*. An n-dimensional vector is essentially an element in $\mathbb{R}^n$. An n-dimensional vector in R can be more general in the sense that its entries need not be just numbers.

To generate in R the vector $(1, 2, 4.5)$ and then naming it x use the command

```
x<-c(1,2,4.5)
```

To see what the vector x is type

```
x
```

and then hit RETURN/ENTER.[2]

To see what the k-th entry of x is us the command

```
x[k]
```

The command

```
x[j:k]
```

will generate all the entries of x from the j-the to the k-th. If you want to add an entry to x, say you want to generate the longer vector $(1, 2, 4, 5, 7)$, use the command

```
c(x,7)
```

For long vectors this approach can be time consuming. The process of describing vectors can be accelerated if the entries of the vector x are subject to patterns. For example, the vector of length 22 with all entries equal to the same number, say 1.5, can be generated using the command

```
rep(1.5, 22)
```

To generate the vector listing in increasing order all the integers between -2 and 10 (included) use the command

```
(-2):10
```

To generate the vector named x consisting of 25 equidistant numbers staring at 1 and ending at 7 use the command

```
x<-seq(from=1, to=7, length.out=25)
```

---

[2] You have to do this after every command so I will omit this.

To add all the entries of a vector $x = (x_1, \ldots, x_n)$ use the command

`sum(x)`

To add all the natural numbers from 50 to 200 use the command

`sum(50:200)`

The result is $18,875$.

You can sort the entries of a vector, if they are numerical. For example

```
> z<-c(1,4,3)
> sort(z)
[1] 1 3 4
```

A very convenient feature of working with vectors in R is that the basic algebraic operations involving numbers extend to vectors, component wise. For example, if z is the above vector, and $y = (1, 8, 9)$, then the command `y/z` returns $(1/1, 8/4, 9/3) = (1, 2, 3)$, while `z^2` returns $(1, 16, 9)$. □

Example A.4 (Logical operators). These are operators whose output is a TRUE or FALSE or a vector whose entries are TRUE/FALSE.

For example, the command $2 < 5$ returns TRUE. On the other hand if x is the vector $(2, 3, 7, 8)$, then the command $x < 5$ return

TRUE, TRUE, FALSE, FALSE.

In R the logicals TRUE/FALSE also have arithmetic meaning,

$$\text{TRUE} = 1, \quad \text{FALSE} = 0.$$

The output of $x < 5$ is a vector whose entries are TRUE/FALSE. To see how many of the entries of x are < 5 use the command

`sum(x<5)`

Above $x < 5$ is interpreted as a vector with 0/1-entries. When we add them we count how many are equal to 1 or, equivalently, how many of the entries of x are < 5.

The R language also has two very convenient logical operators **any** and **all**. When we apply **any** to a vector with TRUE/FALSE entries it returns TRUE if at least one of the entries of v are TRUE and returns FALSE otherwise. When we apply **all** to a vector v with TRUE/FALSE entries it returns TRUE if *all* of the entries of v are TRUE and returns FALSE otherwise. □

Example A.5 (Functions in R). One can define and work with functions in R. For example, to define the function

$$f(q) = 1 + 6q + 10q^2(1-q)^4$$

use the command

```
f<-function(q) (1+4*q+10*q^2)*(1-q)^4
```

To find de value of f at $q = 0.73$ use the command

```
f(0.73)
```

To display the values of f at all the points
$$0,\ 0.01,\ 0.02,\ 0.03,\ldots,\ 0.15,\ 0.16$$
use the command

```
x<-seq(from=0, to=0.16, by=0.01)
f(x)
```

To plot the values of f over 100 equidistant points in the interval $[2, 7]$ use the command

```
x<-seq(from=2, to=7, length.out=100)
y<-f(x)
plot(x,y, type="l")
```

Equivalently, there is the simple command curve(-) that allows drawing multiple graphs in the same coordinate system.

```
function1<-function(x){x^2}
function2<-function(x){1-cos(x)}
curve(function1, col=1)
curve(function2, col=2, add=TRUE)
```

Above col stands for "color". When this option is used different graphs are depicted in different colors.

Here is how we define in R the indicator function of the unit disc in the plane
$$I_D(x,y) = \begin{cases} 1, & x^2 + y^2 \leq 1, \\ 0, & x^2 + y^2 > 1. \end{cases}$$

```
indicator<-function(x,y) if(1 >= x^2+y^2)  1 else 0
```

Another possible code that generates this indicator function is

```
indicator<-function(x,y) as.integer(x^2+y^2<= 1)
```

Above, the command **as.integer** converts TRUE/FALSE to 1/0. □

Example A.6 (Samples with replacement). For example, to sample *with replacement* 7 balls from a bin containing balls *labeled* 1 through 23 use the R command

```
sample(1:23,7, replace=TRUE)
```

The result is a 7-dimensional vector whose entries consists of 7 numbers sampled with replacement from the set $\{1,\ldots,23\}$. Similarly, to simulate rolling a fair die 137 times use the command

```
sample(1:6,137, replace=TRUE)
```

Example A.7 (Rolling a die). Let us show how to simulate rolling a die a number n of times and then count how many times we get 6. Suppose $n = 20$. We indicate this using the command

```
n<-20
```

We now roll the die n times and store the results in a vector x

```
x<-sample(1:6, n, replace=TRUE)
```

Next we test which of the entries of x are equal to 6 and store the results of these 20 tests in a vector y

```
y<-x==6
```

The entries of y are True or False, depending on whether the corresponding entry of x was equal to 6 or not. To find how many entries of y are T use the command

```
sum(y)
```

The result is equal to the number of 6s we got during the string of 20 rolls of a fair die.

We can visualize data. Suppose we roll a die a large number $N = 1200$ of times. For each $1 \leq k \leq N$ we denote by $z(k)$ the fraction of the first k rolls when we rolled a 6. For $k \to \infty$ the Law of Large Numbers states that this frequency should approach $\frac{1}{6}$. The vector z can be generated in R using the commands

```
N<-12000
x<-sample(1:6, N, replace=TRUE)
z<-cumsum(x==6)/(1:N)
```

Above, **cumsum** stands for "cumulative sum". The input of this operator is a numerical vector $x = (x_1,\ldots,x_n)$. The output is a numerical vector s of the same dimension, with $s_k = s_1 = \cdots + s_k$. We can visualize the fluctuations of $z(k)$ around the expected value $\frac{1}{6}$ using the R code

```
plot(1:N, z, type="l", xlab="# of rolls",
ylab="average number 6-s")
abline(h=1/6,col="red")
```

Figure A.1 depicts the output.

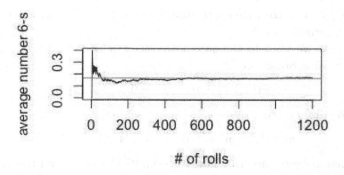

Fig. A.1 Rolling a die.

Example A.8 (Samples without replacement). To sample without replacement 7 balls from an urn containing balls labeled 1 through 23 use the R command

`sample(1:23, 7)`

The number of possible samples above is $(27)_7$ and to compute it use the R command

`prod(21:27)`

□

Example A.9 (Permutations). To sample a random permutation of 7 objects use the R command

`sample(1:7,7)`

To sample 10 random permutations of 7 objects use the R command

`for (i in 1:10 ) print(sample(1:7,7))`

To compute 7! in R use the command

`factorial(7)`

□

Example A.10 (Combinations). Sampling random m-element subsets out of an n-element set possible is possible in R. For example, to sample 4 random subsets with 2 elements out of a 7-element set possible the following command

`replicate(4, sort( sample(1:7, 2) ))`

The sampled sets will appear as columns. To compute $\binom{52}{5}$ in R use the command

`choose(52,5)`

□

Example A.11 (Custom discrete distribution). We can produce custom discrete random variables in R.

Suppose that we want to simulate a discrete random variable X whose values, sorted in increasing order, are

$$x_1 = 0.1, \quad x_2 = 0.2, \quad x_3 = 0.3, \quad x_4 = 0.7.$$

The corresponding probabilities are

$$p_1 = 1/3, \quad p_2 = 1/6, \quad p_3 = 1/4, \quad p_4 = 1/4.$$

The R-commands below describe how to compute the mean and the variance of X and how to sample X.

```
X<-c(0.1,0.2,0.3,0.7) # stores the values  of X in
increasing order.

prob<-c(1/3,1/6,1/4,1/4) # stores the probabilities.
sum(prob) #  This is a  test. If this is 1 prob is a pmf.
# Otherwise check prob.

m<-sum(X*prob) # computes the mean of X and stores in m.
v<-sum((X^2)*prob) -m^2# computes the variance of X
# and stores it in v.

m # produces the value of the mean.

v # produces the variance of X.

sample(X,15, replace=TRUE, prob) # produces 15 random
#samples of X.

cumsum(prob) # computes the values of the cdf of X at
# x_1,x_2,...
```

In R the symbol # indicates a comment. It is only for the programer/user benefit. Anything following a # is not treated by R as a command. □

Example A.12 (Useful discrete distributions). The standard discrete distributions are implemented in R.

The distribution	The R command
The binomial distribution $\text{Bin}(n,p)$	binom(n,p)
The geometric distribution $\text{Geom}(p)$	geom(p)
The negative binomial distribution $\text{NegBin}(k,p)$	nbinom(k,p)
The Poisson distribution $\text{Poi}(\lambda)$	pois(lambda)

A few useful facts 517

The R library however uses rather different conventions

(i) The geometric distribution in R is slightly different from the one described in this book. In R, the range of Geom(p) variable T is $\{0, 1, \dots\}$ and its pmf is $\mathbb{P}[T = n] = p(1-p)^n$. In this book, a geometric random variable has range $\{1, 2, \dots\}$ and its pmf is $\mathbb{P}[T = n] = p(1-p)^{n-1}$; see Example A.14.
(ii) In R the equality **nbinom**(k, p) = n represents the number of *failures* until we register the k-th success; see Example A.15.

The above commands by themselves mean nothing if they are not accompanied by one of the prefixes

- d produces the density or $\boxed{pmf}$.
- p produces the $\boxed{cdf}$.
- r produces random $\boxed{samples}$.
- q produces $\boxed{quantiles}$. □

You can learn more details using R's help function. The examples below describe some concrete situations.

Example A.13 (Binomial). For example, suppose that $X \sim \text{Bin}(10, 0.2)$, i.e., X is the number of successes in a sequence of 10 independent Bernoulli trials with success probability 0.2.
To find the probability $\mathbb{P}(X = 3)$ use the R command

dbinom(3,10,0.2)

If $F_X(x) = \mathbb{P}(X \leq x)$ is the cdf of X, then you can compute $F_X(4)$ using the R command

pbinom(4,10,0.2)

To generate 253 random samples of X use the command

rbinom(253,10,0.2)

To find the 0.8-quantile of X use the R command

qbinom(0.8,10,0.2)

□

Example A.14 (Geometric). Suppose now that $T \sim \text{Geom}(0.2)$ is the waiting time until the first success in a sequence of independent Bernoulli trials with success probability $p = 0.2$.
To find the probability $\mathbb{P}(T = 3)$ use the command

dgeom(3-1,0.2)

To find the probability $\mathbb{P}(T \leq 4)$ use the command

`pgeom(4-1,0.2)`

To generate 253 random samples of T use the command

`1+rgeom(253,0.2)`

To find the 0.8-quantile of T use the R command

`qgeom(0.8,0.2)+1`

□

Example A.15 (Negative Binomial). Suppose that $T \sim \text{NegBin}(8, 0.2)$ is the waiting time for the first 8 successes in a string of Bernoulli trials with success probability.
To find the probability $\mathbb{P}(T = 12)$ use the R command

`dnbinom(12-8,8,0.2)`

You can compute $\mathbb{P}(T \leq 14)$ using the R command

`pnbinom(14-8,8,0.2)`

To generate 253 random samples of T use the command

`8+rnbinom(253,8,0.2)`

To find the 0.8-quantile of T use the R command

`8+qnbinom(0.8,8,0.2)`

□

Example A.16 (Poisson). Suppose that $X \sim \text{Poi}(0.2)$ is a Poisson random variable with parameter $\lambda = 0.2$.
To find the probability $\mathbb{P}(X = 3)$ use the command

`dpois(3,0.2)`

To find the probability $\mathbb{P}(X \leq 4)$ use the command

`ppois(4,0.2)`

To generate 253 random samples of X use the command

`rpois(253,0.2)`

To find the 0.8-quantile of X use the R command

`qpois(0.8,0.2)`

□

Example A.17 (Continuous distributions in R). The continuous distributions $\text{Unif}(a,b)$, $\exp_\lambda$ and $N(\mu, \sigma^2)$ can be simulated in R by invoking

`unif(min=a, max=b)`

`exp(rate=lambda)`

`norm(mean=mu, sd=sigma)`

where sd: = standard deviation.

To invoke the standard normal random variable you could use the shorter command

`norm`

□

As in the case of discrete distributions, we utilize these commands with the prefixes $d-$, $p-$, $q-$ and $r-$ that have the same meaning as in R-Session A.12. Thus $d-$ will generate the pdf, $p-$ the cdf, $r-$ generates a random sample, and $q-$ produces quantiles.

Example A.18. Here are some concrete examples. To find the probability density of $\exp_3$ at $x = 1.7$ use the command

`dexp(1.7, 3)`

To find the probability density of $N(\mu = 5, \sigma^2 = 7)$ at $x = 2.6$ use the command

`dnorm(2.6,5, sqrt(7))`

To produce 1000 samples from $\text{Unif}(3, 13)$ use the command

`runif(1000,3,13)`

□

Example A.19 (Gambler's ruin). Consider two players the first with fortune \$$a$, and the second with fortune \$$b$. Set $N := a + b$. They flip a fair coin. Heads, player 1 gets a dollar from player 2, Tails, player 1 gives a dollar to player 2. The game ends when one of them is ruined. One can simulate this in R using the code

```
r<-function(a,N){
t<-0
x<-a
v<-c(0,N)
while(all(v!=x)){
  f<-sample(0:1,1, replace=TRUE)
```

```
    x<-x+(2*f-1)
    t<-t+1
  }
  y<-c(x,t)
  y
}
```

The output is a two-dimensional vector. Its first entry is the fortune of the first player at the end of the game, while the second entry is duration of the game, i.e., the number of coin flips until one of them is ruined.

To compute the winning probability of the first player and the expected duration of a game we can use the Law of Large Numbers and run a large number G of games

```
empiric_r<-function(G,a,N){
  P<-c()
  T<-c()
  for(i in 1:G){
    P<-c(P,r(a,N)[1])
    T<-c(T,r(a,N)[2])
  }
  c(sum(P==N)/G,sum(T)/G)
}
```

For example if we want to run a number $G = 1200$ of games with the first player's initial fortune $a = 8$ and the combined fortune of the two players is $N = 15$ use the command

```
empiric_r(1200,8,15)
```

The output is a two-dimensional vector. Its first entry describes the fraction of the G games won by the first player, and the second entry is the average duration of these G games.

One can also visualize a game. The code below produces a vector whose entries describe the evolution of the fortune of the first player.

```
rgr<-function(a,N){
  x<-a
  z<-c(a)
  v<-c(0,N)
  while(all(v!=x)){
    f<-sample(0:1,1,replace=TRUE)
    x<-x+(2*f-1)
    z<-c(z,x)
  }
```

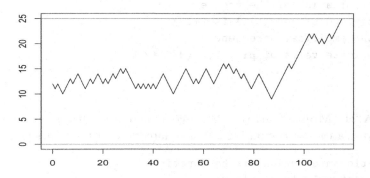

Fig. A.2 *The ruin problem.*

```
  z
}
```

For given values of N and a say, $N = 25$, $a = 12$, one can visualize the evolution of the fortune of the first player using the code below. Its output is a graph similar to the one in Figure A.2.

```
N<-25
a<-12
u<-rgr(a,N)
l<-length(u)-1
plot(0:l, u,type="l", xlab="# of flips",
ylab="the fortune of the first player",ylim=c(0,N))
abline(h=c(0,N),col=c("red","red") )
```

□

Example A.20 (Buffon's needle problem). The R program below uses the Buffon needle problem (see Exercise 1.22) to find an approximation of π.

```
L<-0.7 # L is the length of the needle. It is <1.
N<-1000000 # N is the number of times we throw the needle.
f<-0
#the next loop simulates the tossing of
#N random needles and computes
# the number f  of times they intersect a line

for (i in 1:N){
  y<-runif(1, min=-1/2,max=1/2) #this locates
  # the center of the needle
  t<-runif(1, min=-pi/2,max=pi/2)#this determines
```

```
    #the    inclination of the needle
    if ( abs(y)< 0.5*L*cos(t) ) f<-f+1 }
#f/N  is the empirical frequency
"the approximate value of pi is";   (N/f)*2*L
```

□

Example A.21 (Monte Carlo). The R-command lines below implement the Monte Carlo strategy for computing a double integral over the unit square

```
# Monte Carlo integration of the function f(x,y)
#over the rectangle [a,b] x[c,d]
# First we describe the function
f<- function(x,y) sin(x*y)
# Next, we describe the region of integration [a,b]x[c,d]
a=0
b=1
c=0
d=1
# Finally, we decide the number N  of sample points in
# the region of integration
N=100000
#S will store the integral
S=0
for (i in 1:N){
  x<- runif(1,a,b) #we sample a point uniformly  in [a,b]
  y<- runif(1,c,d) #we sample a point uniformly  in [c,d]
  S<-S+f(x[1],y[1])
}
'the integral is'; (b-a)*(d-c)*S/N
```

The next code describes a Monte-Carlo computation of the area of the unit circle.

```
nsim<-1000000#nsim is the number of simulations
x<-runif(nsim,-1,1)#we choose nsim uniform samples
#in the interval (-1,1) on the x axis
y<-runif(nsim,-1,1)#we choose nsim uniform samples
#in the interval (-1,1) on the y axis
area<-4*sum(x^2+y^2<1)/nsim
"the area of the unit circle is very likely"; area
```

□

Example A.22. Suppose that we have a probability distribution `prob` on the alphabet $\{1, 2, \ldots, L\}$. One experiment consists of sampling the alphabet according to the distribution `prob` until we first observe the given word (or pattern) `patt`. The following R-routine performs m such experiments and returns an m-dimensional vector `f` whose components are the cumulative means of the waiting times

$$f_k = \frac{1}{k} \sum_{j=1}^{k} T_j, \ \ k = 1, \ldots, m,$$

where T_j is the time to observe the pattern in the j-th experiment.

```
Tpattern<-function(patt, prob, m, L){
  k<-length(patt)
  T<-c()
  for (i in 1:m){
    x<-sample(1:L,k,replace=TRUE, prob)
    n<-k
    while ( all(x[(n-k+1):n]==patt)==0 ){
      x<-c(x, sample(1:L,1,replace=TRUE, prob) )
      n<-n+1
    }
    T<-c(T,n)
  }
  f<-cumsum(T)/(1:m)
  f
}
```

If `prob` is the uniform distribution use the faster routine

```
Tpatt_unif<-function(patt, m, L){
  k<-length(patt)
  T<-c()
  for (i in 1:m){
    x<-sample(1:L,k,replace=TRUE)
    n<-k
    while ( all(x[(n-k+1):n]==patt)==0 ){
      x<-c(x, sample(1:L,1,replace=TRUE) )
      n<-n+1
    }
    T<-c(T,n)
  }
  f<-cumsum(T)/(1:m)
  f
}
```

In the uniform case, the expected waiting time to observe the pattern `patt` can be determined using routine below that relies on the identity (3.1.11) in Example 3.31.

```
tau<-function(patt,L){
  n<-length(patt)
  m<-n-1
  t<-2^n
  for (i in 1:m){
      j<-n-i
       k<-i+1
       t<-t+ any(patt[1:j]==patt[k:n])*L^(n-i)
      }
    t
}
```

Bibliography

D. Aldous: *Exchangeability and related topics*, École d'été de Probabilités des Saint Fleur XIII-1983, pp. 2–199, Lect. Notes in Math vol. 1117, Springer Verlag, 1985.

D. Aldous: *Probability Theory*, Course notes, Spring 2017.
https://www.stat.berkeley.edu/~aldous/205B/chewi_notes.pdf

V. I. Arnold, A. Avez: *Ergodic Problems of Classical Mechanics*, Addison Wesley, 1968.

R. B. Ash: *Probability and Measure Theory*, (with contributions from C. Doléans-Dade), 2nd Edition, Academic Press, 2000.

S. Asmunssen: *Applied Probability and Queues*, 2nd Edition, Stoch. Modelling and Appl. Probab., vol. 51, Springer Verlag, 2003.

K. B. Athreya, P. E. Ney: *Branching Processes*, Springer Verlag, 1972.

S. Banach: *Über die Bairésche Kategorie gewisser Functionenmengen*, Studia Mathematica, 3(1931), 174–179.

P. Bamberg, S. Sternberg: *A Course in Mathematics for Students of Physics*, vol. 2, Cambridge University Press, 1990.

R. N. Bhattacharya, E. C. Waymire: *Stochastic Processes with Applications*, SIAM, 2009.

R. N. Bhattacharya, E. C. Waymire: *A Basic Course in Probability Theory*, 2nd Edition, Springer Verlag, 2016.

P. Billingsley: *Ergodic Theory and Information*, John Wiley & Sons, 1965.

P. Billinglsley: *Convergence of Probability Measures*, 2nd Edition, John Wiley & Sons, 1999.

N. H. Bingham: *Fluctuation theory for the Ehrenfest Urn*, Adv. Appl. Prob. **23**(1991), 598–611.

D. Blackwell, D. Freedman: *The tail σ-field of a Markov chain and a theorem of Orey*, Ann. Math. Statist., **35**(1964), 1291–1295.

V. I. Bogachev: *Measure Theory. Vol. 1*, Springer Verlag, 2007.

R. Bott: *On induced representations*, in volume *The Mathematical Heritage of Hermann Weyl*, Proc. Symp. Pure Math., vol. 48, Amer. Math. Soc., 1988

S. Boucheron, G. Lugosi, P. Massart: *Concentration Inequalities. A Nonasymptotic Theory of Independence*, Oxford University Press, 2013.

N. Bourbaki: *General Topology*, Part 2, Hermann, 1966.

L. Breiman: *Probability*, SIAM, 1992.

P. Brémaud: *Markov Chains, Gibbs Fields, Monte Carlo Simulations and Queues*, Springer Verlag, 1999.

P. Brémaud: *Probability Theory and Stochastic Processes*, Springer Verlag, 2020.
H. Brezis: *Functional Analysis, Sobolev Spaces and Partial Differential Equations*, Universitext, Springer Verlag, 2011.
J. Bricmont: *Making Sense of Statistical Mechanics*, Springer Verlag, 2022.
S. A. Broughton, K. Bryan: *Discrete Fourier Analysis and Wavelets. Applications to Signal and Image Processing*, Second Edition, John Wiley & Sons, 2018.
J. C. Butcher: *Numerical Methods for Ordinary Differential Equations*, 3rd Edition, John Wiley & Sons, 2016.
I. Chavel: *Eigenvalues in Riemann Geometry*, Academic Press, 1984.
J. Cheeger: *A lower bound for the smallest eigenvalue of the Laplacian*, in Gunning (ed.) Problems in Analysis, pp. 199–205, Princeton University Press, 1970.
Y. S. Chow, H. Robbins, D. Siegel: *Great Expectations: The Theory of Optimal Stopping*, Houghton Mifflin Co., 1971.
Y. S. Chow, H. Teicher: *Probability Theory. Independence, Interchangeability, Martingales*, 3rd Edition, Springer Verlag, 1997.
K. L. Chung, J. L. Doob: *Fields, Optionality and Measurability*, Amer. J. Math., **87**(1965), 397–424.
K. L. Chung, F. AitSahlia: *Elementary probability theory: with stochastic processes and an introduction to mathematical finance*, Springer Verlag, 2003.
V. Chvátal, D. Sankoff: *Longest common subsequences of two random sequences*, J. Appl. Prob. **12**(1975), 306–315.
E. Çinlar: *Probability and Stochastics*, Graduate Texts in Math., vol. 261 Springer Verlag, 2011.
E. G. Coffman Jr., G. S. Lueker: *Probabilistic Analysis of Packing and Partitioning Algorithms*, John Wiley & Sons, 1991.
D. L. Cohn: *Measure Theory*, 2nd Edition, Birkhäuser, 2013.
I. P. Cornfeld, S. V. Fomin, Ya. G. Sinai: *Ergodic Theory*, Springer Verlag, 1982.
T. M. Cover, M. Thomas, J. A. Thomas: *Elements of Information Theory*, Wiley-Interscience, 2006.
T. M. Davies: *The Boof of R*, No Starch Press, 2015, https://nostarch.com/bookofr
C. Dellacherie, P.-A. Meyer: *Probabilities and Potential*, vol. A, North Holland Mathematical Studies, vol. 29, Hermann Paris, 1978.
C. Dellacherie, P.-A. Meyer: *Probabilities and Potential*, vol. C, North Holland, 1988.
P. Diaconis: *Group Representations in Probability and Statistics*, Institute of Mathematical Statistics, 1988.
P. Diaconis: *The Markov chain Monte Carlo revolution*, Bull. Amer. Math. Soc., **46**(2009), 179–205.
P. Diaconis, R. Griffiths: *Exchangeable pairs of Bernoulli random variables, Krawtchouk polynomials, and Ehrenfest urns*, Aust. N. Z. J. Stat. **1**(2012), 81–101.
P. Diaconis, R. Griffiths: *An Introduction to multivariate Krawtchouk polynomials and their polynomials*, arXiv: 1309.0112, J. Stat. Plann. Inference, **154**(2014), 39–53.
P. Diaconis, B. Skyrmis: *Ten Great Ideas About Chance*, Princeton University Press, 2018.
P. Diaconis, D. Stroock: *Geometric bounds for eigenvalues of Markov chains*, Ann. Appl. Prob., **1**(1991), 31–61.
J. L. Doob: *Stochastic Processes*, John Wiley & Sons, 1953.

P. G. Doyle, J. L. Snell: *Random Walks and Electrical Networks*, MAA, 1984.

L. E. Dubins, L. J. Savage: *How to Gamble if you Must. Inequalities for Stochastic Processes*.

R. M. Dudley: *Real Analysis and Probability*, Cambridge University Press, 2004.

R. M. Dudley: *Uniform Central Limit theorems*, Cambridge University Press, 1999.

N. Dunford, J. T. Schwartz: *Linear Operators. Part I: General Theory*, John Wiley & Sons, 1957.

R. Durrett: *Probability. Theory and Examples*, 5th Edition, Cambridge University Press, 2019.

A. Dvoretzky, P. Erdös, S. Kakutani: *Nonincrease everywhere of the Brownian motion process*, 1961 Proc. 4th Berkeley Sympos. Math. Statist. and Prob., Vol. II pp. 103–116 Univ. California Press, Berkeley, Calif.

P. Erdös, A. Rényi: *On a classical problem of probability theory*, Magyar Tudományos Akadémia Matematikai Kutató Intézetének Közleményei, **6**(1961), 215–220,

G. Fayolle, V. A. Malyshev, M. V. Menshikov: *Topics in the Constructive Theory of Countable Markov Chains*, Cambridge University, Press, 1995.

W. Feller: *The Kolmogorov-Smirnov theorems for empirical distributions*, Ann. Math. Statistics, **19**(1948), 177–189.

W. Feller: *An Introduction to Probability Theory and its Applications*, Volume 1, 3rd Edition, John Wiley & Sons, 1970.

W. Feller: *An Introduction to Probability Theory and its Applications*, Volume 2, 2nd Edition, John Wiley & Sons, 1970.

L. Floridi: *Information Theory. A Very Short Introduction*, Oxford University Press, 2010.

D. Foata, A. Fuchs: *Processus Stochastiques. Processus de Poisson, Chaînes de Markov et Martingales*, 2nd edition, Dunod, 1998.

T. Frankel: *The Geometry of Physics*, 3rd Edition, Cambridge University Press, 2011.

D. Freedman: *Brownian Motion and Diffusion*, Springer Verlag, 1983.

B. Friestedt, L. Gray: *A Modern Approach to Probability Theory*, Birkhäuser, 1997.

F. R. Gantmacher: *Theory of Matrices*, vol. 2, AMS, Chelsea Publishing, 2000.

M. Gardner: *Time Travel and Other Mathematical Bewilderments*, W. H. Freeman & Co., 1988.

A. Garsia: *A simple proof of E. Hopf's maximal ergodic theorem*, J. Math. Mech., **14**(1965), 381–382.

I. M. Gelfand, N. Ya. Vilenkin: *Generalized Functions. Volume 4. Applications of Harmonic Analysis*, Academic Press, 1964.

J. P. Gilbert, F. Mosteller: *Recognizing the maximum of a sequence*, J. Amer. Stat. Assoc., **61**(1966), 35–73.

E. Giné, R. Nickl: *Mathematical Foundations of Infinite Dimensional Statistical Models*, Cambridge University Press, 2016.

J. Gleick: *The Information. A History. A Theory. A Flood*, Pantheon Books, 2011.

B. V. Gnedenko, A. N. Kolmogorov: *Limit Distributions for Sums of Independent Random Variables*, Addison Wesley, 1968.

A. Grigoryan: *Introduction to Analysis on Graphs*, University Lect. Series, Amer. Math. Soc., 2018.

G. R. Grimmett: *Probability on Graphs. Random Processes on Graphs and Lattices*, Cambridge University Press, 2011.

G. R. Grimmett, D. R. Stirzaker: *Probability and Stochastic Processes*, 4th Edition, Oxford University Press, 2020.

L. J. Guibas, A. M. Odlyzko: *String overlaps, pattern matching and nontransitive games*, J. Copmb. Th. Series A, **30**(1981), 183–208.

O. Häggström: *Finite Markov Chains and Algorithmic Applications*, Cambridge University Press, 2002.

P. R. Halmos: *Lectures on Ergodic Theory*, Dover, 2017.

G. H. Hardy: *Divergent Series*, Oxford University Press, 1949.

T. E. Harris: *The Theory of Branching Processes*, Springer Verlag, 1963.

J. Hawkins: *Ergodic Dynamics. From Basic Theory to Applications*, Springer Verlag, 2021.

B. Hayes: *The first links in a Markov chain*, American Scientist, **101**(2013), No. 2, p. 92. https://www.americanscientist.org/article/first-links-in-the-markov-chain

M. Jerrum, M. Sinclair: *Approximate Counting, Uniform Generation and Rapidly Mixing of Markov Chains*, Information and Computation, **82**(1989), 93–133.

M. Kac: *Random walk and the theory of Brownian motion*, Amer. Math. Monthly, **54**(1947), 369–391.

O. Kallenberg: *Foundations of Modern Probability*, 3rd Edition, Springer Verlag, 2021.

S. Karlin, H. M. Taylor: *A First Course in Stochastic Processes*, 2nd Edition, Academic Press, 1975.

J. G. Kemeny, J. L. Snell: *Finite Markov Chains*, with a new appendix "Generalization of a fundamental matrix", Springer Verlag, 1983.

J. H. B. Kemperman: *The Passage Problem for a Stationary Markov Chain*, The University of Chicago Press, 1961.

H. Kesten, P. Ney, F. Spitzer: *The Galton-Watson process with mean one and finite variance*, Th. Prob. Appl., **11**(1966), 513–540.

J. M. Keynes: *A Treatise on Probability*, MacMillan and Co. London, 1921.
Available at Project Guttenberg http://www.gutenberg.org/ebooks/32625.

J. F. C. Kingman: *Uses of exchangeability*, Ann. Prob., **6**(1978), 183–197.

A. Klenke: *Probability Theory. A Comprehensive Course*, Universitext, Springer Verlag, 2008.

U. Krengel: *Ergodic Theorems. With a supplement by Antoine Brunel*, Walter de Guyter, 1985.

A. N. Kolmogorov: *Grundbegriffe der Wahrscheinlichkeitsrechnung*, Springer 1933. English translation *Foundations of the Theory of Probability*, Chelsea 1950.

A. N. Kolmogorov: *The theory of transmission of information*, the volume *Selected Works of A. N. Kolmogorov. Volume III. Information Theory and the Theory of Algorithms*, pp. 6–33, Kluwer Academic Publishers, 1993.

E. Kowalski: *An Introduction to Expander Graphs*, Société Mathématiques de France, 2019.

L. Kuipers, H. Niederreiter: *Uniform Distribution of Sequences*, John Wiley & Sons, 1974. Dover reprint 2006.

K. Kuratowski, A. Mostowski: *Set Theory. With an Introduction To Descriptive Set Theory*, North Hollan Publishing Co., 1976.

S. Lang: *Linear Algebra*, 3rd Edition, Springer Verlag, 1987.

N. N. Lebedev: *Special Functions and Their Applications*, Dover, 1972.

M. Ledoux, M. Talagrand: *Probability in Banach Spaces*, Springer Verlag, 1991.

J.-F. Le Gall: *Intégration, Probabilités et Processus Aléatoares*,
https://www.math.u-psud.fr/~jflegall/IPPA2.pdf

J.-F. Le Gall: *Brownian Motion, Martingales, and Stochastic Calculus*, Graduate Texts in Math., vol. 274, Springer Verlag 2016.

D. A. Levin, Y. Perez, E. L. Wilmer: *Markov Chains and Mixing Times*, Amer. Math. Soc., 2009.

P. Lévy: *Théorie de l'Addition des Variables Aléatoires*, Gauthier-Villars, 1937.

P. Lévy: *Processus Stochastiques et Mouvement Brownien*, Gauthier Villars, 1965.

S.-Y. R. Li: *A martingale approach to the study of occurrence of sequence patterns in repeated experiments*, Ann. Prob. **8**(1980), 1171–1176.

A. Lubotzky: *Expander graphs in pure and applied mathematics*, Bull. A.M.S., **49**(2012), pp. 113–162.

T. Lyons: *A simple criterion of transience of a reversible Markov chain*, Ann. Prob., **11**(1983), 393–402.

T. Lyons, Y. Peres: *Probability on Trees and Networks*, Cambridge University Press, 2017.

M. Loève: *Probability Theory*, vol. I, 4th Edition, Graduate Texts in Math. no. 45, Springer Verlag, 1977.

D. J. MacKay: *Information Theory, Inference and Learning Algorithms*, Cambridge University Press, 18th printing, 2017.

R. Mansuy: *The origins of the word "martingale"* Electronic Journl for History of Probability and Statistics, vol. 5, Fasc. 1, (2009), 1–9.
http://www.jehps.net/juin2009.html

J. Matoušek: *Lectures on Discrete Geometry*, Graduate Texts in Math. no. 212, Springer Verlag, 2002.

Mazurkewicz: *Sur les fonctions non dérivables*, Studia Mathematica, **3**(1931), 92–94.

M. McCaffrey: *Markov Chains: A Random Walk Through Particles, Cryptography, Websites and Card Shuffling*, Senior Thesis, University of Notre Dame, 2017,
https://www3.nd.edu/~lnicolae/Thesis_v3.pdf.

M. Mitzenmacher, E. Upfal: *Probability and Computing. Randomized Algorithms and Probability Analysis*, 7th printing, Cambridge University Press, 2013.

P. Mörters, U. Peres: *Brownian Motion*, Cambridge University Press, 2010.

C. St. J. A. Nash-Williams: *Random walks and electric currents in networks*, Proc. Cambridge. Phil. Soc., **55**(1959), 181–194.

D. J. Newmann, L. Shepp: *The double dixie-cup problem*, Amer. Math. Monthly, **67**(1960), 58–61.

L. I. Nicolaescu: *Introduction to Real Analysis*, World Scientific, 2020.

L. I. Nicolaescu: *Lectures on the Geometry of Manifolds*, 3rd Edition, World Scientific, 2021.

J. R. Norris: *Markov Chains*, Cambridge University Press, 1997.

R. E. A. C. Paley, N. Wiener, A. Zygmund, *Note on random functions*, Math. Z., 1933.

J. L. Palacios: *Fluctuation theory for the Ehrenfest urn via electric networks*, Adv. Appl. Prob., **25**(1993), 472–476.

K. R. Parthasarathy: *Probability Measures on Metric Spaces*, Academic Press, 1967.

V. V. Petrov: *Limit Theorems of Probability Theory. Sequences of Independent Random Variables*, Oxford University Press, 1995.

I. Pinelis: *Martingales converging in probability and not as*, MathOverflow, https://mathoverflow.net/a/410350/20302

D. Pollard: *Convergence of Stochastic Processes*, Springer Verlag, 1984.

D. Pollard: *Empirical Processes: Theory and Applications*, NSF-CBMS Regional Conference Series in Probability and Statistics, vol. 2, 1990.

M. Pollicott, M. Yuri: *Dynamical Systems and Ergodic Theory*, Cambridge University Press, 1998.

S. L. Resnick: *Adventures in Stochastic processes*, Birkhäuser, 2002.

F. Riesz: *Sur la théorie ergodique*, Comment. Math. Helv. **17**(1944), 221–239.

R. T. Rockefellar: *Convex Analysis*, Princeton University Press, 1997.

L. C. G. Rogers, D. Williams: *Diffusions, Markov Processes and Martingales. Volume 1. Foundations*, Cambridge University Press, 2000.

R. L. Schilling, L. Partzch: *Brownian Motion. An Introduction to Stochastic Processes*, DeGruyter, 2012.

K. Schmüdgen: *The Moment Problem*, Springer Verlag, 2017.

D. Serre: *Matrices. Theory and Applications*, 2nd Edition, Grad. Texts Math., vol. 216, Springer Verlag, 2010.

S. Shalev-Shwartz, S. Ben-David: *Understanding Machine Learning. From Theory to Algorithms*, Cambridge University Press, 2014.

A. N. Shiryaev: *Probability*, 2nd Edition, Springer Verlag, 1996.

A. Sinclair: *Algorithms for Random Generation and Counting: A Markov Chain Approach*, Progress in Theoretical Com. Sci., Springer Verlag, 1993.

P. M. Soardi: *Potential Theory of Infinite Networks*, Lect. Notes. Math., vol. 1590, Springer Verlag, 1994.

R. Stanley: *Enumerative Combinatorics. vol. 1*, 2nd Edition, Cambridge University Press, 2012.

J. M. Steele: *Probability Theory and Combinatorial Optimization*, CBMS-NSF Regional Conf. Series in Appl. Math., SIAM, 1997.

J. M. Steele: *Stochastic Calculus and Financial Applications*, Springer Verlag, 2001.

J. M. Stoyanov: *Counterexamples in Probability*, 3rd Edition, Dover, 2013.

L. Takáks: *On an urn problem of Paul and Tatian Ehrenfest*, Math. Proc. Camb. Phil. Soc., **86**(1979), 127–130.

M. Taylor: *Measure Theory and Integration*, Grad. Studies in Math., Amer. Math. Soc., 2006.

C. B. Thomas: *Representation Theory of Finite and Lie Groups*, World Scientific, 2004.

A. W. van der Vaart, J. A. Wellner: *Weak Convergence and Empirical Processes. With Applications to Statistics*, Springer Verlag, 1996.

V. N. Vapnik, A. Ja. Cervonenkis: *On the uniform convergence of relative frequencies to their probabilities*, Theor. Probab. Appl., **16**(1971), 264–240.

V. N. Vapnik: *The Nature of Statistical Learning Theory*, 2nd Edition, Springer Verlag, 2000.

R. S. Varadhan: *Probability*, Courant Lect. Notes in Math., Amer. Math. Soc., 2001.

M. Viana, K. Oliveira: *Foundations of Ergodic Theory*, Cambridge University Press, 2016.

R. von Mises: *Probability, Statistics and Truth*, 2nd Edition, Dover, 1981.

M. J. Wainright: *High Dimensional Statistics. A Non-Asymptotic Point of View*, Cambridge University Press, 2019.

H. Weyl: *Über die Gleichverteilung von Zahlen mod Eins*, Math. Ann., **77**(1916), 313–352.
E. T. Whittaker, G. N. Watson: *A Course in Modern Analysis*, 4th Edition, Cambridge University Press, 1950.
H. S. Wilf: *Generatingfunctionology*, Academic Press, 1994.
D. Williams: *Probability with Martingales*, Cambridge University Press, 1991.

Index

A^c, xi
$B(a,b)$, 143
$B(x,y)$, 507
$B_{a,b}(x)$, 143
D_Γ, 335
$D_n[f]$, 10
$F_\# \mu$, 15
H_Γ, 336
H_x, 385
L-function, 333
$L^p(\Omega, \mathcal{S}, \mu)$, 37
$N(\mu, \sigma^2)$, 65
N_x, 387
PG_X, 142
R-function, 333
T_x^k, 387
T_A, 384, 385
T_x, 384
$X \sim Y$, 42
$X \sim \mu$, 42
$X \stackrel{d}{=} Y$, 42
$X_\bullet^T$, 270
X_T, 269, 337
$[f]_n$, 10
$\#S$, xi
Ber(p), 15
Bin(n,p), 16, 54
Couple(μ, ν), 404
Elem$(\Omega, \mathcal{S})$, 9, 31
Ent$_2$, 164
Gamma(ν, λ), 67
$\Gamma(x)$, 507
Geom(p), 55
HGeom(w, b, n), 58
$\Lambda(\mathcal{C})$, 5

Markov$(\mathcal{X}, \mu, Q)$, 369
Meas(X), 170
NegBin(k, p), 56
$\Phi(x)$, 66
Φ_X, 179
Poi(λ), 59
Prob(X), 170
Prob$(\Omega, \mathcal{S})$, 13
Proc$(\mathcal{F}_\bullet)$, 333
Proc$(\mathcal{F}_{\text{prog}})$, 333
Proc$_{\text{prog}}$, 333
$\mathcal{R}_{\text{eff}}(x_+, S_-)$, 421
Stud$_p$, 144
Unif(a,b), 64
$\|f\|_p$, 38
a.s., 14
$\mathbb{E}[X]$, 42
$\mathbb{E}[X \| \mathcal{F}]$, 94
$\mathbb{E}[Y \| X = x]$, 94
$\mathbb{E}[Y \| X]$, 94
$\mathbb{E}_\mathbb{P}[X]$, 42
$\mathbb{G}(\sigma^2)$, 194
$\boldsymbol{\Gamma}_{\mu, \sigma^2}$, 65
$\mathbb{H}_\varphi$, 48
$\mathbb{I}_n$, xi
$\mathbb{M}_X(t)$, 51
$\mathbb{P}[F \| \mathcal{S}]$, 114
$\mathbb{P}[S \| \mathcal{F}]$, 98
$\mathbb{T}$-filtration
  see filtration, 260
$\beta_{a,b}(x)$, 68
γ_{μ, σ^2}, 65
$\bigvee_{i \in I} \mathcal{S}_i$, 3
$\boldsymbol{\lambda}_n$, xi
$\boldsymbol{\omega}_n$, xi, 508

$C(X_\bullet)$, 266
σ_{n-1}, xi, 508
$\chi(r), \mathscr{X}$, 293
$\chi^2(n)$, 67
$\mathrm{Cov}[X,Y]$, 50
$\mathrm{dens}(\mathcal{F})$, 210
$\dim_{VC}(\mathcal{F})$, 210
$\mathcal{B}_T$, xi
$\mathcal{B}_X$, 4
$\mathcal{C}_T$, 128
$\mathcal{F} \perp_\mathcal{G} \mathcal{H}$, 110
$\mathcal{F}_T$, 271, 334
$\mathcal{F}_{\mathrm{prog}}$, 333
$\mathcal{F}_{t+}$, 335
$\mathcal{L}^0(\Omega,\mathcal{S})$, 8
$\mathcal{L}^0(\Omega,\mathcal{S})_*$, 69
$\mathcal{L}^0(\mathcal{S})$, 8
$\mathcal{L}^1(\Omega,\mathcal{S},\mu)$, 32
$\mathcal{L}^\infty(\Omega,\mathcal{S})$, 8
$\mathcal{L}^0_+(\Omega,\mathcal{S})$, 8
$\mathcal{L}^1_+(\Omega,\mathcal{S},\mu)$, 32
$\mathcal{N}_\mu$, 16
$\mathcal{O}_T$, 478
$\mathcal{S}|_X$, 4
$\mathcal{S}^\mu$, 16
$\mathcal{S}_1 \vee \mathcal{S}_2$, 3
$\mathcal{S}_\infty$, 326
ess sup, 38
$\frac{d\nu}{d\mu}$, 37
$\mathfrak{S}_n$, xi
$\hat{X}_T$, 301
i.i.d., 132
i.o., 83
λ-system, 5
$\langle X \rangle$, 267
$\lceil x \rceil$, xi
$\lfloor x \rfloor$, xi
$\perp Y$, 21
$\mu * \nu$, 77
$\mu \overset{\lambda}{\leftrightsquigarrow} \nu$, 404
$\mu_0 \otimes \mu_1$, 70
$\nu \ll \mu$, 37
π-system, 5
π^x, 396
$\rho[X,Y]$, 50
σ-additive, 13
σ-algebra, 2
  Borel, 4
  complete, 16
  completion of a, 16
  trace of a, 4
$\sigma(X_i, i \in I)$, 11
$\sigma[X]$, 49
$\overset{\mathrm{a.s.}}{\to}$, 83
$\overset{d}{\to}$, 170
$\overset{p}{\to}$, 85
2^X, xi
2^T_0, 127
2^X_0, xi
$\mathrm{Var}[X]$, 49
$|S|$, xi
φ-entropy, 48
$\{F \in S\}$, 3
$]]S,T]]$, 270
d_v, 403
f^+, 9
f^-, 9
$g_\nu(x;\lambda)$, 66
$u \sim v$, 410
$x_0 \leftrightarrow x_1$, 378
$x_0 \to x_1$, 378

absolutely continuous, 37
absorbing state, 380
acceptance-rejection, 237, 459
accessible states, 378
AEP, 166, 167
algebra, 2
  Bernoulli, 3
alphabet, 163
  entropy of the, 164
André's reflection trick, 29, 136, 344, 356
aperiodic, 382
arcsine distribution, 69
Arnold's cat map, 476, 500

baker's transform, 504
ballot problem, 27, 274
Banch space, 38
Bernoulli
  algebra, 3
  random variable, 54
  trial, 54
Bernoulli shift, 478
Beta distribution, 68, 120, 144, 332
Beta function, 68, 332, 507
  incomplete, 143
bin packing, 283
binomial distribution, 16
bit, 164

Borel
  σ-algebra, 4
  set, 4
  subset, 4
Borel-Cantelli, see lemma
boundary operator, 415
bounded difference, 284
branching process, 263, 300, 321, 377
Brownian event, 222
Brownian motion, 225, 229
  quadratic variation, 233
  started at 0, 231
  strong Markov property, 341
Brownian motion reflection principle, 343

Cèsaro convergent, 158, 495
Cèsaro means, 158, 490
Cayley graph, 442
cdf, 19, 20, 41
chains, 414
chamber, 3, 92
character, 443, 499
characteristic function, 179
Cheeger constant, 455
Chernoff
  bound, 189
  method, 189, 280
closed set, 380
  irreducible, 380
coboundary operator, 417
cochain, 417
code
  binary, 168
  instantaneous, 168
  Shannon, 168
  uniquely decodable, 168
coercive, 294
communication class, 379
compensator, 266
conditional
  expectation, 93, 94, 136
  independence, 110
  index, 135
  probability, 25, 98, 114
conditional distribution, 115
conductance, 414, 455
convergence
  L^p, 86
  almost sure, 83
  in p-mean, 86

  in distribution, 170
  in law, 170
  in probability, 85, 296
convex function, 47, 190
  conjugate of, 191
convolution, 77
correlation, 50
coupling, 404
  time, 405
coupon collector problem, 57, 244, 308
covariance, 50
covariance form, 218
covariance kernel, 220
cumulant, 189
cumulative distribution function, 19
current, 415
cutting
  see electric network, 426
cylinder, 128

début time, 335
derangement, 63
derangements problem, 63
detailed balance equations, 393, 451
dipole, 425
Dirac measure, 15
Dirichlet form, 451
disintegration, 120
  kernel, 120
distribution
  Beta, 68, 120, 144, 332
  binomial, 16
  chi-squared, 67
  compound Poisson, 249
  Erlang, 67
  exponential, 74, 79, 243
  Gamma, 66, 144
  geometric, 55, 73, 243
  hypergeometric, 58
  negative binomial, 55, 56
  of stochastic process, 125, 217
  Poisson, 80
  Student, 144
distribution Bernoulli, 15
distribution function, 19
Doeblin condition, 472
Doob
  conditions, 271, 278
Doob decomposition, 266, 358
downcrossing, 286

effective
  conductance, 421
  resistance, 421
Ehrenfest urn, 375, 383, 394, 433, 445, 450, 473
electric network, 414
  cutting, 426
  shorting, 426
empirical
  distribution, 200
  process, 204
empirical gap, 119
entropy, 48, 164, 252, 472
  information, 164, 193
  rate, 472
  relative, 252
  Shannon, 164, 193
equidistribution, 492
ergodic map, 480
ess sup, 38
Euler means, 406
event, 14
  almost sure, 14
  exchangeable, 326
  improbable, 14
  permutable, 326
exchangeable, 325
  sequence, 325, 331, 477
expectation, 42
exponential martingale, 346
extinction
  event, 291, 300
  probability, 291

Fatou's lemma, 35
filtration, 260
  complete, 335
  right-continuous, 335
  usual, 335
flow
  Kirchhoff, 422
formula
  Bayes', 30
  Fourier inversion, 443
  Stirling, 390, 392
  Stirling's, 508
  Viète, 246
  Wald, 137, 305, 306
formula Stirling, 244
Fourier transform, 179, 443

function
  Beta, 507
  convex, 47
  elementary, 9
  Gamma, 507
  strictly convex, 47

Galton-Watson process, 263, 290, 300
gambler's ruin, 307, 375, 390, 519
Gamma function, 66, 507
Gaussian
  Hilbert space, 224
  measure, 65, 218, 254
    covariance form, 218
  process, 220
    centered, 220
  random function, 222
  random variables, 65
  regression, 256
  vector, 218, 254, 255
  white noise, 223, 364
gaussian
  measure
    centered, 219
graph, 265
  locally finite, 265
  random walk on, 265

Haar
  basis, 498
  functions, 498, 503
Hamming distance, 285, 445
harmonic function, 265, 408, 420
Hermite polynomials, 139, 251, 346
hitting time, 269, 336, 385, 408, 420
HMC, 369
  Laplacian of, 408
  reversible, 393
  time reversed, 393
hypothesis class, 212
  PAC learnable, 213

independence
  conditional, 110
independency, 21
independent
  events, 21
  families, 21
  random variables, 21
indicator function, xi

inequality
    Azuma, 209, 280, 282, 284
    Bonferroni, 60
        motivic, 60
    Cauchy-Schwartz, 90
    Chebyshev, 49
    Doob's L^p, 318, 348
    Doob's maximal, 317, 347
    Doob's upcrossing, 287, 289, 349
    Gibbs, 165, 193
    Hölder, 38, 191
    Hoeffding, 195, 208, 280, 285
    Jensen, 47, 164
    Kolmogorov's maximal, 153
    Markov, 34
    McDiarmid, 285
    Mills ratio, 66, 140, 231
    Minkowski, 38
infinitely divisible
    distribution, 249
    random variable, 249
integrable, 32
invariance principle, 232
invariant
    distribution, 392, 397
    function, 478
    measure, 392
    set, 478
irreducible
    HMC, 380
    set, 380

joint probability distribution, 76

kernel, 111
    disintegration, 120
    Markovian, 112
    probability, 112
    pullback, 112
    push-forward by, 112
Kirchhoff current, 418
    potential of, 418
Kirchhoff's laws, 416, 418
Kolmogorov sequence, 481
Koopman operator, 483
Kullback-Leibler divergence, 192, 253

L-process, 333
Lévy's martingale, 409
Laplacian, 408, 451

law
    of rare events, 63
    of total probability, 26
law of total probability, 122
lazy chain, 406, 460
Lebesgue
    measurable, 72
    measure, 72
Lebesgue integral, 32
Lebesgue measure, 19
lemma
    'sooner-rather-than-later', 276, 307
    Borel-Cantelli, 83–85, 157, 227, 290, 338
        first, 84
        second, 84, 356, 358
    Fatou, 36, 178, 179, 289, 296, 303, 323, 349, 411
    Fekete, 89, 283
    Hoeffding, 196, 281
    Kronecker, 358
    Kronecker's, 158
    maximal, 488
    Sauer, 210
    Scheffé's, 298
likelihood, 30
likelihood ratio, 320
log-normal distribution, 142
logistic map, 503
longest common subsequence problem, 88
Lusin space, *see* space
Lyapunov function, 407
    coercive, 410

map
    measurable, 6
Markob property
    strong, 389
Markov
    chain, 368
        aperiodic, 383, 406
        irreducible, 380
        null recurrent, 397
        positively recurrent, 397, 400, 411
        recurrent, 388, 410, 435
        reversible, 393, 413, 441
        transient, 388, 409, 435
    path space, 371
Markov property, 368, 373, 385, 395
    strong, 385, 387–389, 400, 401
martingale, 260–264, 266, 268, 270, 272,

275, 277, 280, 284, 290, 306, 307, 309, 319, 345, 408, 409
  L^p, 319
  L^p-bounded, 319
  backwards, 322
  bounded L^p, 319
  closed, 261, 282
  component, 266
  De Moivre, 262, 308
  discrete time, 260
  Doob, 261, 282
  exponential, 346
  quadratic variation, 267
matrix
  matrix, 438
  primitive, 439
mean, 42
measurable
  map, 6
  set, 2
  space, 2
    isomorphism, 6
measure, 13
  σ-finite, 13
  Borel, 16
  Dirac, 15
  finite, 13
  inner regular, 129
  Lebesgue, 19
  Lebesgue-Stieltjes, 74
  outer regular, 129
  probability, 13
  pushforward of a, 15
  Radon, 130, 176
  regular, 129
  signed, 404
  uniform, 15, 16
measure preserving, 475
measured space, 14
memoryless property, 74
Metropolis
  algorithm, 459
  chain, 459
mgf, 51
Mills ratio, 66, 140, 231
mixture, 77, 113, 121, 325
mixtures, 149
moment generating function, 51
monotone class, 10
Monte-Carlo method, 162

motion
  Brownian, 229
  Brownian standard, 225
  pre-Brownian, 217

negligible, 16
noisy dynamical system, 378
null recurrent, 397

Ohm's law, 418
optimal gambling strategy, 311
optimal stopping, 103
orbit, 478, 492
order statistics, 146, 247
Orlicz function, 253

paradox
  waiting time, 80
partition, 3
  chamber of, 3
path space, 477
period, 382
persistent state, 386
pgf, 44, 53, 142, 361
pmf, 42
Poincaré phenomenon, 198, 253
Poisson approximation, 63
Poisson process, 80
poissonization, 63
Polya's urn, 137, 263, 331
positively recurrent, 397
posterior, 30
predictable, 266
predictor, 98
premeasure, 18
  σ-finite, 18
principle
  Dirichlet, 424
  inclusion-exclusion, 59
  Raleigh, 425, 426
  reflection, 344
  Thompson, 424
prior, 30
probability
  generating function, 44, 53, 142
  measure, 13
    Euclidean, 40
  space, 14
probability distribution
  continuous, 64

joint, 76
problem
    ballot, 27, 274
    Banach's matchbox, 137
    bin packing, 283
    birthday, 138, 244
    Buffon, 138, 521
    coupon collector, 57, 135, 244, 308
    derangements, 63
    gambler's ruin, *see* gambler's ruin
    longest common subsequence, 88, 281
    occupancy, 243
    Polya's urn, 137, 263, 331
    secretary, 106
process
    branching, 263, 300, 321
    empirical, 204
    exchangeable, 325
    Galton-Watson, *see* Galton-Watson process
    independent increments, 345
    L-, 333
    measurable, 333
    Poisson, 80, 147, 148, 249, 306, 346, 363, 464
    predictable, 266, 268, 358
    progressive, 333
    R-, 333
    renewal, 82
    separable, 347, 349
    stochastic, 123
        distribution, 125, 217
        path, 223
    stopped, 270
product formula, 25
projective family, 128
pushforward, 15, 16

quadratic variation, 233, 267, 363
    optional, 354
    predictable, 354
quantile, 20, 201
quasi-invariant
    function, 479
    set, 479
quasi-mixing, 501

R-process, 333
Rademacher
    complexity, 209

function, 503
    random variable, 27, 154, 206, 214
    symmetrization, 206
random
    measure, 112, 325
    variable, 14
    walk, 27, 261, 262, 265, 292, 357, 360, 369, 376, 383, 390, 393
        on groups, 442
        standard, 27
random variable, 14
    Bernoulli, 54
    binomial, 54
    cdf of, 41
    discrete, 42
        probability mass function, 42
    distribution of, 16, 41
    expectation, 42
    exponential, 67, 74
    finite, 14
    Gamma, 67
    Gaussian, 65
    geometric, 55
    hypergeometric, 58
    law of, 41
    mean, 42
    negative binomial, 56
    normal, 65
    Poisson, 59
    probability distribution of, 41
    Raleigh, 244
    standard normal, 66
    subgaussian, 194
    uniform, 64
random vector, 76
    probability distribution of, 76
recurrence class, 388
recurrent state, 386
Reflection Principle, 343, 344
regular version, 114, 115
return time, 384, 420
reversible, 393, 413, 450

sample range, 119
sample space, 14
secretary problem, 106
separating collection, 170
shift, 478
    Bernoulli, 478
shorting, *see* electric network

sieve, 62, 63
sigma-algebra, 2, 3
SLE, 441, 444, 451
SLLN, 151, 156, 161, 200, 203, 204, 321, 330, 338, 401
space
  Lusin, 115, 116, 128
  measurable, 2
  measured, 14
  Polish, 116
  standard measurable, 116
stable sequence, 241
standard deviation, 49
state
  accessible, 378
  aperiodic, 382
  period of, 382
  persistent, 386
  recurrent, 386
  transient, 386
stationary
  distribution, 392
  measure, 392
stationary sequence, 477
Stieltjes
  measure, 20
Stirling's formula, 390, 392, 508
stochastic integral
  discrete, 268, 358
stochastic matrix, 369
stochastic process, 123, 260
  indistinguishable, 226
  stochastically equivalent, 226
  version of, 226
stopped process, 270
stopping time, 103, 269, 384
  optimal, 104
strictly convex function, 47
strong Markov property, 341
subadditivity, 89
subgaussian, see random variable
submartingale, 260, 266, 268, 270, 272, 287, 345
  backwards, 322
  discrete time, 260
superadditivity, 89
superharmonic function, 408
supermartingale, 260, 312, 345, 355, 407
  backwards, 322
  discrete time, 260

survival function, 74

tail
  algebra, 24, 326, 479
  events, 24
tail-algebra, 481
temporal average, 484
tent map, 477, 498, 503
theorem
  L^p-martingale convergence, 319
  asymptotic equipartition property, 165
  Backwards Martingale Convergence, 325, 482
  Birkhoff's ergodic, 487
  Blumenthal's 0-1 law, 339
  Bochner, 183
  Bounded Convergence, 91, 297
  Carathéodory Extension, 18, 129
  Cayley-Hamilton, 441
  central limit, 186
  début, 336
  de Finetti, 326, 330
  Dominated Convergence, 35, 39, 86, 90, 91, 94, 171, 174, 179, 184, 293, 297, 313, 320, 342, 352
  Donsker, 232
  Doob's regularization, 347
  Dynkin, 11, 94
  Dynkin's $\pi - \lambda$, 5, 7, 23
  ergodic, 402, 491
  Fubini-Tonelli, 70, 295
  Glivenko-Cantelli, 202
  Hewitt-Savage 0-1 law, 330
  Ionescu-Tulcea, 132
  Kolmogorov existence, 54, 128, 370
  Kolmogorov one series, 152, 223, 320
  Kolmogorov's 0-1, 24, 84, 152, 299, 492
  Kolmogorov-Smirnov, 203
  Lévy's 0-1, 299, 493
  Lévy's continuity, 183
  Lévy's equivalence, 236
  Lévy's forgery, 366
  Lindeberg, 188
  Lindenstrauss-Johnson, 199
  mapping, 172
  mean ergodic, 484
  Monotone Class, 10, 70, 71, 94, 113, 114, 372
  Monotone Convergence, 32, 35, 36, 46, 71, 83, 97, 113, 320, 349, 352

Optional Sampling, 272, 275, 278, 290, 302–305, 307, 310, 317
Optional Stopping, 270, 292, 302
Perron-Frobenius, 439
portmanteau, 172
Radon–Nicodym, 98
Radon–Nikodym, 37
Raleigh, 426
Riesz Representation, 40, 117
Slutsky, 175, 216, 252
Strong Law of Large Numbers, 157, 321, 330, 492
submartingale convergence, 289, 292, 409, 411
Tikhonov's compactness, 131
Wald's formula, 305
weak law of large numbers, 160
Weyl's equidistribution, 494
theorem Kolmogorov continuity, 226
tight family, 248
time
    hitting, 269
    optional, 334
    stopping, 269, 334
transience class, 388
transient state, 386
transition matrix, 369
    n-th step, 370
    locally finite, 407
tree, 436
    radially symmetric, 436

UI, 293, 294, 296, 298, 302–305, 307, 311, 319
uniform integrability, 293
unimodality, 62
union bound, 62
upcrossing, 286
    number, 286
usual conditions, 335, 338, 347

vague convergence, 170
Vapnik-Chervonenkis, *see* VC
variance, 49
variation distance, 403
VC
    dimension, 210
    family, 210

waiting time, 80
walk, 379
weak convergence, 170
weakly mixing, 500
weight function, 15
Wiener
    integral, 225, 257, 364
    measure, 232
    process, 225
WLLN, 160

zero-one
    algebra, 25, 480, 481, 493
    event, 25, 480

Printed in the United States
by Baker & Taylor Publisher Services